McGraw Hill Circuit Encyclopedia and Troubleshooting Guide

Vol. 1

John D. Lenk

McGraw-Hill, Inc.
New York St. Louis San Francisco Auckland Bogotá
Caracas Lisbon London Madrid Mexico City Milan
Montreal New Delhi San Juan Singapore
Sydney Tokyo Toronto

Library of Congress Cataloging-in-Publication Data

Lenk, John D.
McGraw-Hill circuit encyclopedia and troubleshooting guide, Vol. 1 / by John D. Lenk.
p. cm.
Includes index.
ISBN 0-07-037603-4
1. Electronic circuits. 2. Electronic circuits—Testing.
I. Title.
TK7867.T463 1993 92-333275
621.3815—dc20 CIP

4 5 6 7 8 9 0 DOH/DOH 9 9 8 7 6 5 4

ISBN 0-07-037603-4

The sponsoring editor for this book was Daniel A. Gonneau, the editor was Andrew Yoder, the designer was Jaclyn J. Boone, and the production supervisor was Katherine G. Brown. This book was set in Times Roman. It was composed by the McGraw-Hill Publishing Company Professional and Reference Division composition unit.

Printed and bound by R. R. Donnelley of Harrisonburg.

Dedication

Greetings from the Villa Buttercup!
To my wonderful wife Irene,
Thank you for being by my side all these years!
To my lovely family, Karen, Tom, Brandon, and Justin.
And to our Lambie and Suzzie, be happy wherever you are!
To my special readers, may good fortune
find your doorways to good health and happy things.
Thank you for buying my books!
And a special thanks to Dan Gonneau, Stephen Fitzgerald,
and Robert McGraw of McGraw-Hill/TAB
for making me a best seller again!
This is book number 76.
Abundance!

Acknowledgments

Many professionals have contributed to this book. I gratefully acknowledge the tremendous effort needed to make this book. Such a comprehensive work is impossible for one person and I thank all who contributed, both directly and indirectly.

I give special thanks to the following: Syd Coppersmith of Dallas Semiconductor, Rosie Hinojosa and Kellie Garcia of EXAR Corporation, Jeff Salter of GEC Plessey, Linda daCosta of Harris Semiconductor, Ron Denchfield of Linear Technology Corporation, David Fullagar of Maxim, Linda Capcara of Motorola Inc., Andrew Jenkins of National Semiconductor Corporation, Antonio Ortiz of Optical Electronics Incorporated, Lorraine Jenkins of Raytheon Company Semiconductor Division, Ed Oxner and Robert Decker of Siliconix Incorporated, Amy Sullivan of Texas Instruments, and Fred Swymer of Unitrode (Microsemi) Corporation.

I also thank Joseph A. Labok of Los Angeles Valley College for help and encouragement throughout the years.

Very special thanks to Dan Gonneau, Stephen Fitzgerald, Robert McGraw, Nancy Young, Barbara McCann, Kimberly Martin, Wayne Smith, Charles Love, Peggy Lamb, Thomas Kowalczyk, Suzanne Babeuf, Nancy Rosenblum, Kathy Greene, Kriss Helman, Susan Wahlman, Joanne Bishop, Bob Ostrander, Andrew Yoder and Jeanne Glasser of the McGraw/TAB organization for having that much confidence in the author. I recognize that all books are a team effort and am thankful that I am working with the first team!

And to my wife Irene, my research analyst and agent, I extend my thanks. Without her help, this book could not have been written.

Contents

Introduction

When you have finished this encyclopedia, you should be able to recognize well over 700 circuits that are commonly used in all phases of electronics. You will also understand how the circuits operate and where they fit into electronic equipment and systems. This information alone makes the book an excellent one-stop source or reference for anyone (student, experimenter, technician, or designer) who is involved with electronic circuits.

However, this book is much more than just a mere collection of circuits and descriptions. First, all circuits are grouped by function. With each functional group, the book has a practical guide to test and troubleshooting for that type of circuit. Thus, if you are building any particular circuit, and the circuit fails to perform as desired, you are given a specific troubleshooting approach to locate circuit problems.

Second, you can put the circuits to work "as is" because actual circuits with proven component values are given in full detail. On those circuits where performance (such as frequency range, power outputs bandwidth, etc.) depend on circuit values, you are given information as to how circuit values can be selected to meet a certain performance goal. This information is particularly useful to the student or experimenter, but it can also be an effective time-saver for the designer.

Notice that most electronic equipment applications use combinations of basic circuits, so a desired circuit could be in any of several different chapters. For example, an audio oscillator circuit could appear in chapter 1 (audio) or chapter 5 (oscillators).

If you do not locate the circuit you want on the first try, use the Index at the back of the book. Here, the circuits are indexed under the different names by which they are known or could possibly be classified. Hundreds of cross references in the Index will aid you in this search.

Circuit sources and addresses

The source for each circuit is given at the end of the circuit description. This includes the publication title and date, as well as the page number or numbers, for the particular circuit. This makes it possible for the reader to contact the original source for further information on the particular circuit or circuit components. To this end, the complete mailing address for each source is included in this section. When writing, give the complete information, including publication date and page numbers. Notice that all circuit diagrams have been reproduced directly from the original source, without redrawing, by permission of the publisher in each case.

Dallas Semiconductor
4401 South Beltwood Parkway
Dallas, TX 75244-3292

EXAR Corporation
2222 Qume Drive
PO Box 49007
San Jose, CA 95161-9007

GEC Plessey Semiconductors
Cheney Manor, Swindon, Wiltshire
United Kingdom SN2 2QW

Harris Semiconductor
PO Box 883
Melbourne, FL 32902

Intel Corporation
3065 Bowers Avenue
Santa Clara, CA 95051

Linear Technology Corporation
1630 McCarthy Boulevard
Milpitas, CA 95035-7487

Maxim Integrated Products
120 San Gabriel Drive
Sunnyvale, CA 94086

Motorola, Inc.
Semiconductor Products Sector
Public Relations Department
3102 North 56th Street
Phoenix, AZ 85018

National Semiconductor Corporation
2900 Semiconductor Drive
PO Box 58090
Santa Clara, CA 95052-8090

Optical Electronics Inc
PO Box 11140
Tucson, AZ 85734

Philips
2001 West Blue Heron Boulevard
Riviera Beach, FL 33404

Raytheon Company
Semiconductor Division
350 Ellis Street
Mountain View, CA 94039-7016

Signetics Corporation
811 East Arques Avenue
Sunnyvale, CA 94086

Siliconix Incorporated
2201 Laurelwood Road
Santa Clara, CA 95054

Texas Instruments
PO Box 1443
Houston, TX 77001

Unitrode Corporation
(Microsemi Corporation)
580 Pleasant Street
Watertown, MA 02172

Unitrode Corporation
8 Suburban Park Drive
Billerica, MA 01821

Substitution and cross-reference tables

Substitutions can often be made for the semiconductor and IC types that are specified on the circuit diagrams. Newer components, not available when the original source was published, might actually improve the performance of the particular circuit. Electrical characteristics, terminal connections, and such critical ratings as voltage, current, frequency, and duty cycle must, of course, be taken into account if experimenting without referring to substitution guides.

Semiconductor and IC substitution guides can usually be purchased at electronic parts-supply stores. In the absence of any substitution guides, the following cross-reference tables will help in locating possible substitute ICs.

Raytheon Linear Integrated Circuits, 1989, Pages 1-2 through 1-9.

General Cross References

INDUSTRY TYPE	RAYTHEON DIRECT REPLACEMENT	RAYTHEON FUNCTIONAL REPLACEMENT
ADVFC32		RC4153
ADOP07	OP-07	
ADOP27	OP-27	
ADOP37	OP-37	
ADREF01	REF-01	
ADREF02	REF-02	
AD101	LM101	
AD558		DAC-4888
AD565	DAC-8565	
AD581		REF-01
AD586		REF-02
AD647		RC4207
AD654		RC4152
AD707		RC4077
AD708		RC4277
AD741	RC741	
AD767		DAC-4881
AM686		RC4805
AM6012	DAC-6012	
CA124	LM124	
CA324	LM324	
CA139	LM139	
CA339	LM339	
CA741	RC741	
CS3842		RC4190
CMP-04		LM139
CMP-05		RC4805
DAC-08	DAC-08	
DAC-10	DAC-10	
DAC-80		DAC-4881
DAC-100		DAC-10
DAC-312	DAC-6012	
DAC0800	DAC-08	
DAC0801	DAC-08	
DAC0830		DAC-4888
DAC-888		DAC-4888
DAC1208		DAC-4881
DAC1218		DAC-6012
DAC1219		DAC-6012
DAC1230		DAC-4881
DAC8222		DAC-4881
HA-OP27	OP-07	
HA-OP27	OP-27	
HA-OP37	OP-37	
HA-3182	RC3182	
HA-4741	RC4741	
HA-5147		OP-47
HSOP07	OP-07	
HSOP27	OP-27	
HSOP37	OP-37	

INDUSTRY TYPE	RAYTHEON DIRECT REPLACEMENT	RAYTHEON FUNCTIONAL REPLACEMENT
ICL7660		RC4391
ICL7680		RC4190
ICL8013		RC4200
LF155	LF155	
LF156	LF156	
LF157	LF157	
LH2101	LH2101	
LH2108	LH2108	
LH2111	LH2111	
LM101	LM101	
LM111	LM111	
LM108	LM108	
LM124	LM124	
LM148	LM148	
LM324	LM324	
LM331		RC4152
LM348	LM348	
LM368-5.0		REF-02
LM368-10		REF-01
LM369		REF-01
LM607		RC4077
LM741	RC741	
LM833	RC5532	
LM1458		RC4558
LM1851	LM1851	
LM1851		RC4145
LM2900	LM2900	
LM2901		LM339
LM2902		LM324
LM3900	LM3900	
LP165	LP165	
LP365	LP365	
LT-1001	LT-1001	
LT-1012	LT-1012	
LT-1012		RC4097
LT-1019		REF-01
LT-1019		REF-02
LT-1024		RC4207
LT-1028		OP-37
LT-1054		RC4391
LT-1070		RC4190
LT-1084		RC4292
MAX400		RC4077
MAX630	RC4193	
MAX630		RC4190
MAX634	RC4391	
MC1741	RC741	
MC1747	RC747	
MC3403	RC3403	
MC4558	RC4558	

General Cross References (Continued)

INDUSTRY TYPE	RAYTHEON DIRECT REPLACEMENT	RAYTHEON FUNCTIONAL REPLACEMENT
MC4741	RC4741	
MPREF01	REF-01	
MPREF02	REF-02	
MPOP07	OP-07	
MPOP27	OP-27	
MPOP37	OP-37	
MP108	LM108	
MP155	LM155	
MP156	LM156	
MP157	LM157	
NE5532	RC5532	
NE5534	RC5534	
OPA156		LM156
OPA27		OP-27
OPA37		OP-37
OP-02		RC741
OP-04		RC747
OP-07	OP-07	
OP-14		RC4558
OP-16		LF156
OP-27	OP-27	
OP-37	OP-37	
OP-77	OP-77	
OP-97		RC4097
OP-200		RC4207, RC4277
OP-207		RC4207
OP-227		RC4227
OP-270		RC4227
PM-108	LM108	
PM-139	LM139	
PM-148	LM148	
PM-155	LM155	
PM-156	LM156	
PM-157	LM157	
PM-339	LM339	
PM-348	LM348	
PM-741	RC741	
PM-747	RC747	
RC4136	RC4136	
RC4151	RC4151	
RC4152	RC4152	
RC4558	RC4558	
RC4559	RC4559	
REF-01	REF-01	
REF-02	REF-02	
REF-05		REF-02
REF-10		REF-01
SE5534		RC5534
SG101	LM101	
SG124	LM124	

INDUSTRY TYPE	RAYTHEON DIRECT REPLACEMENT	RAYTHEON FUNCTIONAL REPLACEMENT
SG741	RC741	
SI-9100		RC4292
SSM-2134		RC5534
TA7504	RC741	
TA75339	LM339	
TL494		RC4190
TL496		RC4190
TL497		RC4190
TL510		RC4805
TSC9400		RC4151
TSC9401		RC4151
TSC9402		RC4151
UC1842		RC4292
VFC-32		RC4153
XR-2207	XR-2207	
XR-2208		RC4200
XR-2211	XR-2211	
XR-3403	RC3403	
XR-4136	RC4136	
XR-4194	RC4194	
XR-4195	RC4195	
XR-5532	RC5532	
XR-5534	RC5534	
µA101	LM101	
µA108	LM108	
µA111	LM111	
µA124	LM124	
µA139	LM139	
µA148	LM148	
µA324	LM324	
µA339	LM339	
µA348	LM348	
µA741	RC741	
µA747	RC747	

Precision Operational Amplifier Cross Reference

ANALOG DEV.	RAYTHEON	PACKAGE
AD OP-07AH	*OP-07AT	TO-99
AD OP-07AH/883	*OP-07AT/883B	TO-99
AD OP-07CN	*OP-07CN	PLASTIC
AD OP-07CR	*OP-07CM	SO-8
AD OP-07Q/883	*OP-07D/883B	CERAMIC
AD OP-07DN	*OP-07DN	PLASTIC
AD OP-07EN	*OP-07EN	PLASTIC
AD OP-07H	*OP-07T	TO-99
AD OP-07H/883	*OP-07T/883B	TO-99
AD OP-07Q	*OP-07D	CERAMIC
AD OP-07AQ	*OP-07AD	CERAMIC
AD OP-07AQ/883B	*OP-07AD/883B	CERAMIC
AD OP-27AH	OP-27AT	TO-99
AD OP-27AH/883	OP-27AT/883B	TO-99
AD OP-27AQ	OP-27AD	CERAMIC
AD OP-27AQ/883	OP-27AD/883B	CERAMIC
AD OP-27BH	OP-27BT	TO-99
AD OP-27BH/883	OP-27BT/883B	TO-99
AD OP-27BQ	OP-27BD	CERAMIC
AD OP-27BQ/883	OP-27BD/883B	CERAMIC
AD OP-27CH	OP-27CT	TO-99
AD OP-27CH/883	OP-27CT/883B	TO-99
AD OP-27CQ	OP-27CD	CERAMIC
AD OP-27CQ/883	OP-27CD/883B	CERAMIC
AD OP-27EN	OP-27EN	PLASTIC
AD OP-27FN	OP-27FN	PLASTIC
AD OP-27GN	OP-27GN	PLASTIC
AD OP-37AE	OP-37AL	LCC
AD OP-37AE/883	OP-37AL/883B	LCC
AD OP-37AH	OP-37AT	TO-99

ANALOG DEV.	RAYTHEON	PACKAGE
AD OP-37AH/883	OP-37AT/883B	TO-99
AD OP-37AQ	OP-37AD	CERAMIC
AD OP-37AQ/883	OP-37AD/883B	CERAMIC
AD OP-37BH	OP-37BT	TO-99
AD OP-37BH/883	OP-37BT/883B	TO-99
AD OP-37BQ	OP-37BD	CERAMIC
AD OP-37BQ/883	OP-37BD/883B	CERAMIC
AD OP-37CH	OP-37CT	TO-99
AD OP-37CH/883	OP-37CT/883B	TO-99
AD OP-37CQ	OP-37CD	CERAMIC
AD OP-37CQ/883	OP-37CD/883B	CERAMIC
AD OP-37EN	OP-37EN	PLASTIC
AD OP-37FN	OP-37FN	PLASTIC
AD OP-37GN	OP-37GN	PLASTIC
AD707AQ	*RC4077FD	CERAMIC
AD707CH	*RM4077AT	TO-99
AD707CH/883	*RM4077AT/883B	TO-99
AD707CQ	*RM4077AD	CERAMIC
AD707CQ/883	*RM4077AD/883B	CERAMIC
AD707JN	*RC4077FN	PLASTIC
AD707JR	*RC4077FM	SO-8
AD707KN	*RC4077EN	PLASTIC
AD707KR	*RC4077EM	SO-8
AD707SH	*RC4077AT	TO-99
AD707SH/883B	*RC4077AT/883B	TO-99
AD707SQ	*RC4077AD	CERAMIC
AD707SQ/883	*RC4077AD/883B	CERAMIC
AD707TH	*RC4077AT	TO-99
AD707TH/883B	*RC4077AT/883B	TO-99
AD707TQ	*RC4077AD	CERAMIC
AD707TQ/883	*RC4077AD/883B	CERAMIC

BURR BROWN	RAYTHEON	PACKAGE
OPA27AJ/883	*OP-27AT/883B	TO-99
OPA27BJ/883	*OP-27BT/883B	TO-99
OPA27CJ	*OP-27CT/883B	TO-99
OPA27AJ	*OP-27AT	TO-99
OPA27AZ	*OP-27AD	CERAMIC
OPA27BJ	*OP-27BT	TO-99
OPA27BZ	*OP-27BD	CERAMIC
OPA27CJ	*OP-27CT	TO-99
OPA27CZ	*OP-27CD	CERAMIC
OPA27EP	*OP-27EN	PLASTIC
OPA27FP	*OP-27FN	PLASTIC
OPA27GP	*OP-27GN	PLASTIC
OPA27GU	*OP-27GM	SO-8
OPA27GZ	*OP-27GD	CERAMIC
OPA27AZ/883	*OP-27AD/883B	CERAMIC
OPA27BZ/883	*OP-27BD/883B	CERAMIC
OPA27CZ/883	*OP-27CD/883B	CERAMIC

BURR BROWN	RAYTHEON	PACKAGE
OPA37AJ	*OP-37AT	TO-99
OPA37AJ/883	*OP-37AT/883B	TO-99
OPA37AZ	*OP-37AD	CERAMIC
OPA37AZ/883	*OP-37AD/883B	CERAMIC
OPA37BJ	*OP-37BT	TO-99
OPA37BJ/883	*OP-37BT/883B	TO-99
OPA37BZ	*OP-37BD	CERAMIC
OPA37BZ/883	*OP37-BD/883B	CERAMIC
OPA37CJ	*OP-37CT	TO-99
OPA37CJ/883	*OP-37CT/883B	TO-99
OPA37CJ/883	*OP-37CD/883B	CERAMIC
OPA37CZ	*OP-37CD	CERAMIC
OPA37EP	*OP-37EN	PLASTIC
OPA37FP	*OP-37FN	PLASTIC
OPA37GP	*OP-37GN	PLASTIC
OPA37GU	*OP-27GM	SO-8

* Denotes functionally equivalent types.

Precision Operational Amplifier Cross Reference (Continued)

LTC	RAYTHEON	PACKAGE
OP-07AH	OP-07AT	TO-99
OP-07AH/883B	OP-07AT/883B	TO-99
OP-07AJ8	OP-07AD	CERAMIC
OP-07AJ8/883B	OP-07AD/883B	CERAMIC
OP-07CN8	OP-07CN	PLASTIC
OP-07CS8	OP-07CM	SO-8
OP-07EN8	OP-07EN	PLASTIC
OP-07H	OP-07T	TO-99
OP-07H/883B	OP-07T/883B	TO-99
OP-07J8	OP-07D	CERAMIC
OP-07J8/883B	OP-07D/883B	CERAMIC
OP-27AH	OP-27AT	TO-99
OP-27AH/883B	OP-27AT/883B	TO-99
OP-27AJ8	OP-27AD	CERAMIC
OP-27AJ8/883B	OP-27AD/883B	CERAMIC
OP-27CH	OP-27CT	TO-99
OP-27CH/883B	OP-27CT/883B	TO-99
OP-27CJ8	OP-27CD	CERAMIC
OP-27CJ8/883B	OP-27CD/883B	CERAMIC
OP-27EN8	OP-27EN	PLASTIC
OP-27GN8	OP-27GN	PLASTIC
OP-37AH	OP-37AT	TO-99
OP-37AH/883B	OP-37AT/883B	TO-99
OP-37AJ8	OP-37AD	CERAMIC
OP-37AJ8/883B	OP-37AD/883B	CERAMIC
OP-37CH	OP-37CT	TO-99
OP-37CH/883B	OP-37CT/883B	TO-99
OP-37CJ8	OP-37CD	CERAMIC
OP-37CJ8/883B	OP-37CD/883B	CERAMIC
OP-37EN8	OP-37EN	PLASTIC
OP-37GN8	OP-37GN	PLASTIC
LM108AH	LM108AT	TO-99
LM108AH/883B	LM108AT/883B	TO-99
LM108AJ8/883B	LM108AD/883B	CERAMIC
LM108H	LM108T	TO-99
LM108H/883B	LM108T/883B	TO-99
LM108J8/883B	LM108D/883B	CERAMIC
LT1001ACH	LT-1001ACT	TO-99
LT1001ACN8	LT-1001ACN	PLASTIC
LT1001AMH/883B	LT-1001AMT/883B	TO-99
LT1001AMJ8	LT-1001AMD	CERAMIC
LT1001AMJ8/883	LT-1001AMD/883B	CERAMIC
LT1001CH	LT-1001CT	TO-99
LT1001CN8	LT-1001CN	PLASTIC
LT1001CS8	LT-1001CM	SO-8
LT1001MH	LT-1001MT	TO-99
LT1001MH/883B	LT-1001MT/883B	TO-99
LT1001MJ8	LT-1001MD	CERAMIC
LT1001MJ8/883B	LT-1001MD/883B	CERAMIC
OP-227EN	*RC4227FN	PLASTIC
OP-227GN	*RC4227GN	PLASTIC
OP-227AJ	*RM4227BD	CERAMIC
OP-227AJ/883B	*RM4227BD/883B	CERAMIC

*Denotes functionally equivalent types.
NOTE: LTC OP-227 contains two die in a 14-pin package.
Raytheon's 4227 is a monolithic IC in an 8-pin package.

Precision Operational Amplifier Cross Reference (Continued)

PMI	RAYTHEON	PACKAGE
OP07AJ	OP-07AT	TO-99
OP07AJ/883	OP-07AT/883B	TO-99
OP07AZ	OP-07AD	CERAMIC
OP07AZ/883	OP-07AD/883B	CERAMIC
OP07CP	OP-07CN	PLASTIC
OP07CS	OP-07CM	SO-8
OP07DP	OP-07DN	PLASTIC
OP07DS	OP-07DM	SO-8
OP07EP	OP-07EN	PLASTIC
OP07J	OP-07T	TO-99
OP07J/883	OP-07T/883B	TO-99
OP07RC/883	OP-07L/883B	LCC
OP07Z	OP-07D	CERAMIC
OP07Z/883	OP-07D/883B	CERAMIC
OP27AJ	OP-27AT	TO-99
OP27AJ/883	OP-27AT/883B	TO-99
OP27AZ	OP-27AD	CERAMIC
OP27AZ/883	OP-27AD/883B	CERAMIC
OP27BJ	OP-27BT	TO-99
OP27BJ/883	OP-27BT/883B	TO-99
OP27BRC/883	OP-27BL/883B	LCC
OP27BZ	OP-27BD	CERAMIC
OP27BZ/883	OP-27BD/883B	CERAMIC
OP27CJ	OP-27CT	TO-99
OP27CJ/883	OP-27CT/883B	TO-99
OP27CZ	OP-27CD	CERAMIC
OP27CZ/883	OP-27CD/883B	CERAMIC
OP27EP	OP-27EN	PLASTIC
OP27FP	OP-27FN	PLASTIC
OP27FS	OP-27FM	SO-8
OP27GS	OP-27GM	SO-8
OP27GP	OP-27GN	PLASTIC
OP37AJ	OP-37AT	TO-99
OP37AJ/883	OP-37AT/883B	TO-99
OP37AZ	OP-37AD	CERAMIC
OP37AZ/883	OP-37AD/883B	CERAMIC
OP37BJ	OP-37BT	TO-99
OP37BJ/883	OP-37BT/883B	TO-99
OP37BRC/883	OP-37BL/883B	LCC
OP37BZ	OP-37BD	CERAMIC
OP37BZ/883	OP-37BD/883B	CERAMIC
OP37CJ	OP-37CT	TO-99
OP37CJ/883	OP-37CT/883B	TO-99
OP37CZ	OP-37CD	CERAMIC
OP37CZ/883	OP-37CD/883B	CERAMIC
OP37EP	OP-37EN	PLASTIC
OP37FP	OP-37FN	PLASTIC

PMI	RAYTHEON	PACKAGE
OP77AJ	OP-77AT	TO-99
OP77AJ/883	OP-77AT/883B	TO-99
OP77AZ	OP-77AD	CERAMIC
OP77AZ/883	OP-77AD/883B	CERAMIC
OP77BJ	OP-77BT	TO-99
OP77BJ/883	OP-77BT/883B	TO-99
OP77BRC/883	OP-77BL/883B	LCC
OP77BZ	OP-77BD	CERAMIC
OP77BZ/883	OP-77BD/883B	CERAMIC
OP77EP	OP-77EN	PLASTIC
OP77FP	OP-77FN	PLASTIC
OP77FS	OP-77FM	SO-8
OP77GP	OP-77GN	PLASTIC
OP77GS	OP-77GM	SO-8
PM108AZ	LM108AD	CERAMIC
PM108AZ/883	LM108AD/883B	CERAMIC
PM108AJ	LM108AT	TO-99
PM108AJ/883	LM108AT/883B	TO-99
PM108ARC	LM108AL	LCC
PM108ARC/883	LM108AL/883B	LCC
PM108DZ	LM108D	CERAMIC
PM108DZ/883	LM108D/883B	CERAMIC
PM108J	LM108T	TO-99
PM108J/883	LM108T/883B	TO-99
PM2108AQ	LH2108AD	CERAMIC
PM2108AQ/883	LH2108AD/883B	CERAMIC
PM2108Q	LH2108D	CERAMIC
PM2108Q/883	LH2108D/883B	CERAMIC
OP207AY/883	*RM4207BD/883B	CERAMIC
OP207AY	*RM4207BD	CERAMIC
OP227AY	*RM4227BD	CERAMIC
OP227AY/883	*RM4227BD/883B	CERAMIC
OP227BY/883	*RM4227BD/883B	CERAMIC
OP227GY	*RC4227GN	PLASTIC

* Denotes functionally equivalent types.
NOTE: PMI"s OP207/227 contains two die in a 14-pin package.
Raytheon's 4207/4227 is a monolithic IC in an 8-pin package.

General Purpose Operational Amplifier Cross Reference

Raytheon	PMI	FSC	AMD	Motorola	National	RCA	Signetics	T.I.
LH2101A			LH2101A		LH2101A		LH2101A	
LH2111			LH2111		LH2111			
LM101A		μA101A	LM101A	LM101A	LM101A	CA101A	LM101A	
LM111		μA111	LM111	LM111	LM111	CA111	LM111	
LM124		μA124	LM124	LM124	LM124	CA124	LM124	LM124
LM139	PM139	μA139	LM139	LM139	LM139	CA139	LM139	LM139
LM148	PM148	μA148	LM148		LM148		LM148	
LM301A		μA301A	LM301A	LM301A	LM301A	CA301A	LM301A	LM301A
LM324		μA324	LM324	LM324	LM324	CA324	LM324	LM324
LM339	PM339	μA339	LM339	LM339	LM339	CA339	LM339	LM339
LM348		μA348	LM348		LM348		LM348	LM348
LM2900		μA2900			LM2900			
LM3900		μA3900			LM3900			LM3900
RC3403A		μA3403		MC3403				MC3403
RC4136	OP-09	μA4136						RC4136
RC4156		μA148*		MC4741	LM348*			LM348*
RC4157		μA148/348*		MC4741*	LM348*			LM348*
RC4558		μA4558		MC4558				RC4558
RC4559		μA4558*		MC4558*				RC4559
RC4741N				MC3-4741-5				
RM4741D				MC1-4741-2				
RC5532							NE5532	NE5532
RC5532A							NE5532A	NE5532A
RC5534							NE5534	NE5534
RC5534A							NE5534A	NE5534A
RC741	OP-02	μA741		MC1741	LM741	CA741	CA741	
RC747	OP-04	μA747		MC1747	LM747	CA747	CA747	
RC747S	OP-04	μA747			LM747			

*Functional Equivalent

Data Conversion Cross Reference

Raythen	PMI	AMD	Motorola	NSC	Devices	Analog Power	Micro-Datel
DAC-08AD	DAC-08AQ	AMDAC-08AQ	MC1408L8	DAC-08AQ	AD-1508-9D	MP-7523*	DAC-IC8BC*
DAC-08D	DAC-08Q	AMDAC-08Q		DAC-08Q	AD-1508-9D	MP-7523*	DAC-IC8BC*
DAC-08ED	DAC-08EQ	AMDAC-08EQ		DAC-08EQ	AD-1408-8D	MP-7523*	DAC-IC8UP*
DAC-08EN	DAC-08EP	AMDAC-08EN		DAC-08EP			DAC-IC8UP*
DAC-08CN	DAC-08CP	AMDAC-08CN	MC1408P6	DAC-08CP			DAC-IC8UP*
DAC-10BD	DAC-10BX			DAC-1020 LD*	AD7520/ 30/33*	MP-7520/ 30/33*	DAC- HF10BMM*
DAC-10CD	DAC-10CX			DAC-1021/ 22LD8*	AD7520/ 30/33*	MP-7520/ 30/33*	DAC- HF10BMM*
DAC-10FD	DAC-10FX			DAC-1020 LCN*	AD7520/ 30/33*	MP-7520 30/33*	DAC- HF10BMC*
DAC-10GD	DAC-10GX			DAC-1021/ 22LCN*	AD7520/ 30/33*	MP-7520/ 30/33*	DAC-HF10BMC*
DAC- 6012AMD		AM6012ADM		DAC-1220 LD*	AD6012ADM	MP-7531/ 41*	DAC-HF12BMM*
DAC- 6012MD	DAC-312 BR*	AM6012DM		DAC-1221/ 22LD*	AD6012DM	MP-7531/ 41*	DAC- HF12BMM*
DAC- 6012ACN		AM6012ADC		DAC-1220 LCN*	AD6012ADC	MP-7531/ 41*	DAC- HF12BMC*
DAC- 6012CN	DAC-312FR*	AM6012DC		DAC-1221/ 22LCN*	AD6012DC	MP-7531/ 41*	DAC- HF12BMC*
DAC-8565DS*				MC3412L	DAC-1208 AD-I*	AD565JD/ BIN	
DAC-8565JS*				MC3412L	DAC-1280 HCD-I*	AD565JD/ BIN	
DAC-8565SS*					DAC-1280 HCD-I*	AD565SD/ BIN	

*Functional Equivalent

Special Functions Cross Reference

Raytheon	Teledyne	Analog Devices	EXAR	Motorola	Datel	Burr Brown
RC4151	4780*	AD451*	XR4151		VFQ-1C*	VFC-32KF*
RC4152	4781*	AD452*	XR4151*		VFQ-2C*	VFC-42BP*
RC4153	4782*	AD537*			VFQ-3C*	VFC-52BP*
RC4200/A		AD539*		MC1494*		4202K* & 4205K*
XR2207			XR2207			
XR2211			XR2211			
RC4444				MC3416		

*Functional Equivalent

Voltage Regulator and Voltage Reference Cross Reference

Raytheon	EXAR	Maxim	T.I.	Analog Devices	Motorola	NSC
REF-01	REF-01		MP-5501	AD581*	MC1504AU10*	LH0070-0*
REF-01A	REF-01A		MP-5501A	AD581*		LH0070-1*
REF-01C	REF-01C		MP-5501C	AD581*	MC1404U10*	LH0070-2*
REF-01D	REF-01D		MP-5501D	AD581*	MC1404U10*	
REF-01E	REF-01E		MP-5501E	AD581*		
REF-01H	REF-01H		MP-5501H	AD581*	MC1404AU10*	
REF-02	REF-02		MP-5502		MC1504AU5*	LM136-5.0*
REF-02A	REF-02A		MP-5502A			LM136A-5.0*
REF-02C	REF-02C		MP-5502C		MC1404U5*	LM336-5.0*
REF-02D	REF-02D		MP-5502D		MC1404U5*	LM336-5.0*
REF-02E	REF-02E		MP-5502E			LM336A-5.0*
REF-02H	REF-02H		MP-5502H		MC1404AU5*	
RC4190		MAX630*				
RC4193		MAX630*				
RC4391		MAX634*				
RC4194	XR4194CN					
RC4195	XR4195CP				MC1468/ MC1568*	LM325/326*

*Functional Equivalent

IC packages and pin connections

Not all circuits give power connections and pin locations for ICs, but this information can be obtained from manufacturers' data sheets. Also, looking through other circuits might turn up another diagram on which the desired connections are shown for the same IC.

The following diagrams show some typical pin connections. Notice that the functions shown in the following diagrams apply only to a specific IC, and are included to show the normal pin-numbering sequence only. As shown, numbering normally starts with 1 (looking from the top) for the first pin *counterclockwise* from the notched (or otherwise marked) end, and continues in sequence. The highest number is next to the notch (or mark) on the other side of the IC.

Notice that these guides show all of the common pin-connection configurations, including metal can, DIP (dual-in-line), SO DIP, LCC (leadless chip carrier), multipin DIP, and surface mount.

(Raytheon Linear Integrated Circuits, 1989, p. 4–98, and Harris Semiconductors, Data Acquisition, 1991, p. 2–131)

Connection Information

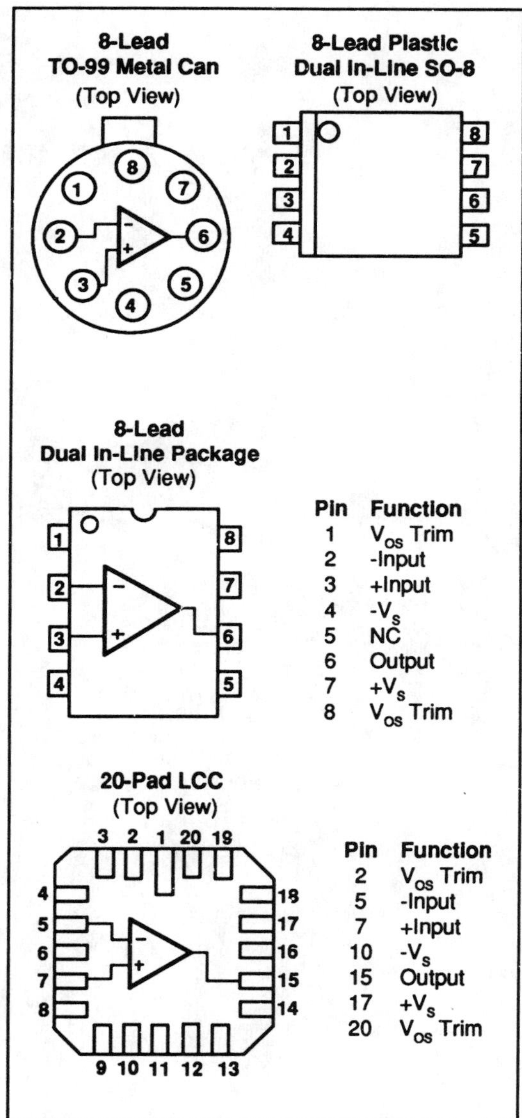

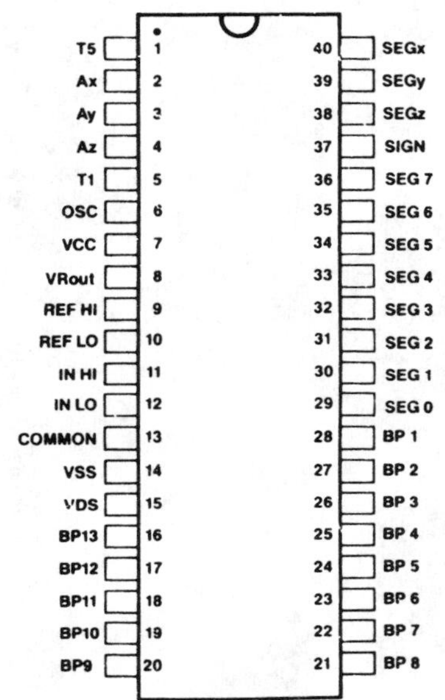

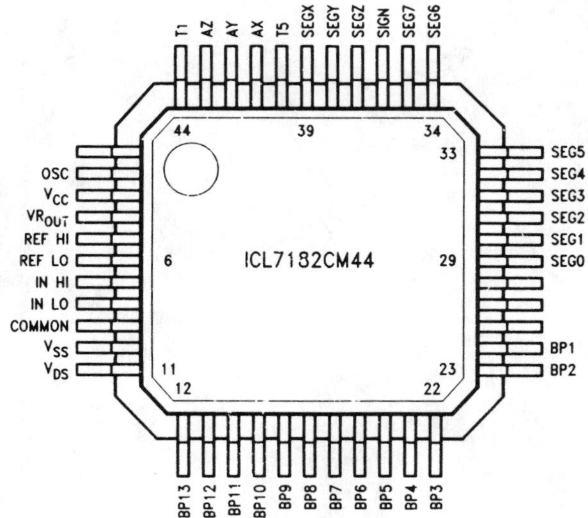

Abbreviations and reference symbols

Most non-US electronics manufacturers and publications use some different reference symbols for electronic components, as well as a different abbreviation system for values or units of measure.

For example, on Fig. 2-33A, notice the resistor symbol (a rectangular box) at pin 2 of the SL6442. Also notice the capacitor at the same pin. One half of the symbol is solid and the other half is open. This is used whenever *polarity* must be observed when connecting a capacitor into the circuit. When polarity is of no particular concern, both halves of the symbol are solid, as shown for the capacitor at pin 5 of the SL6442.

Also, the abbreviations for component values are simplified. Thus, μ after a capacitor value represents μF, n is nF and p is pF. (The Japanese often go one step further and use a lower-case u instead of the micro symbol.)

With resistor values, k is thousands of ohms, M is megohms, and the absence of a unit of measure is ohms (the ohms symbol is omitted to save time). If μ, n, and p are used with an inductance or coil value, they represent μH, nH and pH (such as the 18 nH and 82 nH coils at pin 6 of the SL6442 in Fig. 2-33A).

For a decimal value, the letter for the unit of measure is sometimes placed at the location of the decimal point. Thus, 3k3 is 3.3 kilohms, or 3,300 ohms. 2M2 is 2.2 megohms, 7μ7 is 7.7 μF, 0μ1 is 0.1 μF, and 3n7 is 3.7 nF.

Finding circuits

The circuits are arranged by circuit type or function, with each group assigned to a separate chapter. For example, chapter 1 contains audio circuits, chapter 2 contains RF circuits, and so on. The table of contents and circuit titles list each chapter in order, as well as the title of each circuit in that chapter.

To find a particular circuit, start by noting the chapter in which the circuit is likely to appear. Then, look for a title that best describes the circuit you want. For example, if you are interested in audio amplifiers, start by looking in chapter 1. If you want discrete-component audio amplifiers, note that the titles for Figs. 1-15 through 1-21 indicate discrete-component circuits. If you need to find an IC audio amplifier, note that the title for Fig. 1-45 is a line-operate IC audio circuit.

If you want to test or troubleshoot a particular circuit start with the following table of contents, and find the chapter for that circuit group. For example, if you want to test the audio circuits of Figs. 1-15 through 1-21, or of Fig. 1-45, note that such tests are described in the first section of chapter 1, including a discussion of test equipment and procedures. If the circuit fails to perform properly, note that the second section of chapter 1 describes audio-circuit troubleshooting, including some step-by-step examples.

List of figures

1

Audio, ultrasonic, and direct-current circuits

This chapter is devoted to audio, ultrasonic, and direct-current circuits, including dc bridges. It is assumed that you are already familiar with audio basics (amplifier principles, bias operating points, etc.), practical considerations (heatsinks, power dissipation, component-mounting techniques), simplified audio design (frequency limitations), and basic audio test/troubleshooting. If not, read *Lenk's Audio Handbook*, McGraw-Hill, 1991. However, the following paragraphs summarize both test and troubleshooting of audio circuits. This information is included so that even those readers not familiar with electronic procedures can both test the circuits described here, and localize problems if the circuits fail to perform as shown.

Audio circuit tests

This section covers the basic tests for audio circuits, as well as ultrasonic and direct-current circuits. The section starts with a review of typical audio test equipment, and then goes on to describe test procedures that can be applied to the circuits of this chapter, using the actual circuit test results where practical. If the circuits pass these basic tests, the circuits can be put to immediate use. If not, the tests provide a starting point for the troubleshooting procedures that are described in the next section.

Typical audio test equipment

Test/troubleshooting for the circuits in this chapter can be performed using meters, generators, scopes, power supplies, and assorted clips, patch cords, and so on. So, if you have a good set of test equipment that is suitable for other electronic work, you can probably get by. A possible exception is a distortion meter. Also, here are some points that you should consider when selecting and using audio test equipment.

Matching test equipment to the circuits

No matter what test instrument is involved, try to match the capabilities of the test equipment to the circuit. For example, if pulses, square waves, or complex waves are to be measured, a peak-to-peak meter can possibly provide meaningful indications, but a scope is the logical instrument.

Audio voltmeters

In addition to making routine voltage and resistance checks, the main functions for voltmeters in audio work are to measure frequency response and to trace audio signals from input to output. Many technicians prefer scopes for these procedures. The reasoning is that scopes also show distortion of the waveforms during measurement or signal tracing. Other technicians prefer the simplicity of a meter, particularly in such procedures as voltage-gain and power-gain measurements.

It is possible that you can get by with any ac meter (even a multimeter, analog or digital) for all audio work. However, for accurate measurements, I recommend a wideband ac meter, preferably a dual-channel model. The dual-channel feature makes it possible to monitor both channels of a stereo circuit simultaneously. This is particularly important for stereo frequency-response and cross-talk measurements.

Typical audio-voltmeter characteristics The typical audio voltmeter has ranges (at least 10, possibly 12) that cover audio voltages from about 0.2 mV to 100 V. The usual wideband frequency permits measurements from 5 Hz to 1 MHz. High sensitivity permits measurements down to 30 μV (and lower in some meters).

Audio meters have two voltage scales: a dB scale and a dBm scale (0 dBm = 1 mW across 600 Ω). These scales are used for making relative measurements (gain, attenuation, etc.). Typical ranges are −90 to +40 for dB and −90 to +42 for dBm.

Most audio meters have built-in low-distortion amplifiers that are used to drive the meter movements. This allows the meter to be used as a calibrated, high-gain preamp. Typically, the output is calibrated at 1 Vrms for a full-scale reading.

Generally, audio meters use absolute mean-value (average) sensing, but they are calibrated to read the rms value of a sine-wave voltage. The input voltage is capacitively coupled, which permits measurement of an ac signal superimposed on

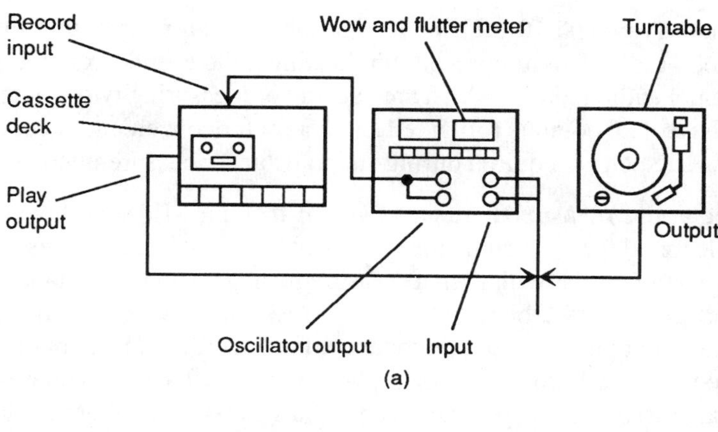

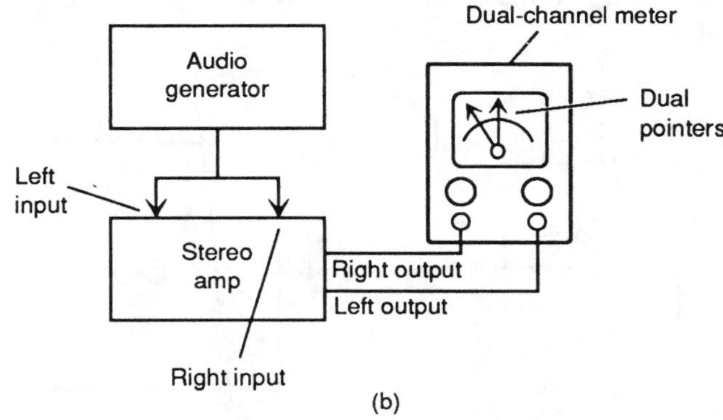

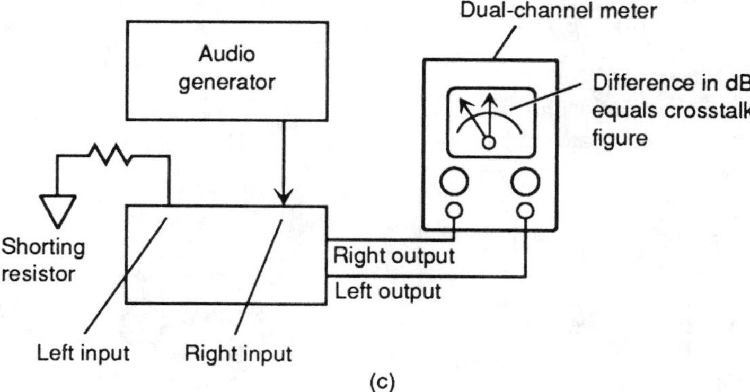

Fig. 1-A Basic test connections for stereo-response measurements.

a dc voltage. A typical 10-MΩ input impedance on all ranges with low shunting capacitance, assures minimum circuit loading to the audio circuit under test. Dual-channel audio meters, which are capable of measuring two voltages simultaneously, have dual pointers (often red and black) for convenient direct comparison of two levels (such as required during stereo balance measurements).

Basic audio measurements and signal tracing Present-day audio meters are suitable for all forms of audio measurement and signal tracing. It is assumed that you are already familiar with such tests as frequency response, voltage gain, power output and gain, power bandwidth, load sensitivity, feedback, dynamic output impedance, and dynamic input impedance. However, for those totally unfamiliar with these tests, the following paragraphs cover some basic procedures.

It is also assumed that you can trace audio signals from input to output through audio circuits, using a meter or scope. If not, you might have a problem

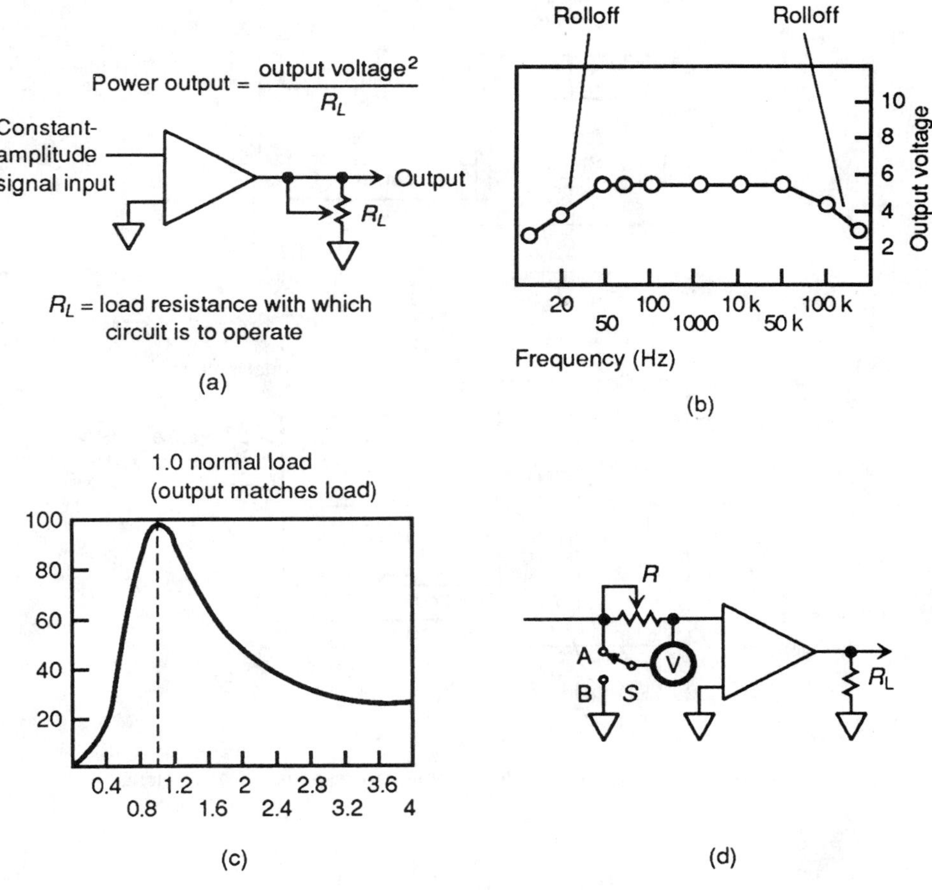

Fig. 1-B Basic audio-circuit test connections.

troubleshooting audio circuits! Before you panic, a step-by-step procedure for basic audio-circuit troubleshooting is included in this chapter.

Stereo measurements A meter with two input channels applied to a dual-pointer scale is more useful for direct-comparison measurements than two separate voltmeters. An excellent example is the measurement of left- and right-channel characteristics of stereo circuits. To show you the advantages of a dual-channel, wide-band meter, the typical stereo-crosstalk and stereo-response measurement (with a dual-channel meter) is covered here.

Figure 1-B shows the basic test connections for stereo-response measurement. With equal audio signals applied to the left- and right-channel inputs of the stereo circuit, set the controls to the desired range.

It might be necessary to set a ground-mode switch for minimum hum. For a stereo circuit (Fig. 1-A) that has no connection common to the left and right channels, this usually means setting the ground-mode switches to open. In most other cases (where there is some connection common to both channels), use the ground position of the ground-mode switch.

With the connections made and controls set, vary the frequency of the audio generator. Notice any difference in frequency response between the left and right channels, as indicated by unequal deflection of the two pointers. Even those not thoroughly familiar with frequency-response measurements will see how much simpler a dual-pointer meter is to use (than two separate meters).

Figure 1-C shows the basic test connections for stereo-crosstalk measurements. With an audio signal applied to one channel and the other channel shorted, set the controls to the desired range. Again, it might be necessary to set a ground-mode switch for minimum hum. Notice that the value of the shorting resistor should be equal to the input impedance of the amplifier.

With the connections made and controls set, notice the difference in dB between the signal level measured on the channel with the signal input. This is the *crosstalk figure*. *Crosstalk* is generally specified in dB (such as −40 dB).

For a complete crosstalk measurement, reverse the input connections to the stereo circuits and repeat the measurement. There should be no substantial difference in the meter readings with the input connections reversed. Make certain that the signal level, frequency, and all other input connections are identical when the connections are reversed.

Distortion meters

If you are already in audio and stereo service work, you probably have *distortion meters* (and know how to use them effectively). There are two types of distortion measurement: *harmonic* and *intermodulation*. No particular meter is described here. Instead, descriptions are included of how harmonic and intermodulation distortion measurements are made.

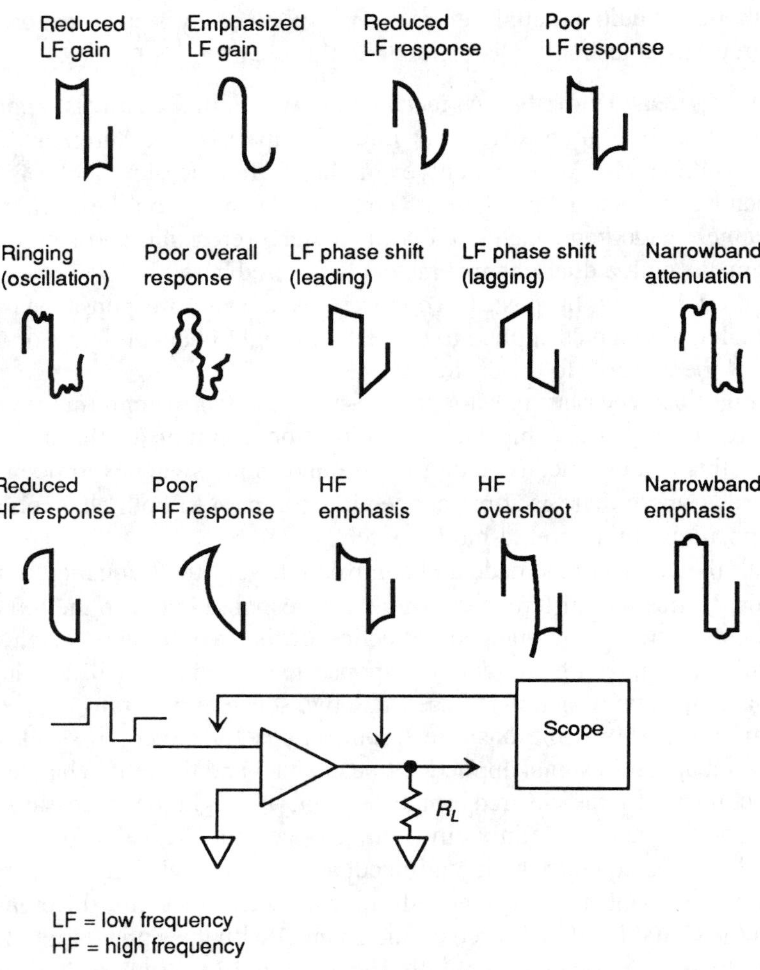

Fig. 1-C Basic square-wave distortion analysis.

Scopes

The *scopes* used for audio work should have the same characteristics as for other electronic circuits. If you have a good scope for TV and VCR work, use that scope for all audio-circuit measurements and troubleshooting. If you are considering a new scope, remember that a dual-channel instrument permits you to monitor both channels of a stereo circuit (as is the case with a dual-channel voltmeter). A scope has the advantage over a meter in that the scope can display such common audio-circuit conditions as distortion, hum, ripple, and oscillation (described in the following paragraphs). However, the meter is easier to read.

Basic audio-circuit tests

Most of the circuits described in this chapter contain audio amplifiers of some kind. For that reason, it is essential that you understand how to make basic audio-amplifier tests and measurements. The following paragraphs describe the basics of such procedures and are included here for those who are totally unfamiliar with audio work. Adjustments for specific audio circuits are given in the related circuit description.

Frequency response

The *frequency response* of an audio circuit can be measured with a generator and a meter or scope. When a meter is used, the generator is tuned to various frequencies, and the resultant output response is measured at each frequency. The results are then plotted in the form of a graph or response curve (Fig. 1-B). The procedure is essentially the same when a scope is used to measure audio-circuit response. The scope gives the added benefit of a visual analysis of distortion, as discussed throughout the following paragraphs.

The basic procedure for measurement of frequency response (with either meter or scope) is to apply a constant-amplitude signal while monitoring the circuit output. The input signal is varied in frequency (but not in amplitude) across the entire operating range of the circuit. The voltage output at various frequencies across the range is plotted on a graph:

1. Connect the equipment (Fig. 1-B).
2. Initially, set the generator frequency to the low end of the range. Then, set the generator amplitude to the desired input level.
3. In the absence of a realistic test input voltage, set the generator output to an arbitrary value. A simple method of finding a satisfactory input level is to monitor the circuit output (with a meter or scope) and increase the generator amplitude at the circuit's center frequency (or at 1 kHz) until the circuit is overdrive. This point is indicated when further increases in generator output do not cause further increases in meter reading (or the output waveform peaks begin to flatten on the scope display). Set the generator output just below this point. Then, return the meter or scope or monitor the generator voltage (at the circuit input) and measure the voltage. Keep the generator at this voltage throughout the test.
4. If the circuit is provided with any operating or adjustment controls (volume, loudness, gain treble, balance, and so on), set the controls to some arbitrary point when making the initial frequency-response measurement. The response measurements can then be repeated at different control settings, if desired.

5. Record the circuit-output voltage on the graph. Without changing the generator amplitude, increase the generator frequency by some fixed amount, and record the new circuit-output voltage. The amount of frequency increase between each measurement is an arbitrary matter. Use an increase of 10 Hz, where rolloff occurs and 100 Hz at the middle frequencies.
6. Repeat the process; check and record the circuit-output voltage at each of the check points. Figure 1-41B shows the frequency response for the amplifier of Fig. 1-41A.
7. After the initial frequency-response check, the effect of operating or adjustment controls should be checked. Volume, loudness, and gain controls should have the same effect all across the frequency range. Treble and bass controls might have some effect on all frequencies. However, a treble control should have the greatest effect at the high end, whereas bass controls should have a greatest effect at the low end.
8. Notice that the generator output can vary with changes in frequency (a fact that is often overlooked in making frequency-response tests during troubleshooting). Even precision lab generators can vary in output with changes in frequency, and thus produce a considerable error. The generator output should be monitored after each change in frequency (many generators have a built-in output meter). Then, if necessary, the generator-output amplitude can be reset to the correct value. It is more important for the generator amplitude to remain constant, rather than set at some specific value when making a frequency-response check.

Voltage gain

Voltage gain in an audio amplifier is measured in the same way as frequency response. The ratio of output voltage to input voltage (at any given frequency or across the entire frequency range) is the voltage gain. Because the input voltage (generator output) is held constant for a frequency-response test, a voltage-gain curve should be identical to a frequency-response curve (such as shown in Fig. 1-41B).

Power output and power gain

The power output of an audio amplifier is found by noting the output voltage across the load resistance (Fig. 1-B) at any frequency or across the entire frequency range. The power output is calculated as shown in Fig. 1-B. For example, if the output voltage is 10 V across a load resistance of 50 Ω, the power output is $10^2/50 = 2$ W.

Never use a wire-wound component (or any component that has reactance) for the load resistance. Reactance changes with frequency, and causes the load to change. Use a composition resistor or a potentiometer for the load.

To find the power gain of an amplifier, it is necessary to find both the input and the output power. Input power is found in the same way as output power, except that the input impedance must be known (or calculated). Calculating input impedance is not always practical in some circuits—especially in designs where input impedance depends on transistor gain. (The procedure for finding dynamic input impedance is described later in this chapter.) With input power known (or estimated), the power gain is the ratio of output power to input power.

Input sensitivity In some audio-amplifier circuits, the input-sensitivity specification is used. Input-sensitivity specifications require a minimum power output with a given voltage input (such as 100-W output with a 1-V input).

Power bandwidth

Many audio circuits include a power-bandwidth specification. Such specifications require that the audio circuit deliver a given power output across a given frequency range. For example, a circuit might produce full-power output up to 20 kHz—even though the frequency response is flat up to 100 kHz. That is, voltage (without a load) remains constant up to 100 kHz, whereas power output (across a normal load) remains constant up to 20 kHz. Figure 1-15F shows a typical power-bandwidth curve or graph.

Load sensitivity

An audio-amplifier circuit is sensitive to changes in load. This is particularly true of power amplifiers (including power-amplifier ICs). An amplifier produces maximum power when the output impedance is the same as the load impedance.

The test circuit for load-sensitivity measurement is the same as the circuit for frequency response (Fig. 1-Ba), except that the load resistance is variable. Again, never use a wire-wound load resistance. The reactance can result in considerable error.

Measure the power output at various load-impedance and output-impedance ratios. That is, set the load resistance to various values (including a value equal to the true amplifier-output impedance) and notice the voltage and/or power gain at each setting. Repeat the test at various frequencies.

Figure 1-Bc shows a generalized load-sensitivity response curve, and Fig. 1-15E shows the curve for an actual circuit. Notice that if the load is twice the output impedance (as indicated by a 2:1 ratio, or a normalized load impedance of 2, Fig. 1-15E), the output power is reduced to about 50 percent.

Dynamic output impedance

The load-sensitivity test can be reversed to find the dynamic output impedance of an audio circuit. The connections (Fig. 1-Ba) and the procedures are the same, except that the load resistance is varied until maximum output power is found. Power is removed, the load resistance is disconnected from the circuit, and the resistance is measured with an ohmmeter. This resistance is equal to the dynamic output impedance of the circuit (but only at the measurement frequency). The test can be repeated across the entire frequency range, if desired.

Dynamic input impedance

Use the circuit in Fig. 1-Bd to find the dynamic input impedance of an audio amplifier. The test conditions are identical to those for frequency response, power output, and so on. Move switch S between points A and B, while adjusting resistance *R* until the voltage reading is the same in both positions of S. Disconnect R and measure resistance *R*. This resistance is equal to the dynamic impedance of the amplifier input.

Accuracy of the impedance measurement depends on the accuracy with which resistance *R* is measured. Again, a noninductive (not wire-wound) resistance must be used for *R*. The impedance found by this method applies only to the frequency used during the test.

Sine-wave analysis

All amplifiers are subject to possible distortion. That is, the output signal might not be identical to the input signal. Theoretically, the output should be identical to the input, except for amplitude.

Some troubleshooting techniques are based on analyzing the waveshape of signals that pass through an amplifier to determine possible distortion. If distortion (or an abnormal amount of distortion) is present, the circuit is then checked further by the usual troubleshooting methods (localization, voltage measurement, etc.).

Amplifier distortion can be checked by sine-wave analysis. The procedures are the same as those used for signal tracing. The primary concern in distortion analysis is deviation of the output (amplifier stage) from the input waveform. If there is no change (except in amplitude), there is no distortion. If there is a change in waveform, the nature of the change will often reveal the cause of distortion. For example, the presence of second or third harmonics can distort the fundamental signal.

In practical troubleshooting, analyzing sine waves to pinpoint amplifier problems that produce distortion is a difficult job that requires considerable experience. Unless distortion is severe, it might pass unnoticed.

Sine waves are best used where harmonic-distortion or intermodulation-distortion meters are combined with the scope for distortion analysis. If a scope is used alone, square waves provide the best basis for distortion analysis (the reverse is true for frequency-response and power measurements).

Square-wave analysis

The procedure for checking distortion with square waves is essentially the same as that used with sine waves. Distortion analysis is more effective with square waves because of the high odd-harmonic content in square waves (and because it is easier to see a deviation from a straight line with sharp corners than from a curving line).

Square waves are introduced into the circuit input, and the output is monitored with a scope (Fig. 1-C). The primary concern is deviation of the output waveform from the input waveform (which is also monitored on the scope). If the scope has a dual-trace feature, the input and output can be monitored simultaneously. If there is change in waveform, the nature of the change will often reveal the cause of the distortion.

Notice that the drawings of Fig. 1-C are generalized and that same waveform can be produced by different causes. For example, poor LF (low frequency) response appears to be the same as HF (high-frequency emphasis). Figure 1-D shows the waveforms that are produced by an actual circuit. Notice that the output (trace B) does a good job of following the input (trace A) at a gain of −1. That is, the output is inverted from the input, and there is no gain, or unity gain. Also notice that there is some slight reduced high-frequency response in output trace B, but not the exaggerated response of Fig. 1-C.

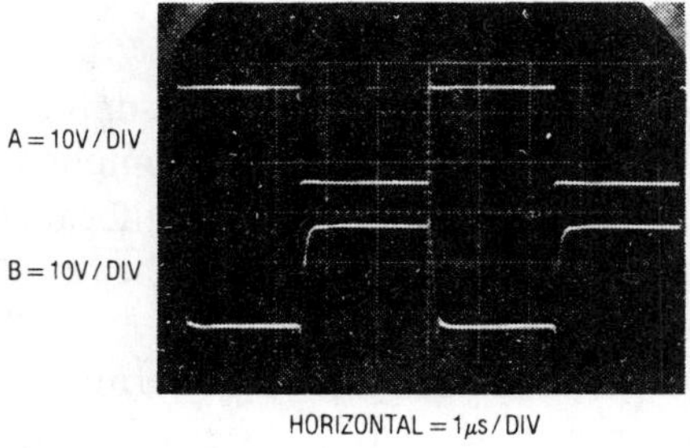

Fig. 1-D
Audio-circuit response to square waves.
Linear Technology Corporation, Linear Applications Handbook, 1991, p. AN4-3.

The third, fifth, seventh, and ninth harmonics of a clean square wave are emphasized. If an amplifier passes a given frequency and produces a clean square-wave output, it is reasonable to assume that the frequency response is good up to at least 9 times the square-wave frequency.

Harmonic distortion

No matter what amplifier circuit is used or how well the circuit is designed, there is the possibility of odd or even harmonics being present with the fundamental. These harmonics combine with the fundamental and produce distortion, as is the case when any two signals are combined. The effects of second- and third-harmonic distortion are shown in Fig. 1-E.

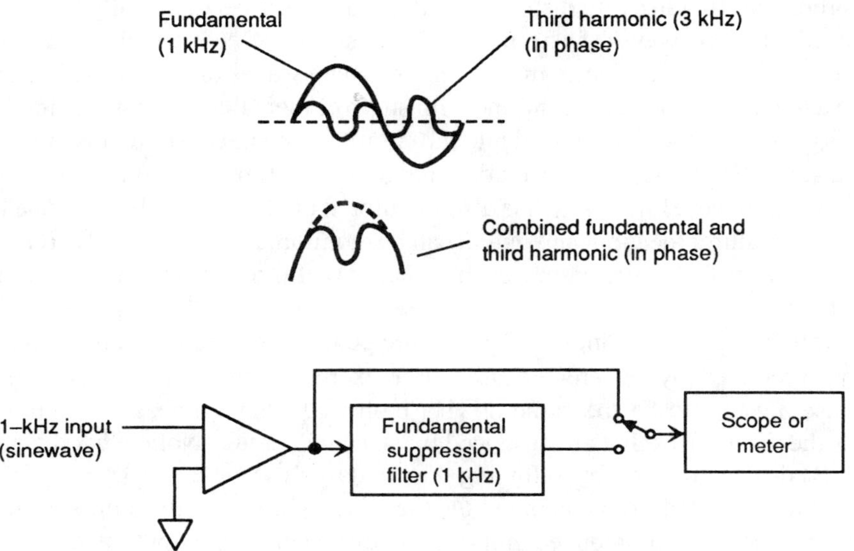

Fig. 1-E Basic harmonic-distortion analysis.

Commercial harmonic-distortion meters operate on the fundamental-suppression principle. A sine wave is applied to the amplifier input, and the output is measured on the scope (Fig. 1-E). The output is then applied through a filter that suppresses the fundamental frequency. Any output from the filter is then the result of harmonics.

The output is also displayed on the scope. Some commercial harmonic-distortion meters use a built-in meter instead of, or in addition to, an external scope. When the scope is used, the frequency of the filter-output signal is checked to determine harmonic content. For example, if the input is 1 kHz and the output (after filtering) is 3 kHz, third-harmonic distortion is indicated. Reduce the scope horizontal sweep down so that you can see one input cycle. If there are three cycles at the output for the same time period as one input cycle, this indicates third-harmonic distortion.

The percentage of harmonic distortion is also determined by this method. For example, if the output is 100 mV without the filter and 3 mV with the filter, a 3-percent harmonic distortion is indicated. Figure 1-44B shows the total harmonic distortion (or THD) of the circuit in Fig. 1-44A. Notice that THD varies with the power output of the circuit over a wide frequency range. Also notice that THD depends on load.

On some commercial harmonic-distortion meters, the filter is tunable; therefore, the amplifier can be tested over a wide range of fundamental frequencies. On other instruments, the filter is fixed in frequency, but it can be detuned slightly to produce a sharp null.

Intermodulation distortion

When two signals of different frequencies are mixed in an amplifier, it is possible that the lower-frequency signal will modulate the amplitude of the higher-frequency signal. This produces a form of distortion that is known as *intermodulation distortion*.

Commercial intermodulation-distortion meters consist of a signal generator and a high-pass filter (Fig. 1-F). The generator portion of the meter produces a higher-frequency signal (usually about 7 kHz) that is modulated by a low-frequency signal (usually 60 Hz). The mixed signals are applied to the circuit input. The

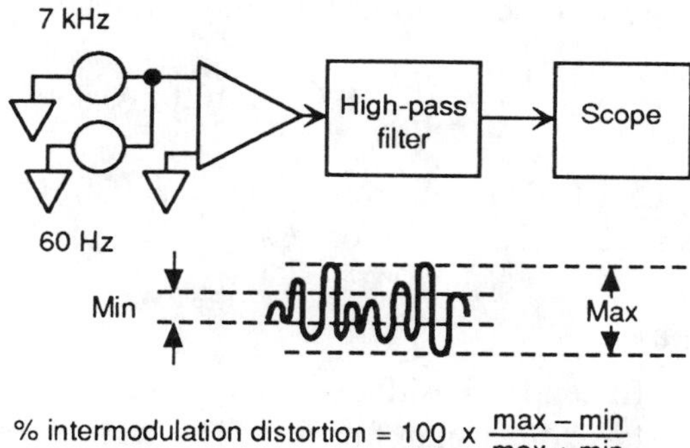

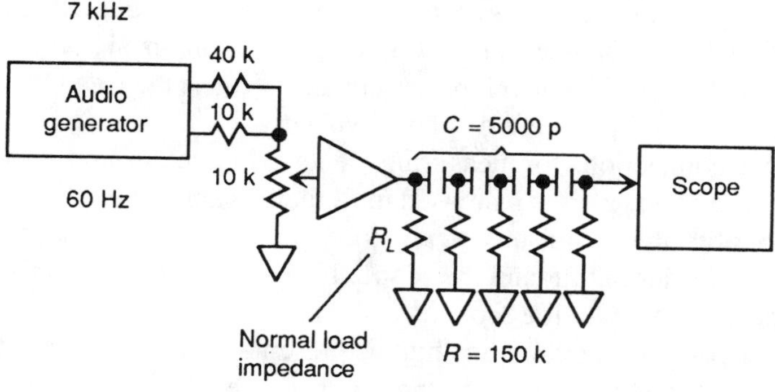

Fig. 1-F Basic intermodulation-distortion analysis.

circuit output is connected through a high-pass filter to a scope. The high-pass filter removes the low-frequency (60 Hz) signal. The only signal that appears on the scope should be the 7-kHz signal. If any 60-Hz signal is present on the scope display, the 60-Hz signal is being passed through as modulation on the 7-kHz signal.

Figure 1-F also shows an intermodulation test circuit that can be fabricated in the shop. Notice that the high-pass filter is designed to pass signals that are about 200 Hz and above. The purpose of the 40- and 10-kΩ resistors is to set the 60-Hz signal at 4 times the amplitude of the 7-kHz signal. Some audio generators provide a line-frequency output, at 60 Hz, that can be used as the low-frequency modulation source.

If the shop circuit (Fig. 1-F) is used instead of a commercial meter, set the generator line-frequency (60 Hz) output to 2 V (if adjustable) or to some value that does not overdrive the circuit that is being tested. Then, set the generator output (7 kHz) to 2 V (or to the same value as the 60-Hz output).

Calculate the percentage of intermodulation distortion using the equation shown in Fig. 1-F. For example, if the maximum output (shown on the scope) is 1 V and the minimum is 0.9 V, the percentage of intermodulation distortion (listed as IM or IMD on most specification sheets) is approximately:

$$\frac{1.0 - 0.9}{1.0 + 0.9} \quad 0.05 \times 100 = 5\%.$$

Notice that a 5% IMD is quite high (Fig. 1-20C). Many home-entertainment audio amplifiers have IMDs of 0.09 or less.

Background noise

If the scope is sufficiently sensitive, it can be used to check and measure the background-noise level of an amplifier as well as to check for the presence of hum, oscillation, and the like. The scope should be capable of a measurable deflection with an input below 1 mV (and considerably less if an IC amplifier is involved).

The basic procedure consists of measuring amplifier output with the volume or gain controls (if any) at maximum, but without an input signal. The scope is superior to a meter for noise-level measurements because the frequency and nature of the noise (or other signal) are displayed visually.

The basic connections for measuring the level of background noise are shown in Fig. 1-G. The scope gain is increased until there is a noise or "hash" indication.

It is possible that a noise indication can be caused by pickup in the leads between the amplifier output and the scope. If in doubt, disconnect the leads from the amplifier, but not from the scope.

If you suspect that a 60-Hz line hum is present in the amplifier output (picked up from the power supply or any other source), set the scope sync control (or whatever other control is required to synchronize the scope trace to the line

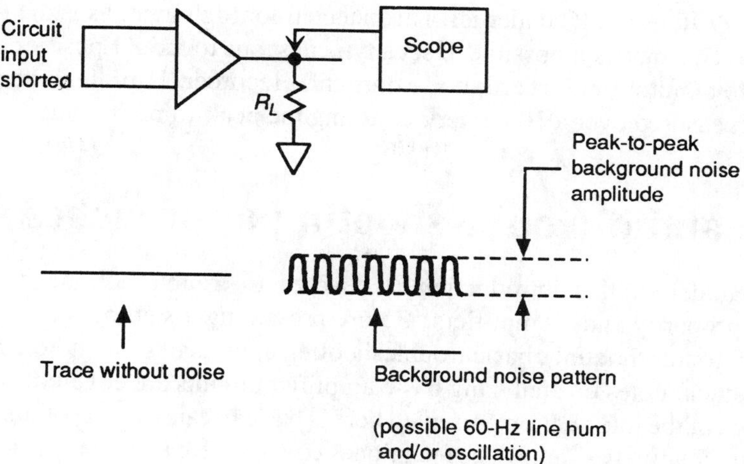

Fig. 1-G Basic audio-circuit background-nose test connections.

frequency) to line. If a stationary signal pattern appears, the signal is the result of line hum getting into the circuit.

If a signal appears that is not at the line frequency, the signal can be the result of oscillation in the amplifier or stray pickup. Short the amplifier input terminals. If the signal remains, suspect oscillation in the amplifier circuits.

With present-day IC amplifiers, the internal or background noise is considerably less than 1 mV , and it is impossible to measure—even with a sensitive scope. It is necessary to use a circuit that amplifies the output of the IC under test before the output is applied to the scope. Figure 1-H is such a circuit (and is used for noise tests

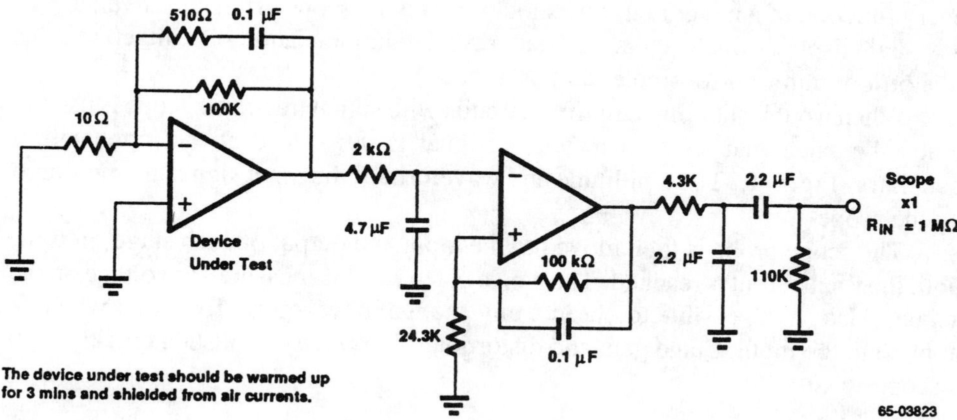

Fig. 1-H 0.1- to 10-Hz noise test circuit. Raytheon Linear Integrated Circuits, 1989, p. 4-116.

of an OP-77 IC). The IC under test is connected for high gain, as is the following amplifier. This makes it possible to set a typical scope to the X1 position. It is also possible to monitor (and record) noise on a chart recorder (Figs. 1-9B and 1-10B). Noise is measured over a 10-s interval, noting the peak-to-peak value.

Basic audio troubleshooting approaches

The remainder of this introduction is devoted to troubleshooting basic audio circuits, primarily audio amplifiers. Before presenting a step-by-step example of amplifier troubleshooting, basic troubleshooting approaches for audio circuits and some practical notes on analyzing basic amplifier circuits are covered.

Much of the information presented here is basic because most solid-state audio circuits are relatively simple. The techniques covered here are of the most benefit to those readers who are totally unfamiliar with audio troubleshooting. These techniques serve as a basis for understanding the more complex IC (or combination discrete and IC) circuits that are covered in the remaining chapters of this book.

Signal tracing

The basic troubleshooting approach for an amplifier involves *signal tracing*. The input and output waveforms of each stage are monitored on the scope or meter. Any stage that shows an abnormal waveform (in amplitude, waveshape, and so on) or the absence of an output (with a known-good input signal) points to a defect in that stage. Voltage and/or resistance measurements on all elements of the transistor (or IC) are then used to pinpoint the problem.

A scope is the most logical instrument for checking amplifier circuits (both complete amplifier systems or a single amplifier stage). The scope can duplicate every function of a meter in troubleshooting, and the scope offers the advantage of a visual display. Such a display can reveal common audio-circuit conditions (distortion, hum, noise, ripple, and oscillation).

When troubleshooting amplifier circuits with signal tracing, a scope is used in much the same manner as a meter. A signal is introduced at the input with a generator (Fig. 1-I). The amplitude and waveform of the input signal are measured on the scope.

The scope probe is then moved to the input and output of each stage, in turn, until the final output is reached. The gain of each stage is measured as voltage on the scope. Also, it is possible to observe any change in waveform from that which is applied to the input. Stage gain and distortion (if any) are established quickly with a scope.

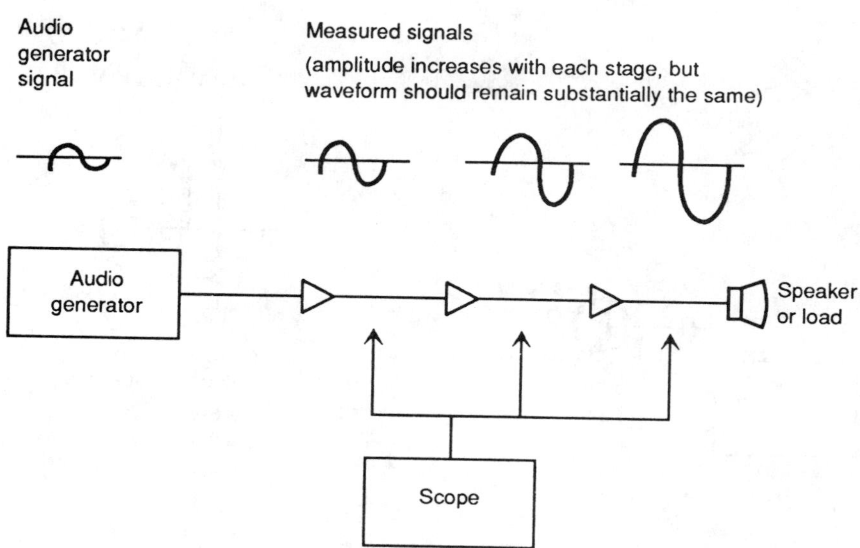

Fig. 1-I Basic audio-circuit signal tracing.

Measuring gain

Take care when measuring the gain of amplifier stages (especially in a circuit where there is feedback). For example, in Fig. 1-Ja, if you measure the signal at the base of Q1, the base-to-ground voltage is not the same as the input voltage.

To get the correct value of gain, connect the low side of the measuring device (meter or scope) to the emitter and the other lead (high side) to the base (Fig. 1-Ja). In effect, measure the signal that appears across the base-emitter junction. This measurement includes the effect of the feedback signal.

As a general safety precaution, never connect the ground lead of a meter or scope to the transistor base unless the lead connects back to an insulated inner chassis on the meter or scope. Large ground-loop currents (between the measuring device and the circuit being checked) can flow through the base-emitter junction and possibly burn out the transistor. This can usually be eliminated by an isolation transformer.

Low-gain problems

Low gain in a feedback amplifier can result in distortion. If gain is normal in a feedback amplifier, some distortion can be overcome. With low gain, the feedback might not be able to bring the distortion within limits. Of course, low gain by itself is sufficient cause to troubleshoot a circuit (with or without feedback).

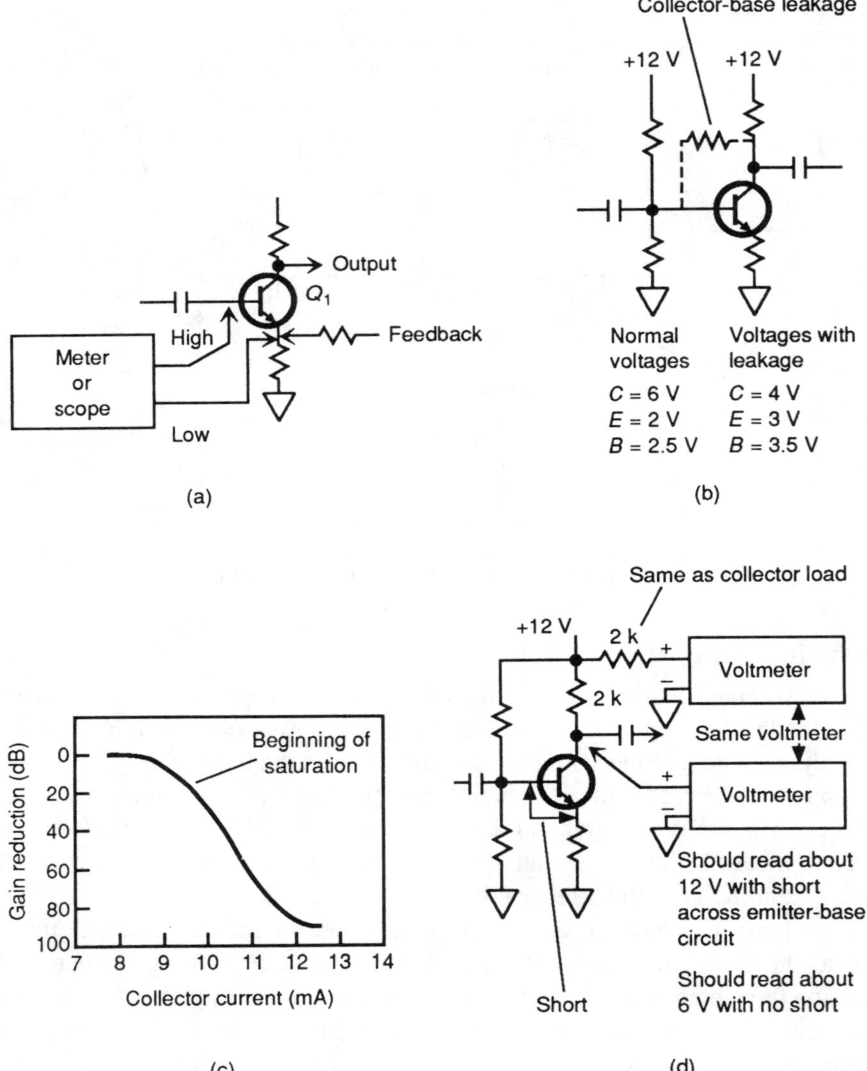

Fig. 1-J Basic audio-circuit troubleshooting techniques.

Most feedback amplifiers have a very high open-loop gain that is set to some specific value by the ratio of resistor values (feedback-resistor value to input load-resistor value). If the closed-loop gain is low in an experimental circuit, this usually indicates that the resistance values are incorrect. In an existing equipment circuit, the problem is usually where the open-loop gain has failed far enough so that the resistors no longer determine the gain.

In troubleshooting such a situation, if waveforms indicate low gain and transistor (or IC) voltages appear normal, try replacing the transistors (or ICs). Never overlook the possibility that the emitter-bypass capacitors might be open or badly leaking. If the capacitors are leaking (acting as a resistance in parallel with the emitter resistor), there is considerable negative feedback and little gain. Of course, a completely shorted emitter-bypass capacitor produces an abnormal voltage indication at the transistor emitter, typically zero volts or ground.

Distortion problems

Distortion can be caused by improper bias, overdriving (too much gain), or underdriving (too little gain, preventing the feedback signal from countering the distortion). One problem that is often overlooked in a feedback amplifier with a pattern of distortion is overdriving that results from transistor leakage. Here's why.

Generally, it is assumed that the collector-base leakage of a transistor reduces gain because the leakage is in opposition to signal-current flow. Although this is true in the case of a single stage, it might not be true when more than one feedback stage is involved. Whenever there is collector-base leakage, the base assumes a voltage nearer to that of the collector (nearer than is the case without leakage). This increases both transistor forward bias and transistor current flow. An increase in the transistor current causes a reduction in input resistance (which might or might not cause a gain reduction, depending on where the transistor is located in the circuit).

If the feedback amplifier is direct coupled, the effects of feedback are increased. This is because the operating point (set by the base bias) of the following stage is changed, which could possibly result in distortion.

Effects of leakage on circuit performance

When there is considerable leakage in a circuit, the gain is reduced to zero, and/or the signal's waveforms are drastically distorted. Such a condition also produces abnormal waveforms and transistor voltage. These indications make troubleshooting relatively easy. The troubleshooting problem becomes really difficult when there is just enough leakage to reduce circuit gain, but not enough to distort the waveform seriously (or to produce transistor voltages that are way off).

Collector-base leakage

Collector-base leakage is the most common form of transistor leakage and produces a classic condition of low gain (in a single stage). When there is any collector-base leakage, the transistor is forward-biased, or the forward bias is increased (Fig. 1-Jb).

Collector-base leakage has the same effect as a resistance between the collector and the base. The base assumes the same polarity as the collector (although at a lower value), and the transistor is forward-biased. If leakage is sufficient, the forward bias can be enough to drive the transistor into or near saturation. When a transistor is operated at or near the saturation point, the gain is reduced (for a single stage), as shown in Fig. 1-Jc.

Capacitor leakage

Capacitor leakage can also cause gain and distortion problems in audio circuits. The effects of leaking capacitors on circuits is discussed in chapter 2.

Checking transistor leakage in-circuit

If the normal operating voltages are not known, as is the case with all experimental circuits, defective transistors can appear to be good because all of the voltage relationships are normal. The collector-base junction is reverse-biased (collector more positive than base for an npn), and the emitter-base junction is forward-biased (emitter less positive than base for an npn).

A simple way to check transistor leakage is shown in Fig. 1-Jd. Measure the collector voltage to ground. Then, short the base to the emitter and remeasure the collector voltage.

If the transistor is not leaking, the base-emitter short turns the transistor off, and the collector voltage rises to the same value as the supply. If there is any leakage, a current path remains (through the emitter resistor, base-emitter short, collector-base leakage path, and collector resistor). There is some voltage drop across the collector resistor, and the collector voltage is at some value lower than the supply.

Notice that most meters draw current, and the current passes through the collector resistor when you measure (Fig. 1-Jd). This can lead to some confusion, particularly if the meter draws heavy current (has a low Ω-per-volt rating). To eliminate any doubt, connect the meter to the supply through a resistor with the same value as the collector resistor. The voltage drop, if any, should be the same as when the transistor collector is measured to ground. If the drop is much different (lower) when the collector is measured, the transistor is leaking.

For example, assume that (Fig. 1-Jd) the collector measures 4 V, with respect to ground. This means that there is an 8-V drop across the collector resistor and a collector current of 4 mA (8/2000 = 0.004). Normally, the collector is operated at about one-half the supply voltage (at about 6 V in this case). Notice that simply because the collector is at 4 V, instead of 6 V, does not make the circuit faulty. Some circuits are designed that way.

In any event, the transistor should be checked for leakage with the emitter-base short test (Fig. 1-Jd). Now assume that the collector voltage rises to 10 V when the base and emitter are shorted (within 2 V of the 12-V supply). This indicates that the transistor is cutting off but there is still some current flow through the resistor, about 1 mA (2/2000 = 0.001).

A current flow of 1 mA is high for a meter. To confirm a leaking transistor, connect the same meter through a 2-kΩ resistor (same as the collector-load resistor) to the 12-V supply (preferably at the same point where the collector resistor connects to the power supply). Now assume that the indication is 11.7 V through the external resistor. This shows that there is some transistor leakage.

The amount of transistor leakage can be estimated as follows: 11.7 − 10 = 1.7-V drop difference, and 1.7/2000 = 0.00085, or 0.85 mA. However, from a practical troubleshooting standpoint, the presence of any current flow with the transistor supposedly cut off is sufficient cause to replace the transistor.

Example of audio circuit troubleshooting

This step-by-step troubleshooting problem involves locating the defective part (or improperly connected wiring) in a combination discrete-IC audio amplifier. Figure 1-21 shows the schematic diagram. This circuit was chosen as an example because it combines both IC and discrete components. The CA3094B is a programmable amplifier (where gain is set by the resistor at pin 5) similar to the OTA ICs (described in chapter 11).

Regardless of the type of trouble symptom, the actual fault can eventually be traced to one or more of the circuit parts (transistors, ICs, diodes, capacitors, etc.), unless, of course, you have wired the parts incorrectly! Even then, the following waveform, voltage, and resistance checks will indicate which branch within the circuit is at fault.

Make the initial check. If you were servicing this circuit in an existing equipment, the first step would be to study the literature and test the circuit to confirm the trouble. In this example, our only "literature" is Fig. 1-21. The circuit description claims an output of 12-W into an 8-Ω load. Although the load is shown as R_L, it can be assumed that the circuit will be used with an 8-Ω speaker. There are no test points, waveforms, the voltage information is incomplete, and there is no resistance-to-ground information. However, with this fragmentary information, you can test the circuit, monitor the signals at various points in the circuit, and localize trouble using the test results.

The first step is to apply a signal at the input and monitor the output. The input can be applied at C1 (as shown). The output is measured at R_L, or at an 8-Ω speaker connected in place of R_L. Use the resistor or the speaker, but never operate the circuit without a load. Q2 and Q3, and possibly Q1, can be destroyed if operated without a load.

To produce a 12-W output across an 8-Ω load, a 10-V signal must be at the speaker, and at the junction of the Q2/Q3 emitters. $E^2/R = W$, $10^2/8 = 12.5$ W. If the circuit has a 40-dB gain (as claimed), 0.1 V (100 mV) at the input should be sufficient to fully drive the speaker, depending on the setting of R1.

Connect an audio generator to the input and set the generator to produce 0.1 V at a frequency of 1 kHz (or some other frequency in the audio range). Set both the

Bass and Treble controls to midrange, and adjust R1 until you get a good tone on the speaker and/or a readable signal at the Q2/Q3 emitter junction. Adjust R1 for a 10-V signal at the speaker or emitter-junctions. The tone will probably burst your eardrums at this point. Adjust R1 until the tone is reasonable, and vary both the Bass and Treble controls. Both tone controls should have some effect on the tone, but the Bass control should have the most control. Change the generator frequency to 10 kHz, and repeat the tone-control test. Now, the Treble control should have the most effect.

If the circuit operates as described thus far, it is reasonable to assume that the circuit is good. Quit while you are ahead! If you have access to distortion meters, check distortion against the performance data on Fig. 1-21. Also check the actual voltage input (at pin 2 of the IC) when a 10-V signal appears at the output, and determine the true amplifier voltage gain (which should be 40 dB).

If the circuit does not operate as described, set R1 and the Bass and Treble controls to midrange, and monitor the signal voltages at pins 2 and 8 of the IC. You can monitor all test points with an ac voltmeter, dc voltmeter with a rectifier probe, or with a scope. I use the scope because any really abnormal distortion at the test points appears on the scope display (as does the voltage).

If there is a signal at pin 2 of the IC, but not at pin 8 (or if the signal at pin 8 shows little gain over the pin-2 signal), the problem is at the IC portion of the circuit. Check all voltages at the IC. You do not know the exact values, but there are some hints. The transformer has a 26.8-V center-tapped secondary, so V+ and V− should be about 12-15 V (and should be substantially the same). In any event, pins 4 and 6 of the IC should be about −12 V, and pin 7 should be about +12 V (although pin 7 will probably be lower than pins 4 and 6 because of the 5600-Ω resistor at pin 7).

If the IC voltages appear to be good, but the gain at pin 8 is low, it is possible that the IC is bad, that the resistor at pin 5 is not of the correct value (this resistor determines IC gain), or that there is too much feedback at pin 3 (from the output through C2 and the tone controls).

If there is a good signal at pin 8 of the IC, but not at the output (speaker or R_L), the problem is at the discrete portion of the circuit (Q1, Q2, and Q3 or the associated circuit parts). Check the collector voltages of Q2 and Q3. These voltages should be about 12 V, and should be substantially the same, except of opposite polarity. Also, the voltages should be substantially the same as at pins 4, 6, and 7 of the IC.

The waveform (signal) and voltage checks that are described here should be sufficient to locate any major defect in the circuit, including bad wiring. Of course, if the circuit operates, but performance is not as claimed, it is possible that the problem is one of poor physical layout, wrong component values, and so on. Also, the basic techniques that are described here can be applied to the other circuits of this chapter.

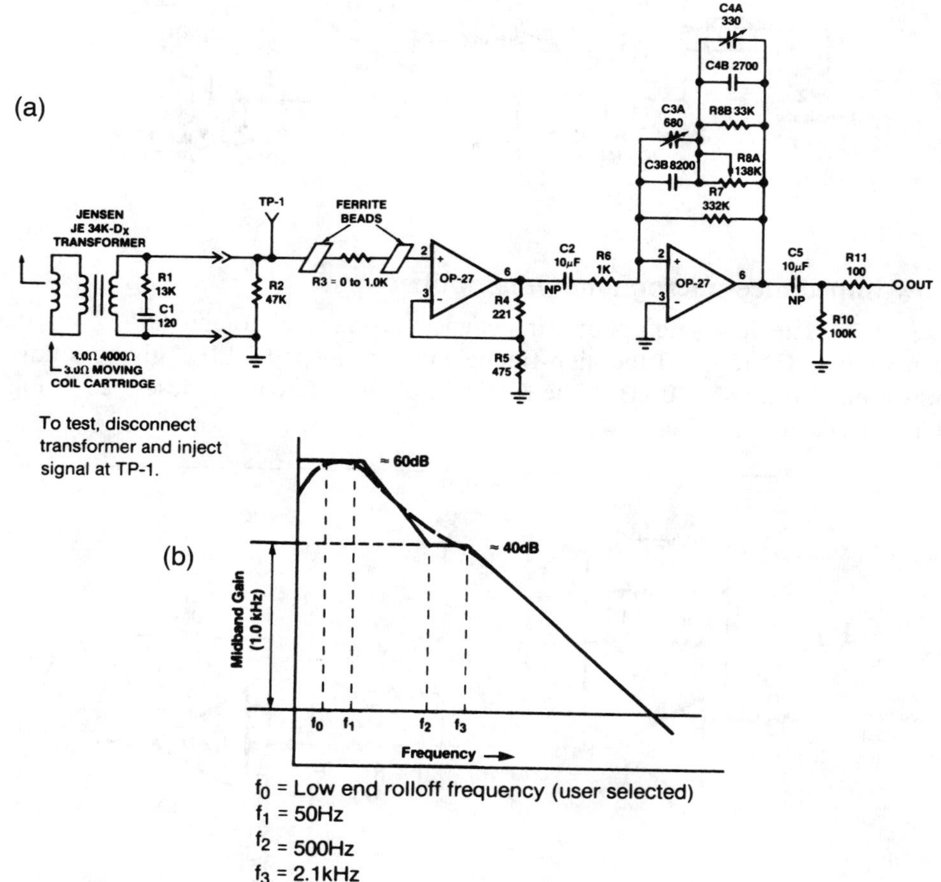

f_0 = Low end rolloff frequency (user selected)

f_1 = 50Hz

f_2 = 500Hz

f_3 = 2.1kHz

RIAA phono preamplifier

Fig. 1-1 The circuit of Fig. 1-1A is adjusted to match a 40-dB RIAA curve (Fig. 1-1B). With break points at 50, 500, and 2100 Hz, the entire curve is fixed by a specified gain at 1 kHz. The circuit is designed for use with newer moving-coil magnetic phono cartridges, which have a sensitivity of 0.1 mV per CM/S (compared to older cartridges with sensitivities of 1 mV per CM/S). The circuit is adjusted by injecting a low-level signal into TP-1, with the transformer disconnected. At 100 Hz, adjust C3A for an output level that is 6 dB lower than the low-frequency output. At 1000 Hz, adjust R8A for an output level 20 dB lower than the low-frequency output. At 21 kHz, adjust C4A for an output that is 40 dB less than the low-frequency output. Raytheon Linear Integrated Circuits, 1989, p. 4-78, 4-79.

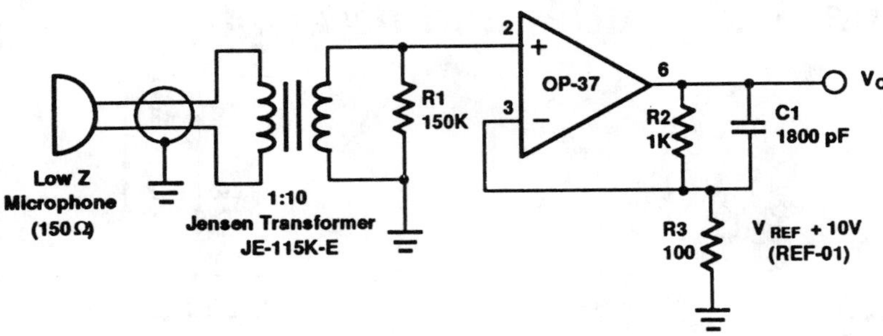

Low-impedance microphone preamplifier

Fig. 1-2 The low microphone impedance is matched to the OP-37 by the transformer. C1 rolls off the high-frequency response at 90 kHz, giving a noise power bandwidth of 140 kHz. The REF-01 voltage reference is described in Fig. 4-1. Raytheon Linear Integrated Circuits, 1989, p. 4-94.

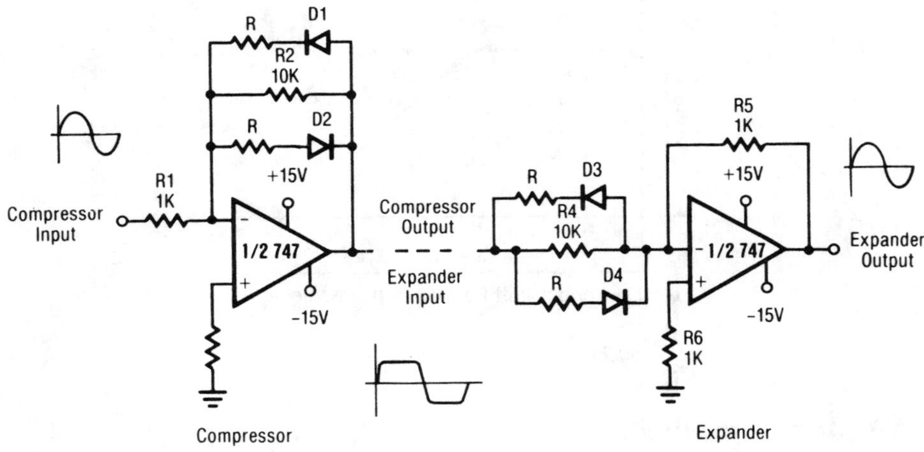

Compressor/expander amplifiers

Fig. 1-3 The maximum compression/expansion ratio of these circuits is set by the ratio of R1/R, as shown. Raytheon Linear Integrated Circuits, 1989, p. 4-149.

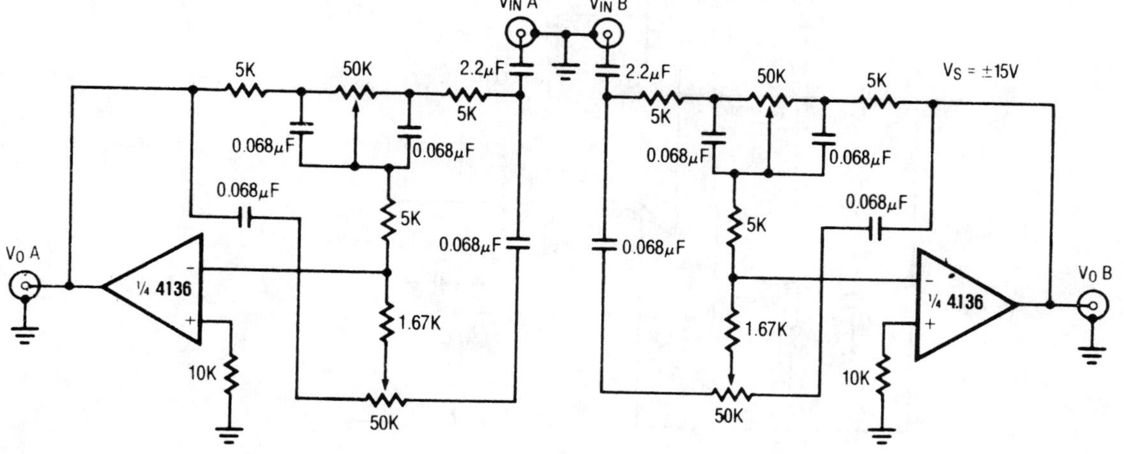

Stereo tone control

Fig. 1-4 This circuit uses only one-half of the 4136 (Fig. 1-5). Raytheon Linear Integrated Circuits, 1989, p. 4-172.

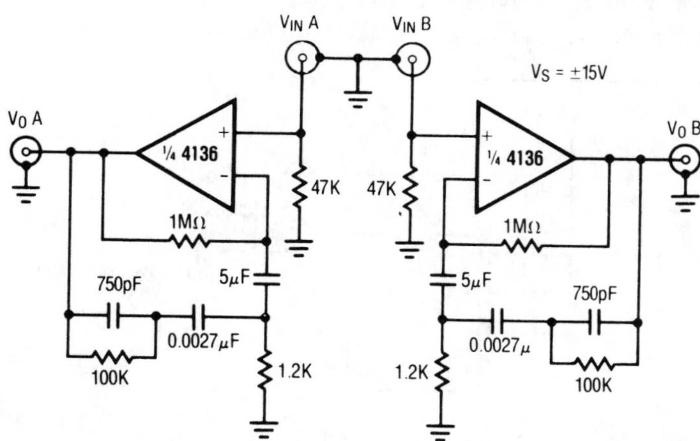

RIAA preamplifier

Fig. 1-5 This circuit also uses only one half of the 4136 (Fig. 1-4). Raytheon Linear Integrated Circuits, 1989, p. 4-173.

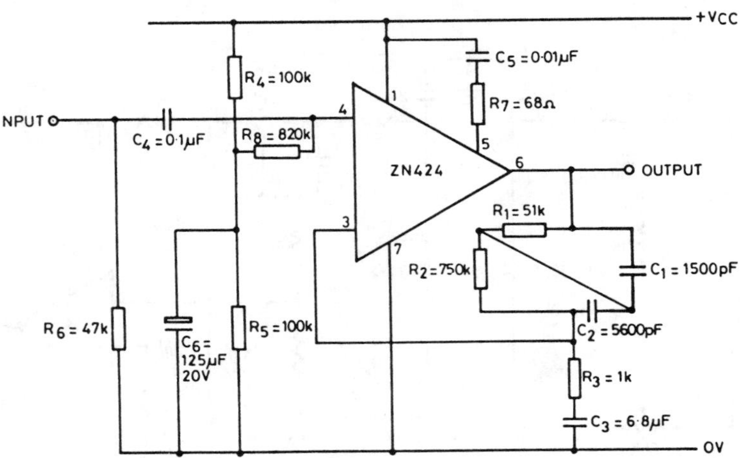

Magnetic cartridge preamplifier

Fig. 1-6 This circuit provides an approximate gain of 50 (34 dB) at 1 kHz with a distortion of 0.004%. If more gain is required, R_3 can be decreased, but C_3 must be increased proportionally to avoid loss of bias. C1, C2, R1, R2 provide RIAA equalization. C3 and R3 provide a rumble filter. The S/N ratio is better than 70 dB below a 5-mV input. GEC Plessey Semiconductors, Professional Products, 1991, p. 1-82.

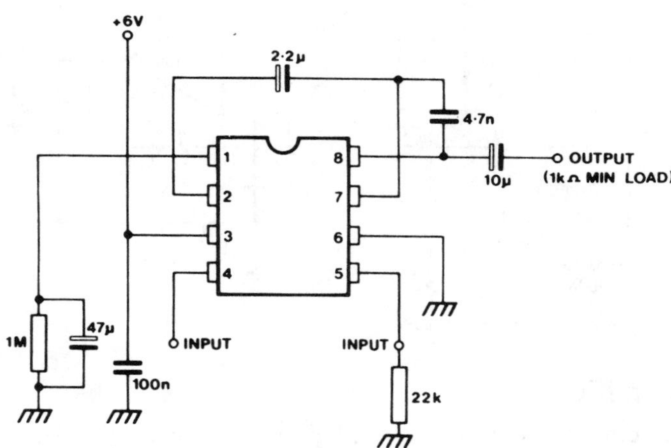

Gain-controlled microphone preamplifier/VOGAD

Fig. 1-7 The SL6270 shown in this figure combines the functions of audio amplifier and voice-operated gain-adjusting device (VOGAD). The SL6270 accepts signals from a low-sensitivity microphone and provides an essentially constant output signal for a 50-dB range of input. The dynamic range, attack time, and decay time are controlled by external components. With the values shown, attack time (time taken for the output to return to within 10% of the original level following a 20-dB increase in input level) is about 20 mS, decay rate is 20 dB/s, voltage gain is 52 dB with 72 μVrms input at pin 4, output level = 90 mVrms with 4 mVrms input at pin 4, and an input impedance of 150 Ω. GEC Plessey Semiconductors, Professional Products, 1991, p. 1-147.

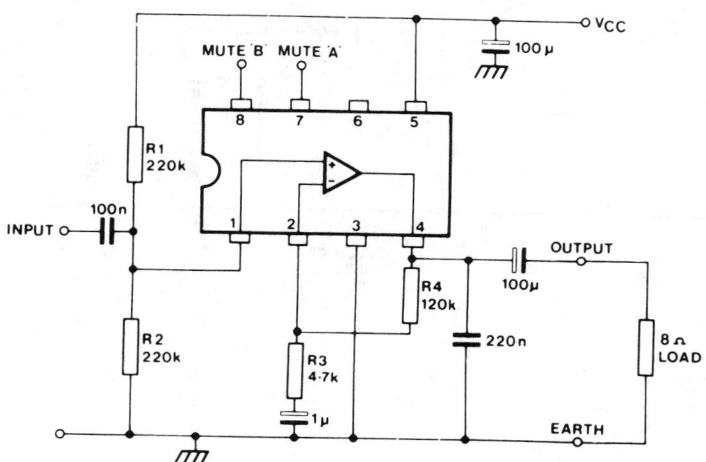

Switchable audio amplifier

Fig. 1-8 The SL6310 shown in this figure is a low-power audio amplifier that can be switched off by applying a mute signal to pins 7 or 8. Despite the low quiescent current consumption of 5 mA (only 0.6 mA when muted), a minimum output power of 400 mW is available into an 8-Ω load from a 9-V supply. With the values shown, the voltage gain is about 25, determined by the ratio of $(R_3 + R_4)/R_3$, with an input impedance of about 100 kΩ, determined by R_1 and R_2. GEC Plessey Semiconductors, Professional Products, 1991, p. 1-149.

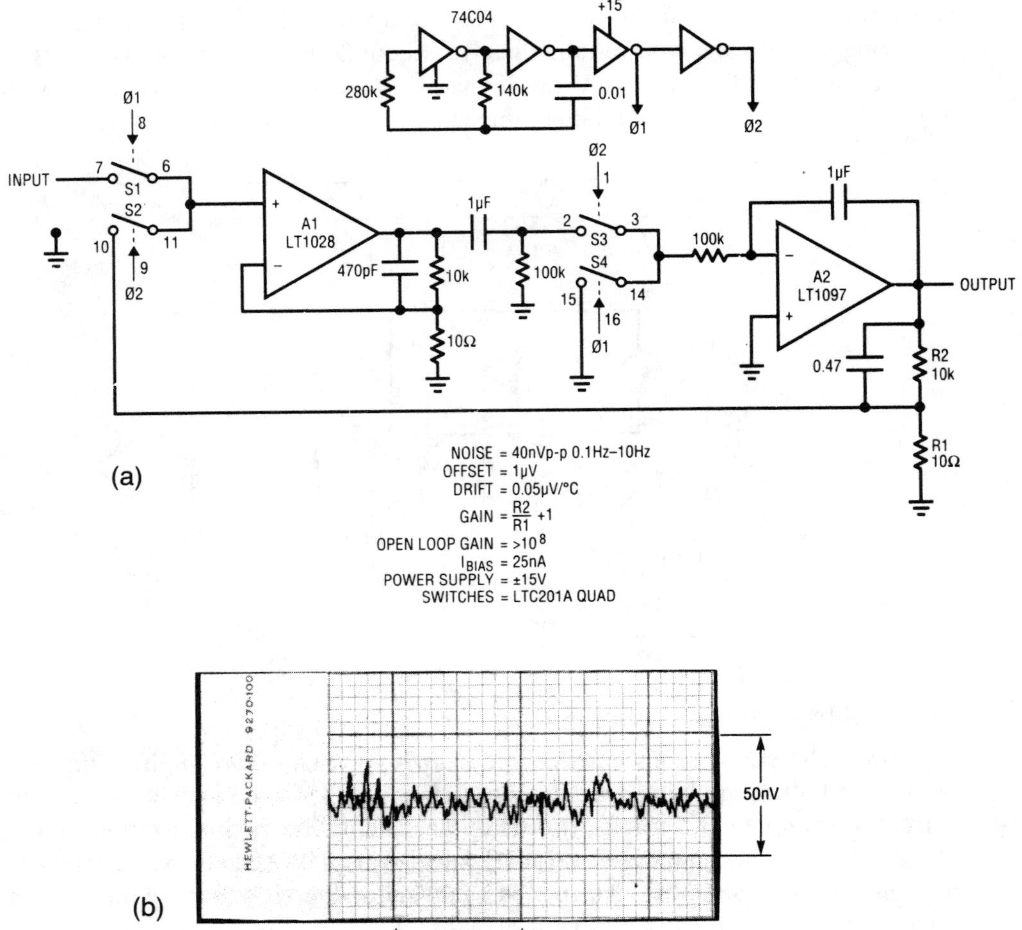

Low noise and drift dc amplifier

Fig. 1-9 This circuit combines the low noise of an LT1028 with a chopper-based carrier-modulation scheme to get a very low-noise, low-drift dc amplifier. As shown in Fig. 1-9B, noise in a 0.1- to 1-Hz bandwidth is less than 40 nV with 0.05 μV/°C drift. In general, to maintain this low-noise performance, the source resistance should be kept below 500 Ω. Such resistance is typical of transducers (strain-gauge bridges, magnetic detectors, etc.). Linear Technology Corporation, 1991, AN45-2.

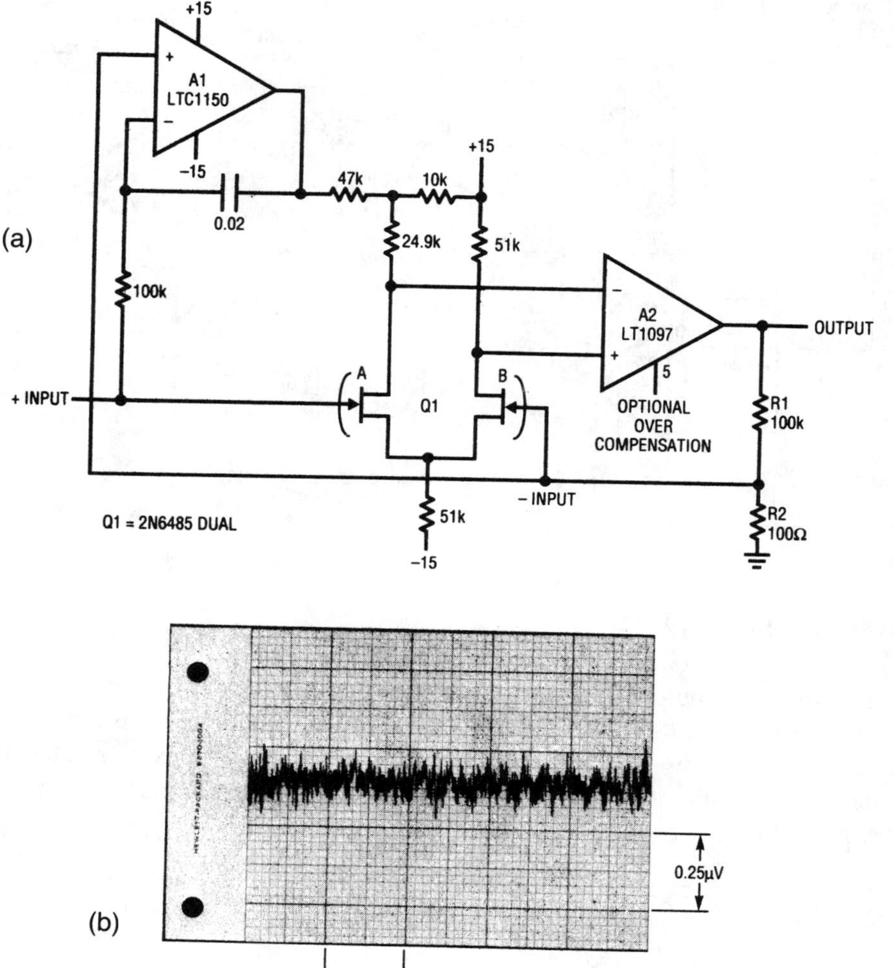

Low noise and drift FET dc amplifier

Fig. 1-10 This circuit combines the low drift of a chopper-stabilized amplifier with a pair of FETs. As shown in Fig. 1-10B, the result is an amplifier with 0.05-μV/°C drift, offset within 5 μV, 50-pA bias current, and 200-nV noise in a 0.1- to 10-Hz bandwidth. Linear Technology Corporation, 1991, AN45-3.

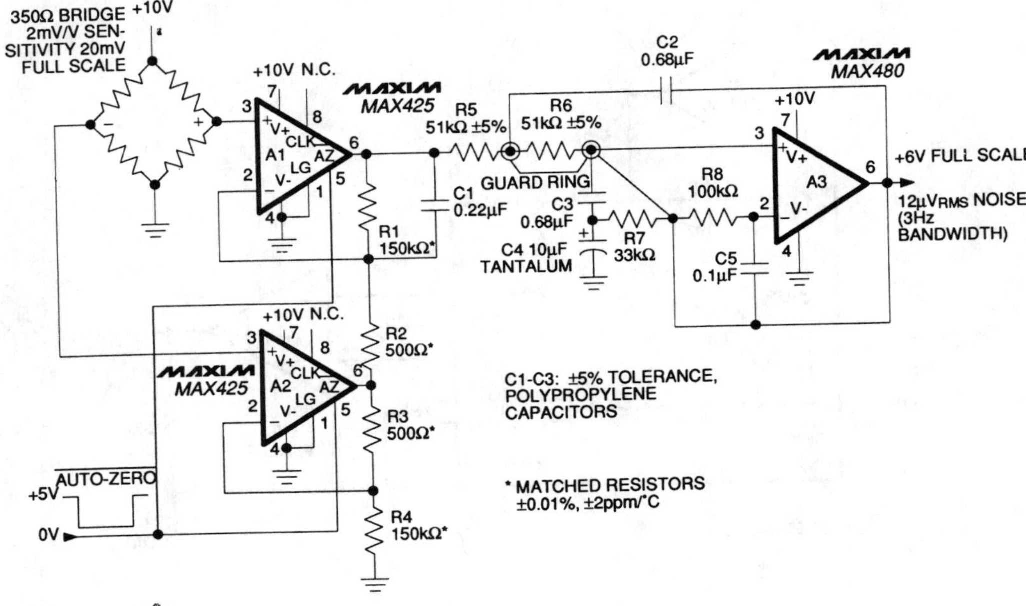

Strain-gauge amplifier

Fig. 1-11 In the circuit, two MAX425s (A1 and A2) amplify the differential 20-mV full-scale output of a strain-gauge bridge. The signal is then filtered and buffered by a MAX480 (A3). The 3-Hz bandwidth of the filter limits output noise to about 12 µVrms (40 nVrms referred to the input). This translates to a signal-to-noise ratio of 114 dB, which allows stable measurements to within a few parts per million. Maxim, 1992, Applications and Product Highlights, p. 5-3.

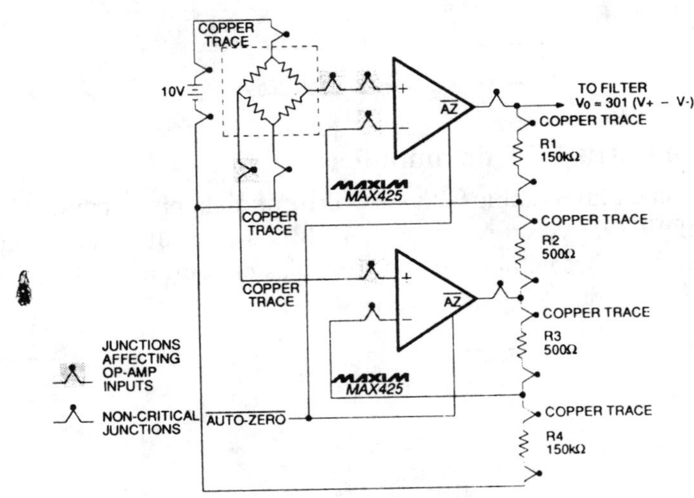

Strain gauge with thermoelectric junction effects

Fig. 1-12 This circuit shows the previous strain-gauge example (Fig. 1-11) with low-level nodes highlighted and shown as thermoelectric junctions. An even number of connections cancels the errors as long as no temperature gradients appear across the circuit. Such gradients are best prevented by symmetric PC layouts with traces sized for matched thermal conductivity. In extreme cases, the circuitry can also be insulated from external heat sources and drafts. Maxim, 1992, Applications and Product Highlights, p. 5-4.

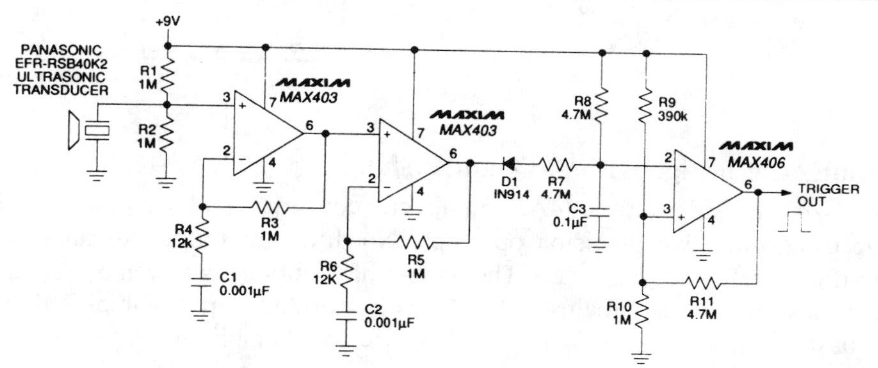

Ultrasonic receiver

Fig. 1-13 This low-power ultrasonic receiver, built around two MAX403s, can operate for several weeks on a 9-V battery. The receiver has high gain at ultrasonic frequencies to detect faint reflections from a separate ultrasonic source (not shown). To prevent false triggering, the 40-kHz received signal passes through R7 to charge C3 after being rectified by D1. The rectified signal is detected by a MAX406 op amp, connected as a comparator, chosen because of its 1-µA supply current and rail-to-rail output swing. Maxim, 1992, Applications and Product Highlights, p. 5-11.

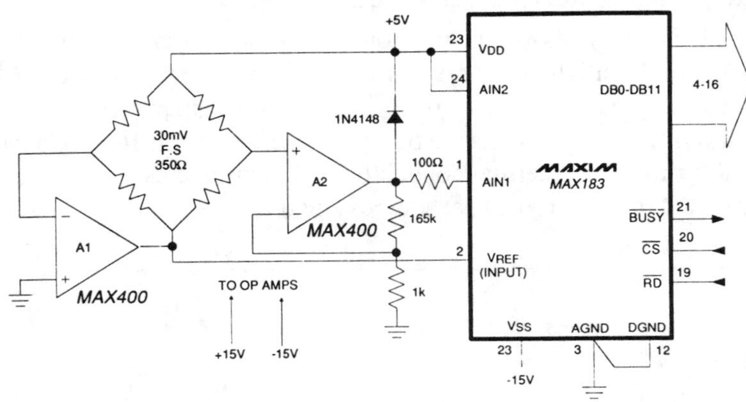

Ratiometric bridge with A/D conversion

Fig. 1-14 In this circuit, a MAX183 directly converts a bridge output to a 12-bit code using only two precision op amps. No differential-input instrumentation amplifier or reference is needed. The bottom of the bridge is driven by A1 so that the A1 inverting input remains at 0 V. The other bridge output is amplified by A2 in a basic noninverting configuration. No differential amplifier is needed because the bridge output to A2 is referenced to 0 V. The A1 output also drives the MAX183 V_{REF} input so that reference changes (V_{DD} is the reference) in the bridge excitation and the A/D track each other. Maxim, 1992, Applications and Product Highlights, p. 8-11.

Audio amplifiers with short-circuit protection

Fig. 1-15 This circuit makes optimum use of economy transistors in discrete-component amplifiers that will operate safely under any usable load conditions, including shorts. Figures 1-15B, 1-15C, and 1-15D show the transistor complement, resistor values/power-supply voltages, and heatsink requirements, respectively, for various power outputs from 35 to 100 W. Input sensitivity is 1 V_{rms} into 10 kΩ for full-rated output power. Frequency response is less than 3-dB rolloff from 10 Hz to 100 kHz, referenced to 1 kHz. Power bandwidth is the full-rated output power ±2 dB from 20 Hz to 20 kHz. Figures 1-15E and 1-15F show the load sensitivity and power bandwidth. Total harmonic distortion (THD) is less than 0.2% at any power level between 100 mW and full-rated output and at any frequency between 20 Hz and 20 kHz. Intermodulation distortion is less than 0.2% at any power level from 100 mW to the full-rated output (60 Hz and 7 kHz mixed 4 to 1). Motorola Applications Literature, 1991, BR135/D REV 7, AN485, p. 2, 3, 4.

Continued

(a)

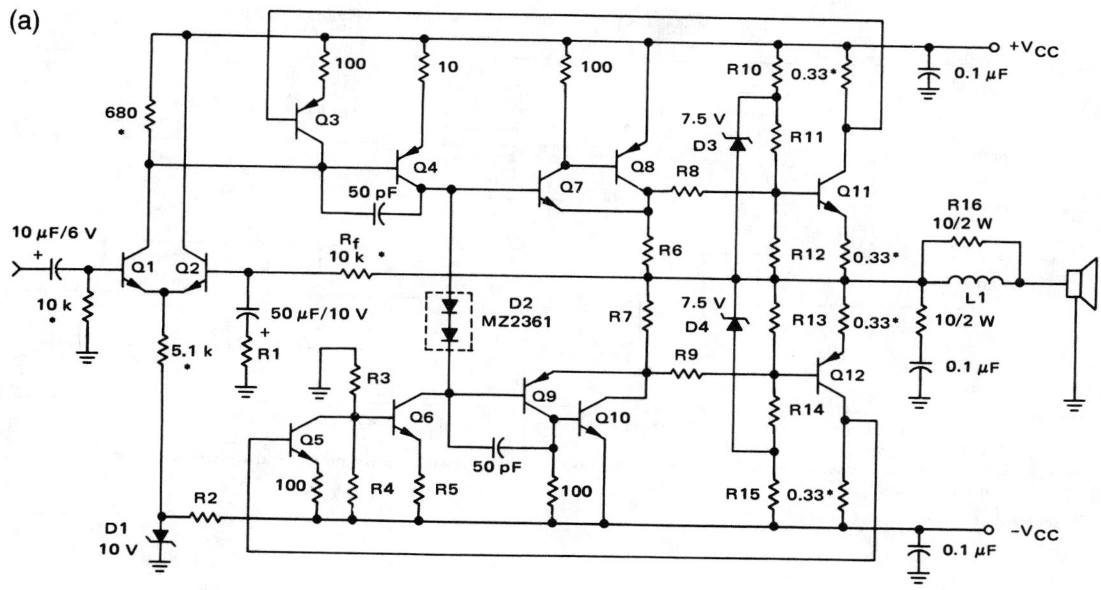

NOTE 1: All of the resistors with the values shown are ±10% tolerance, except where * indicates ±5%.

2: L1 is #20 wire close-wound for the full length of resistor, R16.

(b)

Output Power (Watts-rms)	Load Impedance (Ohms)	R1 ±5%	R2 ±10%	R3 ±5%	R4 ±5%	R5 ±5%	R6, R7 ±5%	R8, R9 ±10%	R10, R15 ±5%	R11, R14 ±5%	R12, R13 ±5%	V_{CC}
35	4	820	2.7 k	18 k	1.2 k	120	0.39	390	2.7 k	1.5 k	470	±21 V
	8	560	3.9 k	22 k	1.2 k	180	0.47*	240	3.0 k	1.2 k	470	±27 V
50	4	680	3.3 k	22 k	1.2 k	100	0.33	360	3.3 k	1.5 k	470	±25 V
	8	470	4.7 k	27 k	1.2 k	150	0.43*	270	3.9 k	1.2 k	470	±32 V
60	4	620	3.9 k	22 k	1.2 k	120	0.33	430	3.9 k	1.5 k	470	±27 V
	8	430	5.6 k	33 k	1.2 k	120	0.39	300	4.7 k	1.2 k	470	±36 V
75	4	560	4.7 k	27 k	1.2 k	91	0.33	620	5.6 k	1.8 k	470	±30 V
	8	390	6.8 k	33 k	1.2 k	150	0.39	390	6.8 k	1.5 k	470	±40 V
100	4	470	5.6 k	33 k	1.2 k	68	0.39	1.0 k	8.2 k	2.2 k	470	±34 V
	8	330	8.2 k	39 k	1.2 k	100	0.39	510	9.1 k	1.8 k	470	±45 V

NOTE: All of the above resistor values are in ohms and are 1/2 W except for R6 and R7.
*R6 and R7 are 5 W resistors except where * indicates 2 W.

Continued

(c)

Minimum Heat Sinking Required for Safe Operation Under Shunted Load at 50°C Ambient Temperature

Output Power (Watts-rms)	Load Impedance (Ohms)	Output Transistor Heat Sink (θ_{CA}) (See Note 1)	Driver Transistor Heat Sink (θ_{CA}) (See Note 2)
35	4	4.2°C/W	None
	8	2.4°C/W	None
50	4	3.0°C/W	60°C/W
	8	2.4°C/W	60°C/W
60	4	2.5°C/W	60°C/W
	8	2.0°C/W	60°C/W
75	4	1.6°C/W	35°C/W
	8	1.6°C/W	70°C/W*
100	4	1.0°C/W	20°C/W
	8	1.0°C/W	50°C/W*

NOTE 1: All of the output transistors are in TO-3 packages with the exception of the MJE2801/2901 (35 W/8 Ω), which are in the Case 90 Thermopad† plastic package.

2: All of the driver transistors are in the plastic Uniwatt† package with the exception of those marked *, which are metal cased TO-5.

†Trademark of Motorola Inc.

(d)

TABLE 1 – Semiconductor Complement

Output Power (Watts-rms)	Load Impedance (Ohms)	Output Transistors		Driver Transistors		Pre-Driver Transistors		Differential Amplifier Transistors (Q1 & Q2)	
		NPN (Q10)	PNP (Q8)	NPN (Q7)	PNP (Q9)	NPN (Q6)	PNP (Q4)	Single Channel	Dual Channel
35	4	MJ2840	MJ2940	MPSU05	MPSU55	MPSA05	MPSA55	MD8001	MFC8000
	8	MJE2801	MJE2901	MPSU05	MPSU55	MPSA06	MPSA56	MD8001	MFC8000
50	4	2N5302	2N4399	MPSU05	MPSU55	MPSA06	MPSA56	MD8001	MFC8000
	8	MJ2841	MJ2941	MPSU06	MPSU56	MPSA06	MPSA56	MD8002	MFC8001
60	4	2N5302	2N4399	MPSU06	MPSU56	MPSA06	MPSA56	MD8001	MFC8000
	8	MJ2841	MJ2941	MPSU06	MPSU56	MPSA06	MPSA56	MD8002	MFC8001
75	4	MJ802	MJ4502	MPSU06	MPSU56	MPSA06	MPSA56	MD8001	MFC8000
	8	MJ802	MJ4502	MM3007	2N5679	MM3007	MM4007	MD8003	MFC8002
100	4	MJ802	MJ4502	MPSU06	MPSU56	MPSU06	MPSU56	MD8002	MFC8001
	8	MJ802	MJ4502	MM3007	2N5679	MM3007	MM4007	MD8003	MFC8002

The following semiconductors are used at all of the power levels:

Q11 – MPSL01
Q5 – MPSA20
Q12 – MPSL51
Q3 – MPSA70

D1 – MZ500-16 or MZ92-20 (See Note 1)
D2 – MZ2361
D3 & D4 – 1N5236B or MZ92-16A (See Note 1)

NOTE 1: For a low-cost zener diode, an emitter-base junction of a silicon transistor can be substituted. A transistor similar to the MPS6512 can be used for the 7.5 V zener. A transistor similar to the Motorola BC317 can be used for the 10 V zener diode.

Continued

(e)

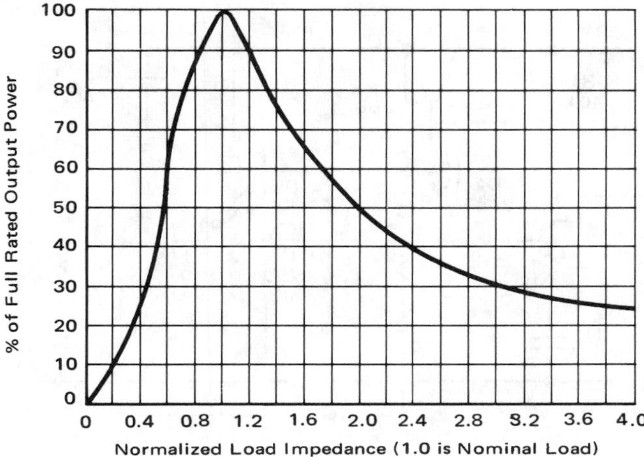

(f)

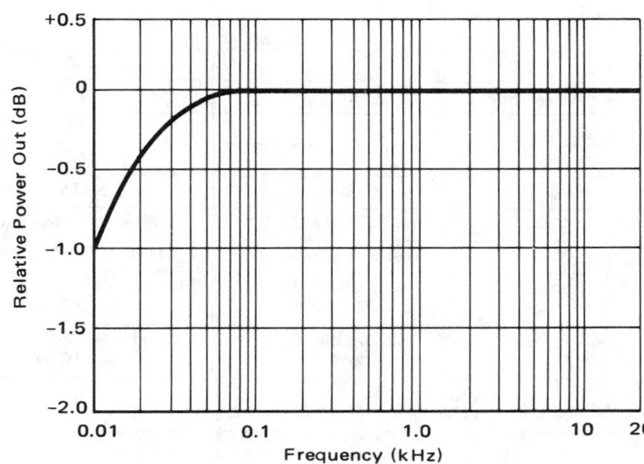

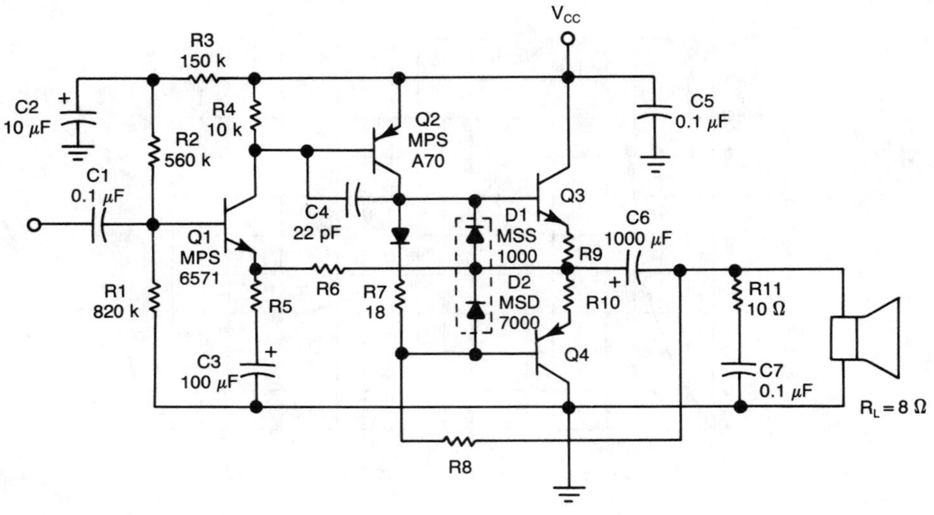

	3 W	5 W
V_{CC}	18 V	22 V
R_S	180	150
R_B	470	390
R9 & R10	0.82	0.56
**Q3	MPSU01 or MJE200	MPSU01 or MJE200
**Q4	MPSU51 or MJE210	MPSU51 or MJE210
Heatsink	with MPSU01/51 27.5°C/W	with MPSU01/51 16.8°C/W
	with MJE200/210 36°C/W	with MJE200/210 19.7°C/W

*Heatsink size calculation is based on a maximum ambient temperature of 50°C and a load phase angle of 60 degrees (see text for method of calculation). Heatsink is for both devices on one sink.

(a)

**Parts in same block are interchangeable.
P.C. board is for MPSU01/51, but can be changed to MJE200/210

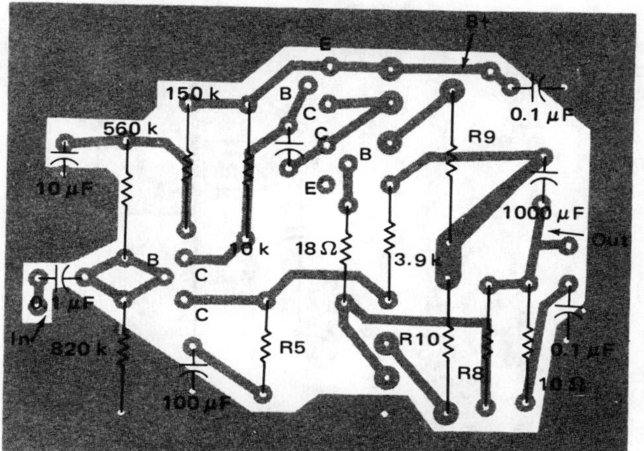

(b)

TABLE B1 – Amplifier Performance Characteristics

Reference Figure 1	3 W 18 Vdc	5 W 22 Vdc
1. Idle Current (normal no-signal)	20 mA	50 mA
2. Current Drain at Rated Power	275 mA	360 mA
3. Typical Input Impedance	280 kohms	280 kohms
4. THD at Rated Output Power 20 Hz or kHz to 20 kHz	< 1%	< 1%
5. IM Distortion at 60 and 7000 Hz 4:1 ratio at Rated Power	< 1%	< 1%
6. –3 dB Bandwidth	20 Hz-290 kHz	20 Hz-325 kHz
7. Typical input sensitivity for rated output power	0.250 VRMS	0.250 VRMS
8. Maximum output power at 5% THD without current limiting	4.2 Watts	7.03 Watts
9. Maximum output power at 5% THD with current limiting	4.06 Watts	6.66 Watts
10. Power Supply Ripple Rejection	34 dB	32 dB
11. Short Circuit Power Supply Current with Current Limiting	750 mA	1 A

(c)

Discrete-component audio amplifier (3 to 5 W)

Fig. 1-16 This circuit provides a 3- or 5-W output, depending on power source and component values, with a pnp driver and overload protection. Figures 1-16B and 1-16C show the PC-board layout and performance characteristics, respectively. Motorola Applications Literature, 1991, BR135/D REV 7, AN484A, p. 17.

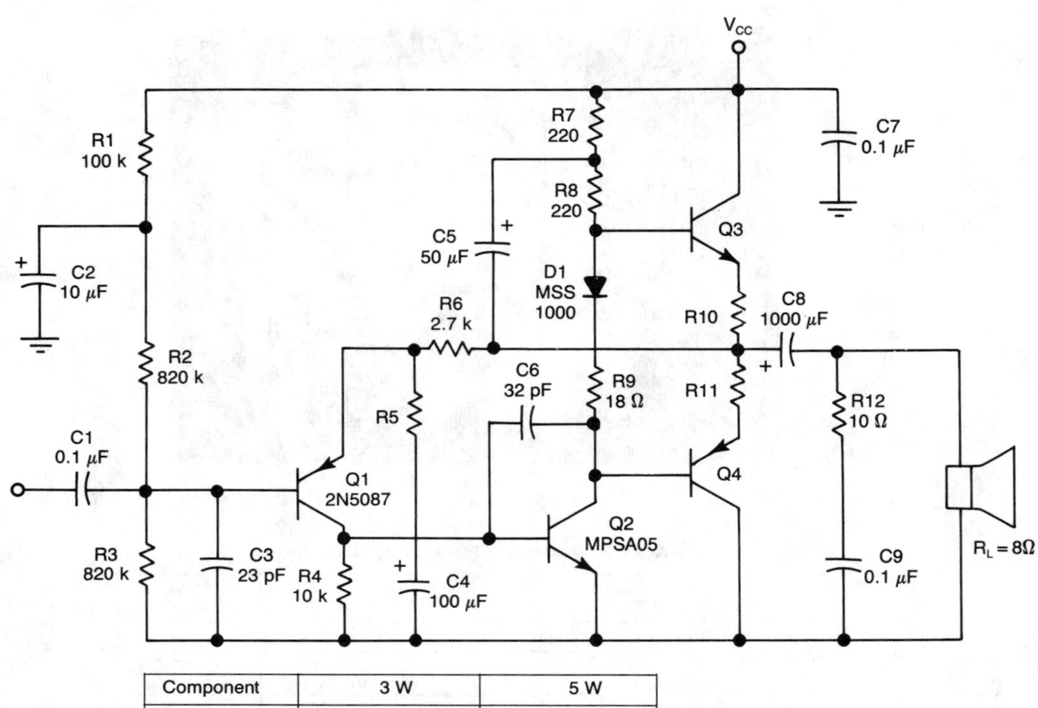

Component	3 W	5 W
V_{CC}	17	22
R_S	120 Ω	100 Ω
R10 & R11	0.82	0.56
Q3*	MPAU01 MJE200	MPSU01 MJE200
Q4*	MPSU51 MJE210	MPSU51 MJE210
**Heatsink	MPSU01/51 27.5°C/W	MPSU01/51 16.8°C/W
**Heatsink	MJE200/210 36°C/W	MJE200/210 19.7°C/W

(a)

*Parts in same block are interchangeable. P.C. board is for MJE200/210, but can be easily changed to MPSU01/51.

**Heatsink size calculation is based on a maximum ambient temperature of 50°C and a load phase angle of 60 degrees (see text for method of calculation) Heatsink is for both devices on one sink.

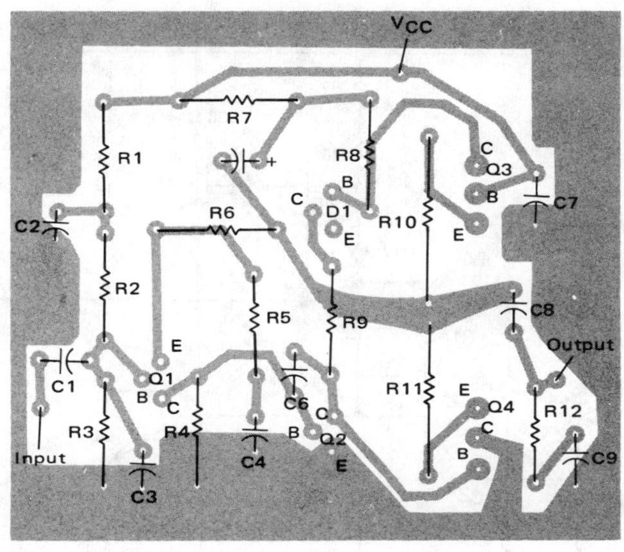

(b)

TABLE B-2 – Amplifier Performance Characteristics

Reference Figure 2	3 W 18 Vdc	5 W 22 Vdc
1. Idle Current (nominal no-signal)	20 mA	54 mA
2. Current Drain at Rated Power	285 mA	365 mA
3. Typical Input Impedance	300 kohms	320 kohms
4. THD at Rated Output Power 20 Hz or 1 kHz 20 kHz	< 1%	< 1%
5. IM Distortion at 60 and 7000 Hz 4:1 ratio at Rated Power	< 1%	< 1%
6. -3 dB Bandwidth	20 Hz-220 kHz	20 Hz-150 kHz
7. Typical input sensitivity for rated output power	0.22 VRMS	0.23 VRMS
8. Maximum output power at 5% THD without current limiting	4.10 Watts	6.8 Watts
9. Maximum output power at 5% THD with current limiting	4.06 Watts	6.65 Watts
10. Power Supply Ripple Rejection	24 dB	36.4 dB
11. Short Circuit Power Supply Current with Current Limiting	800 mA	1 Amp

(c)

Discrete-component audio amplifier (3 to 5 W), npn driver

Fig. 1-17 This circuit provides a 3- or 5-W output, depending on power source and component values, with an npn driver, and overload protection. Figures 1-17B and 1-17C show the PC-board layout and performance characteristics, respectively. Motorola Applications Literature, 1991, BR135/D REV 7, AN484A, p. 18.

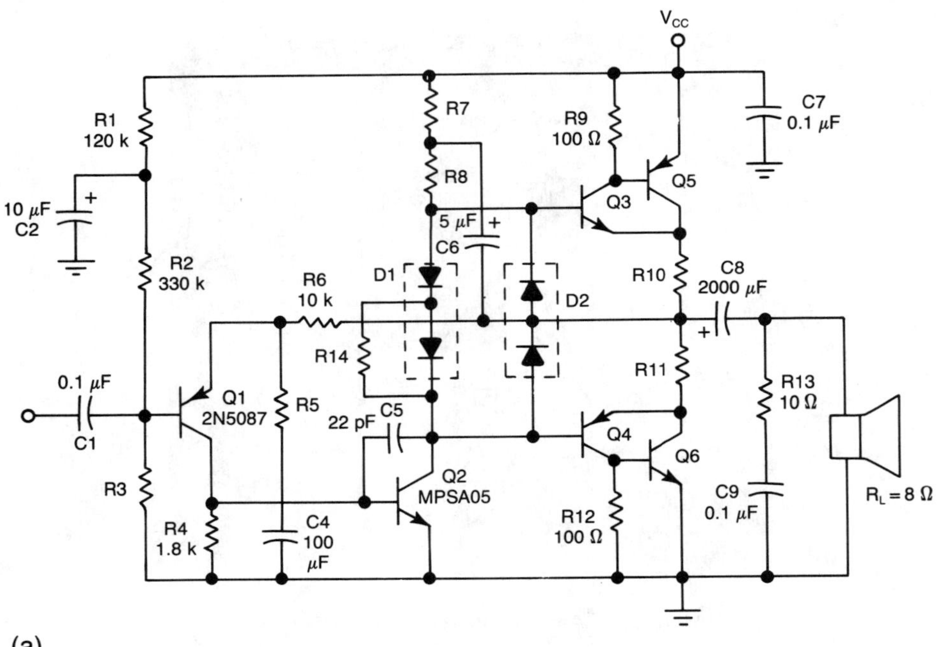

(a)

Components	7 W	10 W	15 W	20 W	25 W	35 W
V_{CC}	25 V	30 V	36 V	43 V	48 V	54 V
R3	560 k	560 k	620 k	620 k	680 k	680 k
R5	180 Ω	180 Ω	120 Ω	120 Ω	120 Ω	120 Ω
R7 & R8	3.9 k	3.9 k	3.9 k	3.9 k	3.9 k	5.6 k
R10 & R11	0.47	0.47	0.47	0.47	0.47	0.33
R14	–	–	–	–	–	470 Ω
Q3	MPSA05	MPSA05	MPSU05 MJE221	MPSU05 MJE224	MPSU05 MJE224	MPSU05 MJE224
Q4	MPSA55	MPSA55	MPSU55 MJE231	MPSU55 MJE234	MPSU55 MJE234	MPSU55 MJE234
Q5	MJE230	MJE230	MJE105	MJE105	MJE2901	MJE2901
Q6	MJE220	MJE220	MJE205	MJE205	MJE2801	MJE2801
D1 & D2	MSD7000	MSD7000	MSD7000	MSD7000	MSD7000	MSD7000
*Heatsink	15.8°C/W	9.8°C/W	6.2°C/W	4.7°C/W	3.4°C/W	2°C/W

*Heatsink size calculation is based on a maximum ambient temperature of 50°C and a load phase angle of 60 degrees (see text for methods of calculation). The heat sink is for both devices on the sink.

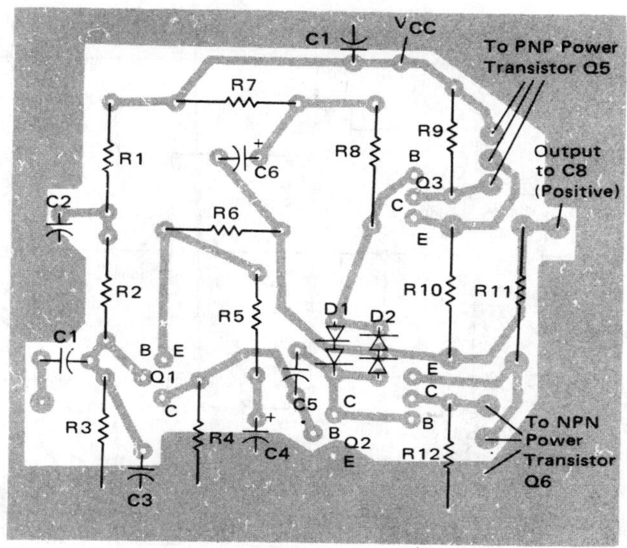

(b)

(c)

TABLE B-3 – Amplifier Performance Characteristics

Reference Figure 3	7 W 26 Vdc	10 W 30 Vdc	15 W 36 Vdc	20 W 43 Vdc	25 W 48 Vdc	35 W 54 Vdc
1. Idle Current (nominal no-signal)	1.6 mA	20 mA	28 mA	65 mA	68 mA	32 mA
2. Current Drain at Rated Power	425 mA	510 mA	610 mA	720 mA	320 mA	940 mA
3. Typical Input Impedance	230 kohms	320 kohms	230 kohms	230 kohms	230 kohms	220 kohms
4. THD at Rated Output Power 20 kHz or 1 kHz to 20 kHz	< 0.5%	< 0.5%	< 0.5%	< 0.5%	< 0.5%	< 0.5%
5. IM Distortion at 60 and 7000 Hz 4:1 ratio at Rated Power	< 1%	< 1%	< 1%	< 1%	< 1%	< 1%
6. -3 dB Bandwidth	16 Hz-300 kHz	15 Hz-380 kHz	16 Hz-250 kHz	16 Hz-250 kHz	16 Hz-275 kHz	16 Hz-230 kHz
7. Typical input sensitivity for rated output power	210 mV	260 mV	200 mV	220 mV	250 mV	280 mV
8. Maximum output power at 5% THD without current limiting	8.8 Watts	12.5 Watts	18 Watts	27.38 Watts	33.6 Watts	45.4 Watts
9. Maximum output power at 5% THD with current limiting	8.85 Watts	12.5 Watts	18 Watts	26.3 Watts	30 Watts	45 Watts
10. Power Supply Ripple Rejection	44.4 dB	42 dB	40 dB	30 dB	36 dB	34 dB
11. Short Circuit Power Supply Current with Current Limiting	1.32 Amps	1.45 Amps	1.5 Amps	1.5 Amps	1.62 Amps	2.2 Amps

Discrete-component audio amplifier (7 to 35 W)

Fig. 1-18 This circuit provides a 7- to 35-W output, depending on power source and component values (as shown), with overload protection. Figures 1-18B and 1-18C show the PC-board layout and performance characteristics, respectively.

Motorola Applications Literature, 1991, BR135/D, REV 7, AN484A, p. 19.

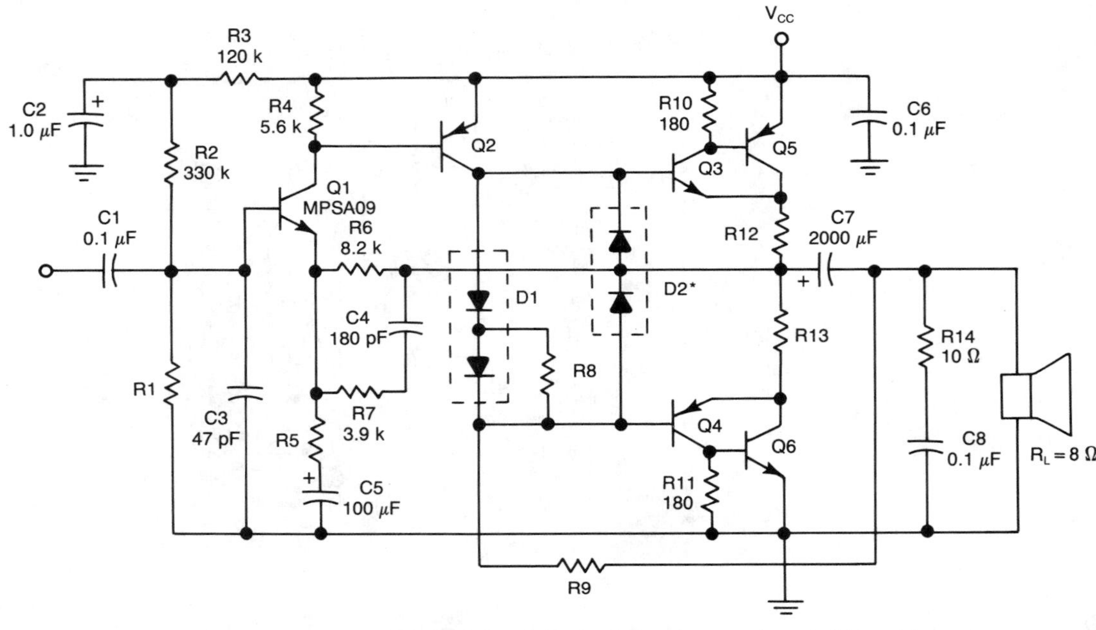

**D2 provides Overload Current Limiting. Provision for D2 was not made on PC Board Layout.

(a)

Component	7 W	10 W	15 W	20 W	25 W	35 W
V_{CC}	25 V	30 V	38 V	46 V	48 V	54 V
R1	560 k	560 k	560 k	620 k	620 k	620 k
R5	100 Ω	82 Ω	100 Ω	100 Ω	150 Ω	180 Ω
R8	Value Selected to Provide 30 mA Collector Current in Q5.					
R9	4.7 k	4.7 k	8.2 k	8.2 k	8.2 k	8.2 k
R12 & R13	0.47	0.47	0.47	0.47	0.33	0.33
Q2	2N5087	2N5087	2N5087	2N5087	MPSA55	MPSA56
Q3	MPSA05	MPSA05	MPSU05 MJE220	MPSU05 MJE224	MPSU05 MJE224	MPSU05 MJE224
Q4	MPSA55	MPSA55	MPSU55 MJE230	MPSU55 MJE234	MPSU55 MJE234	MPSU55 MJE234
Q5	MJE230	MJE230	MJE105	MJE105	MJE2901	MJE2901
Q6	MJE220	MJE220	MJE205	MJE205	MJE2801	MJE2801
D1 & D2	MSD7000	MSD7000	MSD7000	MSD7000	MSD7000	MSD7000
*Heatsink	15.3°C/W	9.8°C/W	6.2°C/W	4.7°C/W	3.4°C/W	2°C/W

*Heatsink size calculation is based on a maximum ambient temperature of 50°C and a load phase angle of 60 degrees (see text for methods of calculation). The heatsink is for both devices on one sink.

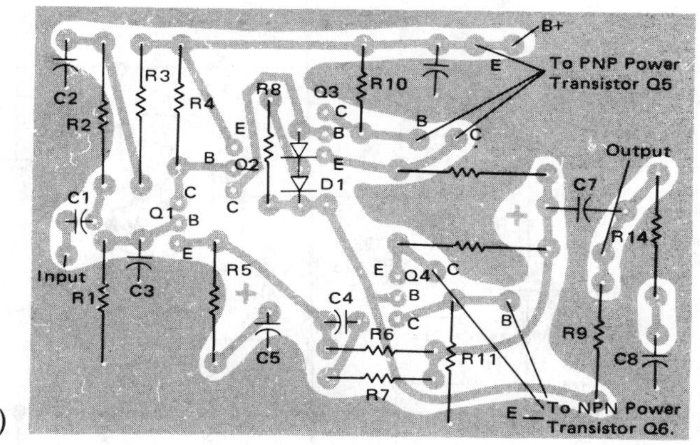

(b)

(c)

TABLE B-4 — Amplifier Performance Characteristics

Reference Figure 4	7 W 25 Vdc	10 W 30 Vdc	15 W 28 Vdc	20 W 46 Vdc	25 W 48 Vdc	35 W 54 Vdc
1. Idle Current (nominal no-signal)	28 mA	40 mA	20 mA	58 mA	20 mA	37 mA
2. Current Drain at Rated Power	440 mA	500 mA	670 mA	720 mA	820 mA	940 mA
3. Typical Input Impedance	230 kohms	230 kohms	210 kohms	220 kohms	220 kohms	210 kohms
4. THD at Rated Output Power 20 Hz or 1 kHz to 20 kHz	$<$0.5%	$<$0.5%	$<$0.5%	$<$0.5%	$<$0.5%	$<$0.5%
5. IM Distortion at 60 and 7000 Hz 4:1 ratio at Rated Power	$<$1%	$<$1%	$<$1%	$<$1%	$<$1%	$<$1%
6. -3 dB Bandwidth	12 Hz-250 kHz	18 Hz-110 kHz	17 Hz-55 kHz	18 Hz-150 kHz	10 Hz-160 kHz	13 Hz-60 kHz
7. Typical input sensitivity for rated output power	220 mV	270 mV	110 mV	120 mV	230 mV	270 mV
8. Maximum output power at 5% THD without current limiting	8.8 Watts	12.5 Watts	21.7 Watts	27 Watts	34 Watts	43 Watts
9. Maximum output power at 5% THD with current limiting	8.75 Watts	11.5 Watts	21.1 Watts	24.5 Watts	34 Watts	43 Watts
10. Power Supply Ripple Rejection	37 dB	26 dB	41.94 dB	33 dB	36.5 dB	38 dB
11. Short Circuit Power Supply Current with Current Limiting	1.2 Amps	1.3 Amps	1.32 Amps	1.4 Amps	2 Amps	2 Amps

Discrete-component audio amplifier (7 to 35 W) pnp driver

Fig. 1-19 This circuit provides a 7- to 35-W output, depending on power source and component values (as shown), with overload protection. Figures 1-19B and 1-19C show the PC-board layout and performance characteristics, respectively.

Motorola Applications Literature, 1991, BR135/D, REV 7, AN484A, p. 20.

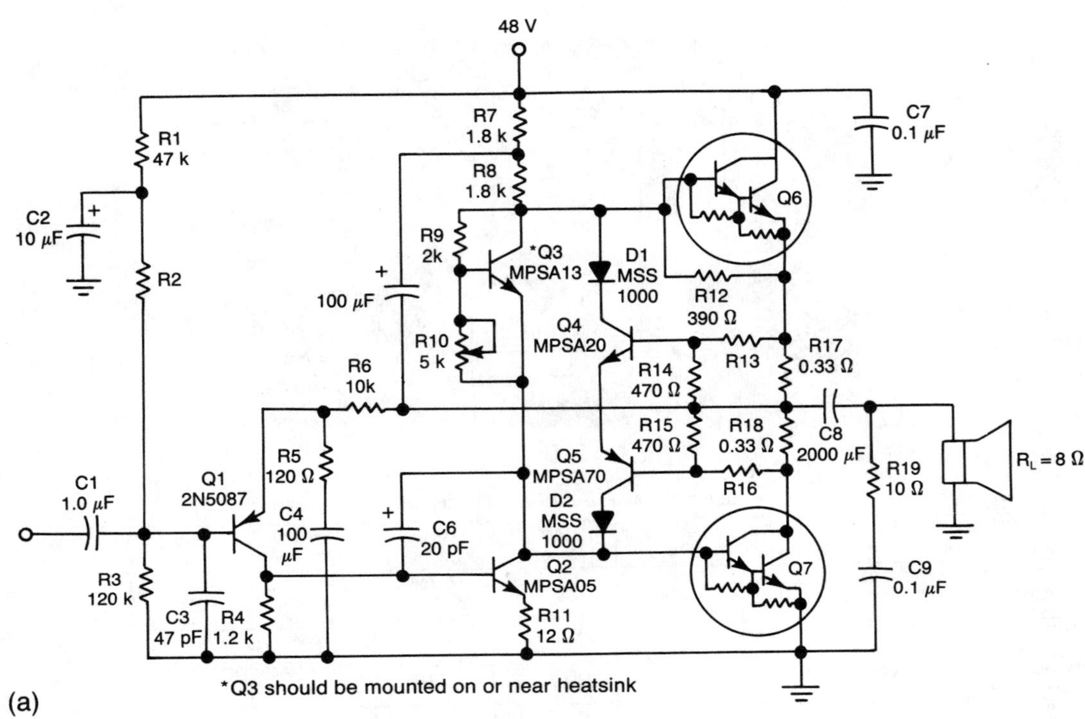

(a)

Component	15 W	20 W	25 W
V_{CC}	42 V	46 V	48 V
R2	220 k	160 k	160 k
R5	120 Ω	180 Ω	180 Ω
R13	180 Ω	220 Ω	220 Ω
R16	180 Ω	220 Ω	220 Ω
Q6	MJE800	MJE1100	MJE1100
Q7	MJE700	MJE1090	MJE1090
*Heatsink	6.5°C/W	4.8°C/W	3.6°C/W
R8	Adjusted for 20-30 mZ Collector Current for Q6		

*Heatsink size calculated is based on a maximum ambient temperature of 50°C and a load phase angle of 60 degrees (see text for method of calculation). The heatsink is for both devices on one sink.

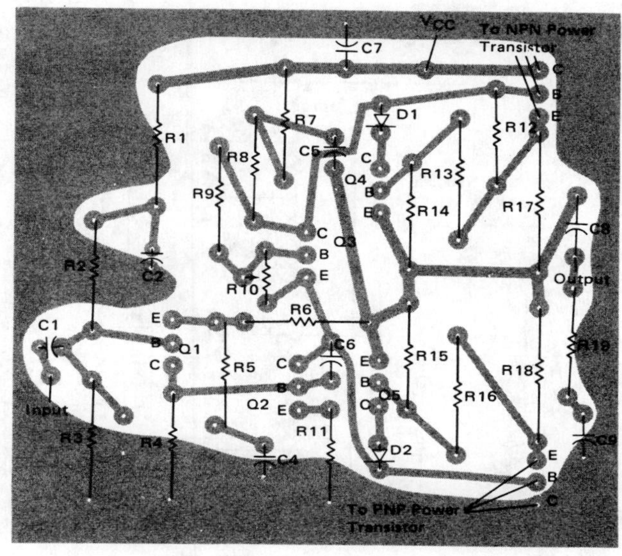

(b)

TABLE B-5 – Amplifier Performance Characteristics

Reference Figure 5	15 W 42 Vdc	20 W 46 Vdc	25 W 48 Vdc
1. Idle Current (nominal no-signal)	20 mA	20 mA	20 mA
2. Current Drain at Rated Power	640 mA	760 mA	800 mA
3. Typical Input Impedance	50 k	56 k	56 k
4. THD at Rated Output Power 20 Hz or 1 kHz to 20 kHz	$<$1%	$<$1%	$<$1%
5. IM Distortion at 60 and 7000 Hz 4:1 ratio at Rated Power	$<$1%	$<$1%	$<$1%
6. −3 dB Bandwidth	23 Hz-170 kHz	27 Hz-180 kHz	28 Hz-190 kHz
7. Typical input sensitivity for rated output power	200 mV	220 mV	230 mV
8. Maximum output power at 5% THD without current limiting	21.2 Watts	28.5 Watts	32.4 Watts
9. Maximum output power at 5% THD with current limiting	21 Watts	28 Watts	30 Watts
10. Power Supply Ripple Rejection	29 dB	34 dB	33 dB
11. Short Circuit Power Supply Current with Current Limiting	1.5 Amps	1.5 Amps	1.5 Amps

(c)

Discrete-component Darlington audio amplifier

Fig. 1-20 This circuit provides a 15-, 20-, or 25-W output, depending on the power source and component values (as shown), with overload protection, using Darlingtons in the output stages. Figures 1-20B and 1-20C show the PC-board layout and performance characteristics, respectively. Motorola Applications Literature, 1991, BR135/D, REV 7, AN484A, p. 21.

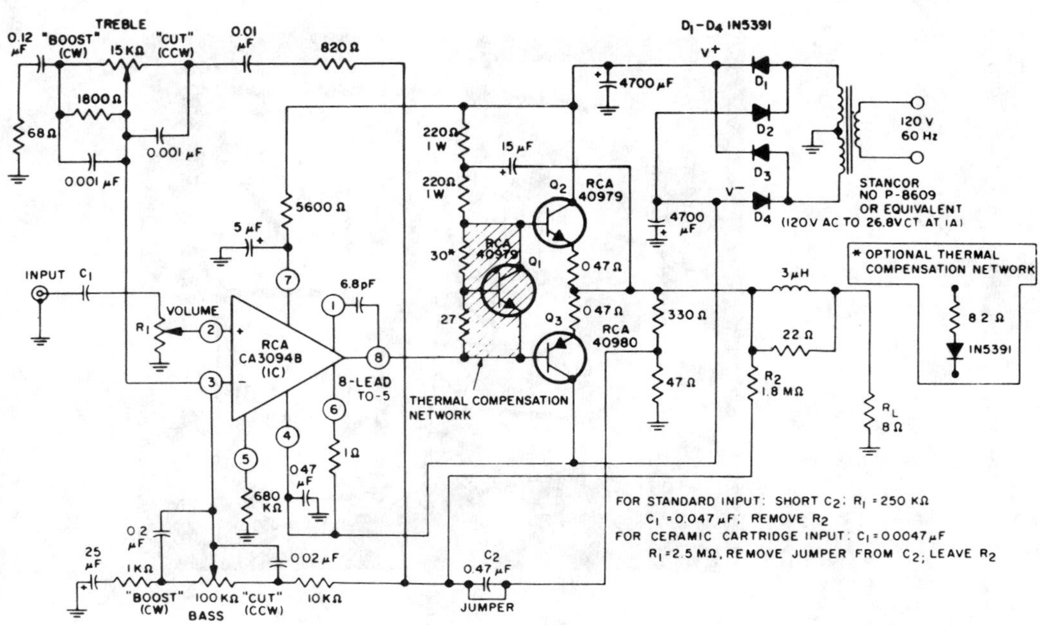

TYPICAL PERFORMANCE DATA
For 12-W Audio Amplifier Circuit

Power Output (8Ω load, Tone Control set at "Flat")		
Music (at 5% THD, regulated supply)	15	W
Continuous (at 0.2% IMD, 60 Hz & 2 kHz mixed in a 4:1 ratio, unregulated supply) See Fig. 8 In ICAN-6048	12	W
Total Harmonic Distoration		
At 1 W, unregulated supply	0.05	%
At 12 W, unregulated supply	0.57	%
Voltage Gain	40	dB
Hum and Noise (Below continuous Power Output)	83	dB
Input Resistance	250	kΩ
Tone Control Range	See Fig. 9 In ICAN-6048	

12-W audio amplifier with discrete output and IC input

Fig. 1-21 This circuit has a true complementary-symmetry output with a programmable IC amplifier input. Harris Semiconductors, Linear & Telecom ICs, 1991, p. 3-81.

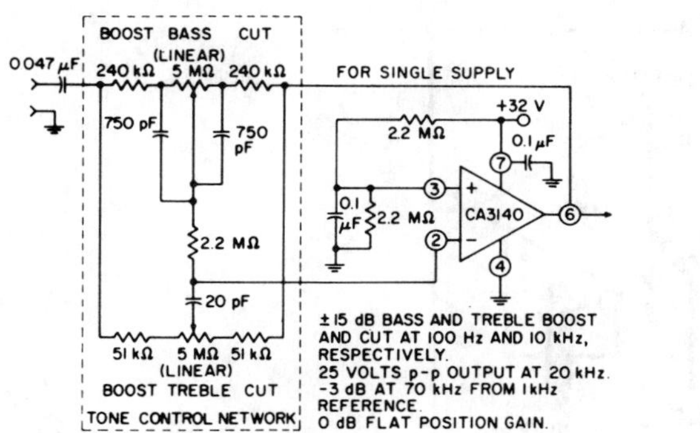

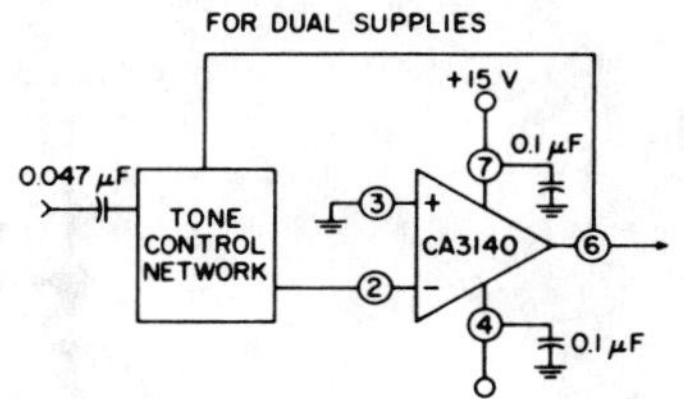

Baxandall tone control

Fig. 1-22 This circuit provides unity gain at midband and uses standard linear potentiometers. Bass/treble, boost, and cut are ±15 dB at 100 Hz and 10 kHz, respectively. Full peak-to-peak output is available up to at least 20 kHz. Amplifier gain is −3 dB down from the "flat" position (70 kHz). Harris Semiconductors, Linear & Telecom ICs, 1991, p. 3-117.

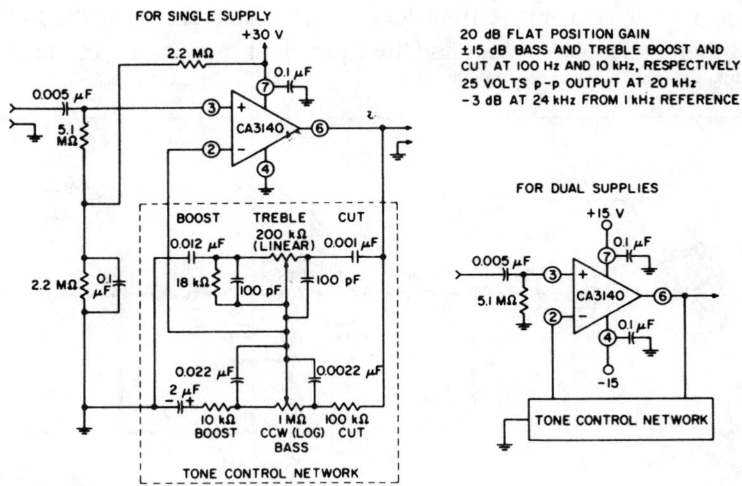

Tone control with 20-dB midband gain

Fig. 1-23 This circuit has boost and cut specifications similar to that of Fig. 1-22. The wideband gain is equal to the ultimate boost or cut, plus one (in this case a gain of 11). For a 20-dB boost and cut, the input loading is essentially equal to the resistance from terminal 3 to ground. Harris Semiconductors, Linear & Telecom ICs, 1991, p. 3-117.

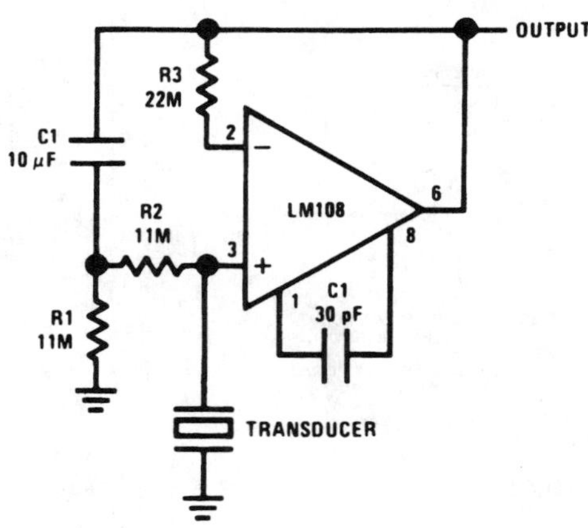

Amplifier for piezoelectric transducers

Fig. 1-24 This circuit shows an LM108 amplifier for high-impedance ac transducers, such as a piezoelectric accelerometer. The circuit has input resistance that is much greater than the dc return resistor values. This is accomplished by bootstrapping the resistors to the output. With such an arrangement, the lower cutoff frequency of a capacitive transducer is determined more by the RC output of R1 and C1 than by resistor values and the equivalent capacitance of the transducer. National Semiconductor, Linear Applications Handbook, 1991, p. 70.

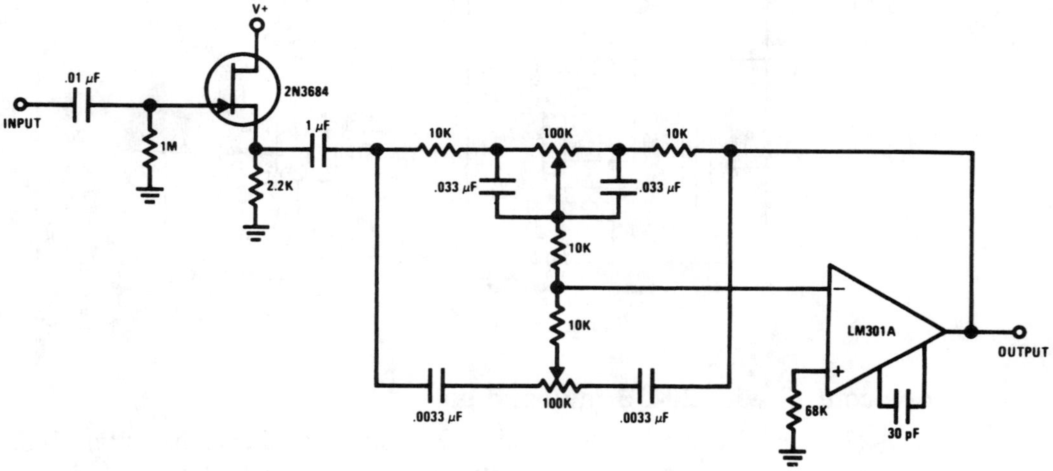

Hi-fi tone control with high-impedance input

Fig. 1-25 The JFET provides both high input impedance and low noise to buffer an op-amp feedback-type tone control. National Semiconductors, Linear Applications Handbook, 1991, p. 107.

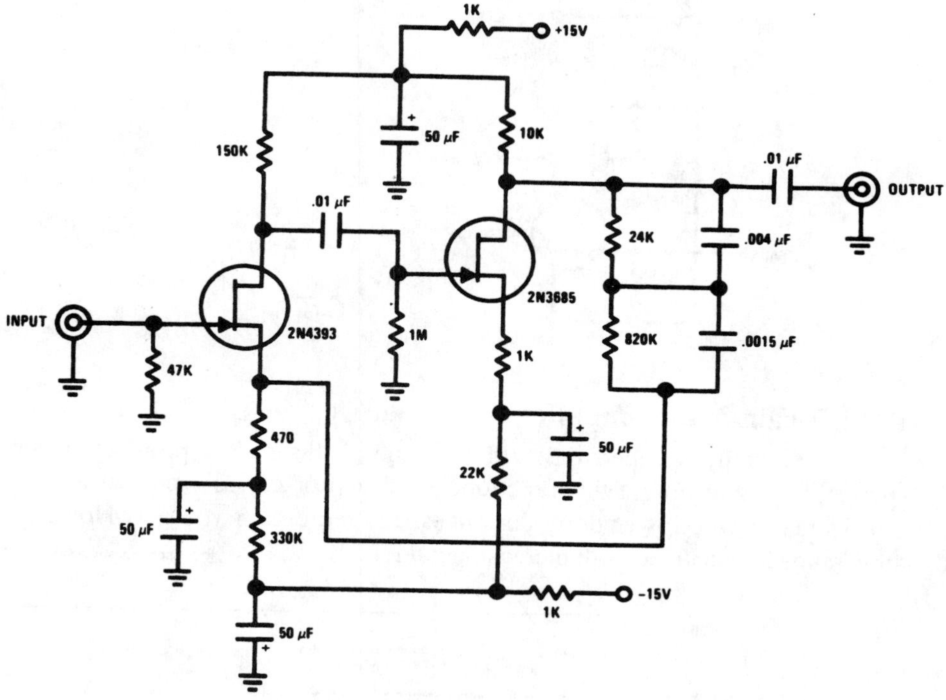

Magnetic-pickup phono preamp

Fig. 1-26 This preamplifier provides proper loading to a reluctance-type phono cartridge. The circuit provides about 25 dB of gain at 1 kHz (2.2-mV input for 100-mV output), and features S + S/N ratio of better than −70 dB (referenced to 10-mV input at 1 kHz). The feedback provides for RIAA equalization with a dynamic range of 84 dB (reference to 1 kHz). National Semiconductor, Linear Applications Handbook, 1991, p. 109.

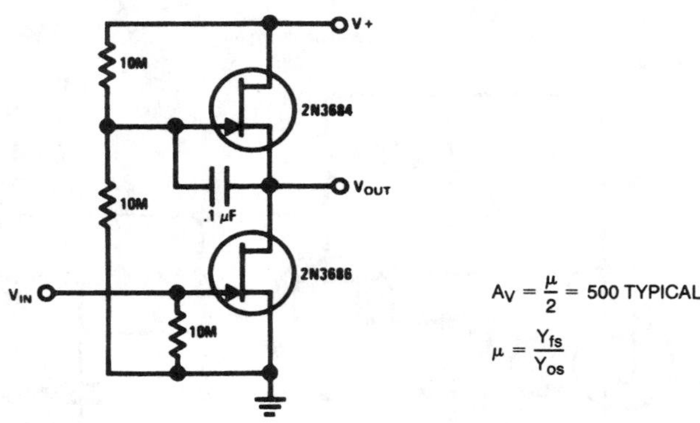

Ultra-high-gain audio amplifier

Fig. 1-27 Sometimes called the "JFET μ amp," this circuit provides a very low-power, high-gain amplifying function. Because μ of a JFET increases as drain current decreases, the lower drain current is, the more gain you get. However, dynamic range is sacrificed with increasing gain. National Semiconductor, Linear Applications Handbook, 1991, p. 110.

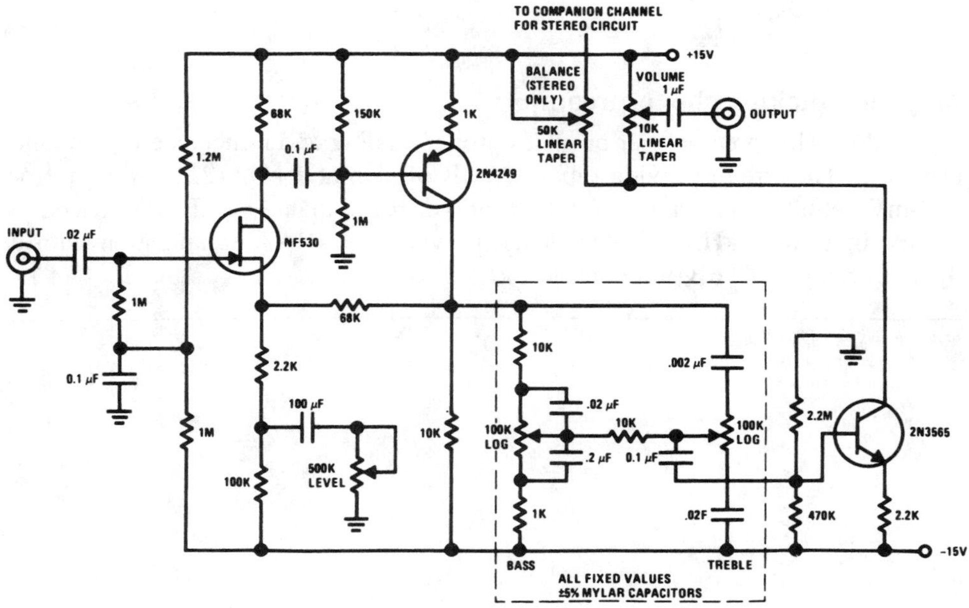

Low-cost high-level preamp and tone control

Fig. 1-28 This circuit uses the JFET to best advantage (low noise with high impedance). All device parameters are non-critical, yet the circuit achieves harmonic distortion of less than 0.05% with a S/N ratio of over 85 dB. The tone controls allow 18-dB of cut and boost. The amplifier has a 1-V output for 100-mV input at maximum level. National Semiconductor, Linear Applications Handbook, 1991, p. 117.

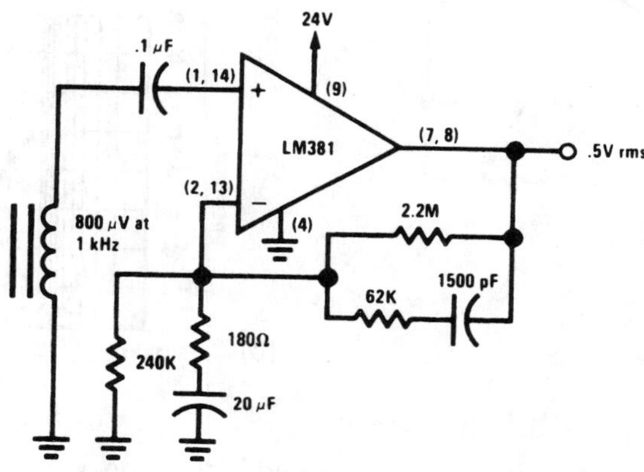

Typical tape playback amplifier

Fig. 1-29 This circuit requires about 5 seconds to turn on (for the gain and supply voltages shown). The turn-on time is approximately

$$t_{ON} = -R_4C_2 \ ln \left(1 - \frac{2.4}{V_{CC}}\right)$$

National Semiconductor, Linear Applications Handbook, 1991, p. 187.

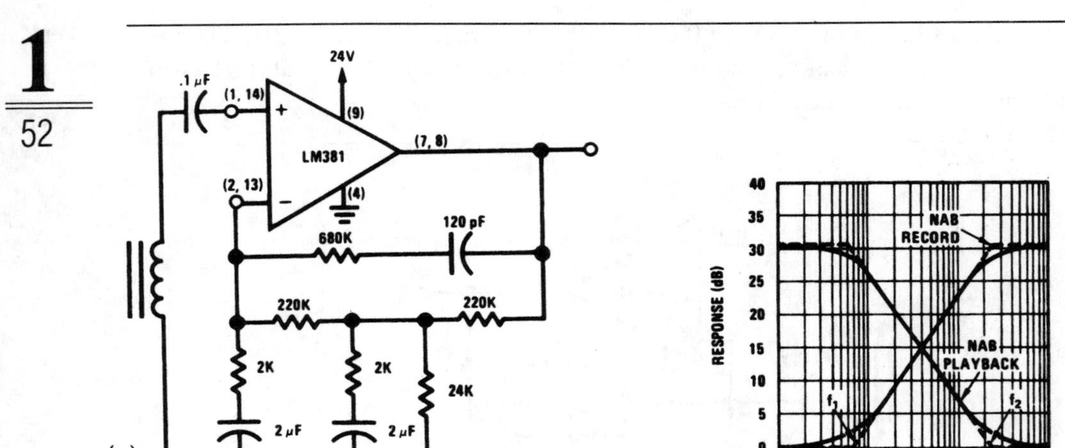

NAB tape preamp with fast turn-on

Fig. 1-30 This circuit requires only 0.1 second for turn-on. Compare to Fig. 1-29. When recording, the frequency response is the complement of the NAB playback equalization, making the composite record and playback response flat (Fig. 1-30B). National Semiconductor, Linear Applications Handbook, 1991, p. 189.

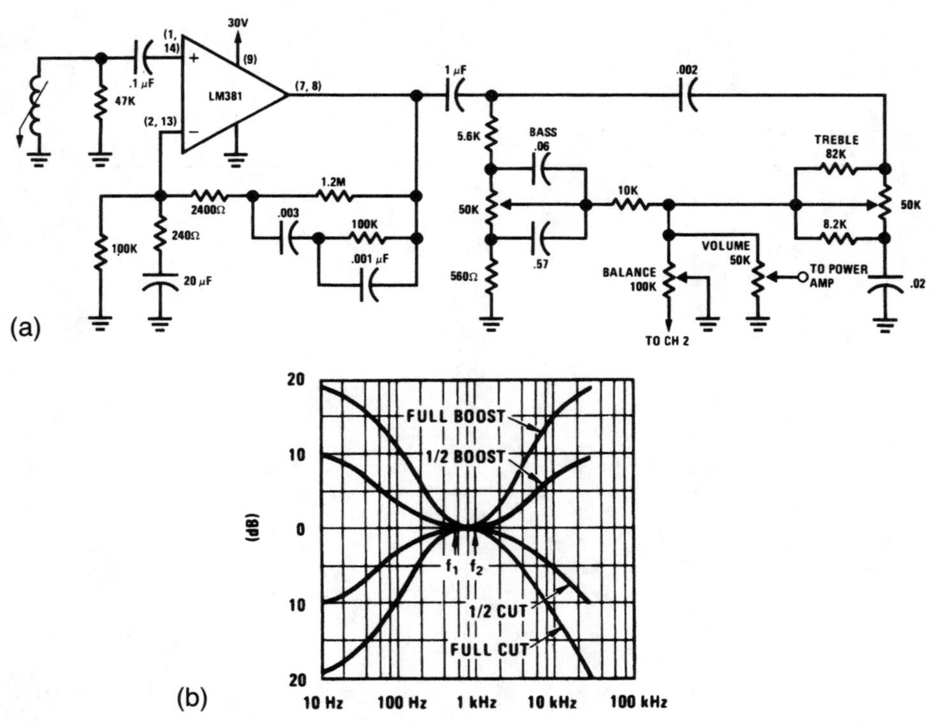

Stereo magnetic-phono preamp and tone controls

Fig. 1-31 This circuit shows a single channel of a stereo phono preamp for magnetic cartridges, including tone, volume, and balance controls. Figure 1-31B shows bass and treble tone-control response for 20-dB boost and cut. National Semiconductor, Linear Applications Handbook 1991, p. 192, 193.

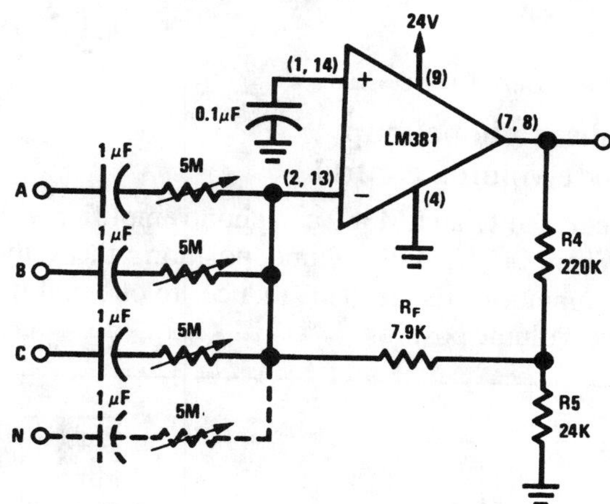

Audio mixer

Fig. 1-32 In many audio applications, it is desirable to provide a mixer to combine or select several inputs. Such applications include public-address systems, where more than one microphone is used, tape recorders, hi-fi phonographs, guitar amplifiers, etc. This circuit provides for mixing of 600-Ω dynamic microphones, each with an output level of 10 mV. The circuit operates from a 24-V supply and delivers 5-V output, with a dynamic range of 80 dB. National Semiconductor, Linear Applications Handbook, 1991, p. 194.

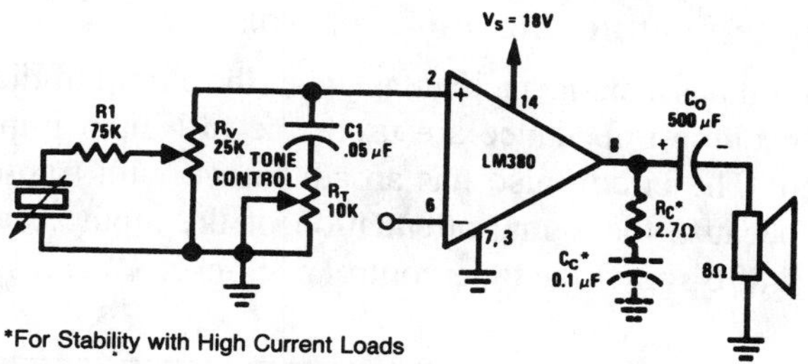

Minimum-component phono amp

Fig. 1-33 This circuit shows an LM380 power-amplifier IC used as a phono amplifier. The circuit has a voltage-divider volume control, with high-frequency rolloff tone control, and provides about 2.5 W to an 8-Ω speaker. National Semiconductor, Linear Applications Handbook, 1991, p. 198.

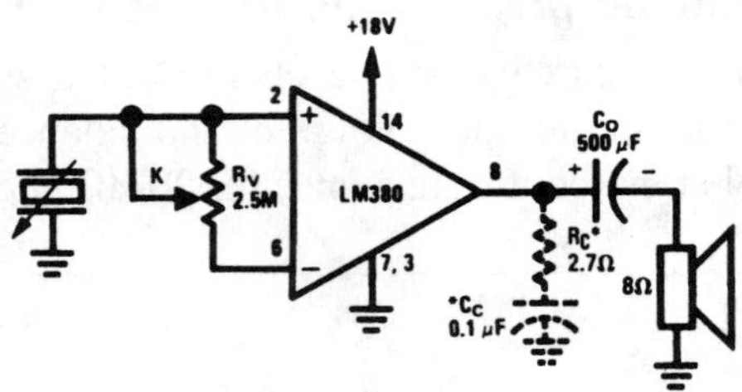

*For Stability with High Current Loads

Minimum-component phono amp with common-mode volume control

Fig. 1-34 This circuit shows an LM380 as a phono amplifier with common-mode volume control. When in the full-volume position, the only source-loading impedance is the IC input impedance. This reduces to one-half the amplifier input impedance at the zero-volume position. National Semiconductor, Linear Applications Handbook, 1991, p. 198.

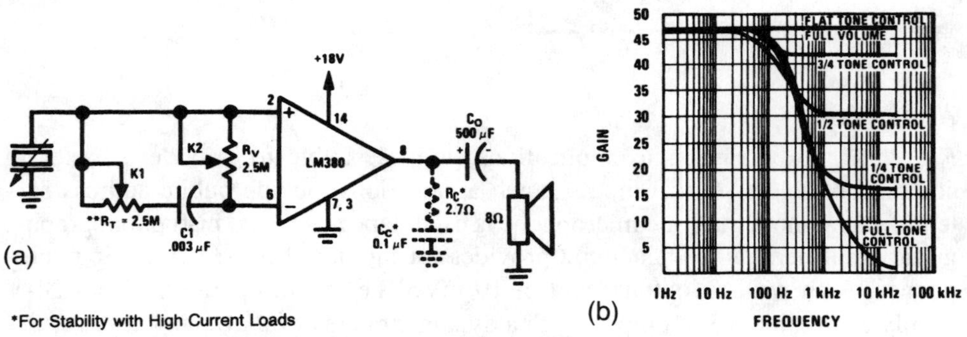

*For Stability with High Current Loads

Minimum-component phono amp with common-mode volume and tone controls

Fig. 1-35 This circuit has a distinct advantage over the circuit of Fig. 1-33. When transducers of high source-impedance are used, the full input impedance of the amplifier is realized. The circuit also has an advantage with transducers of low source impedance because the signal attenuation of the input voltage divider is eliminated. Figure 1-35B shows the tone-control response. National Semiconductor, Linear Applications Handbook, 1991, p. 198.

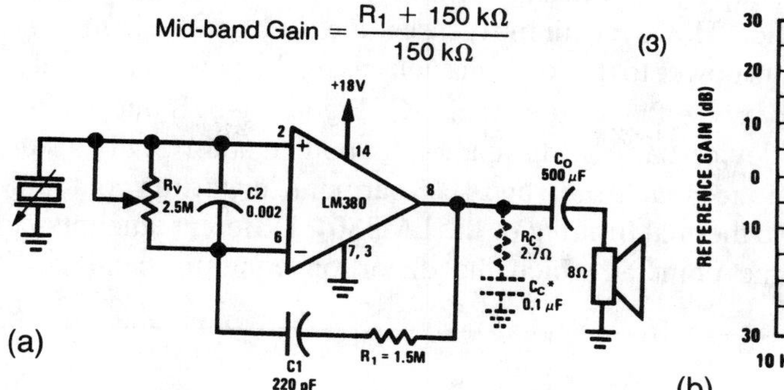

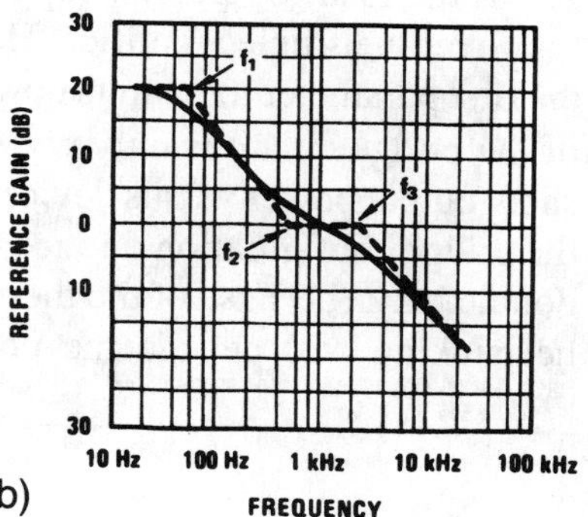

* For Stability with High Current Loads

Minimum-component RIAA phono amp

Fig. 1-36 This circuit shows an LM380 used as a phono amplifier with RIAA playback characteristics. Figure 1-36B shows the playback response. National Semiconductor, Linear Applications Handbook, 1991, p. 198, 199.

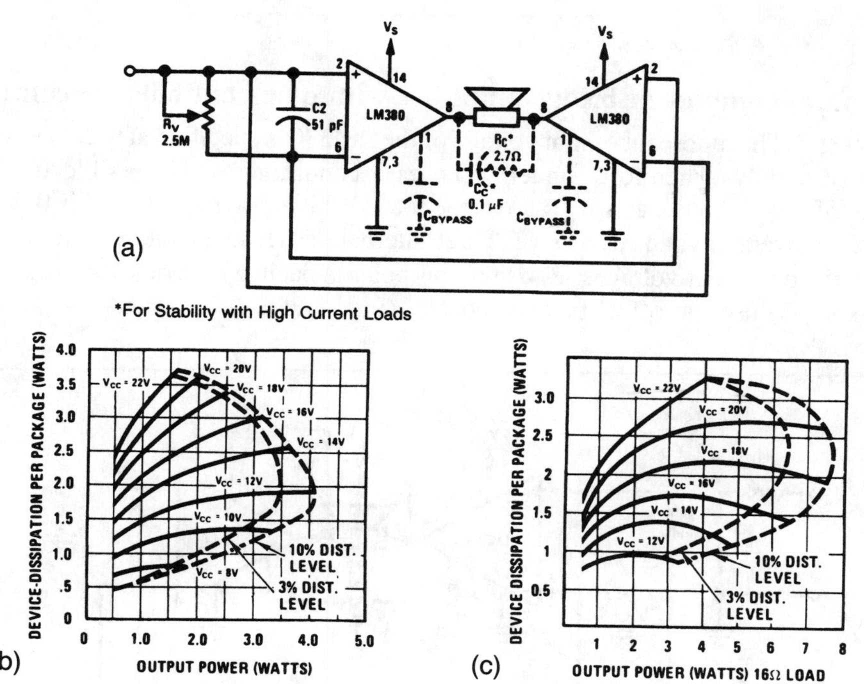

Minimum-component bridge amplifier

Fig. 1-37 This circuit shows two LM380 ICs connected to double the audio output power of a single LM380. The circuit provides twice the voltage swing across the load for a given supply, increasing the power capability by a factor of four over the single amplifier. However, in most cases, the package dissipation is the first parameter that limits power to the load. In such cases, the power capability of the bridge circuit is only twice that of the single IC. Figures 1-37B and 1-37C show output power versus device-package dissipation for both 8- and 16-Ω loads in the bridge configuration. Notice that the 3% and 10% harmonic distortion contours double back (because of the thermal limiting of the LM380). Different amounts of heatsinking will change the point at which the distortion contours bend. National Semiconductor, Linear Applications Handbook, 1991, p. 199.

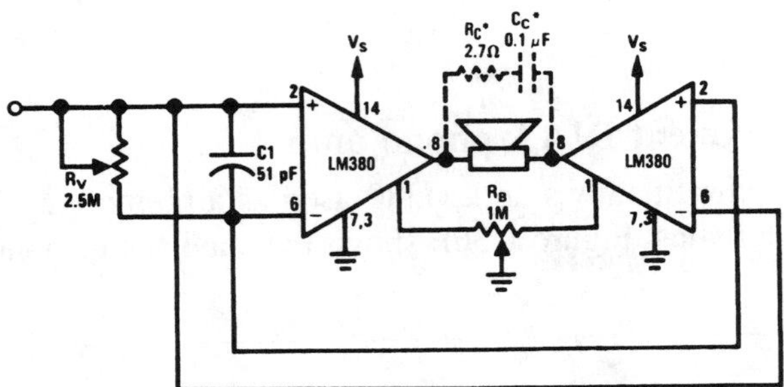

*For Stability with High Current Loads

Minimum-component bridge amplifier with quiescent balance control

Fig. 1-38 The quiescent output voltage of the LM380 is specified at 9 ± 1 V with an 18-V supply. Therefore, under worst-case conditions, it is possible for the circuit of Fig. 1-37 to have 2-V across the load. With an 8-Ω speaker, this 0.25 A of direct current might be excessive. Three alternatives are available: 1) take care to match the quiescent voltages, 2) use a non-polar capacitor in series with the load (speaker), 3) use the offset balance control (Fig. 1-38). National Semiconductor, Linear Applications Handbook, 1991, p. 199.

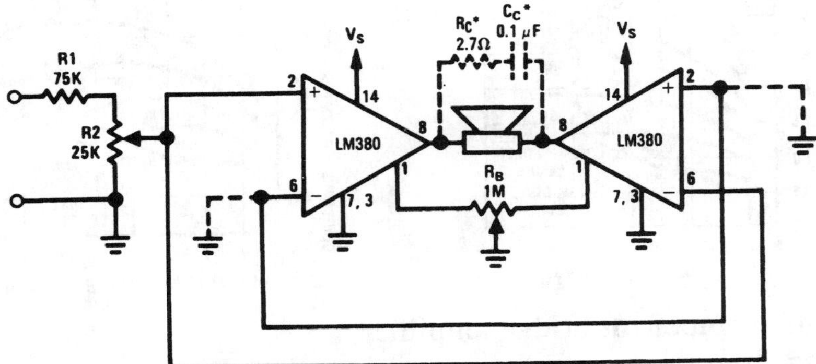

*For Stability with High Current Loads

Minimum-component bridge amplifier with voltage-divider input

Fig. 1-39 This circuit shows a bridge-amplifier configuration with a voltage-divider input (rather than the common-mode input of Fig. 1-37). With the circuit of Fig. 1-39, if the source voltage V_S (pin 14) is more than 3 inches from the power-source filter capacitor, pin 14 should be decoupled with a 1-μF tantalum capacitor. National Semiconductor, Linear Applications Handbook, 1991, p. 200.

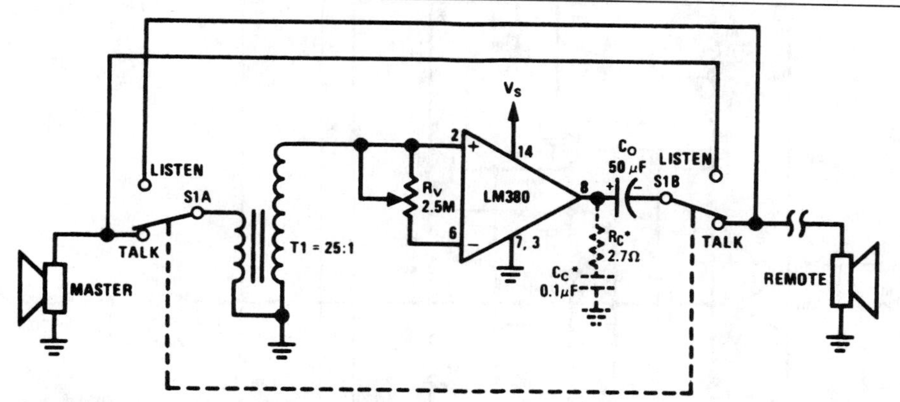

*For Stability with High Current Loads

Minimum-component intercom

Fig. 1-40 With switch S1 in the Talk position, the master-station speaker acts as the microphone. A T1 turns ratio of 25 and an IC gain of 50 allows a maximum loop gain of 1250. R_V provides a "common-mode" volume control. Switching S1 to Listen reverses the role of the master and remote speakers. National Semiconductor, Linear Applications Handbook, 1991, p. 200.

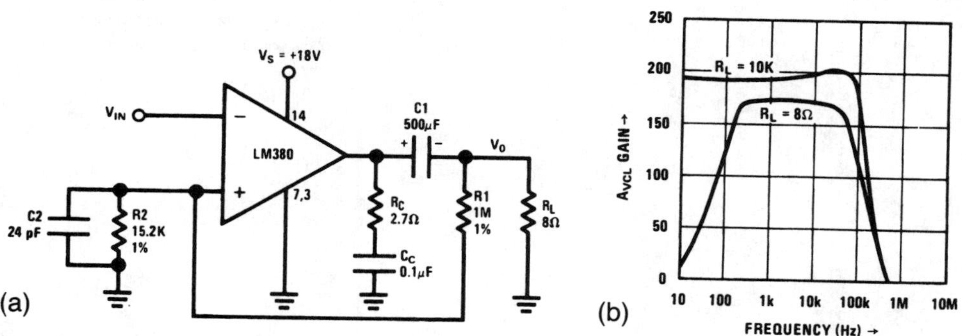

Boosted-gain amplifier

Fig. 1-41 For applications that require gains higher than the internally set gain of 50, it is possible to apply positive feedback around the LM380 for closed-loop gains of up to 300. As shown in Fig. 1-41B, this circuit has a gain of about 200 (well beyond the audio range) when the load is 10 kΩ. When the load is a typical 8-Ω speaker, the circuit provides a gain of over 170 (beyond the audio range). National Semiconductor, Linear Applications Handbook, 1991, p. 201.

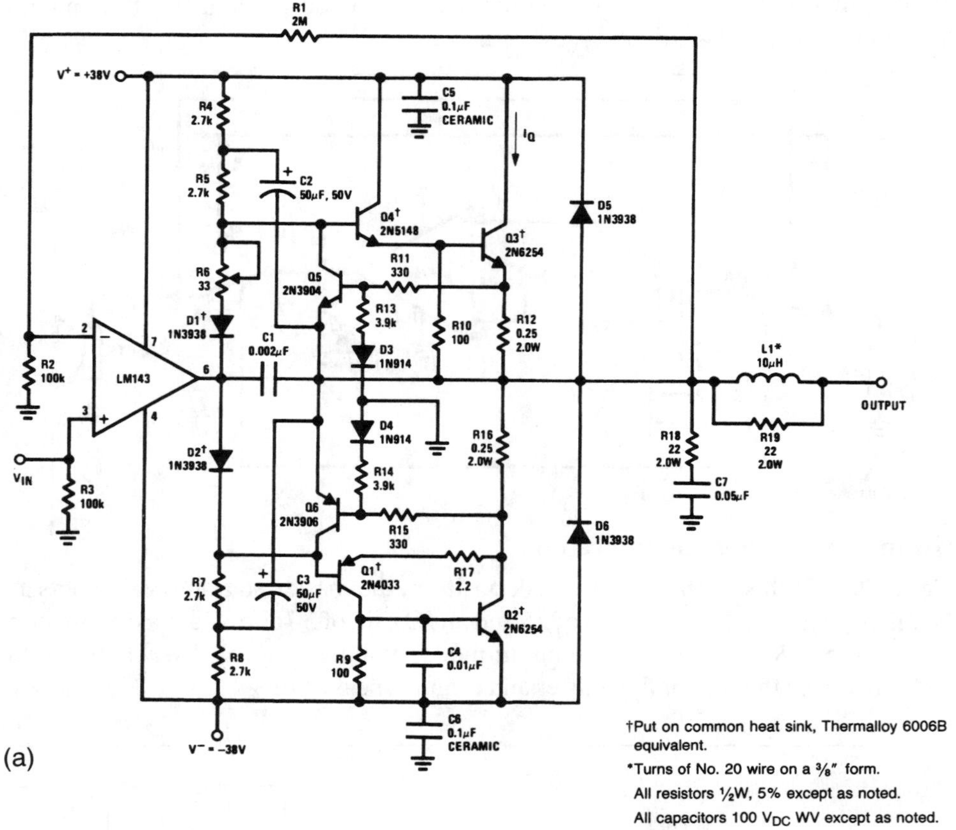

(a)

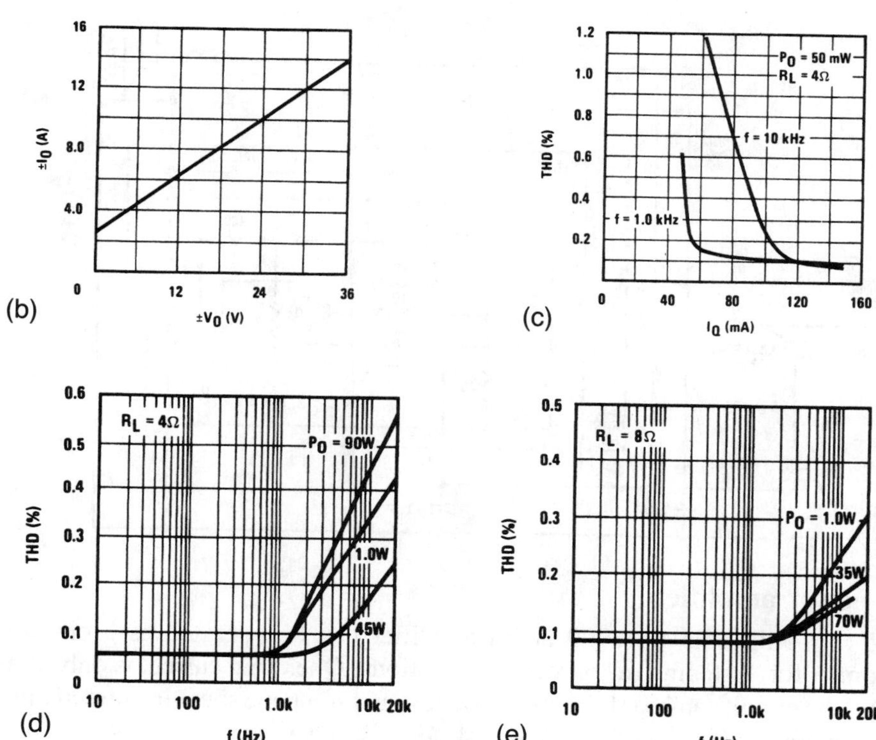

90-W audio power amplifier

Fig. 1-42 This circuit is capable of 90 W into a 4-Ω speaker, or 70 W into an 8-Ω speaker. The circuit features safe area, short-circuit and overload protection, harmonic distortion less than 0.1% at 1.0 kHz, and an all-npn output stage. Figures 1-42B through 1-42E show the circuit performance characteristics. National Semiconductor, Linear Applications Handbook, 1991, p. 383, 384.

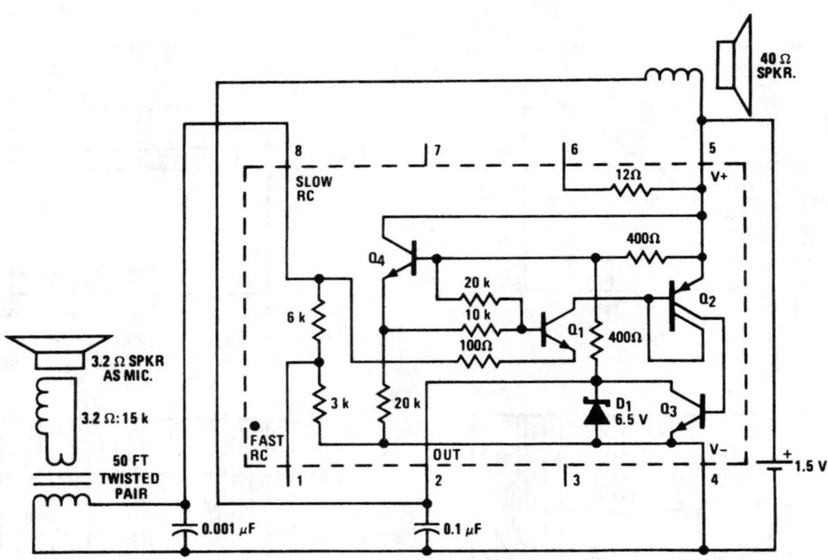

Mini-power amplifier

Fig. 1-43 This little LM3909 audio amplifier can be operated as a one-way intercom or for "listening in" on various situations. Operating current is only 12 to 15 mA. The circuit can hear fairly faint sounds, and someone speaking directly into the microphone generates a full 1.4-V peak-to-peak at the speaker. National Semiconductor, Linear Applications Handbook, 1991, p. 406.

Low-distortion bridge audio power amplifier

Fig. 1-44 This circuit is capable of 40-W into an 8-Ω speaker, or 28-W into a 16-Ω speaker, with a voltage gain of 3 and a full-power bandwidth from dc to 100 kHz. Figures 1-44B and 1-44C show the circuit performance characteristics. National Semiconductor, Linear Applications Handbook, 1991 p. 725, 727.

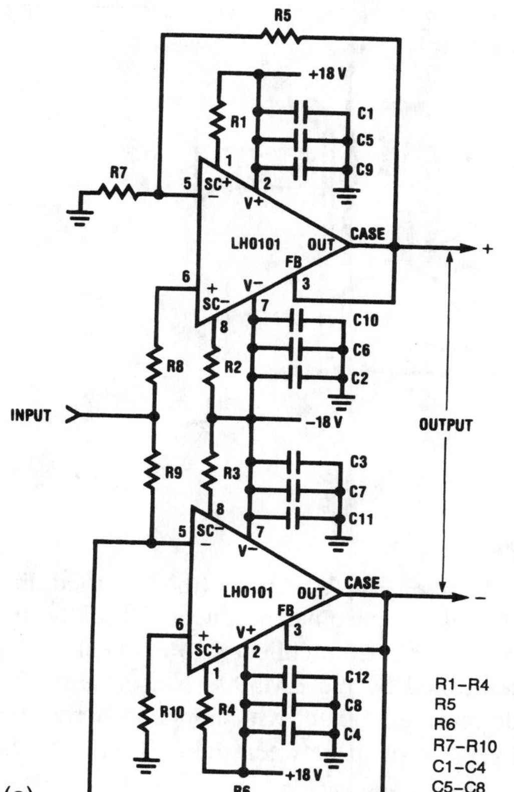

(a)

R1–R4	Current Limit Resistor	0.15Ω, 2W
R5	Feedback Resistor	5 kΩ
R6	Feedback Resistor	15 kΩ
R7–R10	Input Resistors	10 kΩ
C1–C4	Bypass Capacitors	47 μF, 25V Electrolytic
C5–C8	Bypass Capacitors	10 μF, 25V Tantalum
C9–C12	Bypass Capacitors	0.1 μF, 25V Ceramic

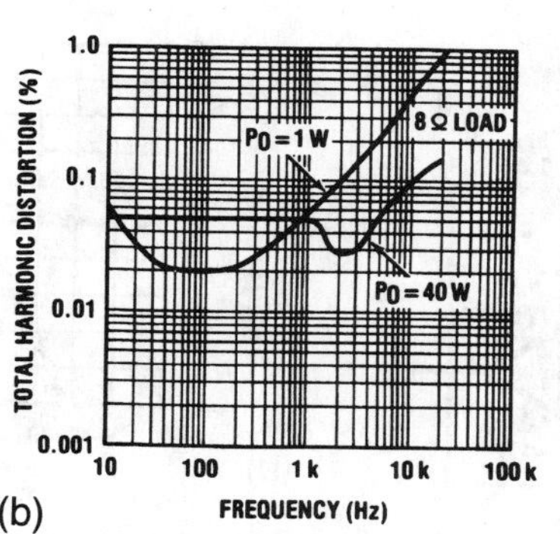

(b)

(c)

TABLE III. Bridge Audio Amplifier Specifications

A_V (Voltage Gain)	3
Z_{IN} (Input Impedance)	10 kΩ
I_q (Quiescent Current)	60 mA
P_O (Output Power)	40 Watts into 8Ω
—RMS Continuous 20 Hz–20 kHz	28 Watts into 16Ω
Full Power Bandwidth	DC to > 100 kHz
THD (Total Harmonic Distortion)	
1W 20–20,000 Hz	<0.8%
40W 20–20,000 Hz	<0.15%
IMD (Intermodulation Distortion)	
1W 60 Hz/100 kHz 4:1	<0.01%
40W 60 Hz/100 kHz 4:1	<0.002%
Peak Output Current	3.125A into 8Ω
Supply Voltage	±18V
Maximum Output Voltage Swing	21.2V rms 30V Peak

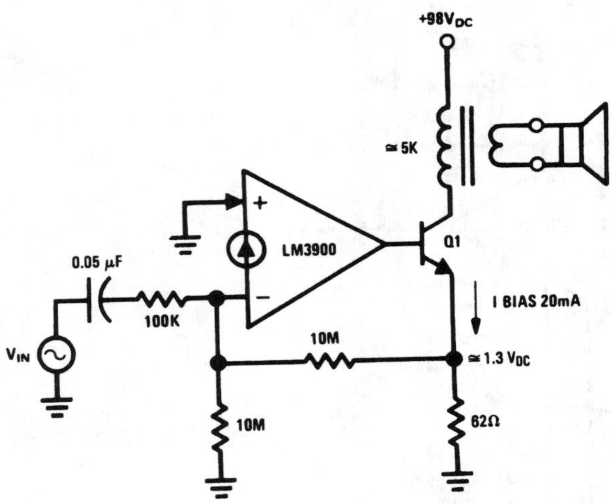

Line-operated IC audio amplifier

Fig. 1-45 This amplifier operates from a +98-V supply (the rectified line voltage). Such amplifiers are often found in consumer products. Virtually any high-voltage transistor (capable of about 100 V from collector to emitter) can be used for Q1, which is biased and controlled by the LM3900 Norton amplifier (chapter 11). The magnitude of the dc biasing voltage, which appears across the emitter resistor of Q1 is controlled by the resistor between the (−) input of the LM3900 and ground. National Semiconductor, Linear Applications Handbook 1991, p. 253.

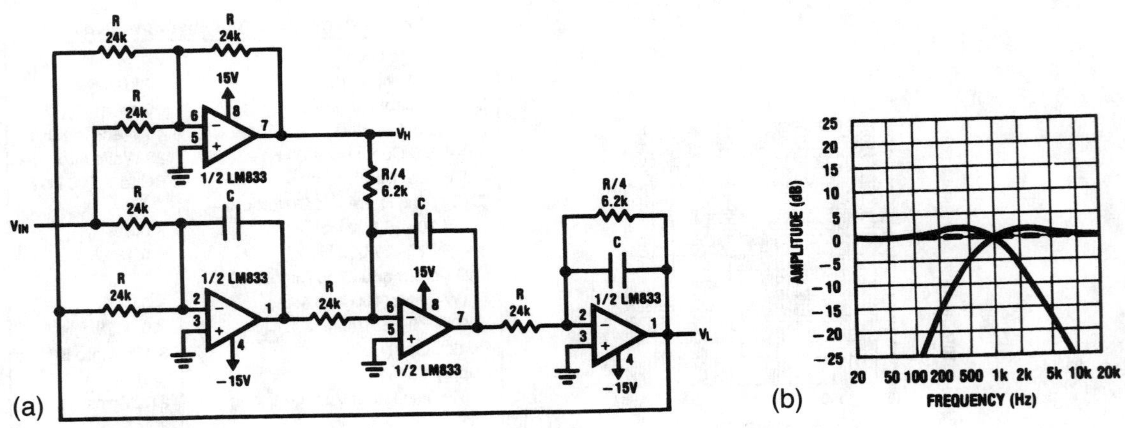

Constant-voltage crossover network

Fig. 1-46 This active crossover network provides 12-dB/octave slopes, where the crossover frequency is determined by: 1/(6.28*RC*). Figure 1-46B shows both the low-pass and high-pass responses, with a crossover frequency of 1 kHz. Notice that the summed response (dashed lines) is perfectly flat. National Semiconductor, Linear Applications Handbook, 1991, p. 943, 944.

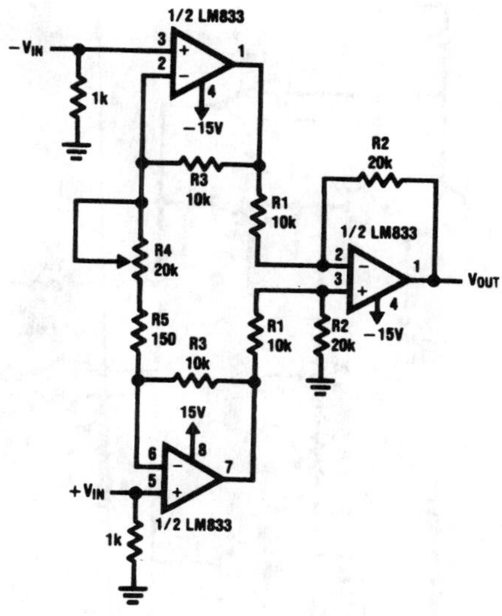

Transformerless microphone preamplifier

Fig. 1-47 This circuit uses two LM833s to amplify the input signal before the common-mode noise is cancelled in the differential amplifier. The equivalent input noise is about 760 nV over a 20-Hz to 20-kHz frequency band (−122 dB referred to 1 V), which is over 26 dB lower than a typical microphone output from the 30-dB SPL ambient noise level in a quiet room. THD is under 0.01% at maximum gain, and 0.002% at minimum gain. R4 adjusts the circuit gain from about 4 to 270. Common-mode noise is about 60 dB (or 44 dB, worst case) when R1, R2, and R3 are of 0.1% tolerance. If better common-mode rejection is needed, one of the R2s can be replaced with an 18-kΩ resistor and a 5-kΩ pot to trim for a better CMRR. If necessary, use 470-pF capacitors between inputs and ground to prevent RF interference from entering the preamp inputs. National Semiconductor, Linear Applications Handbook, 1991, p. 945.

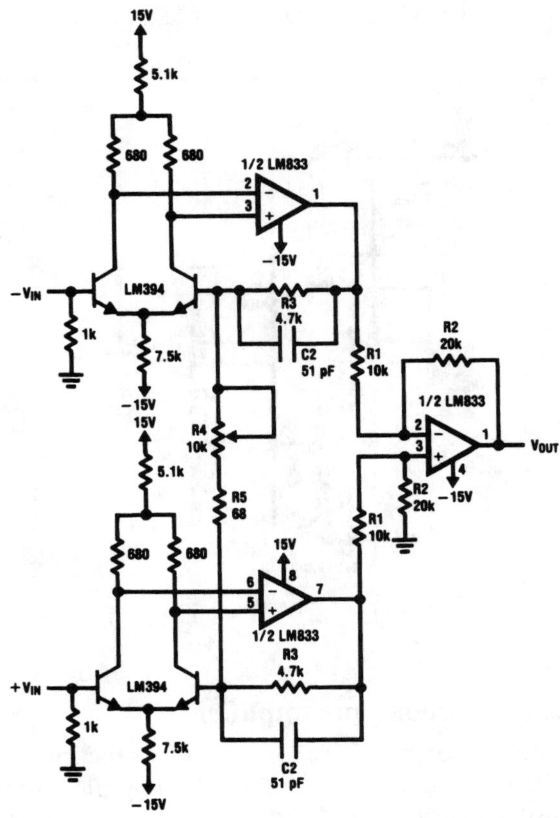

Transformerless microphone preamplifier with low-noise inputs

Fig. 1-48 This circuit is similar to that of Fig. 1-47, except that two LM394s are used as input devices for the LM833 gain stages. The equivalent input noise of the Fig. 1-48 circuit is about 340 nV over a 20-Hz to 20-kHz band (−129 dB referred to 1 V). National Semiconductor, Linear Applications Handbook, 1991, p. 946.

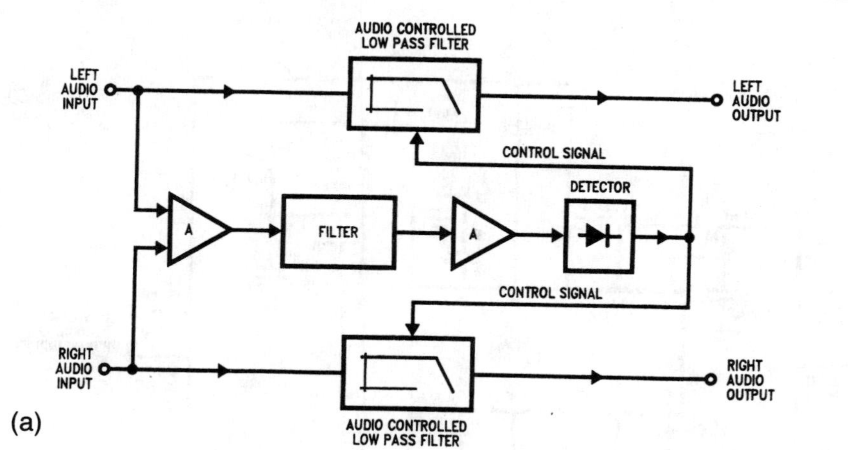

AUDIO CONTROLLED
LOW PASS FILTER
LEFT
AUDIO
INPUT
LEFT
AUDIO
OUTPUT
CONTROL SIGNAL
DETECTOR
A
FILTER
A
CONTROL SIGNAL
RIGHT
AUDIO
INPUT
RIGHT
AUDIO
OUTPUT
AUDIO CONTROLLED
LOW PASS FILTER

(a)

LEFT
OUT
100 kΩ
100 kΩ
LEFT
IN
1 μF
1 μF
0.0033 μF
12
11
13
20 kΩ
20 kΩ
gm
OP AMP
4
3
0.0033 μF
20 kΩ
OP AMP
gm
20 kΩ
2
1 μF
1 μF
12 V
1
100 kΩ
100 kΩ
RIGHT
OUT
RIGHT
IN
14
100 μF
1.2 V
30 kΩ
Σ AMP
A V = 26
27 kΩ
PK DET
1.2 V
700Ω
90Ω
V TO I
CONVERTER
1 kΩ
ON
BYPASS
5
6
8
9
18
7
0.1 μF
.001 μF
.047 μF
1 kΩ
19 kHz
1 μF

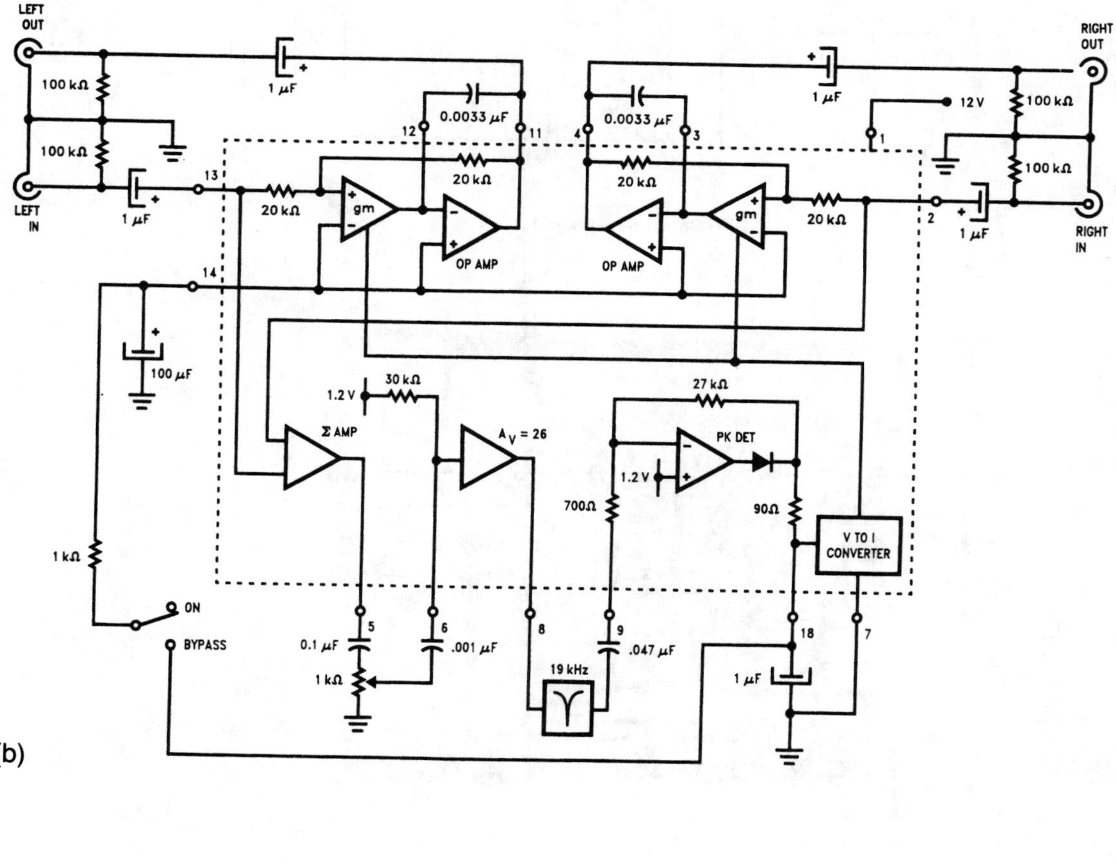

(b)

Continued

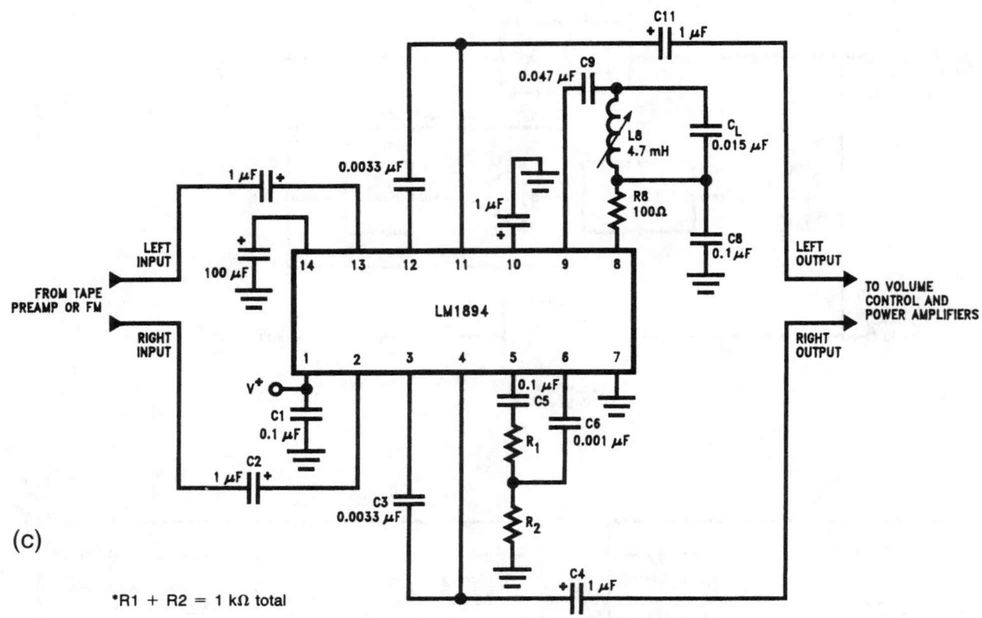

(c)

*R1 + R2 = 1 kΩ total

DNR Component Diagram

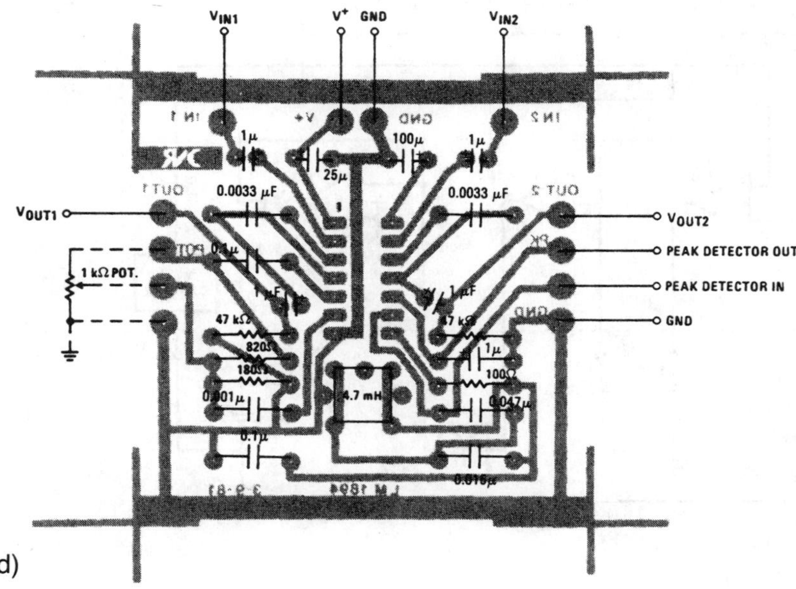

(d)

Noncomplementary audio noise reduction system

Fig. 1-49 Figure 1-49A shows the block diagram, and Figs. 1-49B and 1-49C show the full schematic details of noncomplementary or dynamic noise-reduction (NR) systems. Complementary NR systems require that certain frequencies be emphasized during recording, then de-emphasized during playback in a complementary mode. That is, if highs are emphasized during recording, highs are de-emphasized during playback. These noncomplementary circuits can be applied to any stereo audio system, without regard as to how the material was recorded. Both the left and right audio paths are controlled by signals derived from the mixed audio input (after filtering, amplification, and detection), all of which occurs in the LM1894. The circuits must be placed immediately after the source preamp, and before any circuit that includes user controls (volume, tone, etc.). L8 in Fig. 1-49C is part of a 19-kHz notch filter, and is adjusted to remove the FM pilot signal. R1 and R2 in Fig. 1-49C set overall gain, as does the 1-kΩ pot in Fig. 1-49B. The 1-kΩ pot is adjusted for best noise reduction of a given program material. Figure 1-49D shows the PC layout for the circuit of Fig. 1-49B. National Semiconductor, Linear Applications Handbook, p. 988, 995, 997, 998.

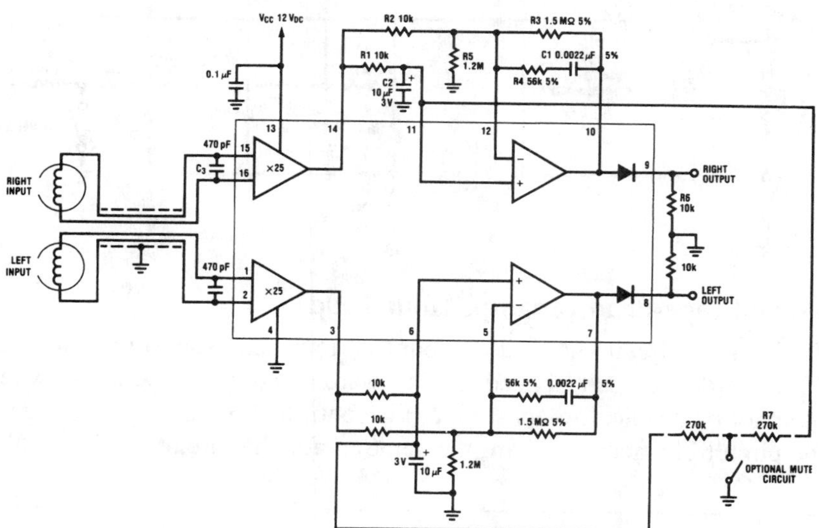

Dc-coupled tape-head preamps (two head)

Fig. 1-50 This circuit shows a dc-coupled, two-head stereo preamp using an LM1897. The circuit characteristics are: a GBW of 76 dB at 20 kHz, an S/N ratio of 62 dB (CCIR/ARM), distortion 0.03% at 1 kHz, power-supply rejection greater than 95 dB, channel separation of 60 dB, and a typical turn-on delay of 0.4 s. National Semiconductor, Linear Applications Handbook, p. 1038.

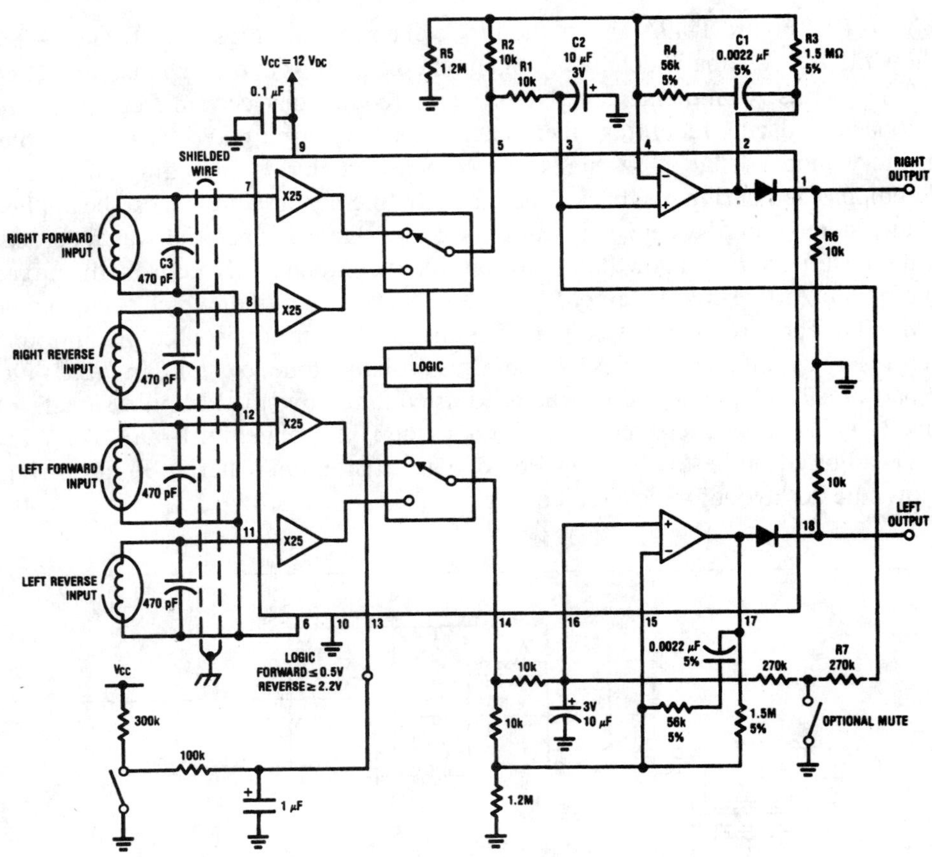

Dc-coupled tape-head preamps (four head)

Fig. 1-51 This circuit shows a dc-coupled, four-head stereo preamp using an LM1937. The desired pair of heads (for forward and reverse operation, where the heads are not mechanically repositioned) is controlled by the logic voltage at pin 13. The circuit characteristics are the same as for the circuit of Fig. 1-50. National Semiconductor, Linear Applications Handbook, p. 1040.

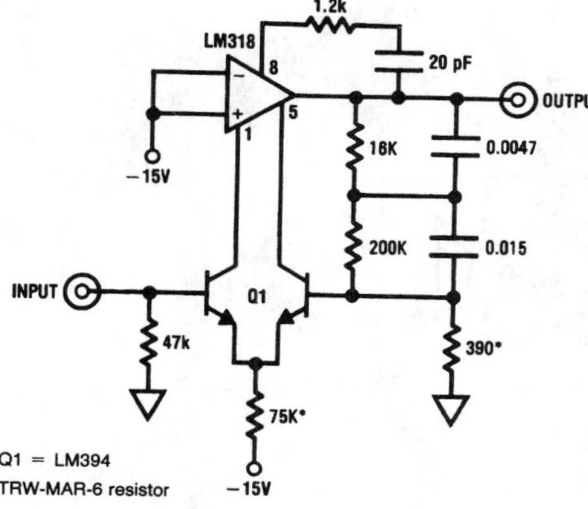

Frequency	Total Harmonic Distortion				
20	<0.002	<0.002	<0.002	<0.002	<0.002
100	<0.002	<0.002	<0.002	<0.002	<0.002
1000	<0.002	<0.002	<0.002	<0.002	<0.002
10000	<0.002	<0.002	<0.002	<0.0025	<0.003
20000	<0.002	<0.002	<0.004	<0.004	<0.007
Output Amplitude (Vrms)	0.03	0.1	0.3	1.0	5.0

Ultra-low-noise RIAA preamp

Fig. 1-52 This circuit uses an LM394 and an LM318 to form an ultra-low noise preamp. Noise referred to a 10-mV input signal is −90 dB down, measuring 0.55 μV and 70 pA in a 20-kHz bandwidth. The noise figure is less than 2 dB when the amplifier is used with standard phono cartridges, which have an equivalent wideband (20 kHz) noise of 0.7 μV. Worst-case dc output offset voltage is about 1 V with a 1-kΩ cartridge. THD for various output amplitudes is shown in the table. National Semiconductor, Linear Applications Handbook, p. 884.

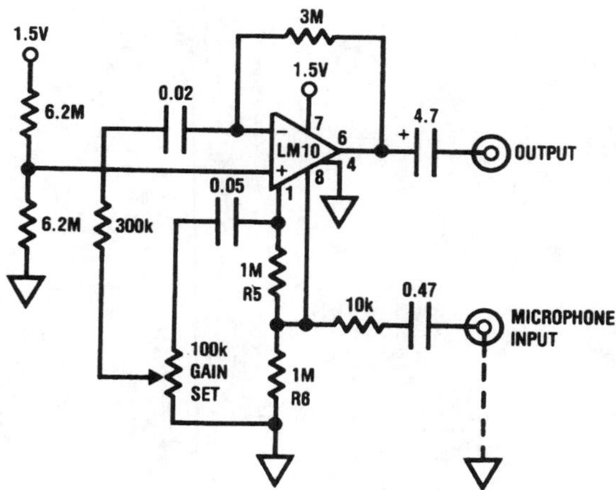

Microphone preamp with low supply voltage

Fig. 1-53 This preamp operates from a single 1.5-V cell, and can be located right at the microphone. The circuit has 60-dB gain with a 10-kHz bandwidth, unloaded, and 5-kHz bandwidth while loaded with 500 Ω. The input impedance is 10 kΩ. National Semiconductor, Linear Applications Handbook, p. 885.

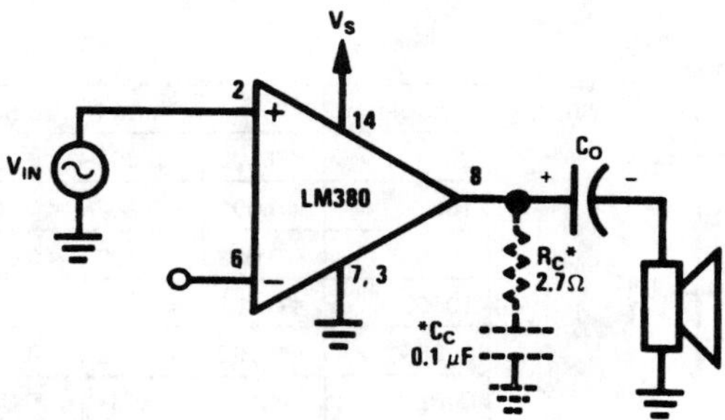

*For Stability With High Current Loads

Minimum-component audio amplifier

Fig. 1-54 Because the LM380 has internal biasing and compensation, this basic audio amplifier requires only an output-coupling capacitor (typically 500 μF). Of course, such an amplifier has no tone- or volume-control (Figs. 1-33 through 1-36). The circuit can supply about 2.5 W into an 8-Ω speaker, with a V_S of 12 V or greater. The R_C and C_C components suppress 5- to 10-MHz oscillation that can occur during the negative swing into a high-current load. Although such oscillation will not pass through the speaker, the radiated signals can produce interference in RF-sensitive equipment. National Semiconductor, Linear Applications Handbook, 1991, p. 197.

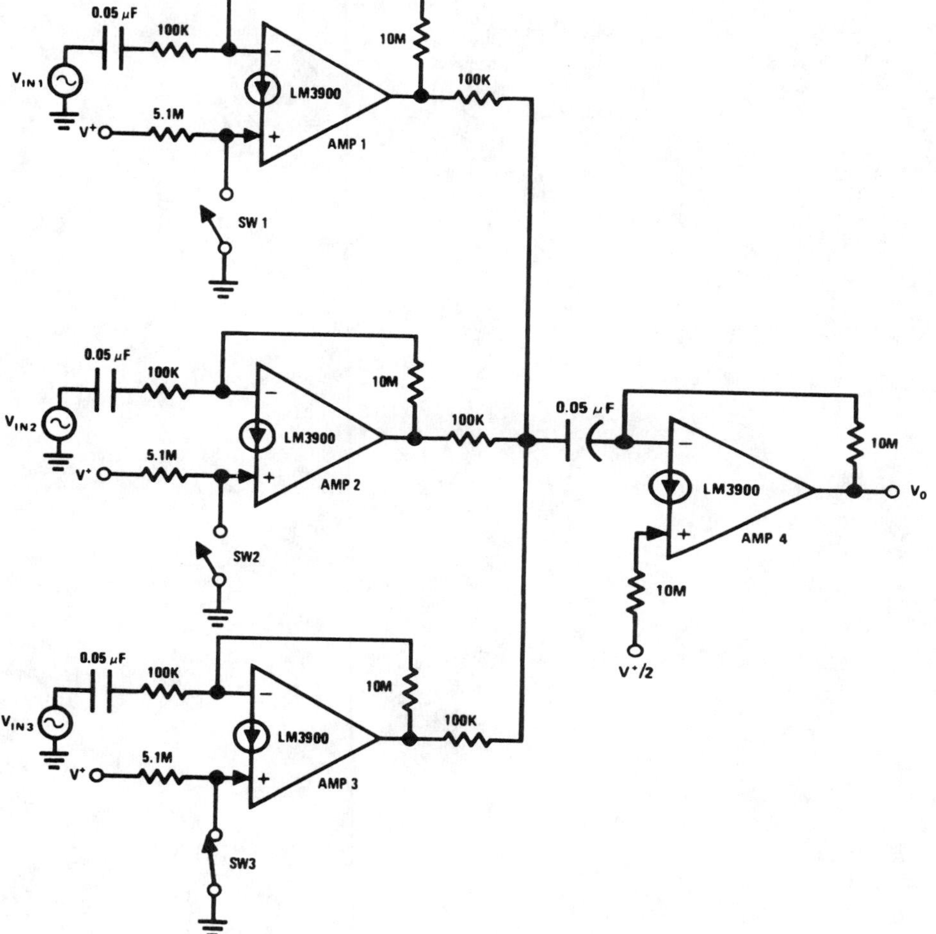

Audio mixer or selector

Fig. 1-55 This circuit uses the four Norton amplifiers of an LM3900 to form an audio-mixing or selection circuit (Norton amplifier circuits are also shown in chapter 11). With this circuit, particular amplifiers can be gated off with dc-control signals applied to the (+) inputs. Amplifier 3 is active, with SW3 closed, while amplifiers 1 and 2 are driven into saturation by the +V input applied through the 5.1-M resistors. National Semiconductor, Linear Applications Handbook, 1991, p. 250.

2

RF and IF circuits

This chapter is devoted to RF and IF circuits, including amplifiers and mixers, that are found in radio receivers and transmitters. Both IC and discrete circuits are included. It is assumed that the reader is already familiar with RF/IF-circuit basics (amplifier principles, bias operating points, etc.), practical considerations (heat-sinks, power dissipation, component-mounting techniques), simplified RF/IF design (frequency limitations) and basic RF/IF test/troubleshooting. If not, read *Lenk's RF Handbook*, McGraw-Hill, 1991. However, the following paragraphs summarize both test and troubleshooting of RF/IF circuits. This information is included so that even those readers not familiar with electronic procedures can both test the circuits described here, and localize problems if the circuits fail to perform as shown.

RF/IF circuit tests

The tests described for audio circuits (chapter 1) can generally be applied to RF/IF circuits. Of course, because of the frequency differences, certain special test equipment and procedures are required for RF/IF. This section concentrates on these differences. Start by reading the first section of chapter 1 if you haven't already. Then, read the following paragraphs and apply the tests to the circuits of this chapter. If the circuits pass these basic tests, the circuits can be used immediately. If not, the tests provide a starting point for the troubleshooting procedures that are described in the next section.

Typical RF/IF test equipment

Test/troubleshooting for the circuits in this chapter can be performed using meters, generators, scopes, frequency counters, power supplies, and assorted clips, patch cords, and so on. So, if you have a good set of test equipment that is suitable for other electronic work, you can probably get by. Possible exceptions are RF generators with internal modulation, RF probes, dummy loads, RF wattmeters, and field-strength meters. Here are some points to be considered when selecting and using RF/IF test equipment.

RF generators

Obviously, the *RF generator* must be capable of operating at frequencies up to and beyond the circuit frequency (the highest circuit frequency in this chapter is 1 GHz, with most circuits operating well below that frequency). For some tests, the generator must have an internal-modulation feature (or be capable of external modulation). A typical RF generator will have 30% internal modulation at either 400 or 1000 Hz. The generator output meter must be capable of indicating signals on the order of a few microvolts to make typical RF circuit checks. Some generators have two meters, one for output amplitude and one for modulation. Other generators use one meter, and select between the two functions with a switch.

RF meters and scopes

The meters used for RF work are essentially the same as for all other electronic circuits. Most tests and troubleshooting can be done with standard multimeters, either digital (most digital meters are electronic meters and thus draw a minimum of current from the circuit) or moving-needle.

Meters can be used to measure both voltages and resistances in RF circuits, as is required for the troubleshooting procedures in this chapter. When used with the appropriate probe, meters can be used to trace signals throughout all RF circuits. With a probe, the meter indicates the presence of a signal in the circuit and the signal amplitude, but not the signal frequency or waveform.

RF signals can be traced with a scope (if the scope is equipped with the proper probe, or has a bandwidth that is greater than the RF signals involved). You can check amplitude, frequency, and waveforms of signals with a scope. However, many RF service technicians do not use scopes extensively, for the following reasons.

The scope can measure signal amplitude, but a meter is easier to read. The same applies to signal frequency. The frequency counter is easier to read, and it is far simpler to measure frequency with a scope, particularly in the typical RF-circuit range. The scope is superior for monitoring waveforms, but most RF signals are sine waves, where waveforms are not critical.

RF, demodulator, and low-capacitance probes

An *RF probe* is required when the signals to be measured are at radio frequencies and are beyond the capabilities of the meter or scope. Always use the probe that is supplied with the meter or scope. If no such probe is available, the circuit shown in Fig. 2-Ab will work with most meters and scopes. That is, the probe will work for testing and troubleshooting, but not for precise voltage measurement.

Demodulator probes are essentially the same as RF probes, but the circuit values and basic functions are somewhat different. When the high-frequency signals contain modulation (which is typical for the modulated RF-carrier signals of most communications equipment), a demodulator probe is more effective for signal tracing. Figure 2-Ad shows a typical demodulator-probe circuit. Notice that the circuit is essentially a half-wave rectifier, with C1 and R2 acting as a filter. The demodulator probe produces both an ac and a dc output. The RF signal is converted into a dc voltage that is approximately equal to the peak value. The modulation voltage (if any) on the RF signal appears as ac at the probe output. In use, the meter is set to dc and the RF signal is measured. Then, the meter is set to ac and the modulating voltage is measured. R1 is adjusted (before use) so that the meter's dc scale reads the correct value.

A *low-capacitance probe* is shown in Fig. 2-Aa. The R_1 and R_2 are selected to form a 10:1 voltage divider between the circuit being tested and the meter or scope input. Such probes serve the dual purpose of capacitance and voltage reduction. When using a low-capacitance probe, remember that voltage indications are one-tenth (or whatever value of attenuation is used) of the actual value. The primary value of low-capacitance probes in RF work is that such probes to not alter the capacitance values of tuned circuits to the same extent as other probes.

RF dummy load

Never adjust an RF circuit without a load connected to the output. This will almost certainly cause damage to the RF circuit. For example, when a transmitter power amplifier is connected to an antenna or load, power is transferred from the final RF-amplifier stage to the antenna or load. Without an antenna or load, the final RF stage must dissipate the full power and will probably be damaged (even with heatsinks). Equally important, you should not make any major adjustments (except for a brief final tune-up) to a transmitter that is connected to a radiating antenna. You will probably cause interference.

These two problems can be overcome by a nonradiating load, commonly called a *dummy load*. A number of commercial dummy loads are available for communications circuit troubleshooting and testing. The RF wattmeters described in this chapter contain dummy loads. It is also possible to make up dummy loads that are suitable for most RF-circuit troubleshooting.

The two generally accepted dummy loads are: the *fixed resistance* and the *lamp*. Remember that these loads are for routine test/troubleshooting, but are not a substitute for an RF wattmeter or special communications test set.

The simplest dummy load is a *fixed-resistor dummy load* that is capable of dissipating the full power output of the RF circuit. The resistor can be connected to

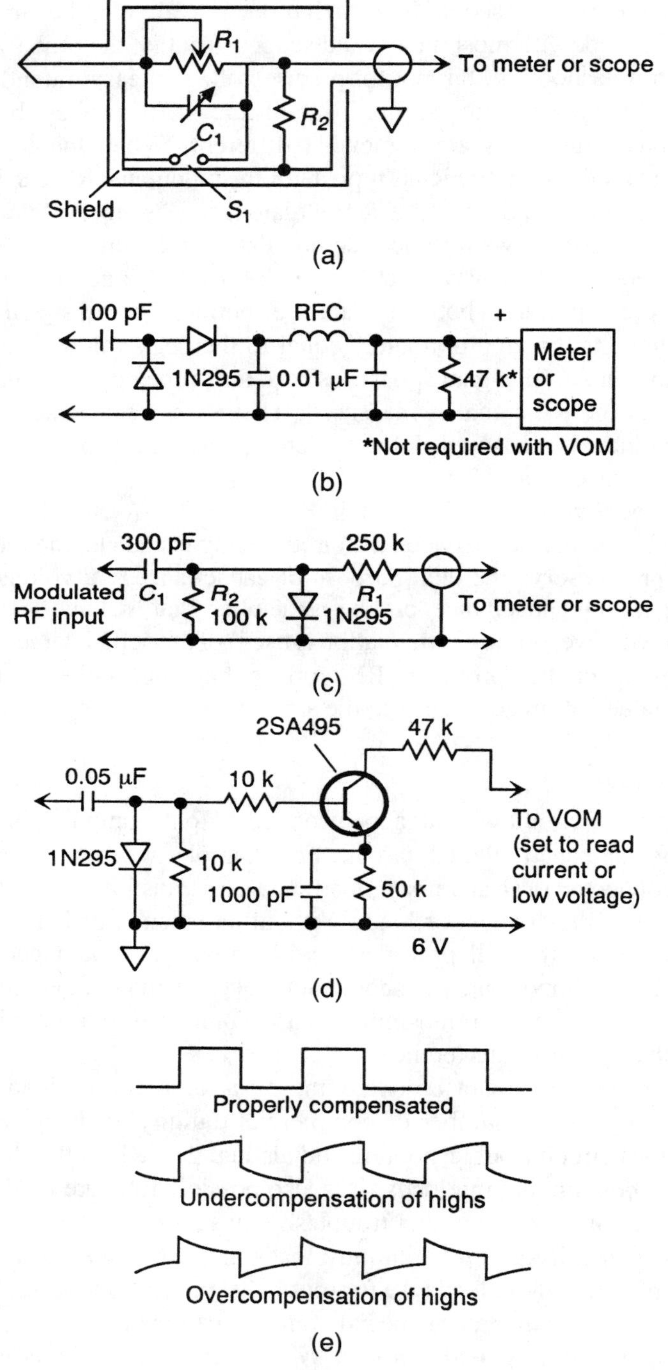

Fig. 2-A RF-probe circuits.

the output of the RF circuit (for example, at the antenna connector of a transmitter) by means of a plug (Fig. 2-Ba).

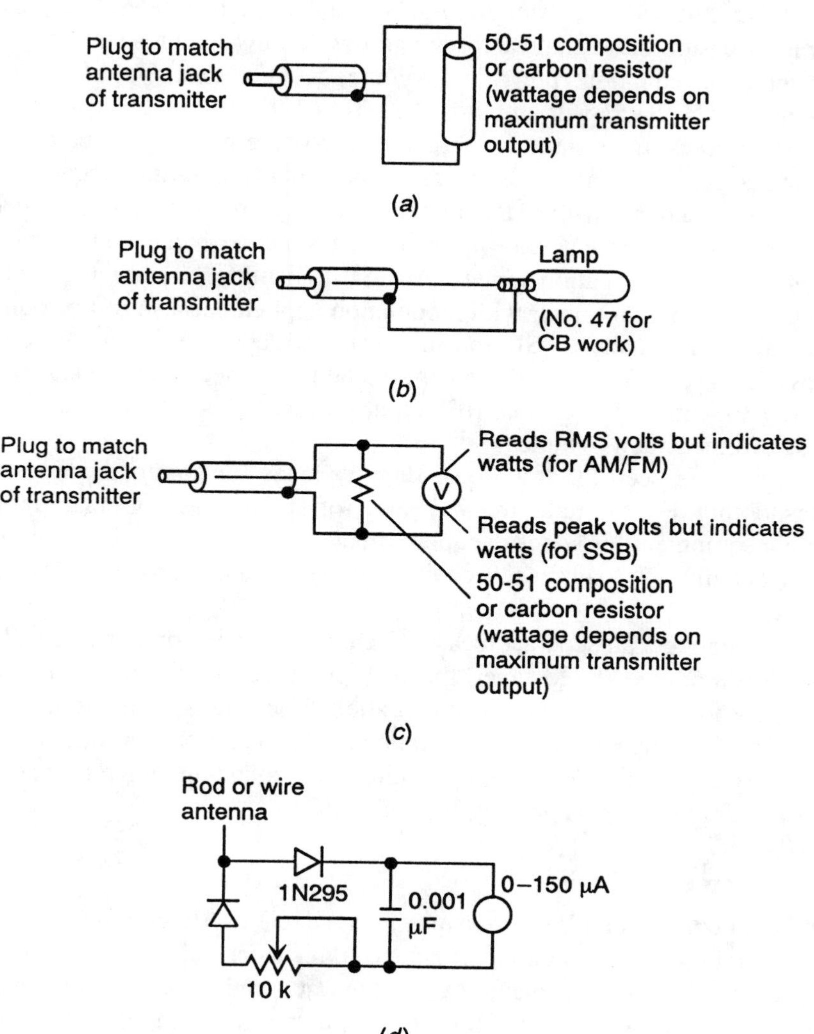

Fig. 2-B RF dummy loads, wattmeters, and RFS meters.

Most transmitters operate with a 50-Ω antenna and lead-in (coax) and thus require a 50-Ω resistor. The nearest standard resistor is 51 Ω. This 1-Ω difference is not critical, but it is essential that the resistor be noninductive (composition or carbon), never wire wound. Wire-wound resistors have some inductance, which changes with frequency and causes the load impedance to change. Always use a resistor with a power rating that is greater than the anticipated maximum output power of the circuit.

It is possible to get an approximate measurement of RF power output from a transmitter circuit with a *resistor dummy load* and a suitable meter. The procedure is simple. Measure the voltage across the 50-Ω dummy-load resistor and find the power with the equation: *power* = *voltage*^{2}/50. For example, if the voltage measured is 14 V, the power output is: $14^2/50 = 3.92$ W.

Certain precautions must be observed. First, the meter must be capable of producing accurate voltage indications at the circuit operating frequency. This usually requires a meter with an RF probe, preferably a probe that is calibrated with the meter. An AM or FM transmitter circuit should be checked with an rms voltmeter with no modulation applied. An SSB-transmitter circuit must be checked with a peak-reading voltmeter with modulation applied because SSB produces no output without modulation. SSB transmitters use wideband power amplifiers (such as shown in Figs. 2-50 and 2-51), where the output is measured as *peak-envelope power (PEP)*, rather than simple RF output power. RF power measurements are discussed further in this chapter.

Lamps have been the traditional dummy loads for communications circuit troubleshooting. For example, the #47 lamp (often found as a pilot lamp for older electronic equipment) provides the approximate impedance and power dissipation required as a dummy load for CB equipment. The connections are shown in Fig. 2-Bb.

You cannot get an accurate measurement of RF power output when a lamp is used as the dummy load. However, the lamp provides an indication of the relative power and shows the presence of modulation. The intensity of the light that is produced by the lamp varies with modulation (more modulation produces a brighter glow), so you can tell at a glance if the circuit is producing an RF carrier (steady glow) and if modulation is present (varying glow).

RF wattmeter

A number of commercial *RF wattmeters* are available for RF circuit test. The basic RF wattmeter consists of a dummy load (fixed resistor) and a meter that measures voltage across the load. The meter reads out in watts (rather than in volts), as shown in Fig. 2-Bc. Simply connect the RF wattmeter to the circuit output (say the antenna connector of a transmitter), key the transmitter, and read the power output on the wattmeter scale.

Figures 2-C and 2-D show typical RF wattmeter connections. Notice that these illustrations show connections for complete transmitters or communications sets, rather than for the power-amplifier circuits of this chapter, and that the circuits are generalized (not for a specific circuit). However, the basic connections can be used for power-output tests and for troubleshooting the circuits that are described here.

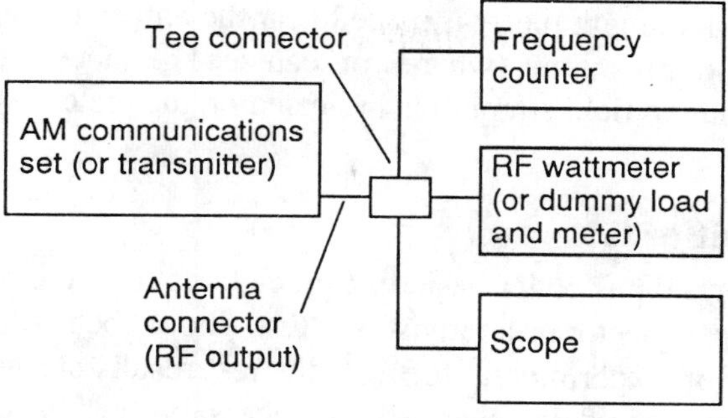

Fig. 2-C AM transmitter test connections.

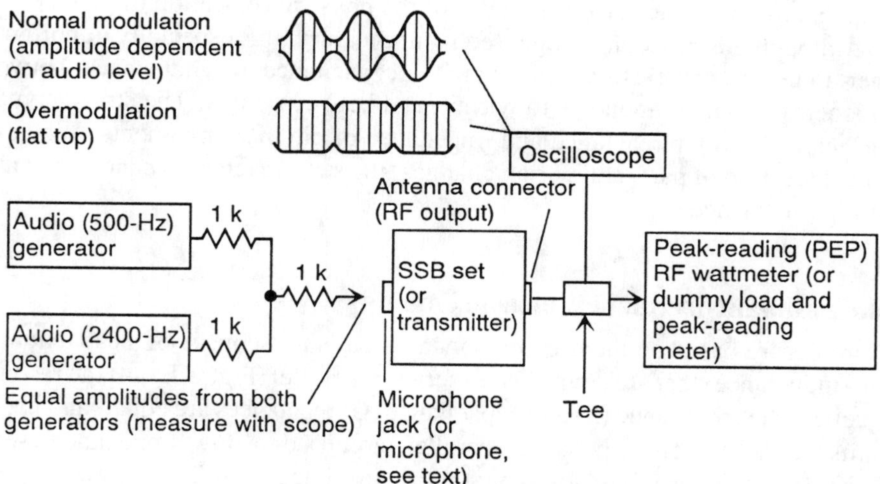

Fig. 2-D SSB transmitter test connections.

Field-strength meter

The two basic types of *field-strength meters* are: the *basic relative field strength (RFS) meter* and the *precision laboratory* or *broadcast-type instrument*. Most RF-circuit tests and troubleshooting can be carried out with simple RFS instruments. An exception is where you must make precision measurements of broadcast antenna radiation patterns.

The purpose of a field-strength meter is to measure the strength of the RF signals radiated by an antenna. This simultaneously tests the RF-circuit output, the antenna, and the lead-in. In the simplest form, a field-strength meter consists of an antenna (a short piece of wire or rod), a potentiometer, diodes, and a microammeter (Fig. 2-Bd). More elaborate RFS meters include a tuned circuit and possibly an amplifier.

In use, a field-strength meter is placed near the antenna at some location that is accessible to the transmitter (where you can see the meter), the transmitter is keyed, and the relative field strength is indicated on the meter.

Basic RF-circuit tests

The following paragraphs cover basic test procedures for RF circuits and components. For the experimenter or hobbyist, the tests should be made when the circuit is first completed in experimental form. If the test results are not as desired, the component values should be changed as necessary (or the circuit should be adjusted) to get the desired results. Also, RF circuits should always be retested in final-form (with all components soldered in place). This shows if there is any change in circuit characteristics because of the physical relocation of parts.

Although this procedure might seem unnecessary, it is especially important at higher radio frequencies. Often, there is capacitance or inductance between components, from components to wiring, and between wires. The stray "components" can add to the reactance and impedance of circuit components. When the physical location of parts and wiring changes, the stray reactances change and alter circuit performance.

Basic resonant-frequency measurements

RF circuits are based on the use of resonant circuits that consist of a capacitor and a coil (inductance) that are connected in series or parallel (Fig. 2-E). At the resonant frequency, the inductance (*L*) and capacitance (*C*) reactances are equal, and the LC circuits act as a high impedance (parallel circuit) or a low impedance (series circuit). In either case, any combination of capacitance and inductance has some resonant frequency.

A meter can be used with a signal generator to find the resonant frequency of either series or parallel LC circuits, using the following procedure.

1. Connect the equipment (Fig. 2-F). Use the connections in Fig. 2-Fa for parallel-resonance, or the connection of Fig. 2-Fb for series-resonance.
2. Adjust the generator output until a convenient midscale indication on the meter. Use an unmodulated signal from the generator.
3. Starting at a frequency well below the lowest possible circuit frequency, slowly increase the generator frequency. If there is no way to judge the approximate resonant frequency, use the lowest generator frequency.
4. If the circuit being tested is parallel-resonant, watch the meter for a maximum (peak) indication.

(a) Resonance and Impedance

Series (zero impedance) 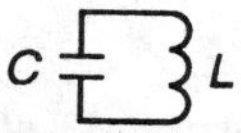   Parallel (infinite impedance)

$$F\,(\text{MHz}) = \frac{0.159}{\sqrt{L\,(\mu\text{H}) \times C\,(\mu\text{F})}} \qquad L\,(\mu\text{H}) = \frac{2.54 \times 10^4}{F\,(\text{kHz})^2 \times C\,(\mu\text{F})}$$

$$C\,(\mu\text{F}) = \frac{2.54 \times 10^4}{F\,(\text{kHz})^2 \times L\,(\mu\text{H})}$$

(b) Capacitive Reactance

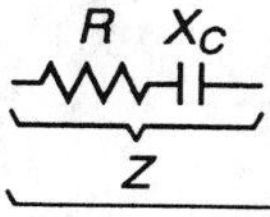

Series:

$$Z = \sqrt{R^2 + X_C^2} \qquad Q = \frac{X_C}{R}$$

$$C = \frac{1}{6.28 F X_C} \qquad F = \frac{1}{6.28 C X_C}$$

Parallel:

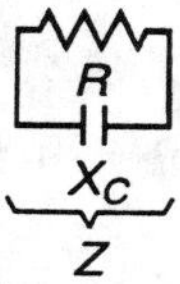

$$Z = \frac{R X_C}{\sqrt{R^2 + X_C^2}} \qquad Q = \frac{R}{X_C} \qquad X_C = \frac{159}{F\,(\text{kHz}) \times C\,(\mu\text{F})}$$

(c) Inductive Reactance

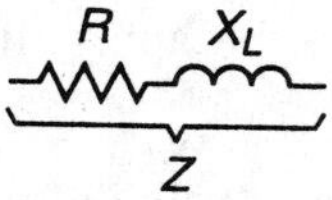

Series:

$$Z = \sqrt{R^2 + X_L^2} \qquad Q = \frac{X_L}{R}$$

$$L = \frac{X_L}{6.28F} \qquad F = \frac{X_L}{6.28L}$$

Parallel:

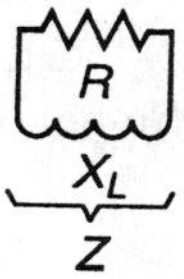

$$Z = \frac{R X_L}{\sqrt{R^2 + X_L^2}} \qquad Q = \frac{R}{X_L} \qquad X_L = 6.28 \times F\,(\text{MHz}) \times L\,(\mu\text{F})$$

Fig. 2-E Resonant-circuit equations: (a) resonance and impedance; (b) capacitive reactance; (c) inductive reactance.

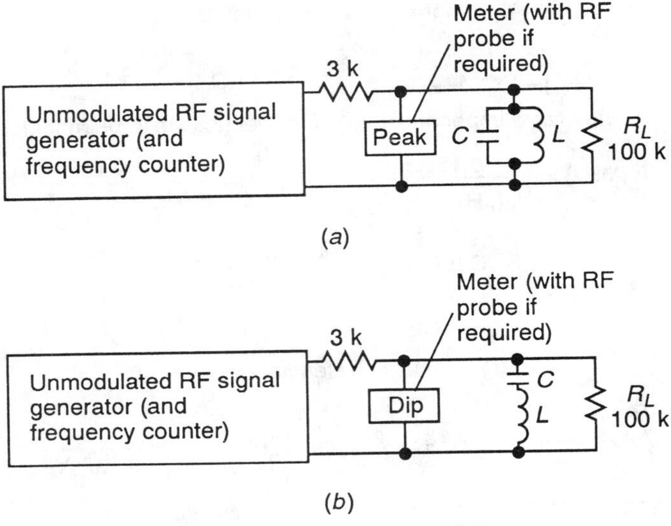

Fig. 2-F Basic RF voltage measurement.

5. If the circuit being tested is series-resonant, watch the meter for a minimum (dip) indication.
6. The resonant frequency of the circuit under test is the one at which there is a maximum (for parallel) or minimum (for series) indication on the meter.
7. There might be peak or dip indications at harmonics of the resonant frequency. Therefore, the test is most efficient when the approximate resonant frequency is known.
8. The value of load resistor R_L is not critical. The load is shunted across the LC circuit to flatten or broaden, the resonant response (to lower the circuit Q) and cause the voltage maximum or minimum to be approached more slowly. A suitable trial value of R_L is 100 kΩ. A lower value of R_L sharpens the resonant response, and a higher value flattens the curve.

Basic coil inductance measurements

A meter can be used with a signal generator and a fixed capacitor (of known value and accuracy) to find the inductance of a coil, using the following procedure.

1. Connect the equipment (Fig. 2-G). Use a capacitor value such as 10 μF, 100 pF, or some other even number to simplify the calculation.
2. Adjust the generator output for a convenient midscale indication on the meter. Use an unmodulated signal output from the generator.

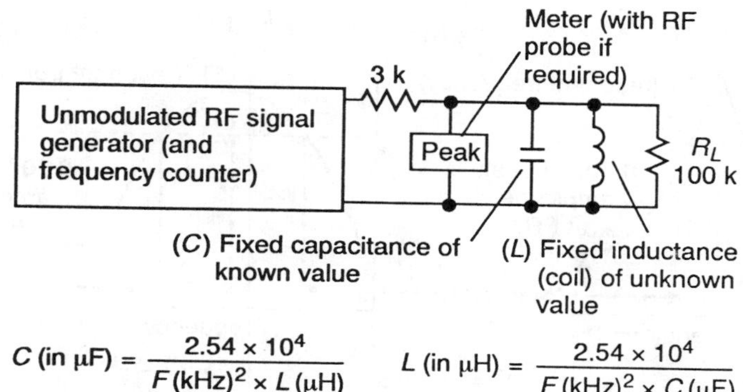

Fig. 2-G Basic coil inductance measurements.

3. Starting at a frequency well below the lowest possible resonant frequency of the inductance-capacitance combination under test, slowly increase the generator frequency. If there is no way to judge the approximate resonant frequency, use the lowest generator frequency.
4. Watch the meter for a maximum (peak) indication. Notice the frequency at which the peak indication occurs. This is the resonant frequency of the circuit.
5. Using the resonant frequency, and the known capacitor value, calculate the unknown inductance using the equation of Fig. 2-F. The procedure can be reversed to find an unknown capacitance value when a known inductance value is available.

RF-coil fabrication Figure 2-H shows the equations that are necessary to calculate the self-inductance of a single-layer, air-core coil (the most common type of coil that is used in RF circuits). Notice that maximum inductance is obtained when the ratio of coil-radius to coil-length is 1.2 (when the length is 0.8 of the radius). RF coils wound for this ratio are the most efficient (maximum inductance for minimum physical size).

Assume that you must design an RF coil with 0.5-μH inductance on a 0.25-in radius (air core, single layer). Using the equations of Fig. 2-Hb, for maximum efficiency, the coil length must be 0.8*R*, 0.2 in. Then:

$$N = \sqrt{\frac{17 \times 0.5}{0.25}} = \sqrt{34} = 5.83 \text{ turns}$$

For practical purposes, use six turns and spread the turns slightly. The additional part of a turn increases inductance, but the spreading decreases inductance. After the coil is made, check the inductance using a fixed capacitor (Fig. 2-G).

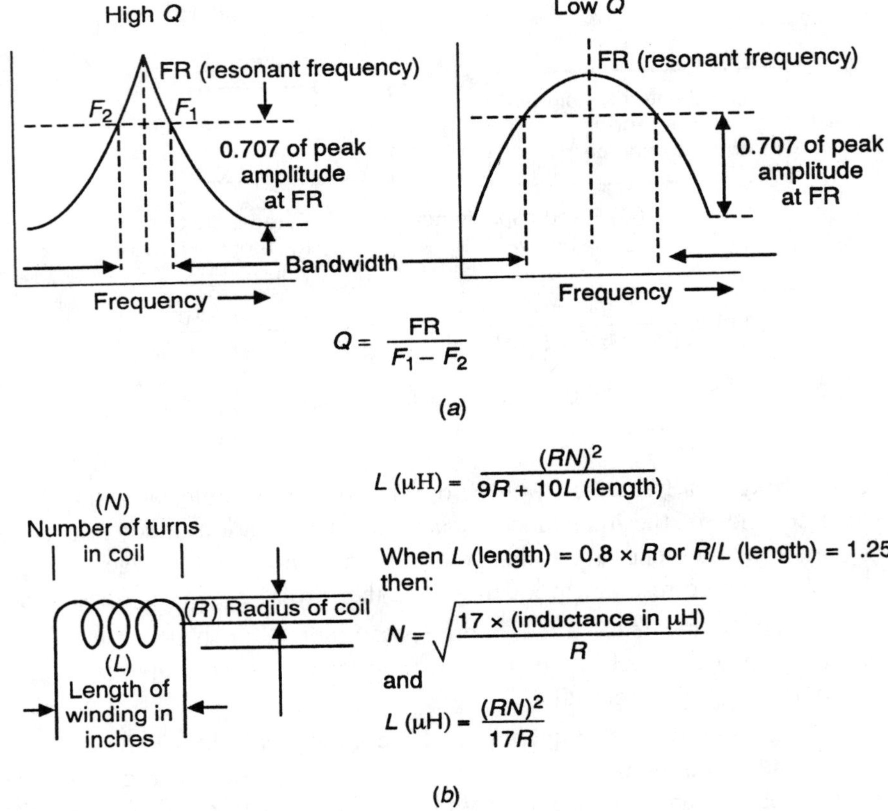

Fig. 2-H *Q* and inductance calculations for RF coils.

Basic coil self-resonance and distributed capacitance measurements

No matter what design or winding method is used, there is some distributed capacitance in any coil. When the distributed capacitance combines with the coil inductance, a resonant circuit is formed. The resonant frequency is usually quite high in relation to the frequency at which the coil is used. However, because self-resonance might be at or near a harmonic of the frequency to be used, the self-resonant effect might limit the usefulness of the coil in LC circuits. Some coils, particularly RF chokes, might have more than one self-resonant frequency.

A meter can be used with a signal generator to find both the self-resonant frequency and the distributed capacitance of a coil, using the following procedure:

1. Connect the equipment (Fig. 2-I). Adjust the generator output for a convenient midscale indication on the meter. Use an unmodulated signal output from the generator.

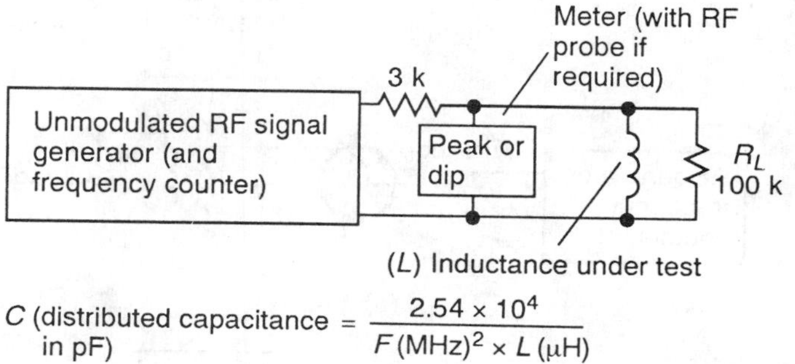

Fig. 2-I Basic coil self-resonance and distributed-capacitance measurements.

2. Tune the generator over the entire frequency range, starting at the lowest frequency. Watch the meter for either peak or dip indications. Either a peak or a dip indicates that the inductance is at a self-resonant point. The generator frequency at that point is the self-resonant frequency. Make certain that peak or dip indications are not the result of changes in generator output level. Even the best lab generators might not produce a flat (constant level) output over the entire frequency range.
3. Because there might be more than one self-resonant point, tune through the entire signal-generator range. Try to cover a frequency range up to at least the third harmonic of the highest frequency that is involved in a resonant-circuit design.
4. Once the resonant frequency is found, calculate the distributed capacitance using the equation in Fig. 2-I. For example, assume that a coil with an inductance of 7 μH is found to be self-resonant at 50 MHz: *C* (distributed capacitance) = (2.54 × 10^4>)/(50^2> × 7) = 1.45 pF.

Basic resonant-circuit *Q* measurements

A resonant circuit has a *Q* (quality) factor. From a practical test standpoint, a resonant circuit with a high *Q* produces a sharp resonance curve (narrow bandwidth), whereas a low *Q* produces a broad resonance curve (wide bandwidth), as shown in Fig. 2-Ha.

The *Q* of a resonant circuit can be measured using a generator and meter (with an RF probe). A high-impedance digital meter generally provides the least loading effect on the circuit and thus provides the most accurate indication.

Figure 2-Ja shows the test circuit in which the generator is connected directly to the input of a complete circuit or stage. Figure 2-Jb shows the indirect method of connecting the generator to the input.

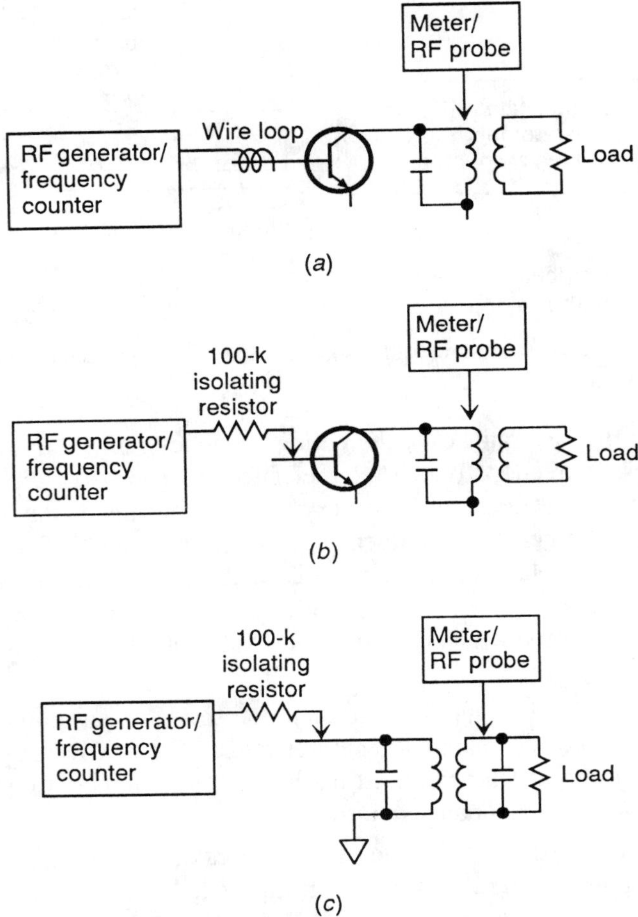

Fig. 2-J Basic resonant-circuit Q measurements.

When the stage, or circuit, has sufficient gain to provide a good reading on the meter a nominal output from the generator, the indirect method (with isolating resistor) is preferred. Any generator has some output impedance (typically 50 Ω). When this resistance is connected directly to the circuit, the circuit Q is lowered, and the response becomes broader (in some cases, the generator output seriously detunes the circuit).

Figure 2-Jc shows the test circuit for a single component (such as an IF transformer). The value of the isolating resistance is not critical and is typically in the range of 100 kΩ. The procedure for determining Q using any of the circuits in Fig. 2-J is:

1. Connect the equipment (Fig. 2-J). Notice that a load is shown in Fig. 2-Jc. When a circuit is normally used with a load, the most realistic Q measurement is made with the circuit terminated in that load value. A fixed resistance can be used to simulate the load. The Q of a resonant circuit often depends on the load value.
2. Tune the generator the circuit resonant frequency. Use an unmodulated output from the generator.
3. Tune the generator frequency for maximum reading on the meter. Note the generator frequency.
4. Tune the generator below resonance until the meter reading is 0.707 times the maximum reading. Note the generator frequency. To make the calculation more convenient, adjust the generator output level so that the meter reading is some even value, such as 1 or 10 V, after the generator is tuned for maximum. This makes it easy to find the 0.707 mark.
5. Tune the generator above resonance until the meter reading is 0.707 times the maximum reading. Note the generator frequency.
6. Calculate the resonant-circuit Q using the equation of Fig. 2-Ha. For example, assume that the maximum meter indication occurs at 455 kHz (F_R), the below-resonance indication is 453 kHz (F_2), and the above resonance indication is 457 kHz (F_1>). Then $Q = 455/(457 - 453) = 113.75$.

Basic resonant-circuit impedance measurements

Any resonant circuit has some impedance at the resonant frequency. The impedance changes as frequency changes. This includes transformers (tuned and untuned), RF tank circuits, and so on. In theory, a series-resonant circuit has zero impedance and a parallel-resonant circuit has infinite impedance, at the resonant frequency. In practical RF circuits, this is impossible because there is always some resistance in the circuit.

It is often convenient to find the impedance of an experimental resonant circuit at a given frequency. Also, it might be necessary to find the impedance of a component in an experimental circuit so that other circuit values can be designed around the impedance. For example, an IF transformer presents an impedance at both the primary and secondary windings. These values might not be specified on the transformer datasheet.

The impedance of a resonant circuit can be measured using a signal generator and a meter with an RF probe (Fig. 2-K). A high-impedance digital meter provides the least loading effect on the circuit and thus produces the most accurate indication.

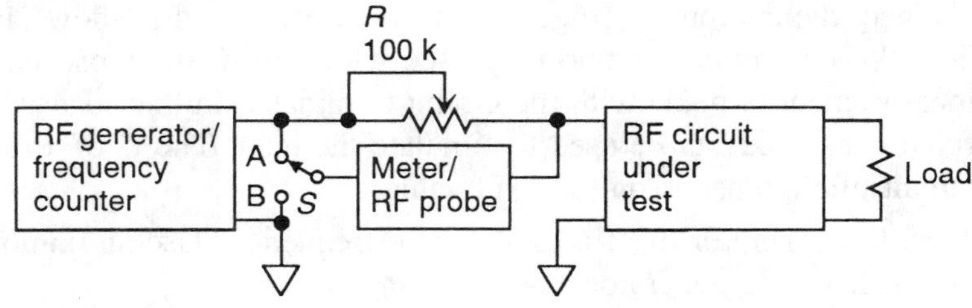

Fig. 2-K Basic resonant-circuit impedance measurements.

If the circuit of a component being measured has both an input and output (such as a transformer), the opposite side or winding must be terminated in the normal load, as shown. If the impedance of a tuned circuit is to be measured, tune the circuit to peak or dip, then measure the impedance at resonance. Once the resonant impedance is found, the generator can be tuned to other frequencies to find the corresponding impedance (if required).

Adjust the generator to the frequency (or frequencies) at which impedance is to be measured. Move switch S back and forth between positions A and B, while adjusting resistance *R* until the voltage reading is the same in both positions of the switch. Disconnect resistor R from the circuit, and measure the dc resistance of *R* with an ohmmeter. The dc resistance, *R*, is then equal to the impedance at the circuit input.

Accuracy of the impedance measurement depends on the accuracy with which the dc resistance is measured. A noninductive resistance must be used. The impedance found by this method applies only to the frequency that is used during the test.

Basic transmitter RF-circuit testing and adjustment

It is possible to test and adjust transmitter RF circuits (such as the amplifiers in Figs. 2-49 through 2-62) using a meter and RF probe. If an RF probe is not available (or as an alternative), it is often possible to use a circuit (such as shown in Fig. 2-La). This circuit is essentially a pickup coil that is placed near the RF-circuit inductance and a rectifier that converts the RF into a dc voltage or measurement on a meter. The basic procedure is:

1. Connect the equipment (Fig. 2-Lb). If the circuit being measured is an RF amplifier, without an oscillator, a drive signal must be supplied by means of a signal generator (as is the case with the amplifiers of Figs. 2-49 through 2-62). Use an unmodulated signal at the correct operating frequency.
2. In turn, connect the meter (through an RF probe or the special circuit in Fig. 2-La) to each stage of the RF circuit. Start with the first stage

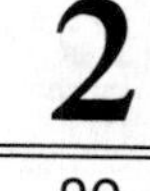

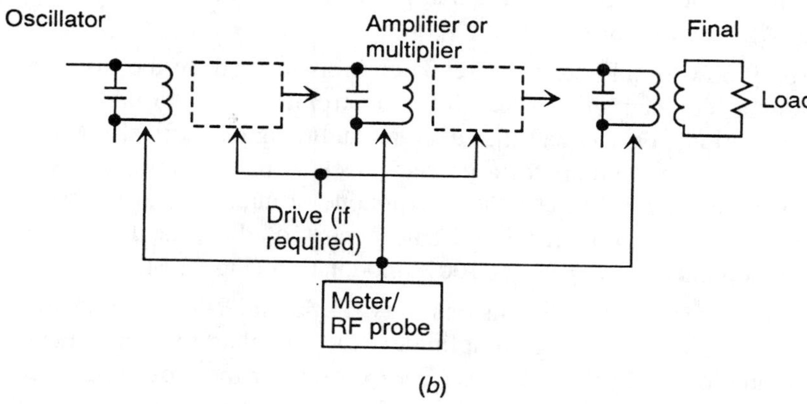

Fig. 2-L Basic transmitter RF-circuit tests and adjustments.

(this is usually the oscillator if the circuit under test is a complete transmitter) and work toward the final (or output) stage, which is usually connected to the transmitting antenna.

3. A voltage indication should be obtained at each stage. Usually, the voltage indication increases with each RF-amplifier stage as you proceed from oscillator to the final amplifier. However, some stages might be frequency multipliers and provide no voltage amplification.
4. If a particular stage is to be tuned, adjust the tuning control for a maximum reading on the meter. If the stage is to be operated with a load (such as the final amplifier into an antenna), the load should be connected or a simulated load (dummy load or RF wattmeter) should be used. As a general rule, a fixed, noninductive resistance provides a good simulated load at frequencies up to about 250 MHz.
5. Notice that this tuning method does not guarantee that each stage is at the desired operating frequency. It is possible to get maximum readings on harmonics. However, if you follow the winding instructions given for the circuits of this chapter, you should have no trouble with harmonics.

Modulation tests

The transmitter amplifier of Fig. 2-52 includes a modulator circuit (Fig. 2-52C). The following procedure can be used as a basic check for both the modulator and amplifier. The same test can be applied to most of the amplifier circuits in Figs. 2-49 through 2-62.

If the vertical-channel response of the scope is capable of handling the RF circuit frequency, the signal can be applied directly through the scope vertical amplifier. The basic test connections are shown in Fig. 2-M. The procedure is:

1. Connect the scope to the antenna jack, or the final RF amplifier (2-M). Use one of the three alternatives.
2. Apply power and adjust the scope controls to produce the displays shown. You can either speak into a microphone (for a rough check of modulation), or you can introduce an audio signal (typically 400 or 1000 Hz) at the input (for a precise check of modulation). In the modulator of Fig. 2-52C, the microphone or audio signal is applied at the 10-μF capacitor. It might be necessary to adjust the 10-kΩ pot of the modulator circuit to get 100%-modulation (Fig. 2-M).
3. Measure the vertical dimensions Figs. 2-Ma and 2-Mb (the crest amplitude and the trough amplitude), and calculate the percentage of modulation using the equation. For example, if the crest amplitude (A) is 63 (63 screen divisions, 6.3 V, and so on) and the trough amplitude (B) is 27, the percentage of modulation is:

$$\frac{63 - 27}{63 + 27} \times 100 = 40\%$$

Make certain to use the same scope scale for both crest (A) and trough (B) measurements, and to use a dummy load or RF wattmeter.

If the scope amplifier is not capable of passing the frequency, the signal can be applied directly to the vertical-deflection plates of the scope display tube, as shown. However, there are two problems with this approach. First, the vertical plates might not be readily accessible. Next, the RF signal might not be of sufficient amplitude to produce measurable deflection of the scope display-tube trace.

Frequency response

If required, the frequency response of amplifiers that are described in this chapter can generally be checked using the same procedures as for the amplifiers of chapter 1. Of course, the operating frequencies are much higher. For example, note the frequency response of the circuits shown in Figs. 2-5C and 2-6C.

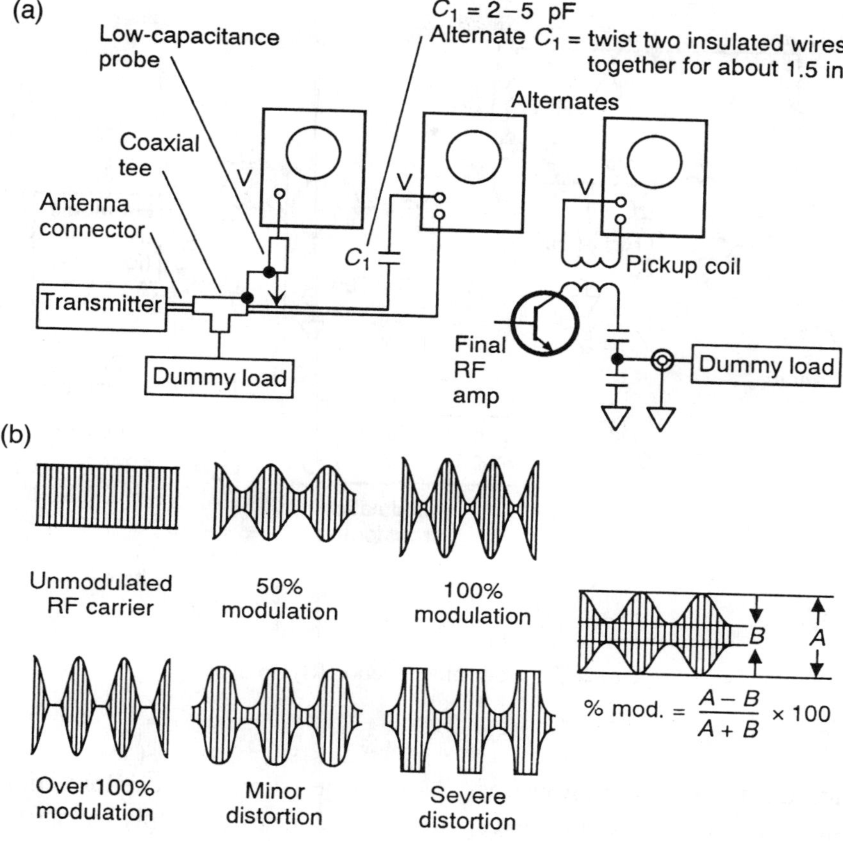

Fig. 2-M Direct measurement of RF-circuit modulation.

Basic RF troubleshooting approaches

The remainder of this introduction is devoted to troubleshooting of basic RF circuits, primarily power amplifiers (such as shown in Figs. 2-49 through 2-62). The basic procedures described in the second section of chapter 1 apply to RF circuits. However, there are certain differences that are covered in the following paragraphs.

General troubleshooting instructions

Figure 2-N shows the schematic of the transmitter circuits for a simple CB set. Notice that the stages of Q2 and Q3 resemble many of the power amplifiers shown in Figs. 2-49 through 2-62. The test points shown have been arbitrarily assigned to

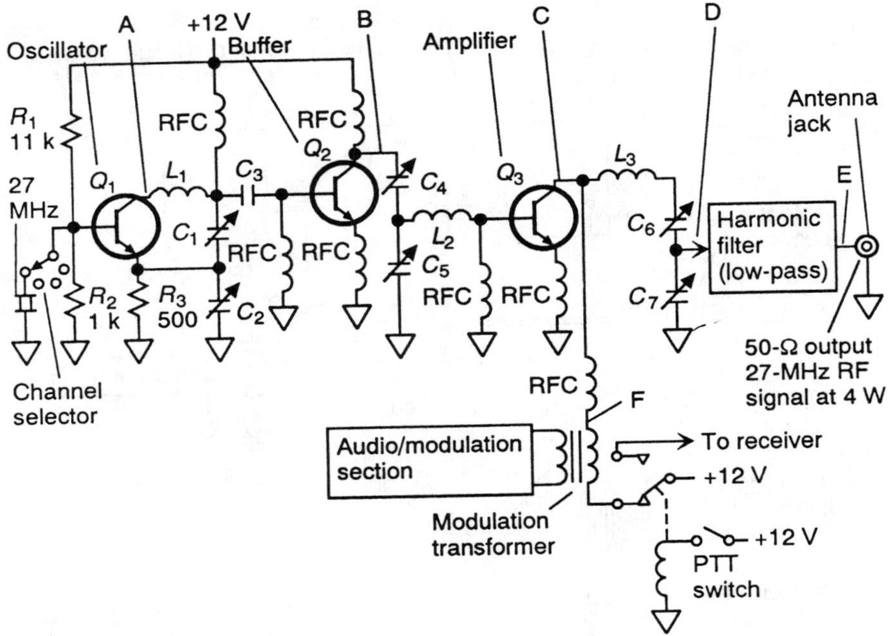

Fig. 2-N Transmitter-section RF circuits.

illustrate the troubleshooting process. No voltage or resistance information (at transistor terminals) is available, except for the +12-V power-supply voltage that is found on the schematic.

Even with the lack of voltage information, you should be able to calculate the voltages at the transistors. The collectors of Q1 through Q3 are all connected to +12 V through RF chokes. Such chokes generally have very little dc resistance and thus produce very little voltage drop (usually a fraction of 1 V). Thus, it is reasonable to assume that the dc voltage at the collectors is about +12 V (or slightly less).

The emitters of Q2 and Q3 are connected to ground through RF chokes. Thus, the dc voltage at the emitters should be zero, and the resistance to ground should be a few Ω, at most. Possibly some dc voltage might be developed across the chokes, but it is not likely. The same is true for the Q2 and Q3 bases.

The base of Q1 has a fixed dc voltage that is applied through the voltage-divider network of R1 and R2. The ratio of R_1 and R_2 indicates that the voltage drop across R2 is about 1 V. Thus, the base of Q1 is about 1 V. Typically, the emitter of Q1 is within 0.5 V of the base (assuming that Q1 is a silicon transistor). The resistance to ground from the emitter of Q1 should be equal to R3 (about 500 Ω).

The schematic shows that the output is 4 W into a 50-Ω antenna. The crystals are in the 27-MHz range, so all three stages are tuned to the channel frequency, and there is no frequency multiplication in any stage.

Capacitors C1, C2, and C4 through C7 are tuning adjustments. There are no other operating or adjustment controls. Once you press the PTT (push to talk) switch, the transmitter section should perform its function and transmit AM-modulated signals to the antenna. With this great wealth of information to draw upon, you are now ready to plunge into troubleshooting.

Determining RF-circuit trouble symptoms

Assume that you have assembled one of the power amplifier circuits in this chapter, and have incorporated the circuit into the transmitter section of a CB set (Fig. 2-N). You decide to test the circuit by applying power and transmitting voice to a friend's CB station across town. (This is a very foolish step! You should test and tune the circuit with a dummy load first, but assume that you are both impatient and overconfident.) As was feared, there is no smoke or flame, but the transmission cannot be heard by any CB stations that are tuned to the same channel. What will you do?

You could start by checking each circuit to determine which one is not operating. This rates a definite no—even though you would eventually locate the faulty circuit. It is not a logical approach and would probably require several unnecessary steps. Try again.

You could perform the tuning procedure to determine which circuit group does not perform properly. You should have tuned the circuit with a dummy load, but, now that you have gone this far, there is a more logical step. Try again.

You should check the output. Notice that I do not tell you how to check the output at this time. With the antenna still connected, you could tune an in-house receiver to the same channel and try to transmit voice. Although this test is quick and easy, it does not provide all the information you need. A weak transmission might be picked up by a nearby receiver, but could not be heard over the normal communications range. Also, the problem could be in the audio and modulation section or the antenna and lead-in might be bad.

At this point, a logical approach is to disconnect the antenna and check the output with a dummy load and meter or scope or RF wattmeter. The wattmeter is best for checking the RF output because a wattmeter provides an immediate answer to the question "is there RF output and is the power correct?" With the dummy load and a meter or scope, you must measure the voltage and calculate the power. For a typical CB set, you will get about 14 V of RF voltage with a 50-Ω antenna or load. Power is equal to E^2/R, or $14^2/50 = 4$ W.

While you have the dummy load connected, it would not hurt to tune the transmitter circuits. Use the procedure described earlier ("Basic transmitter

RF-circuit testing and adjustment"), or as given in the circuit description (such as for the circuit of Fig. 2-40). Notice that many of the circuits in this chapter are broadband or wideband, and are untuned.

Even if you get the correct power output, it is still possible that the transmitter section is not being modulated by the audio and modulation section. Thus, the most positive test of the transmitter output is to monitor the output on a scope, using one of the modulation tests described in this chapter ("Modulation tests"). Assume that you do this, and that there is no RF indication on the scope (no vertical deflection whatsoever) with the transmitter keyed (PTT button pressed) and someone speaking directly into the microphone. In which section of the set do you think the trouble is located?

If you chose the audio and modulation section, you will probably not succeed as a troubleshooter. You could try being a computer consultant. Maybe you are not paying attention. Or maybe you simply do not understand transmitters. The vertical deflection on the scope is produced by the RF signal; the shape of the signal is determined by the modulation. Thus, with no vertical deflection, there is no RF signal.

If you chose the transmitter section, you have made a logical choice. The next step is to isolate the trouble to one of the circuits (oscillator Q1, buffer Q2, or power amplifier Q3) in the transmitter section. What is the next most logical test point?

You could check the power-supply voltage for the transmitter section. This approach does have some merit. Notice that Q1 and Q2 receive collector voltage directly from the +12-V line, whereas Q3 gets collector voltage through a secondary winding on the modulation transformer and through relay contacts (which are operated by the PTT switch). Thus, to make a complete check of the power-supply voltage for the RF circuits, you must measure the voltage at each collector. Of course, if the voltage is good at any collector, the power supply itself is good.

You could check for RF signal at test point A. This is not a bad choice. If there is no RF signal at A, the trouble is traced to Q1. If there is RF at A, you know that oscillator Q1 is operating, but you eliminate only one circuit as a possible trouble area.

You could check for RF signals at test points C, D, and E. This proves very little. In effect, you are checking at these points when you monitor the output signal.

You should check for RF signals at test point B. This is the most logical choice because it is essentially a half-split of the transmitter circuits. As shown in Fig. 2-O, you can check with a meter or scope and an RF probe.

If there is a good RF-signal indication at B, both Q1 and Q2 are operating properly, and the trouble is at Q3 (or possibly to the low-pass TV interference filter (Fig. 2-N). You also must eliminate the power supply (which must be good if Q1 and Q2 are functioning normally). However, do not eliminate the line between the

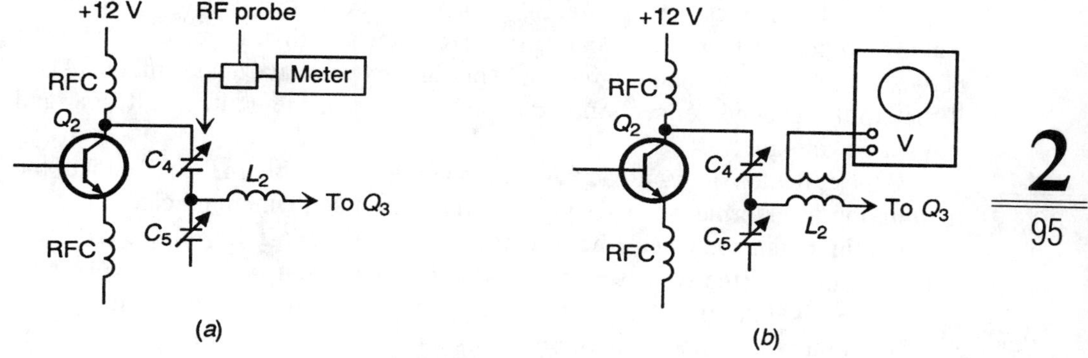

Fig. 2-O Checking for RF at test point B.

power supply and the collector of Q3 (which is a separate path from that between the power supply and the Q1/Q2 collectors).

If there is no RF indication at B, the trouble is traced to Q1 or Q2. The next step is to check for RF at A. If there is a good signal at A, but not B, the problem is in Q2. If there is no signal at A, the problem is in Q1.

Now assume that there is a good RF indication at B. This points to a bad Q3. To confirm this, check for RF signals at C, D, and E on the off chance that there is a good signal at C and D, but not at E (because of a bad low-pass filter).

Locating the specific fault

Now that the trouble is isolated to the circuit, the problem must be located. First, perform a visual inspection of Q3, and the associated wiring (you might have made a mistake). Check that there is no sign of heatsink over heating (you did use a heatsink didn't you?). If there is no apparent trouble with Q3, what is next?

You could make a substitution test of Q3. This is better then trying to test Q3 in-circuit, and you might have to substitute Q3 at some point, but there are more convenient tests.

You could check resistance at all elements of Q3. Although it will probably be necessary to check resistance and continuity before you are through, resistance checks at this time prove to be very little. The resistance to ground at the emitter or base of Q3 should be a few ohms at most. Only a high resistance to ground at the base and emitter (or a very low resistance at the collector) would be significant.

You should check the voltage at all elements of Q3 first. The voltage at the base and emitter should be zero (we are speaking of dc voltage, not RF-signal voltage). The dc voltage at the collector should be about +12 V. If the voltages are all good, you can skip the resistance-to-ground measurements. However, you might still need point-to-point continuity checks if the voltages are abnormal. Examine possible faults that are indicated by abnormal voltages.

Large dc voltages at base or emitter usually indicate that the elements are not making proper contact with ground. For example, a high-resistance solder joint between the Q3 emitter and ground could produce a dc voltage at the emitter. Also, if the emitter-ground connection is completely broken, the emitter will float and show a dc voltage.

If the collector shows no dc voltage, the fault is probably in the RF choke, the modulation-transformer secondary, or the relay contacts, which indicates the need for a continuity check. Check the dc voltage at test point F. If the voltage is correct at F, but not at C (the collector of Q3), the RF choke is at fault. If the voltage is absent at F, the fault is in the transformer winding or in the relay contacts.

To summarize this troubleshooting example, assume that the problem is an open L3 coil winding. The L3 winding is broken from the coil terminal, underneath of which you cannot see the break (that is where they always break). This trouble will not affect dc voltages or resistances. Substituting for Q3 will not cure the problem. Everything appears to be good, but your circuit will not work!

To solve the problem, you must make point-to-point continuity checks through the PC wiring. In this case, if you checked from point C to the top of C6, you would have found the open coil windings.

RF circuits

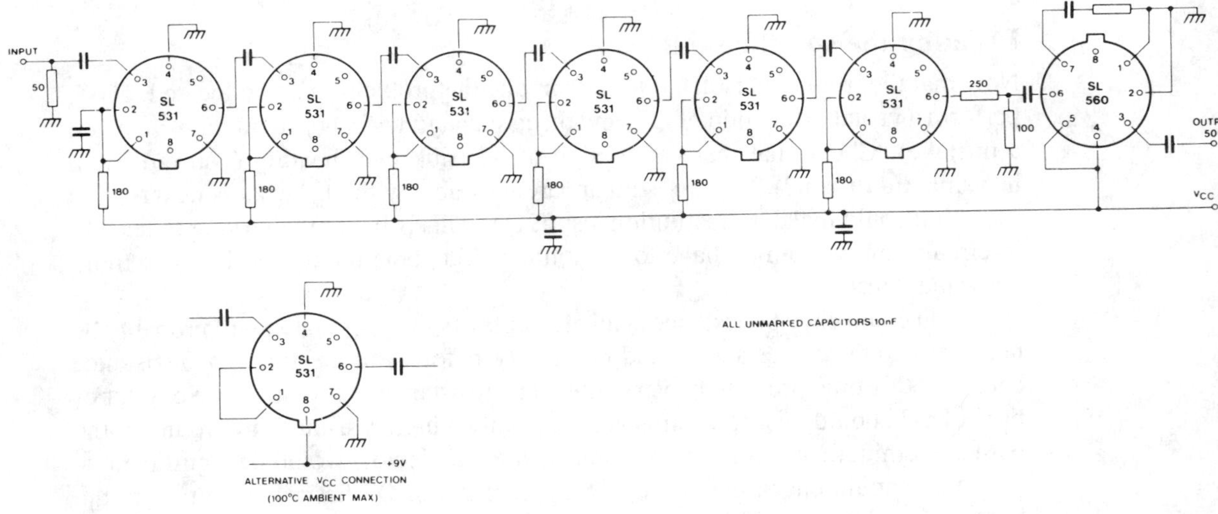

True logarithmic IF amplifier

Fig. 2-1 The SL531 shown here is a wide-band amplifier for use in IF circuits or strips of the true log type (where input and output are at the same frequency and no detection occurs). With the values shown, the frequency range is 10 to 200 MHz.

An SL560 is used as a unity-gain buffer, the output of the log strip being attenuated before the SL560 to give a nominal 0-dBm output into 50 Ω. The low-level gain is 60 dB with an output dynamic range of 20 dB. GEC Plessey Semiconductors, Professional Products, 1991, p. 1-16.

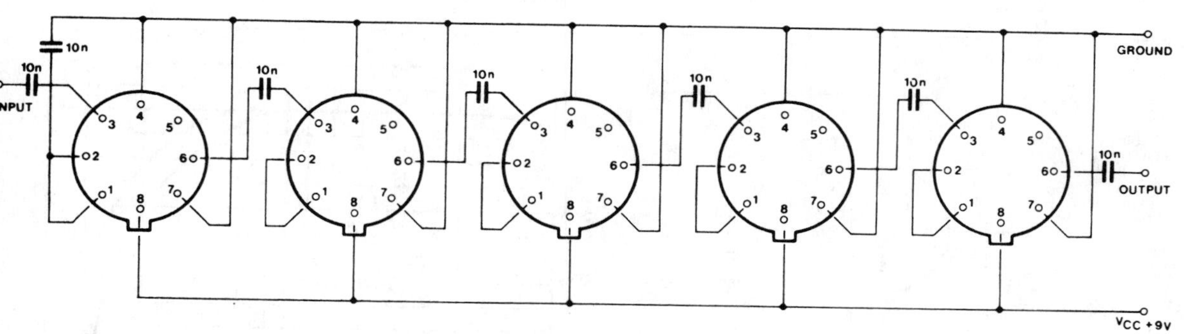

The recommended output buffer amplifier to drive 50Ω loads is the SL560C

Low-phase-shift IF limiter

Fig. 2-2 The SL532 shown here is used for wide-bandwidth limiting IF strips. With the values shown, the input signal for full limiting is 300 μV (rms)(−57 dBm), the output is limited to 1 Vpp, with a phase shift of ±3%. GEC Plessey Semiconductors, Professional Products, 1991, p. 1-19.

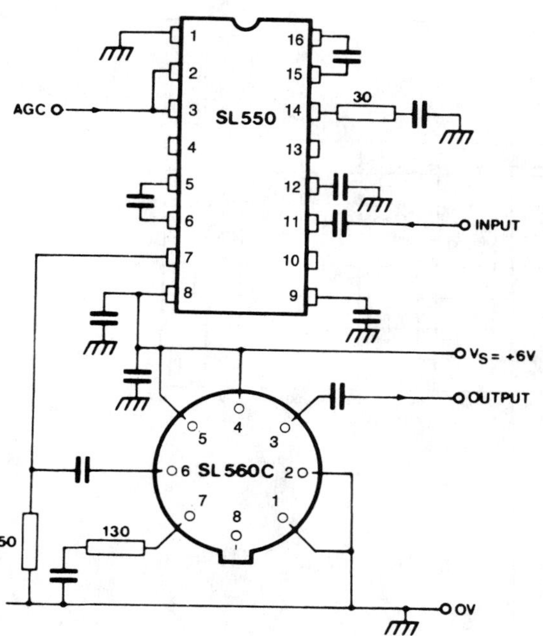

All Capacitors 10nF
Gain 46dB
Noise figure 2.0dB (RS = 200Ω)
Output power +5dBm (R1 = 50Ω)
Frequency response as SL550G
Dynamic range 70dB (1MHz bandwidth)

Low-noise wideband amplifier with AGC

Fig. 2-3 This circuit combines an SL550 (amplifier with external gain control) and an SL560 (low-noise amplifier) to form a very simple wideband amplifier with AGC. GEC Plessey Semiconductors, Professional Products, 1991, p. 1-61.

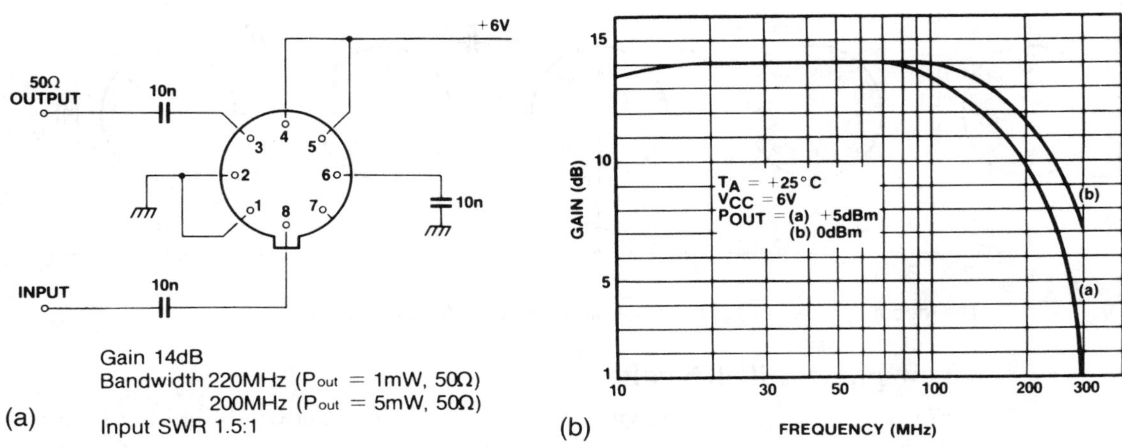

Low-noise line driver

Fig. 2-4 Figure 2-4A shows an SL560 used as a 50-Ω line driver. Figure 2-4B shows the typical frequency response. GEC Plessey Semiconductors, Professional Products, 1991, p. 1-64/1-65.

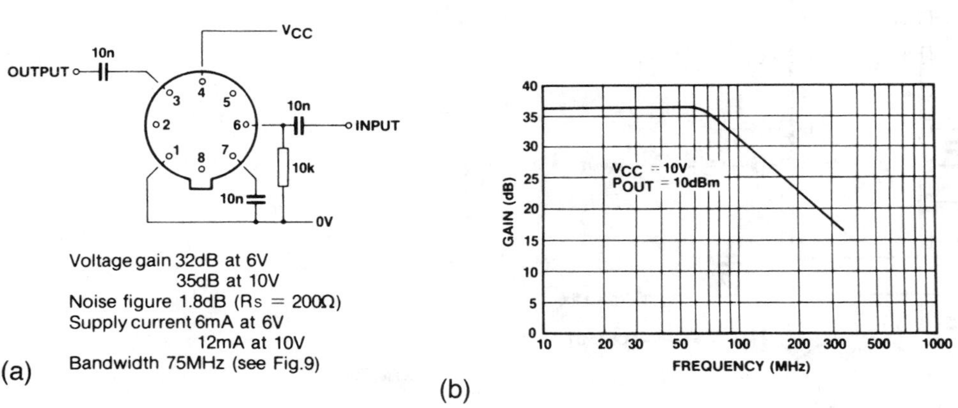

Low-noise preamplifier

Fig. 2-5 Figure 2-5A shows an SL560 used as a preamplifier. Figure 2-5B shows the typical frequency response. GEC Plessey Semiconductors, Professional Products, 1991, p. 1-65.

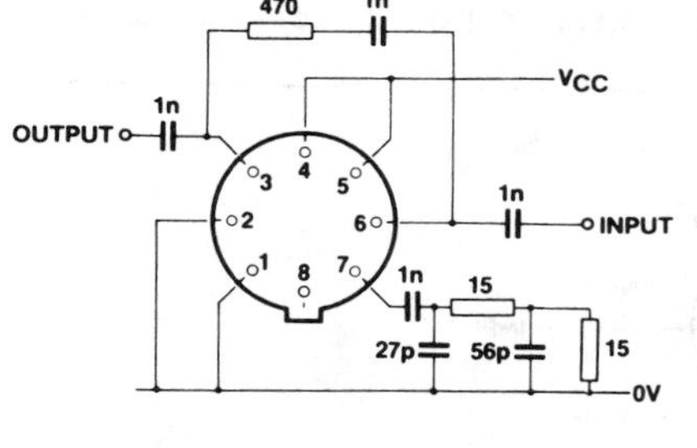

(a) Gain 13dB at Vcc = 9V
-1dB at 6MHz and 300MHz

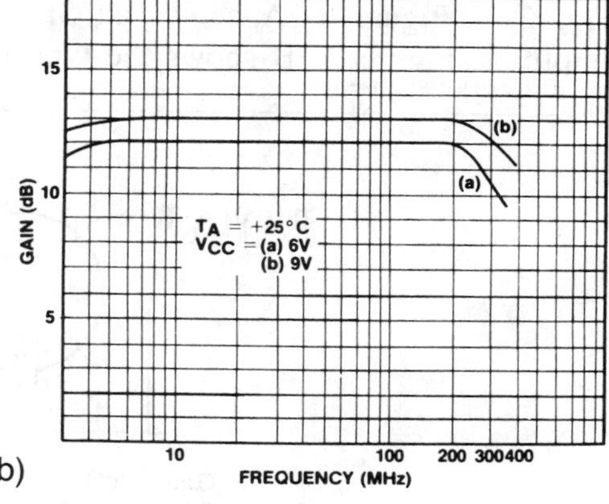

(b)

Low-noise wide-bandwidth amplifier

Fig. 2-6 Figure 2-6A shows an SL560 used as a wide-bandwidth amplifier. Figure 2-6B shows the typical frequency response. GEC Plessey Semiconductors, Professional Products, 1991, p. 1-65.

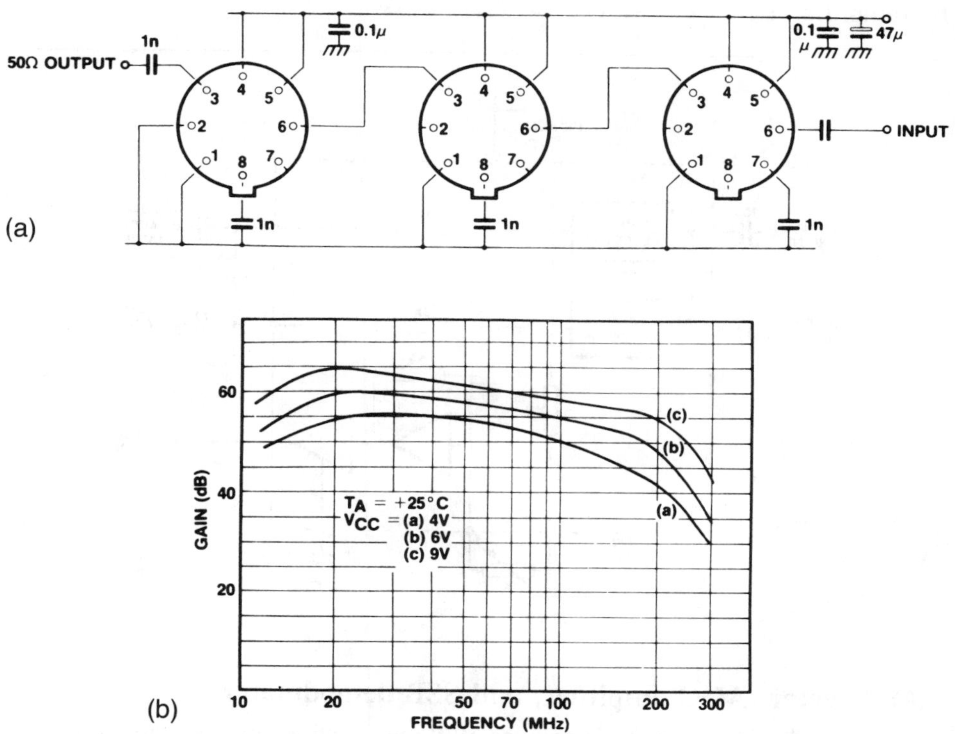

Low-noise high-gain amplifier

Fig. 2-7 Figure 2-7A shows an SL560 used as a direct-coupled high-gain amplifier. Figure 2-7B shows the typical frequency response. GEC Plessey Semiconductors, Professional Products, 1991, p. 1-66.

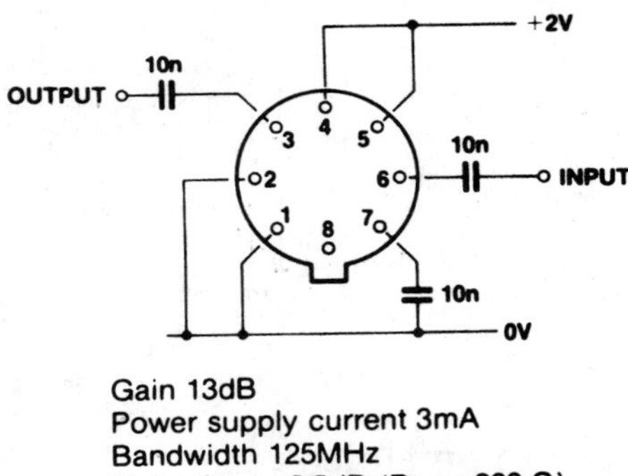

Gain 13dB
Power supply current 3mA
Bandwidth 125MHz
Noise figure 2.5dB (Rs = 200 Ω)

Low-noise low-power amplifier

Fig. 2-8 This circuit shows an SL560 used as a low-noise, low-power consumption amplifier. GEC Plessey Semiconductors, Professional Products, 1991, p. 1-66.

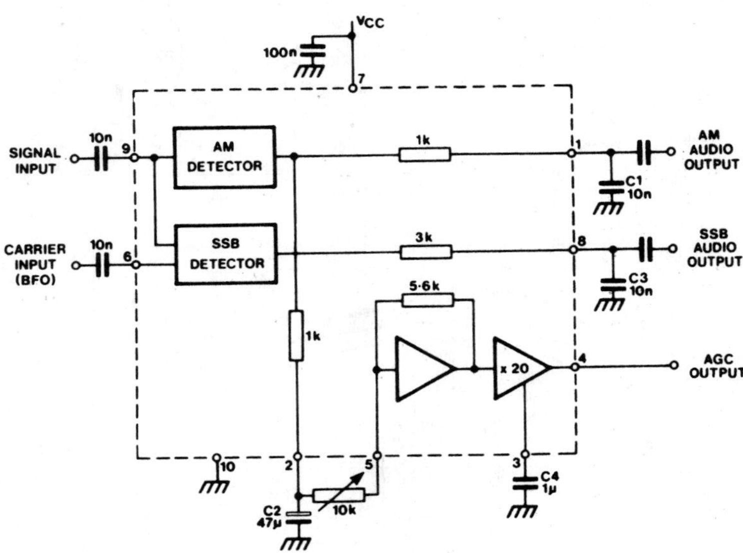

AM detector, AGC amplifier, and SSB demodulator

Fig. 2-9 The SL623 shown provides all three functions on a single chip, and is designed specifically for used in SSB/AM receivers (in conjunction with SL610/

SL611/SL612 RF and IF amplifiers). With the values shown, typical characteristics are: SSB audio output = 30 mV (rms), AM audio output = 55 mV (rms), AGC range = 6 dB (change in input level to increase AGC output voltage from 2.0 to 4.6 V), with an input impedance of 800 Ω at pins 6/9, maximum frequency = 30 MHz, and an AM carrier level down to 100 mV. GEC Plessey Semiconductors, Professional Products, 1991, p. 1-133.

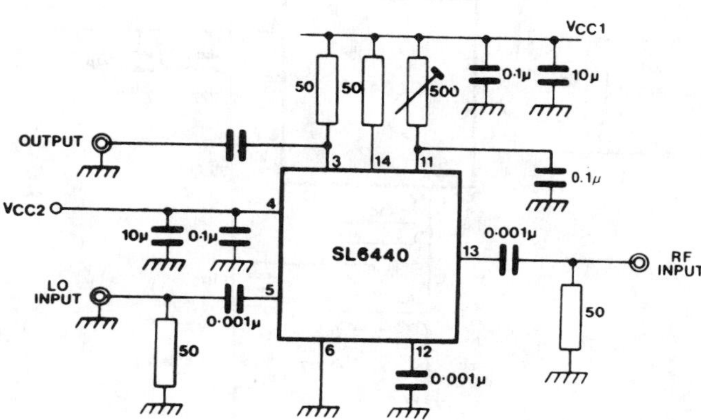

High-level mixer

Fig. 2-10 The SL6440 shown is a double-balanced mixer, for use in RF systems up to 150 MHz. External selection of the dc operating conditions is controlled by the resistor connected between pin 11 (bias) and V_{CC}. Conversion gain for single-ended circuit of Fig. 2-10 is equal to: $GdB = 20\,\text{Log}_{56.61}\, I_P + 0.0785, R_L\, I_P$, where I_P is programmed current at pin 11, and R_L is dc load resistance. GEC Plessey Semiconductors, Professional Products, 1991, p. 1-153.

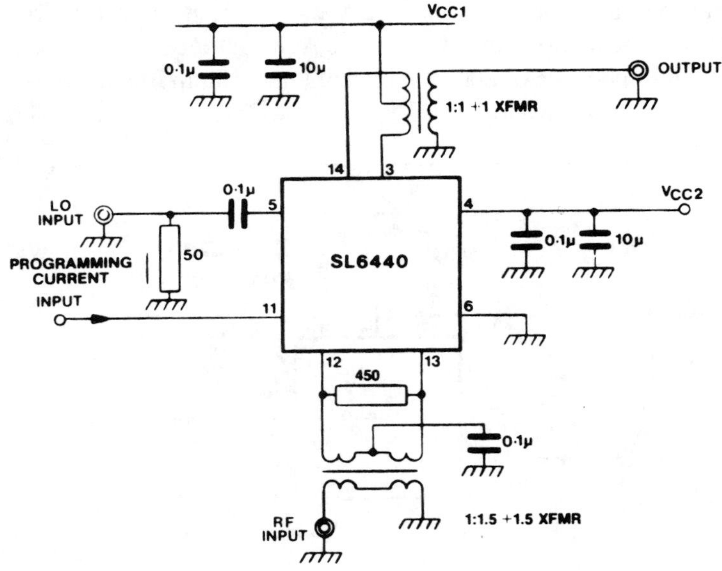

Balanced high-level mixer

Fig. 2-11 This circuit shows the SL6440 (Fig. 2-10) with a balanced input (for improved carrier leak), and balanced output (for increased conversion gain). A lower V_{CC} can be used with this arrangement (for lower device dissipation). Conversion gain for the balanced circuit of Fig. 2-11 is equal to: $Gdb = 20\ \mathrm{Log}_{56.61}\ I_P + 0.0785,\ 2R_L\ I_P$, where I_P is programmed current at pin 11, and R_L is dc load resistance. GEC Plessey Semiconductors, Professional Products, 1991, p. 1-154.

AM IF and detector

Fig. 2-12 The SL6701 shown is a single- or double-conversion IF amplifier and detector for AM-radio applications. Normally, the SL6701 is fed with a first IF signal of 10.7 or 21.4 MHz. There is a mixer for conversion to the first or second IF, a detector, and an AGC generator with optional delayed output. GEC Plessey Semiconductors, Professional Products, 1991, p. 1-168.

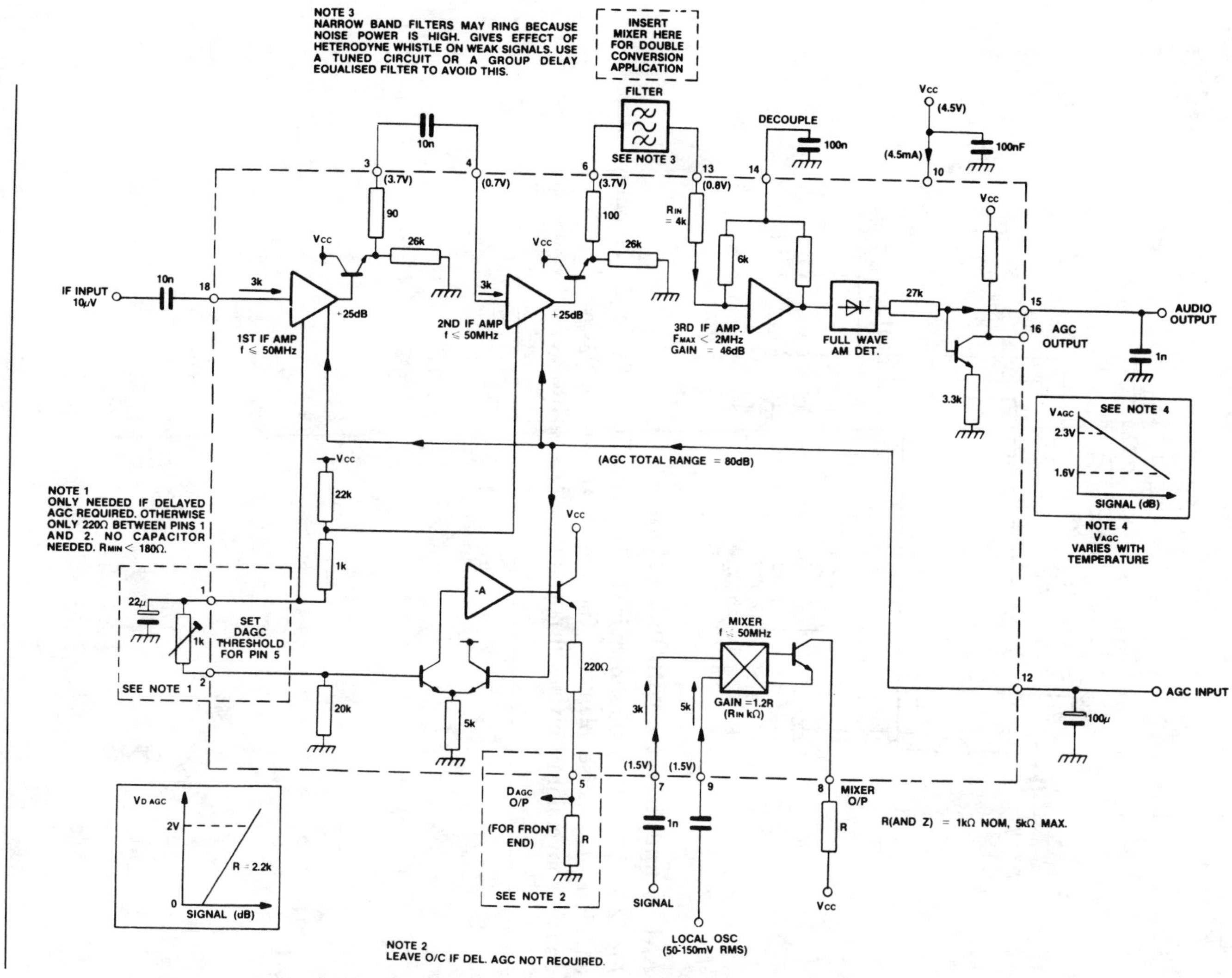
NOTE 3
NARROW BAND FILTERS MAY RING BECAUSE NOISE POWER IS HIGH. GIVES EFFECT OF HETERODYNE WHISTLE ON WEAK SIGNALS. USE A TUNED CIRCUIT OR A GROUP DELAY EQUALISED FILTER TO AVOID THIS.
INSERT MIXER HERE FOR DOUBLE CONVERSION APPLICATION
FILTER
SEE NOTE 3
DECOUPLE
100n
Vcc
(4.5V)
(4.5mA)
100nF
10n
3
(3.7V)
4
(0.7V)
6
(3.7V)
13
(0.8V)
14
10
90
100
RIN = 4k
Vcc
26k
26k
6k
IF INPUT 10µV
10n
18
3k
3k
+25dB
+25dB
1ST IF AMP f ≤ 50MHz
2ND IF AMP f ≤ 50MHz
3RD IF AMP. FMAX < 2MHz GAIN = 46dB
FULL WAVE AM DET.
27k
15
AUDIO OUTPUT
16 AGC OUTPUT
1n
3.3k
VAGC
SEE NOTE 4
2.3V
1.6V
SIGNAL (dB)
NOTE 4
VAGC VARIES WITH TEMPERATURE
(AGC TOTAL RANGE = 80dB)
NOTE 1
ONLY NEEDED IF DELAYED AGC REQUIRED. OTHERWISE ONLY 220Ω BETWEEN PINS 1 AND 2. NO CAPACITOR NEEDED. RMIN < 180Ω.
22k
1k
22µ
1k
1
2
SET DAGC THRESHOLD FOR PIN 5
SEE NOTE 1
-A
220Ω
20k
5k
MIXER f ≤ 50MHz
GAIN = 1.2R (RIN kΩ)
12
AGC INPUT
100µ
3k
5k
(1.5V)
(1.5V)
7
9
8
MIXER O/P
5
DAGC O/P
(FOR FRONT END)
R
SEE NOTE 2
1n
SIGNAL
LOCAL OSC (50-150mV RMS)
R
R(AND Z) = 1kΩ NOM, 5kΩ MAX.
VD AGC
2V
0
R = 2.2k
SIGNAL (dB)
NOTE 2
LEAVE O/C IF DEL. AGC NOT REQUIRED.

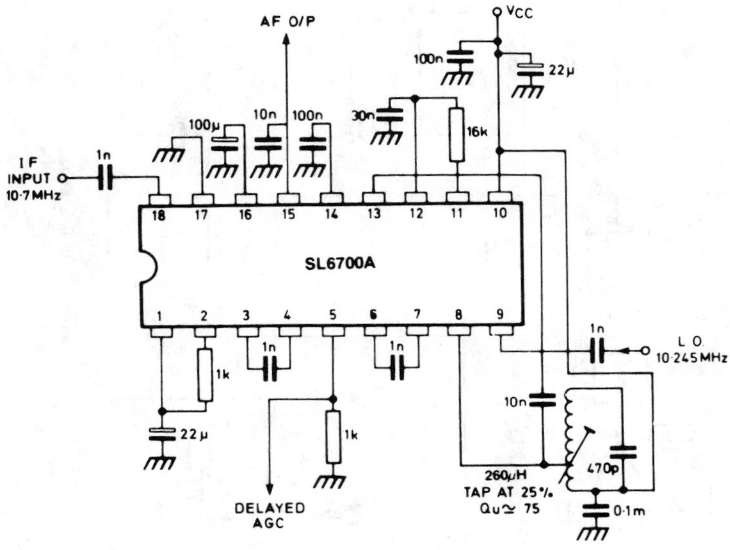

AM IF and detector with noise blanker

Fig. 2-13 The SL6700 shown is similar to the SL6701 (Fig. 2-12), but with the addition of a noise blanker. The circuit shown in Fig. 2-13 is for an AM double-conversion receiver with noise blanker. GEC Plessey Semiconductors, Professional Products, 1991, p. 1-163.

ZN414Z

INPUT
OUTPUT (V_{CC})
GROUND
3 2 1

PLASTIC (TO-92)

(a)

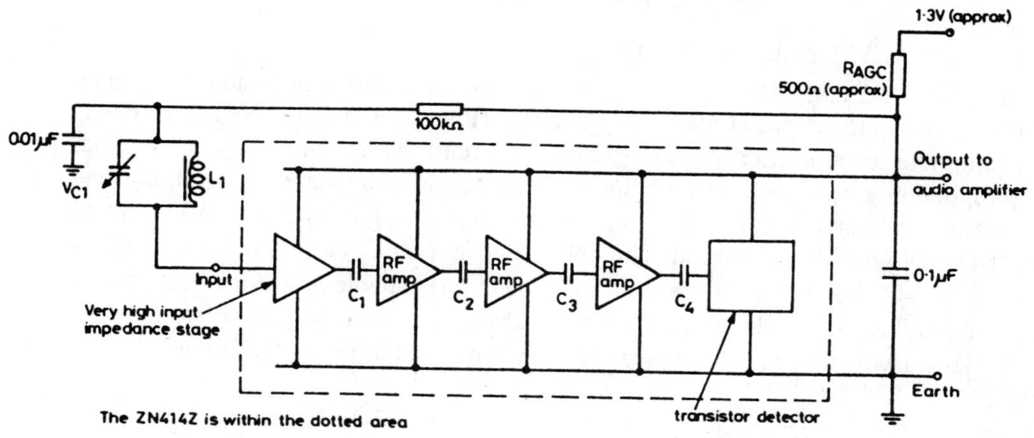

(b)

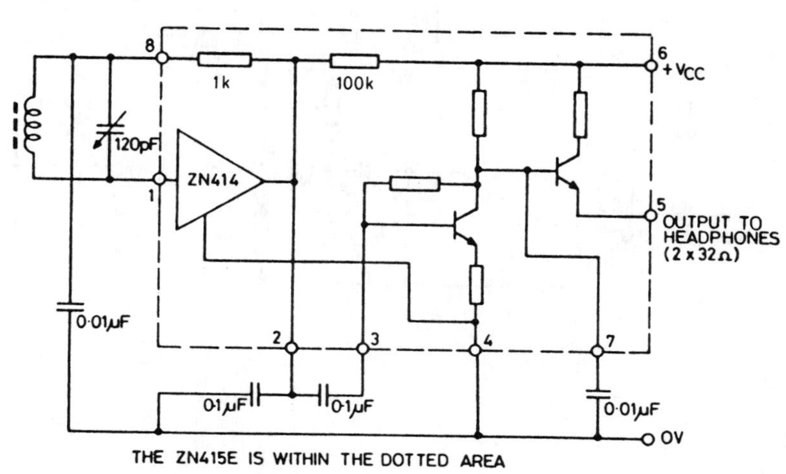

(c)

Continued

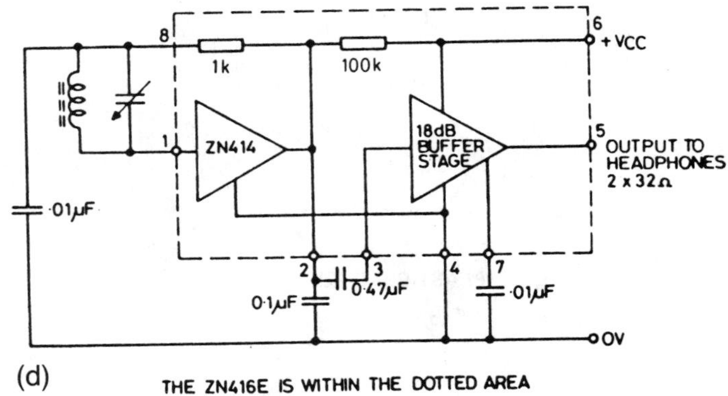

(d) THE ZN416E IS WITHIN THE DOTTED AREA

Miniature AM radio receivers

Fig. 2-14 The ZN414Z (Fig. 2-14A) is a 10-transistor tuned-radio-frequency (TRF) circuit in a 3-pin TO-92 plastic package (Fig. 2-14B). The ZN414Z provides a complete RF amplifier, detector and AGC circuit that requires only six external components to produce an AM tuner. AGC action is adjusted by selecting one external resistor value. Current consumption is low (typically 2 to 4 mA), and no set-up or alignment is required. The ZN415E (Fig. 2-14C) retains all the features of the ZN414Z, but also includes a buffer stage that is capable of driving headphones directly. The ZN416E (Fig. 2-14D) is also a buffered-output version of the ZN414Z, which provides a typical 120 mV (rms)output into a 64-Ω load. GEC Plessey Semiconductors, Professional Products, 1991, p. 1-171, 1-169.

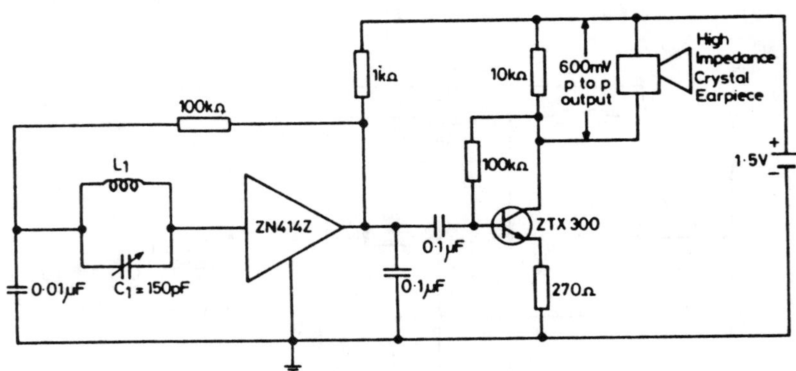

Earphone radio

Fig. 2-15 This circuit shows the ZN414Z (Fig. 2-14A) combined with an external amplifier to drive an inexpensive earpiece. This arrangement is generally cheaper than having the ZN414Z drive an expensive sensitive earpiece directly (to produce an ultra-miniature radio). The arrangement also provides for a volume control, if desired. Substitute a 250-Ω potentiometer in series with a 100-Ω fixed resistor for the 270-Ω emitter resistor. L1 = 80 turns of 0.3-mm diameter enamelled copper wire on a 5-cm or 7.5-cm long ferrite rod. Any value of L_1/C_1 that gives a high Q at the desired frequency can be used. GEC Plessey Semiconductors, Professional Products, 1991, p. 1-177.

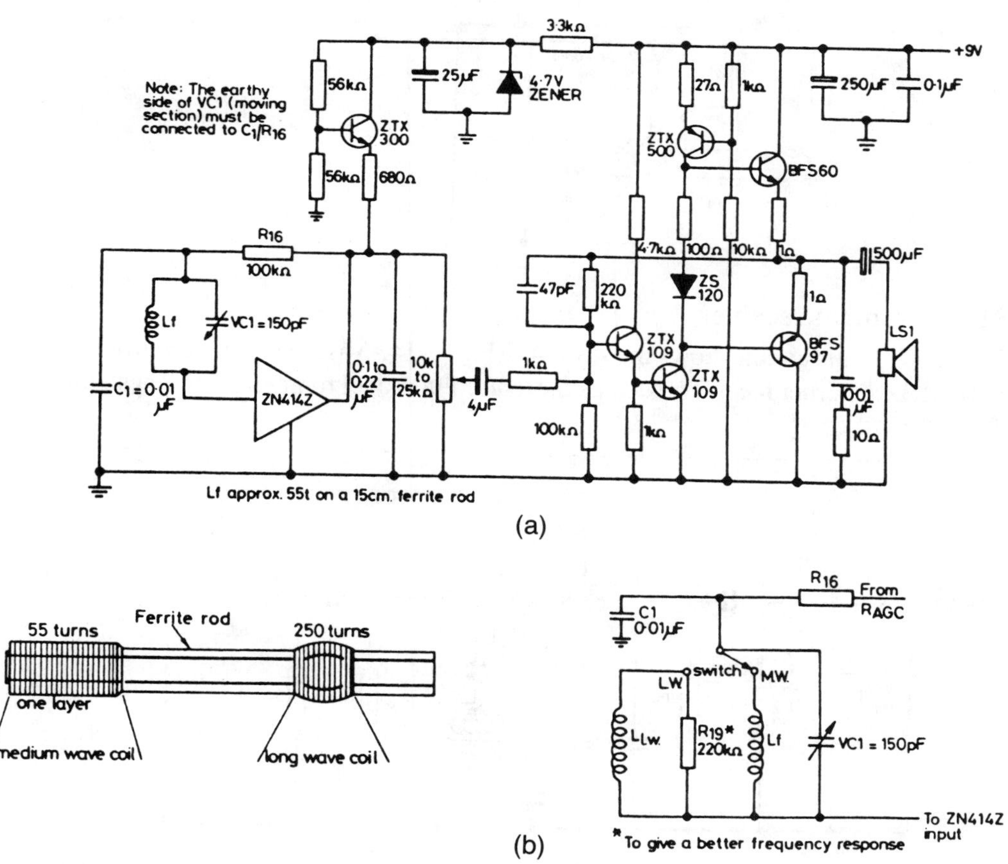

Portable receiver

Fig. 2-16 Figure 2-16A shows the ZN414Z (Fig. 2-14A) combined with external components to form a portable receiver (known as a *Triffid receiver*). Figure 2-16B shows the coil-winding details and waveband-selection circuits. GEC Plessey Semiconductors, Professional Products, 1991, p. 1-177, 1-178.

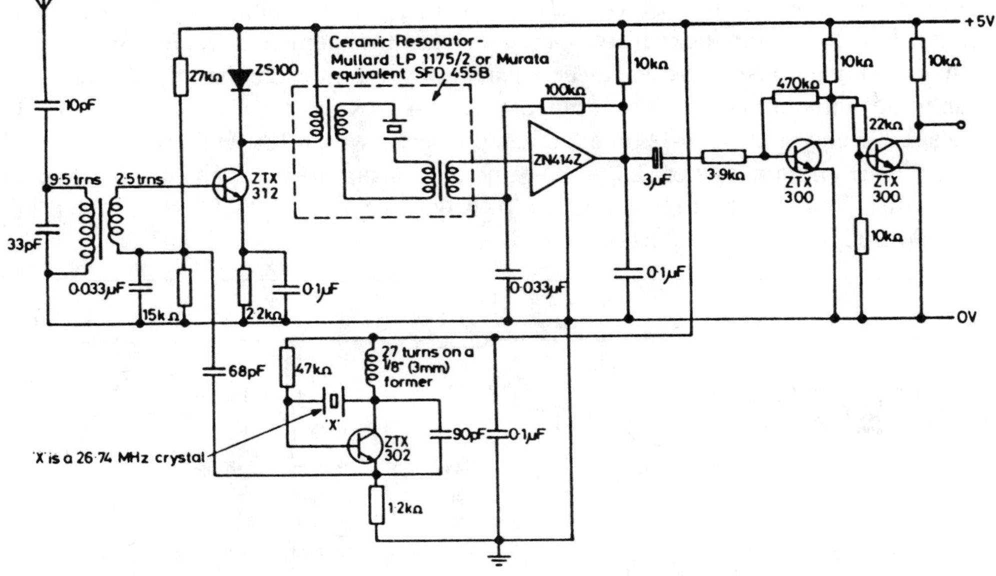

Model-control receiver

Fig. 2-17 This circuit shows the ZN414Z (Fig. 2-14A) used as an IF amplifier for a 27-MHz superhet receiver, such as those used in model remote-control systems.
GEC Plessey Semiconductors, Professional Products, 1991, p. 1-178.

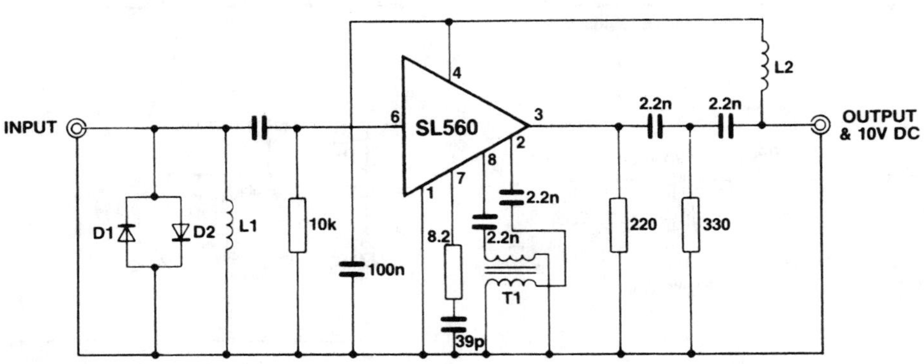

VHF antenna booster

Fig. 2-18 This circuit shows an SL560 amplifier connected as a wideband (40 to 260 MHz) antenna amplifier or booster suitable for any 50-Ω system. D1 and D2 are any general-purpose silicon diodes. L1 is 8 turns of #26, ⅛″ internal diameter. L2 is 20 turns of #26, 3⁄16″ internal diameter. T1 consists of two lengths of #34 wire approximately 6″ long, twisted together (8 twists/inch) and wound on a 6-hole ferrite bead (Mullard FX1898). The circuit is powered through the coaxial cable (+10 V at the coax center, and ground at the coax outer shield), and requires about 30 mA. GEC Plessey Semiconductors, Professional Products, 1991, p. 4-3.

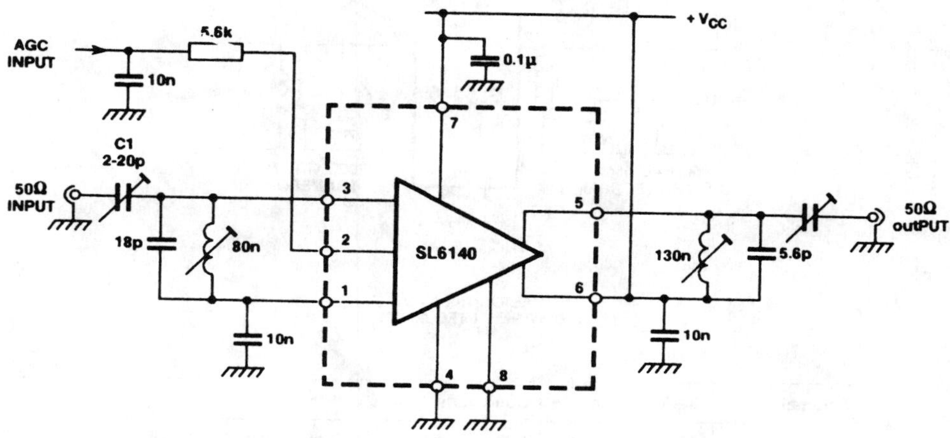

Tuned amplifier (CM pinout)

Fig. 2-19 This circuit shows an SL6140 connected as a single-ended amplifier with tuned input and output networks (Fig. 2-40). The bandwidth and gain depend on the *Q* of the tuned/matching circuits. Using the values shown, the bandwidth is 100 MHz, with a power gain of 35 dB. GEC Plessey Semiconductors, Professional Products, 1991, p. 4-5.

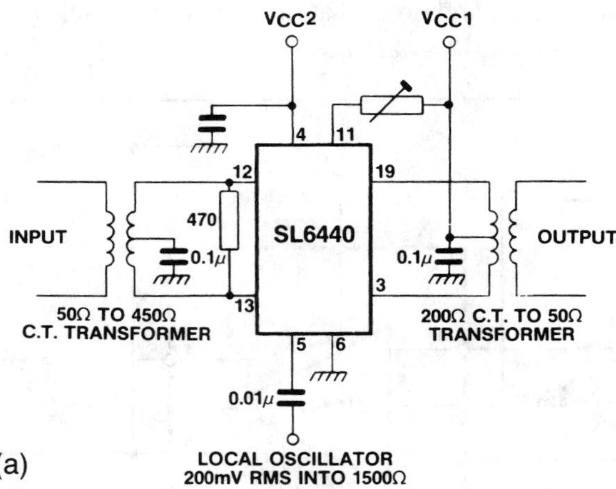

(a)

(b)

Parameter	Performance	Conditions
Sensitivity	15dB S + N/N	1μV EMF, SSB Bandwidth 30MHz, f = 1MHz
IMD 3rd	>-70dB	Input = 142mV EMF each signal 10kHz separation
IMD 2nd	-80dB	Input = 142mV EMF each signal
LO Radiation	-65dBm	Measured in 50Ω at input port
Blocking	100mV EMF	3dB blocking 1μV EMF wanted signal
IF Rejection	30dB	Rejection measured at input port
Input Matching	22dB	Returned loss in a 50Ω system
Gain	10dB	

High-performance mixer

Fig. 2-20 This circuit shows an SL6440 (Fig. 2-10) mixer with transformer coupling. Figure 2-20B shows the circuit characteristics. GEC Plessey Semiconductors, Professional Products, 1991, p. 4-14.

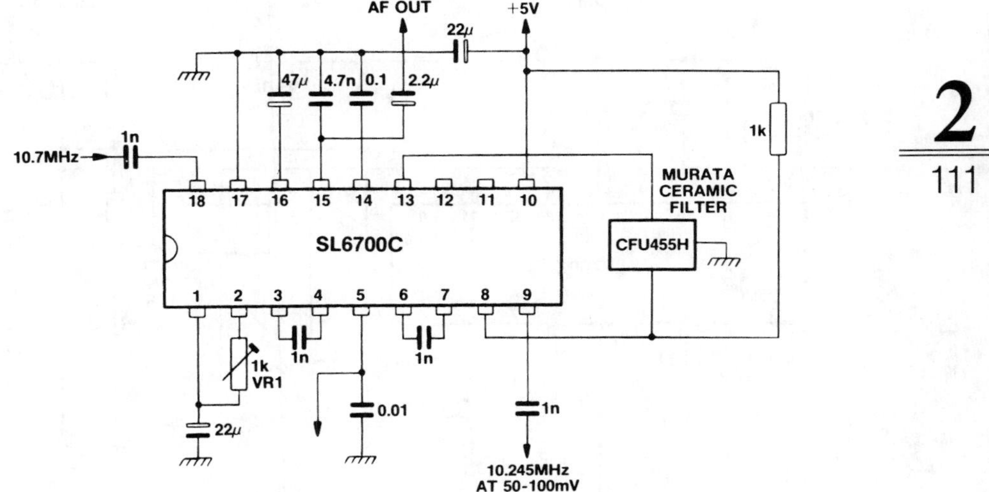

Double-conversion IF strip

Fig. 2-21 This circuit shows an SL6700 (Fig. 2-13) connected to form a double-conversion (10.7 MHz and 455 kHz) IF strip, using a ceramic filter for the 455-kHz network. The sensitivity of this circuit is typically 5 μV (rms), with 30% modulation for a 10-dB S/N ratio. The circuit will accept up to 100 mV (rms) at 80% modulation with distortion below 5%. GEC Plessey Semiconductors, Professional Products, 1991, p. 4-17.

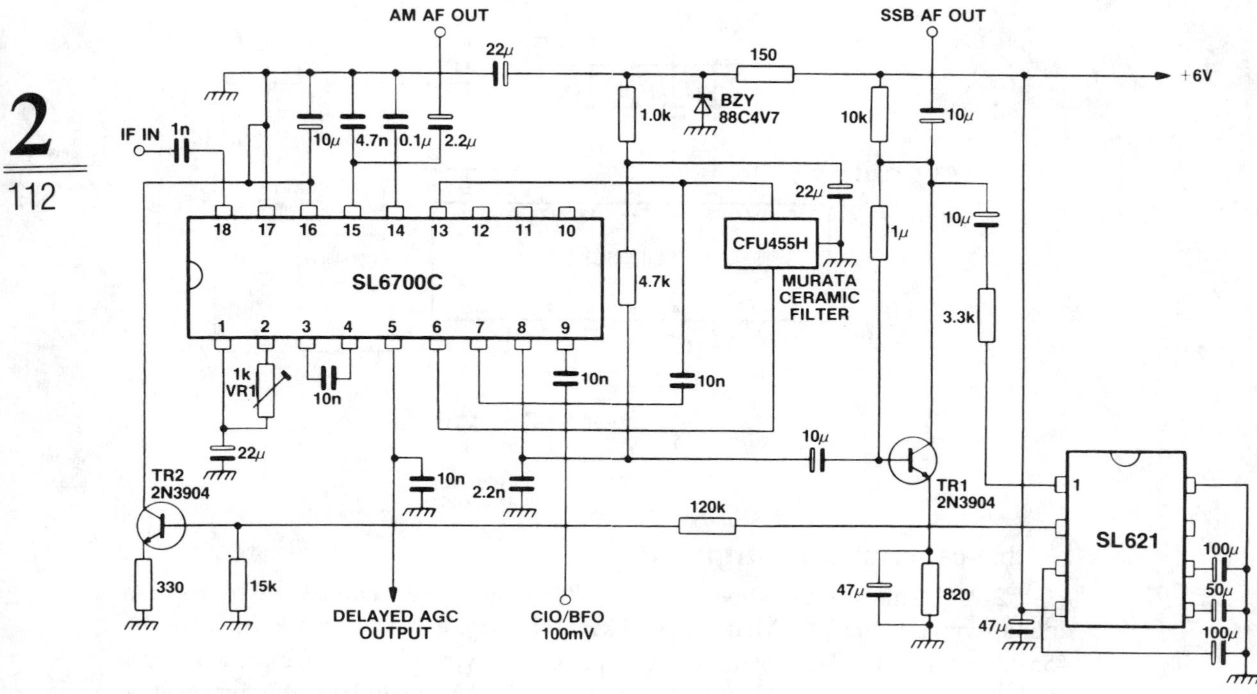

AM/SSB/CW IF strip

Fig. 2-22 This circuit shows an SL6700C (Fig. 2-13) connected to form an AM/SSB/CW IF strip. The SL621 is an AGC generator that is designed specifically for SSB operation. GEC Plessey Semiconductors, Professional Products, 1991, p. 4-19.

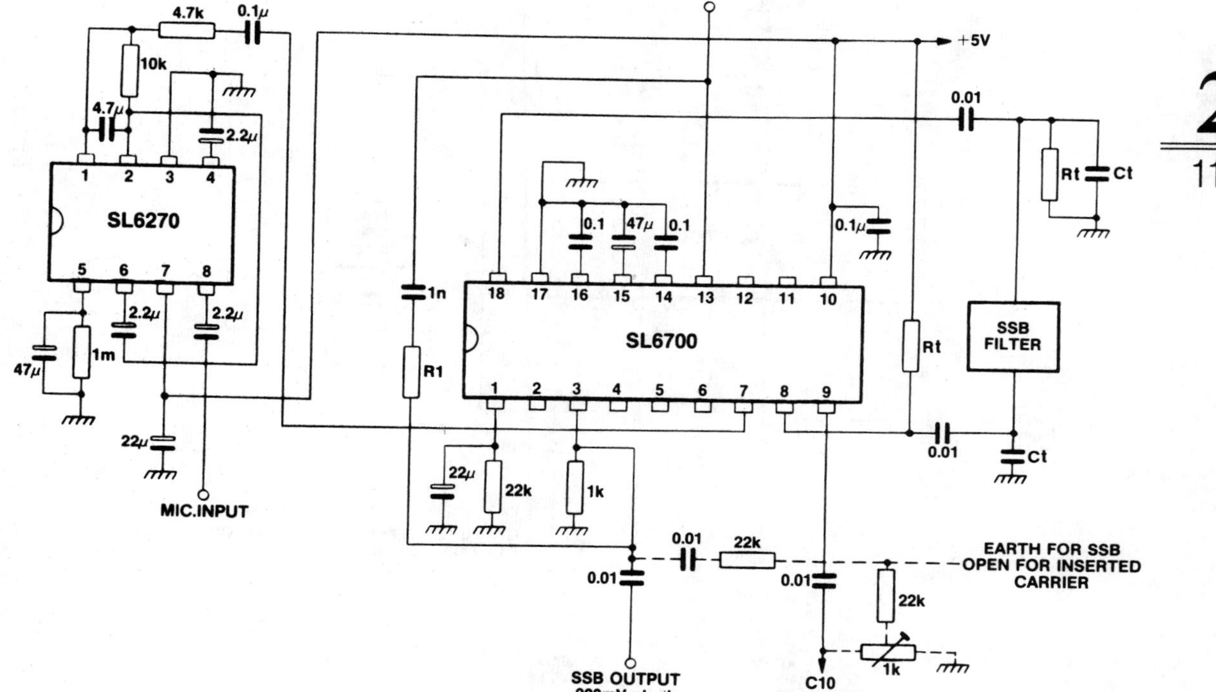

SSB generator

Fig. 2-23 This circuit shows an SL6700 (Fig. 2-13) connected to form a no-adjustment SSB generator. The trim adjustment shown in dashed lines is only used for re-inserting a carrier (not required for basic SSB operation). Notice that the microphone input is applied through an SL6720 (Fig. 1-7). GEC Plessey Semiconductors, Professional Products, 1991, p. 4-22.

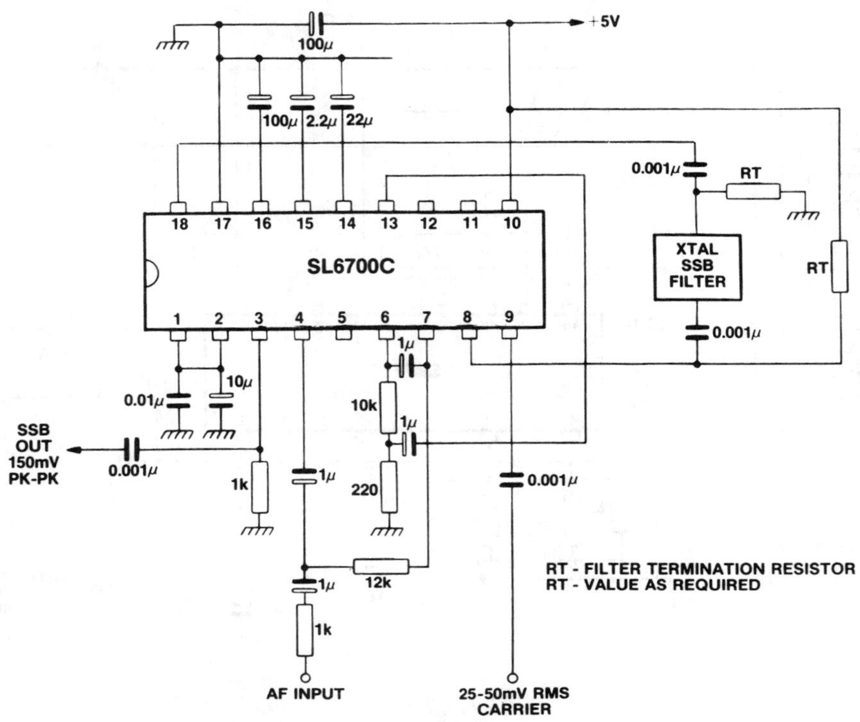

Low-power SSB generator

Fig. 2-24 This circuit shows an SL6700 (Fig. 2-13) connected to form a low-power no-adjustment SSB generator. Notice that the microphone input is applied directly to the SL6700. Although this does not provide the same SSB quality as that of the Fig. 2-23 circuit, the savings in components and lower power consumption make the Fig. 2-24 circuit attractive for hand-held CB, manpacks, etc. GEC Plessey Semiconductors, Professional Products, 1991, p. 4-23.

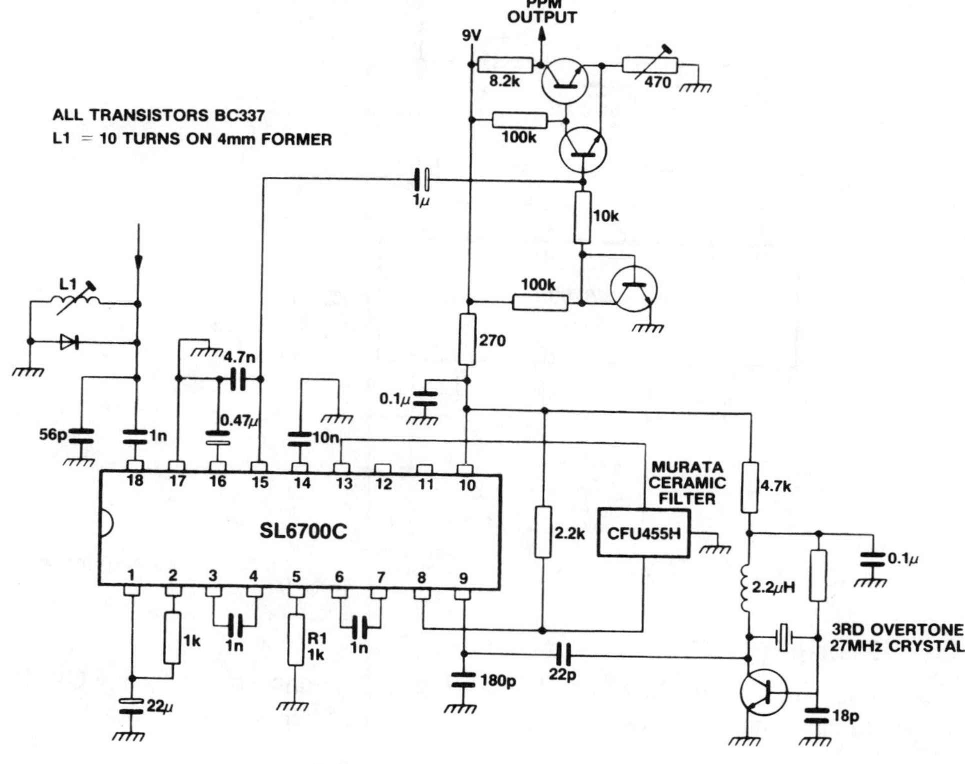

Remote-control receiver

Fig. 2-25 This circuit shows an SL6700 (Fig. 2-13) used as a complete 27-MHz remote-control receiver (such as used for most model controls). GEC Plessey Semiconductors, Professional Products, 1991, p. 4-23.

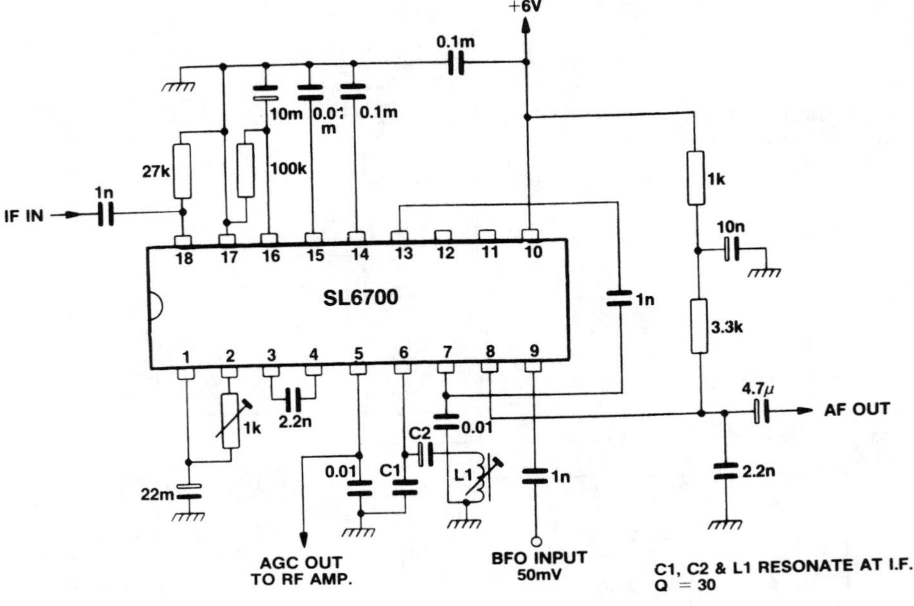

CW IF strip

Fig. 2-26 This circuit shows an SL6700 (Fig. 2-13) connected to form a CW IF strip with minimum component count, and no ceramic filter. GEC Plessey Semiconductors, Professional Products, 1991, p. 4-24.

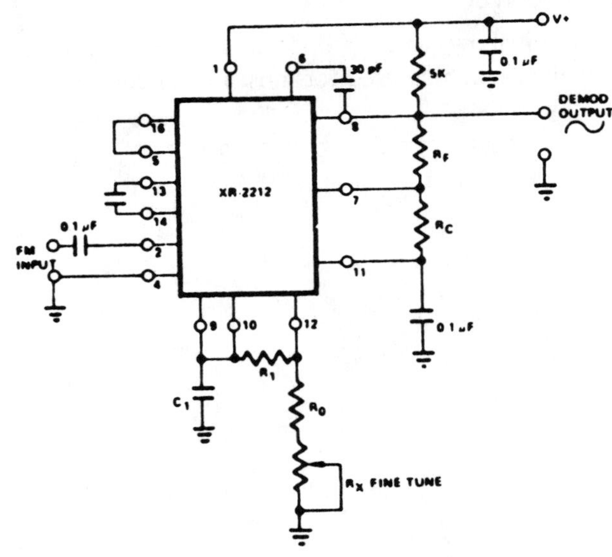

FM demodulator

Fig. 2-27 This circuit shows a precision PLL connected as an FM demodulator. The circuit can be tailored to any FM demodulation application by a choice of external components. For example, to demodulate an FM signal with a 67-kHz carrier and ±5 kHz deviation, use the following components: C_O = 746 pF, R_1 = 89.4 kΩ, C_1 = 186 pF, R_F = 100 kΩ, and R_C = 80.6 kΩ, R_O = 20 kΩ. Use an 18-kΩ fixed resistor and a 5-kΩ potentiometer for R_O and R_X. Fine tune the circuit with R_X. EXAR Corporation Databook, 1990, p. 5-29.

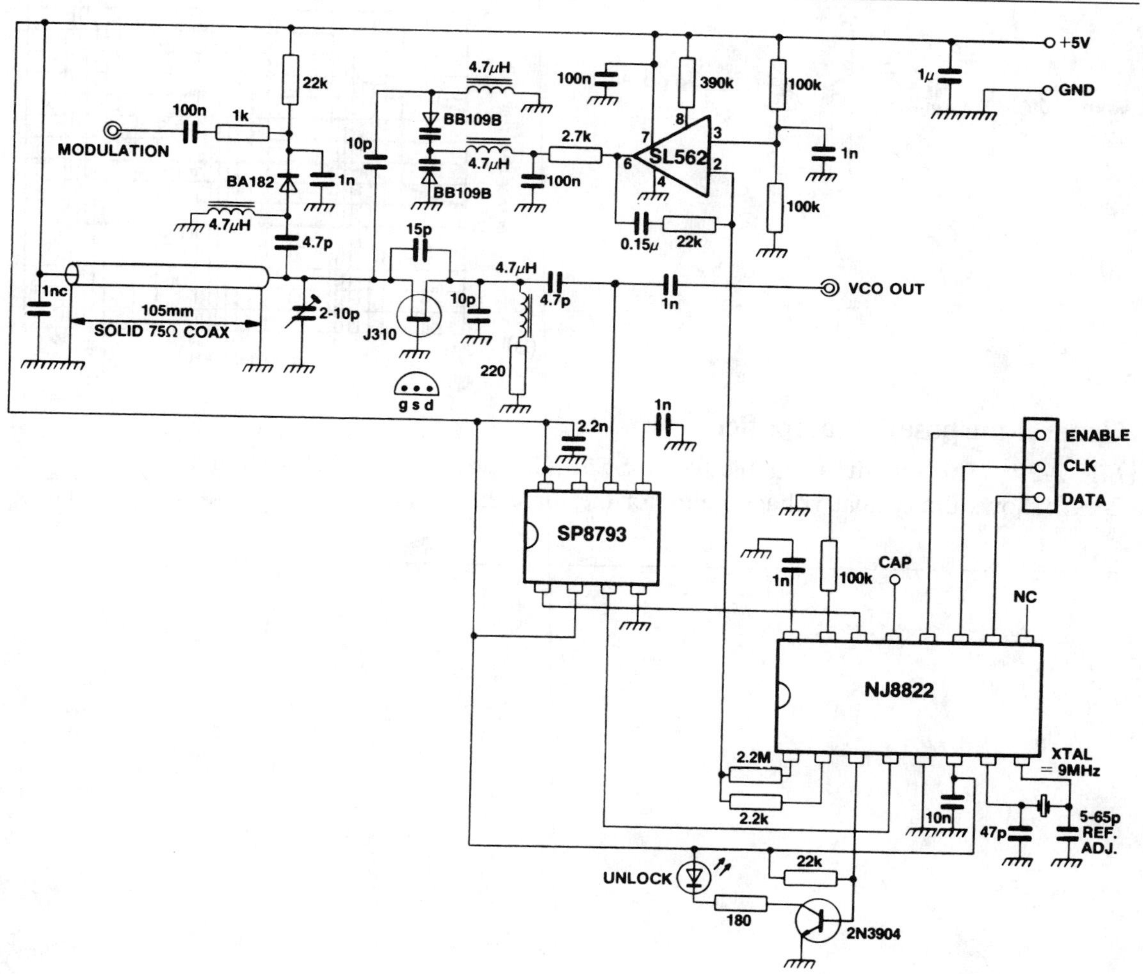

Serially programmable VHF frequency synthesizer

Fig. 2-28 This circuit uses an NJ8822 single-chip synthesizer, an SP8793 dual-modulus prescaler, and an SL562 op amp. The VCO is a JFET oscillator that uses a transmission line as the resonator. This VCO is modulated by applying the audio signal to the cathode of a reverse-biased PIN diode, as shown in the circuit diagram. The loop filter uses the SL562, which (with the values shown) has a loop bandwidth of 60 Hz and a damping factor of 0.6. This filter is followed by a low-pass pole at 3.7 kHz to attenuate the 12.5-kHz reference sidebands. The lock-up time for a 1-MHz change in frequency is 80 ms. The output frequency range is 144 to 146 MHz and the level is +3 dB into 50 Ω. GEC Plessey Semiconductors, Professional Products, 1991, p. 4-46.

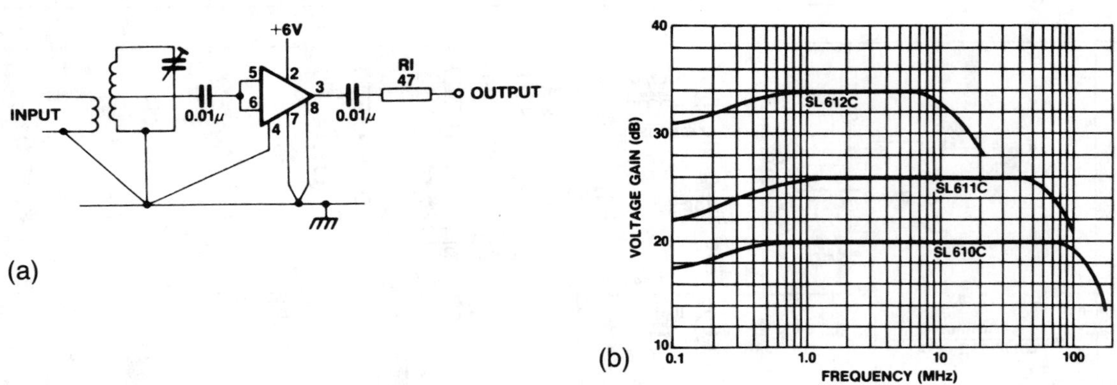

General-purpose RF amplifier

Fig. 2-29 This circuit is suitable for an SL610C, SL611C, or SC612C. Figure 2-29B shows the typical voltage gains for the three amplifiers. GEC Plessey Semiconductors, Professional Products, 1991, p. 1-129.

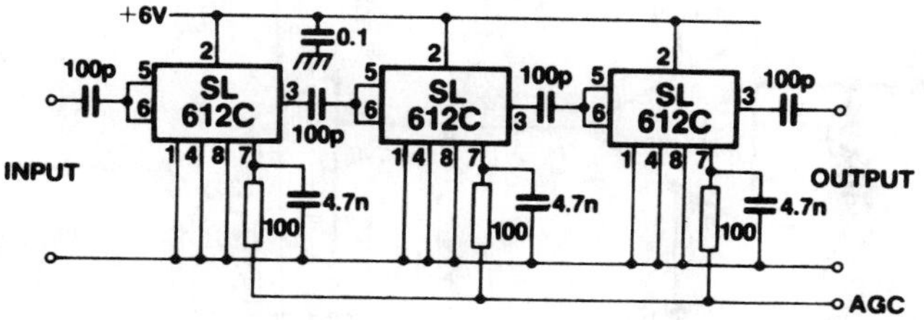

Simple SSB IF amplifier

Fig. 2-30 This circuit uses three SL612C amplifiers, and provides a gain of 100 dB at 9 MHz. GEC Plessey Semiconductors, Professional Products, 1991, p. 1-129.

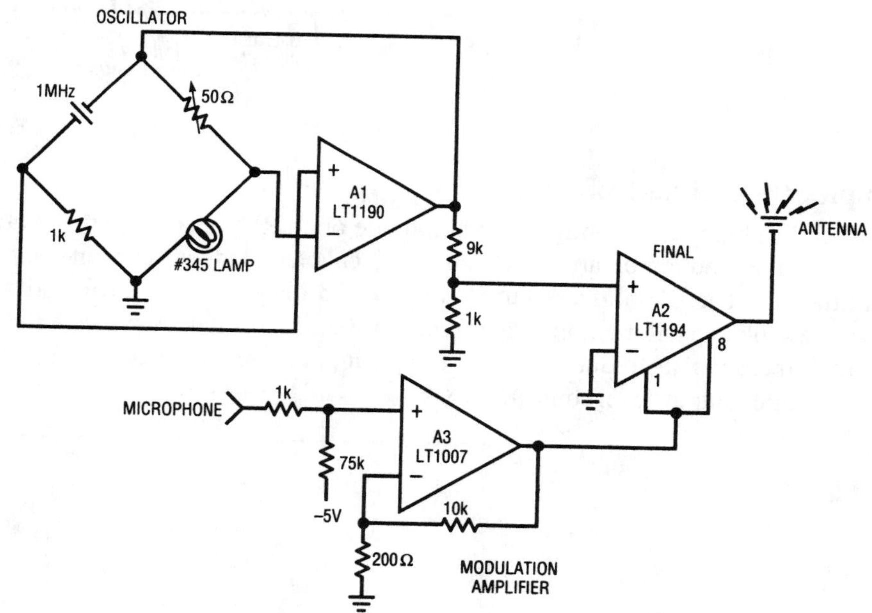

Complete three-chip AM radio station

Fig. 2-31 This circuit is a complete microphone-to-antenna AM radio station, producing a stable 1-Vpp 1-MHz signal at the antenna. Construction and operation of this circuit might require FCC review and/or licensing. The carrier is generated by A1, connected as a quartz-stabilized oscillator, similar to that of Fig. 5-18. The A1 output is applied to modulated RF power-output stage A2. Microphone amplifier A3 supplies bias to pins 1 and 8 of A2, resulting in an amplitude-modulated RF carrier at the A2 output. The circuit is calibrated by adjusting the 50-Ω oscillator potentiometer for a stable 1-Vpp 1-MHz signal at the A1 output. Linear Technology Corporation, 1991, AN47-52.

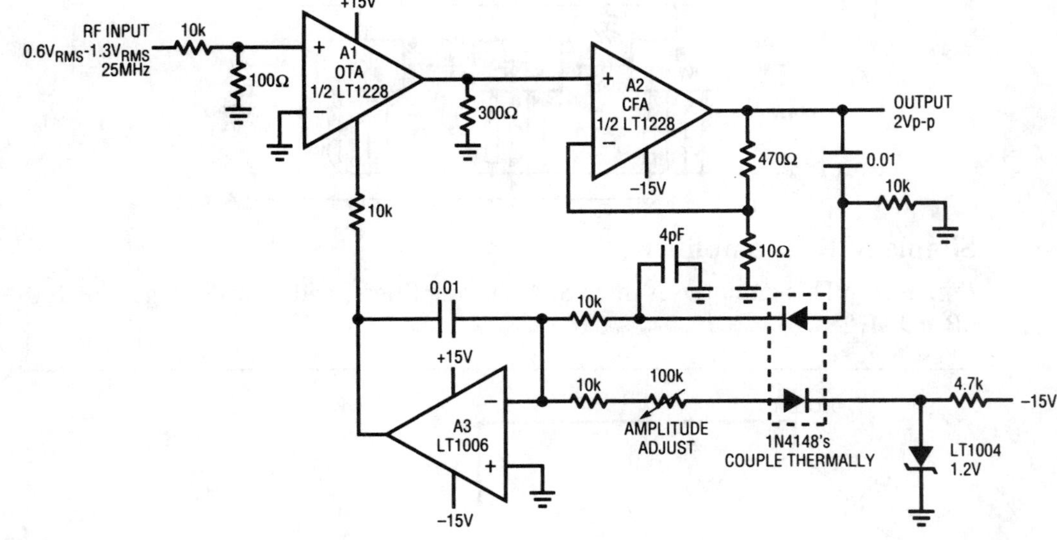

Simple RF-leveling loop

Fig. 2-32 This circuit stabilizes the amplitude of an RF signal against variations in input, time, and temperature. A1 (an OTA, chapter 11) feeds current-feedback amplifier A2. The A2 output is sampled by the A3-biased gain-control configuration. The 4-pF capacitor compensates rectifier diode capacitance, which enhances output flatness versus frequency. The A1 I_{SET} input current controls A1 gain, and allows overall output-level control. Linear Technology Corporation, 1991, AN47-64.

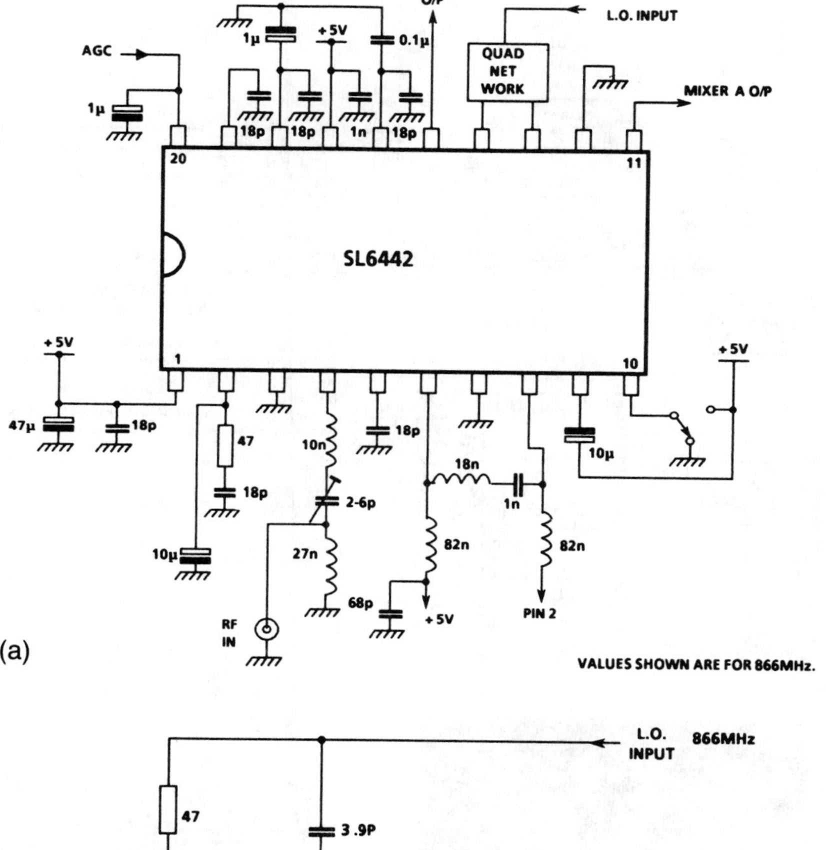

(a)

L.O.
INPUT
866MHz
47
3.9P
47
18P
3.9P
PIN 14
PIN 13

(b)

1-GHz amplifier/mixer

Fig. 2-33 This circuit shows an SL6442 used as an amplifier/mixer suitable for cordless or cellular telephones, pagers, and any low-power receiver operating at frequencies up to 1 GHz. The SL6442 contains a low-noise amplifier with AGC and two mixers for use in *I* and *Q* direct-conversion receivers or image-canceling in superhet receivers, with a single 5-V supply at a typical current of 4 mA. In this circuit, the local-oscillator signal is applied through a quad network (Fig. 2-33B). RF is applied at pin 4, AGC at pin 20, and mixer outputs are taken from pins 11 and 15. GEC Plessey Semiconductors, Personal Communications, 1990, p. 46.

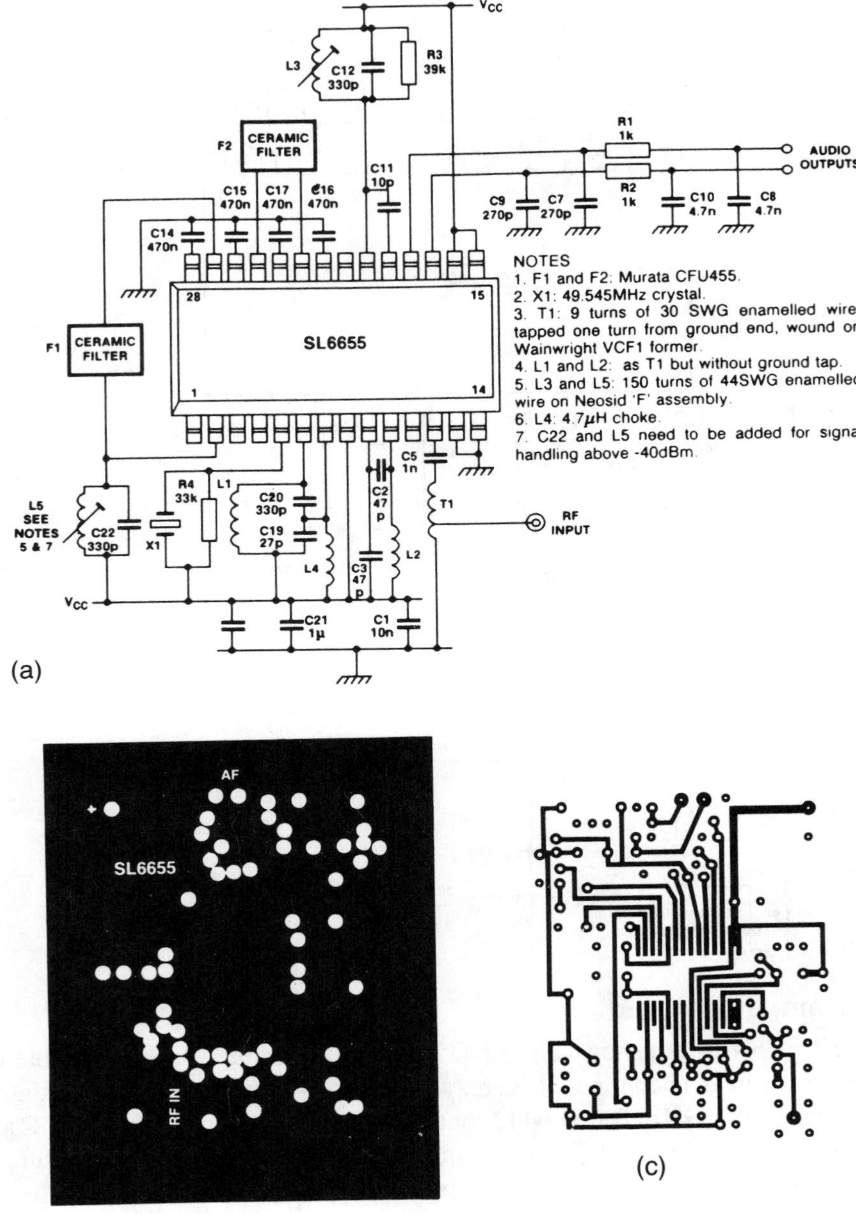
Vcc
L3
C12
330p
R3
39k
F2
CERAMIC
FILTER
C11
10p
R1
1k
AUDIO
OUTPUTS
R2
1k
C15
470n
C17
470n
C16
470n
C9
270p
C7
270p
C10
4.7n
C8
4.7n
C14
470n
28
15
SL6655
1
14
F1
CERAMIC
FILTER
NOTES
1. F1 and F2: Murata CFU455.
2. X1: 49.545MHz crystal.
3. T1: 9 turns of 30 SWG enamelled wire, tapped one turn from ground end, wound on Wainwright VCF1 former.
4. L1 and L2: as T1 but without ground tap.
5. L3 and L5: 150 turns of 44SWG enamelled wire on Neosid 'F' assembly.
6. L4: 4.7μH choke.
7. C22 and L5 need to be added for signal handling above -40dBm.
C5
1n
L5
SEE
NOTES
5 & 7
C22
330p
X1
R4
33k
L1
C20
330p
C19
27p
L4
C2
47
p
C3
47
p
L2
T1
RF
INPUT
Vcc
C21
1μ
C1
10n
(a)
AF
SL6655
RF IN
(b)
(c)

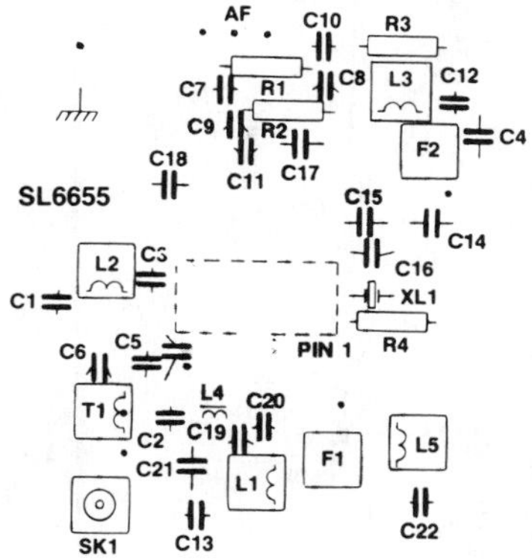

Continued

(d)

Ultra-low-power FM radio receiver

Fig. 2-34 The SL6655 shown in this circuit is a single-chip RF amplifier/mixer/oscillator/IF amplifier/detector, operating with a 0.95- to 5-V supply at a typical current of 1 mA. Typical sensitivity is 250 nV, with a typical audio output of 12 mV (rms). The circuit can operate at frequencies up to 100 MHz. Circuit values for 50-MHz operation are shown. Typical surface-mount construction details are shown in Figs. 2-34B (ground plane), 2-34C (copper track), and 2-34D (component overlay) (all 1:1). GEC Plessey Semiconductors, Personal Communications, 1990, p. 99/100.

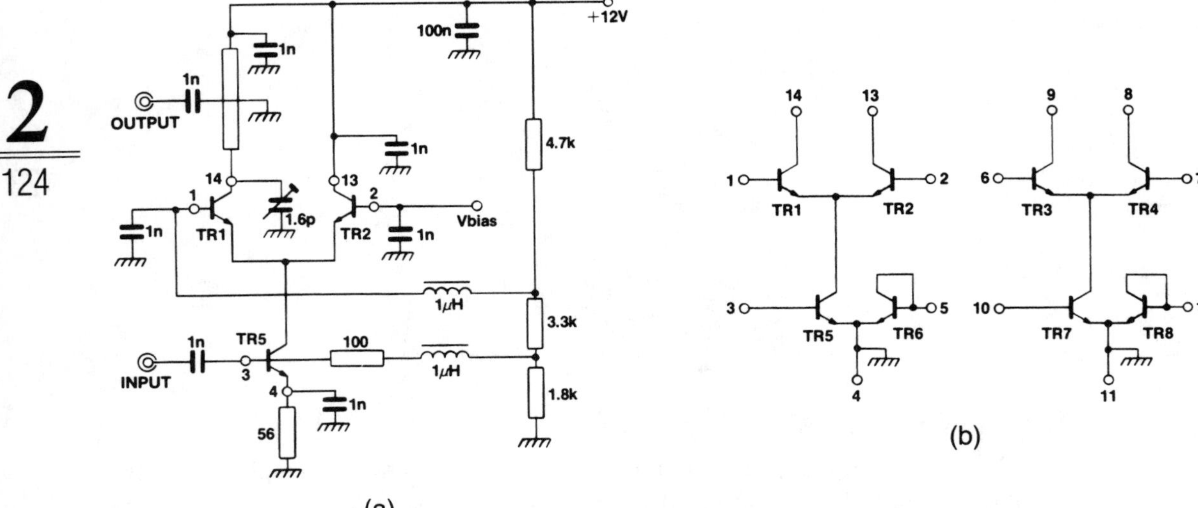

900-MHz gain-controlled amplifier

Fig. 2-35 This circuit shows an SL2365 transistor array connected as a gain-controlled amplifier. The transistor-array connections are shown in Fig. 2-35B. The collector load of TR1 is a transformer that is composed of a 14-mm length of 75-Ω stripline resonated with a 1- to 6-pF variable capacitor. The transformer secondary is a small loop of stiff wire that is grounded at one end and located a few mm above the stripline. The noise figure at full gain is 9 dB. GEC Plessey Semiconductors, Personal Communications, 1990, p. 197.

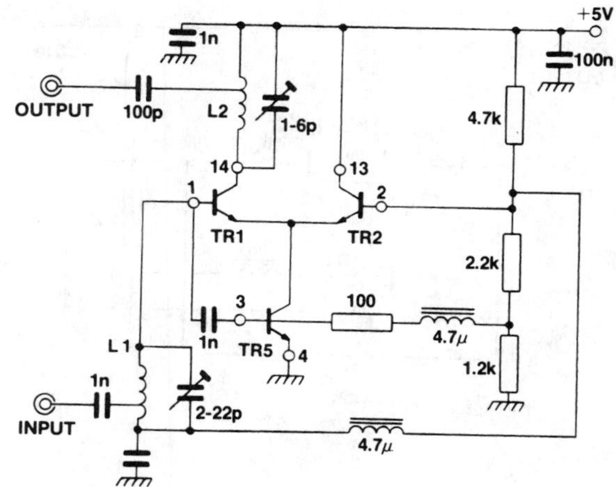

NOTES
.1 = 4 turns 20 swg tinned wire 5mm diameter. Tapped turn from ground end
.2 = 3 turns 20 swg tinned wire 5mm diameter. Tapped turn from 5V rail end

150- to 300-MHz frequency doubler

Fig. 2-36 This circuit shows an SL2365 transistor array (Fig. 2-35B) connected as a frequency doubler, with a gain of +1 dB for a −20 dB input, and an input-frequency rejection of 18 dB. GEC Plessey Semiconductors, Personal Communications, 1990, p. 197.

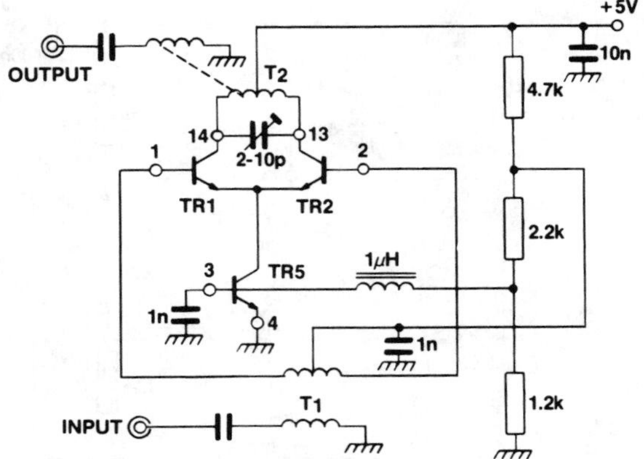

NOTES
T_1 Primary = 4 turns 36 swg enamelled wire wound on ferrite core
T_1 Secondary = 4 turns 36 swg enamelled wire wound on LF ferrite core (twisted pair)
T_2 Primary = 2 turns 20 swg tinned wire 5mm diameter
T_2 Secondary = 4 turns 20 swg tinned wire 5mm diameter, positioned axially relative to T_2 Primary

100- to 300-MHz frequency tripler

Fig. 2-37 This circuit shows an SL2365 transistor array (Fig. 2-35B) connected as a frequency tripler, with a gain of −40 dB, an input-frequency rejection of 30 dB, and second-harmonic rejection of 28 dB. GEC Plessey Semiconductors, Personal Communications, 1990, p. 198.

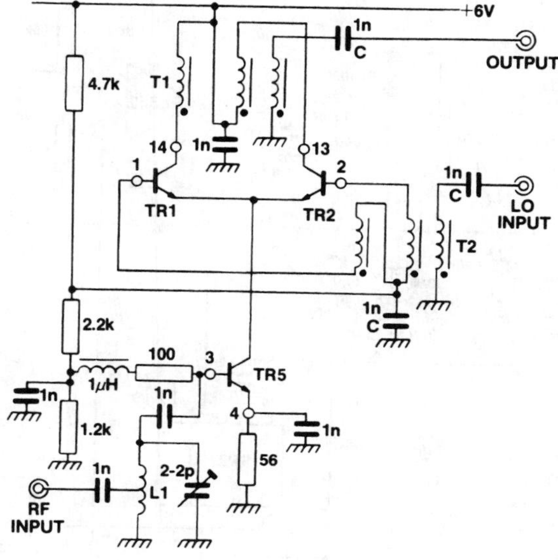

NOTES
L_1 = 4 turns 20 swg tinned wire 5mm diameter. Tapped 1 turn from ground end
T_1 and T_2 = 4 turns of trifilar wound 36 swg enamelled wire wound on LF ferrite core

Single balanced mixer

Fig. 2-38 This circuit shows an SL2365 transistor array (Fig. 2-35B) connected as a balance mixer, with a gain of 13 dB, a third-order incercept of −13 dB, and a noise figure of 10 dB. GEC Plessey Semiconductors, Personal Communications, 1990, p. 198.

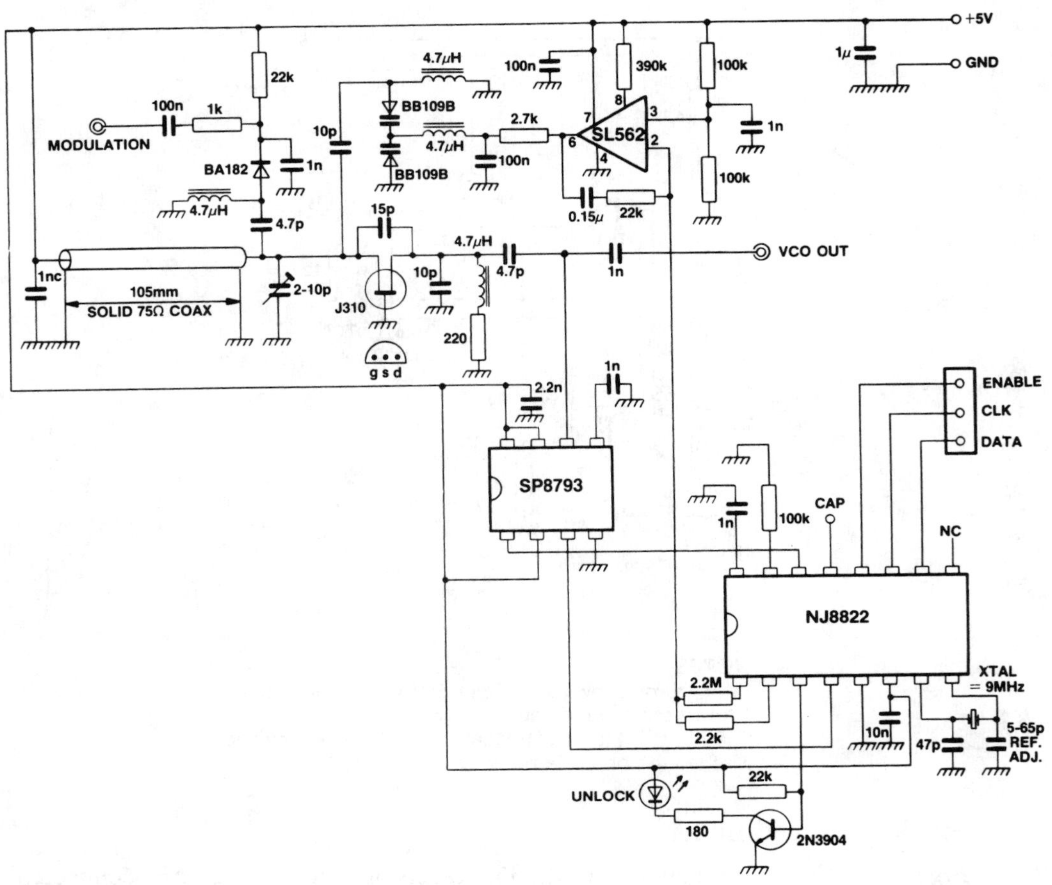

Serially programmable VHF frequency synthesizer

Fig. 2-39 This circuit uses an NJ8822 single-chip synthesizer, an SP8793 dual-modulus prescaler, and an SL562 low-noise op amp to form a frequency synthesizer, with an output range from 144 to 146 MHz at a level of +3 dBm into 50 Ω. The lock-up time for a 1-MHz change in frequency is 80 ms. The NJ8822 is programmed via a serial microprocessor interface. The VCO (using a coax resonator) is modulated by applying audio to the cathode of a reverse-biased PIN diode. The loop filter uses the SL562, which has a loop bandwidth of 60 Hz (with the values shown) and a damping factor of 0.6. The filter is followed by a low-pass pole at 3.7 kHz to attenuate the 12.5-kHz reference sidebands. GEC Plessey Semiconductors, Personal Communications, 1990, p 211.

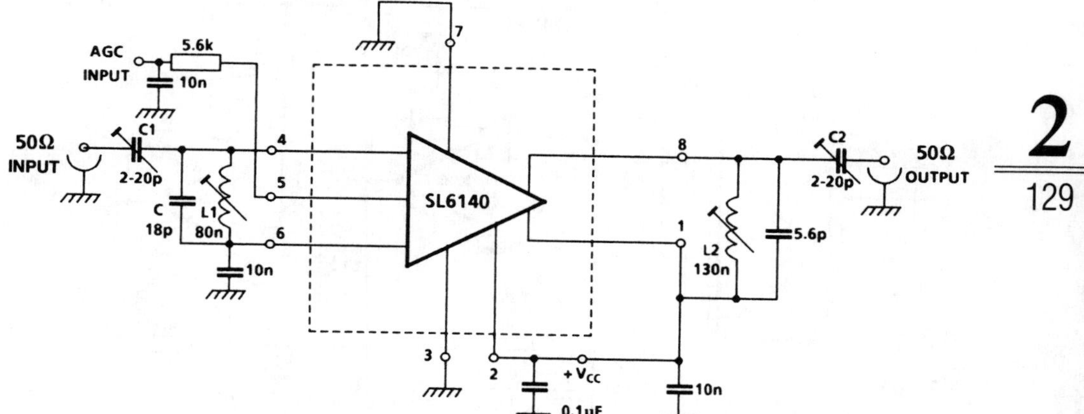

Tuned amplifier (MP pinout)

Fig. 2-40 This circuit shows an SL6140 wideband AGC amplifier used as a 100-MHz tuned amplifier with a power gain of 35 dB (Fig 2-19). The high gain is achieved at the expense of bandwidth by means of the matching networks. L1 and L2 are adjusted to set the tuned frequency. C1 and C2 are adjusted to optimize gain. However, if too high an impedance is "seen" by the input or output, the circuit might oscillate. Adjust C1 and C2 for the minimum bandwidth necessary for the particular application. For best results, use a 50-Ω track ground plane from the matching networks to the 50-Ω source and load. GEC Plessey Semiconductors, Personal Communications, 1990, p. 244.

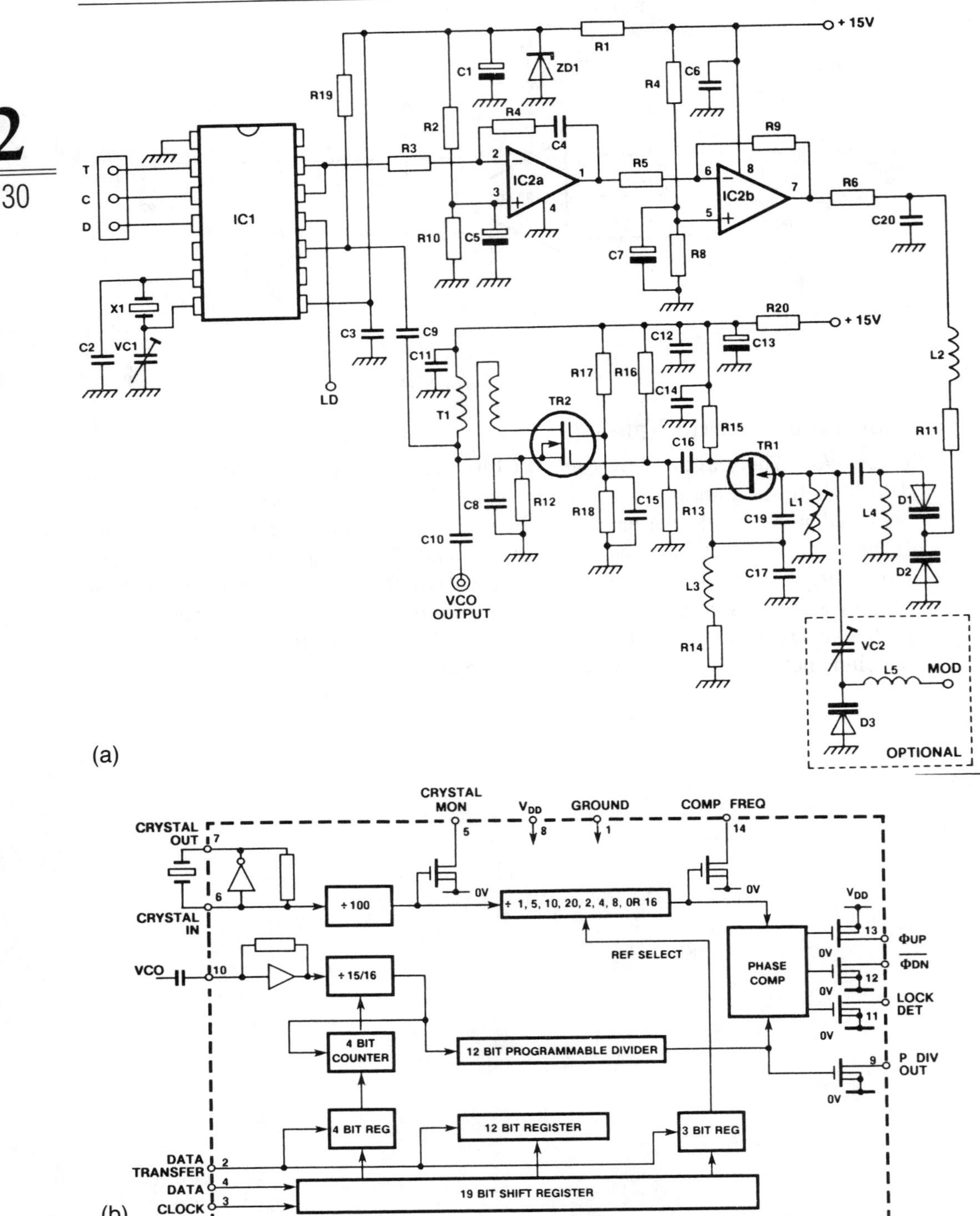
+15V
R1
ZD1
C1
R19
R2
R4
C4
R3
IC2a
R10
C5
R4
C6
R9
R5
IC2b
R6
C20
C7
R8
T
C
D
IC1
X1
C2
VC1
LD
C3
C9
R20
+15V
C12
C13
C11
R17
R16
C14
T1
TR2
R15
C16
TR1
R11
L2
C8
R12
R18
C15
R13
L1
C19
L4
D1
D2
C10
L3
C17
VCO OUTPUT
R14
VC2
L5
MOD
D3
OPTIONAL
(a)
CRYSTAL MON
V_{DD}
GROUND
COMP FREQ
CRYSTAL OUT
CRYSTAL IN
÷100
÷ 1, 5, 10, 20, 2, 4, 8, OR 16
0V
REF SELECT
PHASE COMP
ΦUP
ΦDN
LOCK DET
P DIV OUT
VCO
÷15/16
4 BIT COUNTER
12 BIT PROGRAMMABLE DIVIDER
4 BIT REG
12 BIT REGISTER
3 BIT REG
DATA TRANSFER
DATA
CLOCK
19 BIT SHIFT REGISTER
(b)

Continued

COMPONENTS LIST

R1	1k	C1	10µ	L1	3.5 turns on Toko S18 former (alloy core)
R2	100k	C2	47p	L2	4.7µH
R3	5.6k	C3	100n	L3	1µH
R4	3.3k	C4	1µ (non-polar)	L4	4.7µH
R5	10k	C5	10µ	L5	4.7µH*
R6	22k	C6	100n	T1	3 turns 36 SWG enamelled wire, bifilar wound on B62152A7X1 (LF) 2-hole Ferrite bead
R7	100k	C7	10µ		
R8	100k	C8	1n		
R9	10k	C9	22p	TR1	J310
R10	100k	C10	33p	TR2	BF961
R11	100Ω	C11	1n	D1	BB109B (Varicap)
R12	100Ω	C12	100n	D2	BB109B
R13	27k	C13	33µ	D3	BB109B*
R14	470Ω	C14	10n	ZD1	5.1V Zener diode
R15	100Ω	C15	1n	X1	10MHz series resonant crystal, 50ppm
R16	220k	C16	22p	IC1	NJ88C30
R17	220k	C17	18p	IC2	TL072
R18	100k	C18	4.7p		
R19	22k	C19	3,9p		
R20	100Ω	C20	47n		
		VC1	5-65p		
		VC2	2-10p*		

(c)

*Optional components for modulation input

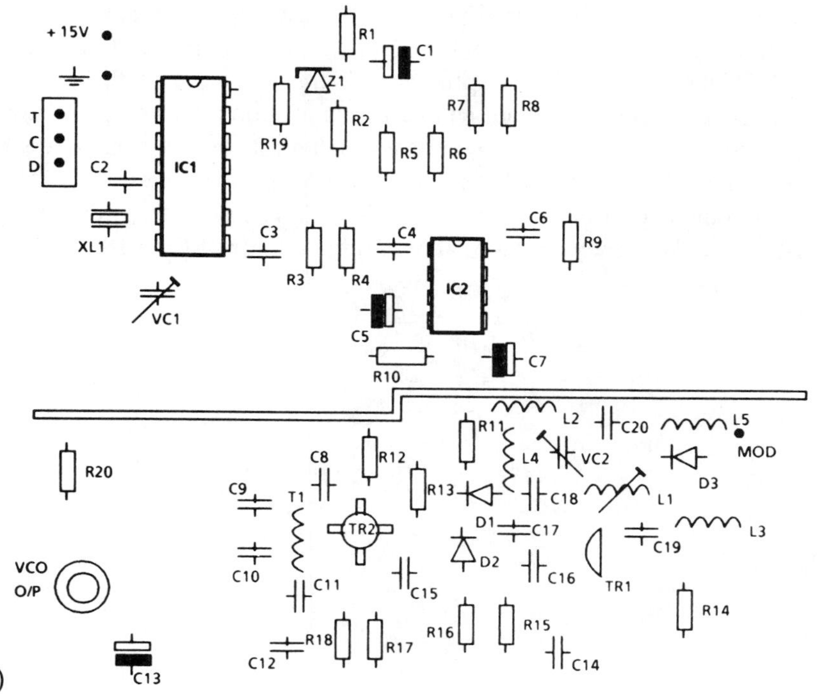

(d)

Continued

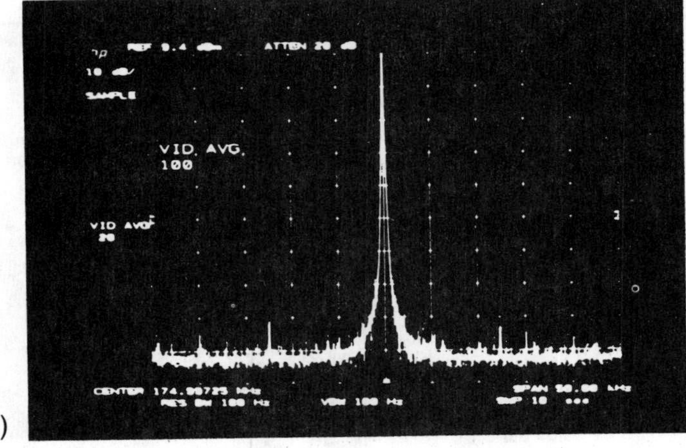

(e)

VHF frequency synthesizer

Fig. 2-41 The NF88C30 shown in this circuit contains all the logic required for a high-band VHF PLL synthesizer. Figure 2-41 shows the internal circuits of the chip. The components list is shown in Fig. 2-41C. As shown in the PC-board layout of Fig. 2-41D, the ground place is split between the VCO and synthesizer-control sections, with dc connections made via narrow tracks on either side of the board. This construction prevents synthesizer currents from causing spurious sidebands in the VCO output. The VCO can be modulated externally by adding L5, D3, and VC2. The output spectrum is shown in Fig. 2-41E. The output power is 9.4 dBm into 50 Ω, with a frequency swing of about 10 MHz (from 170 to 180 MHz in this case). GEC Plessey Semiconductors, Personal Communications, 1990, p. 247, 248, 249.

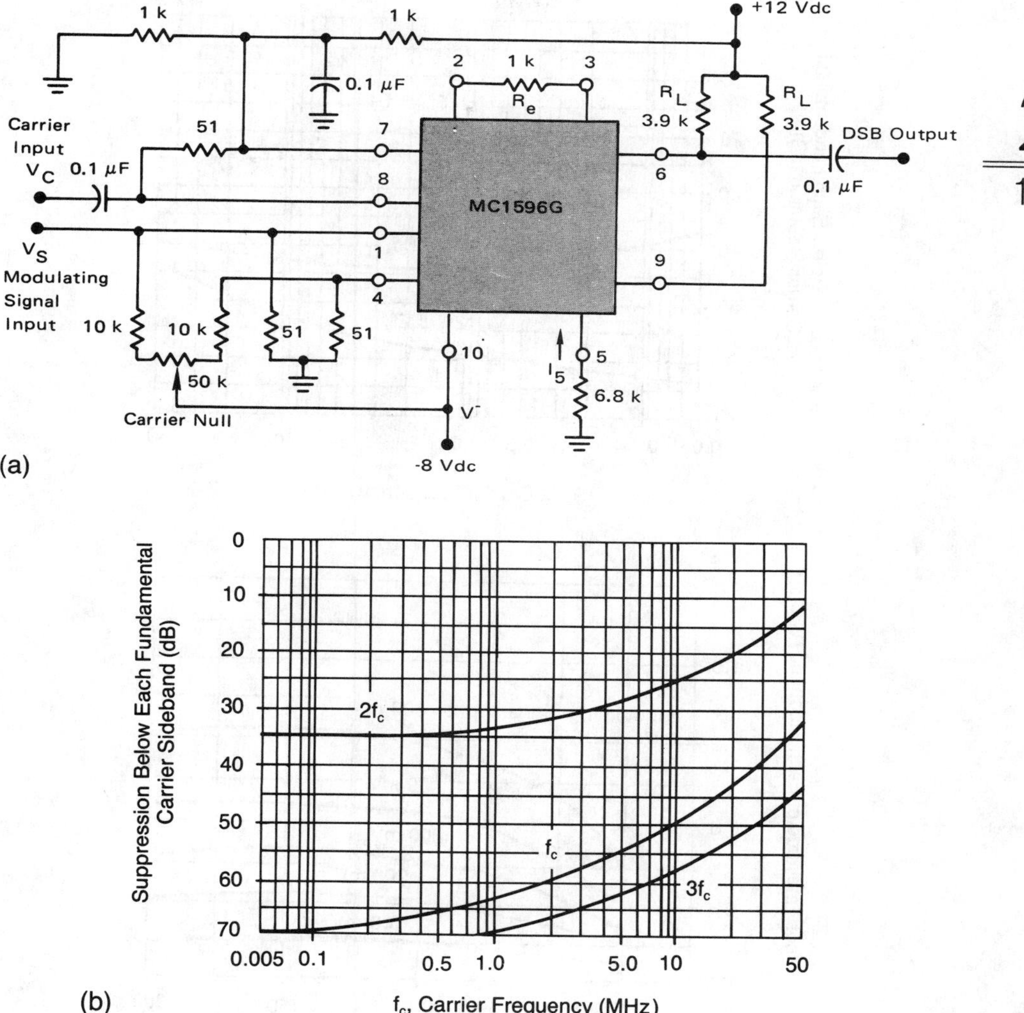
1 k
1 k
+12 Vdc
0.1 μF
2
1 k
3
R_e
R_L
3.9 k
R_L
3.9 k
Carrier
Input
V_C
0.1 μF
51
7
8
MC1596G
6
DSB Output
0.1 μF
V_S
Modulating
Signal
Input
1
4
9
10 k
10 k
51
51
50 k
Carrier Null
10
I_5
5
6.8 k
V⁻
-8 Vdc
(a)
Suppression Below Each Fundamental
Carrier Sideband (dB)
0
10
20
30
40
50
60
70
$2f_c$
f_c
$3f_c$
0.005
0.1
0.5
1.0
5.0
10
50
f_c, Carrier Frequency (MHz)
(b)

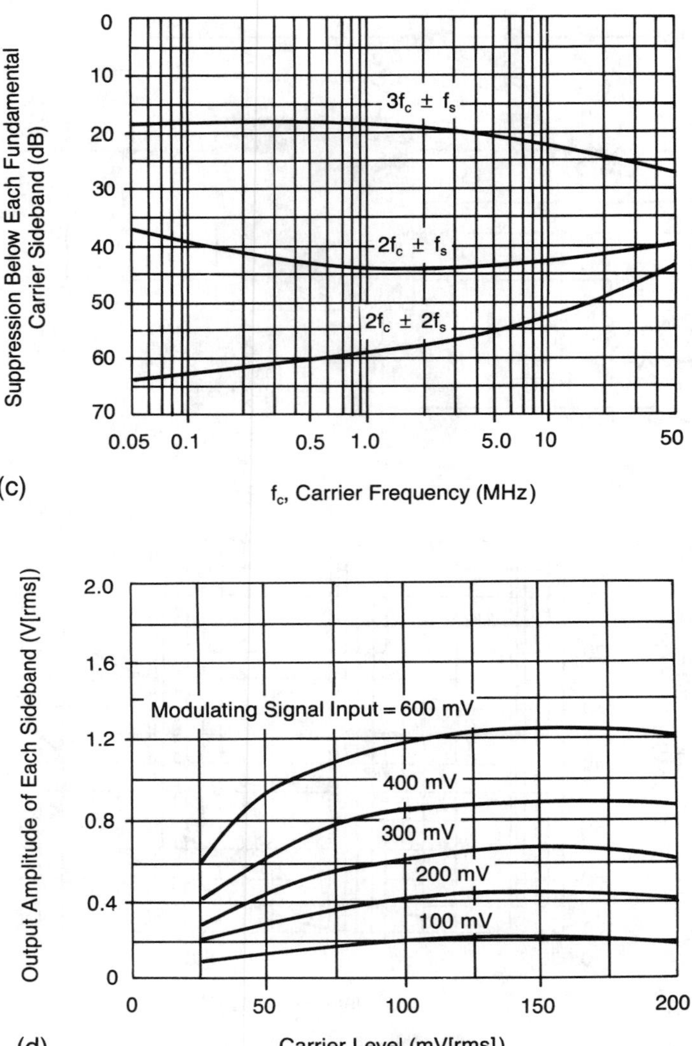

Balanced modulator for SSB operation

Fig. 2-42 This circuit shows an MC1596 operating as a balanced modulator with +12-V and −8-V supplies. Recommended input signal levels are 60 mV (rms) for the carrier and 300 mV (rms) for the maximum modulating signal. Figures 2-42B and 2-42C show the suppression of carrier and sidebands, respectively. Figure 2-42D shows the sideband output levels. Motorola Applications Literature, 1991, BR135/D REV 7, p. AN531, 3, 6.

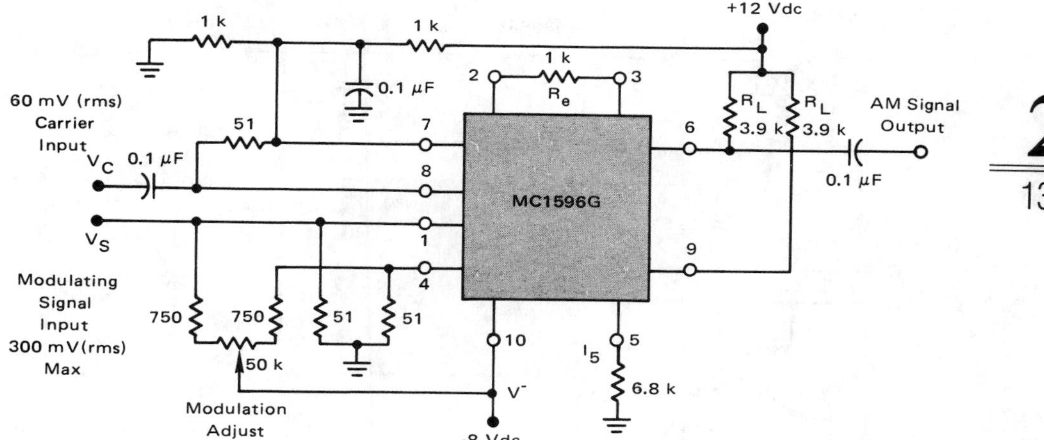

Amplitude modulator

Fig. 2-43 This circuit shows an MC1596 operating as an amplitude modulator with +12- and −8-V supplies. The circuit provides modulation at any percentage from zero to greater than 100%. Motorola Applications Literature, 1991, BR135/D REV 7, AN531, p. 7.

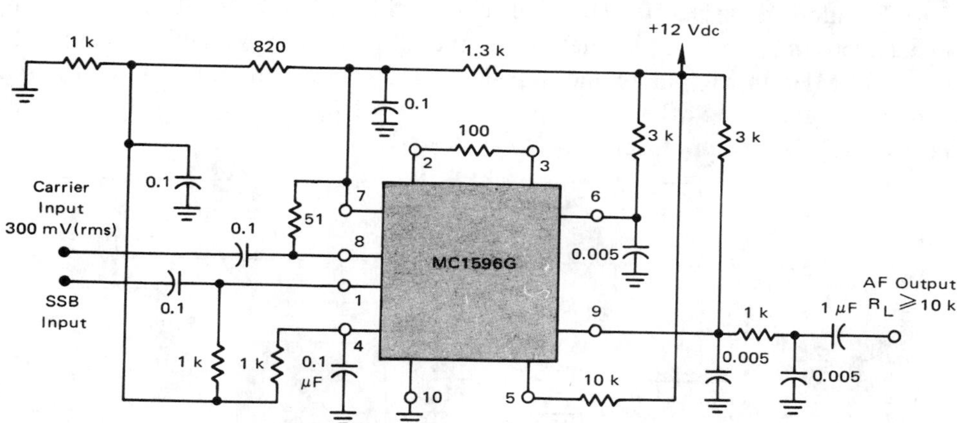

Product detector for SSB operation

Fig. 2-44 This circuit shows an MC1596 operating as a product detector with a single +12-V supply. The circuit can operate up to 100 MHz. When operating in an SSB receiver with 50-Ω input impedance and an 0.5-μV RF input, 12-dB overall gain is required between antenna and the MC1596. Dual outputs (pins 6 and 9) can be used to drive the receiver audio amplifiers, and an AGC circuit, if desired. Motorola Applications Literature, 1991, BR135/D REV 7, AN531, p. 7.

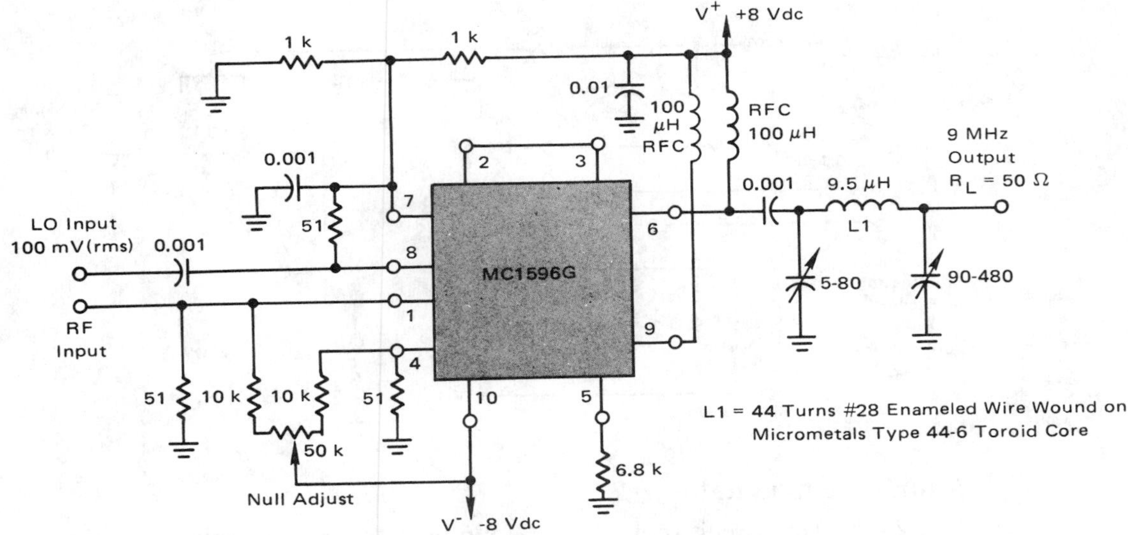

Double-balanced mixer

Fig. 2-45 This circuit shows an MC1596 operating as a high-frequency mixer with a broadband input and a tuned output at 9 MHz. The 3-dB bandwidth of the 9-MHz output tank is 450 kHz. The local-oscillator (LO) signal of 100 mV is injected at pin 8, and the modulated-RF input of about 15 mV is injected at pin 1. For a 30-MHz input (pin 1) and a 39-MHz LO (pin 8), the mixer circuit has a conversion gain of 13 dB and an input sensitivity of 7.5 µV for a 10 dB $(S + N)/N$ ratio in the 9-MHz output signal. Motorola Applications Literature, 1991, BR135/D REV 7, AN531, p. 7.

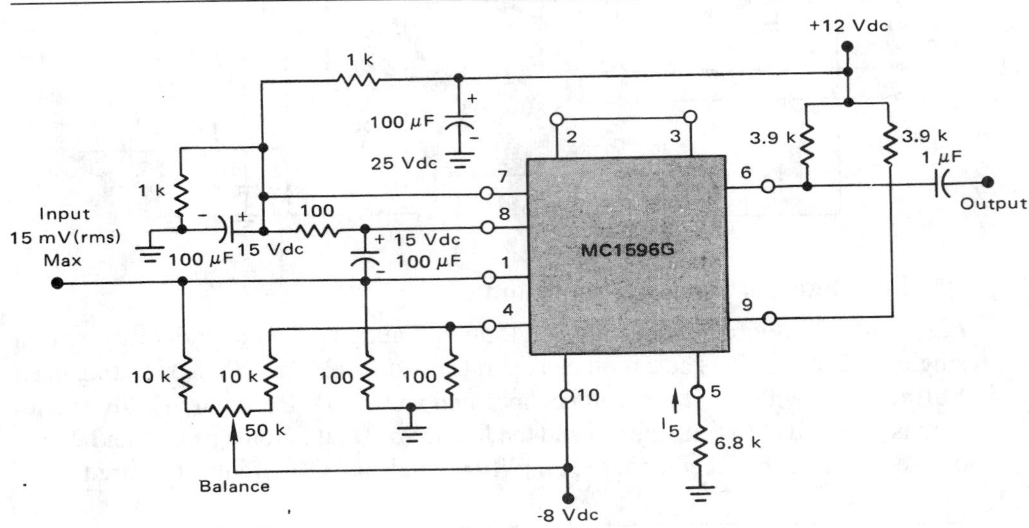

Low-frequency doubler

Fig. 2-46 This circuit shows an MC1596 operating as a frequency doubler in the audio to 1-MHz range, with all spurious outputs greater than 30-dB below the desired $2f_{IN}$-output signals. Motorola Applications Literature, 1991, BR135/D REV 7, AN531, p. 10.

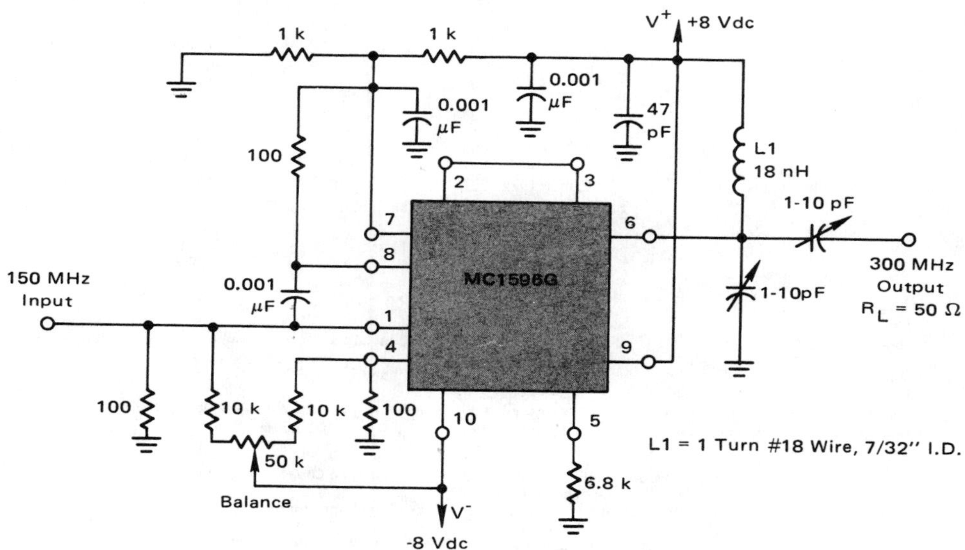

150- to 300-MHz frequency doubler

Fig. 2-47 This circuit shows an MC1596 operating as a 150- to 300-MHz frequency doubler with suitable output filtering. All spurious outputs are 20 dB (or more) below the desired 300-MHz output. Motorola Applications Literature, 1991, BR135/D REV 7, AN531, p. 10.

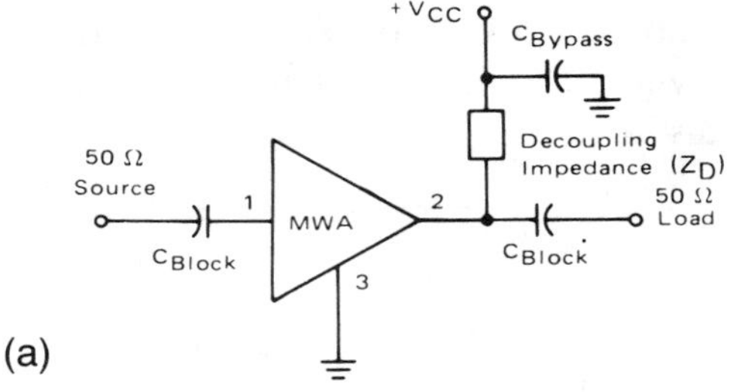

(a)

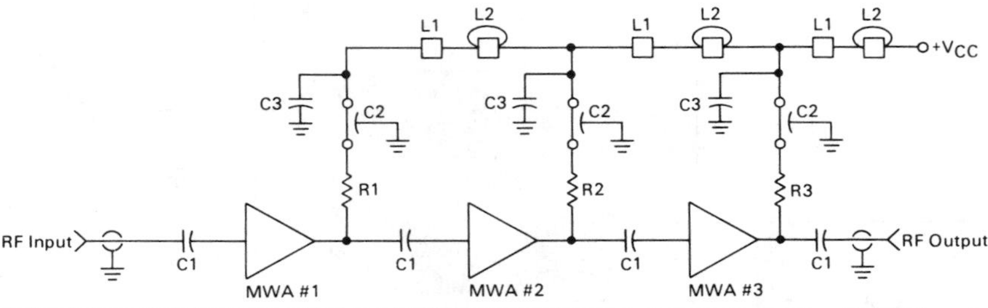

The dc isolation components shown are critical in maintaining good stability in multi-stage designs. Keep Pin #3 (Ground) as short as possible preferably soldering the case to the ground plane for best gain flatness to 1000 MHz.

C1 — For operation to 400 MHz, 1000 pF, 50 mil Chip Capacitor – ATC 50 mil Case (5.0 MHz L.F.)
C1 — For operation to 1000 MHz, 0.018 mF, Chip Capacitor for 0.25 MHz L.F. Cut-Off
C2 — Feedthru Capacitor Centralab SFT-102, 1000 pF or Metuchen 54-794002-681M, 680 pF
C3 — 0.1 μF Sprague 3CZ5U104X0050C5 – 50 Volt
L1 — Ferroxcube Shielding Bead 56-590-65/4A – Single Wire
L2 — Ferroxcube Shielding Bead 56-590-65/4A – 2 Turns #26 AWG

	Cascade 1	Cascade 2
Frequency Range	0.25 to 400 MHz	5.0 to 1000 MHz
Gain	43.5 dB	20.5 dB
Gain Flatness	± 1.0 dB	± 0.75 dB
Input VSWR	2.0:1	2.4:1
Output VSWR	1.2:1	2.1:1
V_{CC} Supply	12 Vdc	33 Vdc
I Supply	44 mAdc	150 mAdc
MWA #1	MWA110	MWA320
MWA #2	MWA110	MWA330
MWA #3	MWA120	MWA330
R1	1000 Ω	1000 Ω
R2	1000 Ω	500 Ω
R3	300 Ω	500 Ω

(b)

Linear amplifier using wideband hybrids

Fig. 2-48 This circuit shows three wideband hybrid amplifiers cascaded to form a linear amplifier. Figure 2-48B shows the hybrid amplifier connections. As shown by the table, the circuit is capable of operation from 0.25 to 400 MHz (or 5 to 1000 MHz), depending on the selection of components and power sources. Motorola RF Device Data, 1991, p. 5-220, 5-221.

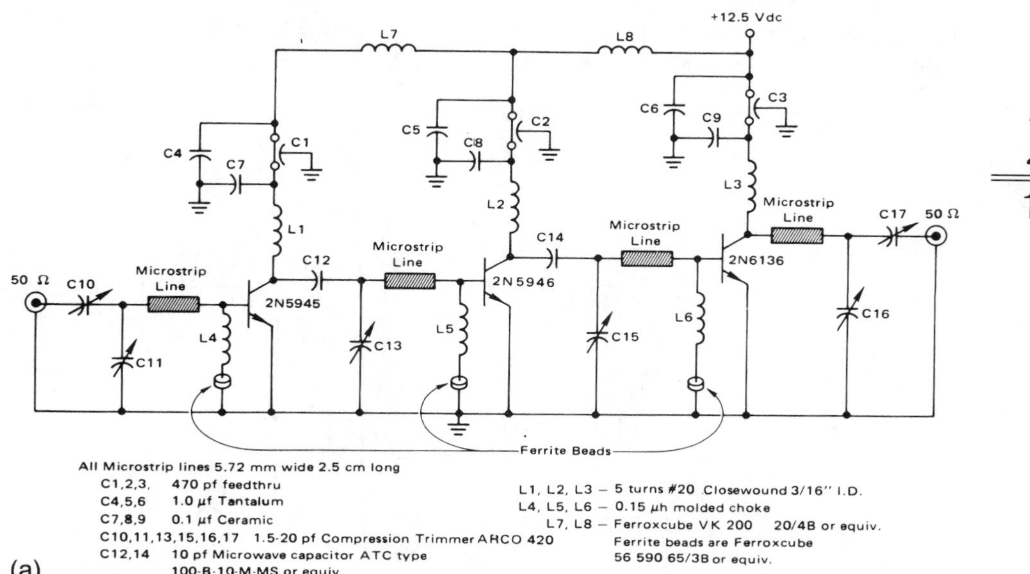

(a)

	450 MHz	480 MHz	512 MHz
Power Gain	18 db	17.2 db	16 db
Bandwidth (−1 db)	5 MHz	6 MHz	8 MHz
Overall Efficiency	44.5%	46.5%	48.5%
Harmonics	All Harmonics Better Than −20 db		
Stability	Amplifier Stable under all Conditions of Drive down to V_{CC} = 5.0 volts		
Power Output	25 w	25 w	25 w
Burnout	No Damage to any Transistor with Load Open & Shorted with 0 to ±180° Phase Angle		

(b)

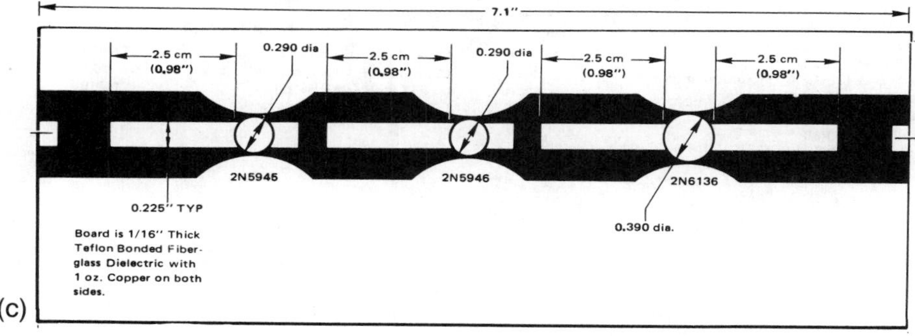

(c)

25-W UHF amplifier using microstrip techniques

Fig. 2-49 This circuit provides 25-W of output power in the 450- to 512-MHz UHF band, and is designed for 12.5-V operation. Figure 2-49B shows typical amplifier performance data. Figure 2-49C shows the microstrip board layout. Those not familiar with microstrip techniques should read the many Motorola publications, such as AN548A and AN555. Motorola RF Device Data, 1991, p. 7-51, 7-54.

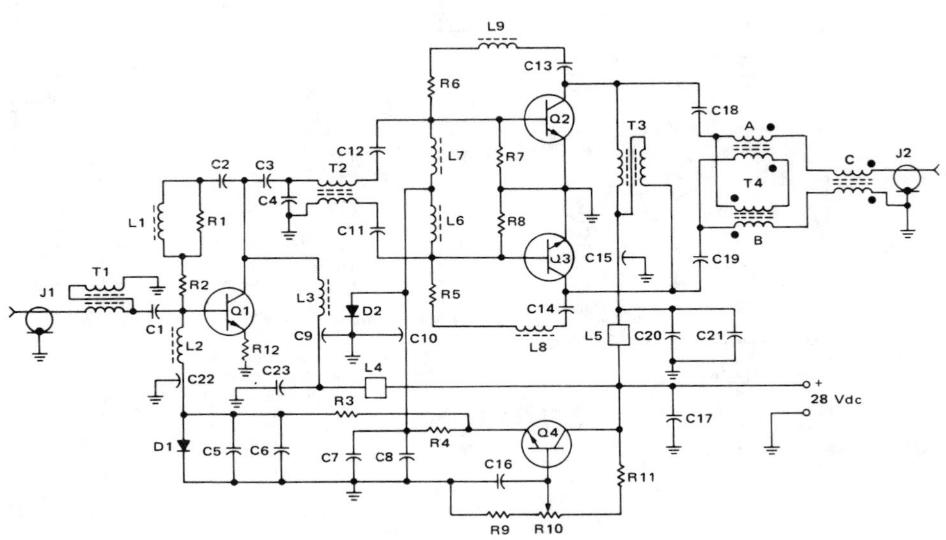

C1 – 0.033 μF mylar
C2, C3 – 0.01 μF mylar
C4 – 620 pF dipped mica
C5, C7, C16 – 0.1 μF ceramic
C6 – 100 μF/15 V electrolytic
C8 – 500 μF/6 V electrolytic
C9, C10, C15, C22 – 1000 pF feed through
C11, C12 – 0.01 μF
C13, C14 – 0.015 μF mylar
C17 – 10 μF/35 V electrolytic
C18, C19, C21 – Two 0.068 μF mylars in parallel
C20 – 0.1 μF disc ceramic
C23 – 0.1 μF disc ceramic
R1 – 220 Ω, 1/4 W carbon
R2 – 47 Ω, 1/2 W carbon
R3 – 820 Ω, 1 W wire W
R4 – 35 Ω, 5 W wire W
R5, R6 – Two 150 Ω, 1/2 W carbon in parallel
R7, R8 – 10 Ω, 1/2 W carbon
R9, R11 – 1 k, 1/2 W carbon
R10 – 1 k, 1/2 W potentiometer
R12 – 0.85 Ω (6 5.1 Ω or 4 3.3 Ω 1/4 W resistors in parallel, divided equally between both emitter leads)

T1 – 4:1 Transformer, 6 turns, 2 twisted pairs of #26 AWG enameled wire (8 twists per inch)
T2 – 1:1 Balun, 6 turns, 2 twisted pairs of #24 AWG enameled wire (6 twists per inch)
T3 – Collector choke, 4 turns, 2 twisted pairs of #22 AWG enameled wire (6 twists per inch)
T4 – 1:4 Transformer Balun, A&B – 5 turns, 2 twisted pairs of #24, C – 8 turns, 1 twisted pair of #24 AWG enameled wire (All windings 6 twists per inch). (T4 – Indiana General F624-19Q1, – All others are Indiana General F627-8Q1 ferrite toroids or equivalent.)

PARTS LIST

L1 – .33 μH, molded choke
L2, L6, L7 – 10 μH, molded choke
L3 – 1.8 μH (Ohmite 2-144)
L4, L5 – 3 ferrite beads each
L8, L9 – .22 μH, molded choke

Q1 – 2N6370
Q2, Q3 – 2N5942
Q4 – 2N5190
D1 – 1N4001
D2 – 1N4997
J1, J2 – BNC connectors

(a)

Continued

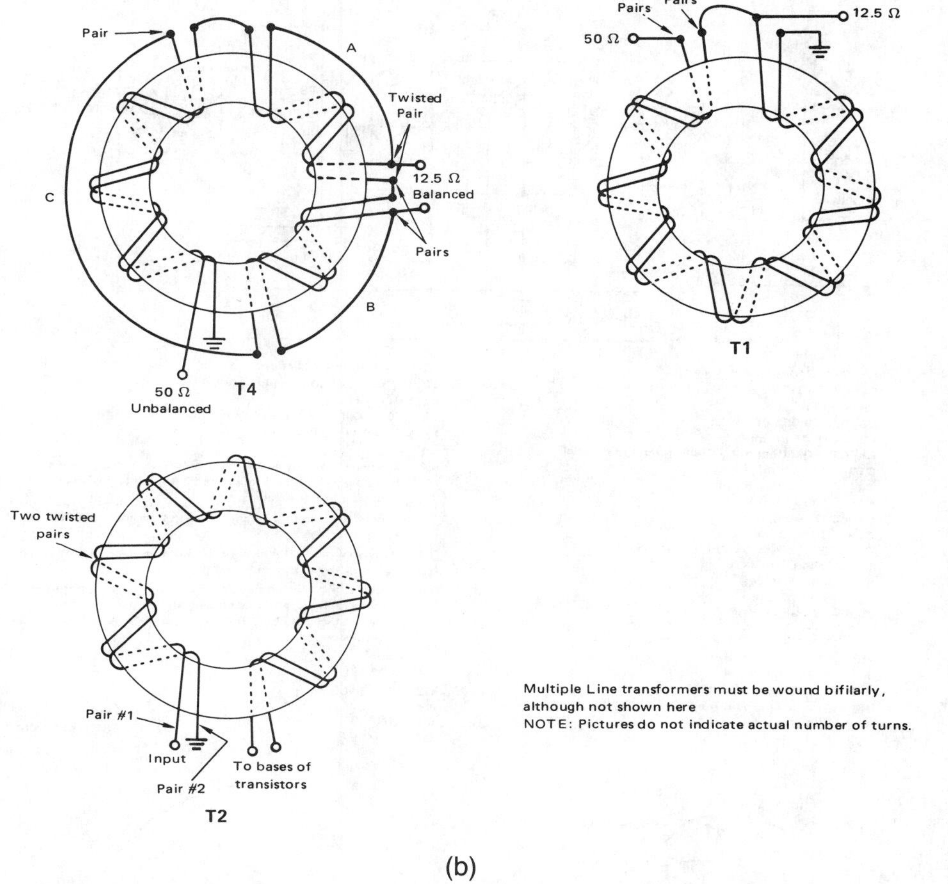

(b)

160-W (PEP) broadband linear amplifier

Fig. 2-50 This circuit provides 160-W (PEP) into a 50-Ω load with an IMD (intermodulation distortion) of −30 dB or better. Figure 2-50B shows the transformer details. Motorola RF Device Data, 1991, p. 7-72, 7-73.

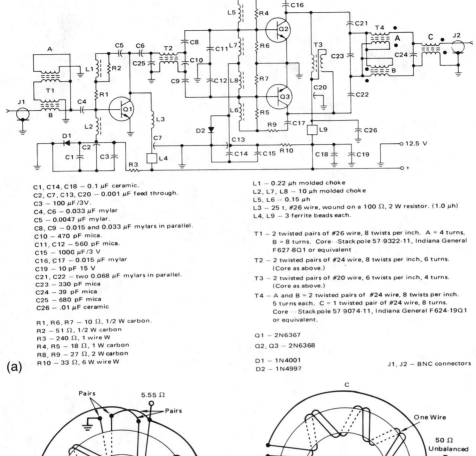

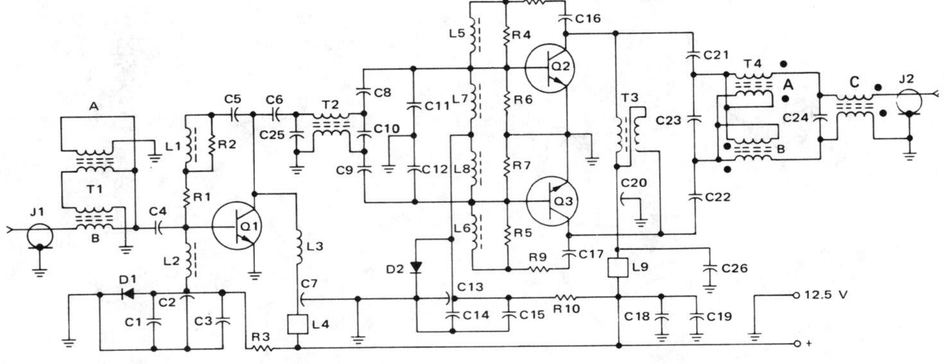

(a)

C1, C14, C18 – 0.1 μF ceramic.
C2, C7, C13, C20 – 0.001 μF feed through.
C3 – 100 μF/3V.
C4, C6 – 0.033 μF mylar
C5 – 0.0047 μF mylar.
C8, C9 – 0.015 and 0.033 μF mylars in parallel.
C10 – 470 pF mica.
C11, C12 – 560 pF mica.
C15 – 1000 μF/3 V
C16, C17 – 0.015 μF mylar
C19 – 10 pF 15 V
C21, C22 – two 0.068 μF mylars in parallel.
C23 – 330 pF mica
C24 – 39 pF mica
C25 – 680 pF mica
C26 – .01 μF ceramic

R1, R6, R7 – 10 Ω, 1/2 W carbon.
R2 – 51 Ω, 1/2 W carbon
R3 – 240 Ω, 1 wire W
R4, R5 – 18 Ω, 1 W carbon
R8, R9 – 27 Ω, 2 W carbon
R10 – 33 Ω, 6 W wire W

L1 – 0.22 μh molded choke
L2, L7, L8 – 10 μh molded choke
L5, L6 – 0.15 μh
L3 – 25 t, #26 wire, wound on a 100 Ω, 2 W resistor. (1.0 μh)
L4, L9 – 3 ferrite beads each.

T1 – 2 twisted pairs of #26 wire, 8 twists per inch. A = 4 turns, B = 8 turns. Core– Stackpole 57-9322-11, Indiana General F627-8Q1 or equivalent

T2 – 2 twisted pairs of #24 wire, 8 twists per inch, 6 turns. (Core as above.)

T3 – 2 twisted pairs of #20 wire, 6 twists per inch, 4 turns. (Core as above.)

T4 – A and B = 2 twisted pairs of #24 wire, 8 twists per inch. 5 turns each. C = 1 twisted pair of #24 wire, 8 turns. Core – Stackpole 57-9074-11, Indiana General F624-19Q1 or equivalent.

Q1 – 2N6367
Q2, Q3 – 2N6368

D1 – 1N4001
D2 – 1N4997

J1, J2 – BNC connectors

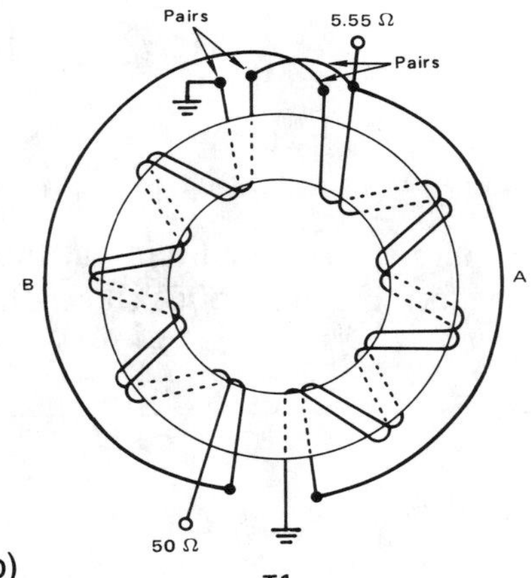

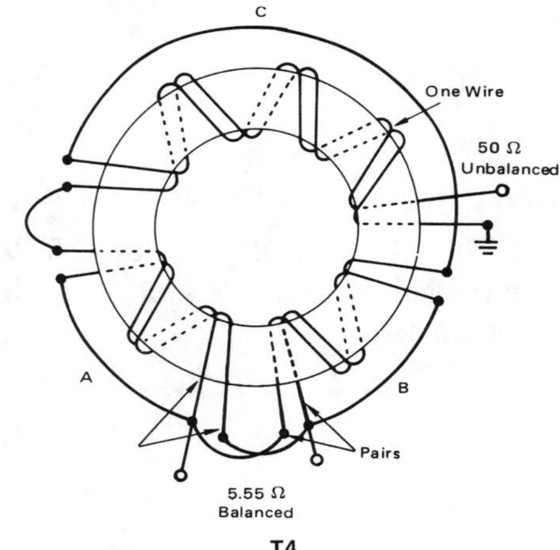

(b)

80-W (PEP) broadband linear amplifier

Fig. 2-51 This circuit provides 80-W (PEP) into a 50-Ω load. Figure 2-51B shows the transformer details. Motorola RF Device Data, 1991, p 7-76, 7-77.

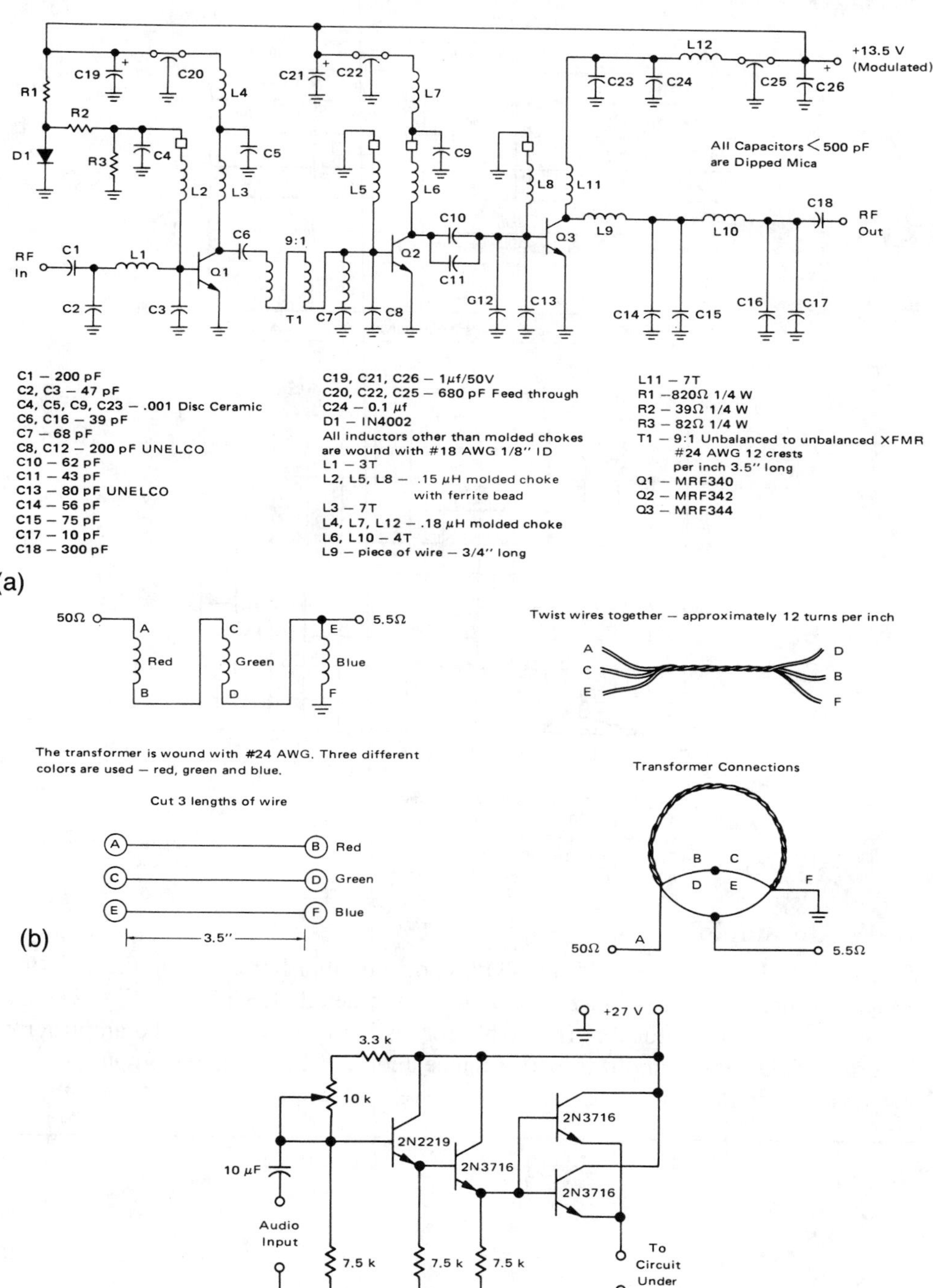
+13.5 V
(Modulated)
All Capacitors < 500 pF
are Dipped Mica
RF In
RF Out
9:1
C1 – 200 pF
C2, C3 – 47 pF
C4, C5, C9, C23 – .001 Disc Ceramic
C6, C16 – 39 pF
C7 – 68 pF
C8, C12 – 200 pF UNELCO
C10 – 62 pF
C11 – 43 pF
C13 – 80 pF UNELCO
C14 – 56 pF
C15 – 75 pF
C17 – 10 pF
C18 – 300 pF
C19, C21, C26 – 1μf/50V
C20, C22, C25 – 680 pF Feed through
C24 – 0.1 μf
D1 – IN4002
All inductors other than molded chokes
are wound with #18 AWG 1/8" ID
L1 – 3T
L2, L5, L8 – .15 μH molded choke
with ferrite bead
L3 – 7T
L4, L7, L12 – .18 μH molded choke
L6, L10 – 4T
L9 – piece of wire – 3/4" long
L11 – 7T
R1 – 820Ω 1/4 W
R2 – 39Ω 1/4 W
R3 – 82Ω 1/4 W
T1 – 9:1 Unbalanced to unbalanced XFMR
#24 AWG 12 crests
per inch 3.5" long
Q1 – MRF340
Q2 – MRF342
Q3 – MRF344
(a)
50Ω
5.5Ω
Red
Green
Blue
Twist wires together – approximately 12 turns per inch
The transformer is wound with #24 AWG. Three different
colors are used – red, green and blue.
Cut 3 lengths of wire
Red
Green
Blue
3.5"
Transformer Connections
(b)
+27 V
3.3 k
10 k
2N2219
2N3716
2N3716
2N3716
10 μF
Audio
Input
7.5 k
7.5 k
7.5 k
To
Circuit
Under
Test
(c)

	118 MHz	127 MHz	136 MHz
P_{in} (mW)	14.0	15.0	18.0
Pcarrier (W)	15.0	15.0	15.0
Total Current (Adc)	2.2	2.0	2.5
Power Supply Voltage (Vdc)	13.5	13.5	13.5
Upward Modulation (%)	89.0	88.0	90.0
Harmonic Rejection (dB) (Relative to Peak Power)			
2f 3f	55.0 58.0	55.0 58.0	52.0 57.0
Load Mismatch	Capable of Operating into 3:1 Load VSWR.		

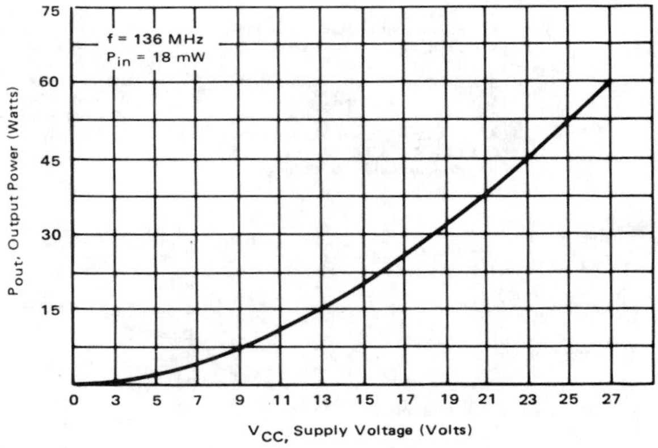

(d)

P_{out} is initially set at the carrier power of 15 watts at 13.5 Vdc, then the supply votlage is varied from 0 to 27 Vdc keeping P_{in} constant. This demonstrates the peak power output capability of the transmitter P.A.

(e)

15-W AM amplifier

Fig. 2-52 This circuit is a 15-W AM broadband amplifier covering the 118- to 136-MHz aircraft band. Figure 2-52B shows the details for T1, and Fig. 2-52C shows a series modulator that is suitable for testing or modulating the amplifier. Figure 2- 52D shows amplifier performance and Fig. 2-52E shows output power versus supply voltage. Motorola RF Device Data, 1991, p. 7-158, 7-162, 7-173, 7-164.

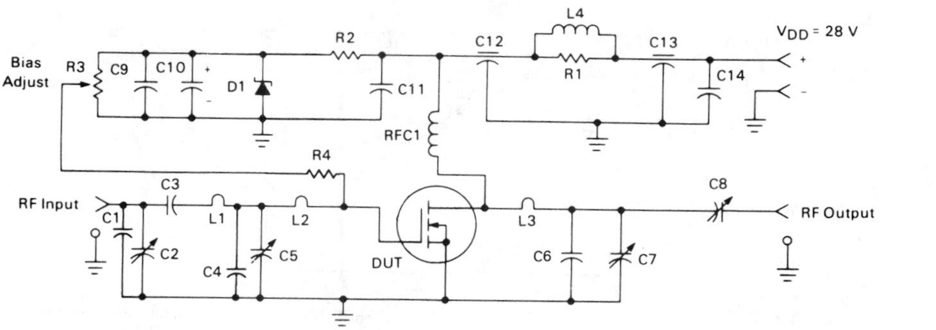

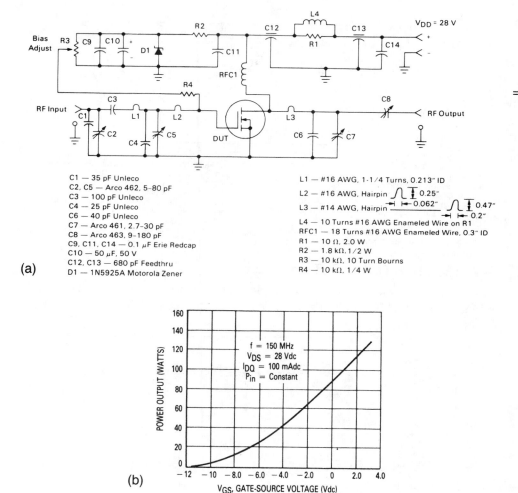

(a)

C1 — 35 pF Unleco
C2, C5 — Arco 462, 5-80 pF
C3 — 100 pF Unleco
C4 — 25 pF Unleco
C6 — 40 pF Unleco
C7 — Arco 461, 2.7-30 pF
C8 — Arco 463, 9-180 pF
C9, C11, C14 — 0.1 μF Erie Redcap
C10 — 50 μF, 50 V
C12, C13 — 680 pF Feedthru
D1 — 1N5925A Motorola Zener

L1 — #16 AWG, 1-1/4 Turns, 0.213" ID
L2 — #16 AWG, Hairpin 0.25" 0.062"
L3 — #14 AWG, Hairpin 0.47" 0.2"
L4 — 10 Turns #16 AWG Enameled Wire on R1
RFC1 — 18 Turns #16 AWG Enameled Wire, 0.3" ID
R1 — 10 Ω, 2.0 W
R2 — 1.8 kΩ, 1/2 W
R3 — 10 kΩ, 10 Turn Bourns
R4 — 10 kΩ, 1/4 W

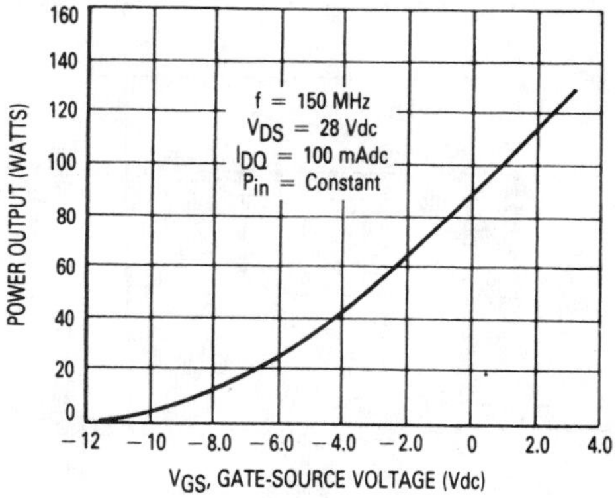

(b)

f = 150 MHz
V_{DS} = 28 Vdc
I_{DQ} = 100 mAdc
P_{in} = Constant

POWER OUTPUT (WATTS)

V_{GS}, GATE-SOURCE VOLTAGE (Vdc)

125-W 150-MHz TMOS FET amplifier

Fig. 2-53 This circuit uses an MRF174 TMOS FET, has a typical gain of 12 dB, and can survive operation into a 30:1 VSWR load at any phase angle with no damage. Notice that the output power can be reduced to less than 1 W continuously by driving the dc gate voltage negative (by adjusting R3). Figure 2-53B shows this performance feature. Motorola RF Device Data, 1991, p. 7-171, 7-172.

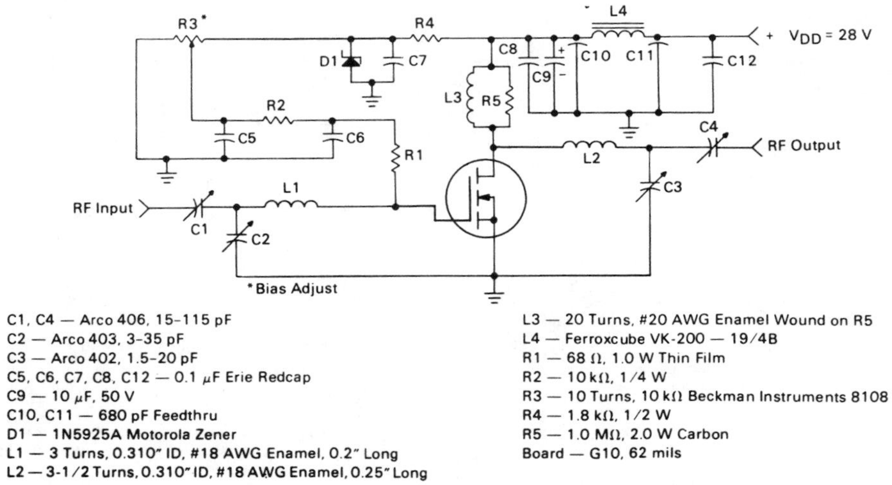

5-W 150-MHz TMOS FET amplifier

Fig. 2-54 This circuit uses an MRF134, has a gain of 14 dB, and has a drain efficiency of 55%. The output amplitude can be adjusted by R3. Motorola RF Device Data, 1991, p. 7-172.

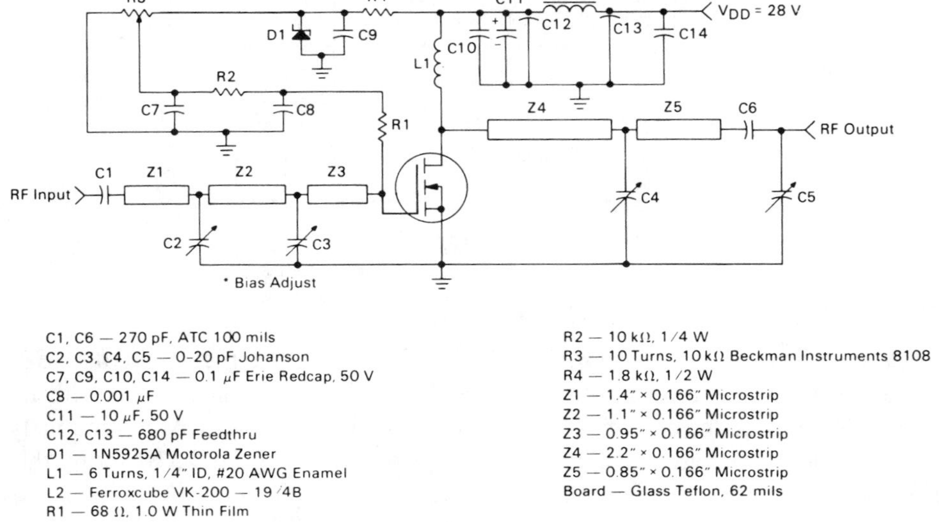

5-W 400-MHz TMOS FET amplifier

Fig. 2-55 This circuit uses an MRF134 and has a gain of 10.5 dB. Motorola RF Device Data, 1991, p. 7-173.

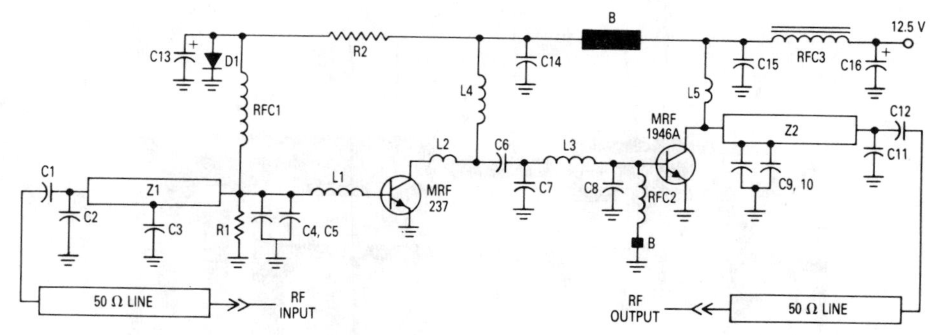

(a)

C1 = 56 pF Dura Mica
C2 = 39 pF Mini-Unelco
C3, C7 = 68 pF Mini-Unelco
C4, C5, C6, C9, C10 = 91 pF Mini-Unelco
C8 = 250 pF Unelco J101
C11 = 36 pF Mini-Unelco
C12 = 43 pF Mini-Unelco
C13 = 1 μF, 25 V Tantalum
C14, C15 = 0.1 μF Mono-Block
C16 = 10 μF 25 V Electrolytic

D1 = Diode, 1N4933 or Equivalent
L1 = Base Lead Cut to 0.4", Formed Into Loop
L2 = Collector Lead Cut To 0.35", Formed Into Loop
L3 = 0.7" #18 AWG Into Loop
L4 = 7 Turns #18 AWG, 1/8" ID
L5 = 3 Turns #16 Enam, 3/16" ID
R1 = 10 Ω, 1/4 W Carbon

R2 = 1500 Ω, 1/2 W (Select For Most Appropriate ICQ)
RFC1 = 10 μH Molded Choke
RFC2 = 0.15 μH Molded Choke
RFC3 = VK200-4B Choke
Z1, Z2 = Printed Line
Z3 = 50 Ohm Printed Line
B = Ferroxcube Ferrite Bead 56-590-65-3B

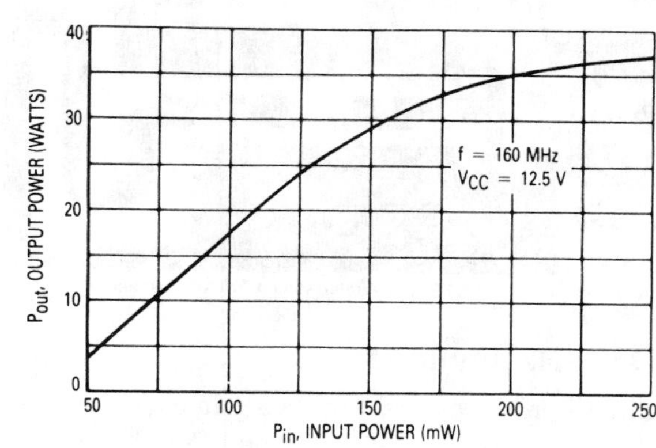

(b)

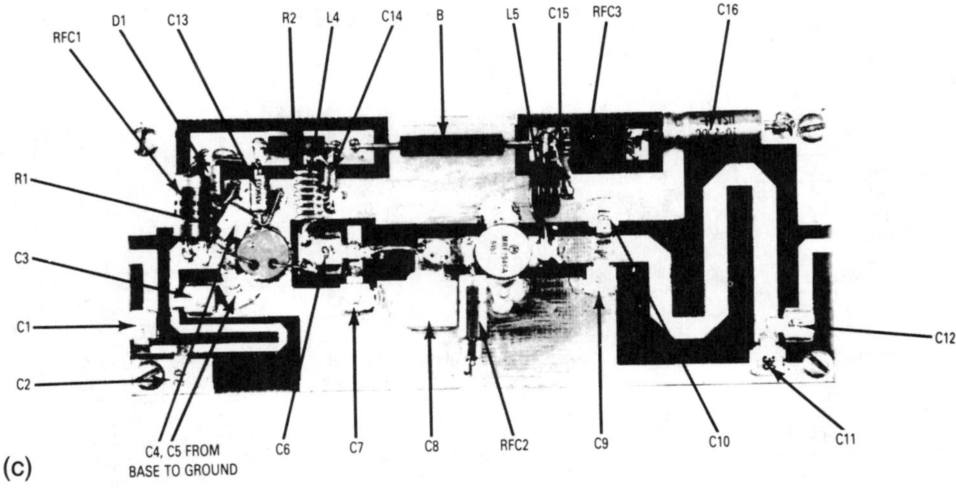

(c)

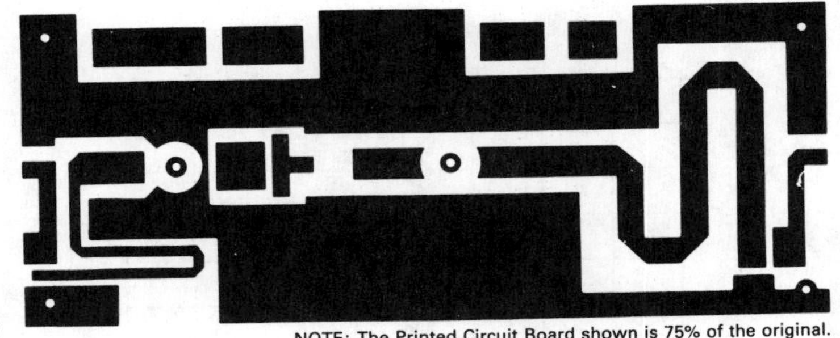

(d) NOTE: The Printed Circuit Board shown is 75% of the original.

30-W 150- to 175-MHz amplifier

Fig. 2-56 This circuit produces an overall gain in excess of 20 dB (Fig. 2-56B). Figure 2-56C shows typical parts placement, and Fig. 2-56D shows the PC-board photomaster (actual dimensions: 5″ × 2″). The amplifier is constructed on 1/16″, double-sided G-10 board with 2-ounce copper cladding. The top and bottom ground planes of the board are connected by wrapping the board edges with thin copper foil (0.002″) and then soldering the foil in place. The MRF237 is inserted into a hole in the board and soldered to the ground plane for heatsinking. The MRF1946A is mounted on a conventional heatsink (an 8-32 stud inserted into an appropriately prepared heatsink). Refer to Motorola AN778 and AN790 for mounting and thermal ratings of power transistors. Motorola RF Device Data, 1991, p. 7-187, 7-188, 7-189.

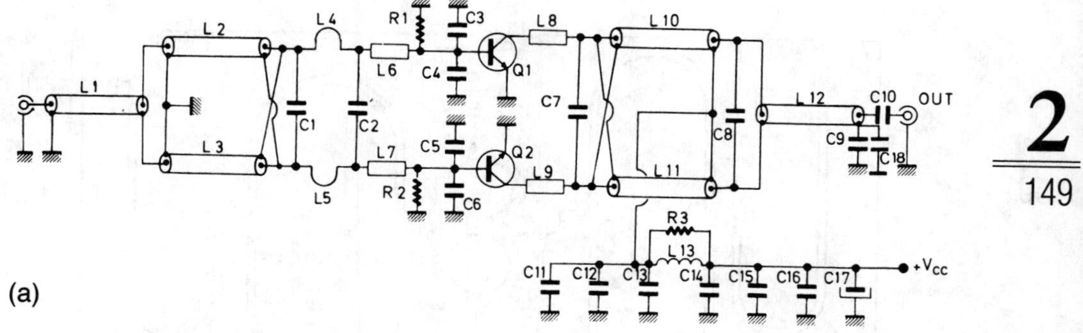

(a)

88-108 MHz; 300 W 28 V

COMPONENTS LIST

C_1	= 120 + 80 pF Chip capacitor ATC 100 B
C_2	= 220 pF Chip capacitor ATC 100 B
C_3, C_4, C_5, C_6	= 470 pF Chip capacitor ATC 100 B
C_7	= 100 pF Chip capacitor ATC 100 B
C_8	= 27 pF Chip capacitor ATC 100 B
C_9, C_{10}, C_{11}, C_{14}	= 1 000 pF Disc capacitor
C_{12}, C_{15}	= 10 nF
C_{13}, C_{16}, C_{18}	= 0,1 μF
C_{17}	= 1 000 μF/63 V Electrolytic

L_1	= 50 Ω coaxial cable ⌀ 3,2 mm (Teflon) L = 110 mm
L_2, L_3	= 25 Ω coaxial cable ⌀ 3,2 mm (Teflon) L = 110 mm
L_4 L_5	= Hair pin : copper foil 18 × 3 mm 0,3 mm thickness
L_6, L_7	= Line on substrate : 15 × 5 mm
L_8, L_9	= Line on substrate : 10 × 5 mm
L_{10}, L_{11}	= 25 Ω coaxial cable ⌀ 5 mm (Teflon) L = 110 mm
L_{12}	= 50 Ω coaxial cable ⌀ 5 mm (Teflon) L = 110 mm
L_{13}	= 15 turns ⌀ 8 mm 1,4 mm wire

R_1, R_2	= 22 Ω 1/2 W
R_3	= 47 Ω 2 W

(b)

Q_1, Q_2	= TP 9383

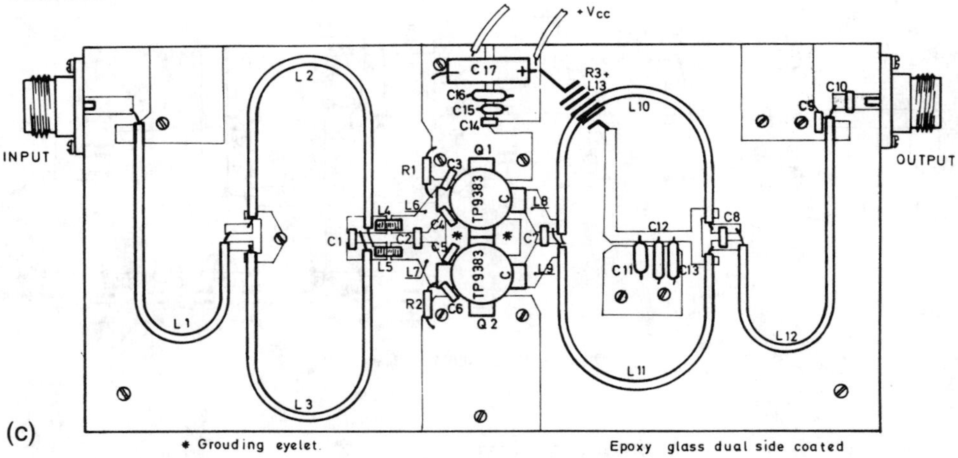

Continued

+Vcc
INPUT
OUTPUT
(c)
* Grouding eyelet.
Epoxy glass dual side coated

300 W PUSH-PULL FM TP 9383

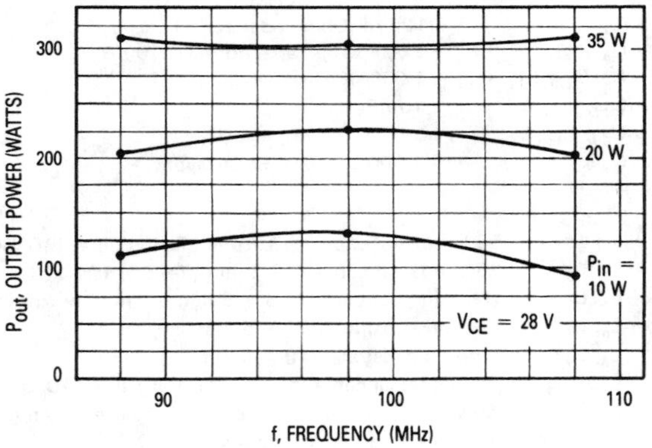

(d)
P_{out}, OUTPUT POWER (WATTS)
f, FREQUENCY (MHz)
35 W
20 W
P_{in} = 10 W
V_{CE} = 28 V

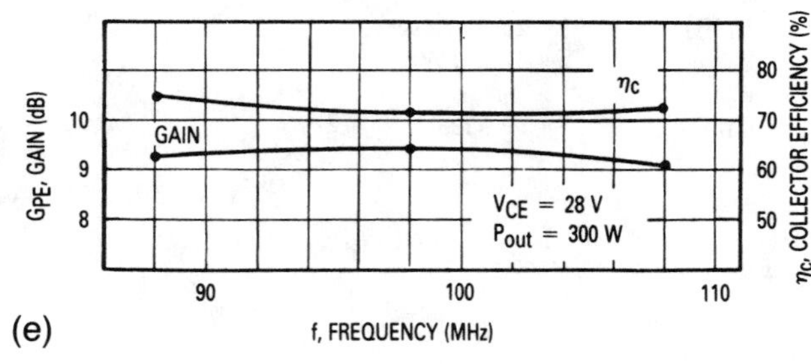

(e)

300-W FM 88- to 108-MHz power amplifier

Fig. 2-57 This circuit has a gain of about 9 dB over the frequency range. The components list, component layout, output power versus input power and frequency, and gain are shown in Figs. 2-57B, 2-57C, 2-57D, and 2-57E, respectively. The actual layout dimensions are: 4.25″ × 2.25″. Motorola RF Device Data, 1991, p. 7-262, 7-263.

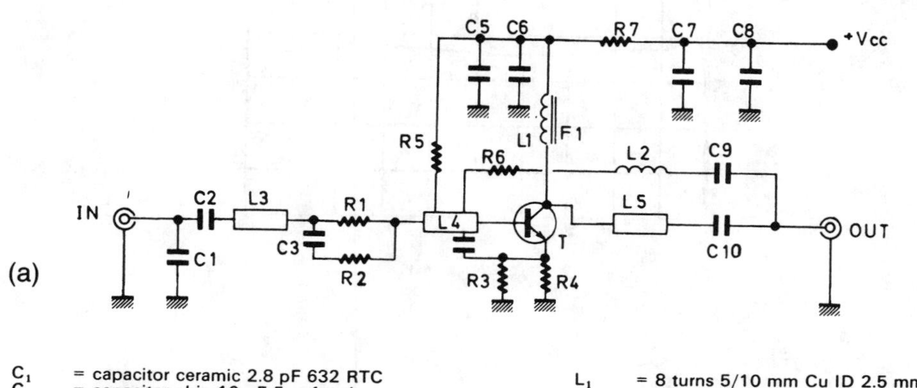

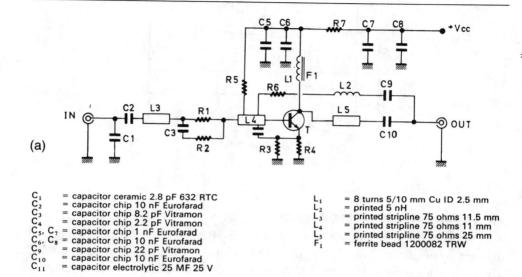

(a)

C_1 = capacitor ceramic 2.8 pF 632 RTC
C_2 = capacitor chip 10 nF Eurofarad
C_3 = capacitor chip 8.2 pF Vitramon
C_4 = capacitor chip 2.2 pF Vitramon
C_5, C_7 = capacitor chip 1 nF Eurofarad
C_6, C_8 = capacitor chip 10 nF Eurofarad
C_9 = capacitor chip 22 pF Vitramon
C_{10} = capacitor chip 10 nF Eurofarad
C_{11} = capacitor electrolytic 25 MF 25 V

L_1 = 8 turns 5/10 mm Cu ID 2.5 mm
L_2 = printed 5 nH
L_3 = printed stripline 75 ohms 11.5 mm
L_4 = printed stripline 75 ohms 11 mm
L_5 = printed stripline 75 ohms 25 mm
F_1 = ferrite bead 1200082 TRW

R_1 = resistor 12 ohms 1/4 W carbon composition
R_2 = resistor 4.7 ohms 1/4 W carbon composition
R_3, R_4 = resistor 10 ohms 1/4 W carbon composition
R_5 = resistor 8.2 kohms 1/4 W carbon composition
R_6 = resistor 240 ohms 1/4 W carbon composition
R_7 = resistor 12 ohms 1/2 W carbon composition

T = transistor TP 3400

(b)

Board Material

Epoxy glass (G 10) 1/16 inch E_R = 4.2

Epoxy glass (G 10), Double Sided

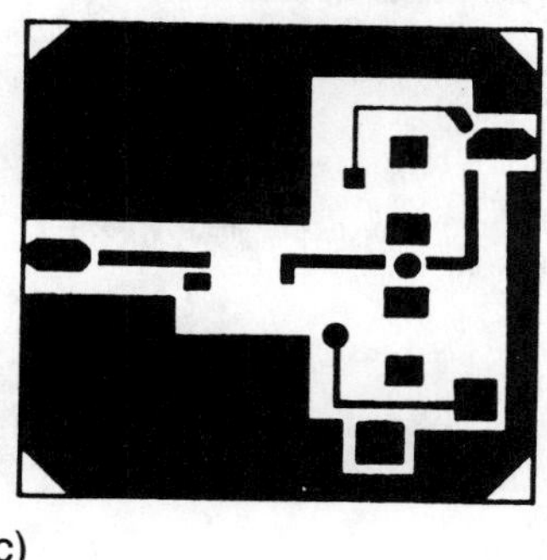

(c)

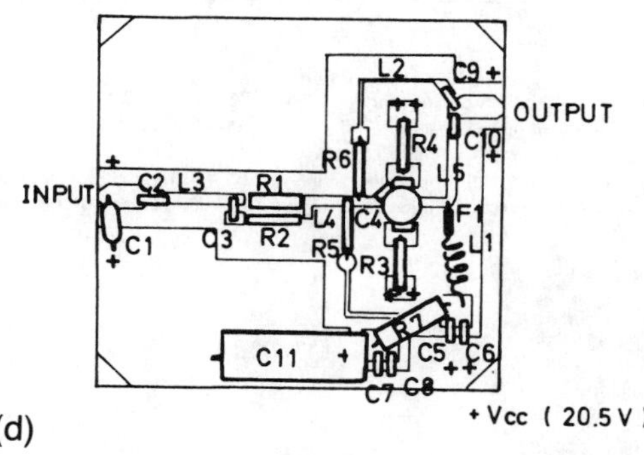

+Vcc (20.5 V)

(d)

+++ FOIL WRAP OR PLATE AROUND PLANE

Titles and descriptions

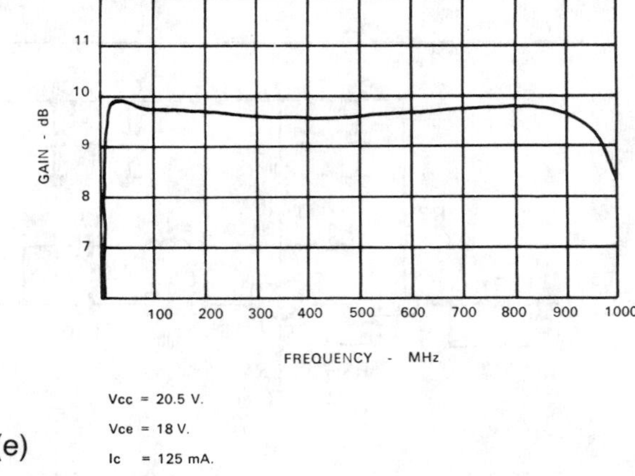

(e)

40- to 900-MHz broadband amplifier

Fig. 2-58 This circuit has a gain of about 9.5 dB over the frequency range, and delivers an output of about 1.2 V into 75 Ω, with an IMD level of about −60 dB. The components lists, PC-board photomaster, components layout, and gain are shown in Figs. 2-58B, 2-58C, 2-58D, and 2-58E, respectively. Motorola RF Device Data, 1991, p. 7-267, 7-268, 7-269.

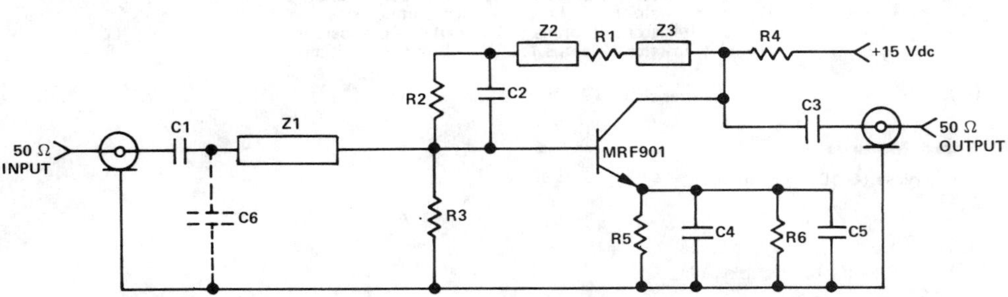

C1–C3 – 2200 pF chip capacitor
C4, C5 – 6.5 pF chip capacitor
C6 – Optional 2.1 pF chip capacitor
Z1 – 0.3″ x 0.125″ microstrip line
Z2 – 0.15″ x 0.125″ microstrip line
Z3 – 0.3″ x 0.125″ microstrip line
R1 – 200 Ω, 1/8″ W, ±5% carbon resistor
R2 – 4.3 kΩ carbon resistor
R3 – 680 Ω carbon resistor
R4 – 560 Ω carbon resistor
R5, R6 – 15 Ω ±5% chip resistor
Substrate – 1 oz. copper, double-sided glass Teflon® board 0.0625″ thick, $\epsilon_r \approx 2.5$
® Registered trademark of DuPont

(a)

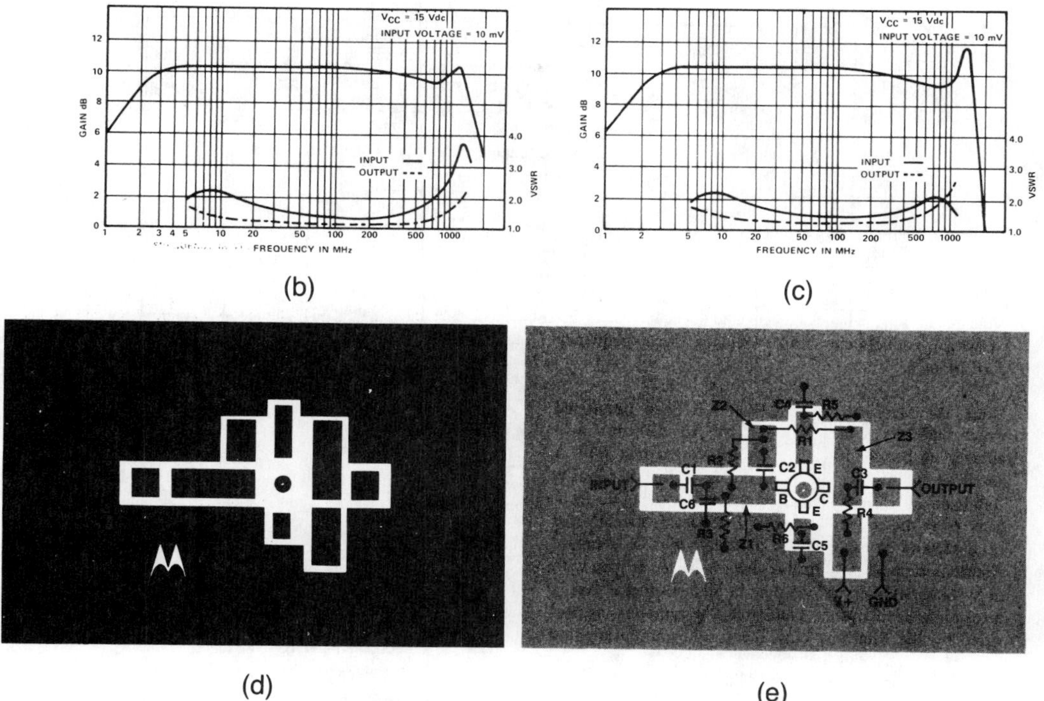

Amplifier with 10-dB gain over nine octaves

Fig. 2-59 This circuit has a gain of about 10 dB from 3 MHz to over 1 GHz (Fig. 2-59B). The gain/VSWR versus frequency, PC-board photomaster, and parts layout are shown in Figs. 2-59C, 2-59D, and 2-59E, respectively. The actual layout dimensions are: 1.8″ × 1.2″. Two type OSM215 50-Ω input and output connectors can be mounted opposite to the component side of the board, if required. Motorola RF Device Data, 1991, p. 7-338.

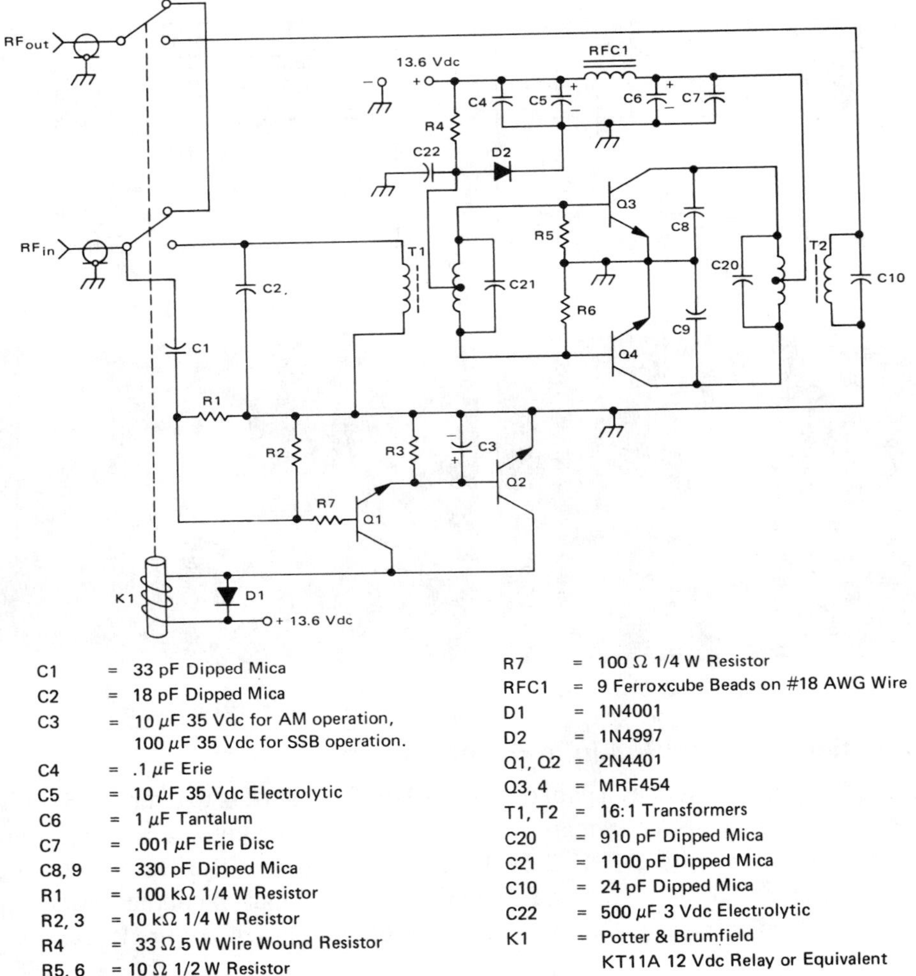

140-W (PEP) 2- to 30-MHz amateur-radio linear amplifier

Fig. 2-60 Both the parts and kits for this amplifier are available from: Communications Concepts, 121 Brown St., Dayton, Ohio 45402, (513) 220-9677. Use of this amplifier is illegal for the class-D citizen band. Motorola RF Device Data, 1991, p. 7-344.

(a)

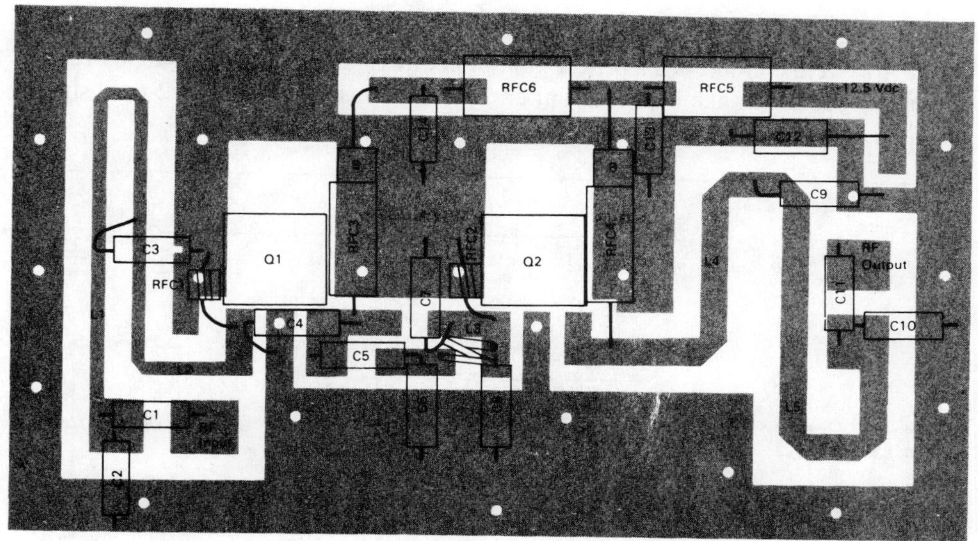

(b)

C1 — 200 pF
C2 — 33 pF
C3 — 47 pF
C4 — 18 pF
C5, C8 — 43 pF
C6 — 12 pF
C7, C9 — 50 pF
C10 — 22 pF
C11 — 100 pF
C12 — 1.0 μF Tantalum
C13, C14 — 0.05 μF Erie Redcap
L1-L5 — Printed Inductor
L3 — 1.25" #18 AWG, 1-1/2 Turns, 9/64 ID
Q1 — MRF260
Q2 — MRF262
RFC1, RFC2 — 2 Turns #26 Enameled on Ferrite Bead Ferroxcube 56-590-65/3B
RFC3 — 10 μH Molded Choke
RFC4 — 0.15 μH Molded Choke
RFC5, RFC6 — VK200-4B
B — Bead, Ferroxcube 56-590-65/3B

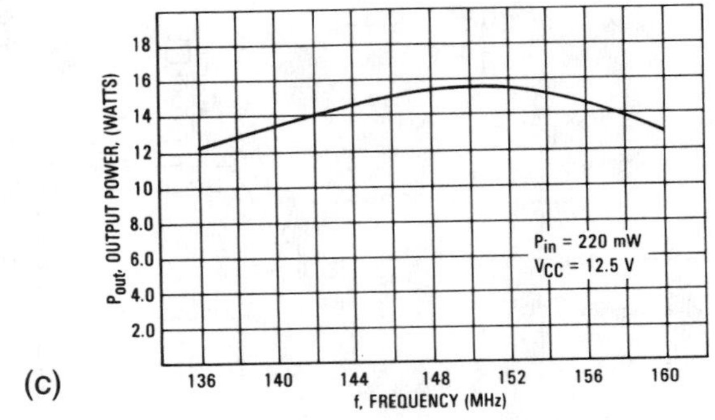

(c)

Low-cost VHF broadband amplifier (136 to 160 MHz)

Fig. 2-61 This circuit has a power gain of about 19 dB over the frequency range. Figure 2-61B shows the board layout and components list, and Fig. 2-61C shows the power output versus frequency. Motorola RF Device Data, 1991, p. 7-362, 7-364.

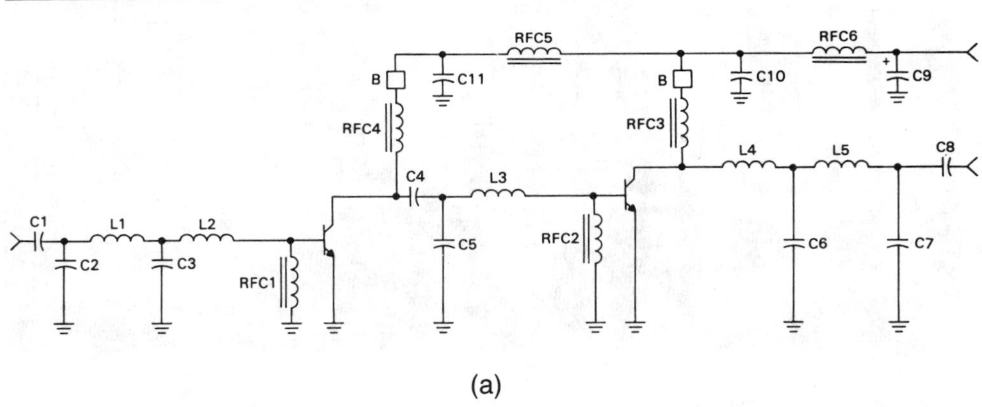

(a)

Continued

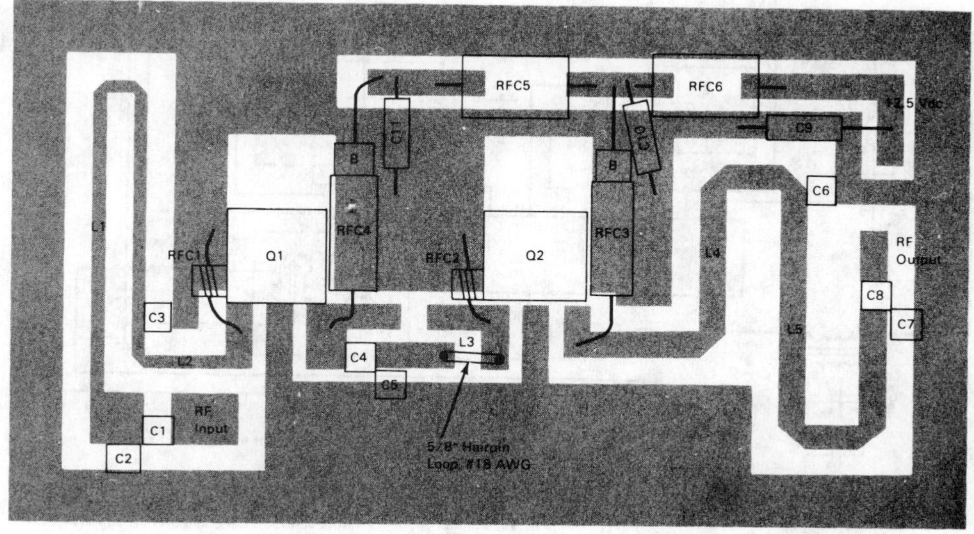

(b)

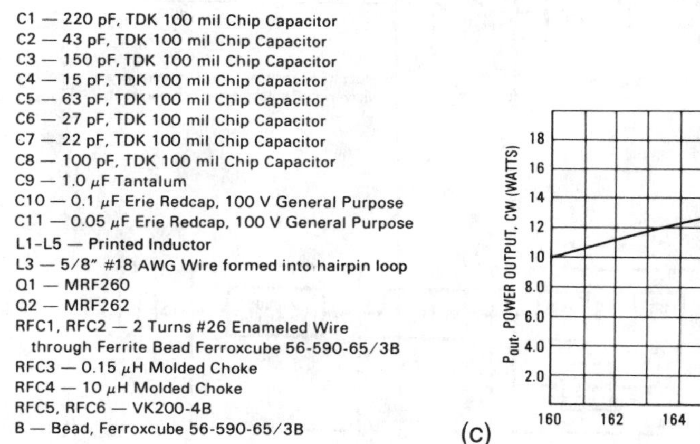
C1 — 220 pF, TDK 100 mil Chip Capacitor
C2 — 43 pF, TDK 100 mil Chip Capacitor
C3 — 150 pF, TDK 100 mil Chip Capacitor
C4 — 15 pF, TDK 100 mil Chip Capacitor
C5 — 63 pF, TDK 100 mil Chip Capacitor
C6 — 27 pF, TDK 100 mil Chip Capacitor
C7 — 22 pF, TDK 100 mil Chip Capacitor
C8 — 100 pF, TDK 100 mil Chip Capacitor
C9 — 1.0 μF Tantalum
C10 — 0.1 μF Erie Redcap, 100 V General Purpose
C11 — 0.05 μF Erie Redcap, 100 V General Purpose
L1-L5 — Printed Inductor
L3 — 5/8" #18 AWG Wire formed into hairpin loop
Q1 — MRF260
Q2 — MRF262
RFC1, RFC2 — 2 Turns #26 Enameled Wire
through Ferrite Bead Ferroxcube 56-590-65/3B
RFC3 — 0.15 μH Molded Choke
RFC4 — 10 μH Molded Choke
RFC5, RFC6 — VK200-4B
B — Bead, Ferroxcube 56-590-65/3B

(c)

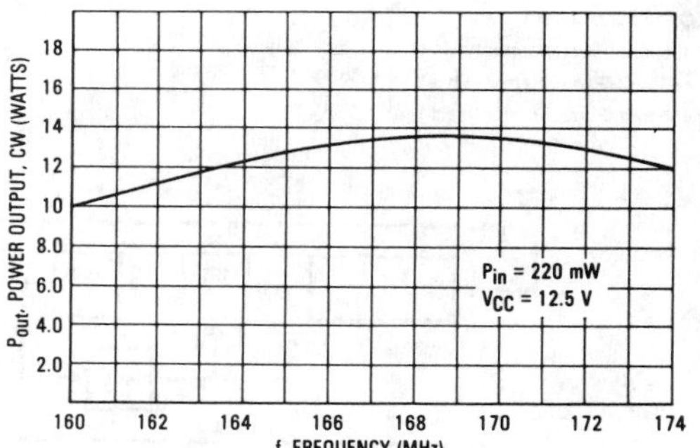

Low-cost VHF broadband amplifier (160 to 174 MHz)

Fig. 2-62 This circuit has a power gain of about 18 dB over the frequency range. Figure 2-62B shows the board layout and components list, and Fig. 2-62C shows the power output versus frequency. Motorola RF Device Data, 1991, p. 7-363, 7-364.

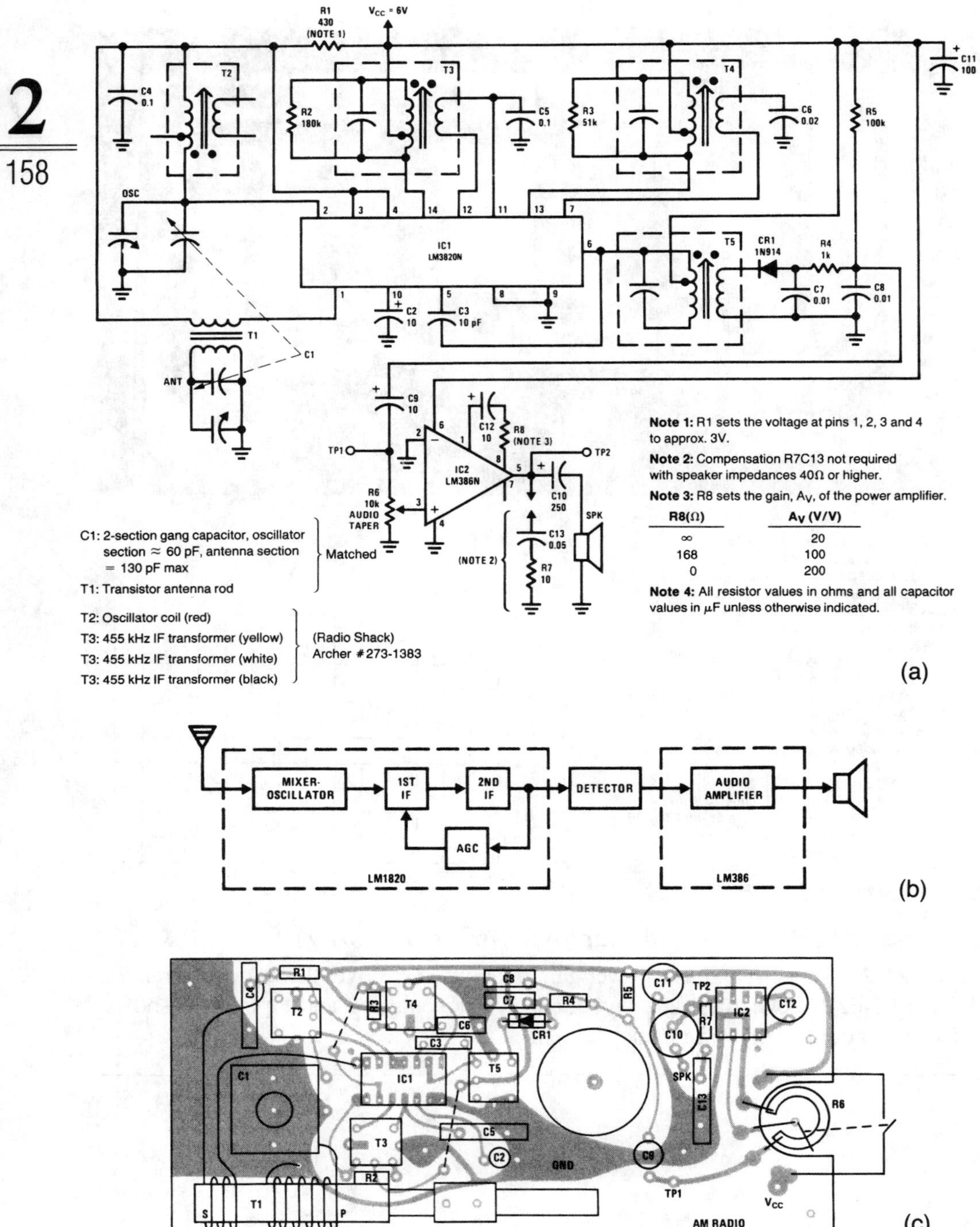
R1
430
(NOTE 1)
V_{CC} = 6V
C11
100
T2
T3
T4
C4
0.1
R2
180k
C5
0.1
R3
51k
C6
0.02
R5
100k
OSC
IC1
LM3820N
T5
CR1
1N914
R4
1k
C7
0.01
C8
0.01
C2
10
C3
10 pF
T1
C1
ANT
C9
10
C12
10
R8
(NOTE 3)
TP1
TP2
IC2
LM386N
C10
250
SPK
R6
10k
AUDIO
TAPER
C13
0.05
(NOTE 2)
R7
10
Note 1: R1 sets the voltage at pins 1, 2, 3 and 4 to approx. 3V.
Note 2: Compensation R7C13 not required with speaker impedances 40Ω or higher.
Note 3: R8 sets the gain, A_V, of the power amplifier.
R8(Ω) | A_V (V/V)
∞ | 20
168 | 100
0 | 200
Note 4: All resistor values in ohms and all capacitor values in μF unless otherwise indicated.
C1: 2-section gang capacitor, oscillator section ≈ 60 pF, antenna section = 130 pF max
T1: Transistor antenna rod
Matched
T2: Oscillator coil (red)
T3: 455 kHz IF transformer (yellow)
T3: 455 kHz IF transformer (white)
T3: 455 kHz IF transformer (black)
(Radio Shack)
Archer #273-1383
(a)
MIXER-OSCILLATOR
1ST IF
2ND IF
AGC
LM1820
DETECTOR
AUDIO AMPLIFIER
LM386
(b)
R1
C4
T2
R3
T4
C8
C7
R4
R5
C11
TP2
IC2
C12
C6
CR1
R7
C10
C3
C1
IC1
T5
SPK
C13
R6
T3
C5
C2
GND
C9
R2
T1
S
P
TP1
V_{CC}
AM RADIO
(c)

Low-cost two-chip AM radio system

Fig. 2-63 This AM radio requires only two ICs and few discrete components. A block diagram of the radio is shown in Fig. 2-63B. The typical PC-board layout is shown in Fig. 2-63C. The mixer-oscillator, two IF stages, and AGC section are contained within the LM3820. The power output of 0.25-W into an 8-Ω speaker is obtained in the LM386. National Semiconductor, Linear Applications Handbook, 1991, p. 1233, 1235.

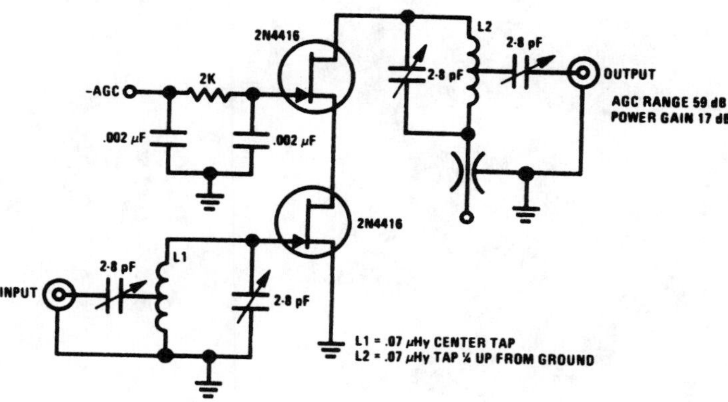

200-MHz cascode amplifier

Fig. 2-64 This JFET cascode circuit features low crossmodulation, large-signal handling ability, no neutralization, and an AGC that is controlled by biasing the upper cascode JFET. The only special requirement of this circuit is that I_{DSS} of the upper JFET must be greater than that of the lower JFET. National Semiconductor, Linear Applications Handbook, 1991, p. 115.

3

Video circuits

This chapter is devoted to video circuits of all types. Both the test and troubleshooting procedures described in chapters 1 and 2 can be applied to the circuits in this chapter. Also, because some of the circuits in this chapter involve the use of op amps, it is recommended that you also read chapter 10 before attempting test or troubleshooting of any video circuits that are described here. It is also recommended that you read *Lenk's Video Handbook*, McGraw-Hill, 1991, for a thorough discussion of video-circuit tests and troubleshooting techniques.

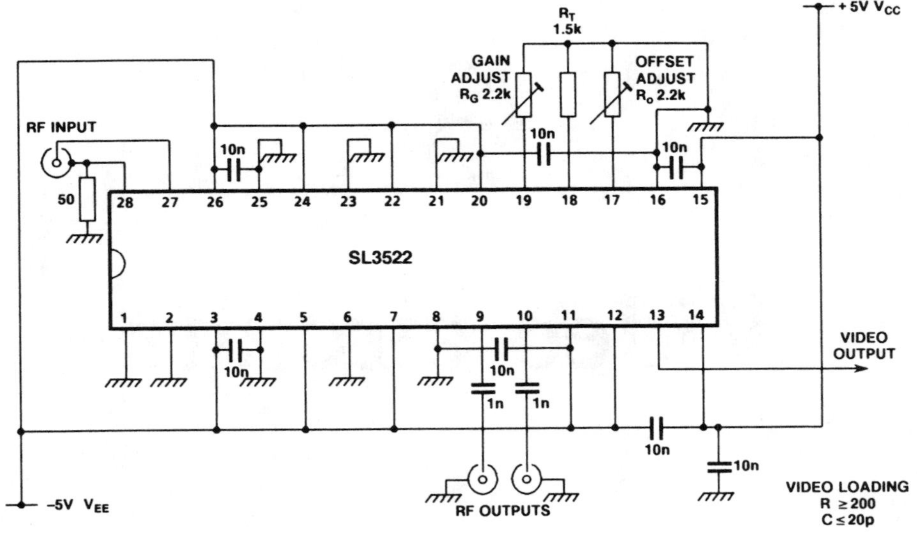

Logarithmic/limiting amplifier

Fig. 3-1 The SL3522 shown in this figure is a 7-stage successive-detection log amp for use in the 100- to 600-MHz range. With the values shown, the frequency range is 500 MHz, with a 40-MHz video bandwidth, and a 450-MHz balanced RF bandwidth. Both video gain and offset are adjustable. Gain has an effect on offset, but not vice versa. Typical dynamic range is 75 dB. GEC Plessey Semiconductors, Professional Products 1991, p. 1-5.

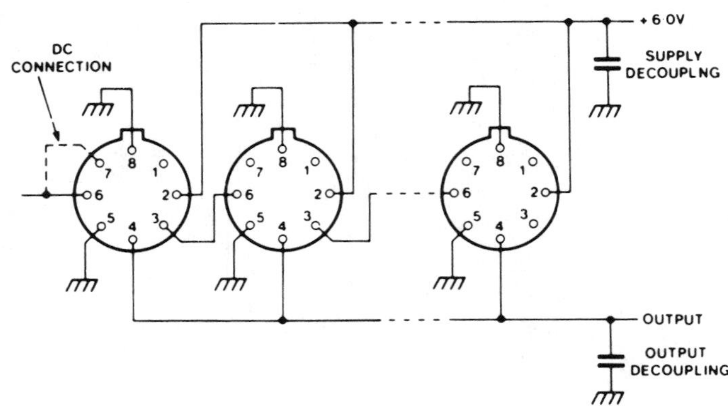

Direct-coupled wideband log amplifier

Fig. 3-2 The SL521 shown in this figure is intended primarily for use in successive-detection log IF strips, operating at center frequencies between 10 and

100 MHz. The SL521 is suitable for direct-coupling, and includes a built-in 500-pF supply-decoupling capacitor. Typical stage voltage-gain is 10 to 12 dB. The RF output is taken from pin 4, and the detected output is taken from pin 3. When more than two untuned stages are used, additional decoupling might be required. The values for decoupling capacitors with untuned cascades are: 3 stages 1 nF, 4 stages 3 nF, 5 stages 10 nF, 6 or more stages 30 nF. GEC Plessey Semiconductors, Professional Products, 1991, P. 1-8.

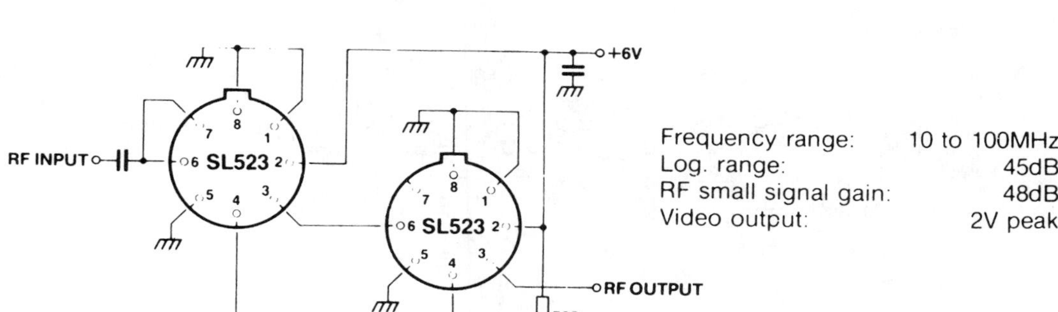

Simple log IF strip

Fig. 3-3 The SL523 shown in this figure is similar to the SL521 shown in Fig. 3-2, except that the SL523 is a dual amplifier and provides an approximate 24-dB gain. Typical value for the supply decoupling capacitor is 3 nF. GEC Plessey Semiconductors, Professional Products, 1991, p. 1-12.

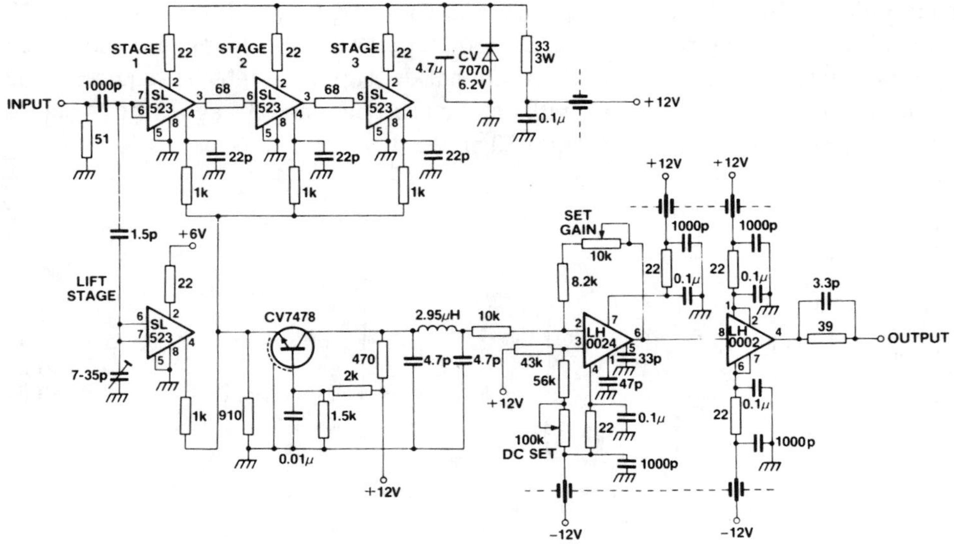

Wideband log amplifier

Fig. 3-4 This circuit consists of six log stages (two stages in each dual SL523) and two "lift" stages, which gives an overall dynamic range of greater than 80 dB. Both the dc level and gain of the video output are adjustable. GEC Plessey Semiconductors, Professional Products, 1991, p. 1-12.

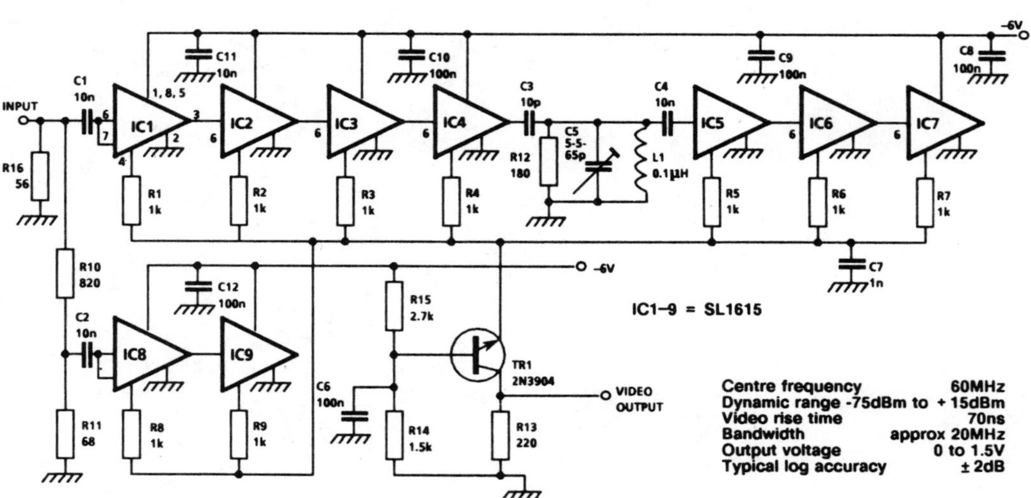

Wideband log IF strip

Fig. 3-5 The SL1615 shown in this figure is a bipolar wideband amplifier for use in successive-detection log IF strips. GEC Plessey Semiconductors, Professional Products, 1991, p. 1-31.

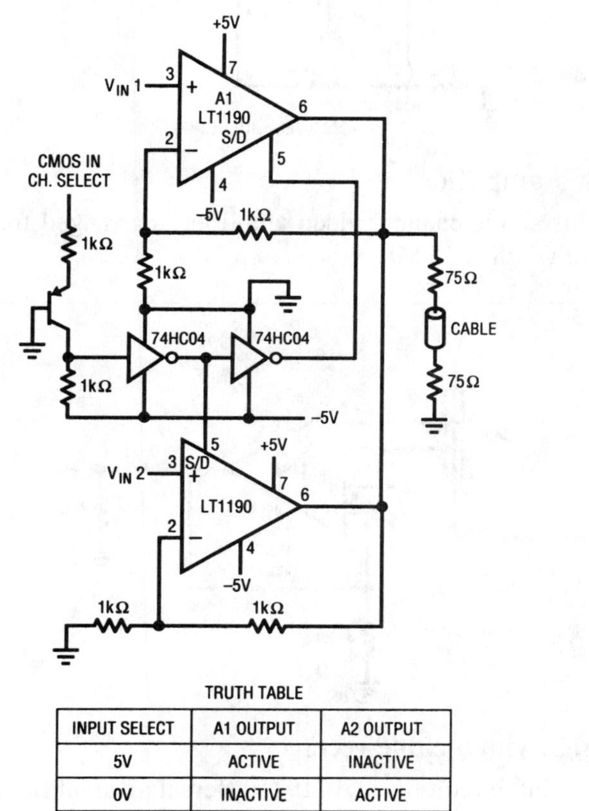

INPUT SELECT	A1 OUTPUT	A2 OUTPUT
5V	ACTIVE	INACTIVE
0V	INACTIVE	ACTIVE

Multiplexed video amplifier

Fig. 3-6 This circuit is a simple way to multiplex two video amplifiers onto a single 75-Ω cable. The appropriate amplifier is activated in accordance with the truth table. Amplifier performance includes 0.02% differential gain error and 0.1° differential phase error. The 75-Ω back termination (looking into the cable) means that the amplifier must swing 2 Vpp to produce 1 Vpp at the cable output. Linear Technology Corporation, 1991, AN47-33.

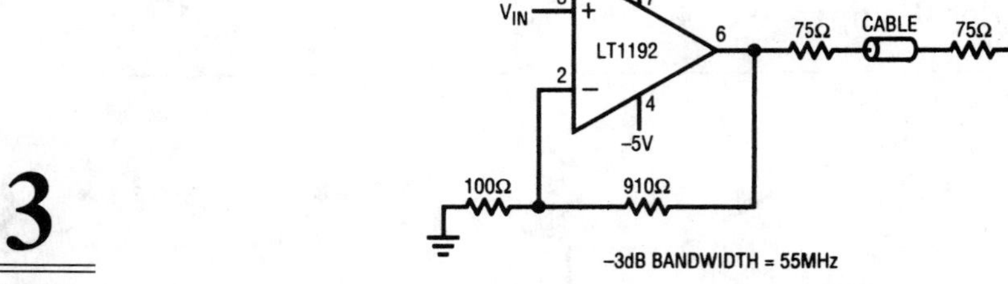

Simple video amplifier

Fig. 3-7 This single-channel video amplifier is arranged for a gain of 10, and delivers a bandwidth of 55 MHz. Linear Technology Corporation, 1991, AN47-33.

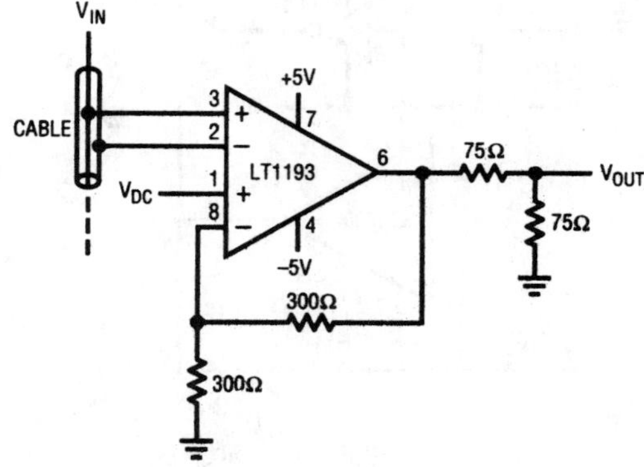

Loop-through video-cable receiver

Fig. 3-8 In this circuit, an LT1193 differential amplifier is placed across a distribution cable to extract the video signal. Common-mode signals are rejected by the LT1193 differential inputs. Differential gain and phase errors measure 0.02% and 0.1°, respectively. A separate input permits dc level adjustment. Linear Technology Corporation, 1991, AN47-33.

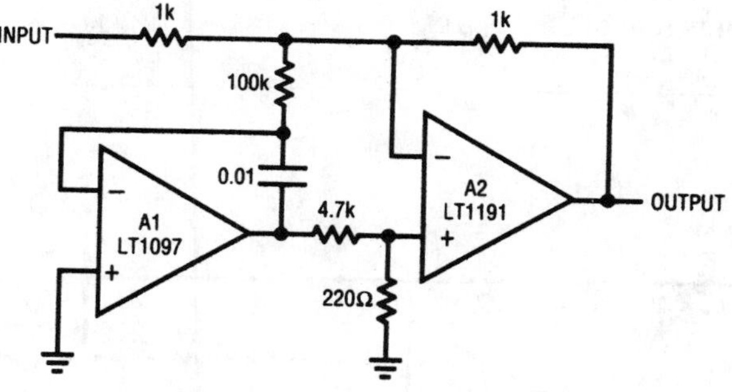

Video-amplifier dc stabilization (summing-point technique)

Fig. 3-9 In this circuit, A2 handles high-frequency inputs while A1 stabilizes the dc operating point. The 4.7 k-220-Ω divider at the A2 input prevents excessive drive during start-up. The circuit combines the A1 35-μV offset and 1.5 V/°C drift with the A2 450 V/μs slew rate and 90-MHz bandwidth. Bias current, dominated by A2, is about 500 nA. Linear Technology Corporation, 1991, AN47-33.

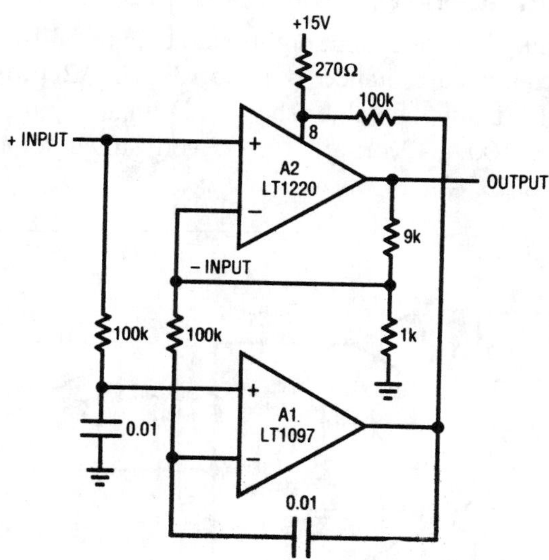

Video-amplifier dc stabilizer (differentially sensed technique)

Fig. 3-10 In this circuit, sensing is done differentially, preserving access to both fast amplifier inputs. A1 measures the dc error at the A2 input, and biases the offset pins to force offset within 50 μV. The offset-pin biasing at A2 is arranged so that A1 is always able to find the servo point. The 0.01-μF capacitors rolls off A1 at the low frequency, and A2 handles the high-frequency signals. The combined characteristics yield: gain-bandwidth of 45 MHz, offset voltage of 50 μV, offset drift of 1 μV/°C, and a slew rate of 250 V/μs. Linear Technology Corporation, 1991, AN47-34.

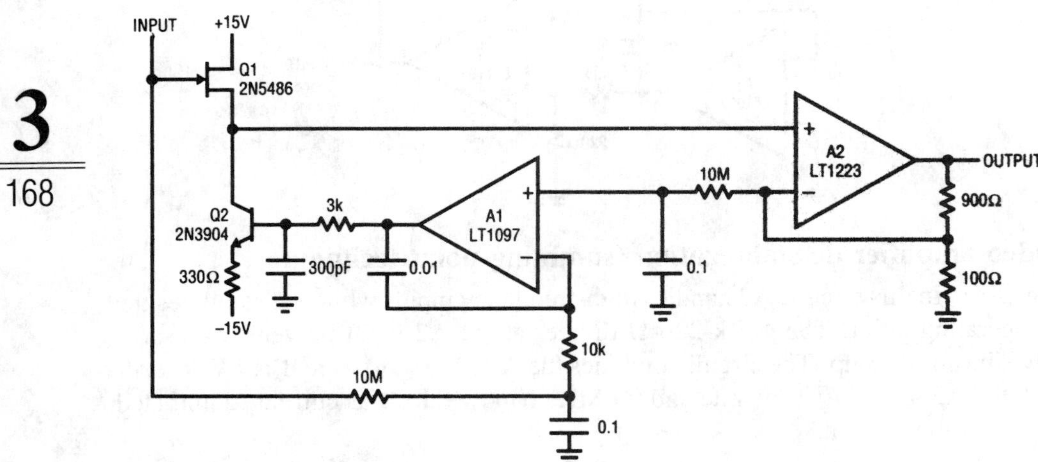

Video-amplifier dc stabilizer (servo-controlled FET)

Fig. 3-11 This circuit is a wideband, highly stable, gain-of-10 amplifier with high input impedance. Input capacitance is about 3 pF. A2 provides a 100-MHz bandwidth gain of 10 (using the values shown). With an input capacitance of 3 pF and bias current of 100 pA, the circuit is well suited to a variety of video applications. Linear Technology Corporation, 1991, AN47-34.

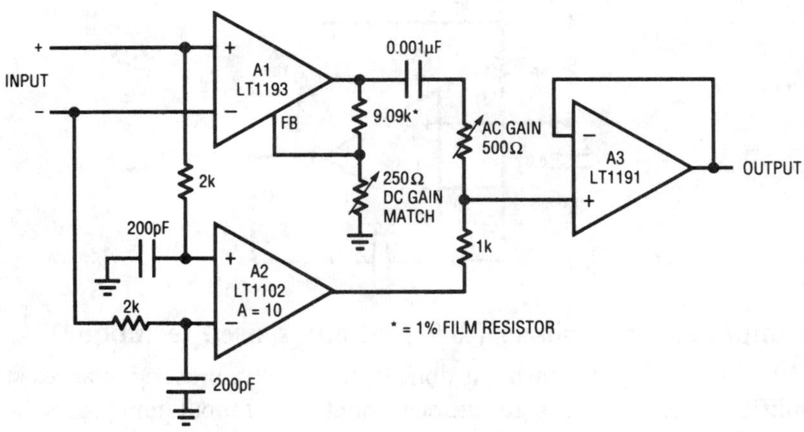

Video-amplifier dc stabilizer (differential inputs)

Fig. 3-12 This circuit shows a way to get full differential inputs with dc-stabilized operation. A1 and A2 both (differentially) sense the input at gains of 10. The A3 output is an undistorted, amplified version (in this case, time 10) of the input. The circuit is adjusted by applying a square wave and adjusting the ac gain for the squarest corners, and the dc gain-match for a flat top. Circuit bandwidth exceeds 35 MHz, slew rate is 450 V/μs, and dc offset is about 200 μV. Linear Technology Corporation, 1991, AN47-35.

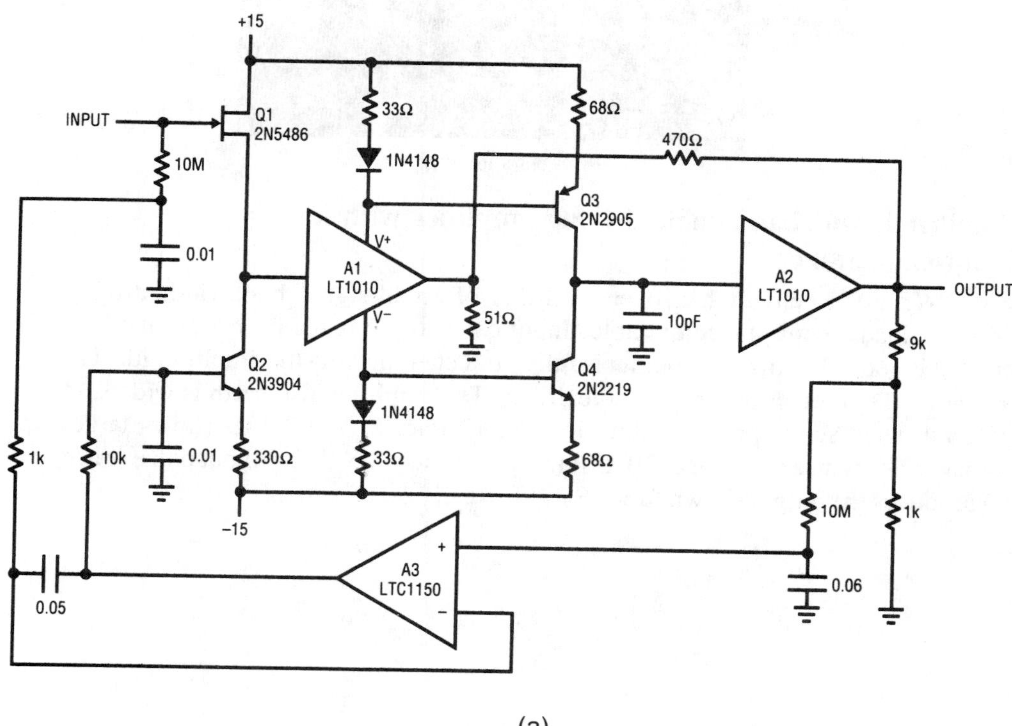

(a)

Continued

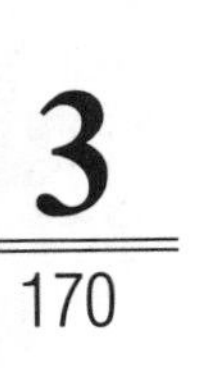

(b)

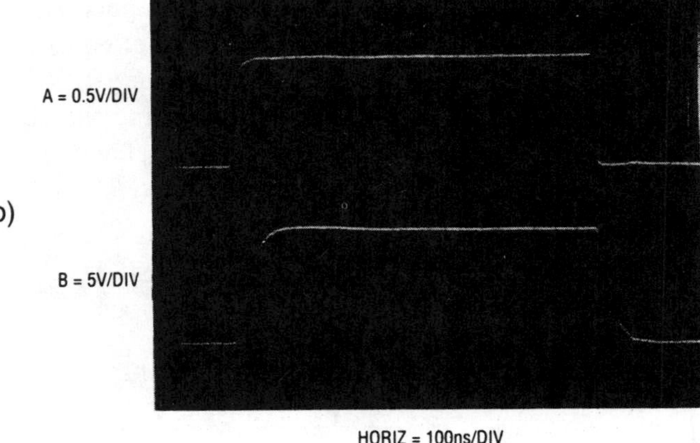

Stabilized, wideband cable-driving amplifier with low input capacitance

Fig. 3-13 This circuit has over 20 MHz of small-signal bandwidth driving 100-mA loads, capacitance or cable. Input capacitance is below 1.5 pF and bias current is about 100 pA. The output is fully protected, making the amplifier ideal as a video A/D input buffer or as a cable driver. The amplifier also permits wideband probing when scope probe loading is not tolerable. Figure 3-13B shows large-signal performance at a gain of 10, driving 10 feet of cable. A fast input pulse (trace A) produces the output shown (trace B). Linear Technology Corporation, 1991, AN45-4.

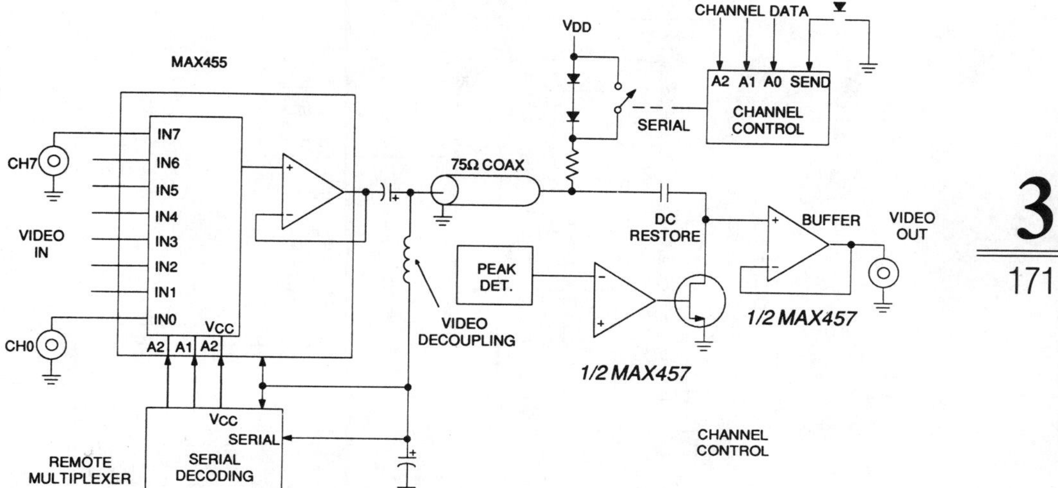

Video, power, and channel-select on a single coax

Fig. 3-14 In this video system, a single coaxial cable carries power to the remote location, selects one of eight video channels, and returns the selected signal. The system can choose one of several remote surveillance-camera signals, for example, and display the picture on a monitor near the channel-select box. Circuit details are shown in Figs. 3-15 and 3-16. Maxim, 1992, Applications and Product Highlights, p. 6-4.

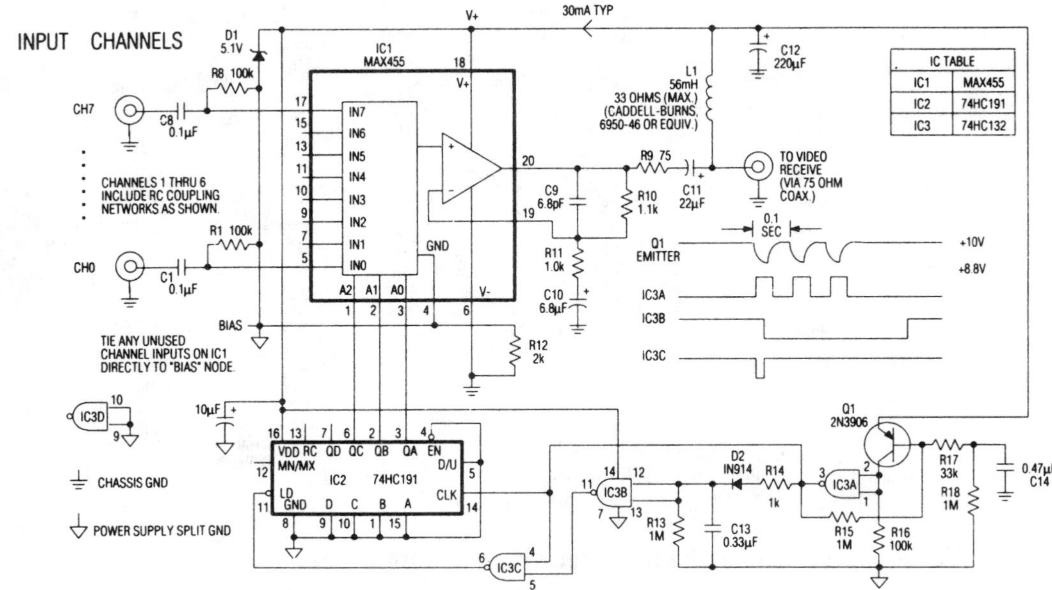

Video-system remote multiplexer box

Fig. 3-15 The heart of the remote multiplexer box in the single-coax video system (Fig. 3-14) is a combination 8-channel multiplexer and amplifier (IC1). C11 couples the MUX baseband video output to the coax, and L1 decouples the video from dc power arriving on the same line. This power (about 30 mA at 10 V) supplies all circuitry in the multiplexer box. Channel-select signals generated at the interface box (1 pulse for channel 0, 8 pulses for channel 7) pulse the 10-V supply to 8.8 V and back at a 10-Hz rate. Q1 and associated components in the remote multiplexer box convert these pulses to 5-V logic levels, which block the 4-bit counter IC2. In turn, IC2 selects the desired multiplexer channel. The first pulse of a burst selects channel 0. Subsequent pulses, arriving before the discharge of timeout network R13/C13 advance IC2 by one count each. Thus, channel 0 appears almost instantly, and channel 7, when selected, appears near the end of a 0.8-s burst. Maxim, 1992, Applications and Product Highlights, p. 6-5.

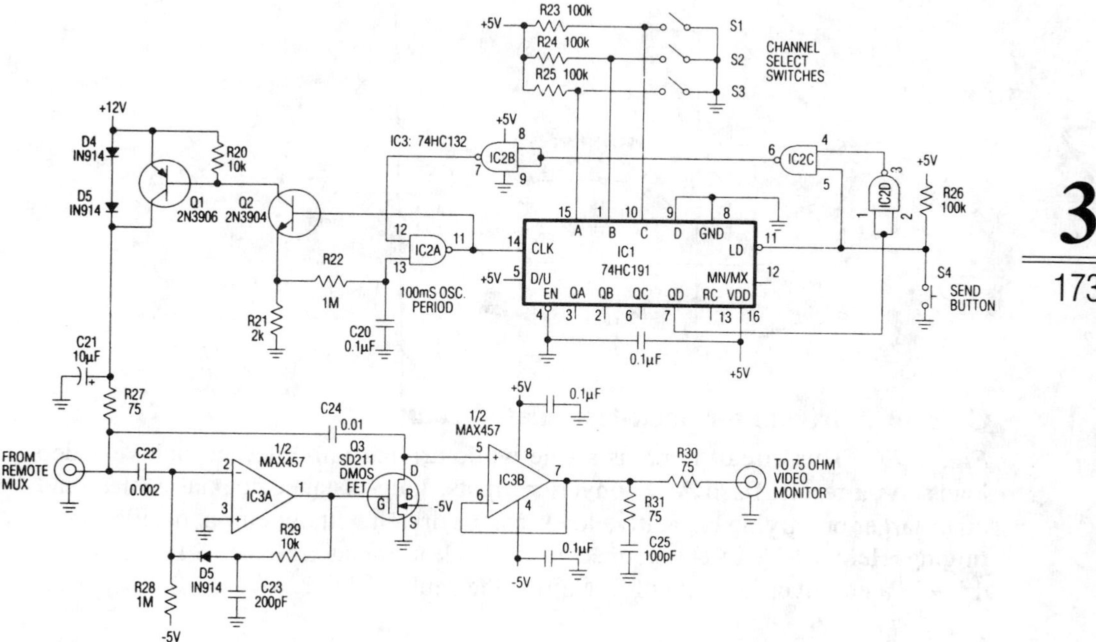

Video-system channel-control box

Fig. 3-16 In the channel control box of the single-coax video system (Fig. 3-14), a desired channel is encoded by three bits, set either by switches (as shown) or by an applied digital input. Momentary depression of the send button triggers down-converter IC1 and gated oscillator IC2A to initiate a channel-selection burst. Supply current flows to the remote multiplexer box through Q1 (normally on and saturated), R27, and the coax center conductor. R27 also terminates the coax via C21. When Q1 turns off momentarily, forward bias across D3 and D4 develops a negative 1.2-V channel-select pulse. This 1.2-V drop in supply voltage does not affect the remote multiplexer video output. Consequently, the video monitor display does not flip during channel changes, provided that the channel signals have common sync timing. The short time constant that is associated with coupling of video to the coax (C11 and R9, R27) enables selection of any channel in less than one second, but also allows the video sync-pulse baseline to shift with picture content. To prevent this shift, peak detector IC3A drives Q3, which applies dc restoration ahead of IC3B. During each negative sync pulse, Q3 turns on just long enough to reclamp the pulse tip at 0 V. Maxim, 1992, Applications and Product Highlights, p. 6-6).

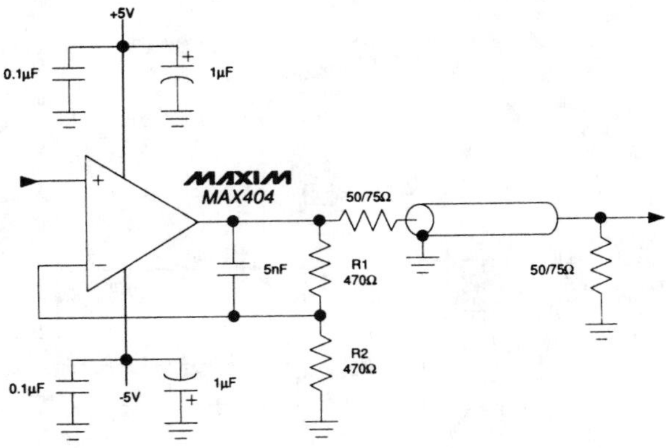

Gain-of-2 drivers for back-terminated coax

Fig. 3-17 This circuit remains stable while driving unlimited capacitive video loads. As a result, flash A/D converter inputs, long-distance coaxial cables, and other larger or varying capacitive loads can be driven without output oscillation or ringing. Here, a MAX404 is connected as a back-terminated 50- or 75-Ω coax cable drive in a noninverting gain of 2. Gain at the cable end is 1. Maxim, 1992, Applications and Product Highlights p. 6-8.

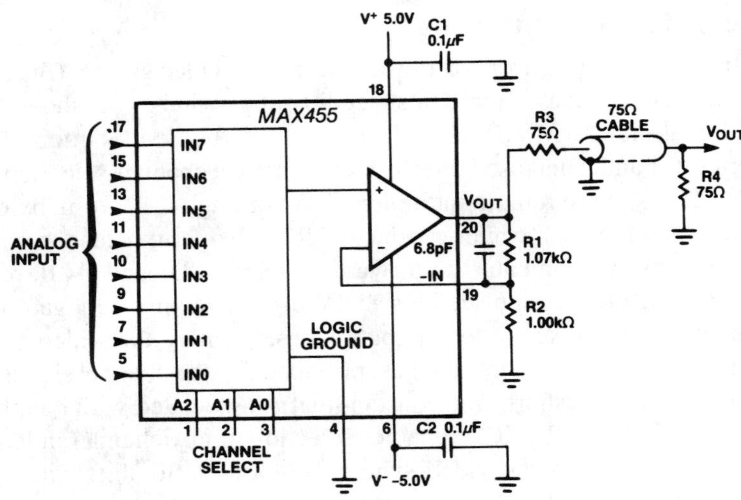

8-channel video mux/amplifier

Fig. 3-18 This circuit provides a 50-MHz gain-bandwidth (unity gain at the cable end) video MUX, which is controlled by a 3-wire channel select. The cable is back-terminated (R3) to match cables and reduce reflections. Maxim, Seminar Applications Book, p. 60.

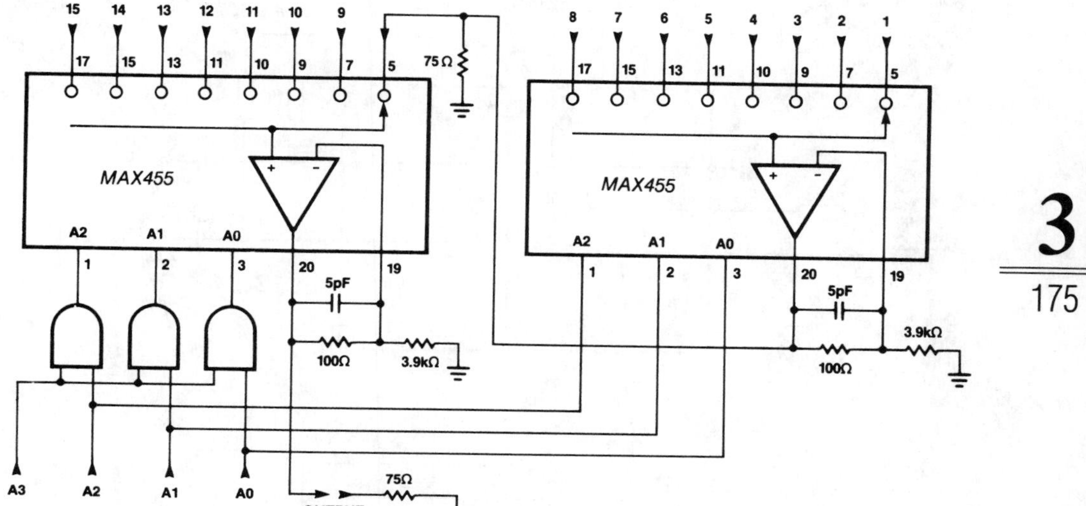

1-of-15 cascaded video mux

Fig. 3-19 In this circuit, two MAX455s are cascaded to form a 1-of-15 video MUX by connecting the output of one MUX to one input of the input channels of a second MUX. Although the two devices are usually close to one another, the output of the first MUX should be terminated to preserve bandwidth. Maxim, Seminar Applications Book, p. 61.

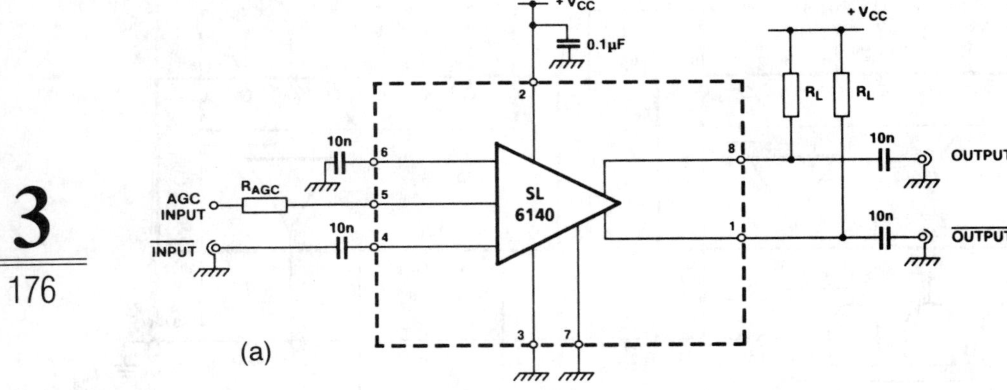
+ V_{CC}
0.1μF
+ V_{CC}
R_L
R_L
10n
10n
OUTPUT
AGC INPUT
R_{AGC}
SL 6140
10n
$\overline{INPUT}$
10n
$\overline{OUTPUT}$
2
6
5
4
8
1
3
7
(a)

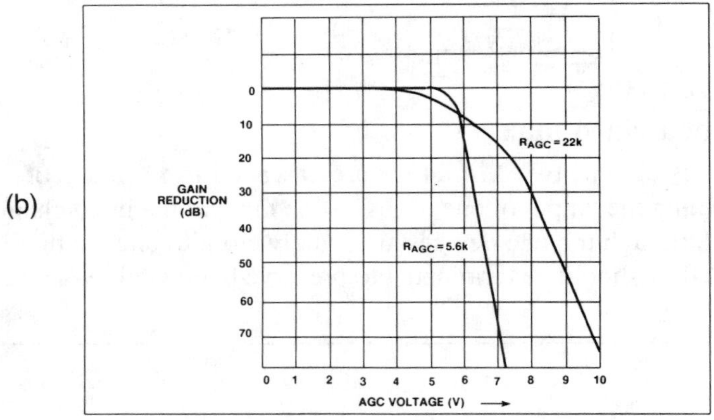
GAIN REDUCTION (dB)
R_{AGC} = 22k
R_{AGC} = 5.6k
AGC VOLTAGE (V)
(b)

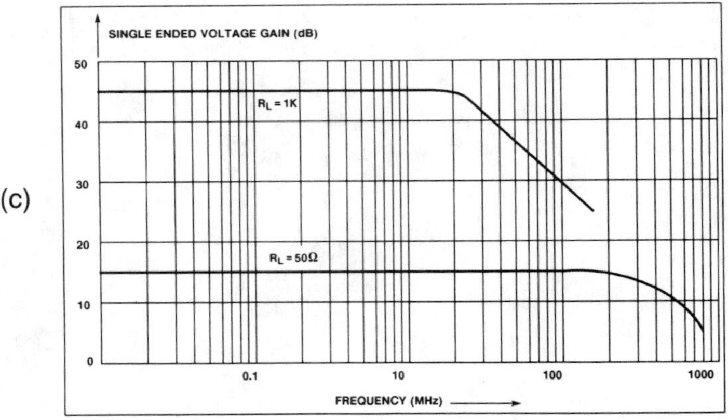
SINGLE ENDED VOLTAGE GAIN (dB)
R_L = 1K
R_L = 50Ω
FREQUENCY (MHz)
(c)

Wideband AGC amplifier

Fig. 3-20 This circuit shows an SL6140 as a high-gain video amplifier, with AGC capable of reducing gain by over 70 dB (Fig. 3-20B). Single-ended voltage gain and bandwidth are set by values of R_L (3-20C). GEC Plessey Semiconductors, Personal Communications, 1990, p. 32/35.

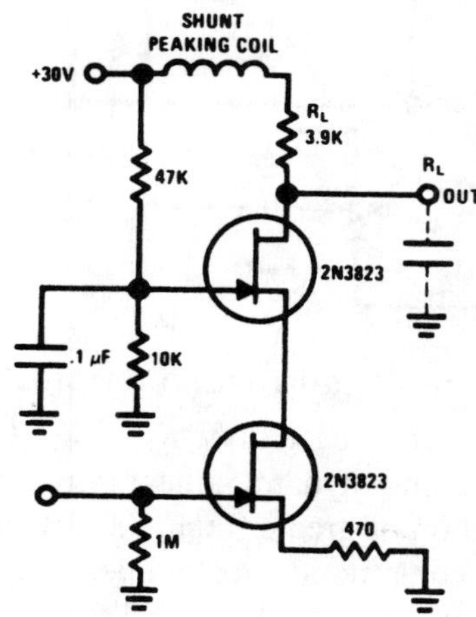

FET cascode video amplifier

Fig. 3-21 This circuit features very low input loading, and reduction of feedback to almost zero. The 2N3823 is used because of the low capacitance and high Y_{fs}. Amplifier bandwidth is limited by R_L and by the load capacitance. National Semiconductor, Linear Applications Handbook, 1991, p. 106.

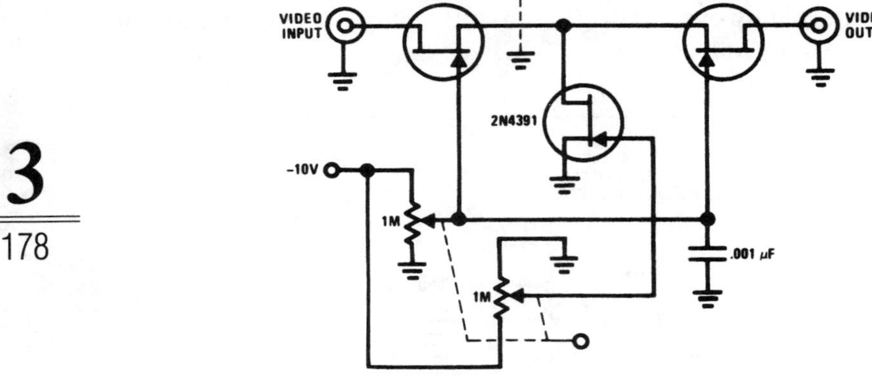

Voltage-controlled, variable-gain video amplifier

Fig. 3-22 The 2N4391 provides a low $R_{DS(ON)}$ (less than 30 Ω). The tee attenuator provides for optimum dynamic linear range for attenuation. If complete turnoff is desired, attenuation of greater than 100 dB can be obtained at 10 MHz, provided that proper RF-construction techniques are used (proper shielding to prevent input passing to output around the FETs). National Semiconductor, Linear Applications Handbook, 1991, p. 110.

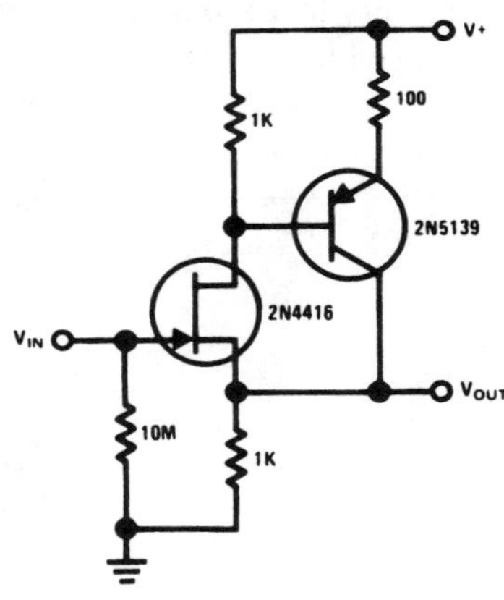

High-impedance low-capacitance video buffer

Fig. 3-23 This compound series-feedback circuit provides high input impedance and stable, wideband unity gain for general-purpose video-buffer applications. National Semiconductor, Linear Applications Handbook, 1991, p. 113.

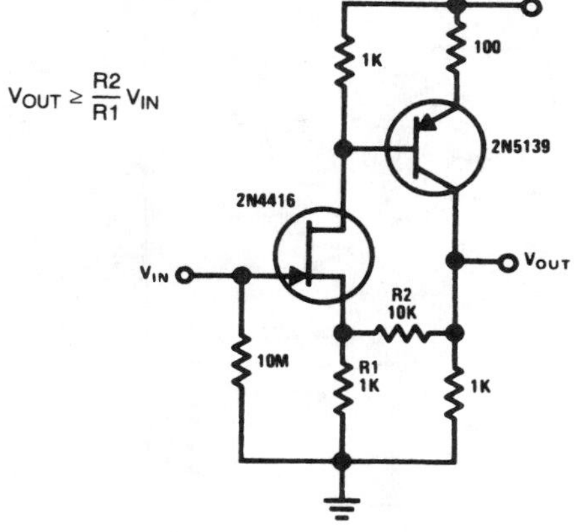

High-impedance low-capacitance video amplifier

Fig. 3-24 This compound series-feedback circuit provides high input impedance and stable, wideband gain for general-purpose video-amplifier applications. National Semiconductor, Linear Applications Handbook, 1991, p. 113.

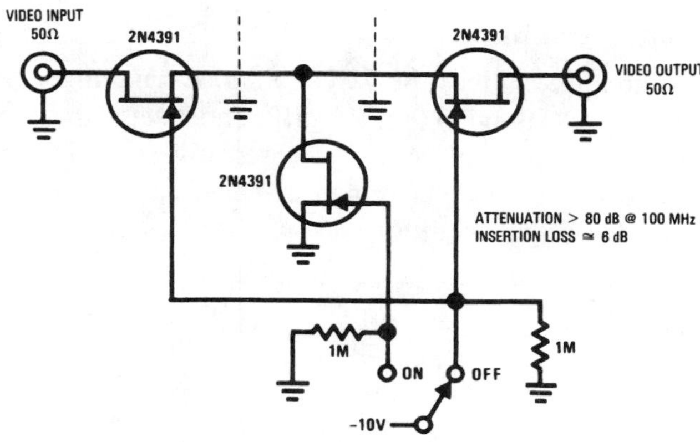

High-frequency video switch

Fig. 3-25 This circuit is similar to that of Fig. 3-22, except that this circuit is used when complete attenuation (rather than variable gain) is required in video applications. The 2N4391 provides a low on-resistance of 30 Ω and a high off-impedance. With proper layout (separating input from output by the maximum allowable distance), and an "ideal" switch, the attenuation and insertion loss shown can be achieved with no difficulty. National Semiconductor, Linear Applications Handbook, 1991, p. 118.

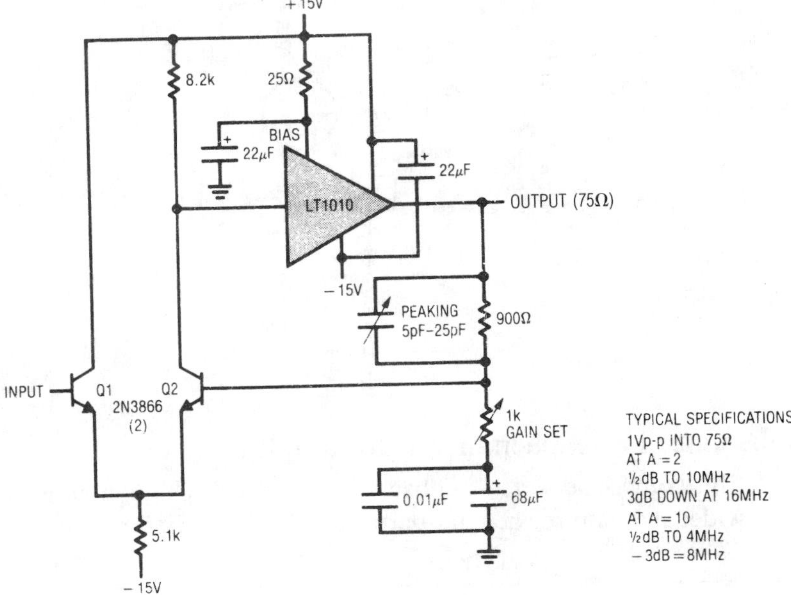

Video line-driving amplifier

Fig. 3-26 This circuit combines the LT1010's load-handling capability with a fast, discrete gain stage. Q1 and Q2 form a differential stage that single-ends into the LT1010. For video applications that are sensitive to NTSC requirements, dropping the bias resistor will aid performance. The peaking adjustment should be optimized under loaded output conditions. Linear Technology, Linear Applications Handbook, 1990, p. AN4-3.

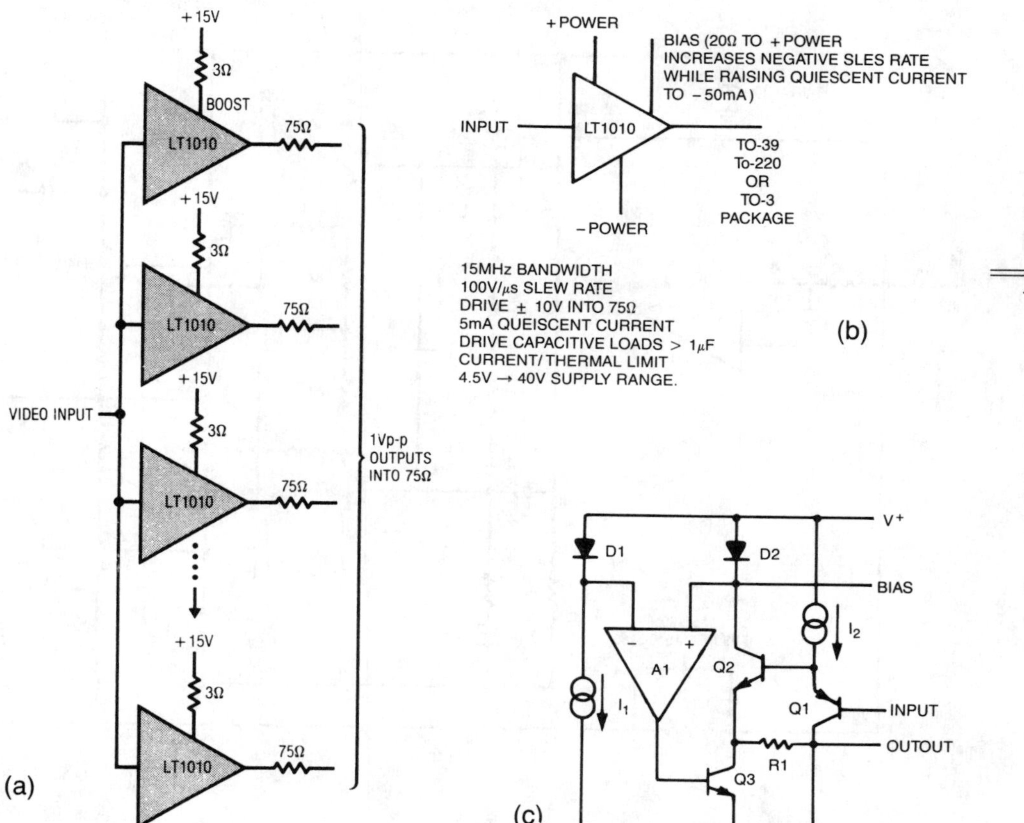

Video distribution amplifier

Fig. 3-27 The resistors in the output lines are included to isolate reflections from unterminated lines. If the line characteristics are known, the resistors can be deleted. To meet NTSC gain-phase requirements, a small-value boost resistor is used. Figures 3-27B and 3-27C show the LT1010 characteristics. Linear Technology, Linear Applications Handbook, 1990, p. AN4-4, AN4-8.

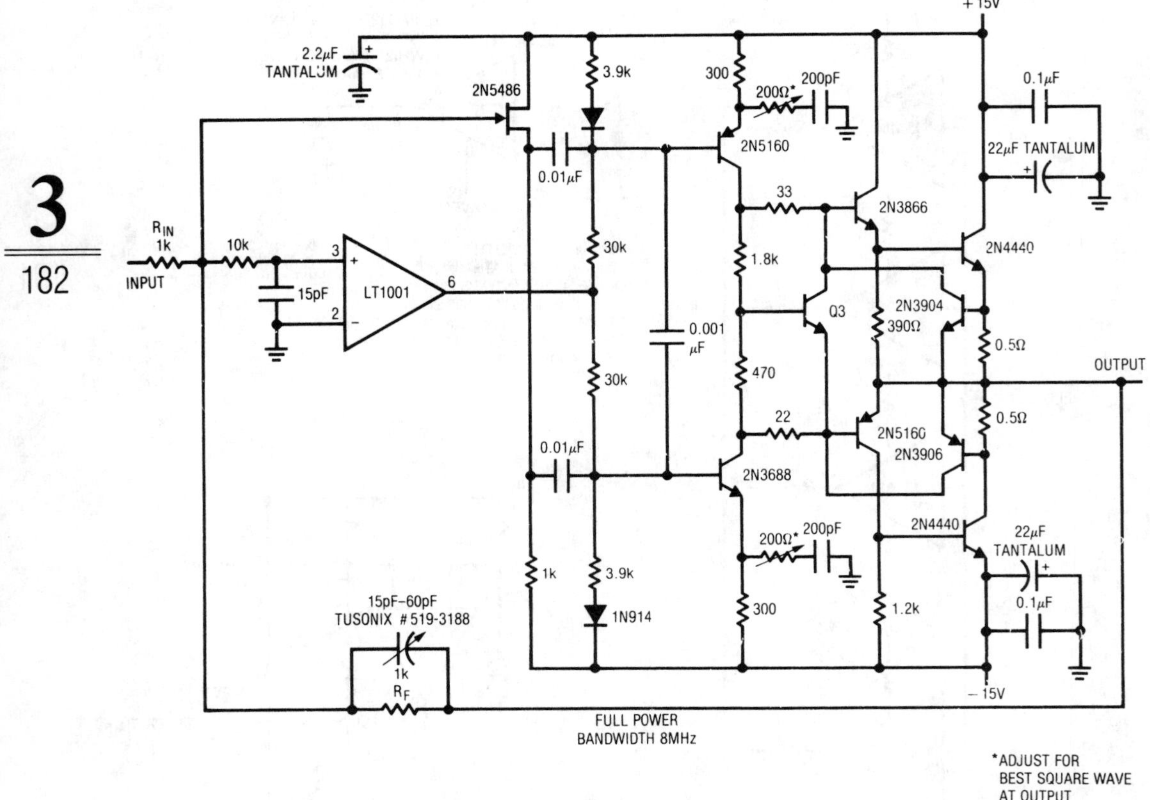

Precision high-speed op amp for video applications

Fig. 3-28 This circuit features a 1500-V/μs slew rate, full output to 8 MHz, ±10-V drive into a 10-Ω load, and is short-circuit protected at ± 1 A. RF layout techniques and a ground plane are mandatory, and the 2N4440s must have heatsinks. The 200-Ω resistors are adjusted for best square-wave output. The 15- to 60-pF peaking-capacitor adjustment should be optimized under loaded output conditions. Linear Technology, Linear Applications Handbook, 1990, p. AN6-7.

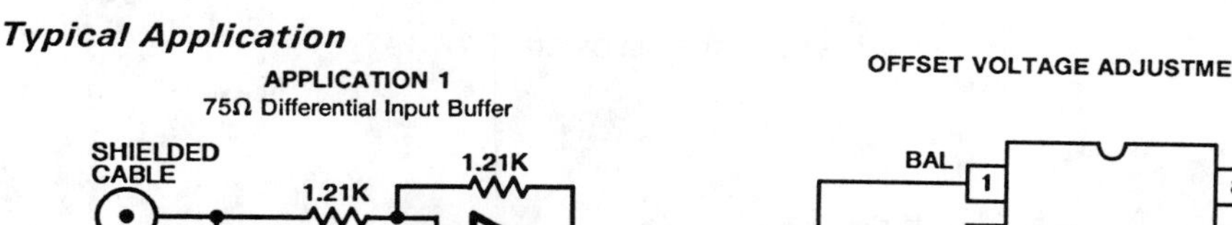
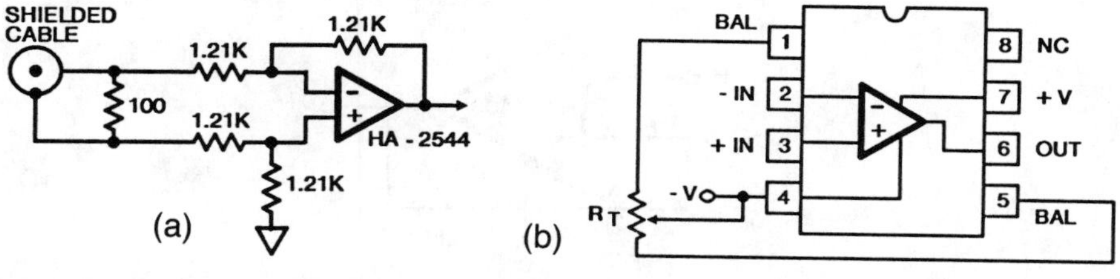

75-Ω differential-input buffer

Fig. 3-29 This circuit uses an HA-2544 video op amp to buffer between a 75-Ω line and other video circuits, with a GBW (gain-bandwidth) up to 50 MHz. Figure 3-29B shows an offset adjustment (if required). A typical range for R_T is 20 M. Harris Semiconductor, Linear & Telecom ICs, 1991, p. 3-307, 3-312.

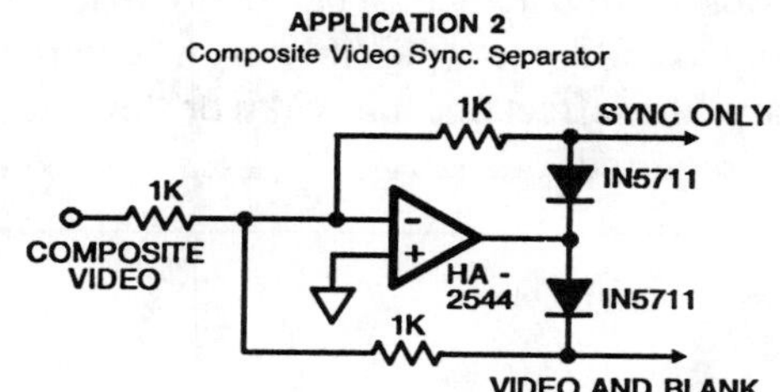

Composite video-sync separator

Fig. 3-30 This circuit uses an HA-2544 video op amp (Fig. 3-29) to separate the TV sync signal from the video and blanking signal with a minimum of external components. Harris Semiconductor, Linear & Telecom ICs, 1991, p. 3-312.

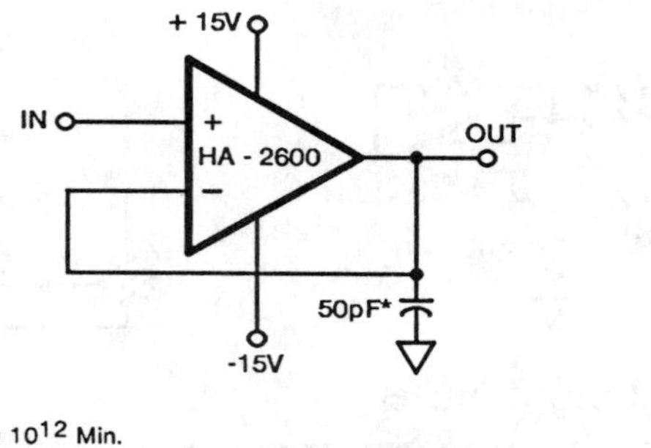

$Z_{IN} = 10^{12}$ Min.
$Z_{OUT} = 0.01$ Max.
Slew Rate = 4V/μs Min.
B.W. = 12MHz Typ.
Output Swing = ±10V Min. to 50kHz

Video voltage follower

Fig. 3-31 This circuit uses an HA-2600 wideband op amp with very high input impedance and low output impedance. The load capacitor is recommended to prevent high-frequency oscillations (possibly from external wiring). Capacitor values up to 100 pF have little effect on bandwidth or slew rate. Figure 3-29B shows an offset adjustment (if required). Harris Semiconductor, Linear & Telecom ICs, 1991, p. 3-324.

Stabilized, ultra-wideband, current-mode-feedback amplifier

Fig. 3-32 When this circuit is constructed using RF layout techniques and a ground plane, the resulting characteristics are suitable for virtually any video or wideband application. For gains from 1 to 20, full-power bandwidth is 25 MHz, with the −3-dB point beyond 110 MHz. Gain is set for 10 (with the values shown) by the 51- and 470-Ω Q3/Q4 emitter resistors. Slew rate exceeds 3000 V/μs. Damping is optimized with the 10-pF trimmer at the Q5/Q6 collectors. To use the circuit, adjust the I_Q level to 80 mA immediately after turn on. Next, set the A2 input resistor divider to a ratio appropriate to the closed-loop gain (1 and 9 kΩ, in this case). Finally, adjust the 10-pF trimmer for best results (preferably with a square-wave input). Notice that this circuit has no output protection. Linear Technology, Linear Applications Handbook, 1990, AN21-8.

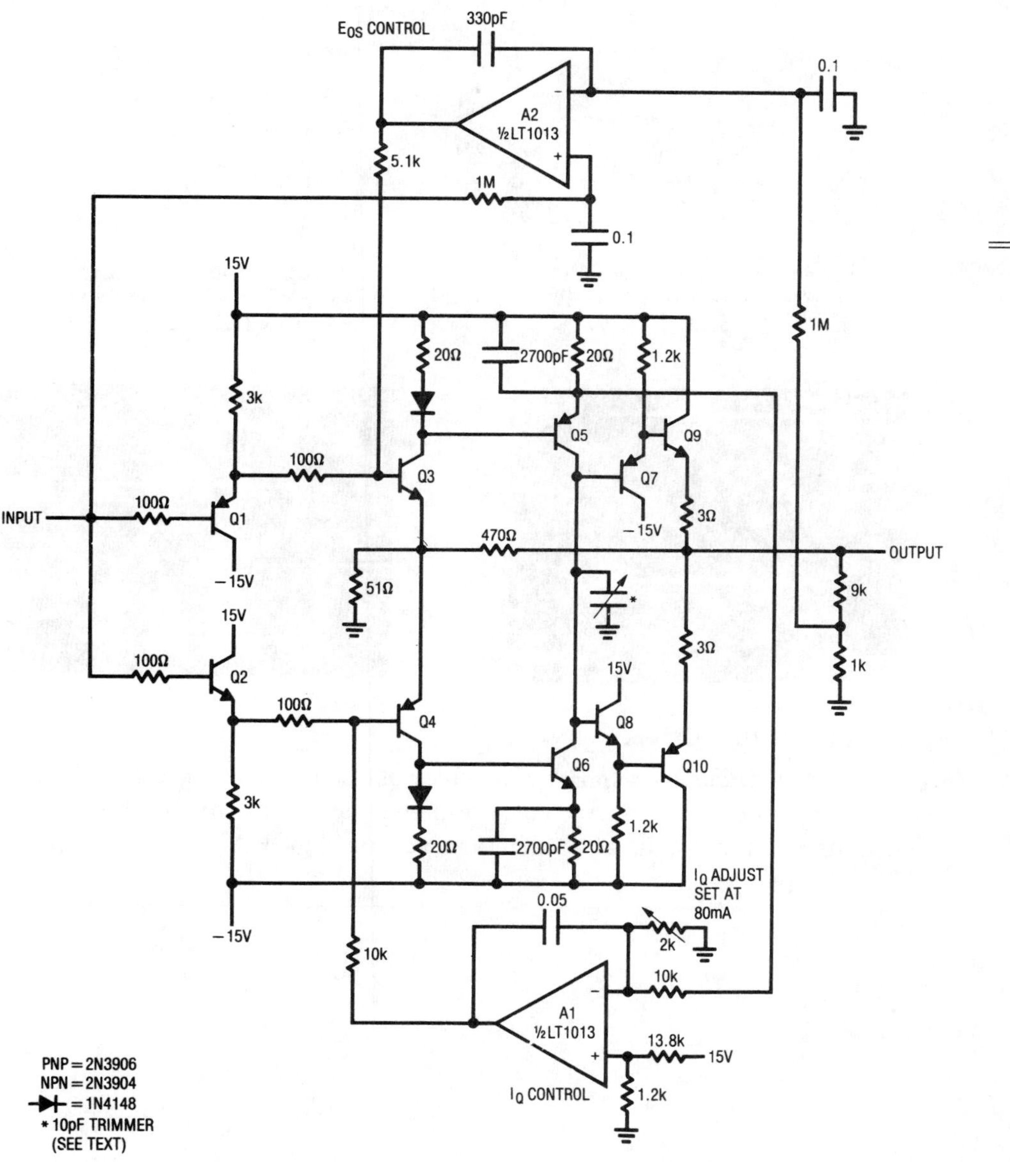

E_OS CONTROL
330pF
A2
½LT1013
5.1k
1M
0.1
0.1
1M
15V
20Ω
2700pF
20Ω
1.2k
3k
Q5
Q9
100Ω
Q3
Q7
INPUT
100Ω
Q1
−15V
3Ω
470Ω
OUTPUT
−15V
51Ω
9k
15V
3Ω
100Ω
Q2
15V
1k
100Ω
Q4
Q8
Q6
Q10
3k
1.2k
20Ω
2700pF
20Ω
I_Q ADJUST
SET AT
80mA
0.05
−15V
2k
10k
10k
A1
½LT1013
13.8k
15V
PNP = 2N3906
NPN = 2N3904
= 1N4148
* 10pF TRIMMER
(SEE TEXT)
I_Q CONTROL
1.2k

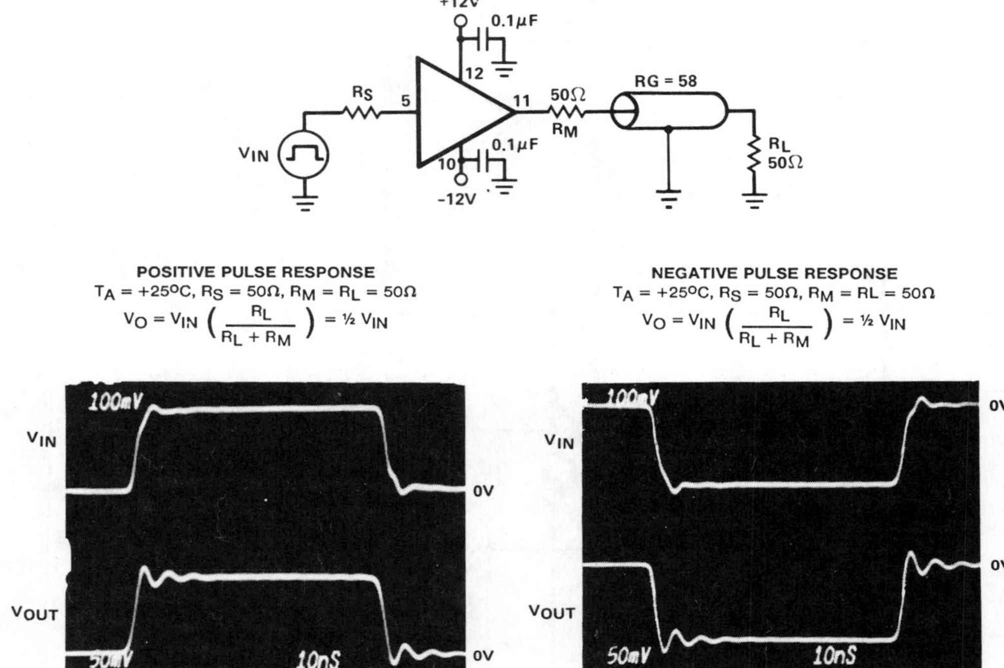

Video coaxial line driver

Fig. 3-33 This circuit uses an HA-5033 video buffer as a coax line driver. Notice that both the positive and negative pulse-response waveforms are also shown. The connections given are for the TO-8 metal-can version of the HA-5033. Harris Semiconductors, Linear & Telecom ICs, 1991, p. 3-379.

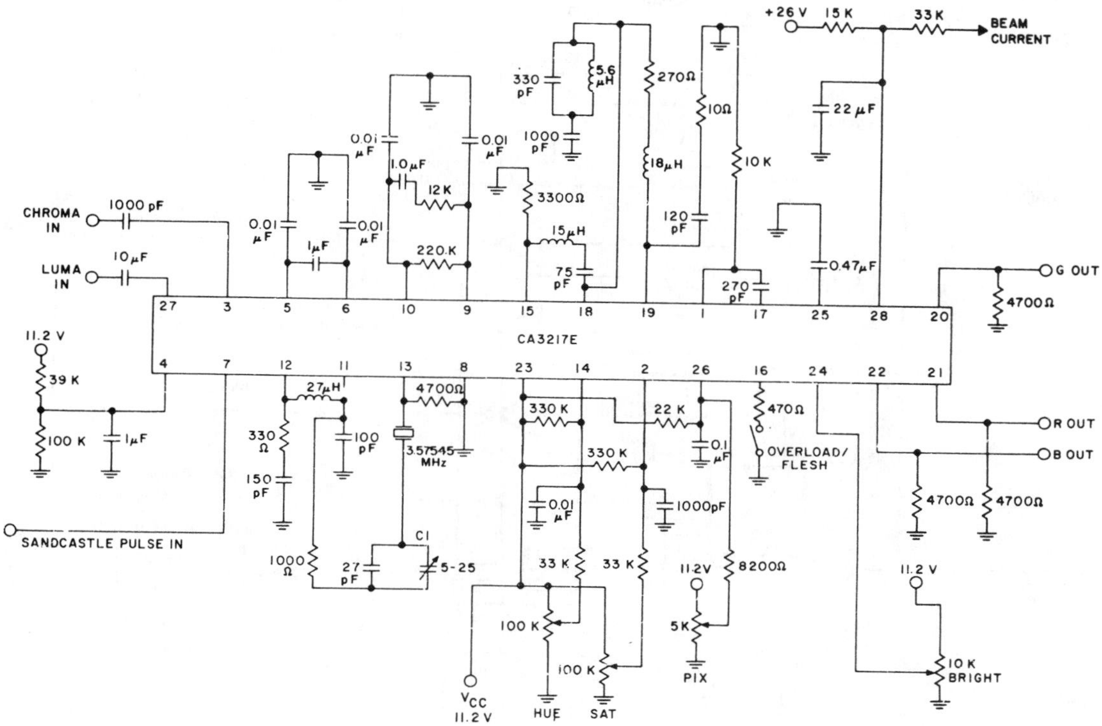

Single-chip TV chroma/luma processor

Fig. 3-34 The CA3217E contains all required circuit functions between the video detector and picture-tube RGB driver stages of a color television receiver. The CA3217E decodes the chroma signals and then produces three different color signals that are internally combined with the luma to develop the RGB signals. The picture saturation, hue, and brightness controls are externally adjustable by the viewers. The AFPC, ACC, dynamic flesh control, beam limiting, and gate black-level (brightness) controls are servo loops that are used to stabilize the RGB output and reduce frequent manual adjustment. The automatic beam-limiter circuit reduces picture contrast and brightness to prevent excessive drive output at the picture tube. Harris Semiconductors, Linear & Telecom ICs, 1991, p. 8-47.

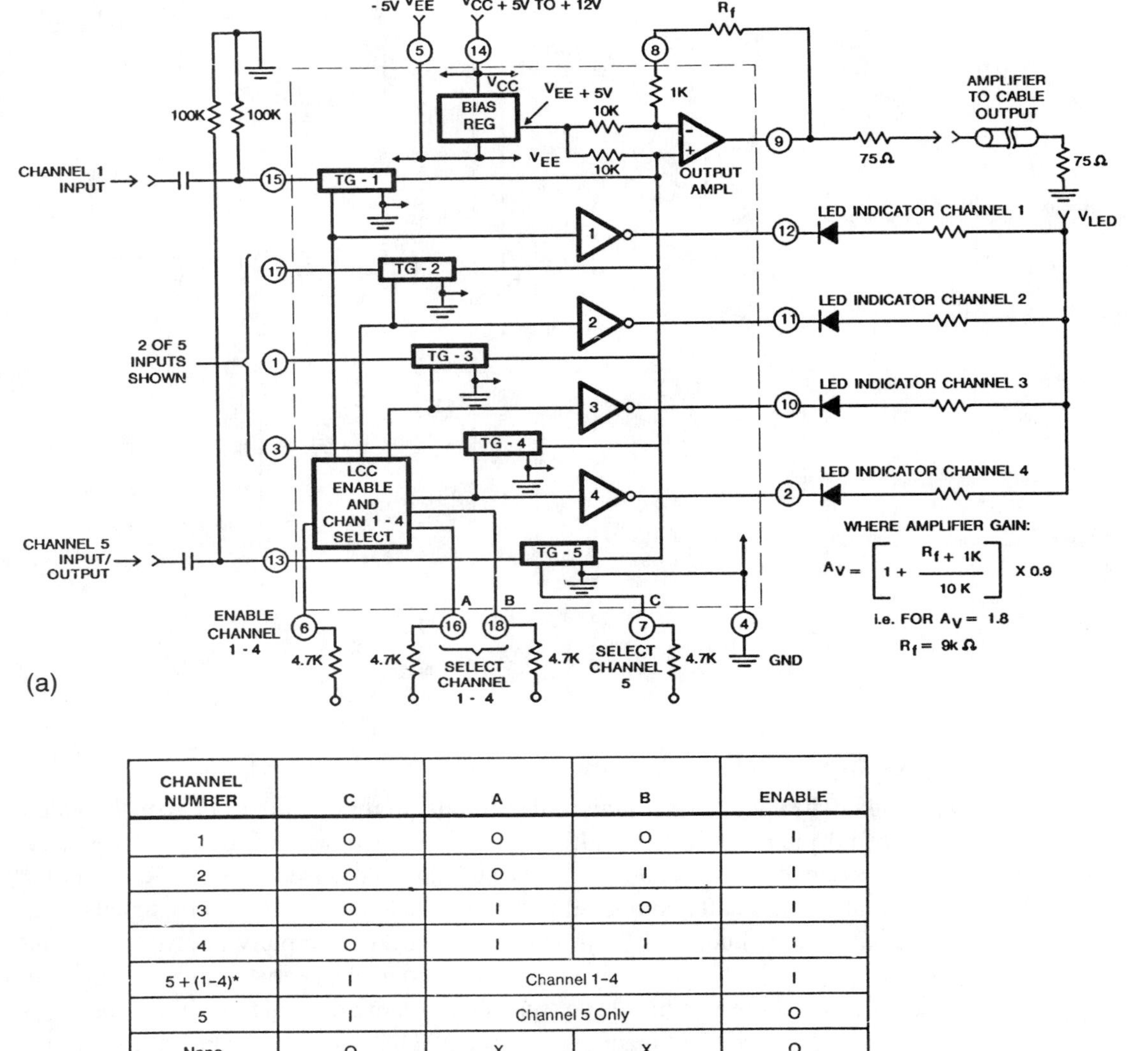

CHANNEL NUMBER	C	A	B	ENABLE
1	O	O	O	I
2	O	O	I	I
3	O	I	O	I
4	O	I	I	I
5 + (1–4)*	I	Channel 1–4		I
5	I	Channel 5 Only		O
None	O	X	X	O

(b) * For Maximum Video Bandwidth, Use Single Channel Selections

Analog video switch and amplifier with direct-coupled output

Fig. 3-35 This circuit shows a CA3256 switch/amplifier connected for a direct-coupled output. One of four channels can be selected in parallel with channel 5. The analog switches of channels 1 to 4 are digitally controlled by logic (Fig. 3-35B). A V_{EE} of −5 V is required. V_{CC} can be from +5 to +12 V. Amplifier gain is determined by the external resistor R_f, as shown by the equation. Harris Semiconductor, Linear & Telecom ICs, 1991, p 8-51, 8-54.

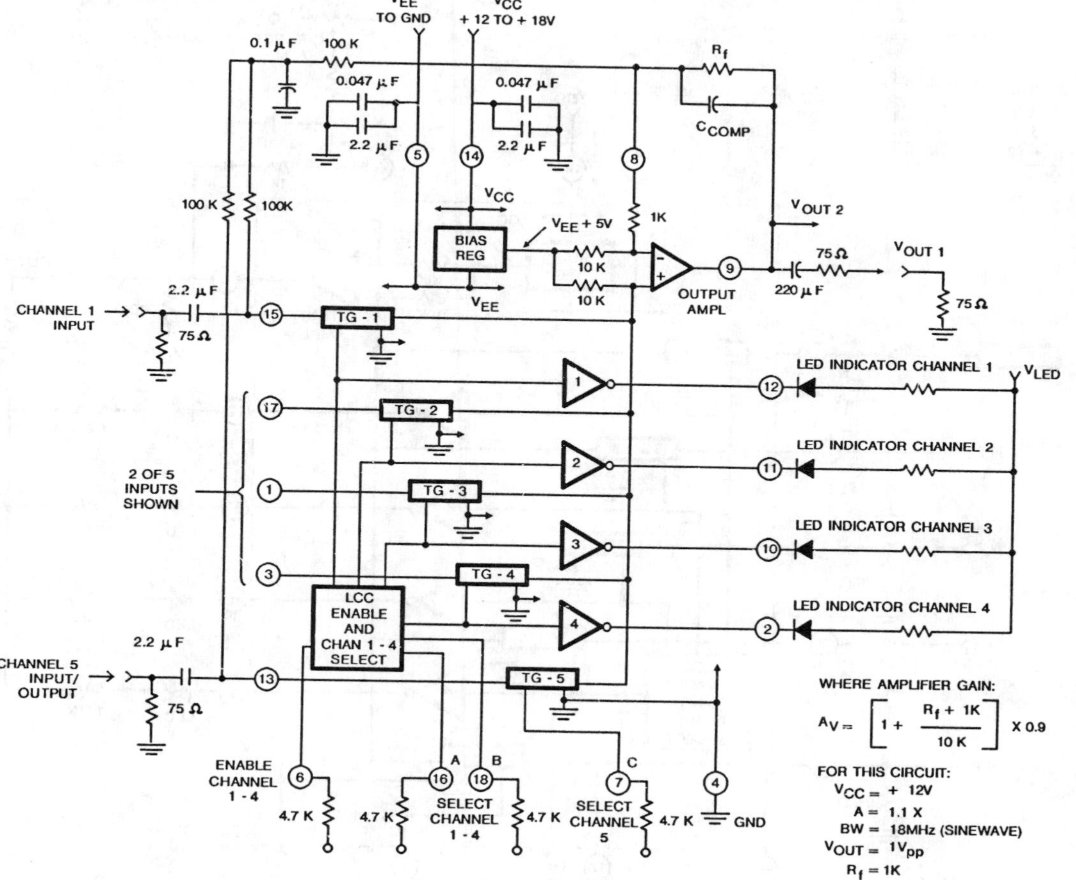

Analog video switch and amplifier with ac-coupled input

Fig. 3-36 This circuit shows a CA3256 switch/amplifier connected for an ac coupled input. One of four channels can be selected in parallel with channel 5. The analog switches of channels 1 to 4 are digitally controlled by logic (Fig. 3-35B). V_{CC} can be from +12 to +18 V. V_{EE} is grounded. Amplifier gain is determined by external resistor R_f, as shown by the equation. A typical value of C_{COMP} is 6 pF. Harris Semiconductor, Linear & Telecom ICs, 1991, p. 8-55.

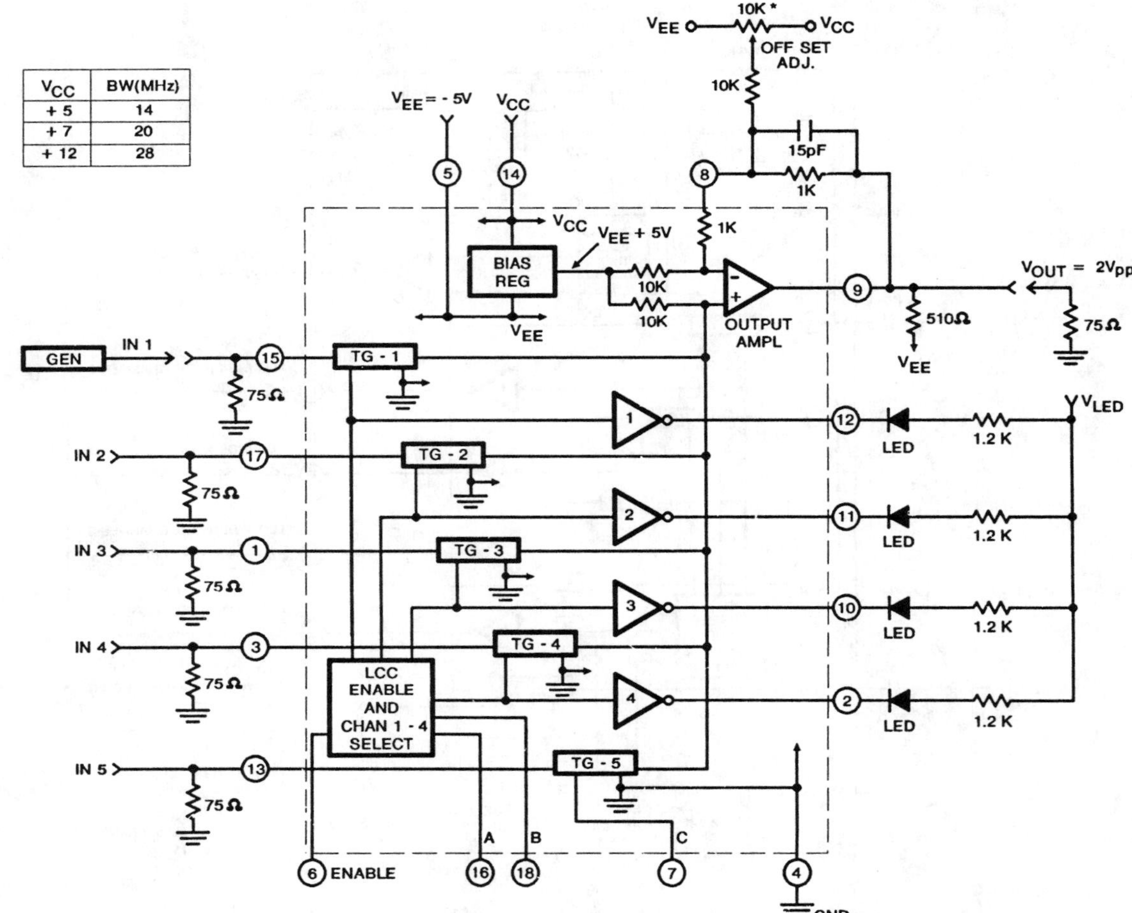

V_{CC}	BW(MHz)
+ 5	14
+ 7	20
+ 12	28

* Adjust offset for V_{DC} at pin 9 equal to zero volts with no AC signal and one channel "ON". Dymanic clamping may be accomplished by error current feedback to pin 8.

Analog video switch and amplifier with direct-coupled input and output

Fig. 3-37 This circuit shows a CA3256 switch/amplifier connected for a direct-coupled input and output. One of four channels can be selected in parallel with channel 5. The analog switches of channels 1 to 4 are digitally controlled by logic (Fig. 3-35B). A V_{EE} of -5 V is required. The peak-to-peak output voltage is fixed by the V_{CC} and V_{EE} range, and is about $+3.6$ to -2.5 V (clipped) for a V_{CC} of $+5$ V and a V_{EE} of -5 V. For a 2-Vpp output, V_{CC} must be $+4$ V, and V_{EE} must be -4 V. This produces a bandwidth of about 10 MHz. Higher V_{CC} produces greater bandwidths, as shown. Harris Semiconductor, Linear & Telecom ICs, 1991, p. 8-56.

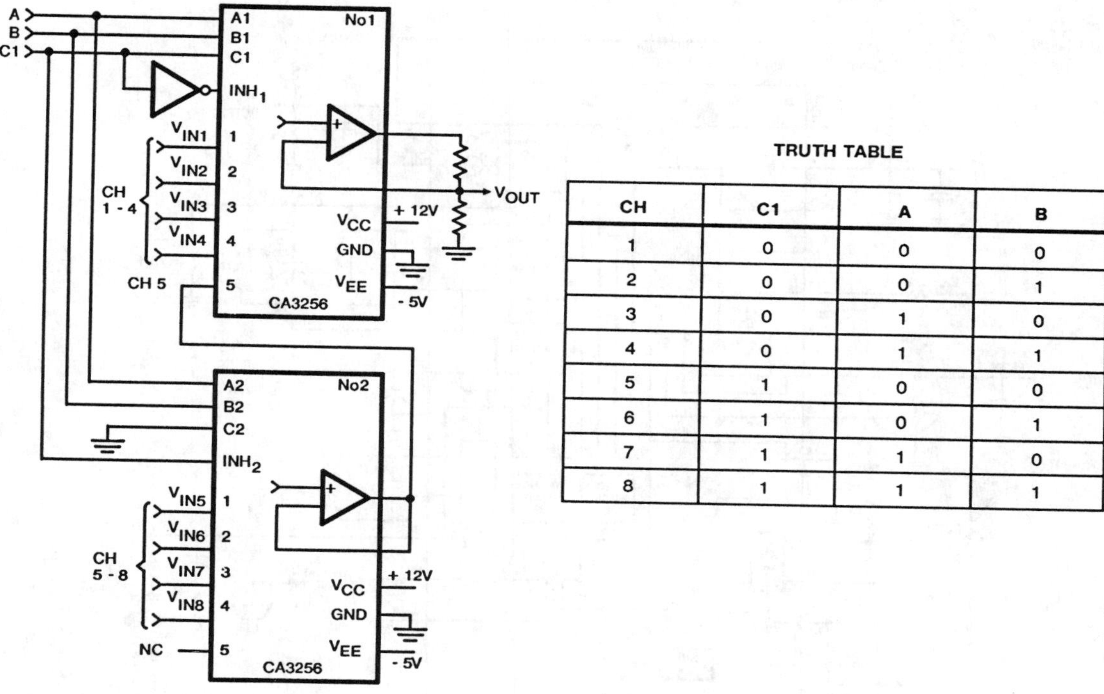

TRUTH TABLE

CH	C1	A	B
1	0	0	0
2	0	0	1
3	0	1	0
4	0	1	1
5	1	0	0
6	1	0	1
7	1	1	0
8	1	1	1

8-to-1 video switch/amplifier

Fig. 3-38 This circuit uses two CA3256 ICs to provide an 8-to-1 video-switch function. The analog switches of channels 1 to 8 are digitally controlled by logic, as shown in the truth table. As in the case of the switch circuits shown in Figs. 3-35 through 3-37, the circuit of Fig. 3-38 has programmable gain for all channels, as well as flexible output-voltage swing. Typically, the output voltage swing can be about 5 Vpp, with a V_{CC} to V_{EE} range of 12 V or greater, and each amplifier can provide a gain of 1 into a 75-Ω load, or a gain of 5 into a 1-kΩ load. Harris Semiconductor, Linear & Telecom ICs, 1991, p. 8-61.

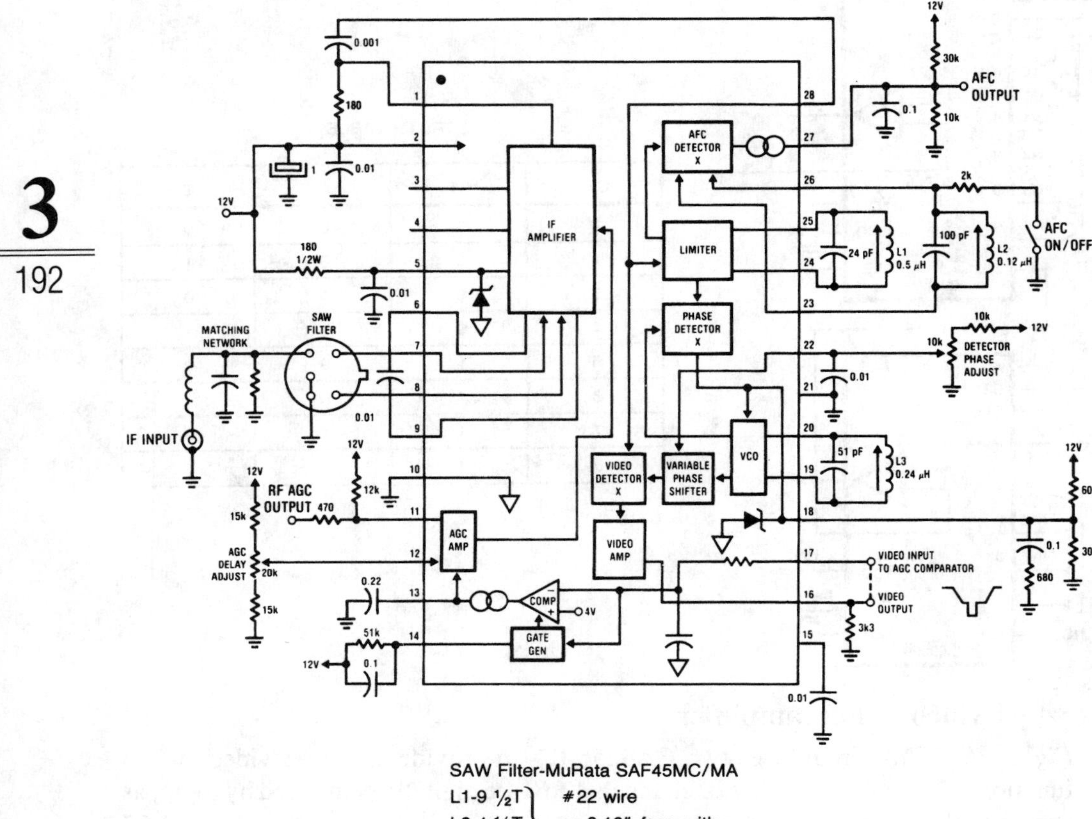

SAW Filter-MuRata SAF45MC/MA

L1-9 ½T } #22 wire
L2-4 ½T } on 3.16″ form with
L3-6 ½T } HF core, shielded

All caps in uF unless noted

(a)

Continued

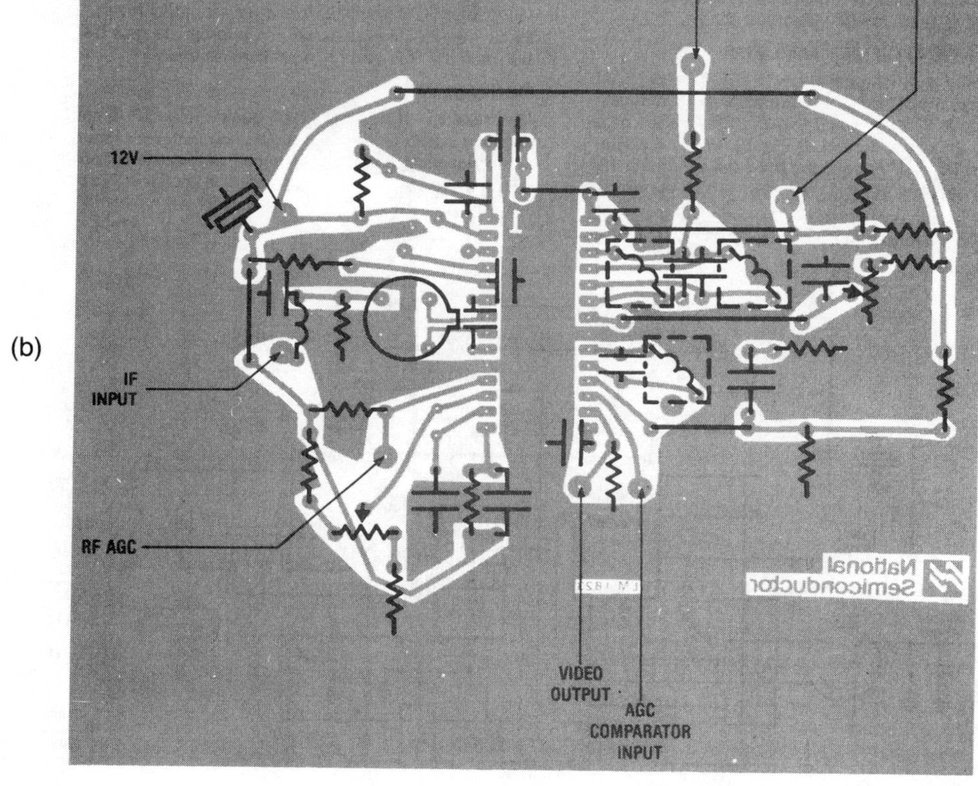

(c)

Maximum system operating frequency	70 MHz	Detector differential phase	1 degree
Typical I.F. amplifier Gain (45.75 MHz)	>60 dB	Noise averaged AGC system	
I.F. amplifier gain control range	55 dB	Internal AGC gated comparator	
True synchronous detector with a PLL		Reverse tuner AGC output	
Detector conversion gain	34 dB	DC controlled video detection phase	
Detector output bandwidth	9 MHz	AFC detector	
Detector differential gain	2%		

TV video IF amplifier and synchronous detector for cable receivers

Fig. 3-39 This circuit shows the external connections for an LM1823, which is a video IF amplifier that is designed to operate at intermediate carrier frequencies up to 70 MHz, and use phase-locked loops (PLLs) for synchronous detection of amplitude modulation on these carrier frequencies. Figure 3-39B shows a typical PC layout, and Fig. 3-39C tabulates the circuit characteristics. National Semiconductor, Linear Applications Handbook, 1991, p. 1007, 1021, 1022.

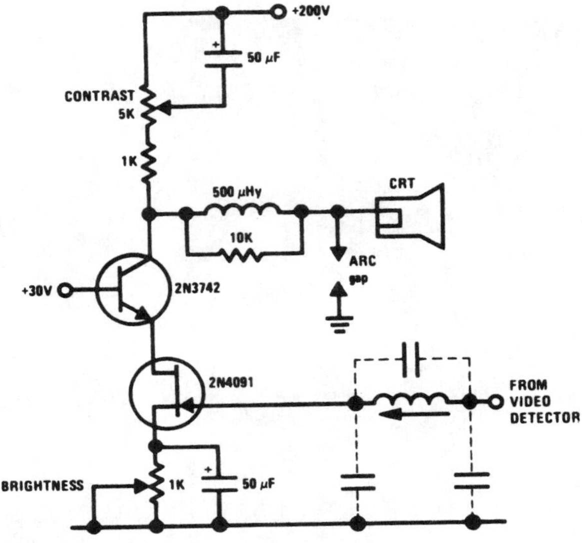

JFET-bipolar cascode video amplifier

Fig. 3-40 This cascode circuit provides full video output for the CRT cathode drive. Gain is about 90. The cascode configuration eliminates Miller-capacitance problems with the JFET, thus allowing direct drive from the video detector. An *m*-derived filter, which uses stray capacitance and a variable inductor, prevents a 4.5-MHz sound frequency from being amplified by the video amplifier. National Semiconductor, Linear Applications Handbook, 1991, p. 112.

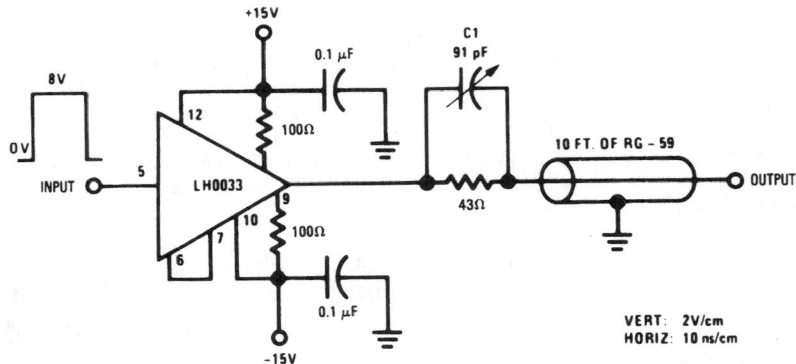

Shield or line driver for high-speed video applications

Fig. 3-41 This circuit uses the low input current, high speed, and high-capacitance drive capabilities of the LH0033 to full advantage for high-speed video applications, such as in automatic test equipment. In this circuit, the LH003 is mounted close to the device under test, and drives the cable/shield, thus allowing higher-speed operation because the device under test does not have to charge the cable. National Semiconductor, Linear Applications Handbook, 1991, p. 158.

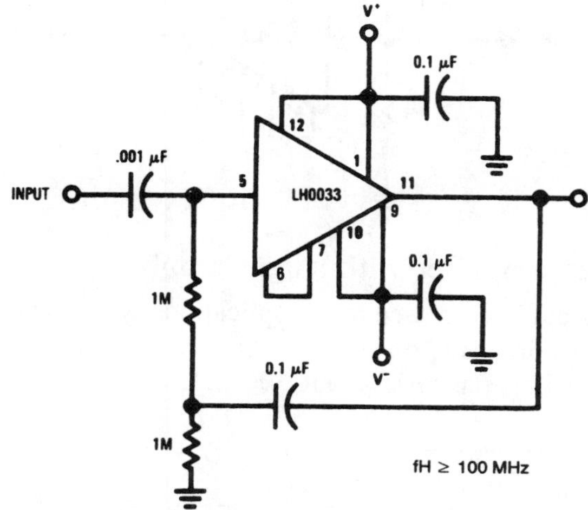

Ac-coupled video amplifier with high input impedance

Fig. 3-42 This circuit can be used in ac applications, such as in video amplifiers and active filters. The circuit uses boot-strapping to achieve input impedances in excess of 10 MΩ. National Semiconductor, Linear Applications Handbook, 1991, p. 160.

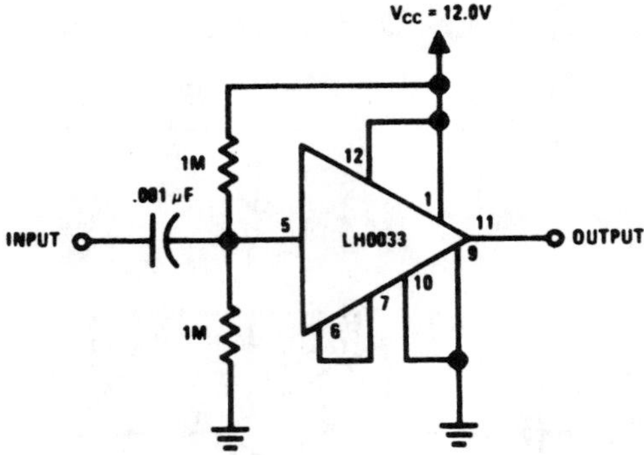

Ac-coupled video amplifier with single supply

Fig. 3-43 This circuit can be used in ac applications, such as video amplifiers and active filters, but requires only one supply. The output swing is over 8 Vpp with a 12-V supply. Input impedance is about 500 kΩ. National Semiconductor, Linear Applications Handbook, 1991, p. 160.

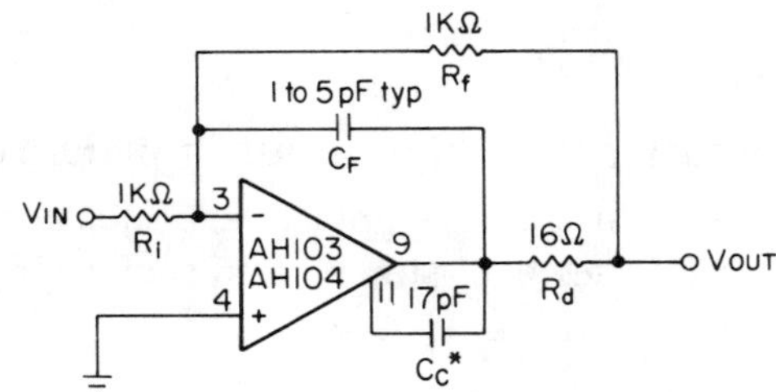

Unity-gain inverting wideband amplifier

Fig. 3-44 This circuit shows the basic connections in a typical unity-gain inverting amplifier. The use of a compensation capacitor C_C is required for the AN104, unless the gain is fairly high. The AN103 only needs additional external compensation in special cases, such as when used as a noninverting unity-gain follower. R_d in the output circuit improves frequency stability. C_F compensates for a pole in the closed-loop transfer function, caused by the input capacitance and feedback resistors. A ground plane is recommended, leads must be short, and R_f should not exceed 5.6 kΩ. Typical bandwidth is 35 MHz, with slew rates of 230 V/µs, into a 330-Ω load. Optical Electronics Incorporated, Product Catalog, 1990-91, p. 45.

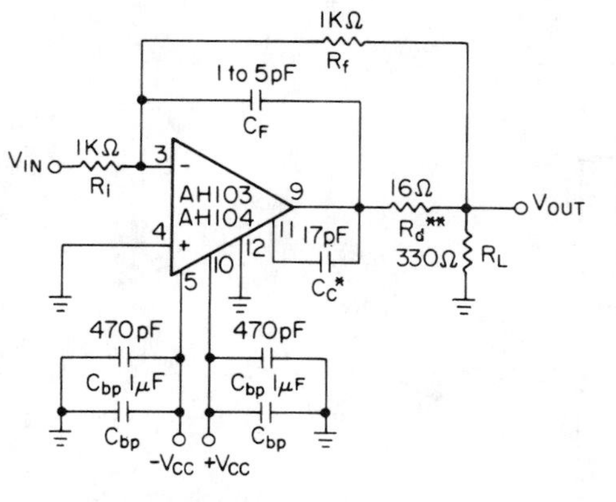

*Not needed for the AH103

**R_d especially critical for capacitive loads. May be unneeded on resistive loads.

(b)

Video pulse amplifier

Fig. 3-45 This circuit shows a video pulse amplifier that uses the basic circuit of Fig. 3-44. Figure 3-45B shows a suggested PC layout. Notice that internal current-limiting resistors provide short-circuit protection to the common line, with a limit of 100-mA maximum. A short to either supply line can destroy the unit. Typical supply voltages are ±20 V. Optical Electronics Incorporated, Product Catalog, 1990-91, p. 45, 46.

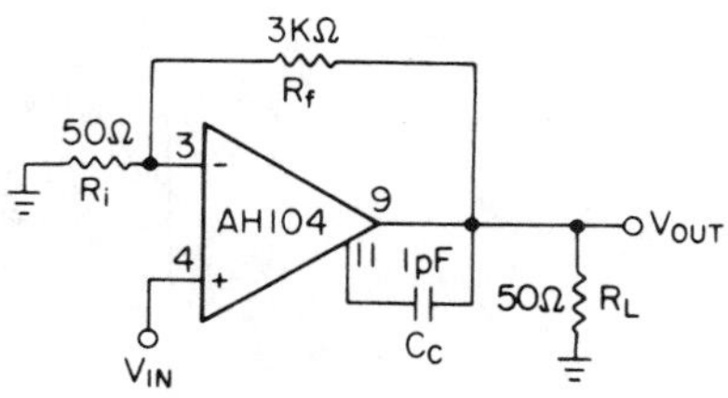

High-gain video amplifier

Fig. 3-46 This circuit shows an AH104 used as a high-gain video amplifier. The high output-current capability makes the AH104 suitable for such high-frequency driver applications. In this circuit, a closed-loop gain of 60 can be obtained to 10 MHz. For lower gains, C_C must be larger to promote stability. The power connections are the same as for Fig. 3-45. The AN103 can be used in this circuit, without C_C, but the internal 20-pF compensation capacitor reduces the gain at any given bandwidth. Optical Electronics Incorporated, Product Catalog, 1990-91, p. 46.

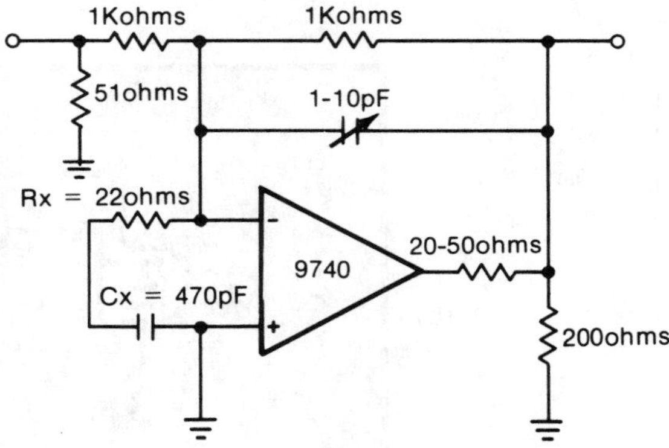

High-speed, fast-settling pulse amplifier

Fig. 3-47 In this circuit, the amplifier is set for unity gain. The input signal is terminated into 51 Ω. This resistance, together with the 470-pF capacitor, compensates for stray capacitances that might appear at the input. With the values shown (and the variable capacitor adjusted to 5 pF), the output is ±10 V swing at ±50-mA drive current, with unity gain up to 100 MHz (assuming a typical ±15-V supply. The pin connections are: 1 output, 10 +V_{CC}, 11-V_{CC}, 16-input, 19 +input, 20 common. Optical Electronics Incorporated, Product Catalog, 1990-91, p. 54.

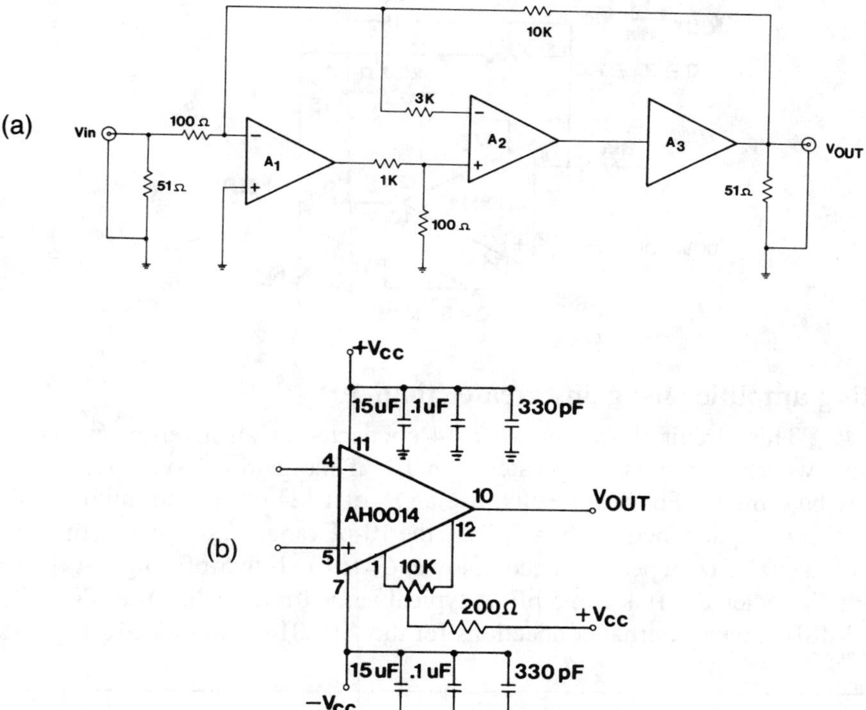

Composite amplifier with current boost

Fig. 3-48 In this circuit, A1 is used to improve the dc characteristics of A2, which, in turn, determines the bandwidth of A3. Using an OP-07 for A1, an AH0014 for A2, and an AH0010 for A3, the circuit is capable of driving a 51-Ω load at ±5 V to 10 MHz. Drift is less than 1 μV/°C with an offset of 20 μV. Figure 3-48B shows the external connections for the AH0014. Optical Electronics Incorporated, Product Catalog, 1990-91, p. 59, 62.

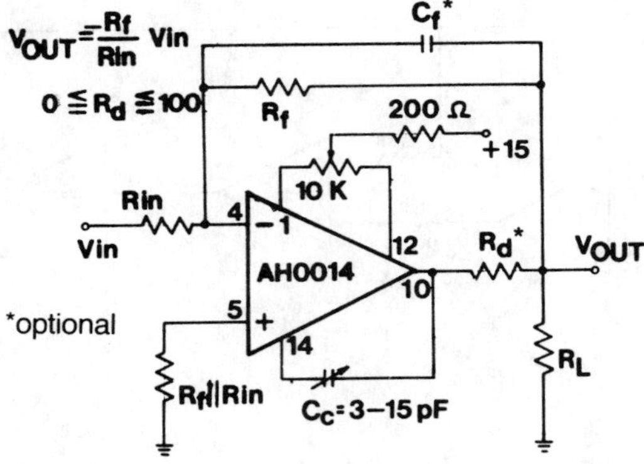

Inverting amplifier for gains greater than 10

Fig. 3-49 This circuit shows an AH0014 connected as an inverting wideband amplifier, where gain must be greater than 10. If the gain is over 100, C_C can generally be omitted. For best results, use an R_f of 1 kΩ or less (to minimize the effects of stray capacitance). When C_C is in the 10-pF range, a stable gain of 20 dB is possible at 10 MHz. If peaking occurs, lower C_C to the 1- to 5-pF range. A typical value for C_F (when used) is 1 to 5 pF. A typical value for R_d (when used) is 33 Ω. Figure 3-48B shows external connections for the AH0014. Optical Electronics Incorporated, Product Catalog, 1990-91, p. 59.

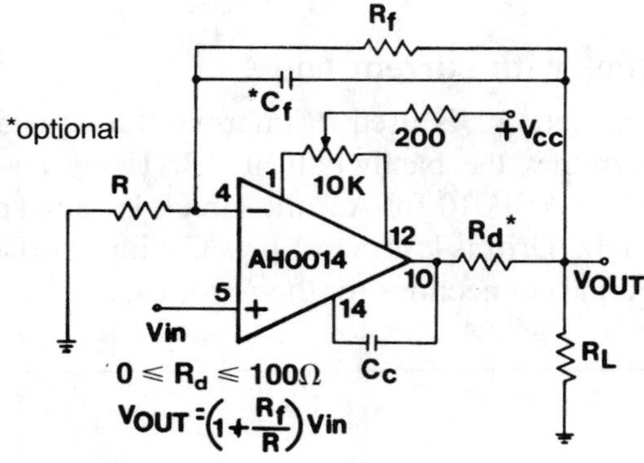

Noninverting amplifier for gains greater than 10

Fig. 3-50 This circuit shows an AH0014 connected as a noninverting wideband amplifier, where gains must be greater than 10. The characteristics are the same as for the circuit of Fig. 3-49. Figure 3-48B shows external connections for the AH0014. Optical Electronics Incorporated, Product Catalog, 1990-91, p. 59.

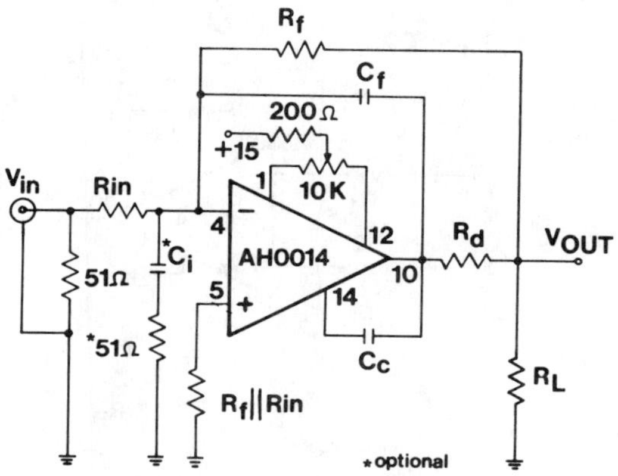

Inverting amplifier for gains less than 10

Fig. 3-51 This circuit shows an AH0014 connected as an inverting wideband amplifier, where gain must be less than 10, but more than 1. The characteristics are the same as for the circuit of Fig. 3-49, but layout is more critical. Keep all leads as short as possible. Solder chip capacitors (bypass capacitors) directly to the chip pins. Use a ground plane. For best results, use an R_i of 1 kΩ or less, and keep C_i in the 15- to 60-pF range (use a variable C_i and adjust for best results. Figure 3-48B shows external connections for the AH0014. Optical Electronics Incorporated, Product Catalog, 1990-91, p. 61.

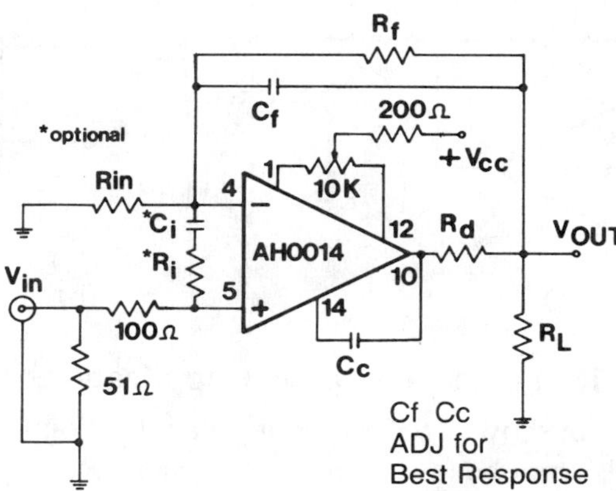

Noninverting amplifier for gains less than 10

Fig. 3-52 This circuit shows an AH0014 connected as a noninverting wideband amplifier, where gains must be less than 10, but more than 1. The characteristics are the same as for the circuits of Figs. 3-49 and 3-51. Figure 3-48B shows external connections for the AH0014. Optical Electronics Incorporated, Product Catalog, 1990-91, p. 61.

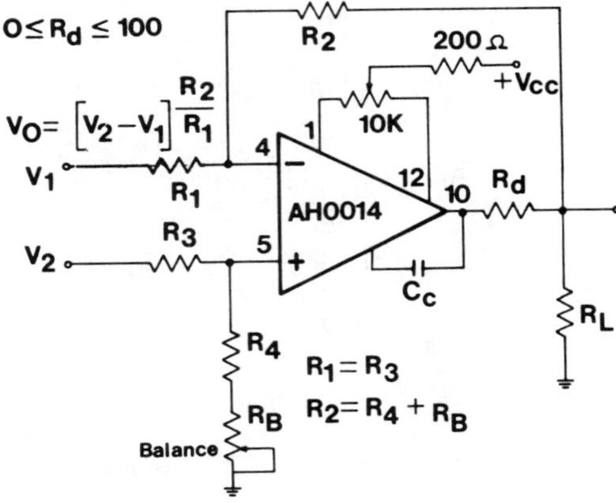

Wideband difference amplifier

Fig. 3-53 This circuit shows an AH0014 connected as a basic differential or instrumentation amplifier with a balance adjustment. The balance resistor is adjusted for best common-mode rejection ratio. Figure 3-48B shows external connections for the AH0014. Optical Electronics Incorporated, Product Catalog, 1990-91, p. 61.

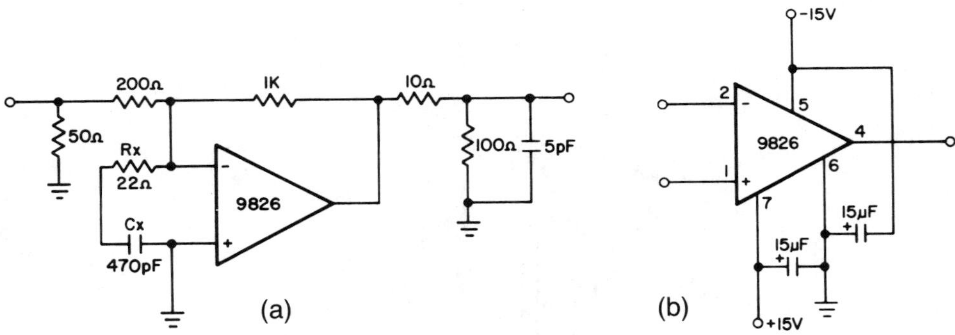

Wideband amplifier for test-equipment applications

Fig. 3-54 This circuit shows a 9826 used in an application where high speed at a relatively low gain is required (a gain of 5 in this case). R_X and C_X form a lead/lag network between the inverting and noninverting inputs. With the values given, this circuit can operate comfortably at frequencies of 20 MHz. If a lower gain is needed (gain less than 5), connect a 1- to 5-pF capacitor between the output and inverting input to aid in suppressing parasitic oscillations. The 100-Ω 5-pF filter at the output has a damping effect to help suppress signal overshoot. Figure 3-54B shows external connections for the 9826. Optical Electronics Incorporated, Product Catalog, 1990-91, p. 72, 74.

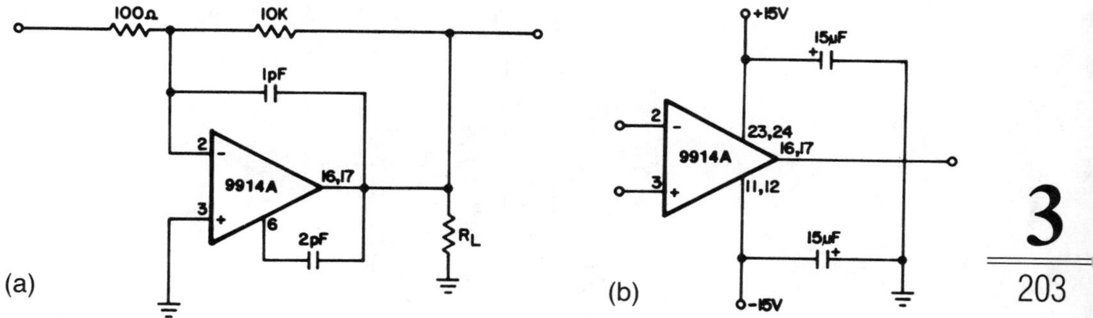

Wideband amplifier for test-equipment applications (high gain)

Fig. 3-55 This circuit shows a 9914A used in an application where high speed at a relatively high gain is required (a gain of 100 in this case). With the values given, this circuit can operate comfortably at frequencies of 20 MHz. If a lower gain is needed, connect a 1- to 5-pF capacitor between the output and inverting input to aid in suppressing parasitic oscillations. Figure 3-55B shows external connections for the 9914A. Optical Electronics Incorporated, Product Catalog, 1990-91, p. 78, 80.

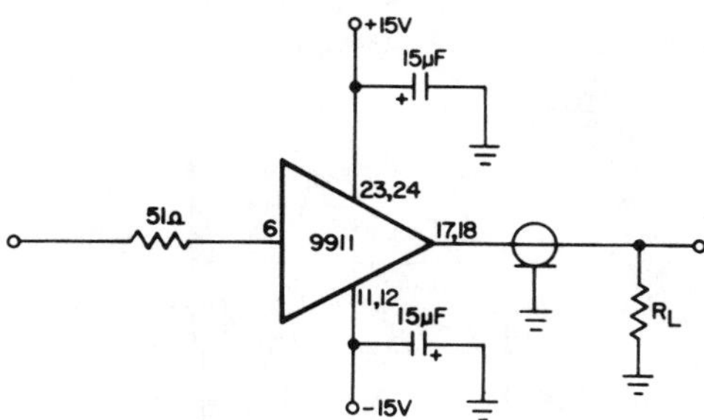

Coaxial line driver

Fig. 3-56 This circuit shows a 9911 used to drive long coaxial lines in a 50- or 75-Ω system, such as those found in test equipment and other applications. The 9911 is a high-current voltage follower and, in this application, is capable of driving long cables (and thus large capacitances) without instabilities. Remember that the 9911 provides 0.96 gain, which can drop to 0.90 when the chip is heavily loaded. R_L should match the cable impedance (typically 50 or 75 Ω). Optical Electronics Incorporated, Product Catalog, 1990-91, p. 103.

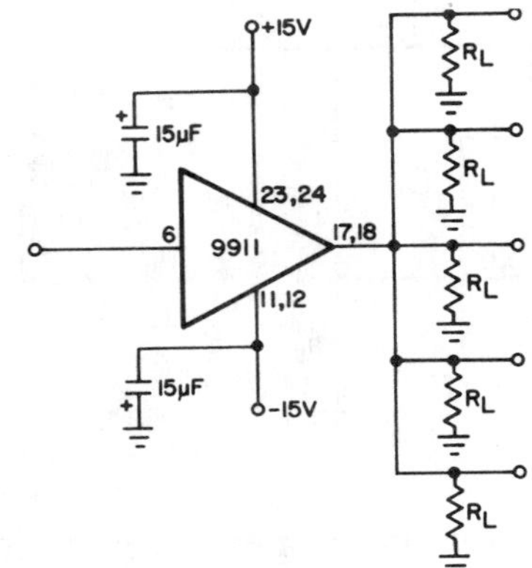

Multiple line driver, buffer

Fig. 3-57 This circuit shows a 9911 used as a driver for multiple lines, which allows a video signal to be routed to 5 (or more) individual directions. A typical application is an apartment building with a common cable system. Although the 9911 is capable of 500-mA drive at a ±10-V output voltage swing, this application usually does not require more than ±1 or ±2 V (for a typical video system). This circuit can be inexpensive because only one voltage-follower/buffer and few external components are needed. Optical Electronics Incorporated, Product Catalog, 1990-91, p. 103.

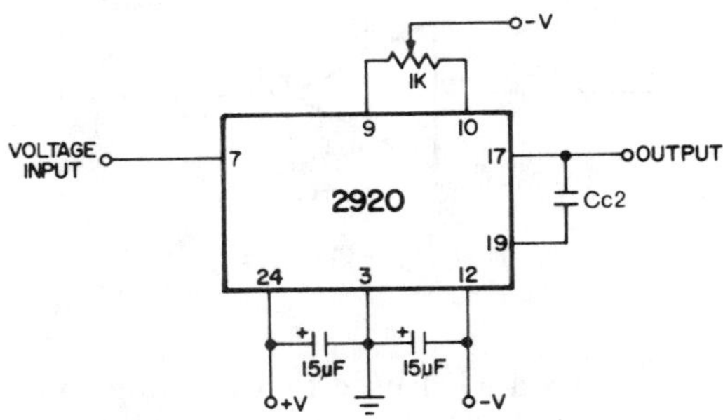

20-MHz video log amplifier

Fig. 3-58 This circuit shows a 2920 in the basic log-amplifier configurations, using a minimum of external components. The offset pot between pins 9 and 10 is mandatory. V_{CC} is typically ±15 V, C_2 is typically between 10 and 50 pF. The circuit has a 20-MHz bandwidth and an 80-dB dynamic range. Optical Electronics Incorporated, Product Catalog, 1990-91, p. 129.

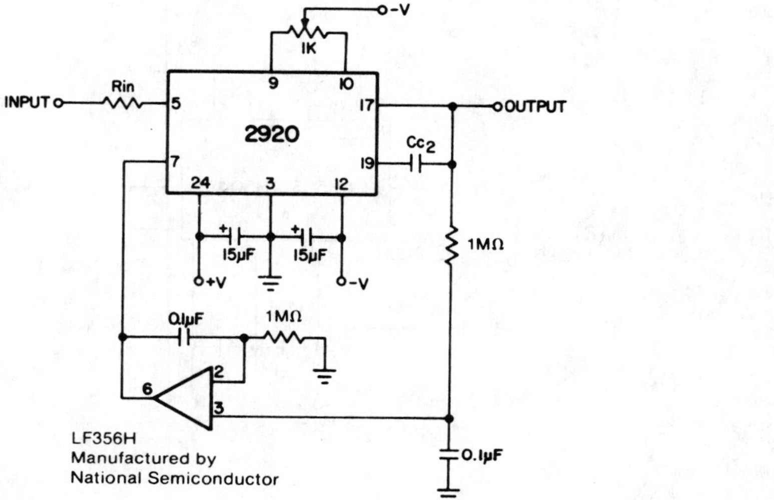

Low duty-cycle log amplifier

Fig. 3-59 This circuit is designed for cases where very accurate offset adjustments must be made. Coarse adjustment is provided by the offset pot at pins 9/10. The output of the 2920 is returned to the input (at pin 7), an LF356H op amp to provide soft offset correction automatically. This circuit is recommended when processing bipolar or low duty-cycle pulse signals. R_{in} should match the source impedance. C_2 is typically between 10 and 50 pF. Optical Electronics Incorporated, Product Catalog, 1990-91, p. 130.

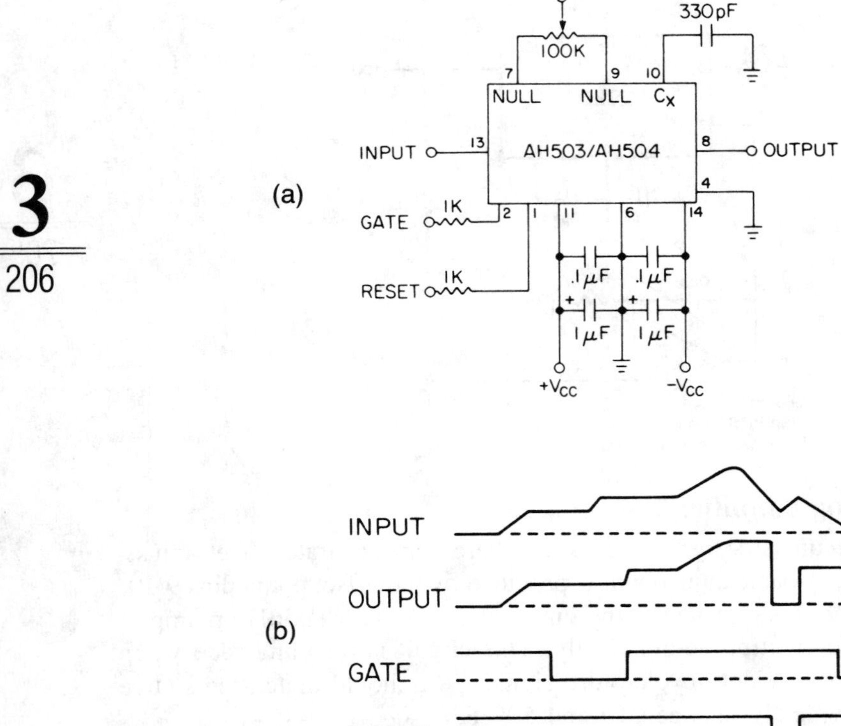

High-speed video-peak detector

Fig. 3-60 This circuit shows an AH503 or AH504 connected as a video-peak sensor or detector, using a minimum of external components. Figure 3-60B shows the related waveforms. The circuit will accommodate a 10-V pulse, and will capture individual 1-V pulses as narrow as 100 ns to within ±10 mV. Delay between input peak and output settling is 500 ns. Optical Electronics Incorporated, Product Catalog, 1990-91, p. 142, 143.

4

Power-supply and regulator circuits

It is assumed that the reader is already with power-supply and regulator basics (such as operation of diodes, switch-mode regulators and supplies, converters, inverters, etc.) and basic power-supply testing/troubleshooting. However, the following paragraphs summarize both the testing and troubleshooting of power-supply/regulator circuits. This information is included so that even those readers who are not familiar with electronic procedures can both test the circuits described here, and localize problems if the circuits fail to perform as shown.

Power-supply/regulator testing

This section describes the basic tests for power-supply and regulator circuits. Both simple tests and more advanced tests are described. If the circuits pass these tests, the circuits can be used immediately. If not, the tests provide a starting point for the troubleshooting procedures that are described in this chapter ("Power supply/regulator troubleshooting").

Basic tests

This section is devoted to simple, practical test procedures that can be applied to a just-completed power supply during design/experimentation (or to a suspected power supply as part of troubleshooting). More advanced tests are covered later in this chapter. However, the following procedures are usually sufficient for practical applications.

The basic function of a power supply is to convert alternating current into direct current (although an inverter does the opposite). In the case of a dc/dc converter, direct current is inverted to direct current, but at a different voltage (usually higher). In any event, the power-supply function can be checked quite simply by measuring the output voltage. However, for a more thorough test of a supply, the output voltage should be measured without a load, with a load, and possibly with a partial load.

If the supply delivers the full-rated output voltage into a full-rated load, the basic supply function is met. In addition, it is often helpful to measure the regulating effect of a power supply, the power-supply internal resistance, and the amplitude of any ripple at the power-supply output. The following paragraphs describe each of these basic tests.

Output tests Figure 4-A is the basic power-supply test circuit. This arrangement permits the power supply to be tested at no load, half load, and full load, depending on the position of S1. With S1 in position 1, no load is on the supply. At positions 2 and 3, there is half load and full load, respectively.

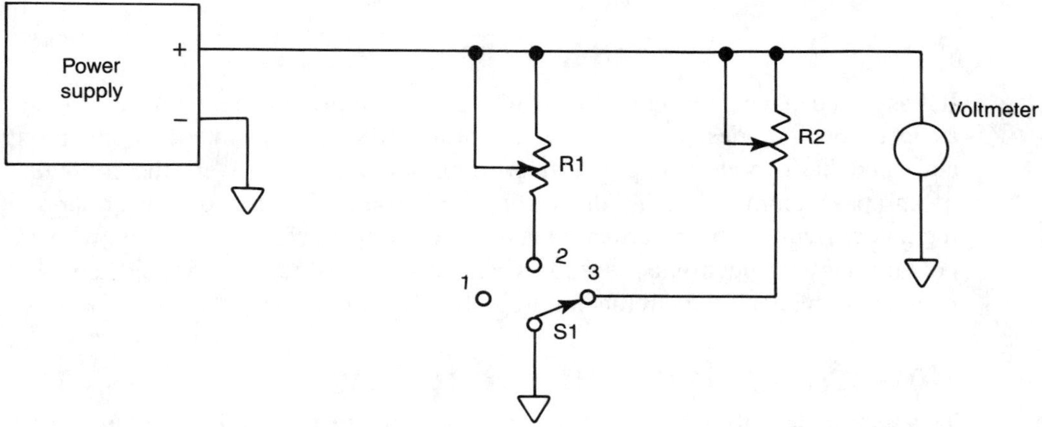

Fig. 4-A A basic power-supply test circuit.

Using Ohm's Law, R = E/1, R_1 and R_2 are chosen on the basis of output voltage and load current (maximum or half load). For example, if the supply is designed for an output of 25 V at 500 mA (full load), the value of R_2 is 25/0.5 = 50 Ω. The value of R_1 is 25/0.25 = 100 Ω. Where more than one supply is to be tested, R1 and R2 should be variable, and adjusted to the correct value before testing (using an ohmmeter with the power removed). The resistors must be noninductive (not wire wound) and must be capable of dissipating the rated power without overheating. For example, using the previous values for R_1 and R_2, the power dissipation of R1 is 25 × 0.5, or 12.5 W (use at least 15 W), and the dissipation for R2 is 25 × 0.25 or 6.25 W (use at least 10 W).

1. Connect the equipment (Fig. 4-A).
2. Set R_1 and R_2 to the correct value.
3. Apply power. Set the input voltage to the correct value. Use the midrange value for the input voltage if a variable transformer or variac is available.
4. Measure the output voltage at each position of S1.
5. Calculate the current at positions 2 and 3 of S1, using Ohm's Law, 1 = E/R. For example, assume that R_1 is 100 Ω, and the meter indicates 22 V at position 2 of S1. The actual load current is 22/100 = 220 mA. If the supply output is 25 V at position 1, and drops to 22 V at position 2, the supply is not producing full output with a load. This is an indication of poor regulation, possibly resulting from poor wiring design (in the case of an experimental supply) or component failure, all of which are covered later in this chapter.

Regulation tests Power supplies can be checked for both output regulation and input regulation. Generally, output regulation is the most important in practical applications.

Output regulation Power-supply output regulation is usually expressed as a percentage, and is determined by the equation:

$$\%\ regulation = \frac{(no\text{-}load\ voltage) - (full\text{-}load\ voltage)}{full\text{-}load\ voltage} \times 100$$

A low percentage-of-regulation figure is desired because this indicates that the output voltage changes very little with load.

1. Connect the equipment (Fig. 4-A).
2. Set R_2 to the correct value.
3. Apply power. Measure the output voltage at position 1 (no load) and position 3 (full load).
4. Using the equation, calculate the percentage of regulation. For example, if the no-load voltage is 26 V, and the full-load voltage is 25 V, the percentage of regulation is (26 − 25)/25, or 4 percent (very poor regulation).
5. Notice that power-supply output regulation is usually poor (high percentage) when the internal resistance is high.

Input regulation Power-supply input regulation is usually expressed as a percentage and represents the maximum allowable output variation (with a given load) for maximum rated input variation.

As an example, if the supply is designed to operate with an ac input from 110 to 120 V, and the dc output is 100 V, the output is measured (1) with an input of 120 V, and (2) with an input of 110 V. If there is no change in output, input regulation is perfect (and probably impossible). If the output varies by 1 V, the output variation is 1 percent (which might or might not be within the allowable maximum).

The actual power-supply input regulation can be measured at full load and/or half load, as desired, using the same test connections as shown in Fig. 4-A. However, the input voltage must be varied from maximum to minimum values (with a variable transformer or variac) and monitored with an accurate voltmeter.

Internal-resistance test Power-supply internal resistance is determined by the equation:

$$Internal\ resistance = \frac{(no\text{-}load\ voltage) - (full\text{-}load\ voltage)}{current} \times 100$$

A low internal resistance is most desirable because this indicates that the output voltage changes very little with load.

1. Connect the equipment (Fig. 4-A).
2. Set R_2 to the correct value.
3. Apply power. Measure the actual output voltage at position 1 (no load) and position 3 (full load).
4. Calculate the actual load current at position 3 (full load). For example, if R_2 is adjusted to 50 Ω, and the output voltage at position 3 is 25 V (as in the preceding example), the actual load current is 25/50 = 500 mA.

With no-load voltage, full-load voltage, and actual current established, find the internal resistance using the equation. For example, with no-load voltage of 26 V, a full-load voltage of 25 V, and a current of 500 mA (0.5 A), the internal resistance is (26 − 25)/0.4 = 2 Ω.

Ripple tests Any power supply, no matter how well regulated or filtered, has some ripple. This ripple can be measured with a meter or scope. Usually, the factor of most concern is the ratio between ripple and full-output voltage. For example, if 3 V of ripple is measured together with a 100-V output, the ratio is 3 to 100 = 3% ripple.

1. Connect the equipment (Fig. 4-A).
2. Set R_2 to the correct value. Ripple is usually measured under full load power.

3. Apply power. Measure the dc output voltage at position 3 (full load).
4. Set the meter to measure alternating current. Any voltage measured under these conditions is ac ripple.
5. Find the percentage of ripple, as a ratio between the two voltages (*ac*/*dc*).
6. One problem often overlooked in measuring ripple is that any ripple voltage is not a pure sine wave. Most meters provide accurate ac voltage indications only for pure sine waves. A more satisfactory method of measuring ripple is with a scope, where the peak value can be measured directly.
7. Adjust the scope controls to produce two or three stationary cycles of ripple on the screen (Fig. 4-B). Notice that a full wave produces two ripple "humps" per cycles, whereas a half-wave supply produces one "hump" per cycle.
8. A study of the ripple waveform can sometimes show defects in an experimental circuit. Here are some examples. If the supply is unbalanced (one rectifier passes more current than the others), the ripple humps are unequal in amplitude. If there is noise or fluctuations in the supply (particularly in zener diodes), the ripple humps will vary in amplitude or shape. If the ripple varies in frequency, the ac source is varying, or (in switching supplies) the switching frequency is varying. If a full-wave supply produces a half-wave output, one rectifier is not passing current.

Advanced tests

The basic tests of the last section are generally sufficient for most practical applications (usually for the experimenter and serious hobbyist). However, many other tests can be used to measure the performance of commercial and lab power supplies. The most common (and most important) of such tests are covered in the following paragraphs.

Figure 4-B shows test connections for measurement of the five most important operating specifications of a power supply: source effect, load effect, PARD, drift, and temperature coefficient. There are additional specifications, such as noise-spike measurement and transient-recovery time measurement. However, these measurements require elaborate test setups, and are generally applied to complete lab or commercial power supplies.

Test equipment All of the tests described here can be performed with only four test instruments: a variable autotransformer, a differential or digital dc voltmeter, a true ac voltmeter, and a scope. Of course, on those supplies that are battery operated, the variable autotransformer can be omitted. When used, make

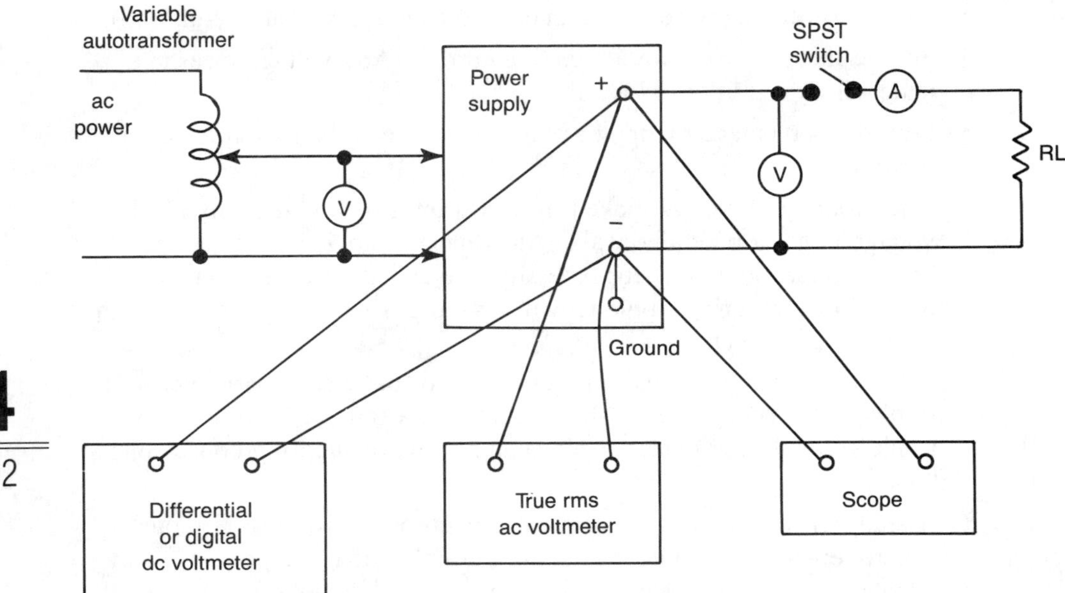

Fig. 4-B Test connections for measurement of source effect, load effect, PARD, drift, and temperature coefficient.

certain that the autotransformer has an adequate current rating. If not, the input voltage applied to the supply might be severely distorted and the rectifying/regulating circuit within the supply might operate improperly.

The dc voltmeter should have a resolution of 1 mV (or better) at voltages up to 1000 V. The scope should have a sensitivity of 100 μV/cm and a bandwidth of 10 MHz (although bandwidth is not critical for most of the supplies in this chapter).

Proper connections For the most accurate measurements, the test connections should be permanent (not clip leads), and should be made on the exact point on the supply terminals. Clip-lead connections often produce measurement errors. Instead of measuring pure supply characteristics, you are measuring supply characteristics plus the resistance between output terminals and point of connection. Even using clip leads to connect the load to the supply terminals can produce a measurement error.

Separate leads All measurement instruments must be connected directly by separate pairs of leads to the monitoring points (Fig. 4-B). This avoids the subtle mutual-coupling effects that can occur between measuring instruments (unless all are returned to the low-impedance terminals of the supply). Twisted pairs or shielded cable should be used to avoid pickup on the measuring leads.

Load resistance Make certain that the load resistance is adequate for the supply and test requirements. Typically, the load resistance and load wattage should permit operation of the supply at maximum rated output voltage and current.

Current limit If the supply has a current limit or adjustment control, set the control well above the maximum output current of the supply. In many supply circuits, the initial regulating action can cause a drop in output voltage, increased ripple, and other performance changes that could make a good supply appear bad.

Pickup and ground-loop effects Always check test-connection setups for possible pickup and/or ground-loop problems (covered further in this chapter). As a simple test, turn off the supply and observe the scope for any unwanted signals (particularly at the line frequency) with the scope leads connected directly on the supply output terminals. Then, connect both scope leads to either terminal (+ or −), whichever is grounded to the chassis or to the common ground. If there is any noise in either test condition, with the supply off, you have possible pickup and/or ground-loop effects.

Ac voltmeter connections Connect the ac voltmeter as close as possible to the input ac terminals of the supply. The voltage indication is then a valid measurement of the supply input, without any error introduced by the drop present in the leads that connect the supply input to the ac line.

Line regulator Do not use any form of line regulator when testing a supply and when using the supply (unless specifically recommended for a particular supply). This is especially true for switching supplies and regulators. A line regulator can change the shape of the output waveform in a switching supply/regulator, thus offsetting any improvement produced by a constant line input to the supply.

Source effect or line regulation

No matter what the test is called, the measurement is made by turning the variable autotransformer throughout the specified range from low-line to high-line, and noting the change in voltage at the supply output terminals. The test is performed with all other test conditions constant. The supply should stay within specifications for any rated output voltage, combined with any rated output current. The extreme source-effect test is with maximum output voltage and maximum output current.

Load effect or load regulation

No matter what the test is called, the measurement is made by closing and opening switch S1 (of Fig. 4-B), and noting the resulting static change in output voltage. The test is performed with all other test conditions constant. The supply should stay within specifications for any rated output voltage, combined with any rated input line voltage. The extreme load-effect test is with maximum output voltage and maximum output current.

Noise and ripple (or PARD)

In many cases, *PARD (periodic and random deviation)* has replaced the terms *noise and ripple*, and represents deviation of the dc output voltage from the average value, over a specified bandwidth, with all other test conditions constant. For example, with Hewlett-Packard lab supplies, PARD is measured in RMS and/or peak-to-peak values over a 20-Hz to 20-MHz bandwidth. Fluctuations below 20 Hz are considered to be drift. Peak-to-peak measurements are of particular importance for applications where noise spikes can be detrimental (such as in digital logic circuits, chapter 6). The RMS measurement is not ideal for noise because output noise spikes of short duration can be present in ripple, but not appreciably increase the RMS value. Always use twisted-pair leads (for single-ended scopes) or shielded two-wire leads (for differential scopes) when making PARD or noise/ripple tests.

Drift (stability)

Drift measurements are made by monitoring the supply output on a differential or digital voltmeter over a stated measurement interval (typically 8 hours, after a 30-minute warmup). In some cases, a strip chart is used to provide a permanent record. A thermometer is placed near the supply to verify that the ambient temperature remains constant during the period of measurement. The supply should be at a location immune from stray air currents (away from open doors or windows and from air- conditioning vents). If practical, place the supply in an oven and hold the temperature constant. A well-regulated supply will drift less during the 8-hour period than during the 30-minute warmup.

Temperature coefficient

Temperature-coefficient measurements are made by placing the supply in an oven and varying the temperature over a given range, following a 30-minute warmup. The supply is allowed to stabilize at each measurement temperature. In the absence of other specifications, the temperature coefficient is the output-voltage change that results from a 5°C change in temperature. The measuring instrument should be placed outside the oven, and must have a long-term stability that is adequate to ensure that any voltmeter drift does not affect measurement accuracy.

Power-supply/regulator troubleshooting

The remainder of the introduction to this chapter is devoted to troubleshooting of power-supply/regulator circuits. In general, most supply problems are the result of wiring mistakes (which you never make), defective (or inadequate) components, and possibly test errors, all of which can be located by basic voltage checks, resistance checks, and point-to-point wiring checks. However, switch-mode supplies and switching regulators present particular problems—especially when an experimental circuit is first tested. The following notes describe some of the most common troubleshooting problems for supplies and regulators.

Supply/regulator test-measurement techniques

The following notes apply specifically when connecting test instruments to an experimental supply or regulator circuit.

Ground loops Figure 4-C shows a typical ground-loop condition. A generator is driving a 5-V signal into 50 Ω on the experimental circuit, which results in 100-mA current. The return path for this current divides between the ground from the signal generator (typically the shield on a BNC cable) and the secondary ground "loop" that is created by the scope probe ground clip (shield), and the two "third-wire" connections on the generator and scope. In this case, assume that 20 mA flows in the parasitic ground loop. If the scope ground lead has a resistance of 0.2 Ω, the scope will show a 4-mV "bogus" signal. The problem gets much worse for higher currents, and for fast-signal edges, where the inductance of the scope probe shield is important. The most practical solution is to use an isolation transformer for the scope. As a quick check, touch the scope probe tip to the probe ground clip, with the clip connected to the experimental-circuit ground. The scope should show a flatline. Any signal displayed on the scope is a ground-loop problem or pickup problem.

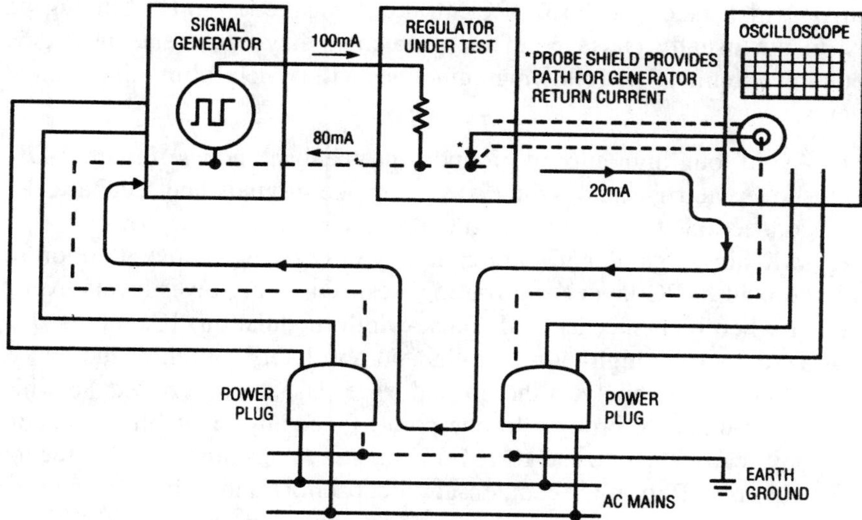

Fig. 4-C Ground-loop errors Linear Technology Corporation, 1991 p. AN44-44.

Scope probe compensation Always check that the scope probe is properly compensated when testing switching supply/regulators. It is especially important for the ac attenuation (on a 10X probe, for example) to match the dc attenuation exactly. If not, low-frequency signals will be distorted and high-frequency signals will have the wrong amplitude. Remember that at typical switch-mode frequencies, the waveshape might look good because the probe acts purely capacitive, so the wrong amplitude might not be immediately obvious.

Ground-clip pickup Do not make any test measurements on a switching regulator with a standard (alligator) ground-clip lead. Replace the alligator clip with a special soldered-in probe terminator, which can be obtained from many probe manufacturers. The standard alligator ground-clip lead can act as an "antenna", and can pick up magnetic and other radiated signals. Make the test described for ground loops if you suspect pickup by the scope probe.

Measure at the component Make all measurements (output voltage, ripple, etc.) at the component, not at a wire that is connected to a component. This is because wires are not shorts. For example, a switching regulator delivering square waves to an output capacitor can generate about 2-V per inch "spikes" in the lead inductance of the capacitor. The further you measure from the capacitor, the greater the spike voltage.

EMI suppression

EMI (electromagnetic interference) is a fact of life with switching regulators. EMI takes two basic forms: *conducted*, which travels down input and output wiring, and *radiated,* which takes the form of electric and magnetic fields. Although these fields do not usually cause regulator problems, they can create problems for surrounding circuits. The following guidelines will be helpful in minimizing EMI problems.

1. Avoid long high-current grounds and feedback nodes. Figure 4-D shows the right and wrong ways to make grounds and feedback connections to switching regulators. Even though low-power switching-regulator ICs are generally easy to use, some attention must be paid to PC layout and routing—especially at power levels over 1 W or when high-speed PWM (pulse-width modulation) ICs are used. Trace out the high-current paths and absolutely minimize their lengths—especially in the ground trace. Use a *star ground*, in which all grounds are brought to one point. Place any input filter capacitor physically close to the IC. Minimize stray capacitance at the feedback (FB) pin. Return all compensation capacitors and bypass capacitors to quiet, well-filtered points (such as an analog ground pin).
2. Use inductors or transformers with good EMI characteristics, such as toroids or pot cores. Avoid rod inductors. If you must use rod

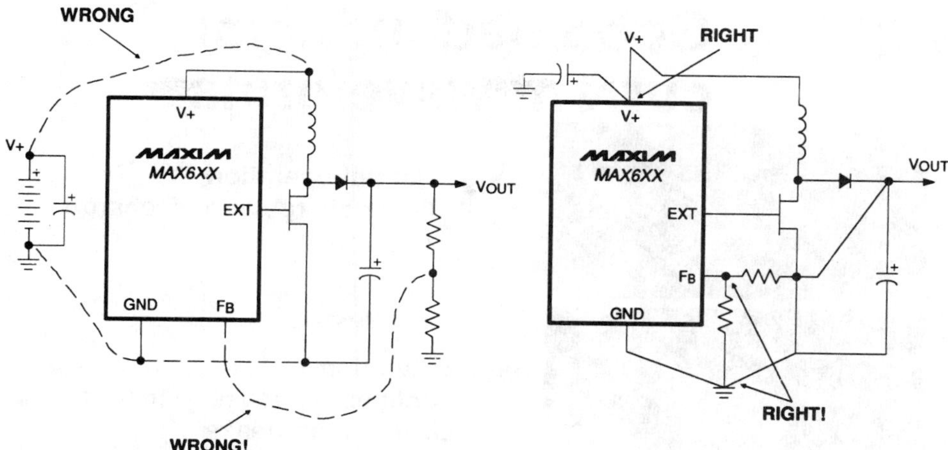

Fig. 4-D Right and wrong ground and feedback connections for switching regulators Maxim Applications and Product Highlights, 1992, p. 4-13.

inductors, keep them in the output filter, where ripple current is low (hopefully). Use the indicator values that are shown in the circuit descriptions. Figure 4-E shows some typical current waveforms that are produced by good and bad inductors. In general, most inductor problems (other than using the incorrect size) can be traced to inadequate saturation (peak current) ratings or excessive dc resistance. If an inductor saturates, its current rises exponentially with time. If there is excessive resistance, a distinct LR characteristic is seen. If the waveform takes small, but strange, bends, the inductor might be producing both effects.

3. Route all traces carrying high ripple current over a ground plane to minimize radiated fields. This includes the catch-diode leads, input and output capacitor leads, snubber leads, inductor leads, IC input and switch pin leads, and input power leads. Keep these leads short and keep the components close to the ground plane.
4. Keep sensitive low-level circuits as far away as possible, and use field-canceling tricks, such as twisted-pair differential lines.
5. In very critical applications, add a "spike killer" bead on the catch diode to suppress high harmonics. This can create higher transient switch voltages at switch turn-off, so check switch waveforms carefully.
6. Add an input filter if radiation from input lines could be a problem. Just a few μH in the input line will allow the regulator input capacitor to swallow nearly all the ripple current that is created at the regulator input.

Good/bad inductor current waveforms

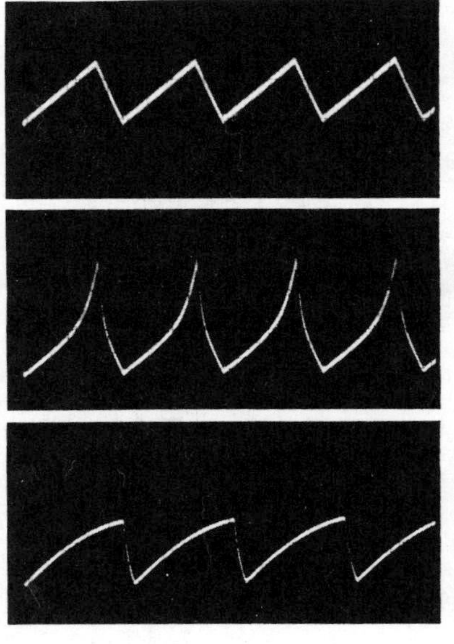

**Good: Normal operation
linear charge and discharge slopes**

**Bad: Saturation—
nonlinear increase in inductor current near peaks**

Bad: Excessive resistance—
1. **High winding resistance**
2. **High transistor RON**
3. **High source resistance**

Fig. 4-E Typical current waveforms produced by good and bad inductors Maximum Applications and Product Highlights, 1992, p. 4-14.

Troubleshooting hints

The following notes apply specifically when troubleshooting an inoperative or poorly performing switch-mode supply or switching regulator, particularly those in experimental form.

If the circuit is totally inoperative, look for such things as transformers wired backwards (always check the polarity dots on transformers), electrolytic capacitors wired backwards (usually you will find this out shortly after power is applied), and IC pins reversed (check the datasheet and follow the wiring that is given on the circuit schematics).

If the input voltage appears to dip (especially at the switching frequency), it is possible that the input leads (from the battery to the switching IC, for example) are too long when connected in experimental form. Switching regulators draw current from the input supply in pulses. Long input wires can cause dips in the input voltage at the switching frequency. Even though the circuit schematic does not show an input capacitor, add a 100-μF or larger input capacitor close to the regulator during the experimental stage.

If the input supply simply will not come up, and switching will not start (with the right components all properly connected), it is possible that the input supply cannot deliver the necessary start-up current. Switching regulators have negative input resistance at start-up, and draw high current. This can latch some input supplies to a low or off condition.

If the supply is operating, but efficiency is low (much more power going into the supply than power coming out), suspect the inductors (or transformers). Core or copper losses might be the problem. Of course, the problem could be an accumulation of all losses (inductor, capacitor, diode, etc.), which results in an inefficient supply/regulator circuit.

If the switch timing varies, check for excessive ripple at the output, and at the V_C and FB pins of the IC. For example, if you monitor the circuit waveforms (such as shown in Fig. 4-12B) and find that switch on time is alternating from cycle to cycle, it is possible the problem is one of excessive ripple. Try connecting a capacitor (about 1000 to 3000 pF) from the V_C and FB pins to ground, and paralleling the output capacitor with a capacitor of the same value. If any of these capacitors eliminate the variation in switch timing, you have located the problem area.

If the IC blows up (with the right components, all properly connected), it is possible that start-up surges are causing momentary large switch voltages. This problem should not occur if you use the circuit values that are shown in the circuit schematics of this chapter.

If the IC runs hot, it is possible that you need a heatsink. For example, a TO-220 package has a thermal resistance of 50°C/W with no heatsink. A 5-V, 3-A output (15 W) with 10% switch loss, will dissipate over 1.5 W in the IC. This means a 75°C temperature rise, or 100°C case temperature at 25°C room ambient (which is hot!). Simply soldering the TO-220 tab to an enlarged copper pad on the PC board will reduce thermal resistance to about 25°C/W.

If you have high output ripple or noise spikes, suspect the output capacitor.

If you have poor load or line regulation, check in the following order: (1) the secondary output filter dc resistance (if it is outside the feedback loop), (2) ground-loop error in the scope, (3) improper connection of output divider resistors to current-carrying lines, (4) excess output ripple.

Power-supply circuits titles and descriptions

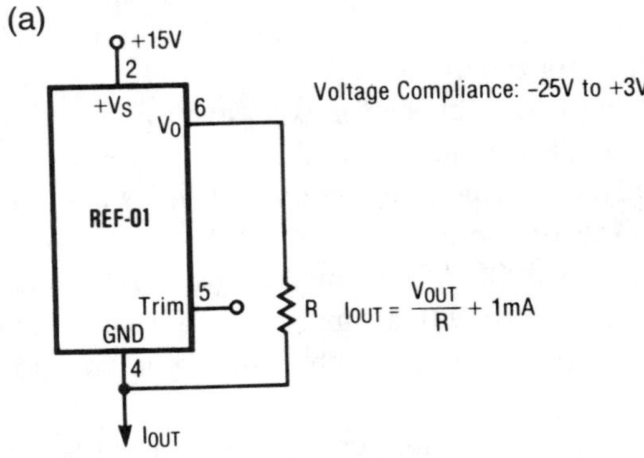

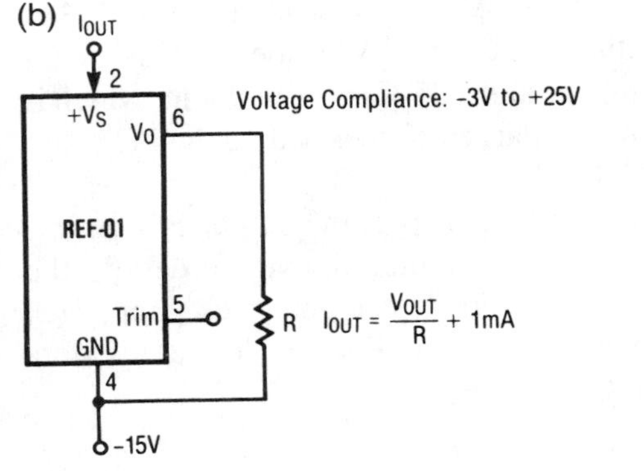

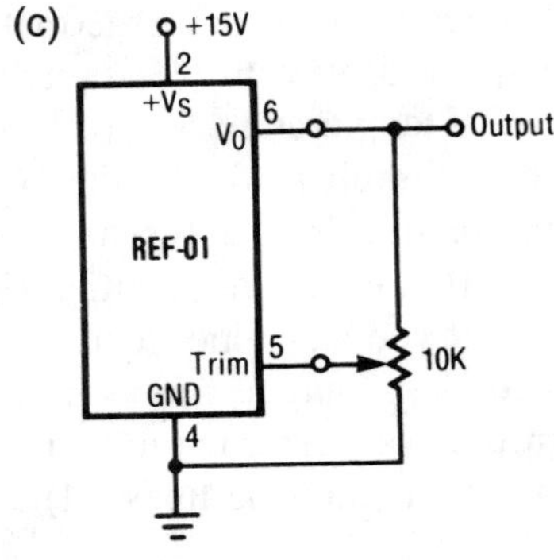

Voltage reference

Fig. 4-1 Figures 4-1A and 4-1B show an REF-01 connected as a current source, and current sink, respectively. The REF-01 trim terminal can be used to adjust the output voltage over a 10-V ±300-mV range (Fig. 4-1C). This permits the output to be set at exactly 10.000 V or to 10.240 V for binary operation. Raytheon Linear Integrated Circuits, 1989, p. 8-6, 8-7).

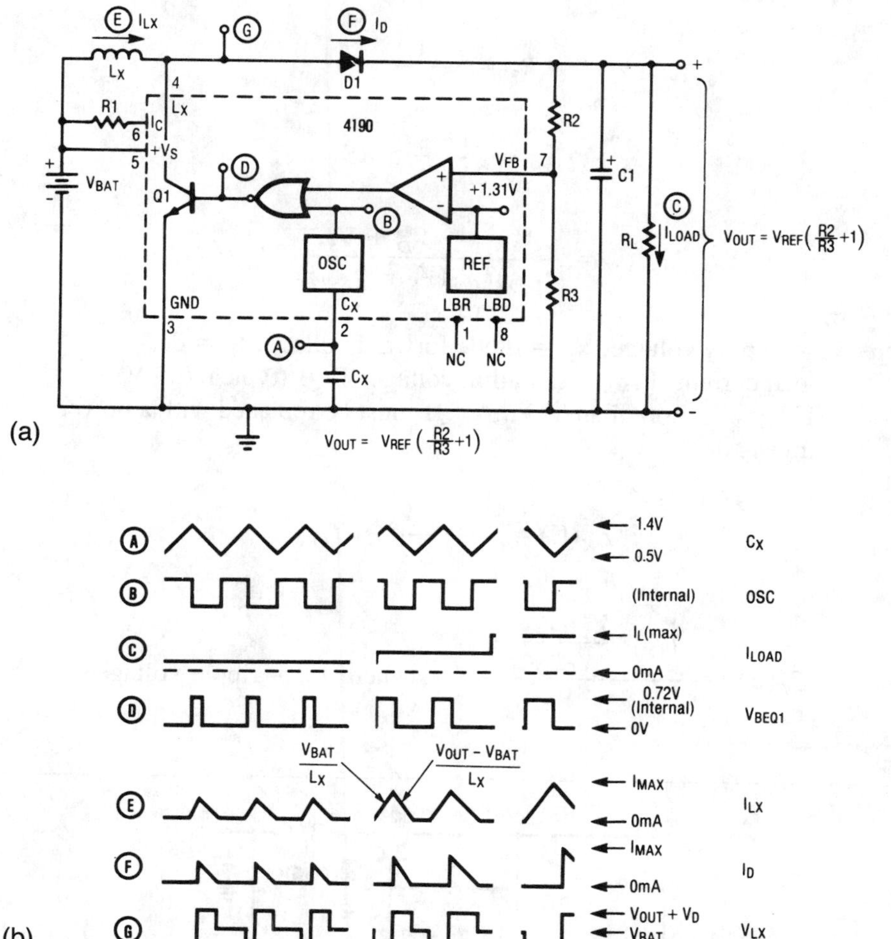

Basic step-up voltage regulator

Fig. 4-2 Figures 4-2A and 4-2B show a basic step-up voltage regulator, and waveforms, respectively. Component values are tailored to circuit requirements, as follows:

$$C_X = \frac{2.4 \times 10^{-6}}{F_O(\text{Hz})} = \text{timing capacitor}$$

$$R_1 = \frac{V_S - 1.2\ \text{V}}{5\ \mu\text{A}}$$

$$R_2 = \frac{V_{\text{OUT}} - 1.31\ \text{V}}{I_A}$$

where: I_A = feedback divider current (typically 50 to 100 μA)

$$R_3 = \frac{1.31\ \text{V}}{I_A}$$

$$T_{ON} = \frac{1}{2F_O} + 5\ \mu\text{S}$$

$$I_{MAX} = \frac{V_{OUT} + V_D + V_S}{(F_O)T_{ON}(V_S - V_{SW})}\ 2I_L$$

where: V_s = supply voltage, V_D = diode forward voltage, I_L = dc load current, V_{SW} = sauration voltage of Q1 (typical 0.5 V), If I_{MAX} is more than 375 mA, Q1 must be replaced with a power transistor.

$$L_X(\text{H}) = \frac{V_S - V_{SW}}{I_{MAX}}\ T_{ON}$$

$$C_F(\mu\text{F}) = \frac{T_{ON}\ \dfrac{V_S I_{MAX}}{V_{OUT}} + I_L}{V_R}, \text{ where } V_R = \text{ripple voltage}$$

Raytheon Linear Integrated Circuits, 1989, p. 9-8.

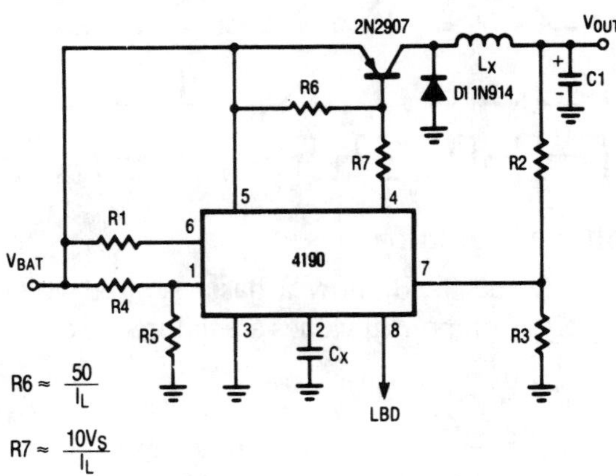

Basic step-down voltage regulator

Fig. 4-3 Figure 4-3 shows a basic step-down voltage regulator, where loads are from 500 mW to 2 W. Component values are tailored to circuit requirements, as described for Fig. 4-2, except as:

$$I_{\text{MAX}} = \frac{2I_L}{(F_O)(T_{ON})\left(\dfrac{V_S - V_{\text{OUT}}}{V_{\text{OUT}} - V_D} + 1\right)}$$

$$L_X = \left(\frac{V_S - V_{\text{OUT}}}{I_{\text{MAX}}}\right) T_{\text{ON}}$$

$$C_F\ (\mu\text{F}) = \frac{T_{\text{ON}}\left(\dfrac{V_S - V_{\text{OUT}}\ I_{\text{MAX}}}{V_{\text{OUT}}} + I_L\right)}{V_R}$$

Raytheon Linear Integrated Circuits, 1989, p. 9-10.

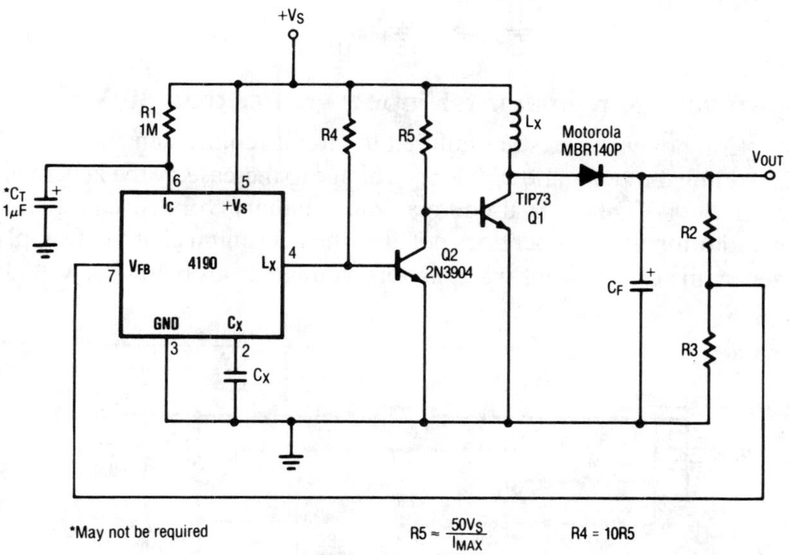

High-power step-up voltage regulator

Fig. 4-4 Figure 4-4 shows a step-up regulator for loads up to 10 W. Component values are tailored to circuit requirements, as described for Fig. 4-2, except for R4 and R5, as shown. Raytheon Linear Integrated Circuits, 1989, p. 9-9.

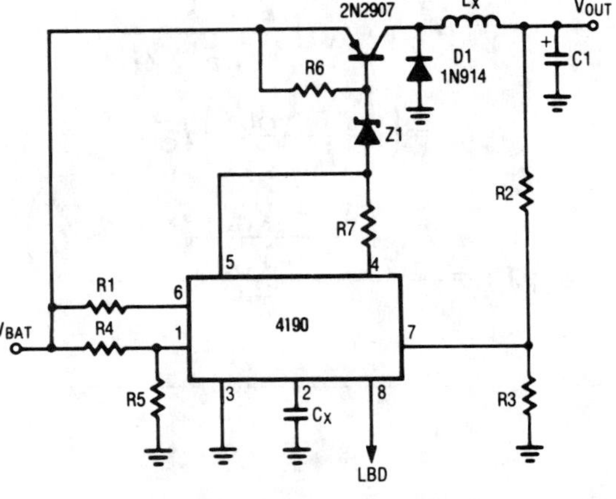

Step-down voltage regulator for inputs greater than 30 V

Fig. 4-5 Component values are tailored to circuit requirements as described for Fig. 4-2. Adding the zener allows battery voltage to increase by the zener value. For example, if a 24-V zener is used, maximum battery voltage can go to 48 V. However, addition of the zener does not alter the maximum charge of supply. With a 24-V zener, the circuit stops when battery voltage drops below 24 V + 2.2 V = 26.2 V. Raytheon Linear Integrated Circuits, 1989, p. 9-10.

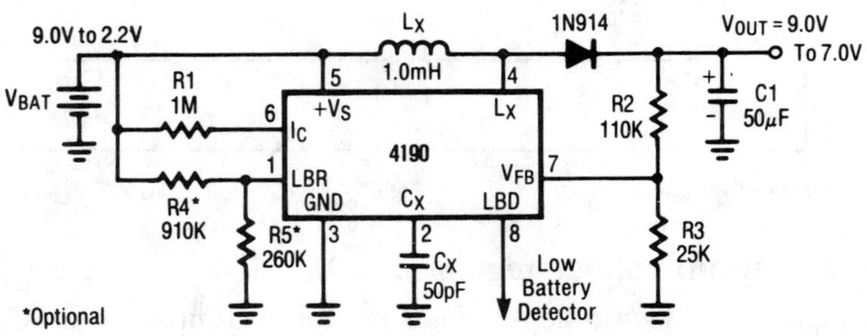

Battery-life extender

Fig. 4-6 This circuit extends the lifetime of a 9-V battery. The regulator remains in a quiescent state (drawing only 215 μA) until the battery voltage decays below 7.5 V, at which time the circuit starts to switch and regulate the output at 7.0 V until the battery falls below 2.2 V. If this circuit is operated at a typical 80% efficiency with an output current of 10 mA, at 5.0-V battery voltage, the average input current is 17.5 mA. Raytheon Linear Integrated Circuits, 1989, p. 9-14.

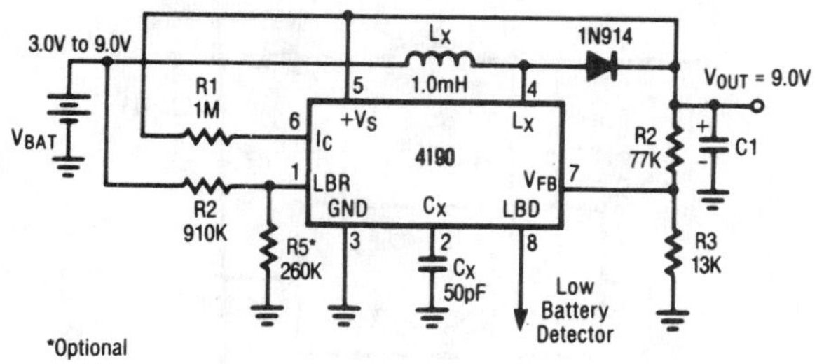

Bootstrap voltage regulator

Fig. 4-7 In this circuit, power to the IC is taken from the output voltage by connecting the $+V_s$ pin and the top of R1 to the output voltage. Notice that the initial battery voltage must be greater than 3.0 V when the circuit is energized. If not, there will not be enough voltage at pin 5 to start the IC. The big advantage of this circuit is the ability to operate down to a discharged battery voltage of 1.0 V. The value of C_1 is determined, as described for Fig. 4-2. Raytheon Linear Integrated Circuits, 1989, p. 9-15.

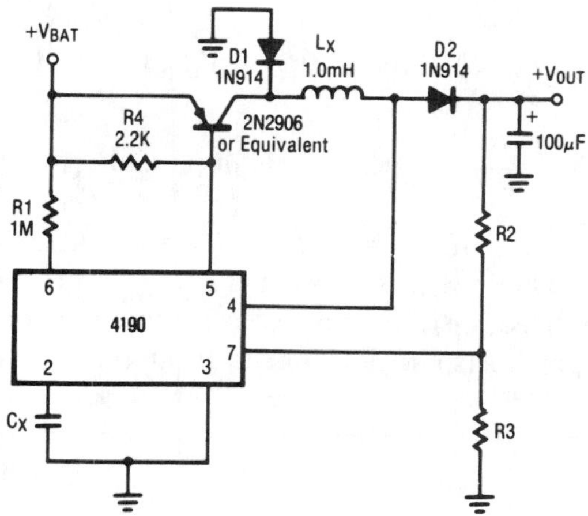

Buck-boost voltage regulator

Fig. 4-8 A disadvantage of the standard step-up and step-down circuits is the limitation of the input voltage range. For a step-up circuit (Fig. 4-2), the battery voltage must always be less than the programmed output voltage, and for a step-down circuit (Fig. 4-3), the battery voltage must always be greater than the output voltage. Figure 4-8 eliminates this disadvantage, and allows a battery voltage above the programmed output voltage to decay to well below the output voltage. The values of R_2 and R_3 are determined, as described for Fig. 4-2. Raytheon Linear Integrated Circuits, 1989, p. 9-15.

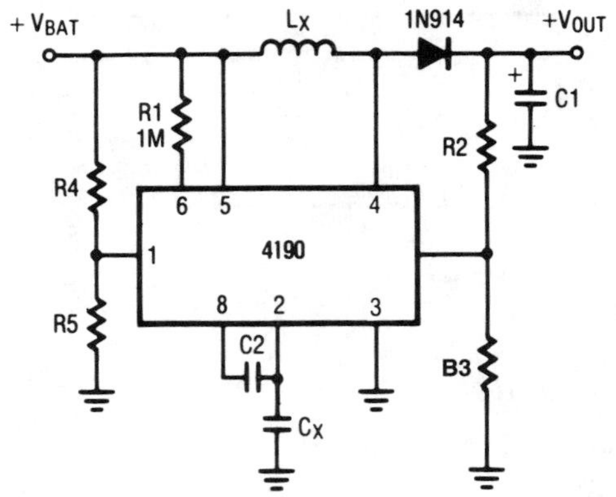

Step-up voltage regulator with voltage-dependent oscillator

Fig. 4-9 The circuit of Fig. 4-9 offers a compromise between load-current capability and output ripple (increased load current typically causes increased ripple). Component values are tailored to circuit requirements, as described for Fig. 4-2, except as follows:

$$\text{Threshold voltage, } V_{TH} = V_{REF}\left(\frac{R_4}{R_5} + 1\right)$$

$$F_O = \frac{2.4 \times 10^{-6}}{C_X + C_2}, \text{ where } C \text{ is in pF and } F_O \text{ is in Hz.}$$

Typical component values are: $R_2 = 330\ \text{k}\Omega$, $R_S = 150\ \text{k}\Omega$, $C_X = 100$ pF, $C_2 = 100$ pF. With these values, V_{TH} is 4-1 V and $F_O = 24$ kHz. If the IC oscillator appears to be destabilized, or if there is excessive low-frequency ripple, look for stray capacitance at pin 7. Also, try a 100-pF to 10-nF capacitor in parallel with R2.

Raytheon Linear Integrated Circuits, 1989, p. 9-16.

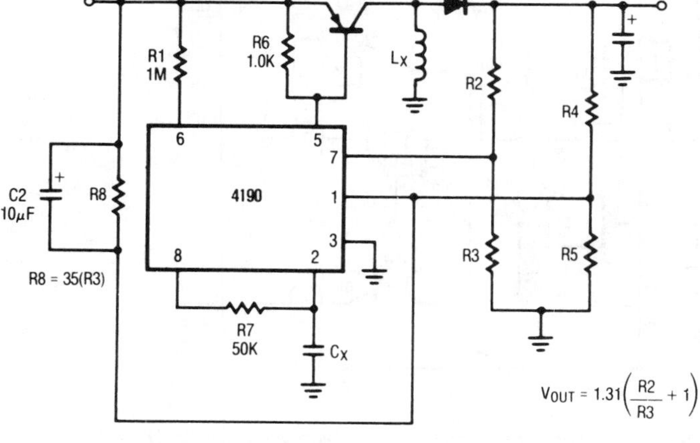

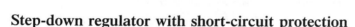

Step-down regulator with short-circuit protection

Fig. 4-10 With this circuit, the low-battery detector (LBD) is connected to sense the output voltage, and shuts off the oscillator by forcing pin 2 low if the output voltage drops. Component values are tailored to circuit requirements, as described in Fig. 4-2, except: choose resistor values so that $R_5 = R_3$ and $R_4 = R_2$, and make R_8 25 to 35 times higher than R_3. Raytheon Linear Integrated Circuits, 1989, p. 9-17.

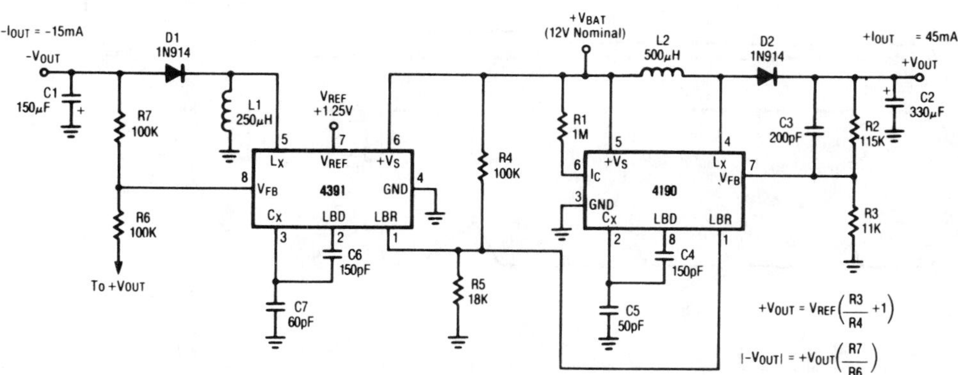

Positive/negative dual-tracking power supply

Fig. 4-11 This circuit uses the 4190 as a step-up regulator and the 4391 as an inverter. The supply is capable of delivering +45 mA (15 mA with regulation) until the battery decays below 5.0 V. Output voltage ripple is under 100 mVpp at ±15-V output. Raytheon Linear Integrated Circuits, 1989, p. 9-18.

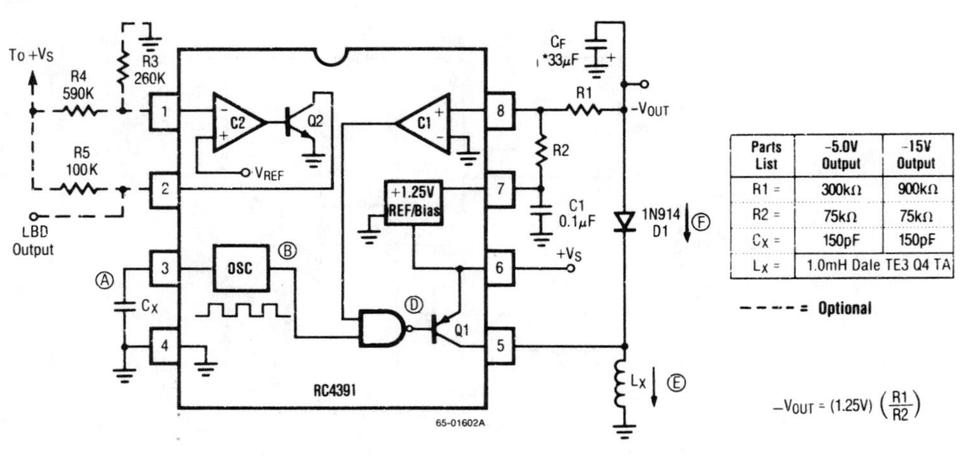

Parts List	-5.0V Output	-15V Output
R1 =	300kΩ	900kΩ
R2 =	75kΩ	75kΩ
C_X =	150pF	150pF
L_X =	1.0mH Dale TE3 Q4 TA	

*Caution: Use current limiting protection circuit for high values of C_F (Fig. 13)

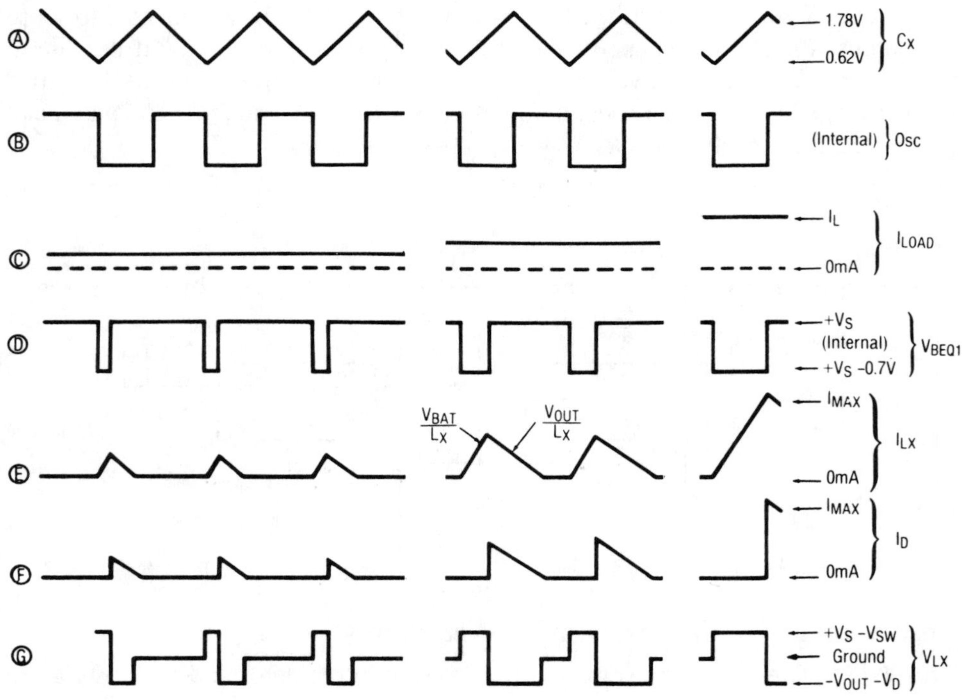

Continued

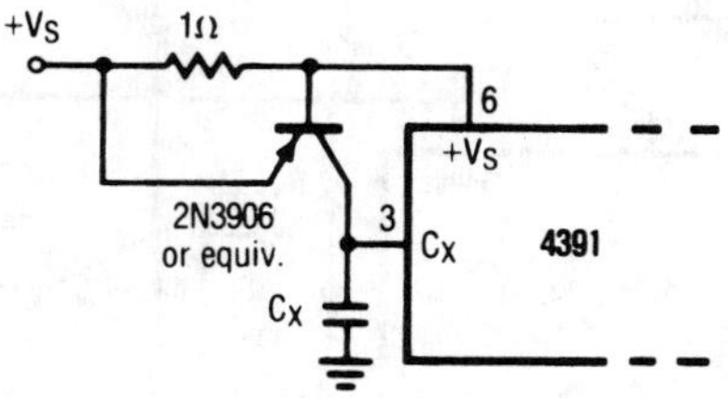

Basic inverting voltage regulator

Fig. 4-12 Figures 4-12A and 4-12B show a basic inverting voltage regulator and waveforms, respectively. The outputs are −5 or −15 V, using the values shown. Other outputs can be selected by changing R1 and R2. It may be necessary to change other circuit values. If high values of C_F are used, a current-limiting protection circuit (Fig. 4-12C) might be required. Raytheon Linear Integrated Circuits, 1989, p. 9-54, 9-55, 9-62.

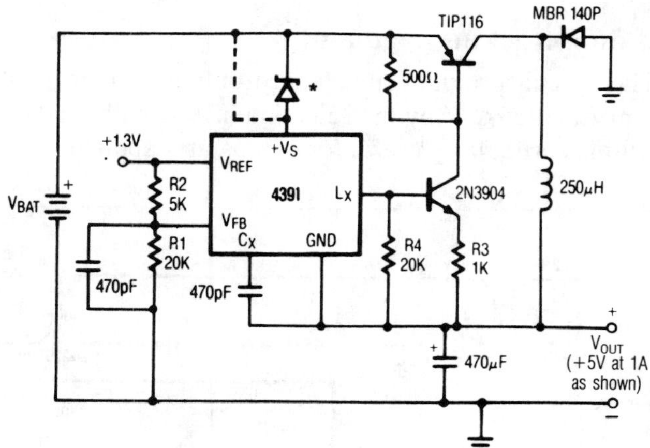

Note: A minimum load ≥ 1mA must be connected.
*Optional — Extends supply voltage range

High-power step-down voltage regulator

Fig. 4-13 This circuit shows a step-down regulator for loads up to 5 W. Notice that a minimum load of at least 1 mA must be connected when the circuit is energized. Raytheon Linear Integrated Circuits, 1989, p. 9-62.

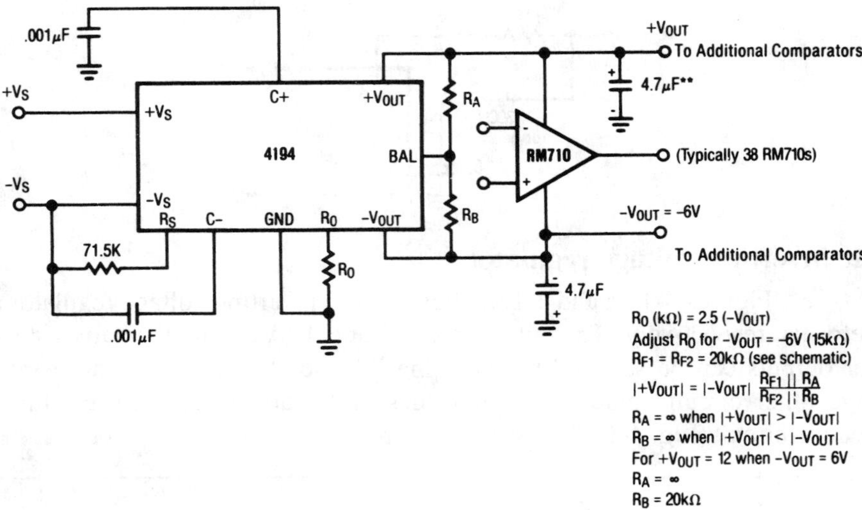

Unbalanced dual-tracking regulator

Fig. 4-14 This circuit provides unbalanced output voltages that can be varied between ±50 mV and ±42 V by the selection of R_O, with load currents of ±200 mV. This circuit is particularly useful for comparator applications. Raytheon Linear Integrated Circuits, 1989, p. 9-68.

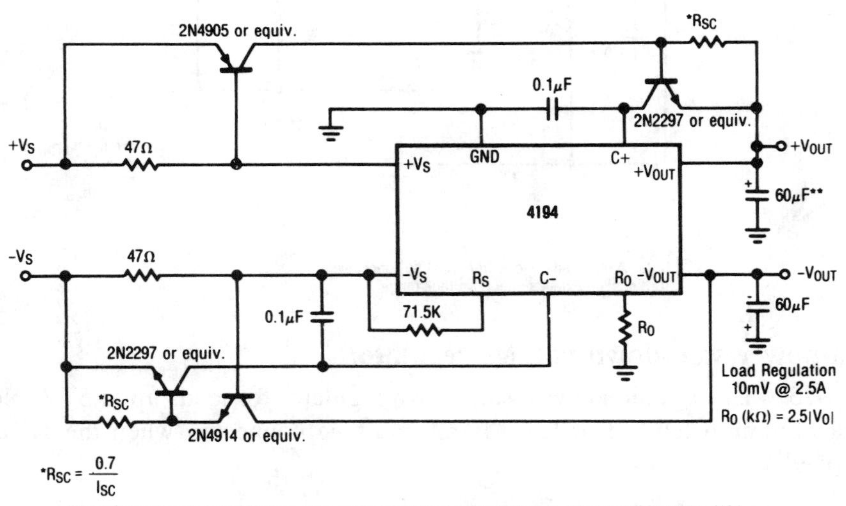

High-output dual-tracking regulator

Fig. 4-15 This circuit provides balanced output voltages with a load regulation of 10 mV at 2.5 A. Raytheon Linear Integrated Circuits, 1989, p. 9-68.

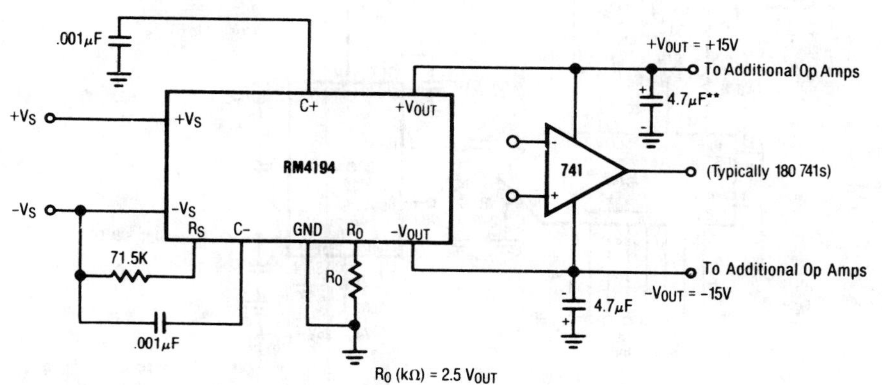

Balanced dual-tracking regulator

Fig. 4-16 This circuit provides balanced output voltages that can be varied between ±50 and ±42 mV by selection of R_O, with load currents of ±200 mV. This circuit is particularly useful for op-amp (chapter 10) applications. Raytheon Linear Integrated Circuits, 1989, p. 9-69.

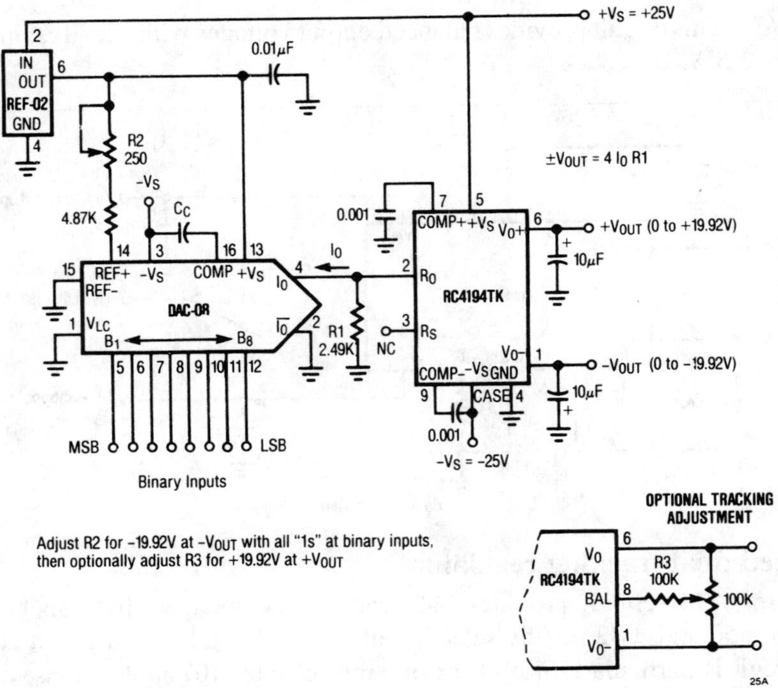

Digitally controlled dual-tracking regulator

Fig. 4-17 This circuit provides balanced output voltages that can be set by binary inputs applied to the DAC. Outputs vary between 0 and ±19.92 V at loads of ±200 mA. Raytheon Linear Integrated Circuits, 1989, p. 9-69.

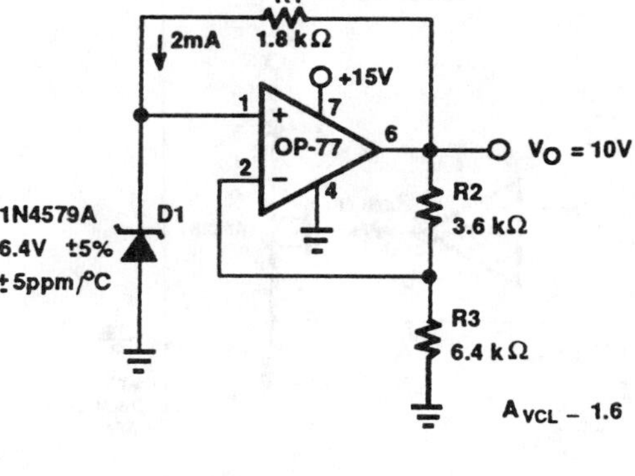

This simple bootstrapped voltage reference provides a precise 10V virtually independent of changes in power supply voltage, ambient temperature and output loading. Correct zener operating current of exactly 2 mA is maintained by R1, a selected 5 ppm/°C resistor, connected to the regulated output. Accuracy is primarily determined by three factors: the 5 ppm/°C temperature coefficient of D1, 1 ppm/°C ratio tracking of R2 and R3, and operational amplifier V_{OS} errors.

V_{OS} errors, amplified by 1.6 (A_{VCL}), appear at the output and can be significant with most monolithic amplifiers. For example: an ordinary amplifier with TCV_{OS} of 5 µV/°C contributes 0.8 ppm/°C of output error while the OP-77, with TCV_{OS} of 0.3 µV/°C, contributes but 0.05 ppm/°C of output error, thus effectively eliminating TCV_{OS} as an error consideration.

High-stability voltage reference

Fig. 4-18 This circuit provides a precision voltage reference without the use of an IC voltage reference. Raytheon Linear Integrated Circuits, 1989, p. 4-117.

Positive Current Sink

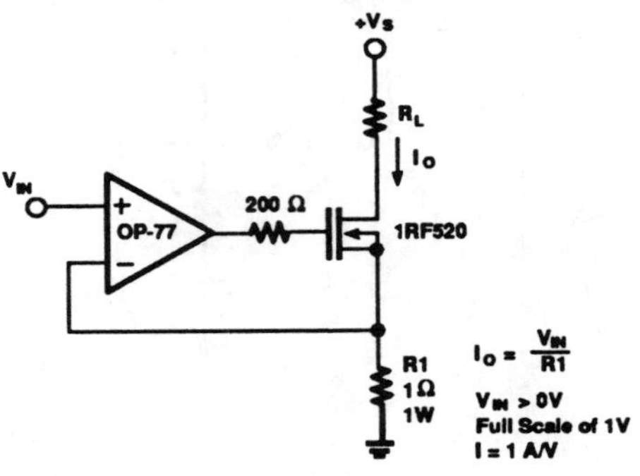

Positive Current Source

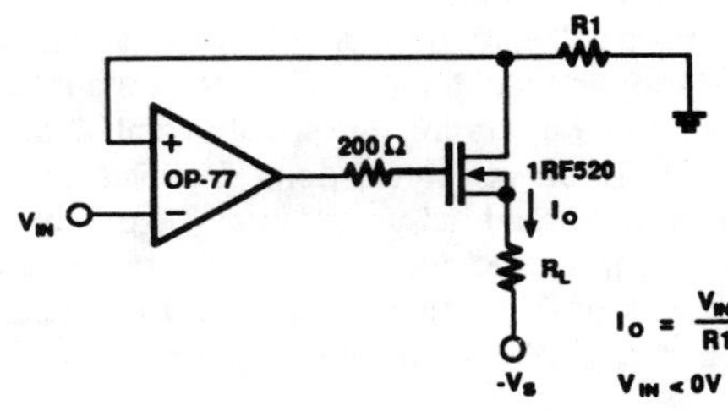

These simple high current sinks require that the load float between the power supply and the sink.

In these circuits, OP-77's high gain, high CMRR, and low TCV_{OS} assure high accuracy.

Precision current sink/source

Fig. 4-19 These circuits provide positive current-sink and current-source capability. Raytheon Linear Integrated Circuits, 1989, p. 4-117.

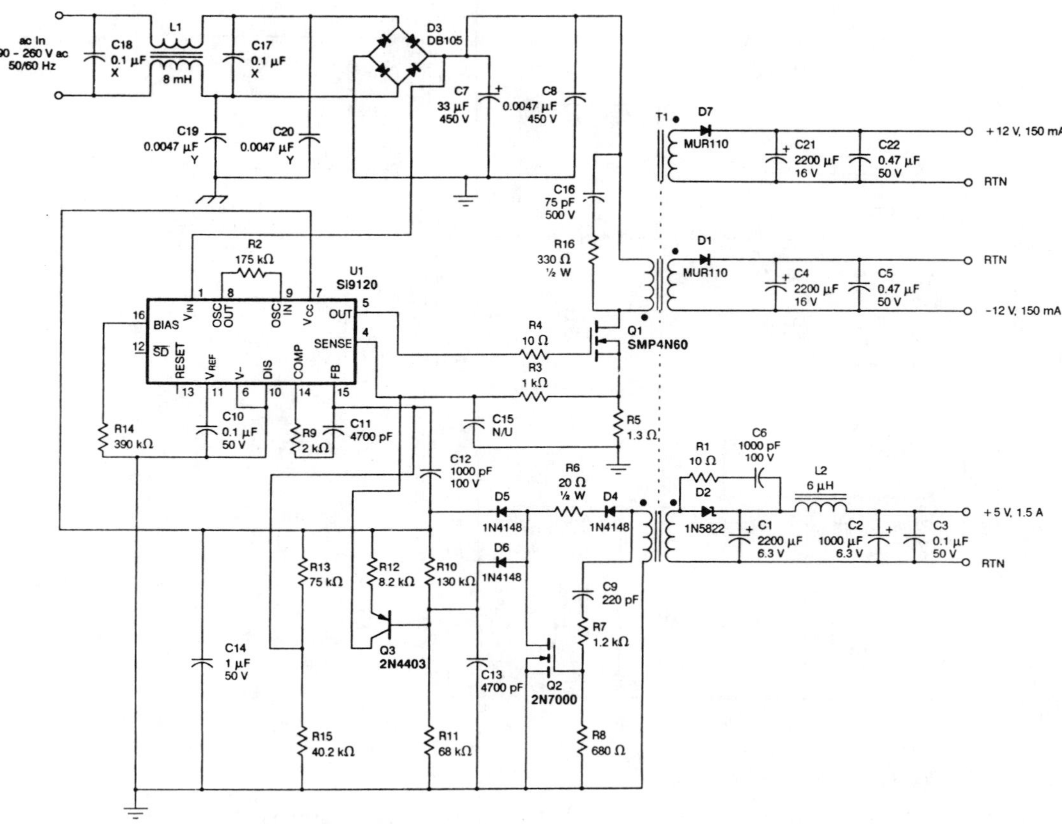

Universal-input power supply

Fig. 4-20 This circuit operates directly from 100-, 110-, and 220-V without the use of selector switches or jumpers (permitting the supply to be plugged into wall outlets anywhere in the world). T1 is a Schott Corp. #67122700, L1 is a Renco 1361- 2 common-mode choke, L2 is any 6-μH 1.5-A inductor, and D3 is a 1-A 600-V bridge. Siliconix Power Products Data Book, 1991, p. 9-2.

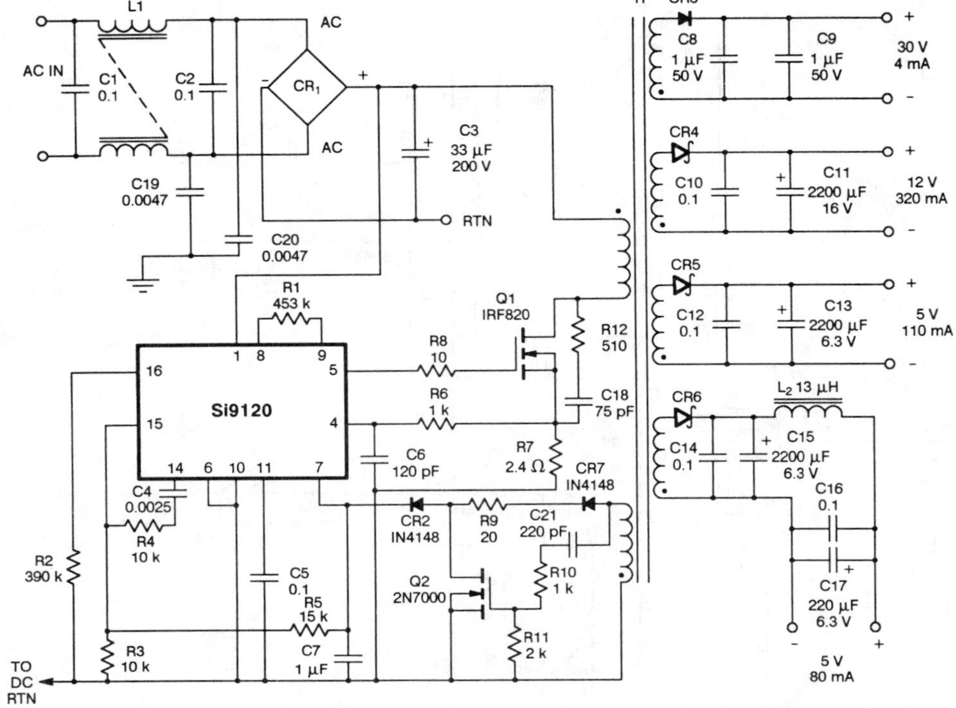

Offline flyback converter

Fig. 4-21 This circuit operates from 90- to 130-V. Regulation is ±5% on all outputs. Ripple is 200 mVp-p on the +30 and +12 outputs; 50 mVp-p on the main +5 output; 15 mVp-p on the auxiliary +5 output. Power consumption is 15 W. T1 is a Schott Corp. #6712244, L1 is any 8 mH common-mode choke, L2 is a Magnetics Inc J40401TC with 6 turns #26 AWG, and CR1 is any 1-A 600-V bridge. Siliconix Power Products Data Book, 1991, p. 9-31.

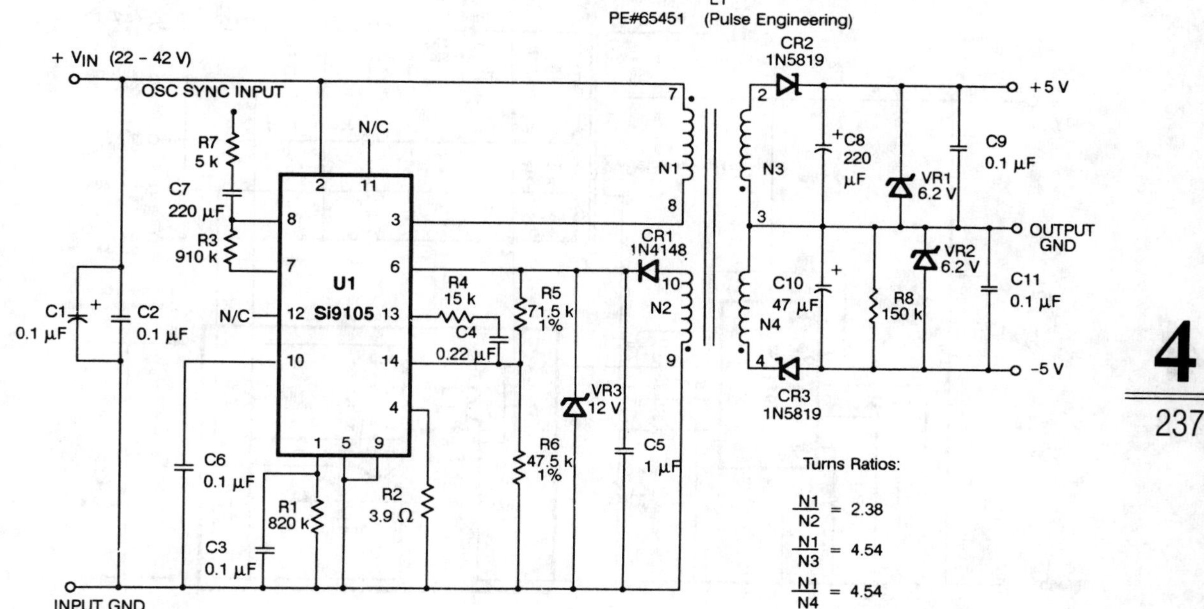

Dc/dc converter

Fig. 4-22 Although this circuit is designed to meet CCITT Standard I.430 for ISDN (Integrated Services Digital Network) terminals, the circuit can be used for any battery-operated ±5-V power-supply application. The input voltage is 22 to 42 V for maximum loads, and 30 to 42 V for minimum loads. Typical minimum loads are 3 mA at +5 V, and 0 mA at −5 V. Typical maximum loads are 100 mA at +5 V and 30 mA at −5 V. Ripple is 100 mV maximum at full load, and 60 mV typical at full load. Regulation is ±5% typical and ±7% worst case. Siliconix Power Products Data Book, 1991, p. 9-56.

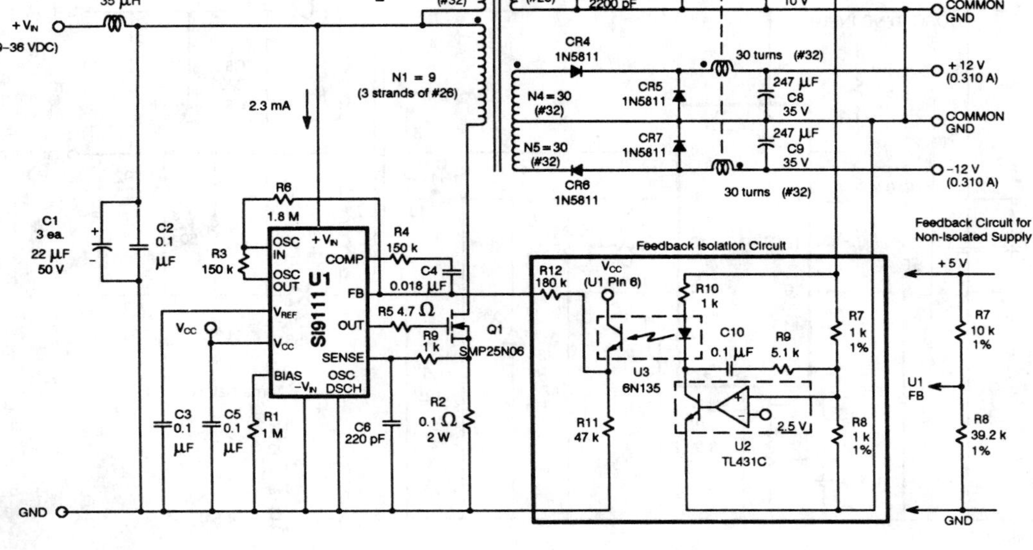

Dc/dc forward converter

Fig. 4-23 The feedback isolation circuit (shown within the box) can be replaced by the simple R7/R8 voltage feedback divider if the supply-output feedback does not need to be isolated from the load. Siliconix Power Products Data Book, 1991, p. 9-66.

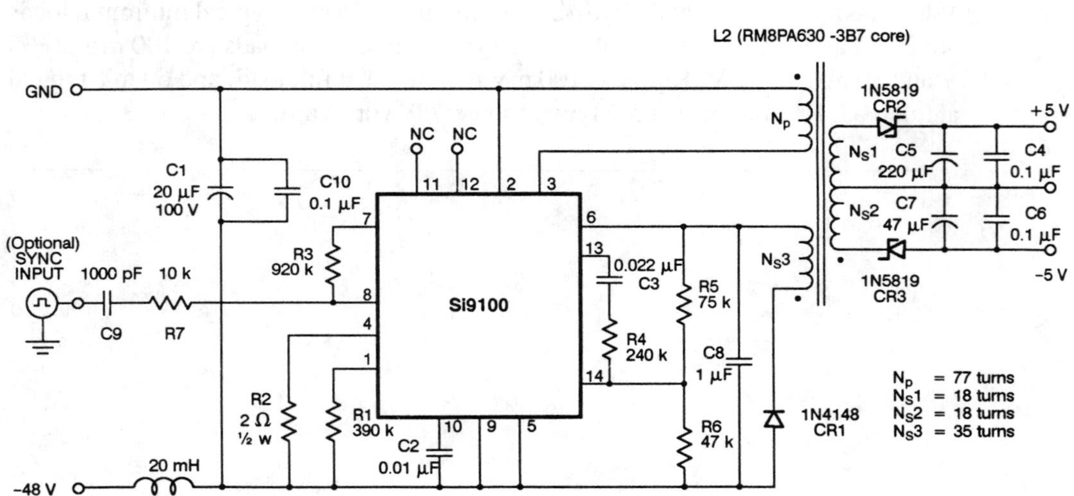

Dc/dc converter

Fig. 4-24 This circuit is similar to that of Fig. 4-22, except that a 3-W Si9100 is used instead of a 1-W Si9105 as the switchmode regulator. Notice that the sync input (at pins 7 and 8) is optional for both circuits. Siliconix Power Products Data Book, 1991, p. 9-82.

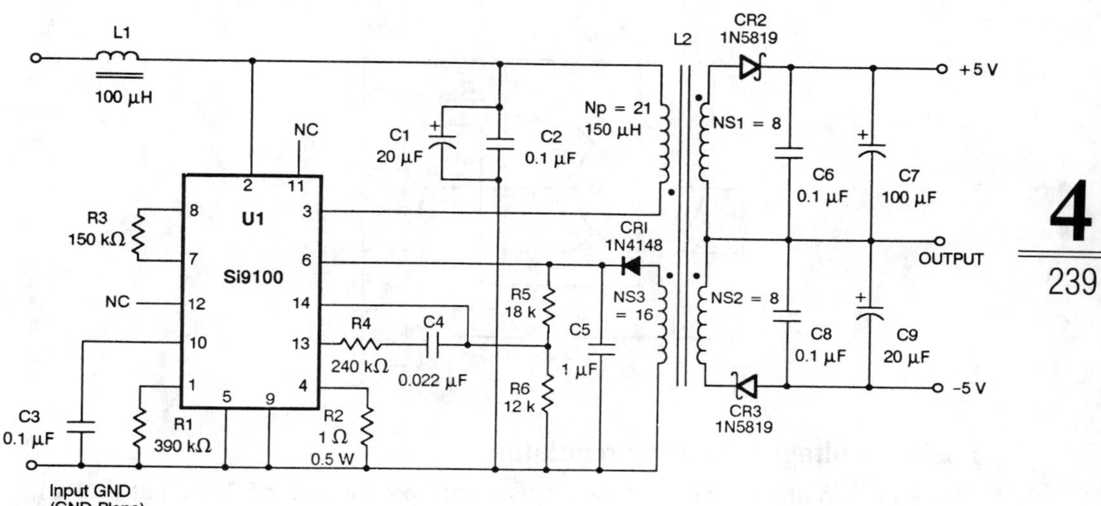

Dc/dc converter (without external sync)

Fig. 4-25 This circuit is similar to that of Figs. 4-22 and 4-24, except that there is no provision for external sync. The input voltage is 15 to 70 V. Typical maximum loads are 167 mA at +5 V, and 33 mA at −5 V. Typical minimum loads are 32 mA at +5 V and 8 mA at −5 V. Maximum ripple is 100 mVpp. Regulation is ±5%. L1 is any 100-µH 75-mA inductor. L2 is GFS Manufacturing #85-787-4. Siliconix Power Products Data Book, 1991, p. 9-85.

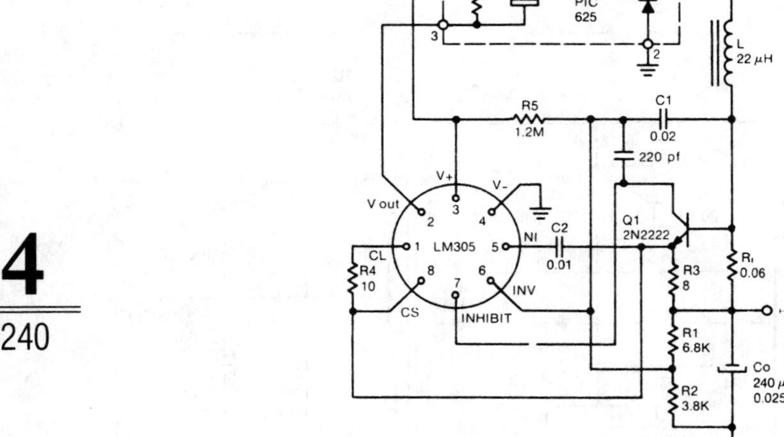

Positive-voltage switching regulator

Fig. 4-26 With the values shown, this circuit produces a ±5-V output (±1%), with a 20- to 40-V input, at currents between 2 and 10 A. Ripple is 2% or 100 mVpp. Unitrode Semiconductor Products DB600, 1990, p. 12-9.

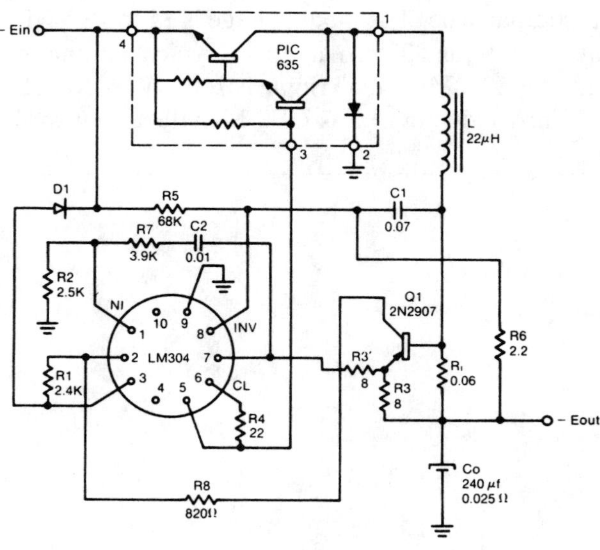

Negative-voltage switching regulator

Fig. 4-27 The input and output characteristics of this circuit are the same as in Fig. 4-26, except that input and output are negative. Unitrode Semiconductor Products DB600, 1990, p. 12-11.

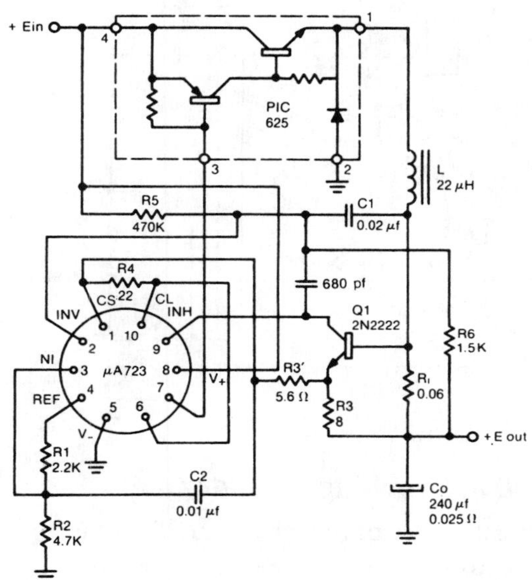

Alternate positive-voltage switching regulator

Fig. 4-28 The input and output characteristics of this circuit are the same as in Fig. 4-26, except that a uA723 is used instead of an LM305. Notice that a 2N2222 is required for Q1. Unitrode Semiconductor Products DB600, 1990, p. 12-11.

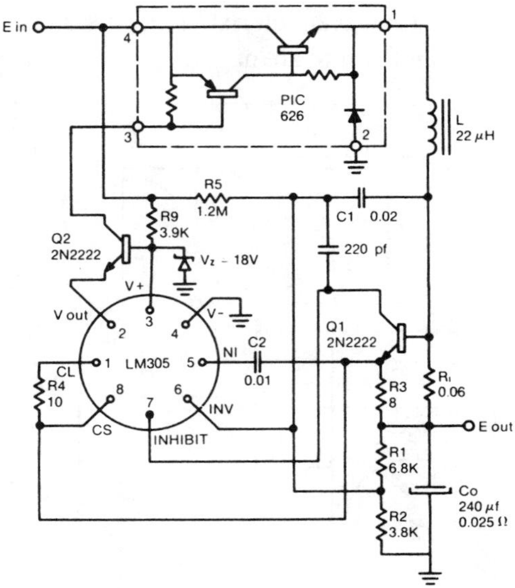

High-voltage positive switching regulator

Fig. 4-29 The characteristics of this circuit are the same as in Fig. 4-26, except that the input can exceed 40 V. Unitrode Semiconductor Products DB600, 1990, p. 12-11.

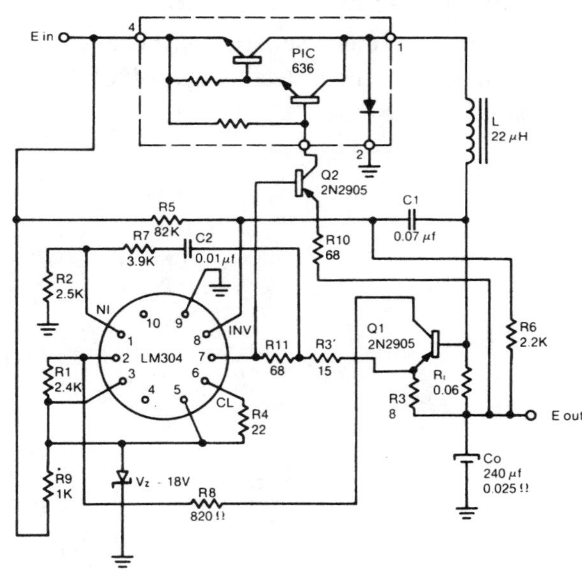

High-voltage negative switching regulator

Fig. 4-30 The characteristics of this circuit are the same as in Fig. 4-26, except that the input can exceed 40 V, and the input/output is negative. Unitrode Semiconductor Products DB600, 1990, p. 12-11.

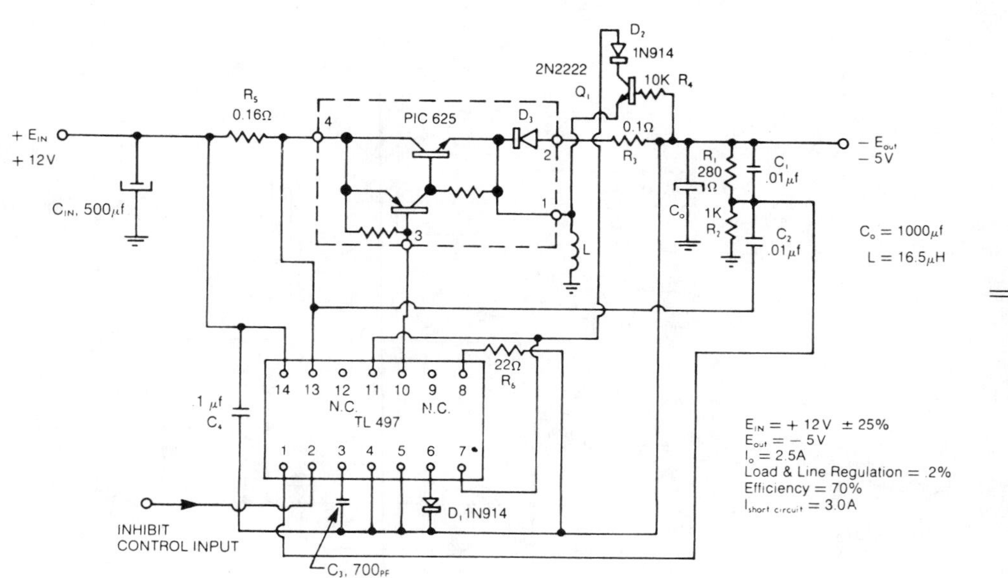

Negative-voltage flyback regulator

Fig. 4-31 This circuit provides a negative output (-5 V) with a positive input (+12 V). Unitrode Semiconductor Products DB600, 1990, p. 12-40.

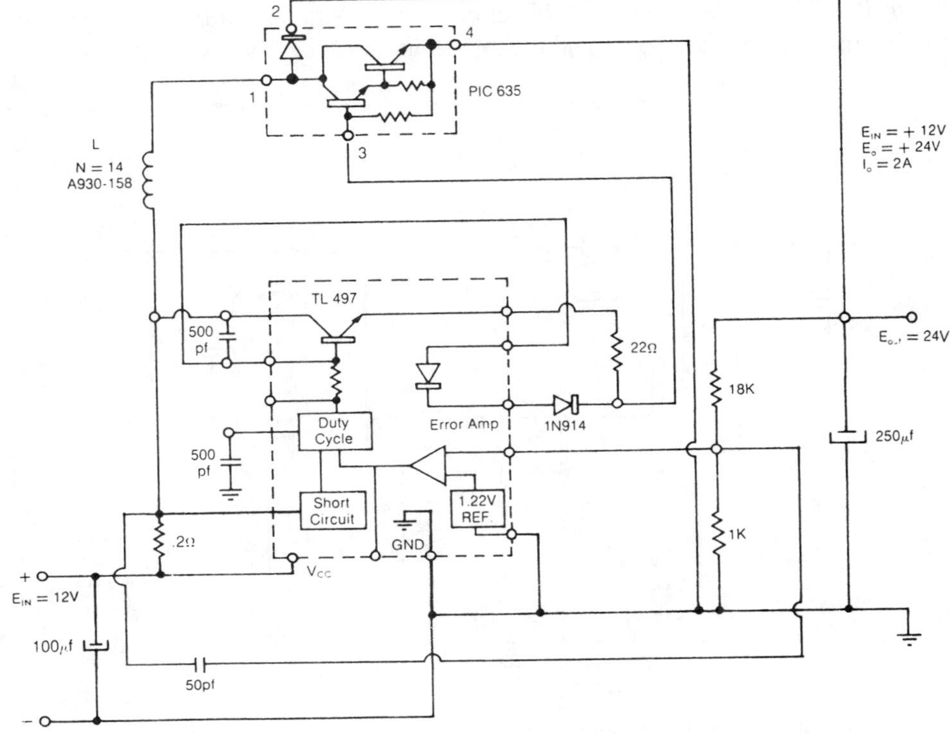

Boost switching regulator

Fig. 4-32 This circuit provides an output (+24 V) that is double the input (+12 V). Unitrode Semiconductor Products DB600, 1990, p. 12-44.

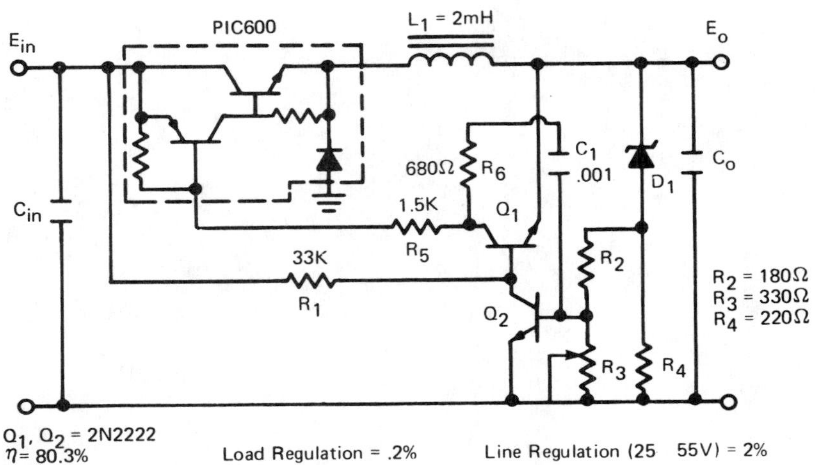

Low-cost buck regulator

Fig. 4-33 This low-cost circuit provides line regulation (at a value set by D1) with a minimum of components. Typical values for C_{in} and C_o are 500 μF. Unitrode Semiconductor Products DB600, 1990, p. 12-64.

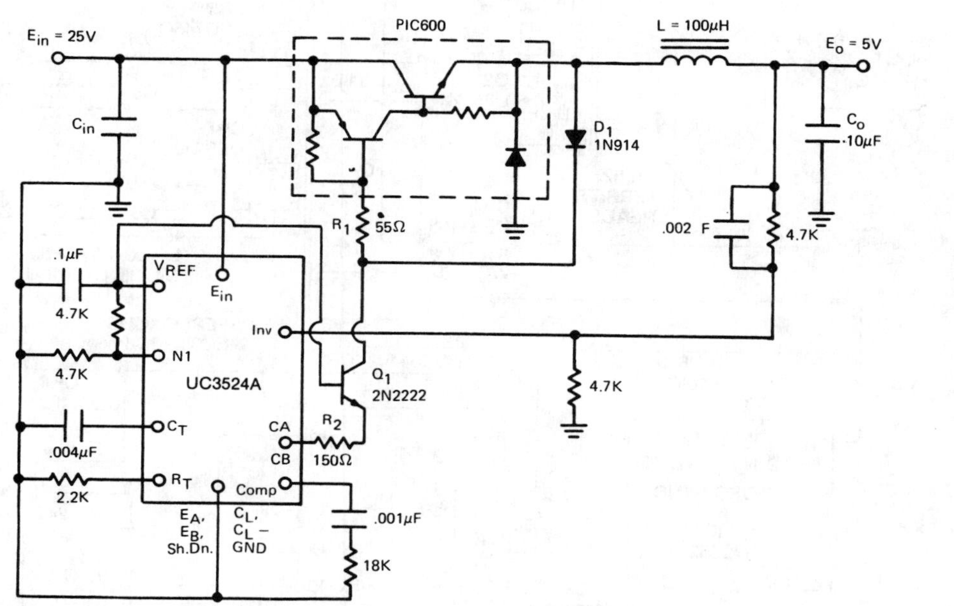

High-frequency switching regulator

Fig. 4-34 Because of their fast switching time, low-voltage hybrid circuits (such as the PIC600) can be operated as high as 250 kHz. The advantages of the higher frequencies are: lower filter cost, reduced size and weight, improved transient response, output ripple less dependent on capacitor ESR, and simpler EMI and RFI filtering. Unitrode Semiconductor Products DB600, 1990, p. 12-64.

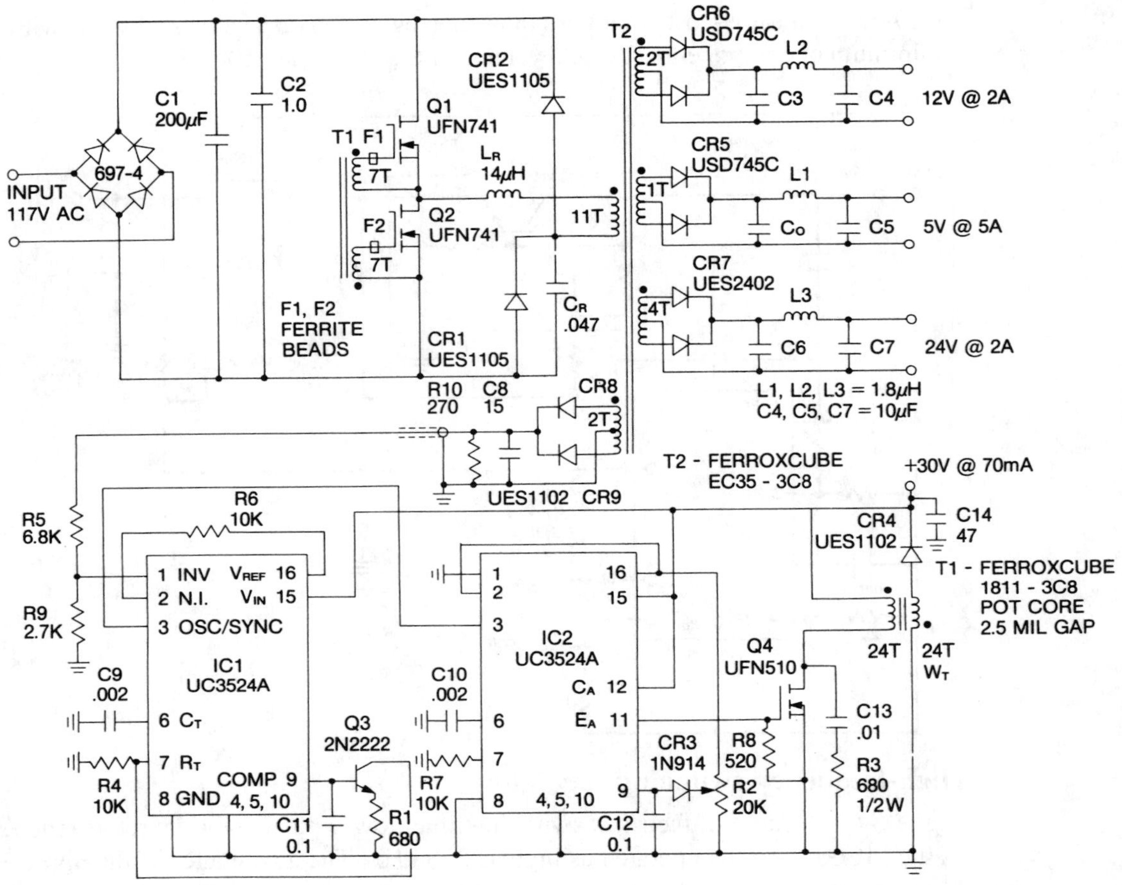

Series-resonant power supply

Fig. 4-35 The advantages of a series-resonant converter, compared to a buck-derived switching regulator, are higher overall efficiency, smaller weight and volume, reduction in EMI, and increased reliability. The disadvantages are: an additional resonant circuit, the rating of the power switch is about 1.4 times higher, and the output filter capacitors must have low ESR and high ripple-current ratings.

Unitrode Semiconductor Products DB600, 1990, p. 12-138.

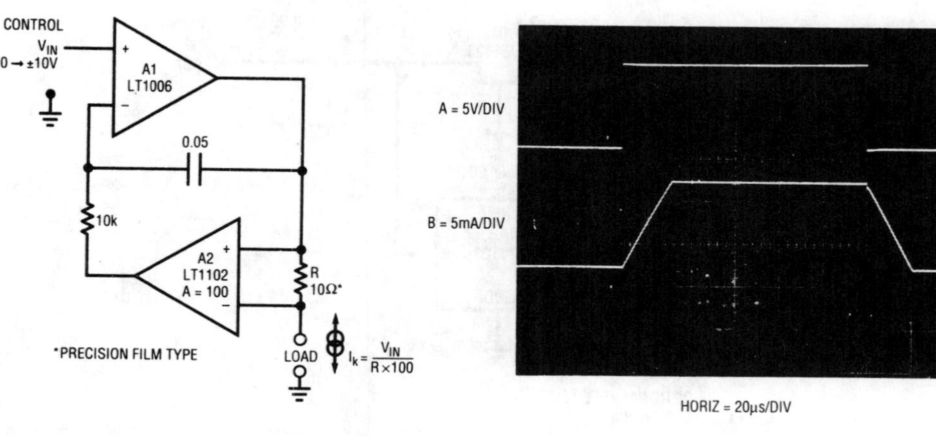

Simple voltage-programmable current source

Fig. 4-36 This circuit produces output current in strict accordance with the sign and magnitude of the control voltage, with no trimming required. Circuit accuracy and stability depend almost entirely on resistor R. Figure 4-36B shows dynamic response for a full-scale input step. Trace A is the voltage-control input and trace B shows the output current. Linear Technology Corporation, 1991, AN45-5.

4

248

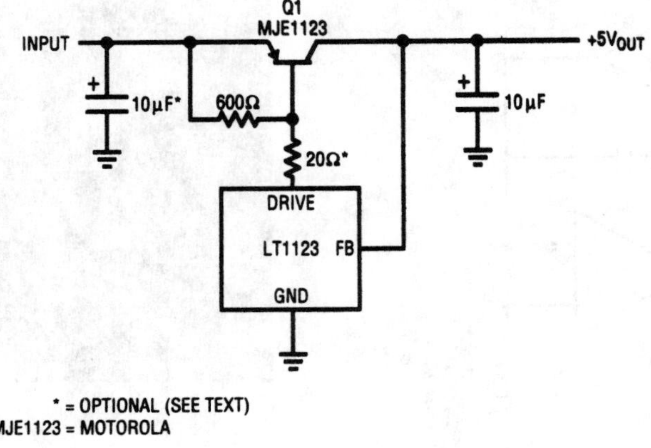

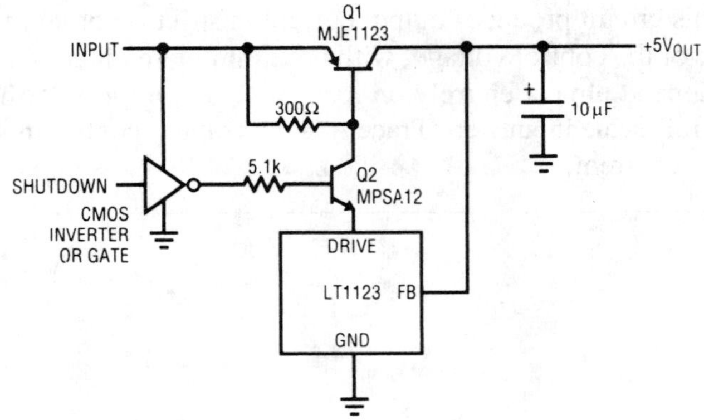

Regulators with ultra-low dropout

Fig. 4-37 Battery life is significantly affected by the dropout performance of linear regulators. These circuits offer lower dropout voltage than any monolithic regulator (below 50 mV at 1 A, increasing to only 450 mA at 5 A). Line and load regulation are within 5 mV, and initial output accuracy is inside 1%. The circuits are fully short-circuit protected, and have no-load quiescent current of 600 µA. Figure 4-37B shows a simple way to add shutdown to the regulator of Fig. 4-37A. A CMOS inverter or gate biases Q2 to control LT1123 bias. When Q2 is driven, the loop functions normally. With Q2 unbiased, the circuit goes into shutdown and pulls no current. Linear Technology Corporation, 1991, AN45-20/21.

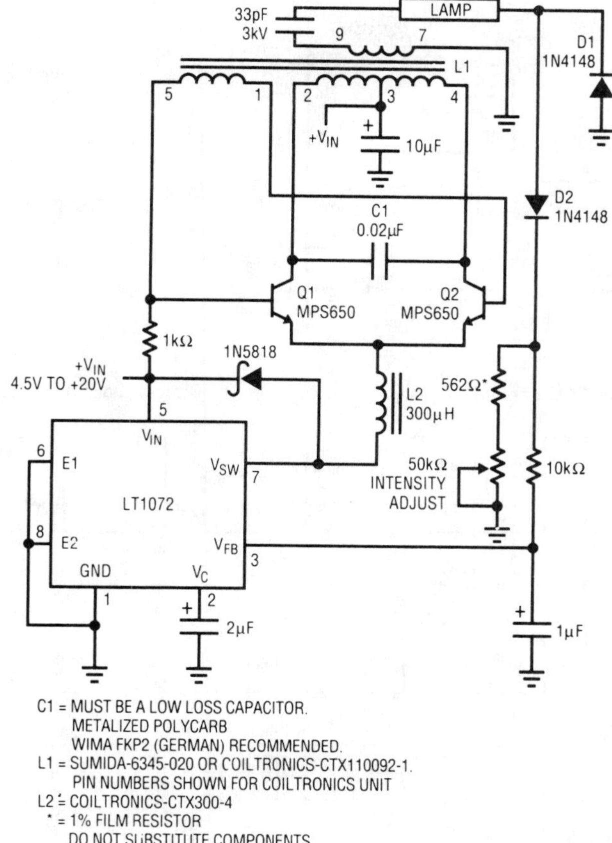

Cold-cathode fluorescent-lamp (CCFL) power supply

Fig. 4-38 CCFLs are often used to backlight the LCD displays of portable computers. Such lamps require a high-voltage sine-wave drive. This circuit provides such drive, and permits lamp intensity to be varied continuously and smoothly from zero to full intensity. Notice that a Tektronix probe type P-6009 (acceptable) or types P6013A and P6015 (preferred) probes must be used to read the L1 output. The vast majority of oscilloscope probes will break down if used for this measurement, unless the probes are rated for wideband high voltage. Linear Technology Corporation, 1991, AN45-22.

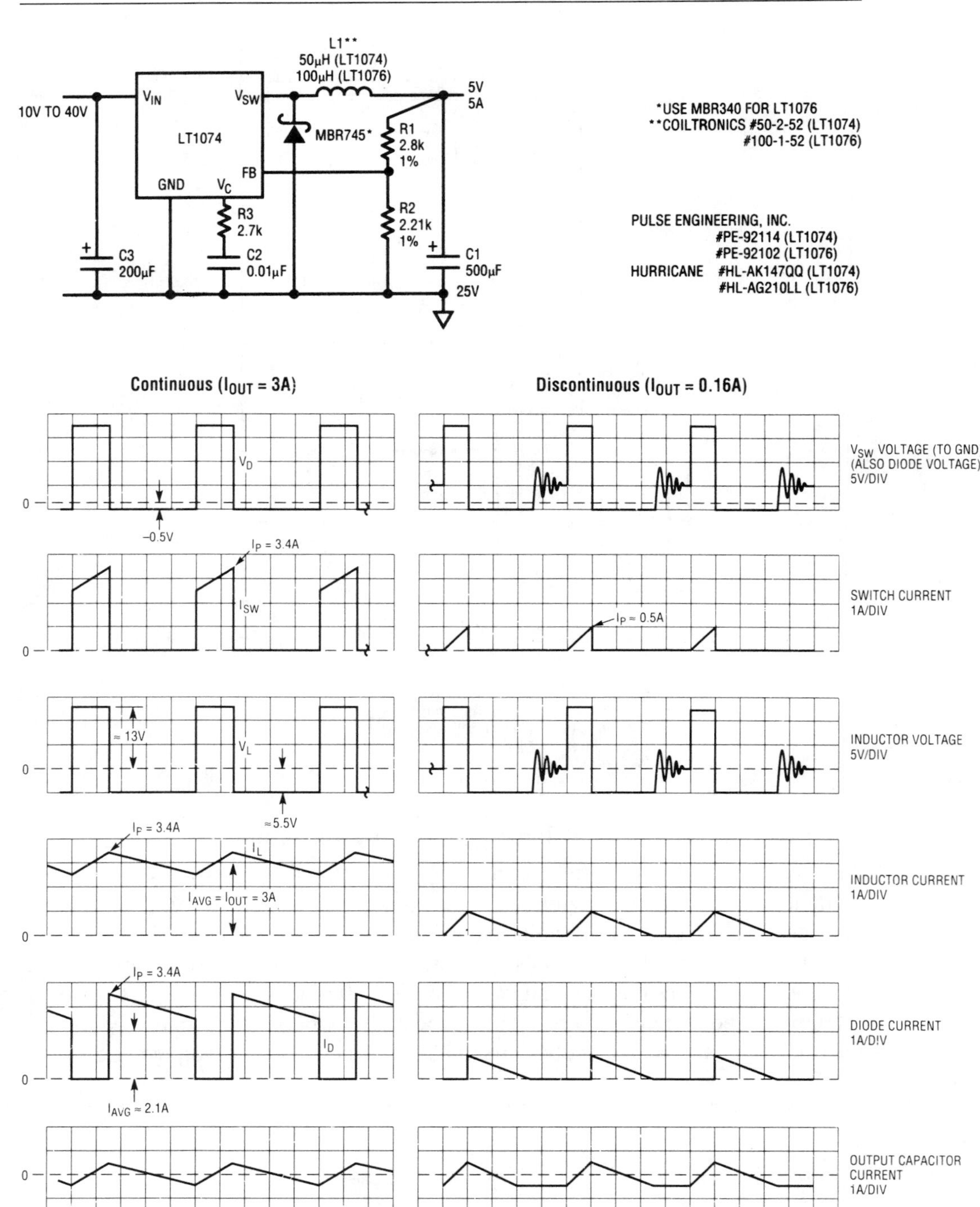
L1**
50μH (LT1074)
100μH (LT1076)
10V TO 40V
VIN
VSW
LT1074
GND
VC
FB
MBR745*
R1
2.8k
1%
R2
2.21k
1%
R3
2.7k
C3
200μF
C2
0.01μF
C1
500μF
25V
5V
5A
*USE MBR340 FOR LT1076
**COILTRONICS #50-2-52 (LT1074)
#100-1-52 (LT1076)
PULSE ENGINEERING, INC.
#PE-92114 (LT1074)
#PE-92102 (LT1076)
HURRICANE #HL-AK147QQ (LT1074)
#HL-AG210LL (LT1076)
Continuous (IOUT = 3A)
Discontinuous (IOUT = 0.16A)
VD
−0.5V
IP = 3.4A
ISW
IP ≈ 0.5A
≈ 13V
VL
≈5.5V
IL
IAVG = IOUT = 3A
ID
IAVG ≈ 2.1A
VSW VOLTAGE (TO GND)
(ALSO DIODE VOLTAGE)
5V/DIV
SWITCH CURRENT
1A/DIV
INDUCTOR VOLTAGE
5V/DIV
INDUCTOR CURRENT
1A/DIV
DIODE CURRENT
1A/DIV
OUTPUT CAPACITOR
CURRENT
1A/DIV
2μs/DIV

Basic positive-buck converter

Fig. 4-39 This circuit is used to convert a larger positive input voltage to a lower positive output, using an LT1074 switching regulator. Typical waveforms are shown in Fig. 4-39B. These waveforms are based on a V_{IN} of 20 V, V_{OUT} of 5 V, and L of 50 μH for both continuous mode (inductor current never drops to zero) with I_{OUT} of 3 A, and discontinuous mode (inductor current drops to zero during a portion of the switching cycle (I_{OUT} = 0.17 A). Linear Technology Corporation, 1991, AN44-18/19.

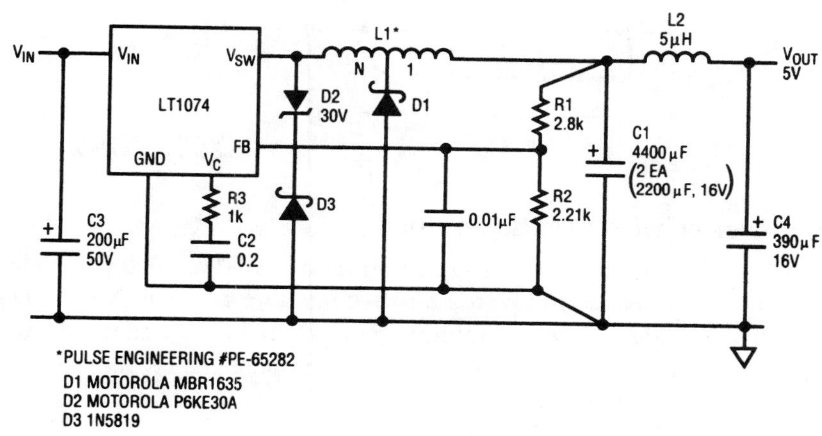

Tapped-inductor buck converter

Fig. 4-40 This circuit uses a tapped inductor to increase the current output capability. The ratio of "input" turns to "output" turns is N. With an N of 3 (three times as many turns to the left of D1 as at the right of D1), a V_{IN} of 20 V, an L of 100 μH, and a V_{OUT} of 5 V, the I_{OUT} is approximately 10 A (double that of Fig. 4-39). Linear Technology Corporation, 1991, AN44-25.

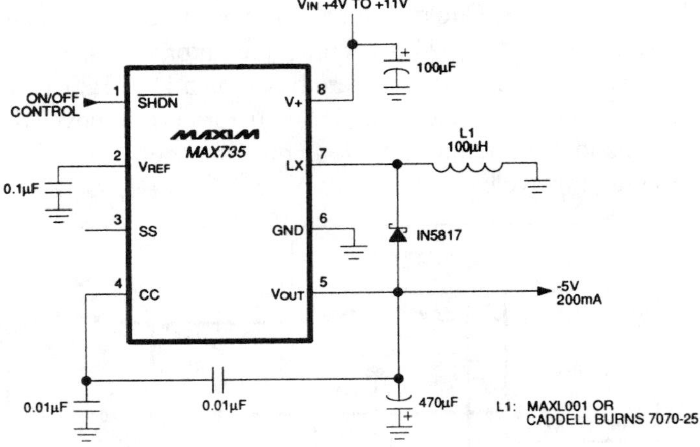

Simple inverter with negative output

Fig. 4-41 In this circuit, a MAX739 current-mode PWM regulator provides all of the active circuitry needed to invert a positive input (+4 to +10 V) to a negative output (−5 V at 200 mA), with a minimum of external components. A switching frequency of 165-kHz allows for small external components. Maxim, 1992, Applications and Product Highlights, p. 4-4.

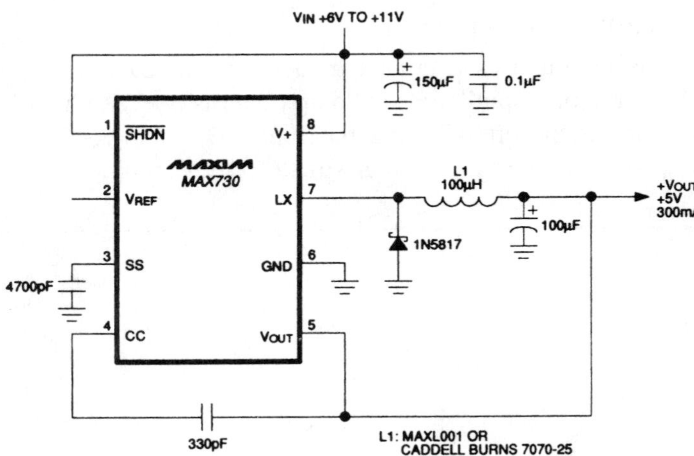

Simple buck regulator

Fig. 4-42 In this circuit, a MAX730 provides a 5-V, 300-mA output for inputs from 6 to 10 V. Maxim, 1992, Applications and Product Highlights, p. 4-6.

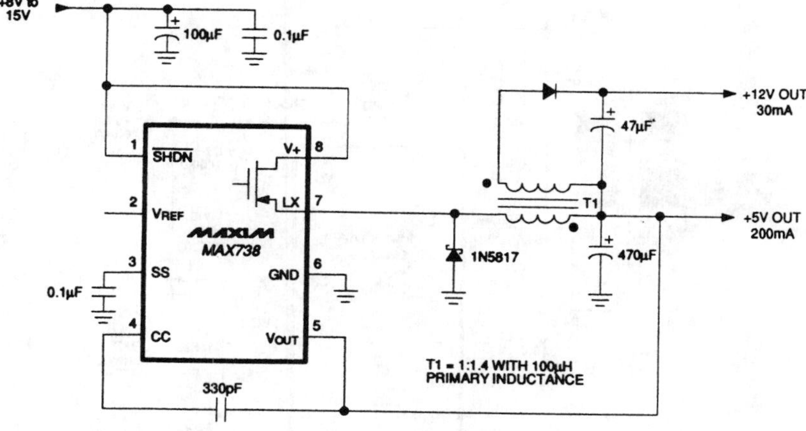

4

253

Simple buck regulator with two outputs

Fig. 4-43 In this circuit, a MAX738 provides a +5-V 200-mA output through the inductor primary, and a +12-V 30-mA output through the inductor secondary for inputs from 8 to 15 V. Maxim, 1992, Applications and Product Highlights, p. 4-6.

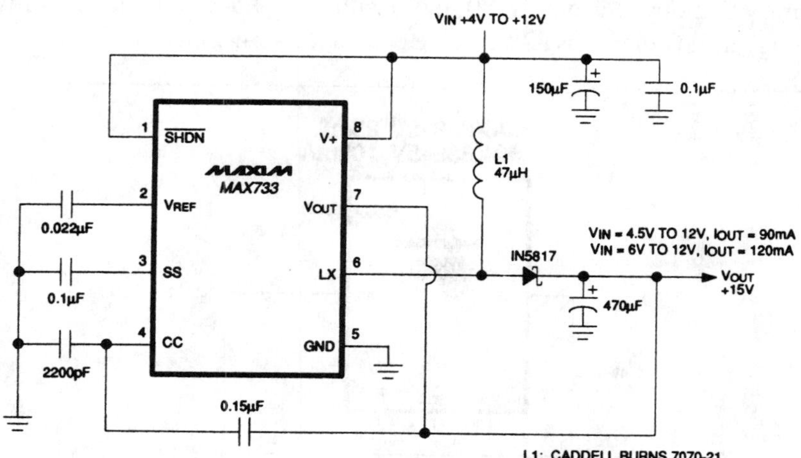

Simple boost regulator

Fig. 4-44 In this circuit, a MAX733 provides a +15-V output at 90 mA (for inputs from +4.5 to +12 V) or at 120 mA (for inputs from +6 to +12 V). Maxim, 1992, Applications and Product Highlights, p. 4-7.

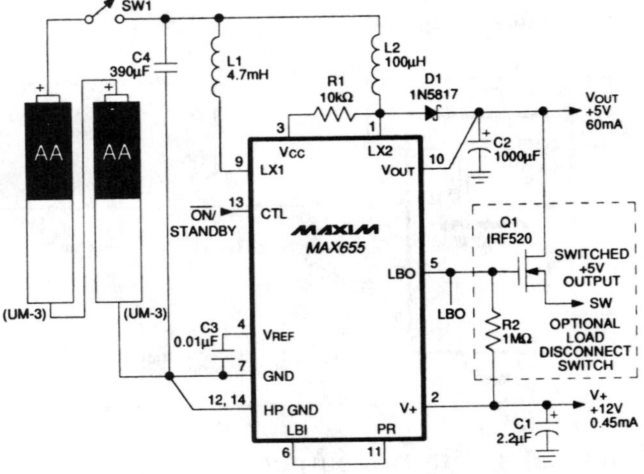

Step-up +5-V output from two AA cells

Fig. 4-45 In this circuit, a MAX655 provides a +5-V 60-mA output from two AA battery cells. The circuit also provides a +12-V 500-µA output. Notice that the low-battery detector (LBO) is used as a level translator for optional high-side switching. Standby current is 80 µA (with the +5-V output still alive and regulating) and efficiency is 82%. Maxim, 1992, Applications and Product Highlights, p. 4-8.

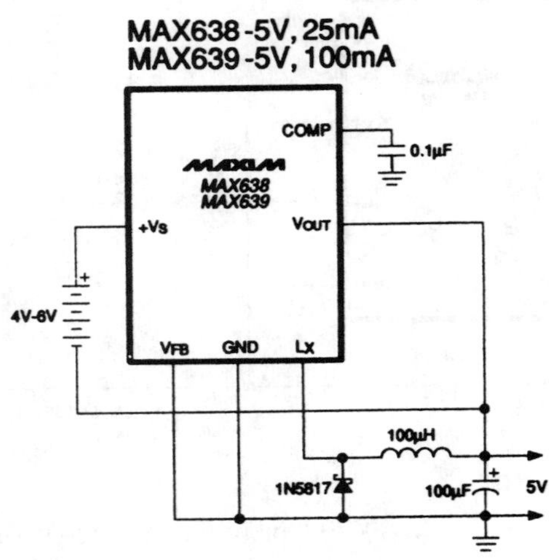

Battery-input buck/boost regulator

Fig. 4-46 This circuit converts an input voltage that can range from above (6 V) and below (4 V) the desired output voltage (5 V). Normally, such a circuit must provide both step-up and step-down actions. However, if the input ground can be floated, as is often the case with a battery, an inverter can be used instead of a transformer or a complicated and lossy step-up/down circuit by fixing the most-negative output voltage at ground. Maxim, 1992, Applications and Product Highlights, p. 4-12.

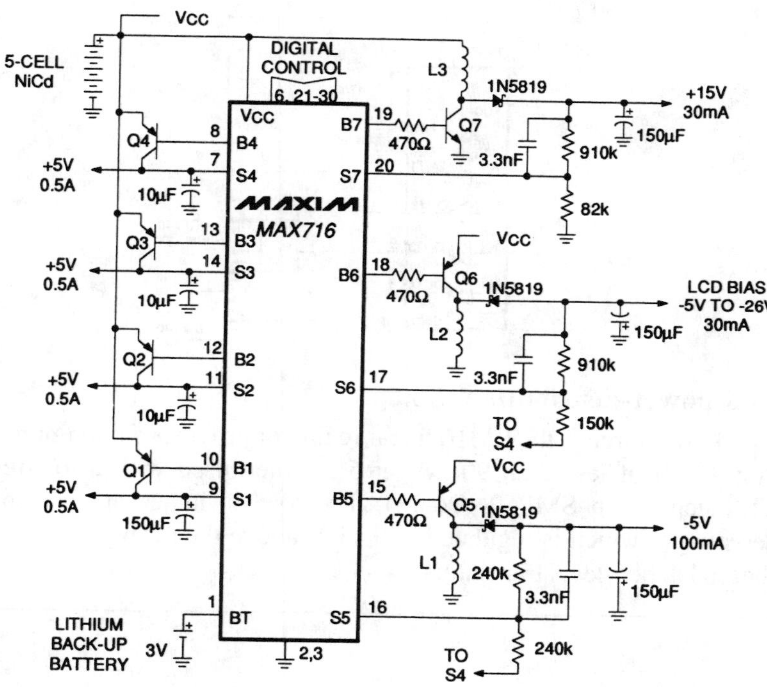

Full-function portable power

Fig. 4-47 This circuit provides a variety of outputs for battery-operated microprocessor systems. All of the outputs are under microprocessor control. A 5-cell NiCd battery is used for the main power source, with a 3-V lithium back-up battery. One of the outputs is negative (and is controlled by a 5-bit on-board D/A converter) for LCD contrast adjustment. Maxim, 1992, Applications and Product Highlights, p. 4-15.

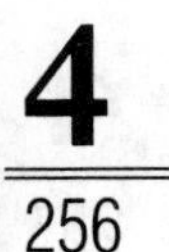

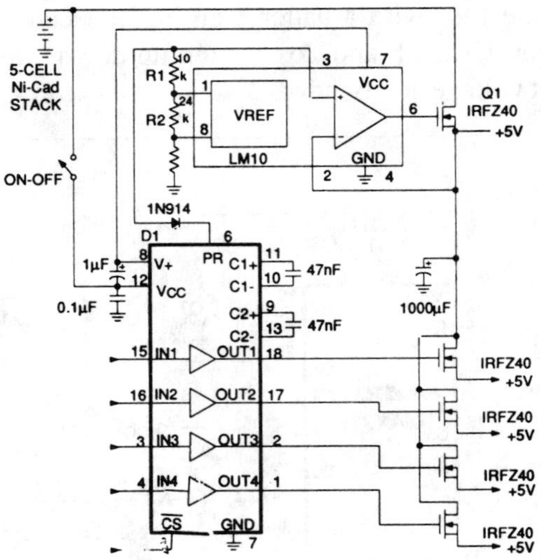

Regulated power-distribution system

Fig. 4-48 In this circuit, the LM10 linear regulator provides a maximum dropout voltage (at 25°C) of less than 90 mV at 5 A when used with a low-resistance MOSFET (such as an SMP60NO6-18). The load switches are driven by the high-side voltage, which is regulated at +11 V above the supply voltage, via the four on-board latched-level translators. Maxim, 1992, Applications and Product Highlights, p. 4-16.

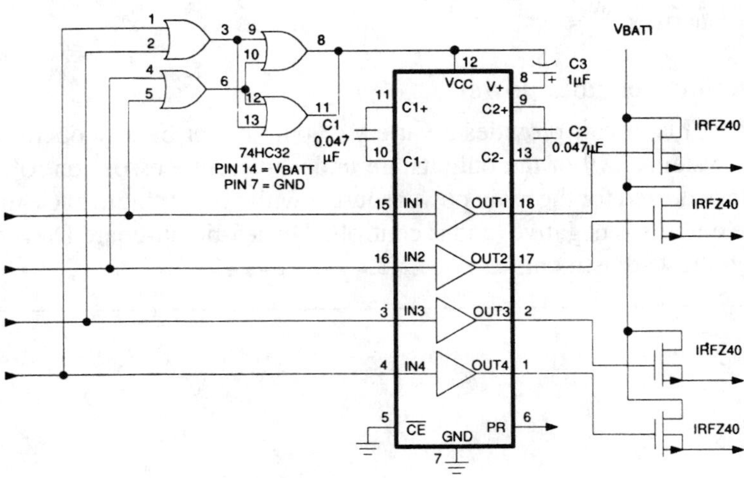

Battery-management circuit with ultra-low standby

Fig. 4-49 This circuit controls four loads via low on-resistance N-channel MOSFETs. If all four switches are programmed off, the power supply input to the MAX620 is disconnected, and the standby supply drain is reduced to 2 μA. Maxim, 1992, Applications and Product Highlights, p. 4-16.

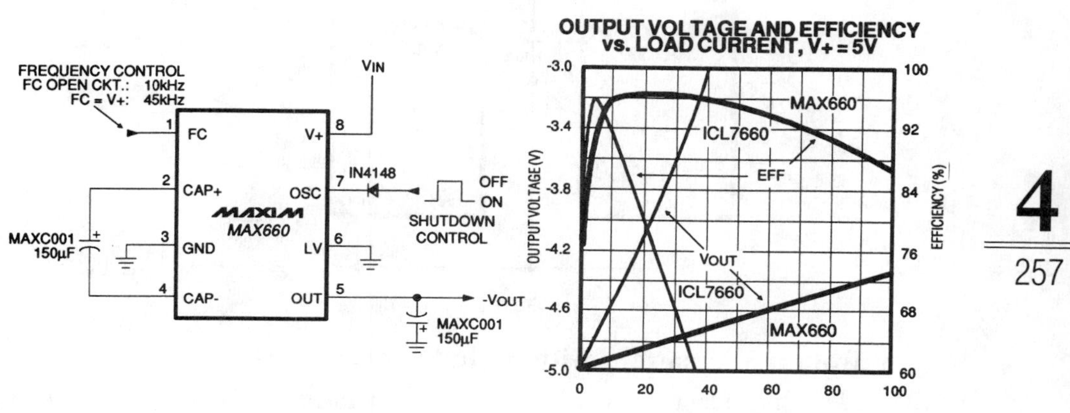

Charge pump with shutdown

Fig. 4-50 In this circuit, a MAX660 charge pump is used as a voltage inverter, creating a negative voltage of approximately equal magnitude to the input voltage. Although the MAX660 has no feedback mechanism, and so is unregulated, the output is a stiff, accurate supply when operated from a regulated input. An output current of 100 mA results in a typical voltage loss of 0.65 V (the drop at 10 mA is less than 100 mV). The FC input selects a 10- or 45-kHz oscillator frequency. Notice that this circuit provides an optional shutdown that disables the internal oscillator, and reduces the supply current to less than 1 μA. Maxim, 1992, Applications and Product Highlights, p. 4-17.

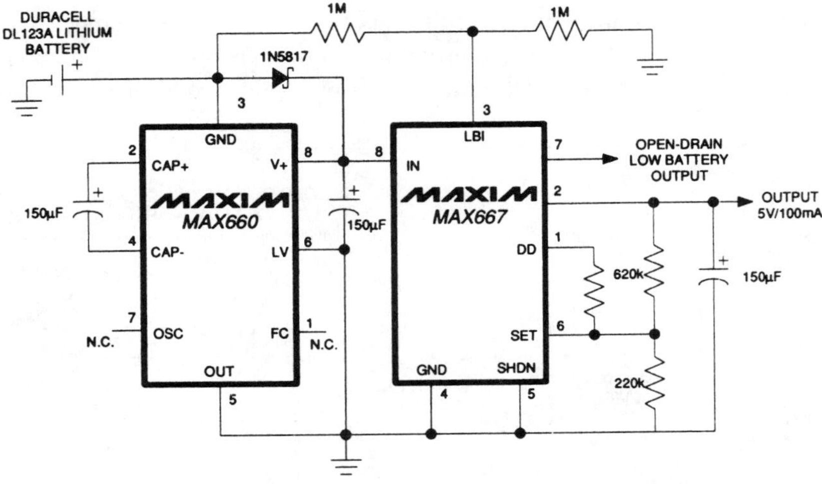

+5 V from a 3-V battery with no inductors

Fig. 4-51 This inductorless dc/dc converter generates a regulated +5 V at up to 100 mA from a single lithium cell. When powered by a DL123A (smaller than an AA cell), the circuit provides 40 mA of load current for 12 hours or more. The MAX660 in a doubler configuration steps up the battery voltage to +6 V, which is then regulated down to +5 V by the MAX667. The surface-mount version of this circuit occupies only 1/2 square inches of PC-board space, because the circuit uses capacitors instead of inductors. The circuit capitalizes on the low dropout voltage of the MAX667 (typically less than 100 mV at 150 mA) and the 100-mA load capability and 95% efficiency of the MAX660. Maxim, 1992, Applications and Product Highlights, p. 4-18.

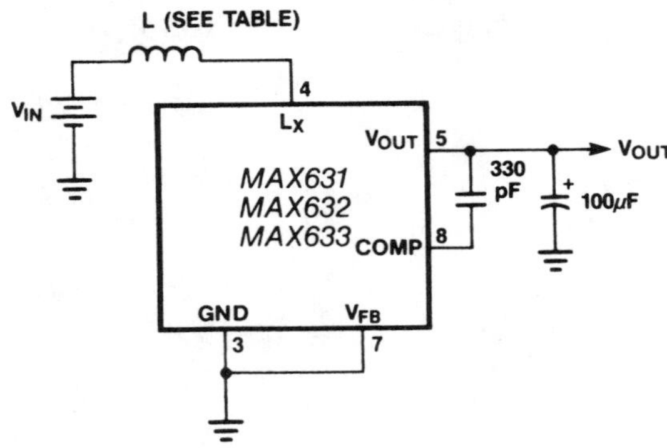

Maxim Part No.	V_{IN} (V)	V_{OUT} (V)	I_{OUT} (mA)	Typ Eff (%)	Part No.*	Inductor (L) µH	Ω
MAX631	2	5	5	78	6860-21	470	0.44
	2	5	10	74	6860-17	220	0.28
	2	5	15	61	6860-13	100	0.1
	3	5	25	82	6860-21	470	0.44
	3	5	40	75	7070-29	220	0.55
MAX632	3	12	5	79	6860-10	330	0.35
	3	12	10	79	7070-28	180	0.48
	5	12	12	88	6860-21	470	0.44
	5	12	25	87	6860-19	330	0.35
MAX633	3	15	5	73	7070-29	220	0.55
	3	15	8	71	7070-27	150	0.43
	5	15	10	85	6860-21	470	0.44
	5	15	15	85	6860-19	330	0.35
	8	15	35	90	6860-21	470	0.44

* Caddell-Burns, NY, (516) 746-2310

Low-power step-up converters

Fig. 4-52 The table of Fig. 4-52B lists the measured efficiency of circuits using the indicated coils. Efficiencies can be improved slightly by placing a Schottky diode (such as the IN5817) in parallel with the internal diode, from pin 4 to 5. The increase in efficiency is most noticeable for the 5-V output circuits. Maxim, Seminar Applications Book, p. 76.

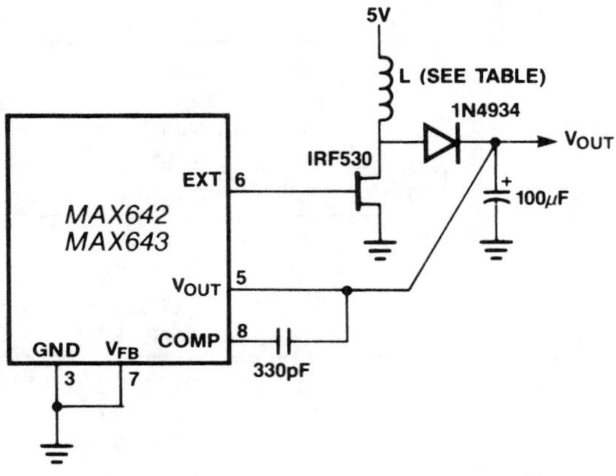

Maxim Part No.	V_{IN} (V)	V_{OUT} (V)	I_{OUT} (mA)	Typ Eff (%)	I_{pk} (A)	Inductor (L) Part No.*	µH	Ω
MAX642	5	12	200	91	1.2	6860-08	39	0.05
	5	12	350	89	2	6860-04	18	0.03
	5	12	550	87	3.5	7200-02	12	0.01
MAX643	5	15	100	92	1.2	6860-08	39	0.05
	5	15	150	89	1.5	6860-06	27	0.04
	5	15	225	89	2	6860-04	18	0.03
	5	15	325	85	3.5	7200-02	12	0.01

* Caddell-Burns, NY, (516) 746-2310

Medium-power step-up converters

Fig. 4-53 The table of Fig. 4-53B lists measured efficiency of circuits using the indicated coils. In selecting coils for circuits that use external MOSFETs, it is important to calculate the peak current and to select a coil that will not saturate at that peak current. Maxim, Seminar Applications Book, p. 77.

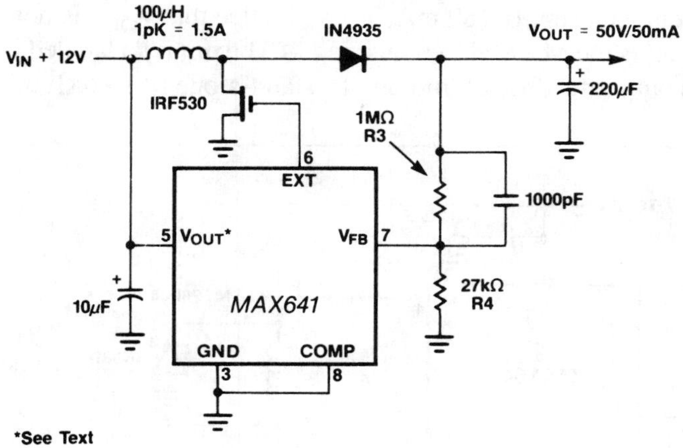

High-voltage step-up converter

Fig. 4-54 In this circuit, a +12-V input is converted to a +50 V at 50 mA by adding an IRF530 N-channel FET, which has a voltage rating of 100 V. The circuit differs from the basic MAX641 hookup in that an external resistor-divider must provide the feedback to the V_{FB} input and that chip power comes from the +12-V input via the V_{OUT} pin. Maxim, Seminar Applications Book, p. 77.

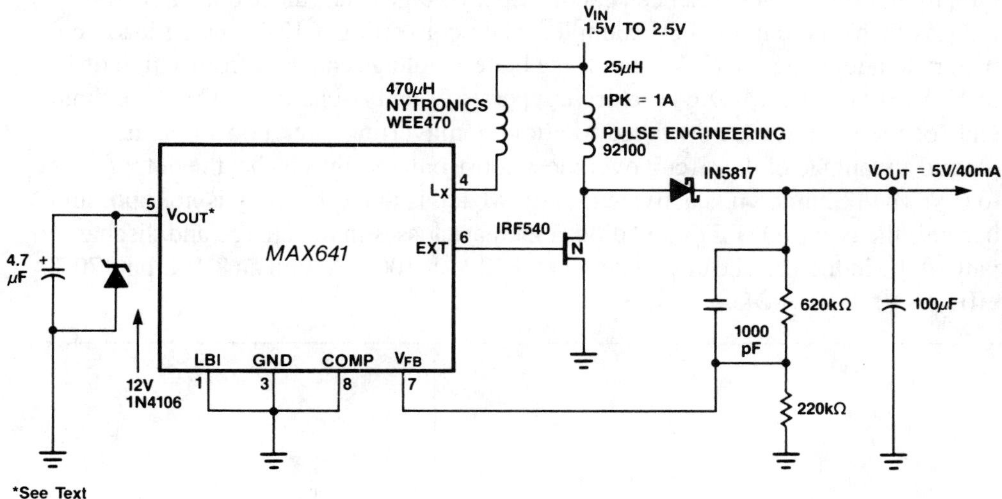

Low-voltage battery to +5 V

Fig. 4-55 By connecting a second inductor to the L_X output of a MAX641 step-up dc/dc converter, the efficiency and power-handling ability can be dramatically improved. This circuit can supply 5 V at 40 mA with only 1.5-V input. With 2.4-V input, the circuit can supply 180 mA at 5 V. Notice that V_{OUT} is not an output (in this circuit), but is the MAX641 voltage input. The pin is so labeled because V_{OUT} is usually the output/feedback terminal of a stand-alone MAX641. Maxim, Seminar Applications Book, p. 78.

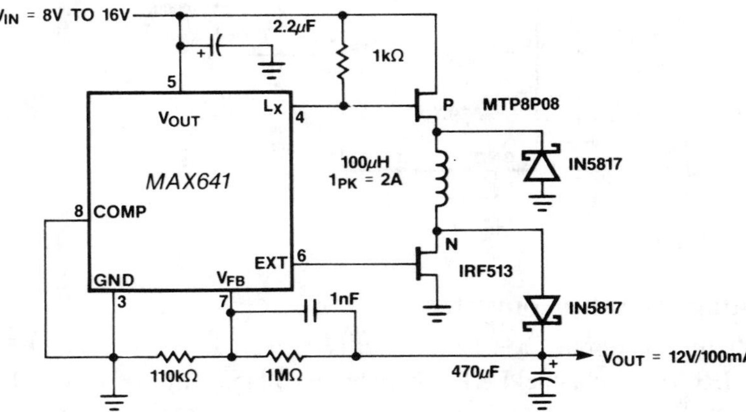

Step-up/down dc/dc converter

Fig. 4-56 Positive-output step-up and step-down dc/dc converters have a common limitation in that neither can handle input voltages that can be both greater than (or less than) the output. For example, when converting a 12-V sealed lead-acid battery to a regulated +12-V output, the battery voltage can vary from a high of 15 to 10 V. By using a MAX641 to drive separate P- and N-channel MOSFETs, both ends of the inductor are switched to allow noninverting buck/boost operation. A second advantage of the circuit over most boost-only designs is that the output goes to 0 V when Shutdown is activated. A drawback is that efficiency is not optimum because the two MOSFETs and diodes increase losses in the charge and discharge path of the inductor. The circuit delivers +12 V at 100 mA with an 8-V input (70% efficiency). Maxim, Seminar Applications Book, p. 78.

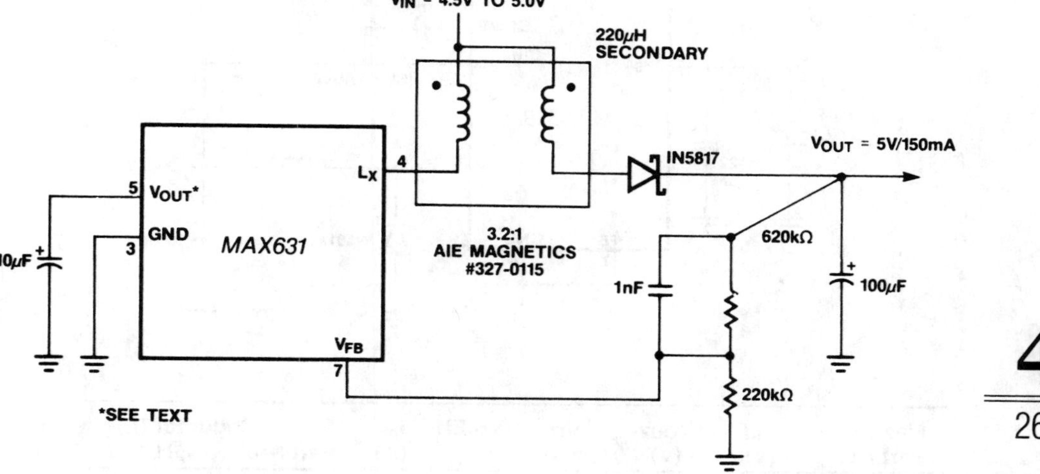

Long-life IR-drop voltage-recover system

Fig. 4-57 This circuit provides a unique solution to a common system-level power-distribution problem. When the supply voltage to a remote board must traverse a long cable, the voltage at the end of the line sometimes drops to unacceptable levels. This circuit solves the problem by taking the reduced voltage at the end of the supply line and boosting it back to +5 V. This can be especially useful in remote-display devices, such as some point-of-sale (POS) terminals, where several meters of cable can separate the terminal from the readout. The 3.2:1 turns ratio of the transformer allows the MAX631 to provide more than the usual output current, without an external MOSFET, at these relatively low operating voltages. Output current is 150 mA at 5 V, with an input of 4 to 5 V. The MAX631 also makes use of the reflected voltage in the transformer primary to generate a higher supply voltage of about +9 V. By operating at 9 V, rather than 5 V, the on resistance of L_X is reduced. Maxim, Seminar Applications Book, p. 79.

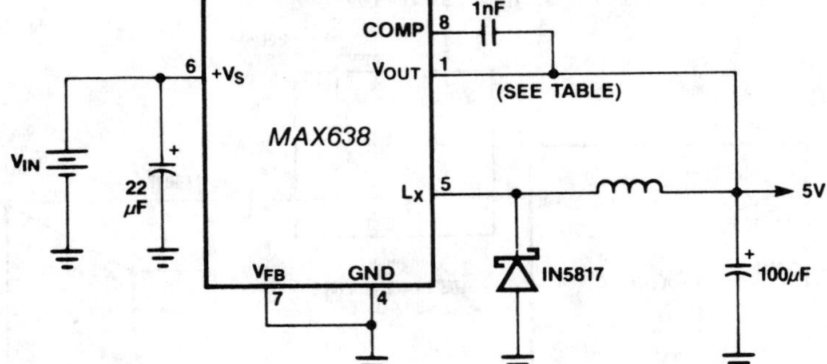

Maxim Part No.	V_{IN} (V)	V_{OUT} (V)	I_{OUT} (mA)	Typ Eff (%)	I_{pk} (A)	Inductor (L) Part No.*	µH	Ω
MAX638	7-9.5	5	35	92	200	7070-27	150	0.4
	8-9.5	5	55	89	200	7070-27	150	0.4
	10-14	5	50	92	300	7070-30	270	0.6
	12	5	60	92	250	7070-30	270	0.6
	12	5	75	89	300	7070-28	180	0.5

* Caddell-Burns, NY, (516) 746-2310

Low-power step-down converters

Fig. 4-58 The table of Fig. 4-58B lists measured efficiency and currents of circuits that use the indicated coils. Maxim, Seminar Applications Book, p. 79.

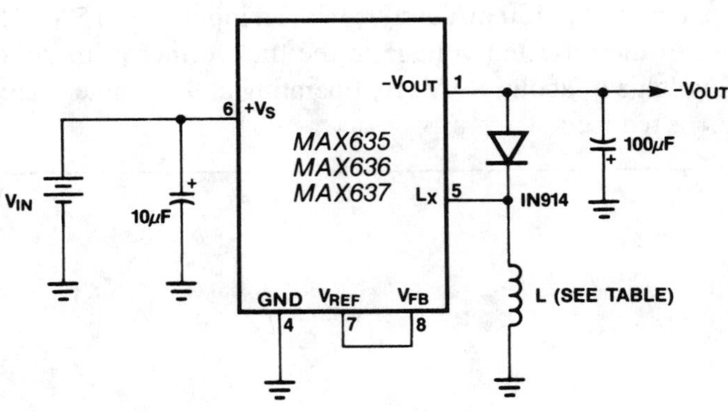

Maxim Part No.	V_{IN} (V)	V_{OUT} (V)	I_{OUT} (mA)	Typ Eff (%)	Inductor (L) Part No.*	μH	Ω
MAX635	+3	-5	5	60	6860-19	330	0.35
	+5	-5	25	76	6860-19	330	0.35
	+9	-5	40	79	6860-19	330	0.35
	+12	-5	45	85	6860-21	470	0.40
	+15	-5	50	90	6860-23	680	0.55
MAX636	+5	-12	12	74	6860-19	330	0.35
	+9	-12	30	84	6860-19	330	0.35
	+12	-12	40	89	6860-21	470	0.40
MAX637	+3	-15	2	65	6860-21	470	0.40
	+5	-15	8	77	6860-19	330	0.35
	+9	-15	25	85	6860-19	330	0.35

* Caddell-Burns, NY, (516) 746-2310

Low-power inverters

Fig. 4-59 The table of Fig. 4-59B lists measured efficiency and currents of circuits that use the indicated coils and corresponding ICs. Maxim, Seminar Applications Book, p. 80.

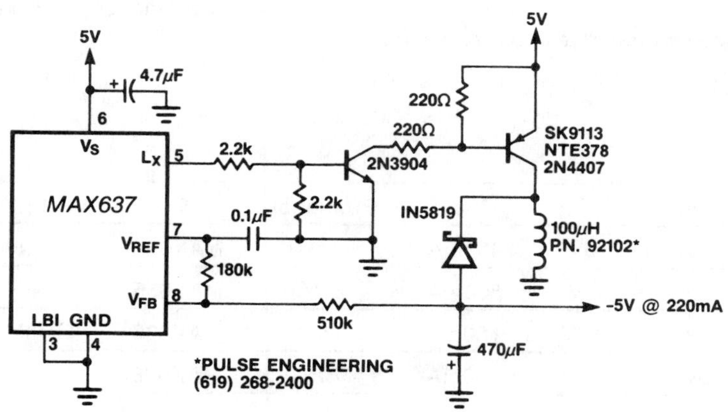

+5- to −5-V at 220 mA (with low power) inverter

Fig. 4-60 This circuit uses two transistors to buffer the L_X output and produce 220-mA of output current. The 2N3904 is used invert the L_X output and drive the power pnp. When using external transistor buffers as shown here, the output voltage is set by the external feedback-resistor network (180 and 510 kΩ). Maxim, Seminar Applications Book, p. 80.

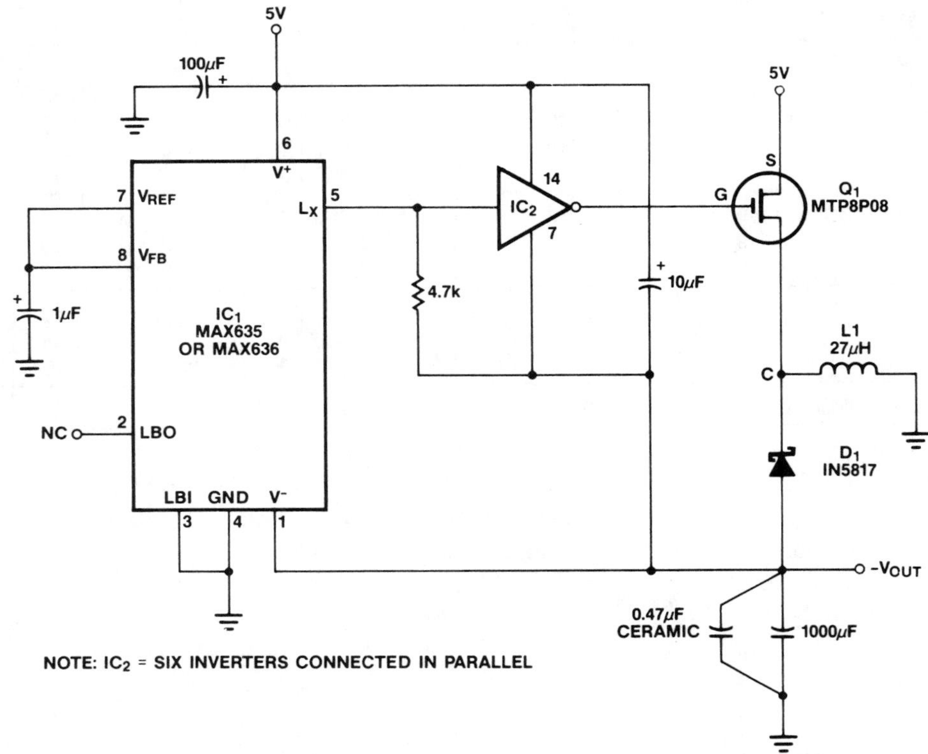

V_{IN}	$-V_{OUT}$	I_{OUT}	Efficiency	IC_1	L_1
5V	-5V	400mA	70%	MAX635	27μH
5V	-5V	500mA	64%	MAX635	18μH
5V	-12V	150mA	75%	MAX636	27μH
5V	-12V	200mA	70%	MAX636	18μH

NOTES:
18μH Coil = Caddell-Burn's (Mineola, NY) Model 6860-04.
27μH Coil = Caddell-Burn's Model 6860-06.

Medium-power inverters

Fig. 4-61 The table of Fig. 4-61B lists measured efficiency and currents of circuits using the indicated coils and corresponding ICs. At startup, the $-V_{OUT}$ is on Schottky diode-drop above ground and the gate drive to the power MOSFET is slightly less than 5 V. The output should be only slightly loaded to ensure startup because the output-power capability of the circuit is very low until $-V_{OUT}$ is about 2 V negative. Maxim, Seminar Applications Book, p. 81.

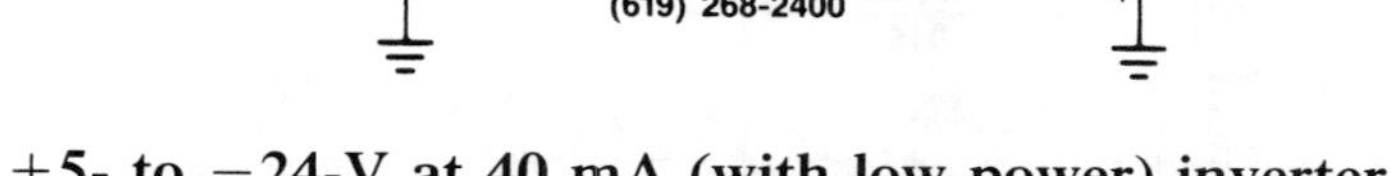

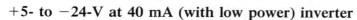

+5- to −24-V at 40 mA (with low power) inverter

Fig. 4-62 This circuit is similar to that of Fig. 4-60 in that the 2N4407 is a buffered replica of the MAX637 L_X output. However, the 2N4407 has a high breakdown voltage and can be used to generate a −24-V output. The −24-V output does not appear directly on any pin of the MAX637 because it is sensed via the 1.5-MΩ external feedback resistor. Maxim, Seminar Applications Book, p. 82.

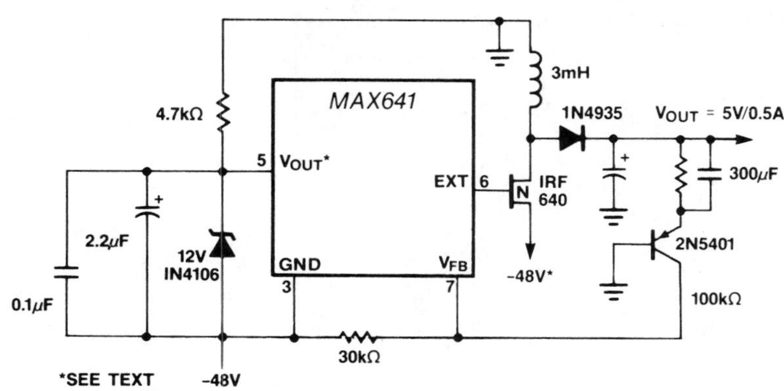

Telecom −48 to 5 V at 0.5 A

Fig. 4-63 The small current consumption of a MAX641 allows the IC to be biased at a −50-V rail with a shunt zener so that the IC can convert −50 to +5 V. This is a common requirement in telecom systems, where logic circuits must be powered from the central-office battery voltage. A small high-voltage pnp level-shifts the feedback signal from the +5-V output down to the pin MAX641, where the ground pin is tied to the 50-V input (instead of true ground). This is done so that EXT can drive the MOSFET to switch the inductor to −50 V, making the circuit operate similar to a step-up dc/dc converter. The 330-pF capacitor provides feed-forward compensation to stabilize the regulator control loop. Maxim, Seminar Applications Book, p. 82.

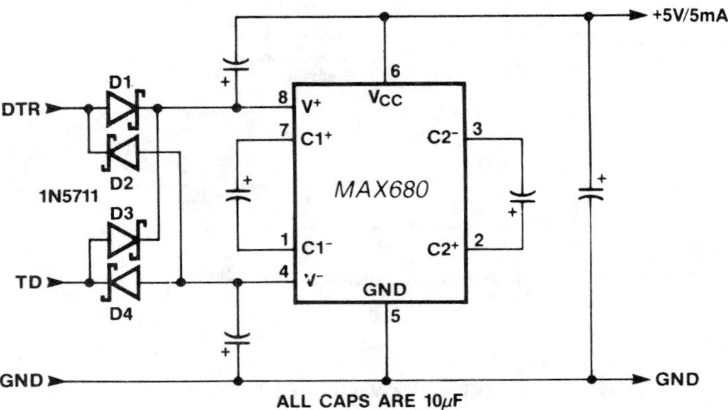

RS-232 line to 5 V

Fig. 4-64 The MAX680 is normally used as a charge-pump voltage converter which converts +5 V to ±10 V. In this circuit, the MAX680 works in reverse, and converts the RS-232 signal levels to a lower voltage. When the RS-232 inputs are driven by 1488 drivers powered by ±12 V, the output voltage varies from 5.3 V open-circuit to 4-5 V with a 5-mA output load. Maxim, Seminar Applications Book, p. 82.

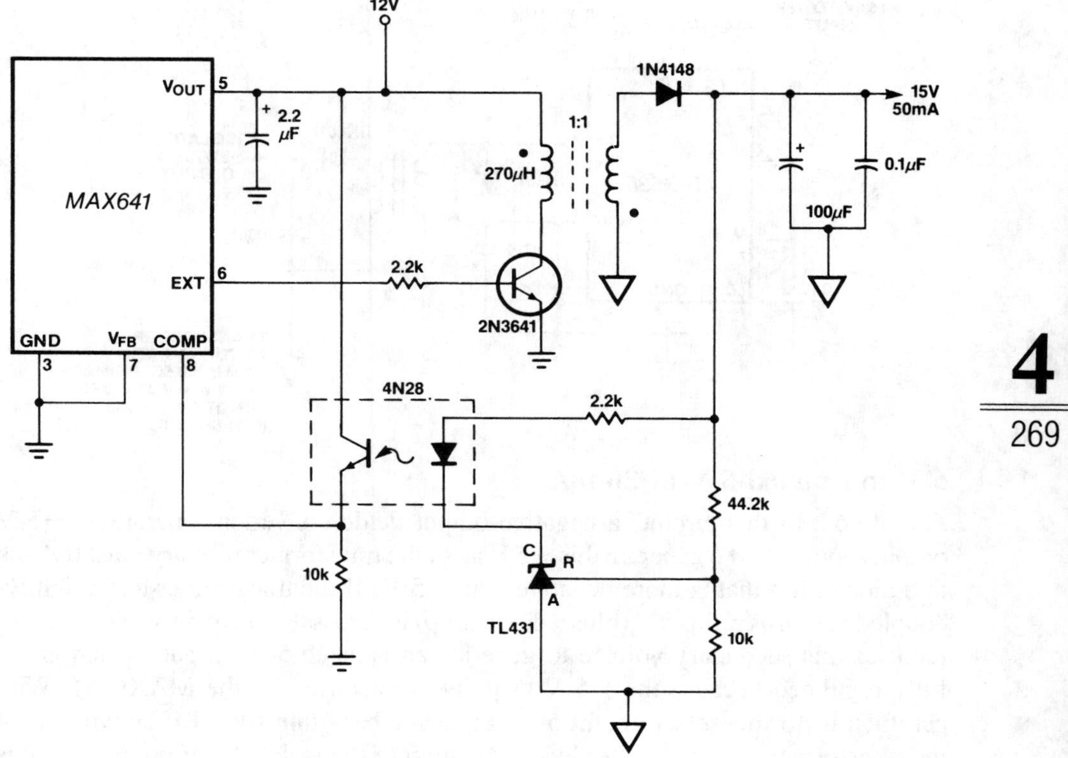

Isolated +15-V dc/dc converter

Fig. 4-65 In this circuit, a TL431 shunt regulator is used to sense the output voltage. The TL431 drives the LED of a 4N28 optocoupler, which provides feedback to the MAX641 while maintaining isolation between the +12-V input and the +15-V output (which is fully regulated with respect to both line and load changes). Maxim, Seminar Applications Book, p. 83.

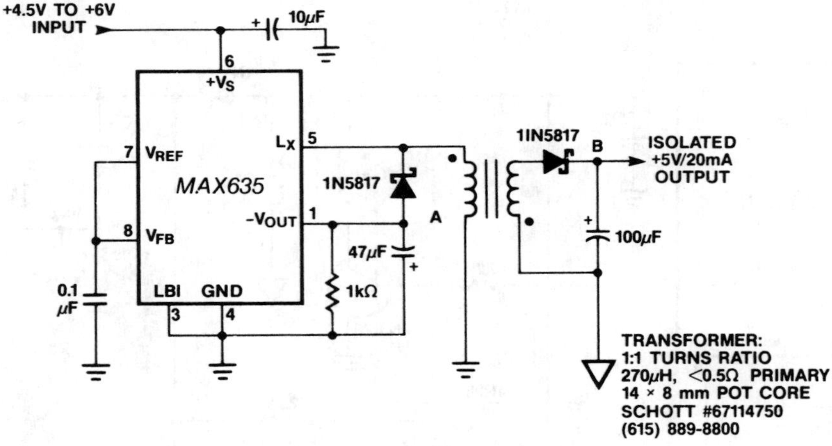

5 V to isolated 5 V at 20 mA

Fig. 4-66 In this circuit, a negative-output dc/dc converter generates a −5-V output at point A. To generate this −5 V at A, the transformer primary must flyback to a diode drop that is more negative than −5 V. If the transformer has a tightly-coupled 1:1 turns ratio, 5 V (plus a diode drop) is across the secondary. The lN5817 rectifies this secondary voltage to generate an isolated 5-V output (which is not fully regulated because the −5 V at point A is sensed by the MAX635). With careful transformer selection, the 5-V output will be within 10%. Bifilar winding of the transformer provides better load regulation of the isolated 5-V output, but this does reduce the isolation by increasing the capacitance between the primary and secondary. The isolation voltage breakdown is determined by the characteristics of the transformer, not the MAX635. Maxim, Seminar Applications Book, p. 83.

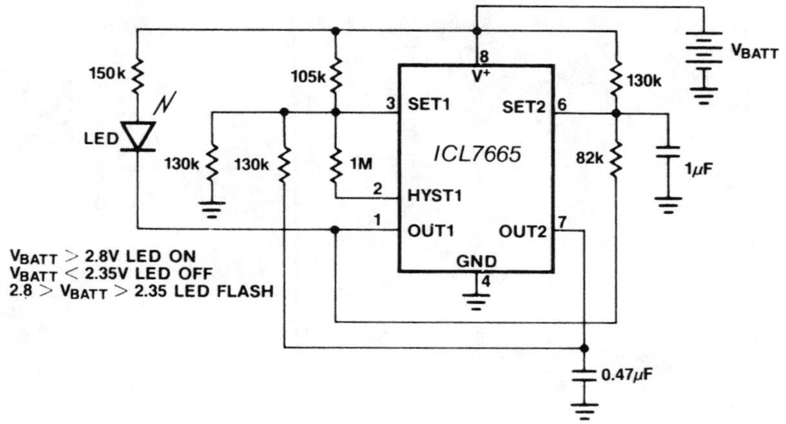

$V_{BATT} > 2.8V$ LED ON
$V_{BATT} < 2.35V$ LED OFF
$2.8 > V_{BATT} > 2.35$ LED FLASH

3-state battery indicator

Fig. 4-67 This circuit shows the state of a battery (typically 3 V) by means of an LED and an ICL7665. The LED remains on if the battery voltage is above 2.8 V, and flashes if the battery voltage is between 2.35 and 2.8 V. The LED remains off if the battery voltage drops below 2.35 V. Maxim, Seminar Applications Book, p. 84.

5

Oscillator and generator circuits

In this chapter, I am assuming that you are already familiar with oscillator basics (such as how discrete-component and IC oscillators operate) and basic oscillator/generator test and troubleshooting. However, the following paragraphs summarize both test and troubleshooting of oscillator/generator circuits. This information is included so that even if you are not familiar with electronic procedures, you can both test the circuits described here and locate problems if the circuits fail to perform as shown.

For our purposes, a circuit is considered to be an oscillator if the primary function is to produce sine waves (such as Fig. 5-1). If the output from the circuit is a pulse, square wave, triangular wave, ramp, etc. (such as Fig. 5-2), the circuit is considered a generator.

A scope is the logical instrument for both testing and troubleshooting generators because the shape of the circuit output is often critical. However, it is convenient to monitor output amplitude with a meter, and a frequency counter is much easier to use when measuring output frequency (although you can measure both amplitude and frequency with a scope).

Oscillator/generator testing and troubleshooting

No matter how complex or simple the circuits of this chapter appear, they are essentially oscillators (or generators) and can be treated as such from a practical test and troubleshooting standpoint. For example, each circuit produces output signals (possibly crystal-controlled, but often where frequency depends on *RC* or *LC* values). The output signals must have a given amplitude and must be at a given frequency (or must be capable of tuning across a given frequency range). In the case of generators, the output must also be of a given shape (square, pulse, etc.). If you measure the signals and find them to be of the correct frequency, amplitude, and shape, the oscillators or generators are good from a troubleshooting standpoint. If not, the test results provide a good starting point for troubleshooting.

Basic oscillator/generator tests

The first step in troubleshooting oscillator/generator circuits is to measure the amplitude, frequency, and shape of the output signals. You can generally skip the waveshape measurements for audio, RF, and video circuits (chapters 1 through 3). Many oscillator/generator circuits have a built-in test point. For example, the sine and cosine outputs are available from the first and second 747s, respectively, in Fig. 5-1. Likewise, the triangular and square-wave outputs are available from the first and second 3403s, respectively, in Fig. 5-2.

In the case of an RF oscillator, the signal can be monitored at the collector or emitter (Fig. 5-Aa). In this case, signal amplitude is monitored with a meter or scope using an RF probe. If you are only interested in the frequency, use a frequency counter.

Oscillator frequency problems

When you measure the oscillator signal, the frequency is (1) right on, (2) slightly off, or (3) way off. If the frequency is right on, leave the oscillator alone (unless the amplitude is low). If the frequency is slightly off, it is possible to correct the problem with adjustment. Many oscillators are adjustable—even those with crystal control. The most precise adjustment is made by monitoring the oscillator signal with a frequency counter and adjusting the circuit for exact frequency. However, it is also possible to adjust an oscillator with a meter or scope.

Generally, when an oscillator circuit is adjusted for maximum signal amplitude, the oscillator is at the crystal frequency. However, it is possible (but not likely) that the oscillator is being tuned for a harmonic (multiple or submultiple) of the crystal frequency. A frequency counter shows this, whereas a meter or scope does not.

If the oscillator frequency is way off, look for a defect, rather than improper adjustment. For example, a coil or transformer might have shorted turns, a transistor or capacitor might be leaking badly, and IC or crystal might be defective, or you might have wired the circuit incorrectly (impossible?).

Measuring the frequency with a scope If you do not have a frequency counter and must measure frequency with a scope, use the following procedure.

Adjust the scope controls so that you can measure the duration of one complete cycle (along the horizontal trace). For example (Fig. 5-22B), trace C; two complete cycles occupy one horizontal division. Because each horizontal division is 50 ns, each cycle is 25 ns (or 25^{-9} s) in duration. Find the reciprocal of 25 ns, or 1/25 = $0.04 \times 10^9 = 40^6$ = 40 MHz.

Oscillator signal-amplitude problems

When you measure the oscillator signal, the amplitude is (1) right on, (2) slightly low, or (3) very low. If the amplitude is right on, leave the oscillator alone (unless the frequency is off). If the amplitude is slightly low, it is sometimes possible to correct the problem with adjustment. Monitor the signal with a meter or scope, and adjust the oscillator for maximum signal amplitude. This also locks the oscillator on the correct frequency (in the case of adjustable crystal-controlled oscillators). If the adjustment does not correct the problem, look for leakage in transistors and capacitors, or look for a possible defective IC.

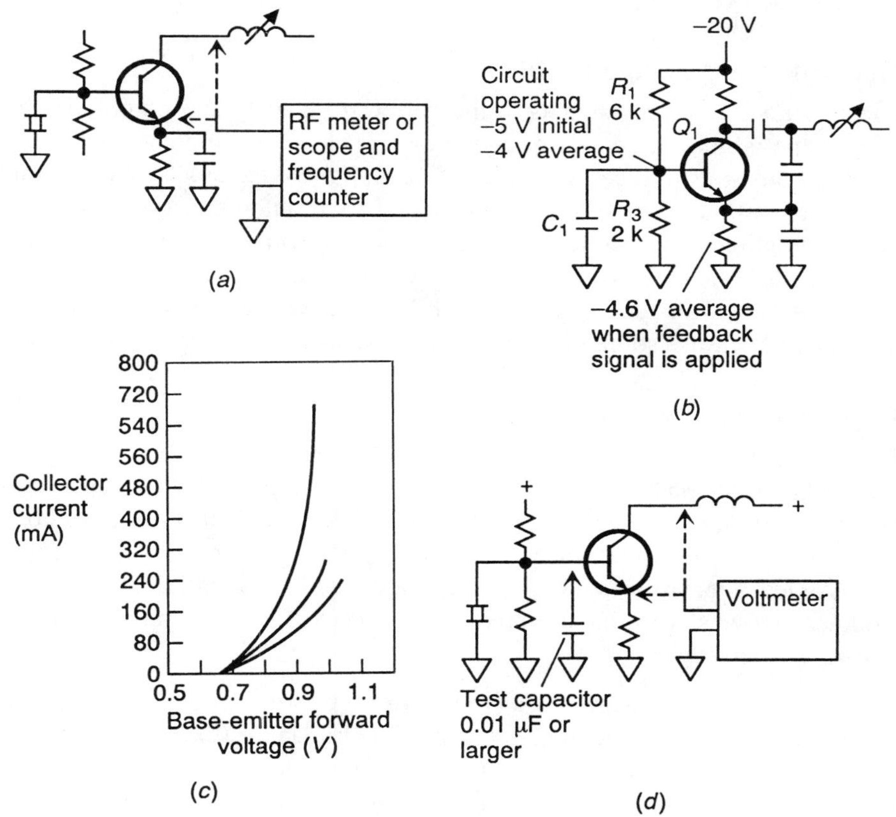

Fig. 5-A Oscillator testing and troubleshooting.

If the amplitude is very low, look for defects (such as low power-supply voltages, badly leaking transistors and/or capacitors, defective ICs, and a shorted coil or transformer turns). Usually, if the signal output is very low, there are other indications, such as abnormal voltages and resistance values.

Determine the output voltage amplitude If you are wondering what output voltage to expect from an oscillator, check the power supply or source voltage (often shown as V_S on the circuit schematics of this chapter). The output voltage will be slightly less than V_S. For example, in Fig. 5-1, both the sine and cosine outputs will be slightly less than 15 V. The same is generally true of the RF oscillator in Fig. 5-Ad. However, in the oscillator of Fig. 5-Ab, the output voltage will be less than 20 V because of the drop across the collector resistor of Q1.

Measuring amplitude with a scope If you want to measure oscillator signal amplitude with a scope, use the following procedure. Adjust the scope controls so that you can measure the amplitude of several cycles (along the vertical scale). For example, as shown in Fig. 5-22B, trace C, the signal occupies one vertical division. Because each vertical division represents 5 V, the signals are 5V in amplitude.

Generator waveshape problems

If a generator produces signals at the correct frequency and amplitude, but the waveshape is not correct, suspect leakage. Usually, leaking capacitors and/or a leaking transistor (in the case of discrete circuits) are the culprit. For example, if either or both the triangle-wave and square-wave outputs in Fig. 5-2 are of the correct amplitude and frequency, but are distorted (square-wave sides not straight, tops not flat, triangle-wave ramps bending in or out, etc.) suspect capacitor C.

Both transistors and capacitors can be checked for leakage, as described in chapter 1. Also, it is sometimes possible to check capacitors in-circuit, as follows.

Capacitor quick checks

During the troubleshooting process, suspected capacitors can be removed from the circuit and tested on bridge-type checkers. This establishes that the capacitor value is correct. With a correct value, it is reasonable to assume that the capacitor is not open, shorted or leaking. From another standpoint, if the capacitor shows no shorts, opens or leakage, it is also fair to say that the capacitor is good. So, from a practical troubleshooting standpoint, a simple test that checks for shorts, opens, or leakage is usually sufficient.

Checking capacitors with circuit voltages As shown in Fig. 5-Ba, this method involves disconnecting one lead of the capacitor (the ground or cold end) and connecting a voltmeter between the disconnected lead and ground. In a good capacitor, there should be a momentary voltage indication (or surge) as the capacitor charges up to the voltage at the hot end.

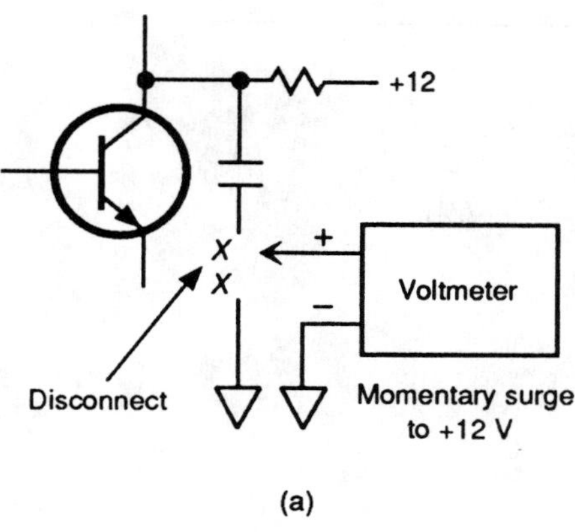

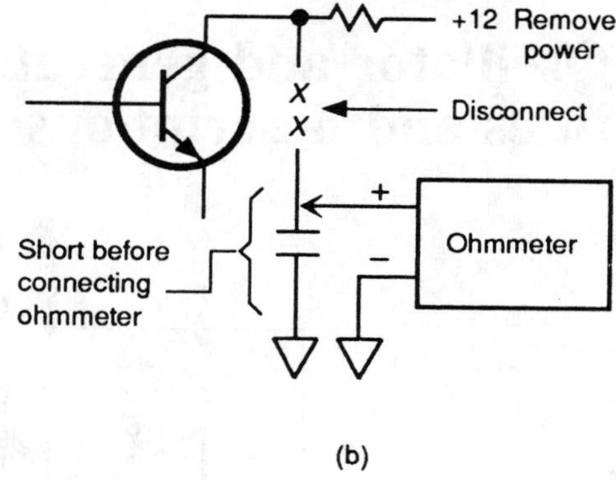

Fig. 5-B RF-circuit capacitor tests.

If the voltage indication remains high, the capacitor is probably shorted. If the voltage indication is steady, but not necessarily high, the capacitor is probably leaking. If there is no voltage indication whatsoever, the capacitor is probably open. Notice that this test is good only where one end of the capacitor is connected to a point in the circuit with a measurable voltage above ground.

Checking capacitors with an ohmmeter As shown in Fig. 5-Bb, this method involves disconnecting one lead of the capacitor (usually the hot end) and connecting an ohmmeter across the capacitor. Make certain that all power is removed from the circuit. As a precaution, short across the capacitor (after the power is removed) to make sure that no charge is retained. In a good capacitor, there should be a momentary resistance indication (or surge) as the capacitor charges up to the voltage of the ohmmeter battery.

If the resistance indication is near zero and remains so, the capacitor is probably shorted. If the resistance indication is steady at some high value, the capacitor is probably leaking. If there is no resistance indication (or surge) whatsoever, the capacitor is probably open.

Oscillator and generator circuits titles and descriptions

5

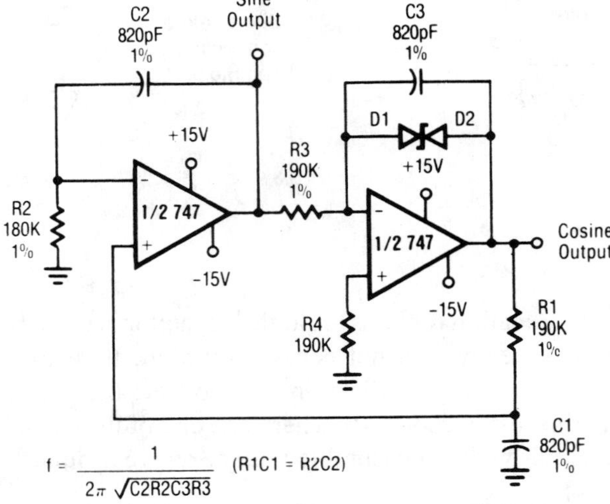

Quadrature oscillator

Fig. 5-1 This circuit uses both sections of a 747 op amp. The frequency of the sine and cosine output is set by the values of R_3, C_3, R_2, and C_2, as shown. Raytheon Linear Integrated Circuits, 1989, p. 4-148.

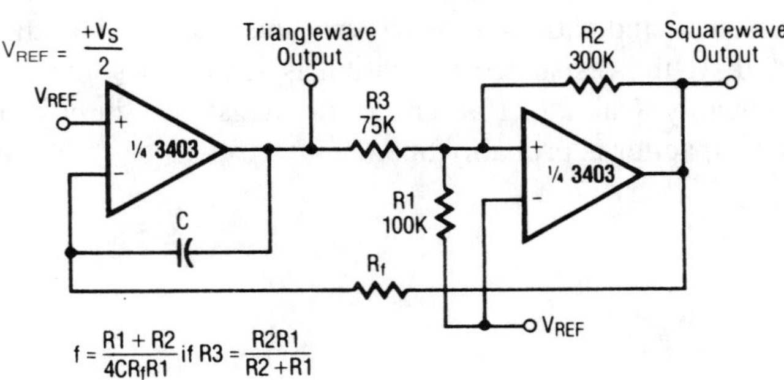

Function generator

Fig. 5-2 This circuit uses two sections of a 3403 op amp. The frequency of the triangle- and square-wave output is set by the values of R_1, R_2, C, and R_f, as shown. Raytheon Linear Integrated Circuits, 1989, p. 4-159.

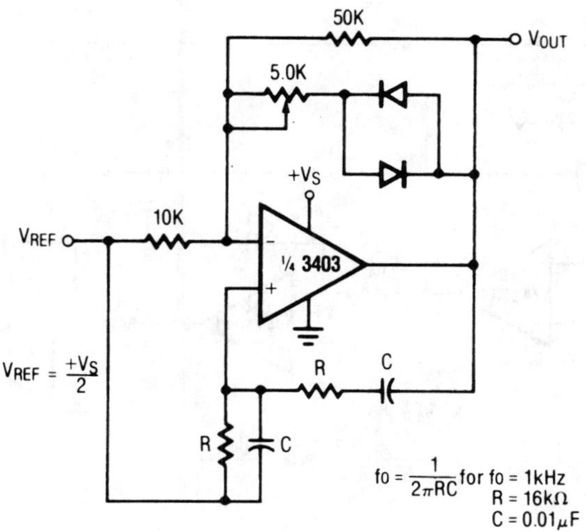

Wien-bridge oscillator

Fig. 5-3 The frequency of V_{OUT} is set by the values of R and C, as shown. Adjust the 5- kΩ potentiometer until the V_{REF} point is one-half of the source voltage, V_S.
Raytheon Linear Integrated Circuits, 1989, p. 4-161.

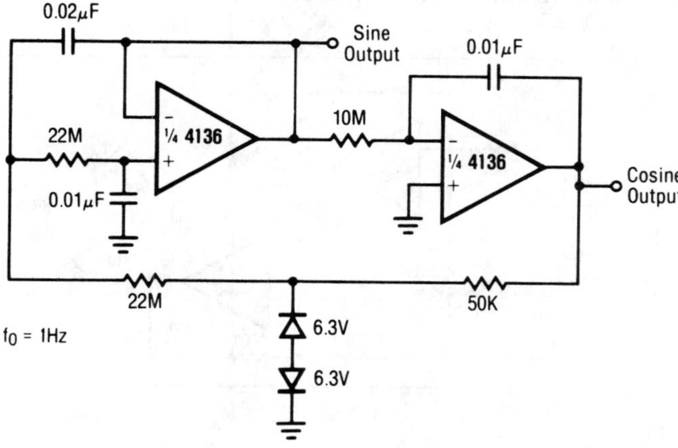

Low-frequency sine-wave generator with quadrature output

Fig. 5-4 This circuit uses two sections of a 4136 op amp. The frequency of both the sine and cosine output is 1 Hz with the values shown. Raytheon Linear Integrated Circuits, 1989, p. 4-173.

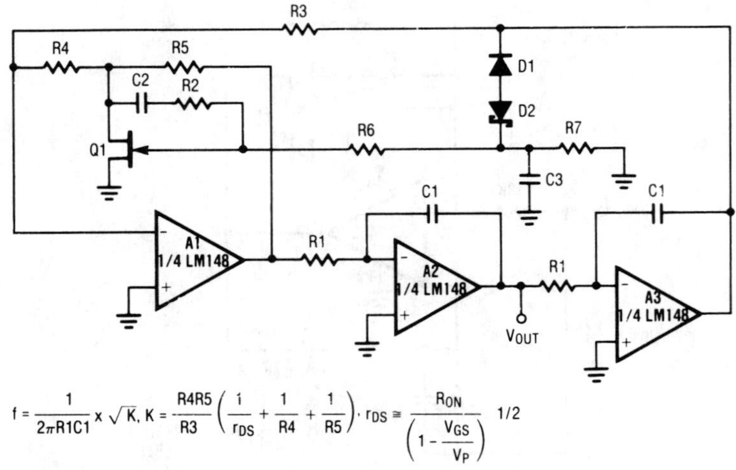

$f_{MAX} = 5.0\text{kHz}$, THD $\leq 0.03\%$
R1 = 100K pot., C1 = 0.0047μF, C2 = 0.01μF, C3 = 0.1μF, R2 = R6 = R7 = 1M, R3 = 5.1K, R4 = 12Ω,
R5 = 240Ω, Q1 = NS5102, D1 = 1N914, D2 = 3.6V avalanche diode (ex. LM103), $V_S = \pm 15V$
A simpler version with some distortion degradation at high frequencies can be made by using A1 as a simple inverting amplifier, and by putting back to back zeners in the feedback loop of A3.

One-decade low-distortion sine-wave generator

Fig. 5-5 This circuit uses three sections of an LM148 op amp. The frequency of V_{OUT} is set by R_1, and is 5 kHz maximum with the values shown. Raytheon Linear Integrated Circuits, 1989, p. 4-263.

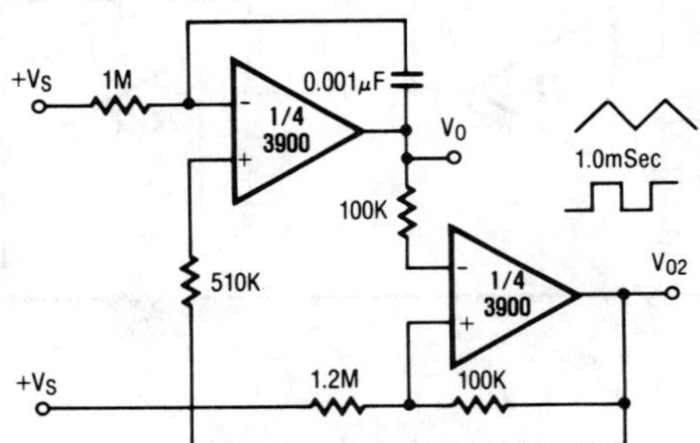

Triangle/square-wave generator

Fig. 5-6 This circuit uses two sections of 3900 op amp. The duration of the triangle-(V_0) and square-wave (V_{02}) is 1.0 mS with the values shown. Raytheon Linear Integrated Circuits, 1989, p. 4-273.

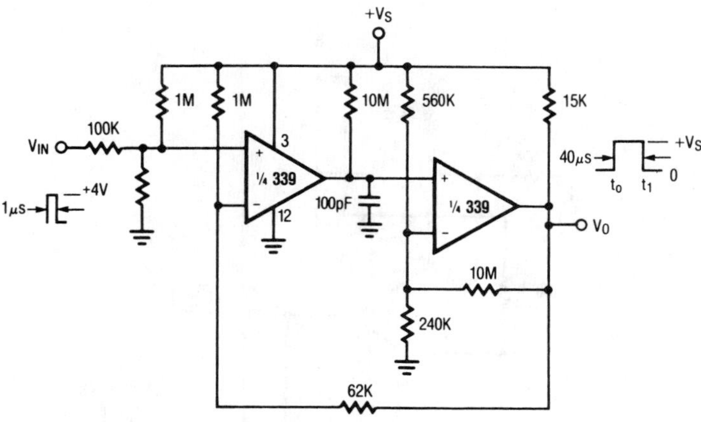

One-shot multivibrator with input lockout

Fig. 5-7 The input threshold of this comparator IC circuit is set by the resistance from the input to ground. This resistance combines with the series 100-kΩ resistor to form an input lockout. For example, if the input-to-ground resistance is also 100 kΩ, the multivibrator remains off until the input exceeds +8 V. Raytheon Linear Integrated Circuits, 1989, p. 5-30.

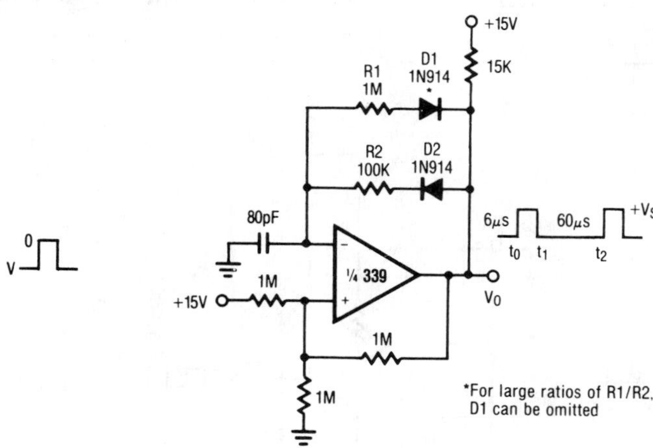

Pulse generator

Fig. 5-8 This circuit uses only one section of a 339 comparator and produces pulses of 6-µs duration with 60-µs spacing, using the values shown. Raytheon Linear Integrated Circuits, 1989, p. 5-31.

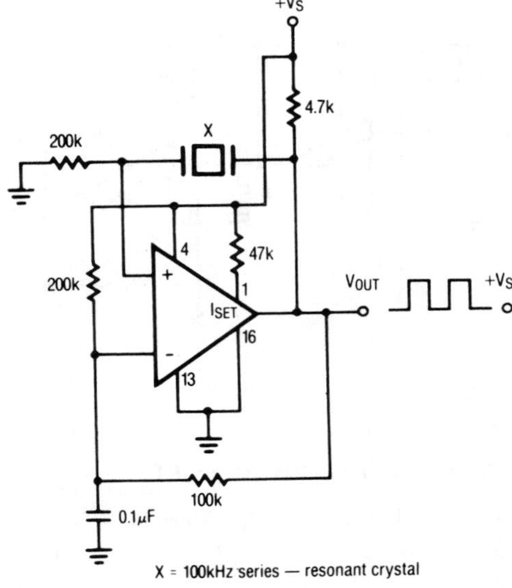

Crystal-controlled oscillator

Fig. 5-9 This circuit uses only one section of an LP165/365 comparator, and produces pulses at 100 kHz using the values shown, with a single supply. Raytheon Linear Integrated Circuits, 1989, p. 5-42.

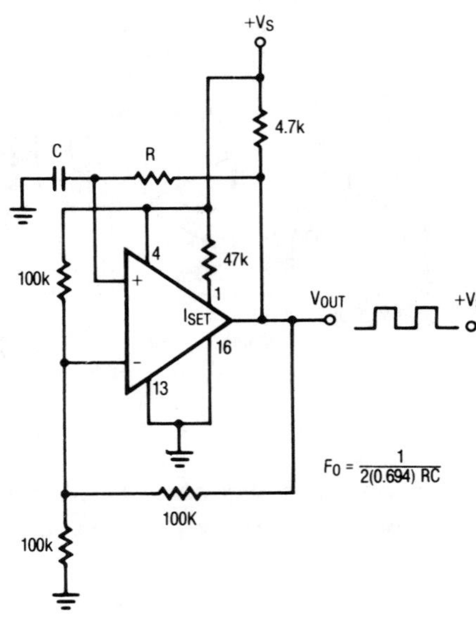

Square-wave oscillator

Fig. 5-10 This circuit uses only one section of an LP165/365 comparator, and produces square waves at a frequency that is determined by the values of R and C, as shown. Raytheon Linear Integrated Circuits, 1989, p. 5-42.

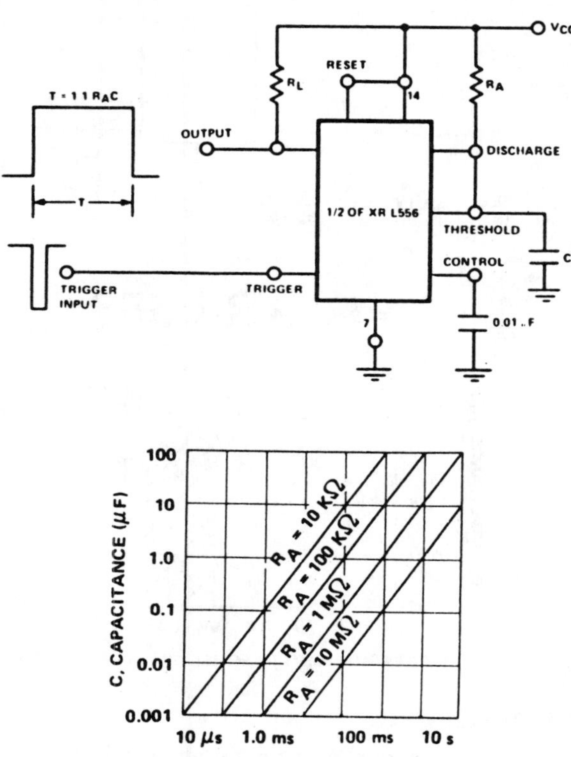

Monostable (one-shot)

Fig. 5-11 This circuit uses one section of an XR-L556 dual timer. Duration of the output pulse is set by C and R_A (Fig. 5-11B). R_L should equal the load being driven. EXAR Corporation Databook, 1990, p. 5-218/5-220.

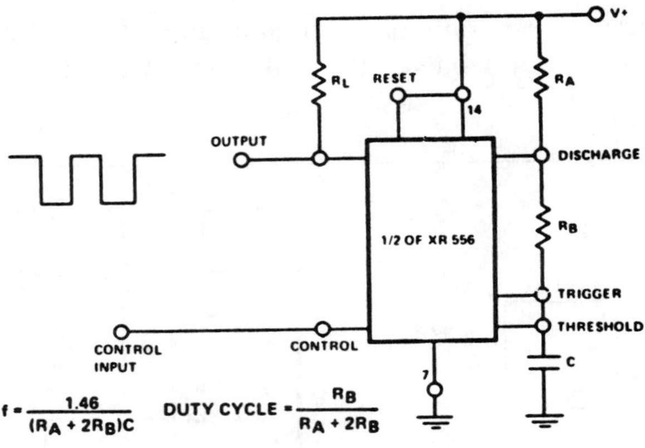

XR-L556

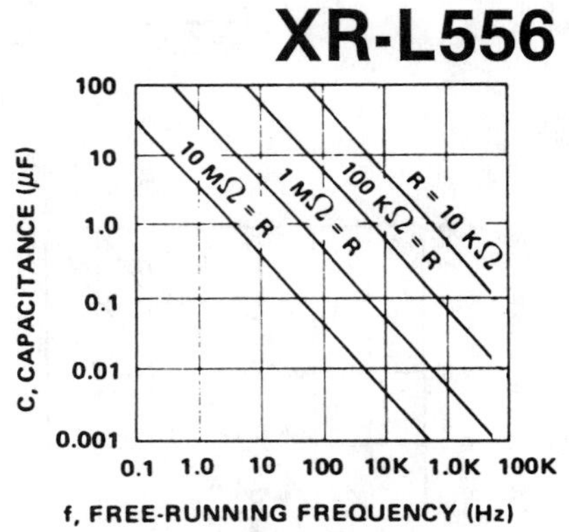

Astable (free running)

Fig. 5-12 This circuit uses one section of an XR-L556 dual timer. Output frequency and duty cycle are set by C and R (Fig. 5-12B). Notice that $R = R_A + 2R_B$ (Fig. 5-12B). EXAR Corporation Databook, 1990, p. 5-218/5-221.

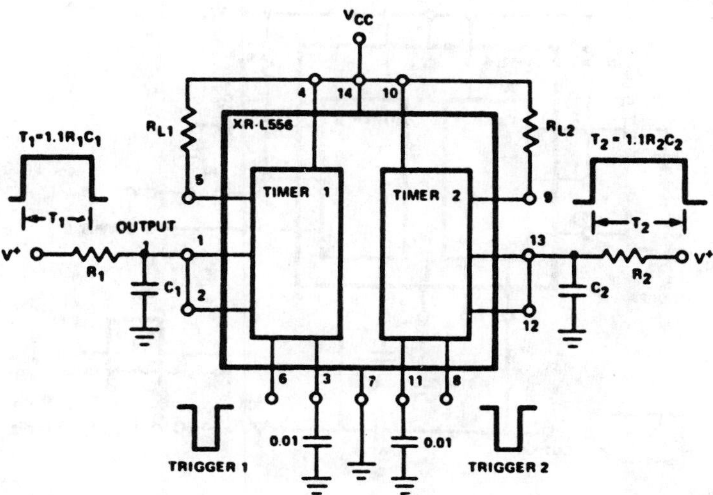

Independent time delay

Fig. 5-13 This circuit uses both sections of an XR-L556 dual timer to produce two independent time delays. Each section is used separately in the monostable mode to produce respective time delays T_1 and T_2. EXAR Corporation Databook, 1990, p. 5-221.

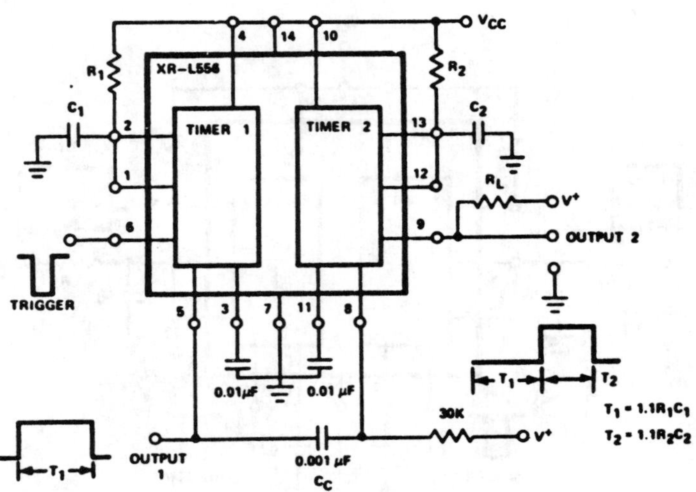

Sequential timing (delayed one-shot)

Fig. 5-14 In this circuit, the output of one timer section triggers the other section so that the output of T_2 is delayed from the initial trigger at pin 6 by a time delay of T_1. Both T_1 and T_2 are set by values of external components, as shown. EXAR Corporation Databook, 1990, P. 5-221.

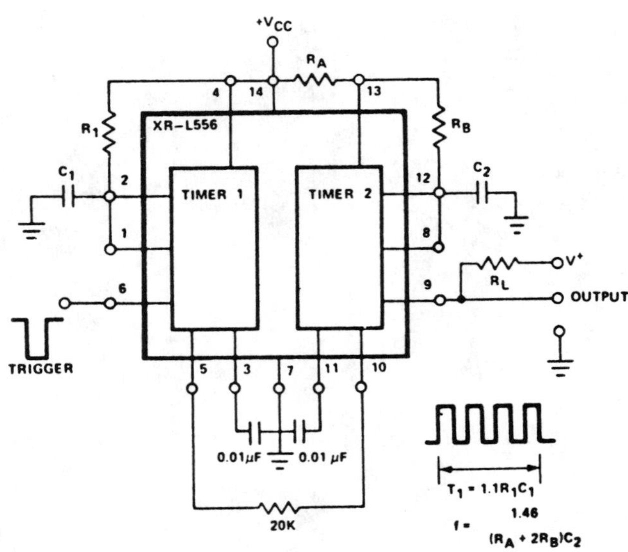

Keyed oscillator

Fig. 5-15 In this circuit, one timer section of an XR-L556 is operated in the free-running mode, and is keyed on and off by the other section. The circuit output (timer 2) appears as a tone burst, whose frequency is set by R_A, R_B, and C_2, and whose duration is set by R_1 and C_1. EXAR Corporation Databook, 1990, p. 5-222.

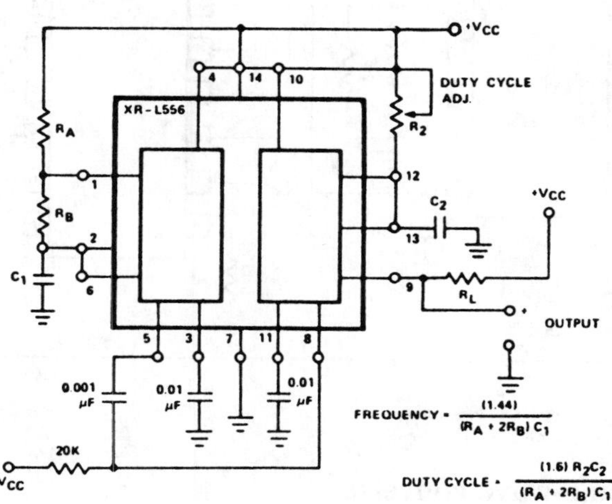

Micropower oscillator with fixed frequency and variable duty cycle

Fig. 5-16 In this circuit, timer 1 is operated in the astable mode, and timer 2 is operated monostable; timer 1 triggers timer 2. The output (pin 9) has the same frequency as timer 1, but with a duty cycle determined by the timing cycle of timer 2. The output duty cycle can be adjusted from 1 to 99% by R2, typically a 10-kΩ potentiometer. EXAR Corporation Databook, 1990, p. 5-222.

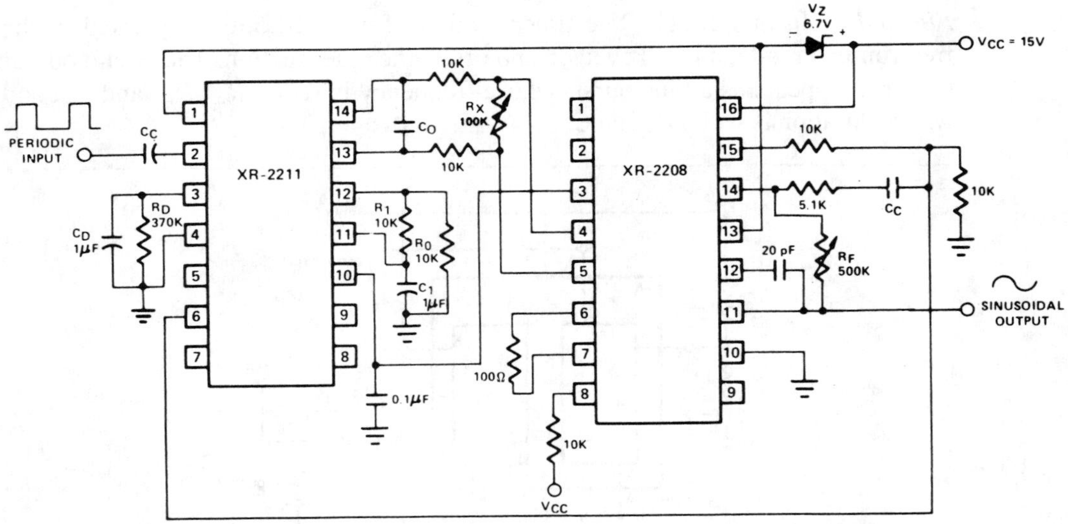

R_X = Distortion Adj. Potentiometer
R_F = Output Amplitude Adj. Pot.

C_C = Coupling Capacitor ($\geqslant 0.1\ \mu F$)

Regenerative sine-wave converter

Fig. 5-17 This circuit uses an XR-2211 PLL and an XR-2208 multiplier to form a universal sine-wave converter. The circuit converts any periodic input-signal waveform to a low-distortion sine-wave, the frequency of which is identical to the repetition rate of the input signal. The center frequency, f_O, is determined by the values of C_O and R_O. With R_O at 10 kΩ, as shown, the value of C_O is determined by: $C_O(\text{in } \mu F) = 100/f_O$ in Hz. Notice that the XR-2211 VCO might not oscillate if R_O is greater than 10 kΩ. The distortion and amplitude of the sine-wave output is adjusted by R_X and R_F, as shown. EXAR Corporation Databook, 1990, p. 5-312.

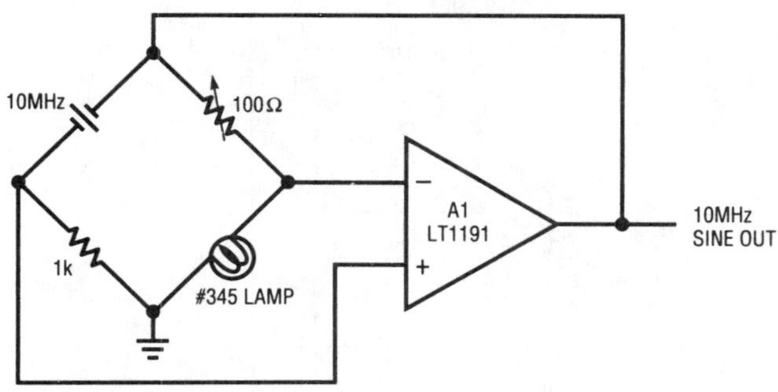

Basic quartz-stabilized oscillator

Fig. 5-18 This circuit uses a crystal within an amplifier's feedback path to create an oscillator. Although shown as 10 MHz, the circuit works well with a wide variety of crystal types over a 100-kHz to 20-MHz range. The use of a lamp to control amplifier gain is a classic technique that was first used in 1938. Linear Technology Corporation, 1991, AN47-49.

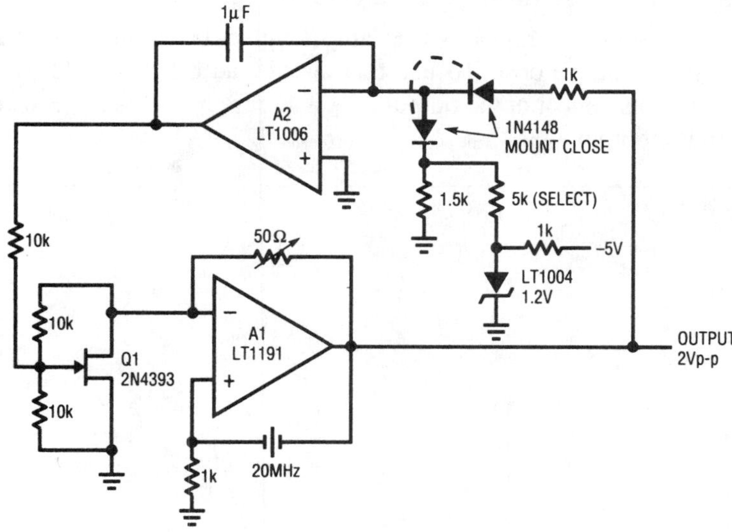

Continued

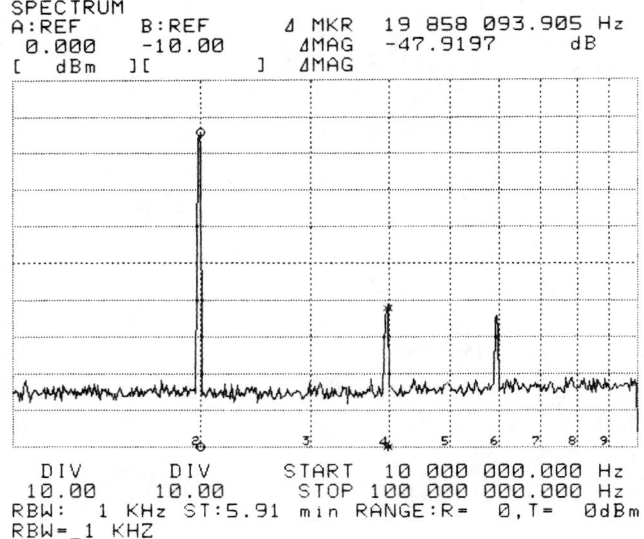

Quartz-stabilized oscillator with electronic gain control

Fig. 5-19 This circuit replaces the lamp (Fig. 5-18) with an electronic gain control or stabilization loop. To use this circuit, adjust the 50-Ω trimmer until 2-Vpp oscillations appear at the output of A1. Figure 5-19B is a spectrum analysis of the oscillator output. Linear Technology Corporation, 1991, AN47-50.

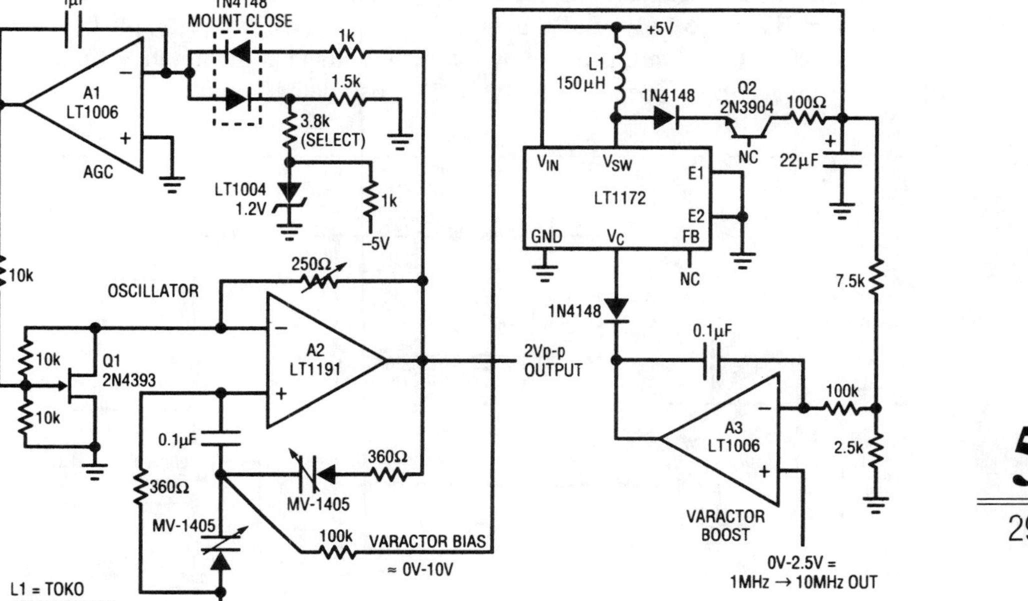

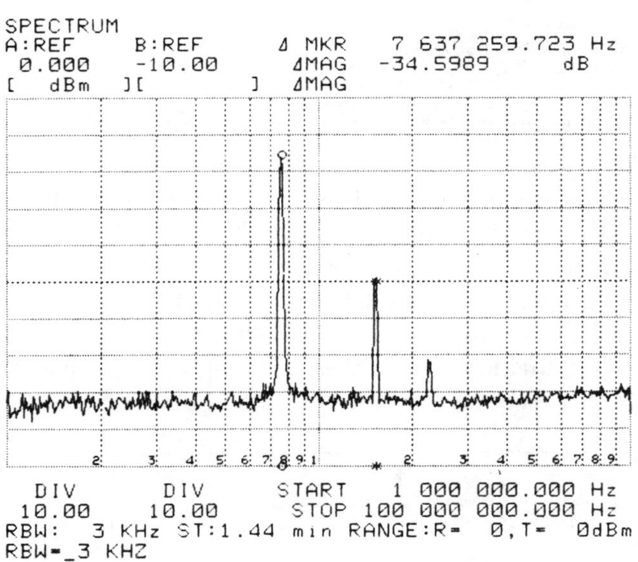

Varactor-tuned Wien-bridge oscillator

Fig. 5-20 This circuit replaces the lamp (Fig. 5-18) and electronic control (Fig. 5-19) with a Wien network to produce a variable-frequency output from 1 to 10

MHz. Usually, Wien circuits require manually adjustable elements (dual pots or two-section variable capacitors) for tuning. In this circuit, the Wien-network resistors are fixed (360 Ω), and the capacitive elements are accomplished with varactor diodes. The circuit is tuned by a control voltage at the noninverting input of A3. Figure 5-20B is a spectrum analysis of the oscillator output. Linear Technology Corporation, 1991, AN47-51.

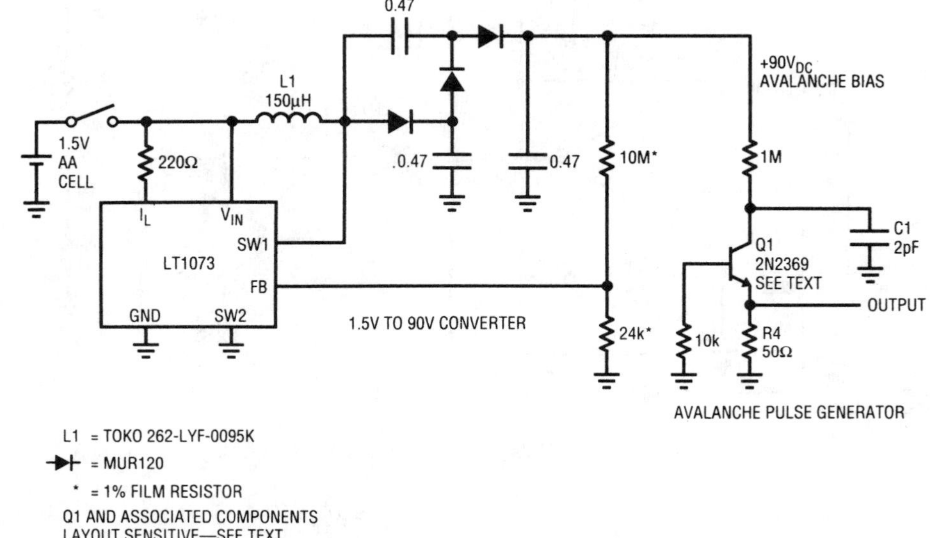

Avalanche pulse generator for rise/fall measurements

Fig. 5-21 This circuit draws only about 5 mA from the 1.5-V battery, and produces a 10-V pulse with a typical rise and fall time of about 350 ps. Q1 might require selection to get the desired avalanche behavior. A sample of 50 Motorola 2N2369s, spread over a 12-year date-code span, yielded 82% (with some in the 220- to 230-ps range). All good 2N2369s switched in less than 650 ps. C1 is selected for a 10-V amplitude output. The value spread is typically 2 to 4 pF. Ground-plane construction with high-speed layout techniques are essential for good results from this circuit. Linear Technology Corporations 1991, AN47-93.

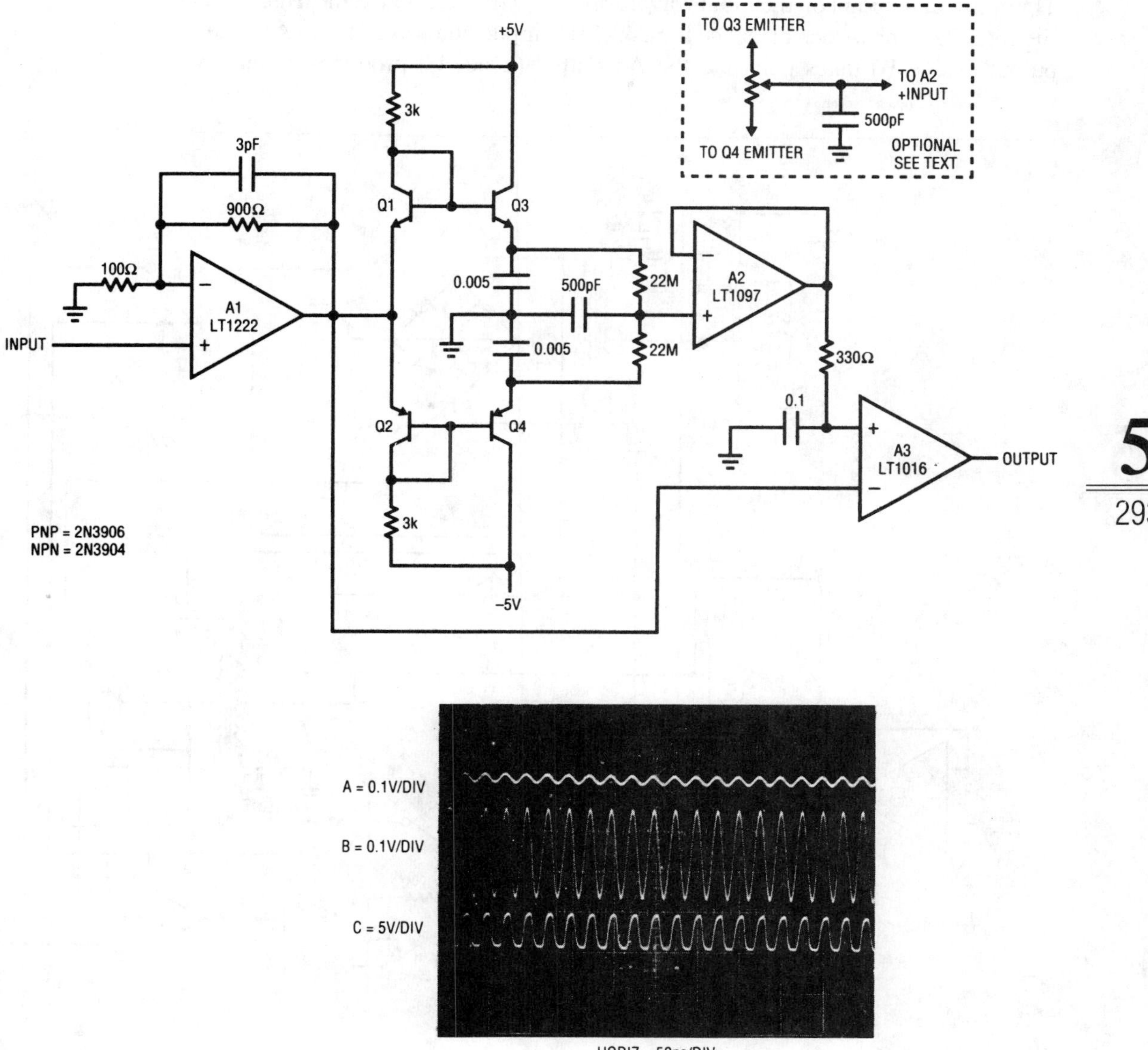

Trigger with adaptive threshold

Fig. 5-22 This circuit triggers from dc to 50 MHz over a 2- to 300-mV input range with no level adjustment. The circuit maintains the A3 output trip-point at one-half the input amplitude, regardless of input magnitude. This ensures reliable automatic triggering over a wide input amplitude range—even for very low-level inputs. As an option, the network shown in dashed lines permits changing the trip threshold. This allows any point on the input waveform edge to be selected as the trigger point. Figure 5-22B shows performance for a 40-MHz input sine wave (Trace A). The A1 output (Trace B) takes gain and the A3 output (Trace C) produces a clean 5-V signal. Linear Technology Corporation, 1991, AN47-59.

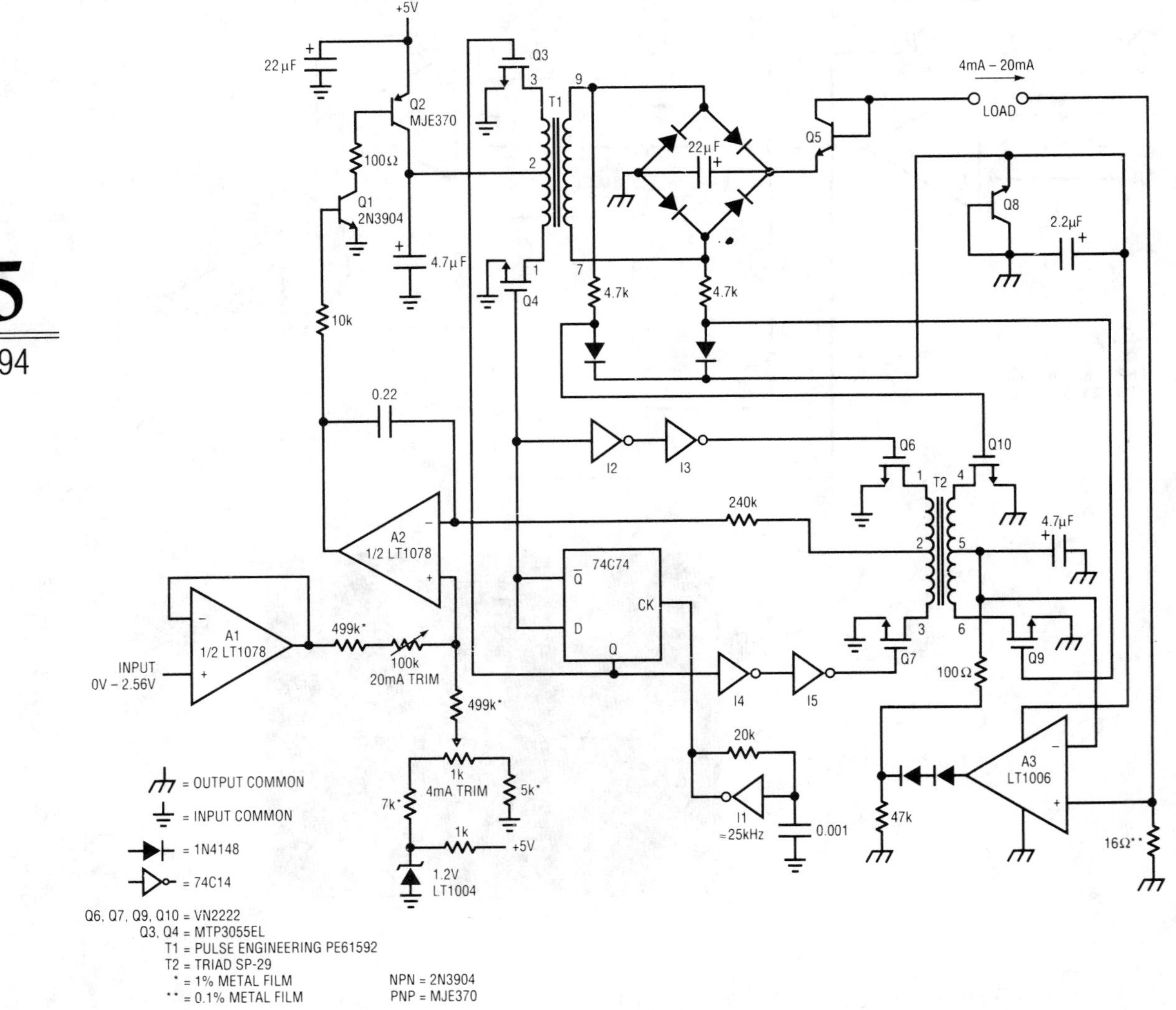

Floating current-loop transmitter

Fig. 5-23 A 4- to 20-mA current-loop transmitter is frequently required in industrial process control. Because of uncertain or dangerous common-mode voltages, it is desirable that the generated current be completely galvanically isolated from the transmitter input. This circuit does this while operating from a single 5-V supply. To calibrate this circuit, apply 0-V input and adjust the 4-mA trim for 4.00-mA output (0.064 V across the 16-Ω shunt). Next, apply 2.56-V input and set the 20-mA trim for 20.00-mA output (0.3200 V across the 16-Ω shunt). Repeat this procedure until both points are fixed. Notice that the 2.56-V input range is directly compatible with D/A converter (chapter 6) outputs, which permits digital control. Linear Technology Corporation, 1991, AN45-6.

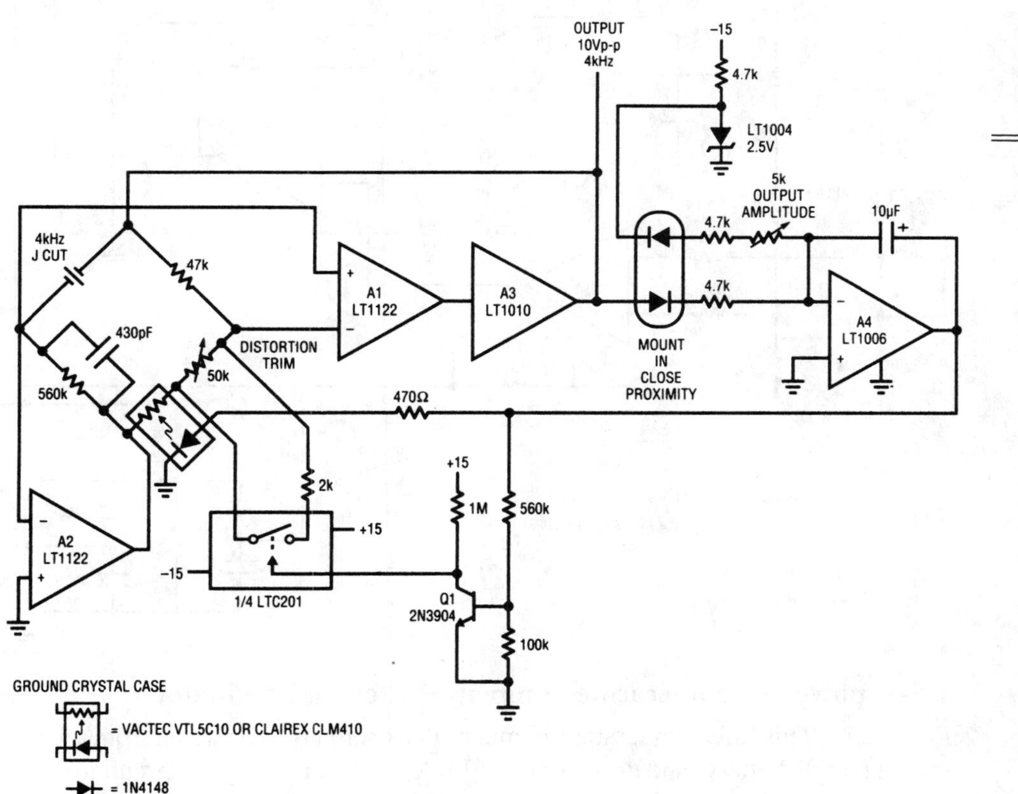

Quartz-stabilized 4-kHz oscillator with 9-ppm distortion

Fig. 5-24 A spectrally pure sine-wave oscillator is required for data converter, filter, and audio testing. This oscillator has less than 9 ppm (0.0009%) distortion in the 10-Vpp output. The circuit is adjusted for minimum distortion by adjusting the

50-kΩ potentiometer while monitoring the A3 output with a distortion analyzer. This trim sets the voltage across the photocell to the optimum value for lowest distortion. Linear Technology Corporation, 1991, AN45-12.

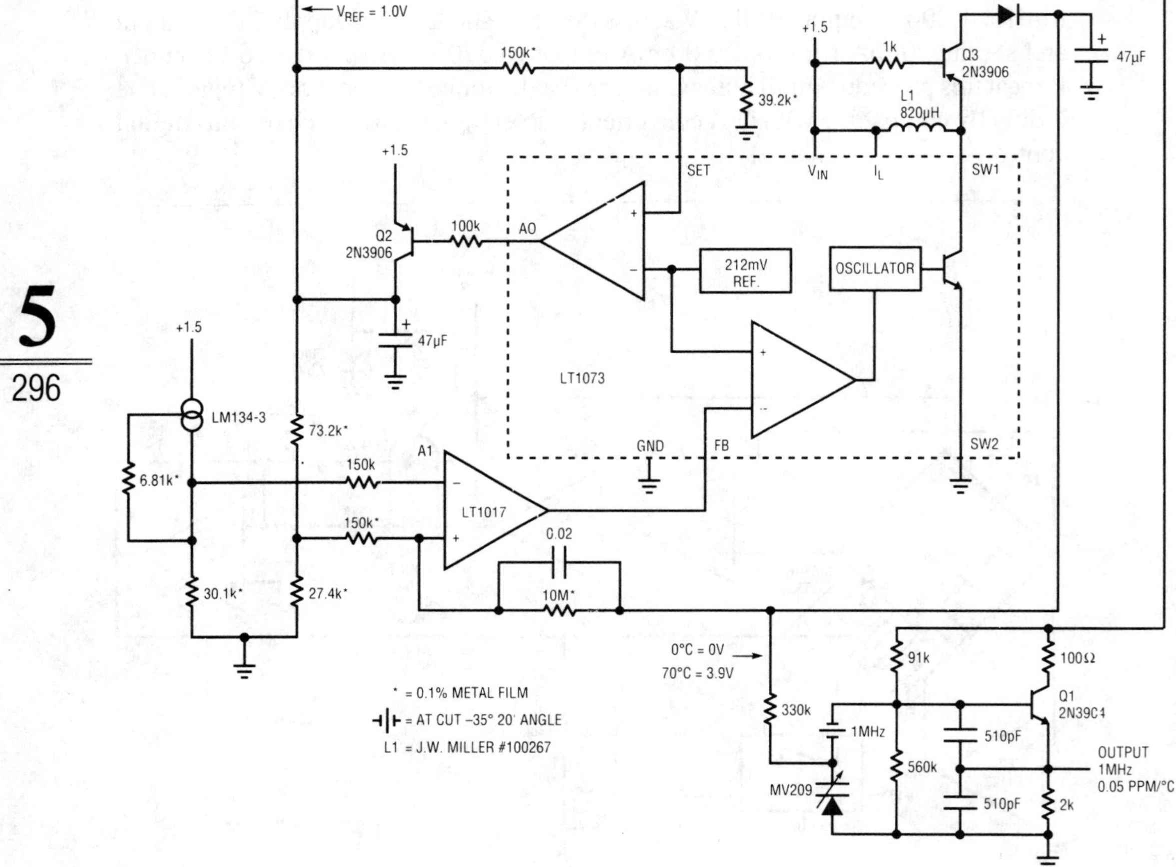

1.5-V powered temperature-compensated crystal oscillator

Fig. 5-25 This fully temperature-compensated oscillator is operated from a 1.5-V source (single battery) and draws only 230-µA current, making the circuit ideal for portable high-accuracy clocks, survival radios, and secure communications. Linear Technology Corporation, 1991, AN45-14.

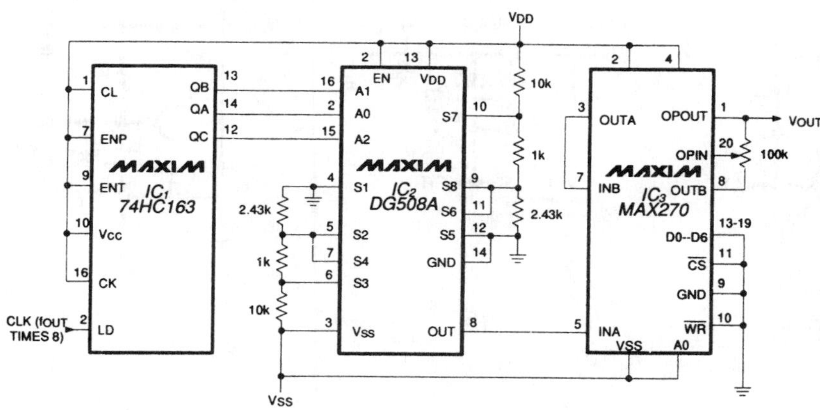

High-purity sine-wave generator

Fig. 5-26 By combining a MAX270 4th-order low-pass filter, a 74HC163 counter, and an 8-channel analog MUX, high-purity 1- to 25-kHz sine waves are generated with a THD of less than −80 dB. The low-pass filter is set to the desired frequency and a clock 8 times the sine frequency is applied to the 74HC163 counter. The MAX270 uncommitted op amp sets the filter output level. Gain accuracy is set by the MAC270 gain at the corner frequency, which is guaranteed between −2.4 and −3.6 dB at 1 kHz. The 100-kΩ potentiometer provides gain control if desired. Two resistor dividers provide the input voltages that are required at the MUX inputs. When the MUX switches through channels 0 to 7, an 8-times oversampled staircase approximation of a sine wave is generated. Compared with a square wave, the oversampled waveform greatly reduces smoothing-filter requirements by pushing the first significant harmonic out to 7 times the fundamental. All higher-order harmonics are filtered to below −80 dB by the MAX270. Maxim, 1992, Applications and Product Highlights, p. 7-7.

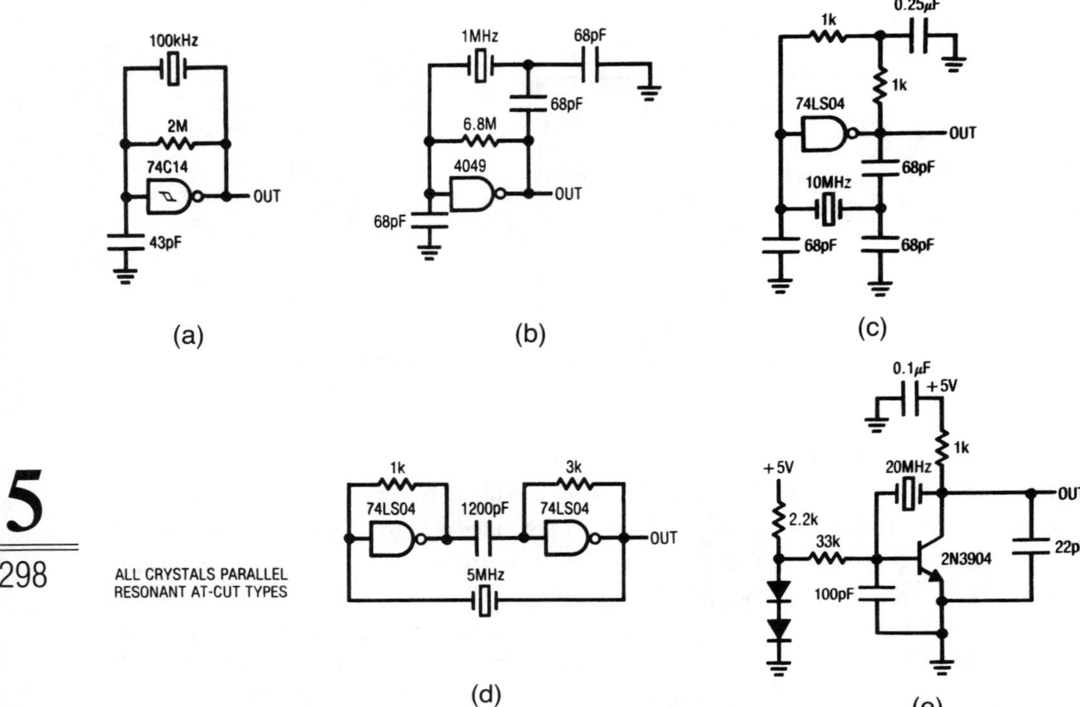

Simple clock oscillators

Fig. 5-27 These circuits show four basic clock oscillators composed of digital IC gates, and a discrete-component oscillator, all of which are well suited as clock sources. Notice that the diodes used in the discrete-component oscillator can be IN4148s or equivalent. Linear Technology Corporation, Linear Applications Handbook, 1990, p. AN12-2.

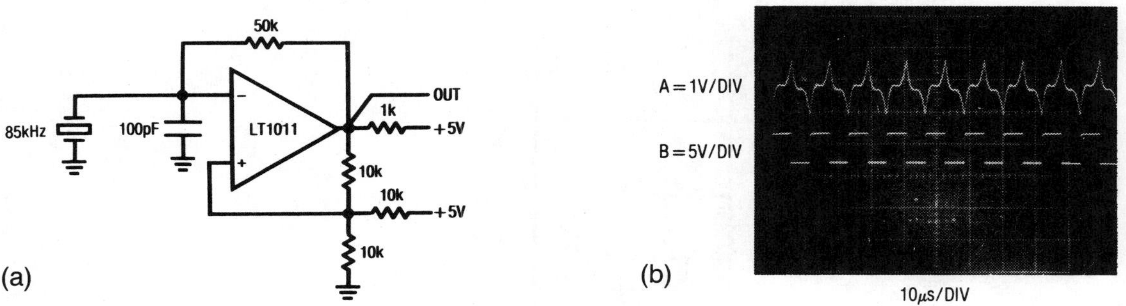

Crystal-stabilized relaxation oscillator

Fig. 5-28 This circuit shows an LT1011 connected as a simple relaxation oscillator with crystal control. Figure 5-28B shows the corresponding waveforms. Waveform A is taken from the inverting (−) input of the LT1011, while waveform B is taken from the LT1011 output. Linear Technology Corporation, Linear Applications Handbook, 1990, p. AN12-3, -2.

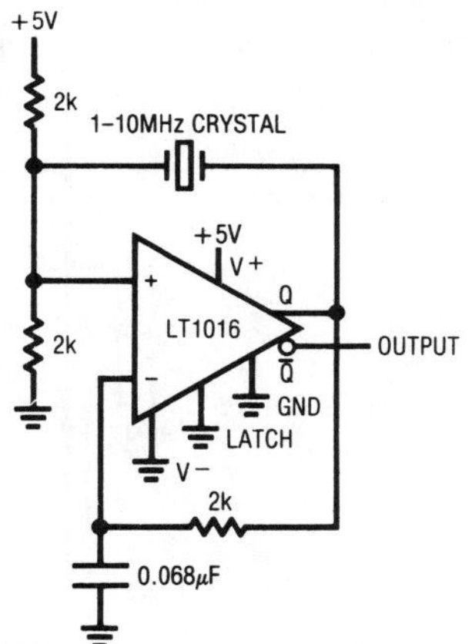

Crystal oscillator (1 to 10 MHz)

Fig. 5-29 This circuit shows an LT1016, which is connected as an oscillator suitable for operation in the 1- to 10-MHz range. Linear Technology Corporation, Linear Applications Handbook, 1990, p. AN12-3.

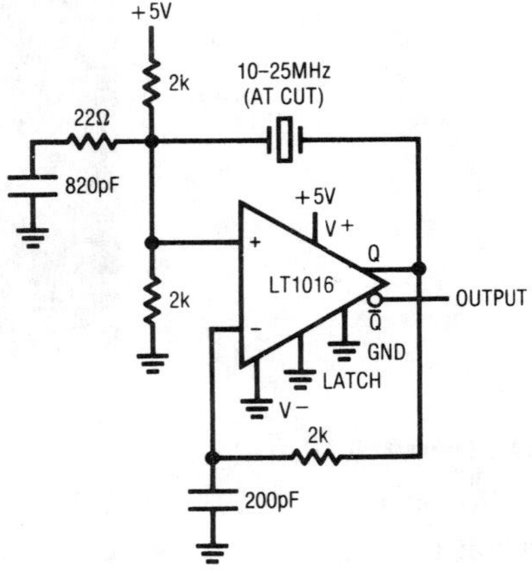

Crystal oscillator (10 to 25 MHz)

Fig. 5-30 This circuit shows an LT1016 connected as an oscillator, which is suitable for operation in the 10- to 25-MHz range. Notice that the damper network (C=820 pF, R=22) prevents the AT-cut crystal from operating in the overtone mode. Linear Technology Corporation, Linear Applications Handbook, 1990, p. AN12-3.

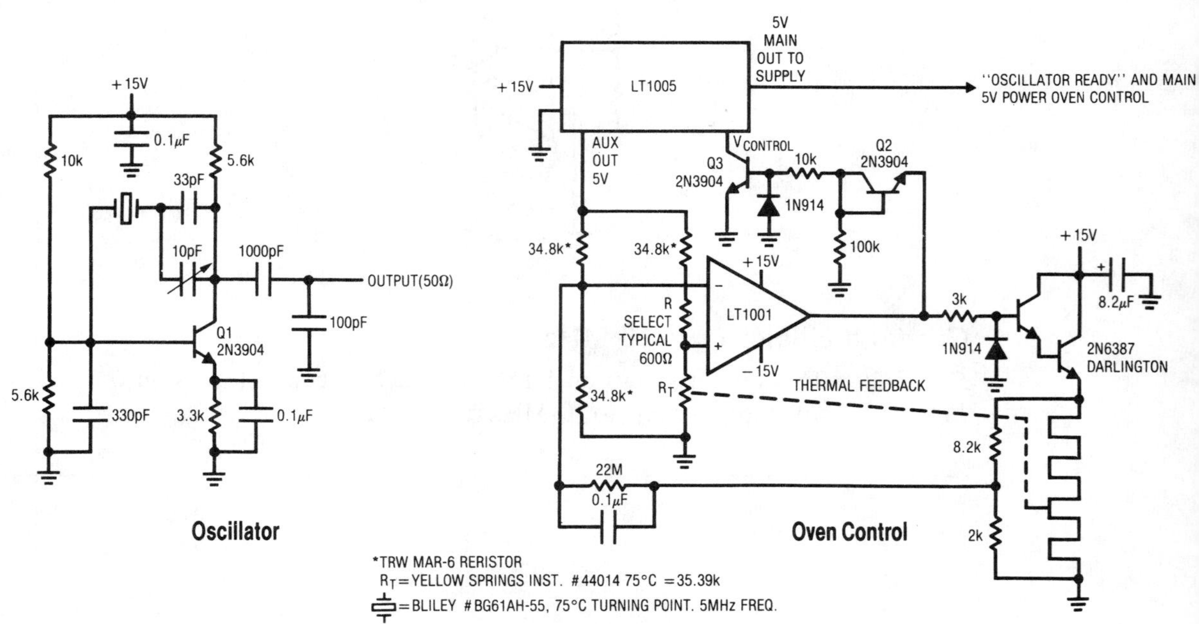

Oven-controlled crystal oscillator

Fig. 5-31 This circuit shows a Pierce-class oscillator with fine-frequency trimmer and an oven-control system. The LT1005 voltage regulator and the LT1001 op amp are used in a precision temperature-servo to control crystal temperature. Linear Technology Corporation, Linear Applications Handbook, 1990, p. AN12-3.

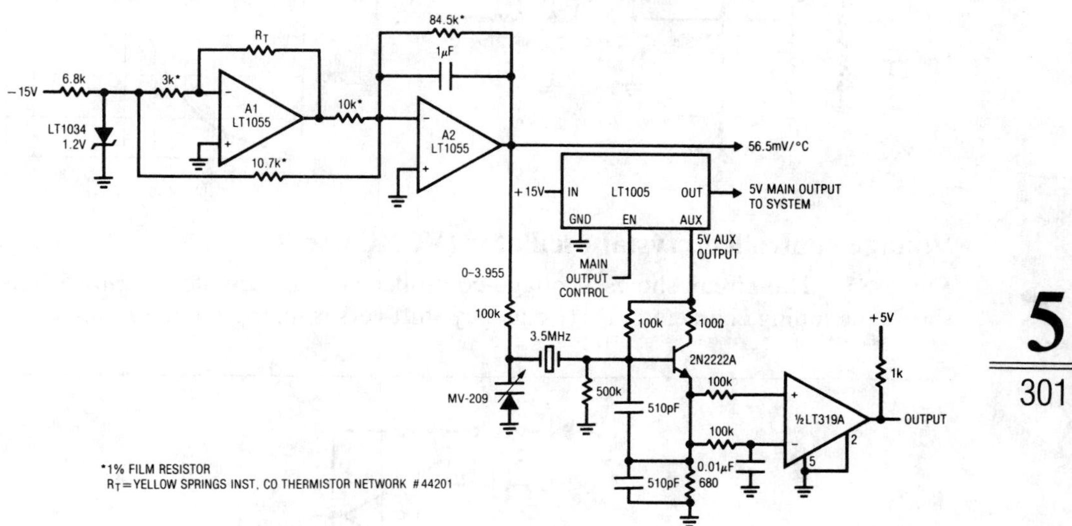

Temperature-compensated crystal oscillator (TCXO)

Fig. 5-32 This circuit shows a temperature-controlled crystal oscillator without an expensive, power-consuming oven. Instead, the crystal frequency is controlled by an MV-209 varactor. In turn, the MV209 is controlled by a linear thermistor in the feedback loop of A1. Linear Technology Corporation, Linear Applications Handbook, 1990, p. AN12-4.

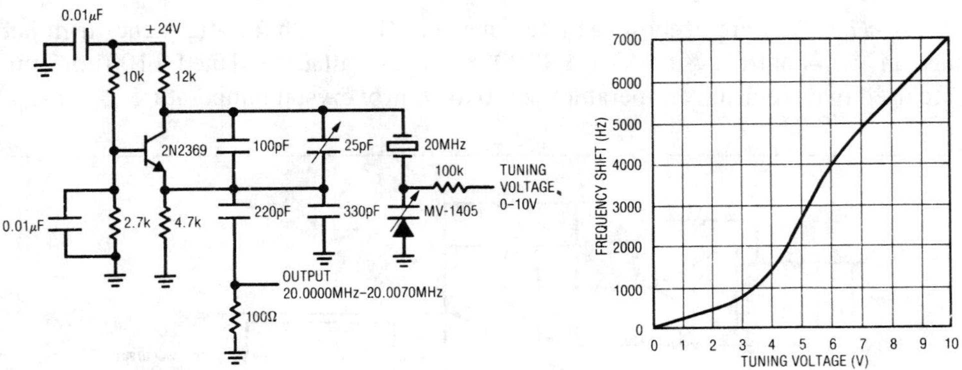

Voltage-controlled crystal oscillator (VCXO)

Fig. 5-33 This circuit shows a voltage-controlled crystal oscillator. Figure 5-33B shows the tuning characteristics (frequency shift versus tuning voltage). Linear Technology Corporation, Linear Applications Handbook, 1990, p. AN12-5.

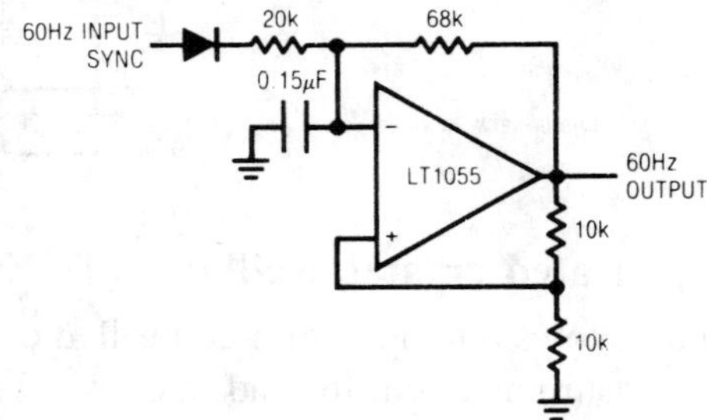

Synchronized oscillator

Fig. 5-34 This circuit shows a line-synchronized oscillator that will not lose lock under noisy line conditions. The circuit is suited to applications that require a reliable 60-Hz line-synchronous clock, and is superior to the usual zero-crossing detectors and simple voltage-level detectors. In this circuit, the basic RC multivibrator is tuned to free-run near 60 Hz, but the line-derived sync input forces the oscillator to lock on the ac line. Linear Technology Corporation, Linear Applications Handbook, 1990, p. AN12-6.

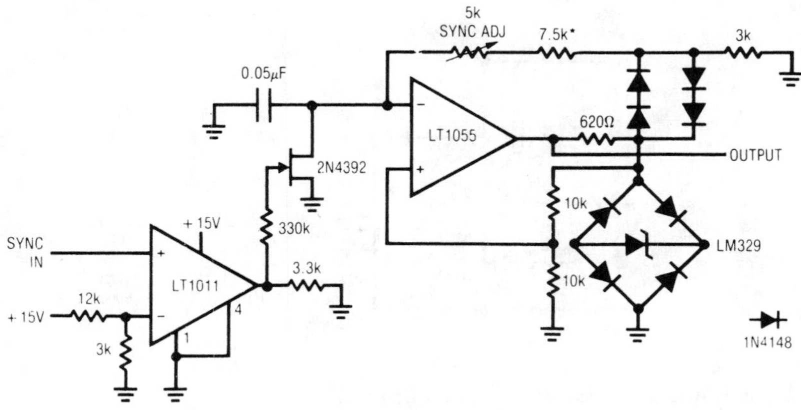

Reset-stabilized oscillator

Fig. 5-35 This circuit shows a synchronous clock generator, where the output locks at a higher frequency than the sync input. The maximum practical output-frequency to sync-frequency ratio is about 50 times. The output frequency is set to a multiple of the input frequency by the 5-kΩ Sync Adj pot. Linear Technology Corporation, Linear Applications Handbook, 1990, p. AN12-6.

5

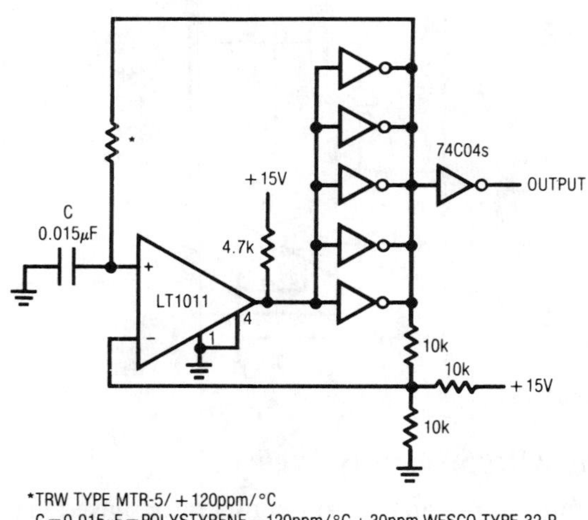

Stable RC oscillator

Fig. 5-36 This circuit shows an RC clock circuit, which depends primarily on the RC elements for stability. The nominal-120-ppm/°C temperature coefficient of the polystyrene capacitor is offset by the opposing positive temperature coefficient of the specified resistor. Linear Technology Corporation, Linear Applications Handbook, 1990, p. AN12-7.

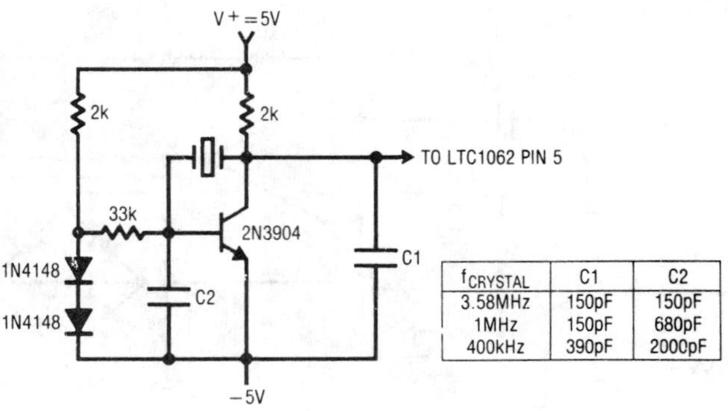

$f_{CRYSTAL}$	C1	C2
3.58MHz	150pF	150pF
1MHz	150pF	680pF
400kHz	390pF	2000pF

Low-cost oscillator with AT-cut crystal

Fig. 5-37 This circuit shows a crystal oscillator that is designed as a clock for an LTC1062 filter (chapter 7). The clock frequency is determined by the crystal and the values of C_1 and C_2, as shown by the table. Linear Technology Corporations Linear Applications Handbook, 1990, p. AN20-12.

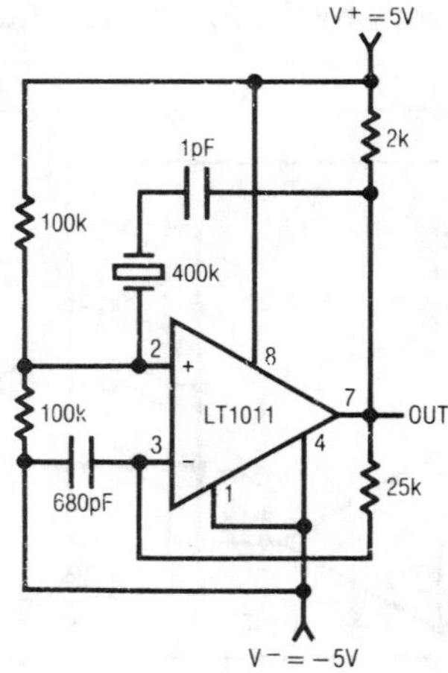

Crystal oscillator with 50% duty cycle

Fig. 5-38 This circuit shows an LT1011 comparator biased in the linear mode and using a crystal to establish the resonant frequency. This circuit can achieve a temperature-independent clock up to a few hundred kHz. Linear Technology Corporation, Linear Applications Handbook, 1990, p. AN20-12.

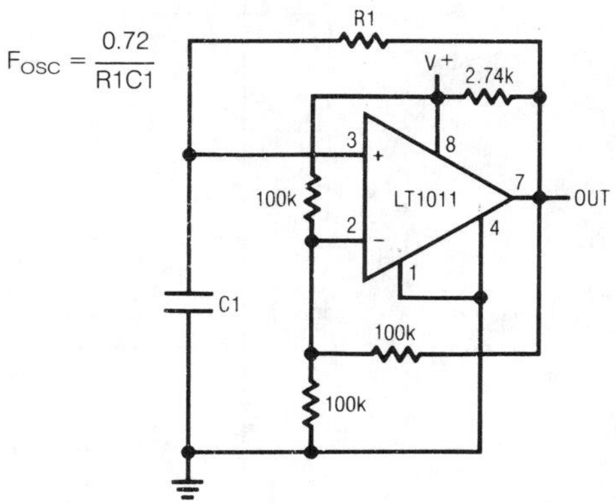

Low-frequency precision RC oscillator

Fig. 5-39 This circuit shows a precision RC oscillator, where the frequency is determined by the equation $0.72/(R_1C_1)$. The circuit is particularly useful for frequencies below 3 kHz. For a bipolar ±5-V output swing, refer the ground connection to −5 V. Linear Technology Corporation, Linear Applications Handbook, 1990, p. AN20-11.

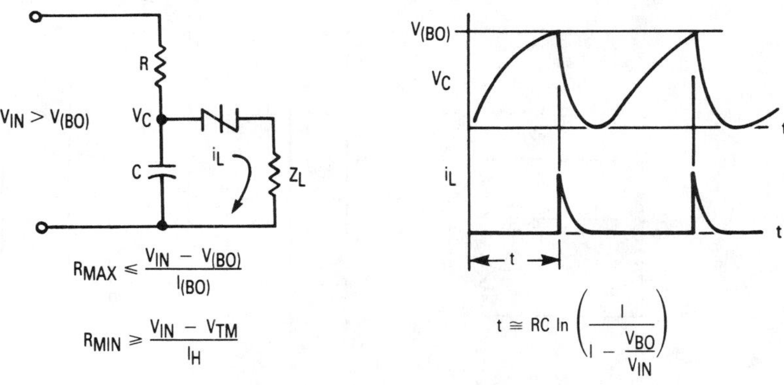

Relaxation oscillator using a SIDAC

Fig. 5-40 This figure shows the circuit and corresponding waveforms, where a SIDAC is used as a relaxation oscillator. Being a negative-resistance device, the SIDAC can be used as a simple relaxation oscillator, where frequency is determined primarily by the RC time constant. Once the capacitor voltage reaches the SIDAC breakover voltage $V_{(BO)}$, the SIDAC fires, and dumps the charged capacitor. Power can be obtained by placing the load in the discharge path. Such a circuit can be used as an xeon flasher (Fig. 8-14). Motorola Thyristor Device Data, 1991, p. 1-4-6.

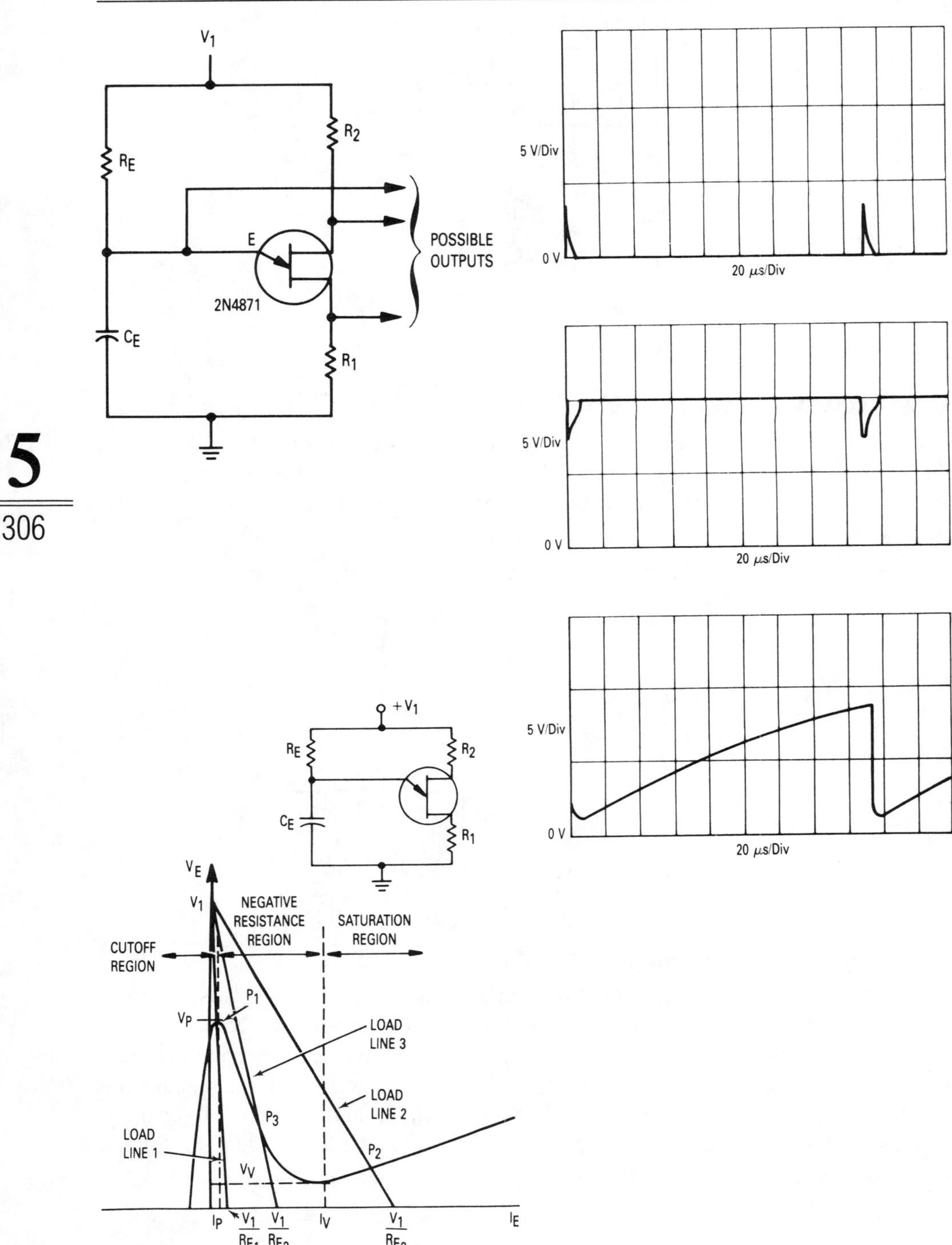

V1
R2
RE
E
POSSIBLE OUTPUTS
2N4871
CE
R1
5 V/Div
0 V
20 μs/Div
5 V/Div
0 V
20 μs/Div
5 V/Div
0 V
20 μs/Div
+V1
RE
R2
CE
R1
VE
V1
NEGATIVE RESISTANCE REGION
SATURATION REGION
CUTOFF REGION
P1
VP
LOAD LINE 3
LOAD LINE 2
P3
LOAD LINE 1
P2
VV
IP
V1/RE1
V1/RE2
IV
V1/RE2
IE

UJT relaxation oscillator

Fig. 5-41 This figure shows the circuit, waveforms, and characteristics, where a UJT is used as a relaxation oscillator. This basic building block is used in most of the UJT timer and control circuits (described in chapter 9). The waveforms and characteristics are shown in Figs. 5-41B and 5-41C, respectively. The waveforms are those obtained when the circuit values are: $R_E = 10\ \text{k}\Omega$, $C_E = 0.01\ \mu\text{F}$, $R_2 = 200\ \Omega$, $R_1 = 47\ \Omega$, and $V_1 = 20$ V. The approximate period of the oscillator pulses can be found by: $R_E C_E \times 1.7$. The value of R_E must be such that the load line (Fig. 5-41C) intersects the emitter characteristic in the negative-resistance region. Motorola Thyristor Device Data, 1991, p. 1-6-37, 38.

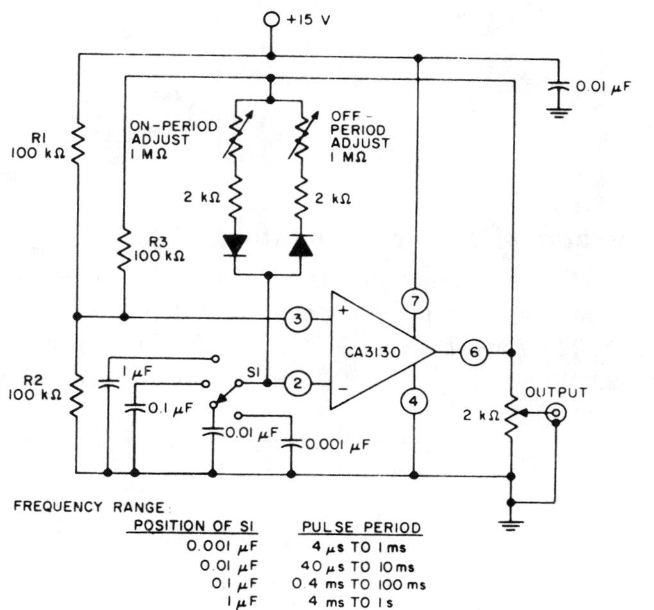

POSITION OF S1	PULSE PERIOD
0.001 μF	4 μs TO 1 ms
0.01 μF	40 μs TO 10 ms
0.1 μF	0.4 ms TO 100 ms
1 μF	4 ms TO 1 s

Pulse generator with independent on and off periods

Fig. 5-42 This circuit is a pulse generator or astable multivibrator, where the on and off periods can be adjusted separately. The pulse rate is selected by the setting of S1, and it remains essentially constant when the on-period and off-period resistors are adjusted. Harris Semiconductor, Linear and Telecom ICs, 1991, p. 3-100.

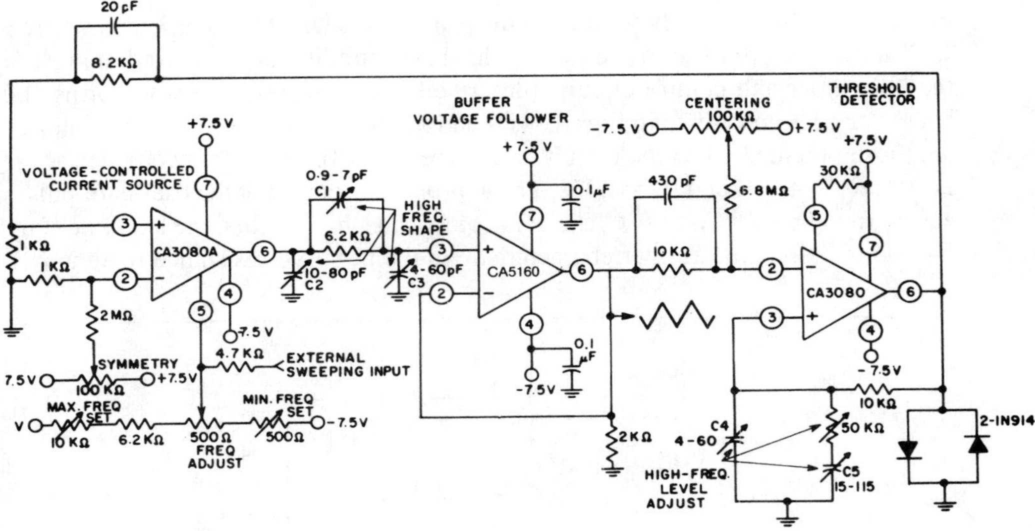

5

308

Function generator with wide tuning range

Fig. 5-43 The adjustment range of this generator is in excess of 1,000,000/1, using a single frequency-adjust potentiometer. C1, C2, and C3 shape the triangular signal between 500 kHz and 1 MHz. C4, C5, and the 50-kΩ trimmer are adjusted to maintain essentially constant (±10%) amplitude up to 1 MHz. Harris Semiconductor, Linear and Telecom ICs, 1991, p. 3-224.

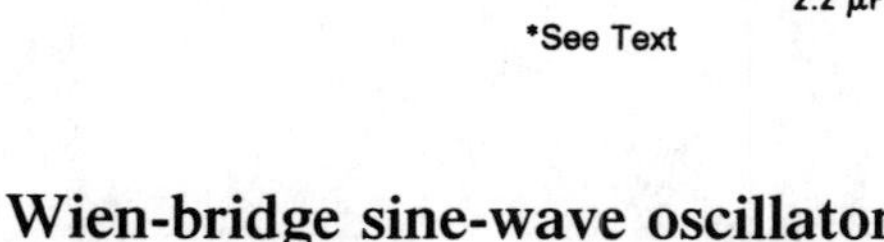

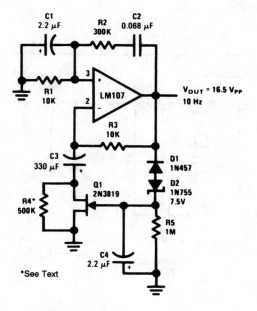

Wien-bridge sine-wave oscillator

Fig. 5-44 This circuit differs from other Wien-bridge oscillators described in this chapter only in the form of negative-feedback stabilization. Negative peaks in excess of −8.5 V cause D1 and D2 to conduct, which charges C4. The C4 charge biases Q1, which determines amplifier gain. C3 provides low-frequency roll-off in the feedback network, and prevents offset voltage/current errors from being multiplied. R5 is chosen to adjust the negative feedback loop so that Q1 is operated at a small negative gate bias. National Semiconductors, Linear Applications Handbook, 1991, p. 28.

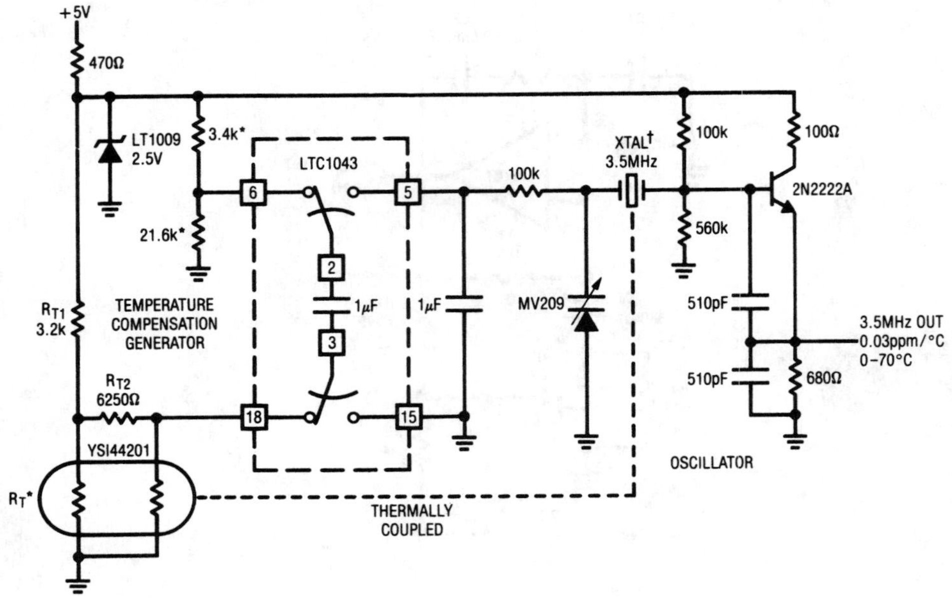

Temperature-compensated crystal oscillator

Fig. 5-45 This circuit uses an LTC1043 to differentiate between a temperature-sensing network and a dc reference. The single-ended output biases a varactor-tuned crystal oscillator to compensate drift. The varactor-crystal network has high dc impedance, which eliminates the need for an output amplifier from the LTC1043. Connect other LTC1043 pins as follows: 4 to +5 V, 16 to ground through a 0.001-μF capacitor, and 17 directly to ground. Linear Technology, Linear Applications Handbook, 1990, p. AN3-15.

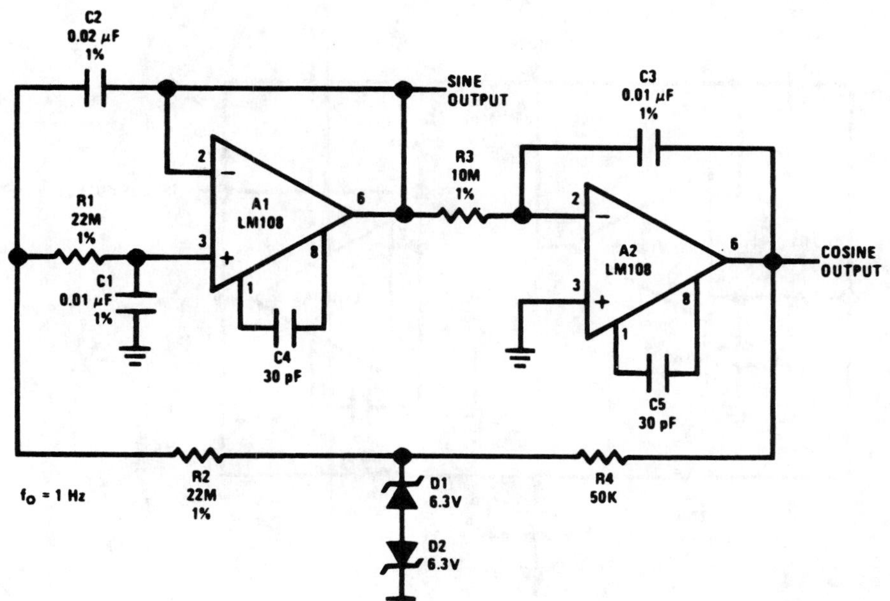

Low-frequency sine-wave generator with quadrature output

Fig. 5-46 This circuit uses two LM108s to provide a both a sine and cosine output at 1 Hz. National Semiconductor, Linear Applications Handbook, 1991, p. 84.

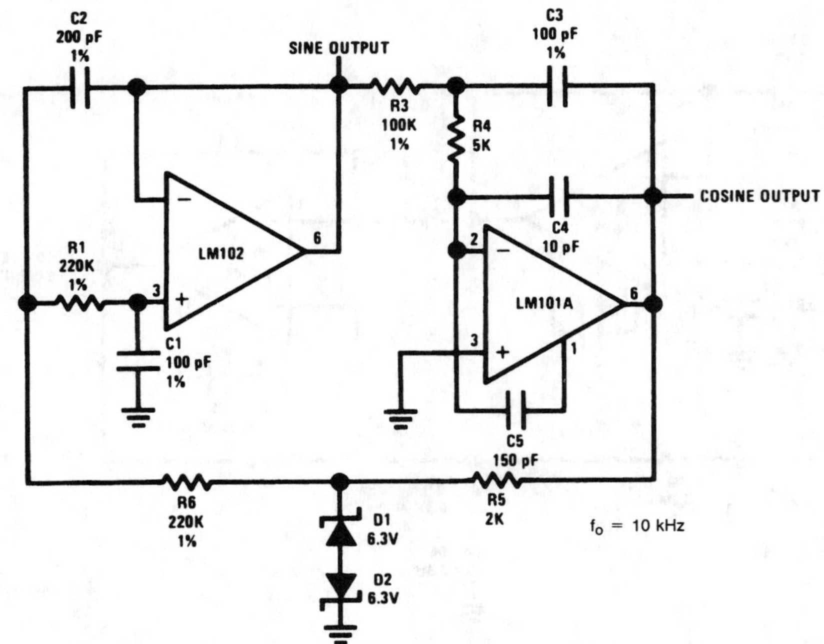

5
312

High-frequency sine-wave generator with quadrature output

Fig. 5-47 This circuit uses an LM102 and an LM101A to provide both a sine and cosine output at 10 kHz. National Semiconductor, Linear Applications Handbook, 1991, p. 84.

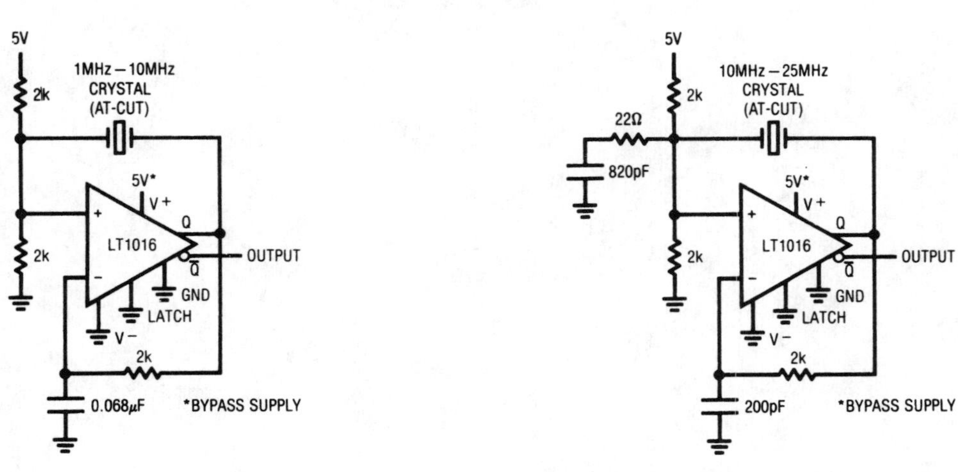

Crystal-oscillator clock circuits for digital systems

Fig. 5-48 These two oscillator circuits cover the frequency range for most digital-system clock requirements. Linear Technology, Linear Applications Handbook, 1990, p. AN31-11.

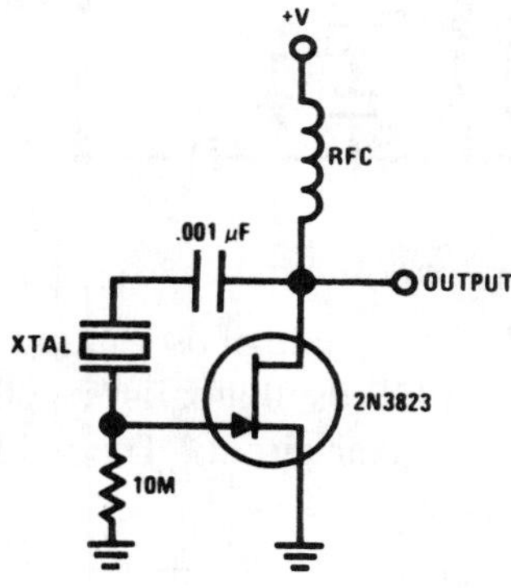

JFET Pierce crystal oscillator

Fig. 5-49 This circuit allows a wide frequency range of crystals to be used without circuit modification. Because the JFET gate does not load the crystal, good *Q* is maintained, which thus ensures good frequency stability. National Semiconductor, Linear Applications Handbook, 1991, p. 106.

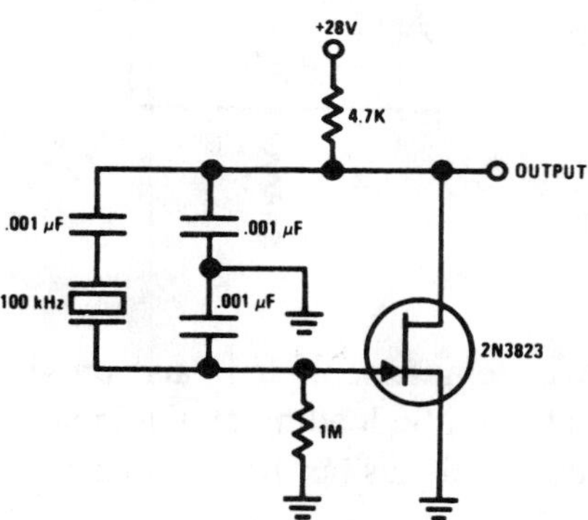

Stable low-frequency crystal oscillator

Fig. 5-50 This Colpitts crystal oscillator is ideal for low-frequency oscillator applications. Excellent stability is assured because the 2N3823 JFET circuit loading does not vary with temperature. National Semiconductor, Linear Applications Handbook, 1991, p. 114.

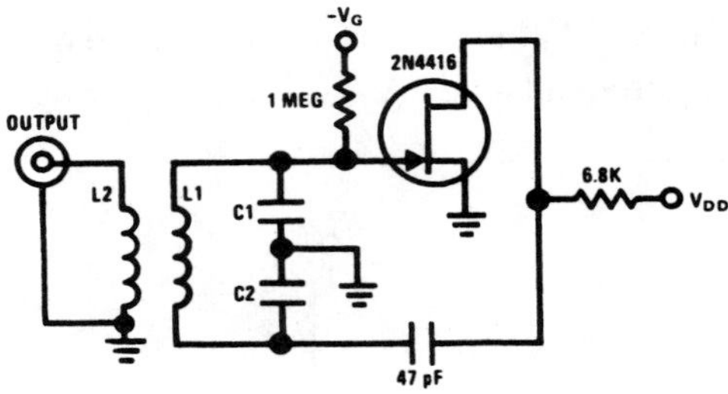

Low-distortion oscillator

Fig. 5-51 The 2N4416 JFET is capable of oscillating in a circuit, where harmonic distortion is very low. The JFET oscillator is excellent when a low harmonic content is required for a good mixer circuit. The values given are for a 20-MHz oscillator. National Semiconductor, Linear Applications Handbook, 1991, p. 114.

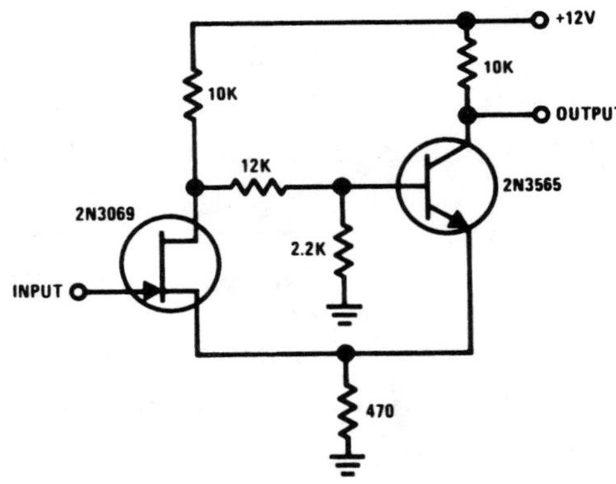

Schmitt trigger

Fig. 5-52 This circuit is "emitter coupled" and provides a simple comparator action. The JFET places very little loading on the input. The 2N3565 bipolar is a high h_{FE} transistor, so the circuit has fast transition action and a distinct hysteresis loop. National Semiconductor, Linear Applications Handbook, 1991, p. 118.

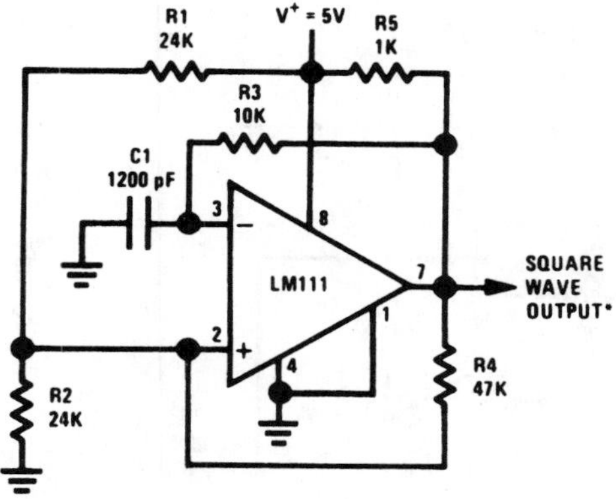

*TTL or DTL Fanout of two.

Free-running multivibrator

Fig. 5-53 This circuit uses an LM111 comparator to produce a 100-kHz square-wave output. The frequency can be changed by varying C1, or by adjusting R1 through R4, while keeping the ratios constant. Because of the low input current of a comparator, large circuit impedances can be used. Thus, low frequencies can be obtained with relatively small capacitor values (compared to most op amp MVs). A 1-Hz output requires a 1-μF capacitor. The speed of the comparator also permits operation above 100 kHz. National Semiconductor, Linear Applications Handbook, 1991, p. 132.

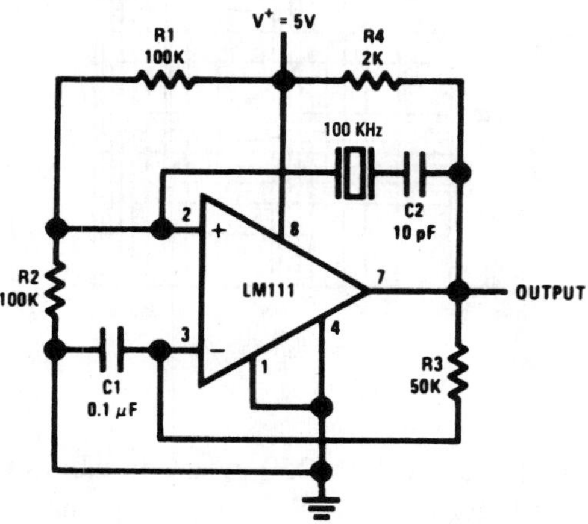

Crystal-controlled comparator oscillator

Fig. 5-54 This circuit also uses an LM111 comparator to produce a 100-kHz output. The circuit is similar to that of Fig. 5-53, except that positive feedback is obtained through the crystal (operating in a series-resonant mode). The high input impedance of the comparator minimizes crystal loading and contributes to frequency stability. National Semiconductor, Linear Applications Handbook, 1991, p. 132.

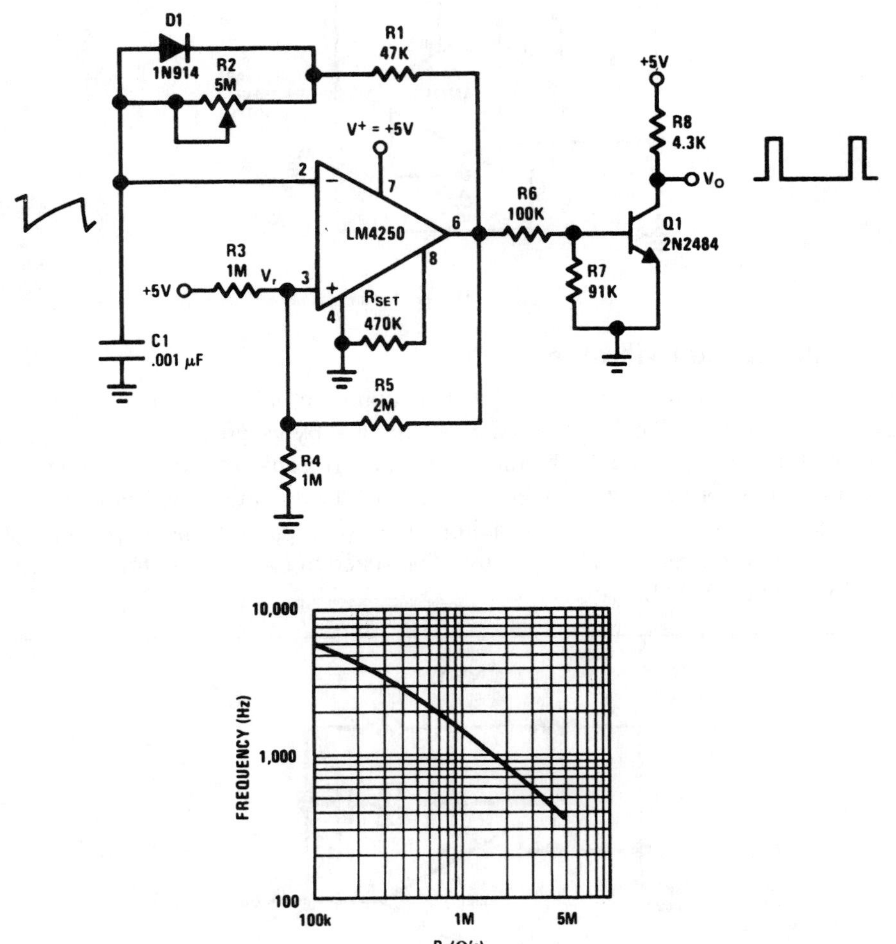

Pulse generator

Fig. 5-55 This circuit uses an LM4250 programmable op amp. Figure 5-55B shows the relationship between the pulse frequency and the setting of R2. The output of Q1 can interface directly with TTL/DTL circuits. National Semiconductor, Linear Applications Handbook, 1991, p. 208.

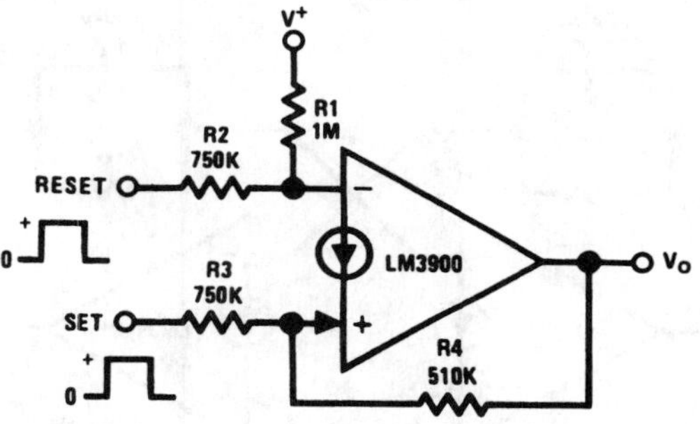

Bi-stable multivibrator

Fig. 5-56 This circuit uses an LM3900 Norton amplifier as a bi-stable MV (or asynchronous RS flip-flop). A positive pulse at the Set input causes the output to go high (approximately V+), and a Reset positive pulse returns the output to essentially 0 V. National Semiconductor, Linear Applications Handbook, 1991, p. 240.

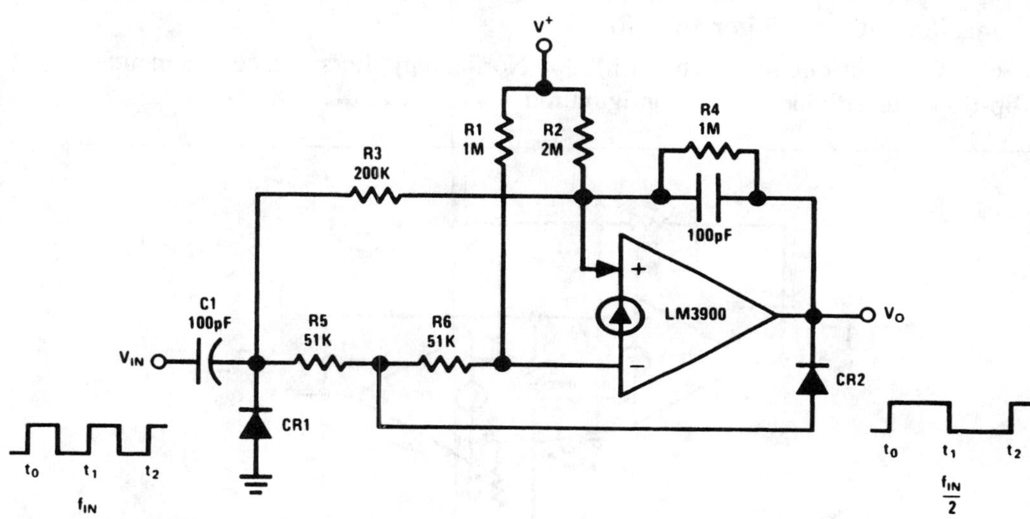

Trigger flip-flop

Fig. 5-57 This circuit uses an LM3900 Norton amplifier as a trigger flip-flop or as a divide-by-two configuration. CR1 and CR2 can be 1N914s or equivalent. National Semiconductor, Linear Applications Handbook, 1991, p. 241.

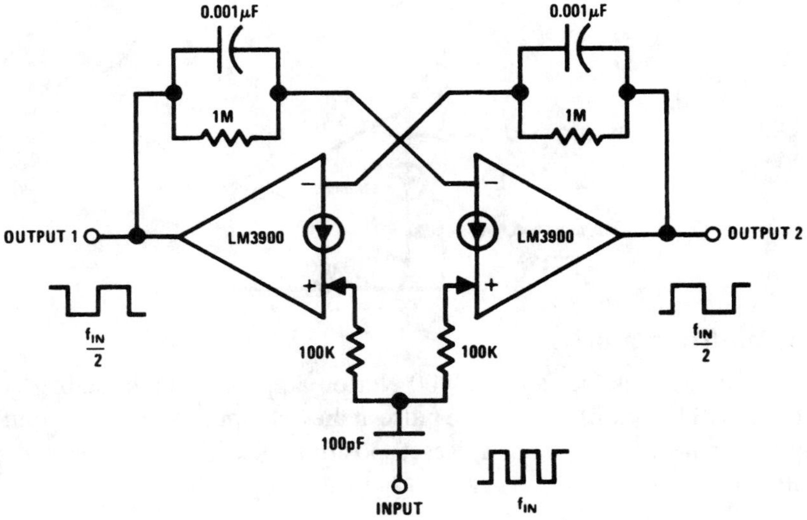

0.001μF
1M
0.001μF
1M
OUTPUT 1
LM3900
LM3900
OUTPUT 2
$\frac{f_{IN}}{2}$
100K
100K
$\frac{f_{IN}}{2}$
100pF
INPUT
f_{IN}

Complementary trigger flip-flop

Fig. 5-58 This circuit uses two LM3900 Norton amplifiers as a complementary flip-flop or as a divide-by-two configuration. National Semiconductor, Linear Applications Handbook, 1991, p. 241.

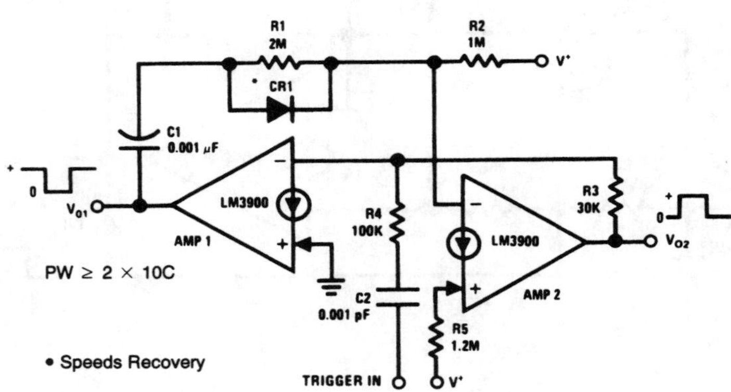

One-shot multivibrator

Fig. 5-59 This circuit uses two LM3900 Norton amplifiers as a one-shot or monostable MV. CR1 can be a lN914 or equivalent. National Semiconductor, Linear Applications Handbook, 1991, p. 242.

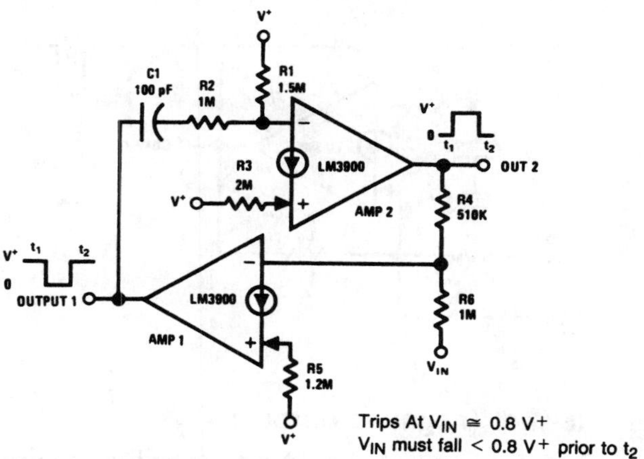

Combined one-shot and comparator

Fig. 5-60 This circuit uses two LM3900 Norton amplifiers to provide a one-shot output pulse only when the input exceeds a given value (0.8 V in this case). National Semiconductor, Linear Applications Handbook, 1991, p. 242.

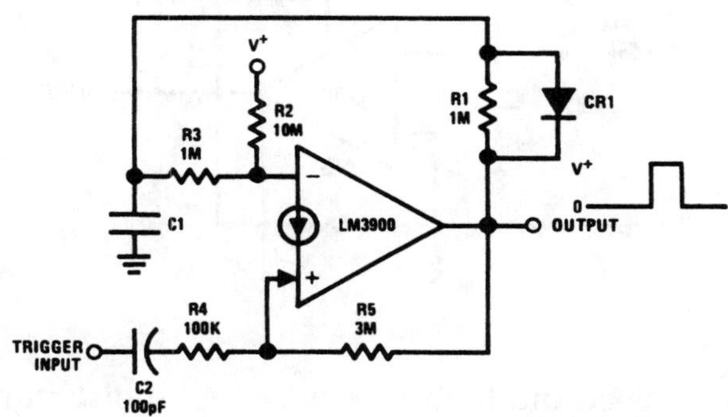

One-amplifier one-shot (positive output)

Fig. 5-61 This circuit uses one LM3900 Norton amplifier to provide a positive one-shot output pulse. CR1 (a lN9I4 or equivalent) permits rapid re-triggering. National Semiconductor, Linear Applications Handbook, 1991, p. 242.

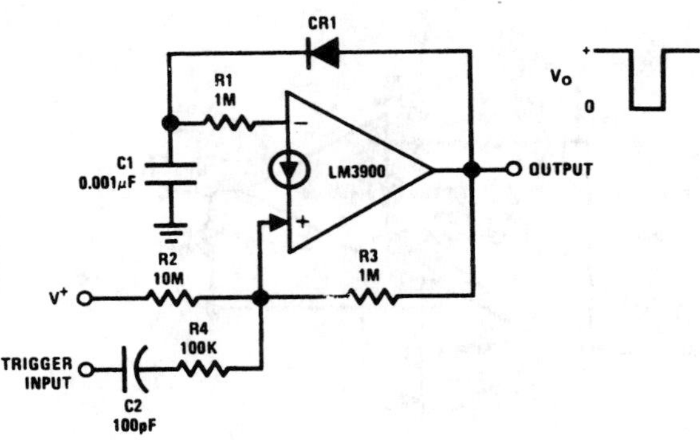

One-amplifier one-shot (negative output)

Fig. 5-62 This circuit uses one LM3900 Norton amplifier to provide a negative one-shot output pulse. CR1 (a IN914 or equivalent) permits rapid re-triggering. National Semiconductor, Linear Applications Handbook, 1991, p. 243.

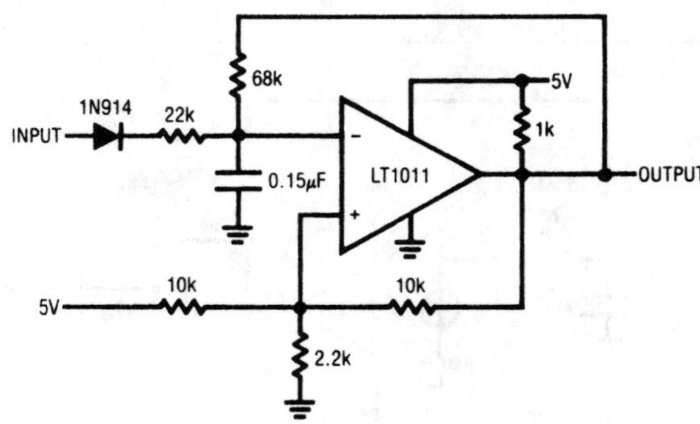

Line-synchronous clock multivibrator with high noise immunity

Fig. 5-63 This circuit uses an LT1011 to provide a line-synchronous clock that will not lose lock under very noisy line conditions. The basic RC multivibrator is tuned to free-run near 60 Hz, but the line-derived synchronizing input forces the oscillator to lock to the line. Noise rejection is obtained from the integrator characteristics of the RC network. Linear Technology, Linear Applications Handbook, 1990, p. AN31-12.

6

Digital circuits

It is assumed that you are already familiar with digital basics (number systems, logic elements, gates, microprocessors, digital test equipment, etc.). If not, read *Lenk's Digital Handbook*, McGraw-Hill, 1992. However, the following paragraphs summarize both test and troubleshooting for digital circuits. This information is included so that even those readers who are not familiar with electronic procedures can both test the circuits in this chapter, and localize problems if the circuits fail to perform as shown.

Digital circuit testing and troubleshooting

Both testing and troubleshooting for the circuits of this chapter can be performed with conventional test equipment (meters, generators, scopes, etc.), covered in previous chapters. However, a logic or digital probe and a digital pulser can make life much easier if you must regularly test and troubleshoot digital devices. So, this section starts with brief descriptions of the probe and pulser. This is followed by testing/troubleshooting for the various types of circuits in this chapter.

Logic or digital probe

Logic probes are used to monitor in-circuit pulse or logic activity. By means of a simple lamp indicator, a logic probe shows you the logic state of the digital signal and allows brief pulses to be detected (that you might miss with a scope). Logic probes detect and indicate high and low (1 or 0) logic levels, as well as intermediate or "bad" logic levels (indicating an open circuit) at the terminals of a logic element (the inputs and outputs of a gate or a digital-to-analog D/A converter).

Not all logic probes have the same functions and you must learn the operating characteristics for your particular probe. For example, on the more sophisticated probes, the indicator lamp can give any of our indications: (1) off, (2), dim (about half brilliance), (3) bright (full brilliance), or (4) flashing on and off.

The lamp is normally in the dim state and must be driven to one of the other three states by voltage levels at the probe tip. The lamp is bright for inputs at or above the 1 state, and off for inputs at or below 0. The lamp is dim for voltages between the 1 and 0 states, and for open circuits. Pulsating inputs cause the lamp to flash at about 10 Hz (regardless of the input pulse rate). The probe is particularly effective when it is used with the logic pulser.

Logic pulser

The hand-held *logic pulser* (similar in appearance to the logic probe) is an in-circuit stimulus device that automatically outputs pulses of the required logic polarity, amplitude, current, and width to drive lines and other test points high and low. A typical pulser also has several pulse burst and stream modes available.

Logic pulsers are compatible with most digital devices. Pulse amplitude depends on the equipment supply voltage, which is also the supply voltage for the pulser. Pulse current and pulse width depend on the load being pulsed. The frequency and number of pulses that are generated by the pulser are controlled by operation of a switch. A flashing LED indicator on the pulser tip indicates the output mode.

The logic pulser forces overriding pulses onto lines or test points, and can be programmed to output single pulses, pulse streams, or bursts. The pulser can be used to force ICs to be enabled or clocked. Also, the circuit inputs can be pulsed while observing the effects on the circuit outputs with a logic probe.

Interface and translator circuits

The circuits shown in Figs. 6-2 through 6-10 can be tested by applying pulses at the input and monitoring the output. This can be done with a generator or pulser at the input and a scope or probe at the output. For example, pulses with ECL levels can be applied at the ECL-gate input and TTL pulses can be monitored at the output (in Fig. 6-6). If TTL pulses are absent at pin 7 of the 4805 comparator, check for pulses at pins 2 and 3. This will isolate the problem to the gate or the comparator.

D/A converters

Digital-to-analog converters (Figs. 6-11 through 6-15) can be tested by applying digital inputs and monitoring the output for corresponding voltages. For example, the B1 through B8 inputs of the DAC-08 can be connected to ground (for a 0) or to +5 V (for a 1), and the output monitored with a voltmeter at pins 2 and 4 in the circuit of Fig. 6-13. If both output voltages are slightly off, suspect the 2.00-mA reference. If one of the output voltages is slightly off, suspect the corresponding 5.000- kΩ resistors. If the output voltages are absent or are way off, suspect the DAC-08.

A/D converters

Analog-to-digital (A/D) converters (Figs. 6-16 and 6-17) can be tested by applying precision voltages at the input and monitoring the output for corresponding digital values. For example, a fixed voltage between 0 and +10 V can be applied to pin 2 of A1, and the corresponding digital value can be read out at pins 5 through 14 of the DAC-10. The pins should go to +5 V for a 1 and to ground or 0 V for a 0 in the circuit of Fig. 6-16. Also notice that there is a serial digital output at pin 2 of the 2504 SAR. This output must be monitored with a scope. The rate at which conversions are done is controlled by the 1- to 2-MHz clock at pin 13 of the 2504. A start-input signal is at pin 14 and a conversion-complete signal is at pin 3. If the output readings are slightly off, try correcting the problem by adjusting R7 for zero, or by trimming the +5-V reference for full-scale output. If the outputs are absent or are way off, suspect the DAC-10 (A1), or the 2504. Also notice that the accuracy of this circuit depends on the precision of R1, R2, and R3.

Bus drivers

Digital *bus drivers* (Fig. 6-18) can be tested by applying pulses at the input and monitoring the output. This can be done with a generator or pulser at the input and a scope or probe at the output. For example, pulses can be applied at pins 4 and 13 of the RM3182 differential driver and pulses can be monitored at pins 6 and 11. If the output pulses are absent, suspect the RM3182. Notice that the frequency limitation of the circuit is set by the values of the capacitors at pins 5 and 12 (a higher frequency requires larger capacitances).

LED and LCD displays

LED and LCD display circuits (Figs. 6-20 through 6-22) can be tested by applying a specific voltage at the input and noting the display readout. For example, in the circuit of Fig. 6-20, the LCD display should follow voltage applied at pins 10 and 11 of the CA3162E. If the display is slightly off, try correcting the problem with adjustment. Adjust the 50-kΩ pot at pins 8 and 9 of the CA3162E for 000 on the display, with pins 10 and 11 shorted. Then, adjust the 10-kΩ pot at pins 7 and 13 for 900 on the display, with 900 mV applied at pins 10 and 11.

If all of the display digits are blank, suspect the LCD backplane oscillator (G1 and G2). Notice that an LED display does not require a backplane drive signal.

If only one of the display digits is bad, check power (pin 16) and ground (pins 7 and 8) of the corresponding CD4056B decoder/driver. Also check the corresponding signal from the CA3163E to pin 1 of the CD4056B. For example, the

MSD signal from pin 4 of the CA3162E is applied to pin 1 of the MSD CD4056B through a 0.047-μF capacitor and inverter G3.

If certain segments of the display are absent on all three digits, check the corresponding inputs at pins 2 through 6 of the CD4056B from pins 1, 2, 15, and 16 of the CA3163E. Notice that the inputs to pin 2 of the CD4056Bs are made through inverter G6 and gates G7, G8, and G9.

Optoisolators

Circuits involving *optoisolators* (Figs. 6-37 through 6-43) can be tested by applying a voltage or signal to the optoisolator and checking the corresponding response. For example, in the circuit of Fig. 6-38, the 2N6071B triac is triggered by a signal from pin 4 of the MOC3011 when a pulse is applied to the buffer input. With the triac triggered on, 115-V power is applied to the load.

If this sequence does not occur, first check that pulses are arriving at pin 2 of the MOC3011. If not, suspect the buffer. Then, check that there is a trigger at pin 4 of MOC3011 when pulses are applied to pin 2. If not, suspect the MOC3011. If you get a trigger at pin 4 of the MOC3011, but power is not applied to the load, suspect the triac.

In the case of an optocoupler that is used as a solid-state relay (Fig. 6-42), simply apply 5 V to the 4N40, and check that the indicator lamp turns on. If not, suspect the 4N40 (or the lamp). Of course, it is possible that one of the three resistors is open or that the capacitor is shorted (or leaking badly), but this is not likely.

Single-chip devices

The *single-chip digital devices* in Figs. 6-44 through 6-58 can best be tested by checking their function in the circuit. For example, the digital power monitor DS1231 (Fig. 6-47) should produce an RST or $\overline{\text{RST}}$ and $\overline{\text{NMI}}$ (nonmaskable interrupt) to a microprocessor when voltage is removed from the input. To test this function, apply +10 V to the voltage sense point of the Fig. 6-47 circuit, and monitor the RST, $\overline{\text{RST}}$ and $\overline{\text{NMI}}$ pins of the DS1231 (make certain that +5 V is applied to V_{CC} and MODE pins). Now, remove the voltage from the voltage sense point, and check that RST, $\overline{\text{RST}}$ and NMI signals appear at the corresponding pins of the DS1231.

The full-duplex circuit of Fig. 6-58 can be tested by applying pulses at both inputs and monitoring for pulses at both outputs. For example, apply pulses to the DIN pin of the DS1275, and check that pulses appear at pin 3 (RXD) of the PC serial port. There should be no substantial difference between the input and output pulses (same amplitude, shape, etc.). Then, apply pulses to pin 2 (TXD) of the PC serial port, and check for pulses at the DOUT pin of the other DS1275. If either receive (signals to the PC port) or transmit (signals from the PC port) functions are abnormal or absent, suspect the corresponding DS1275. Of course, it is possible that the diode or capacitor is leaking or shorted, but this is not likely.

General IC digital troubleshooting tips

The following troubleshooting tips apply to digital circuits, where the majority of components are contained within ICs.

Power and ground connections The first step in tracing problems in a digital circuit with ICs is to check all power and ground connections to the ICs. Many ICs have more than one power and one ground connection. For example, the MAX132 in Fig. 6-30 requires +5 V at pin 24, and −5 V at pin 13. Also, there is a digital ground at pin 12 and an analog ground at pin 16.

Reset and chip-select signals With all power and ground connections confirmed, check that all the ICs receive reset and chip-select signals, as required. For example, the MAX132 in Fig. 6-30 requires a chip-select signal at pin 1 (but no reset signal). On the other hand, the DS1262 in Fig. 6-44 requires a reset signal at pin 15, but no chip-select signal.

Usually, the reset signal is a pulse, whereas the chip-select is a steady signal (but not always). In any event, if a reset line is open, or is shorted to ground, or to power (typically + 5 V or +12 V), the ICs are not reset (or remain locked in the reset condition), no matter what signals are applied. This brings the entire digital operation to a halt. So, if you find a reset pin that is always high, always low, or apparently connected to nothing (floating), check the wiring to that pin carefully. The same applies to chip-select pins, unless the circuit calls for the chip-select to remain high or low.

Clock signals Many digital ICs require clocks. For example, there is a clock at pins 5 and 6 of the MAX132 in Fig. 6-30, and a clock at pin 13 of the DS1262. Generally, the presence of pulse activity on any pins of a digital IC indicates the presence of a clock, but do not count on it. Check directly at the clock pins.

It is possible to measure the presence of a clock signal with a scope or logic probe. However, a frequency counter provides the most accurate measurement. Obviously, if any ICs do not receive required clock signals, the IC cannot function. On the other hand, if the clock is off frequency, all of the ICs might appear to have a clock signal, but the IC function can be impaired. Notice that crystal-controlled clocks do not usually drift off frequency, but can go into some overtone frequency (typically a third overtone).

Input-output signals Once you are certain that all ICs are good and have proper power and ground connections, and that all reset, chip-select, and clock signals are available, the next step is to monitor all input and output signals at each IC. This can be done with either a scope or probe.

Digital circuits titles and descriptions

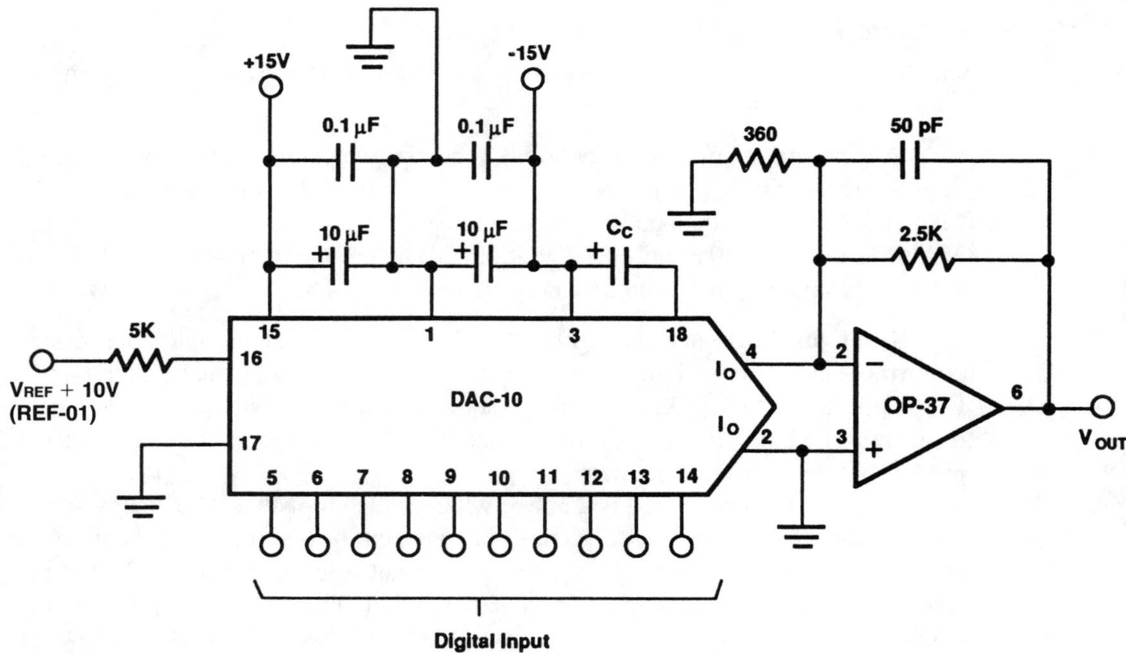

6

326

Basic digital D/A converter

Fig. 6-1 This circuit shows the classic DAC/op-amp combination that produces a digital D/A converter. V_{OUT} is selected, in steps, by the binary signals at the digital input of DAC-10, with the highest V_{OUT} set by V_{REF} at pin 16. V_{REF} is supplied by REF-01 (Fig.4-1). With the circuit assembled using the values shown (0.01 μF is typical for C_C), apply binary inputs (typically $-V_S$ to $-V_S$ plus 36 V) at pins 5 through 14, starting with all zeros (or all 1s). Notice that V_{OUT} varies (in steps) between 0 and a typical 9.9 V. Record actual values. Raytheon Linear Integrated Circuits, 1989, p. 4-95.

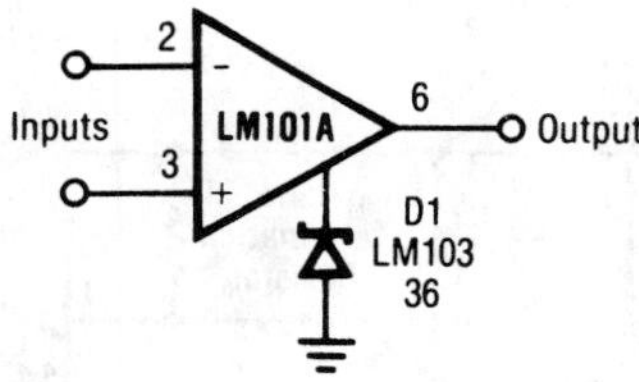

DTL/TTL driver

Fig. 6-2 This circuit uses an LM101A as a voltage comparator to drive DTL/TTL ICs. Raytheon Linear Integrated Circuits, 1989, p. 4-247.

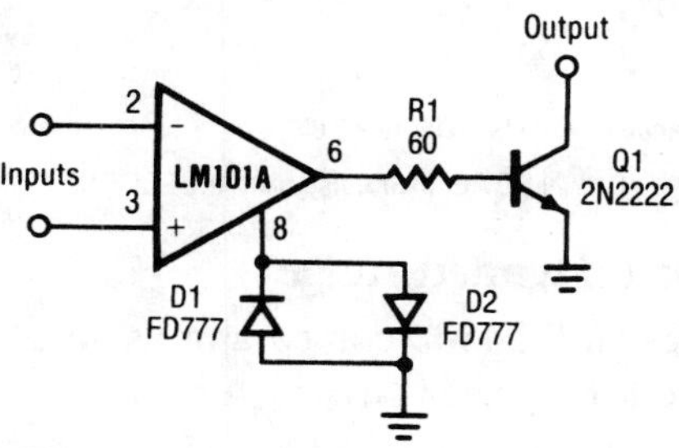

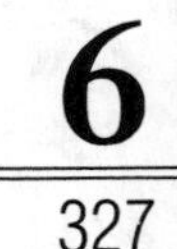

RTL or high-current driver

Fig. 6-3 This circuit uses an LM101A as a voltage comparator to drive RTL or other high-current devices. Raytheon Linear Integrated Circuits, 1989, p. 4-247.

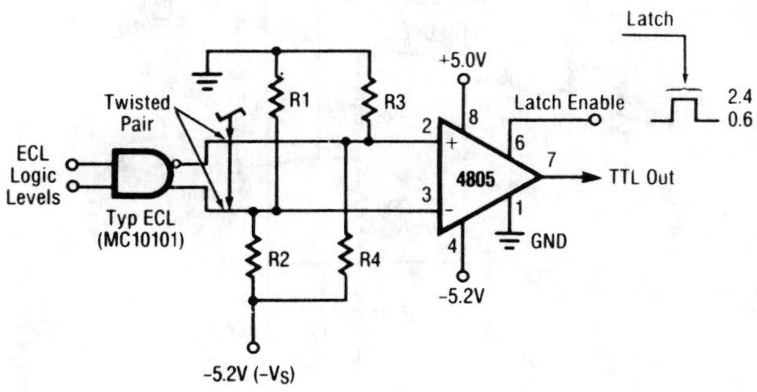

ECL-to-TTL translator

Fig. 6-4 This circuit uses an RC4508 comparator as an ECL-to-TTL translator. Raytheon Linear Integrated Circuits, 1989, p. 5-12.

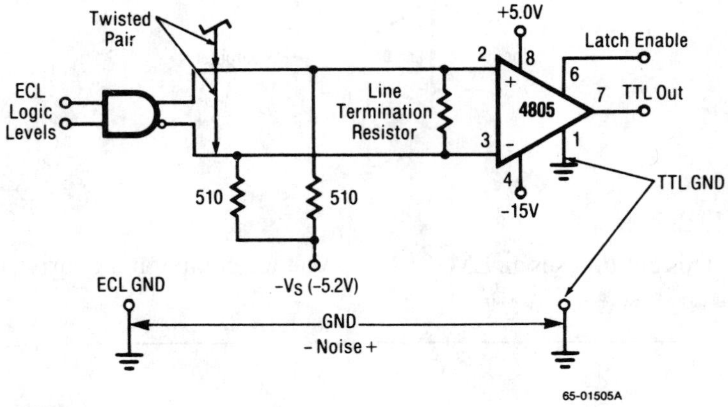

Notes:

1. Common mode range of 4805 is −8.0V to +2.0V.
2. The 4805 can stand −3.0V, +5.0V of GND noise from the ECL GND to the TTL GND.

ECL-to-TTL translator (extended range)

Fig. 6-5 This circuit uses an RC4508 comparator as an ECL-to-TTL translator with extended common-mode range. Raytheon Linear Integrated Circuits, 1989, p. 5-13.

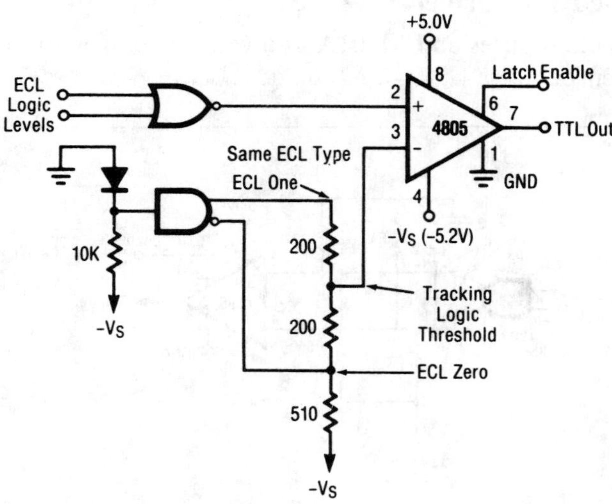

ECL-to-TTL translator (tracking)

Fig. 6-6 This circuit uses an RC4508 comparator as an ECL-(single-ended) to-TTL translator with tracking ECL reference. Raytheon Linear Integrated Circuits, 1989, p. 5-13.

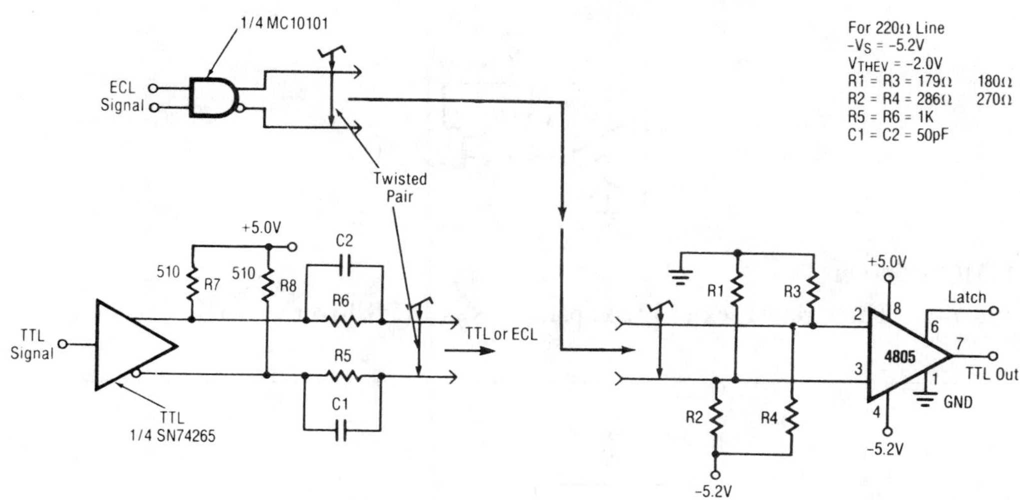

ECL/TTL-to-TTL translator

Fig. 6-7 This circuit adapts either ECL or TTL circuits using an RC4508 comparator. Raytheon Linear Integrated Circuits, 1989, p. 5-14.

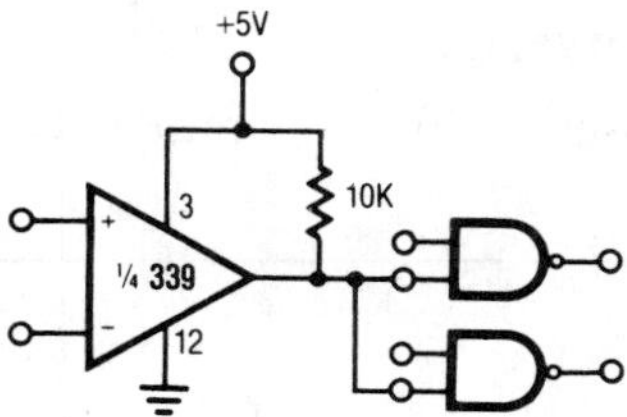

TTL driver

Fig. 6-8 This circuit uses a 339 comparator as a TTL driver. Raytheon Linear Integrated Circuits, 1989, p. 5-30.

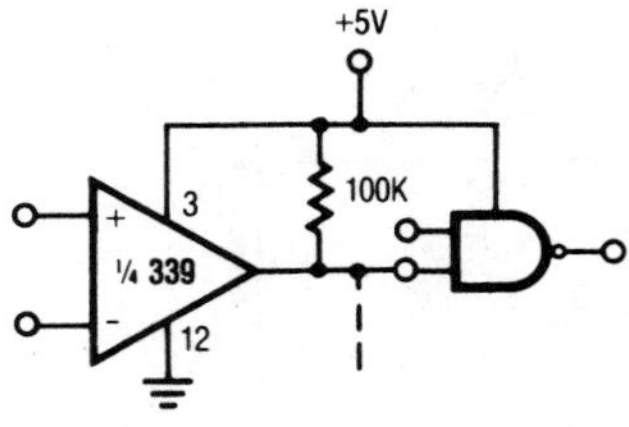

CMOS driver

Fig. 6-9 This circuit uses a 339 comparator as a CMOS driver. Raytheon Linear Integrated Circuits, 1989, p. 5-30.

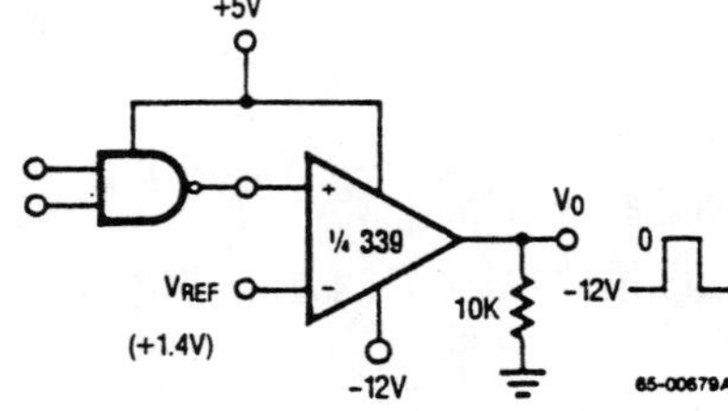

TTL-to-MOS logic converter

Fig. 6-10 This circuit uses a 339 comparator to convert from TTL to MOS logic. Raytheon Linear Integrated Circuits, 1989, p. 5-31.

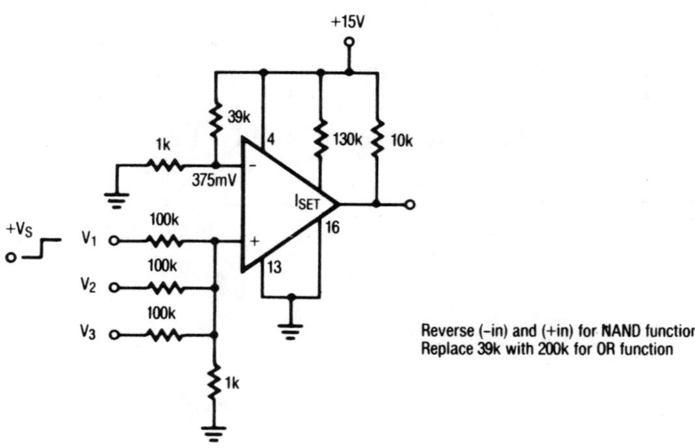

Three-input AND gate

Fig. 6-11 This circuit uses an LP165/365 comparator as a 3-input AND gate. The circuit can be converted to NAND or OR, as shown. Raytheon Linear Integrated Circuits, 1989, p. 5-43.

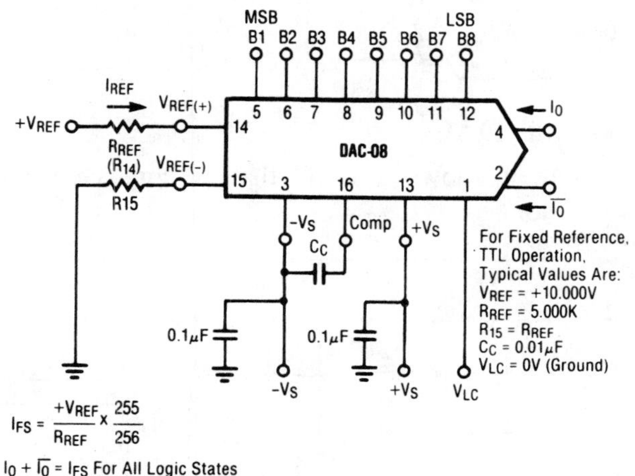

Basic positive-reference DAC

Fig. 6-12 This circuit shows a DAC that is connected for positive-reference D/A-converter operation. Raytheon Linear Integrated Circuits, 1989, p. 6-12.

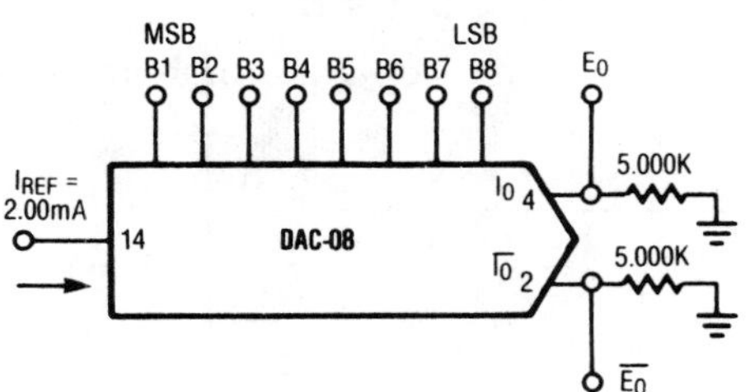

Scale	B1 B2 B3 B4 B5 B6 B7 B8	I_O mA	$\overline{I_O}$ mA	E_O	$\overline{E_O}$
Full Scale	1 1 1 1 1 1 1 1	1.992	0.000	-9.960	-0.000
Half Scale +LSB	1 0 0 0 0 0 0 1	1.008	0.984	-5.040	-4.920
Half Scale	1 0 0 0 0 0 0 0	1.000	0.992	-5.000	-4.960
Half Scale -LSB	0 1 1 1 1 1 1 1	0.992	1.000	-4.960	-5.000
Zero Scale +LSB	0 0 0 0 0 0 0 1	0.008	1.984	-0.040	-9.920
Zero Scale	0 0 0 0 0 0 0 0	0.000	1.992	0.000	-9.960

Basic unipolar negative-reference DAC

Fig. 6-13 This circuit shows a DAC that is connected for unipolar negative-reference D/A-converter operation. Raytheon Linear Integrated Circuits, 1989, p. 6-13.

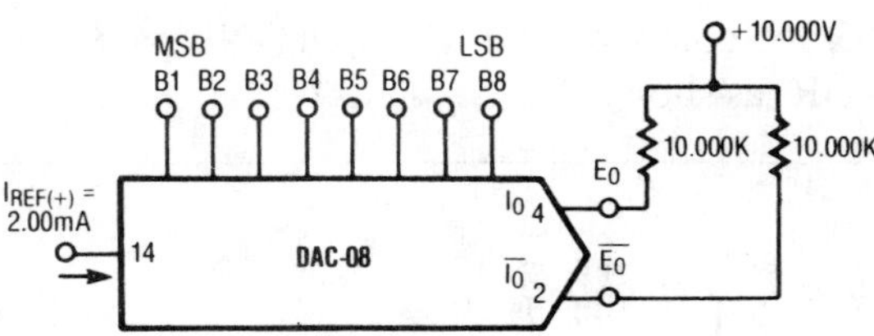

Scale	B1 B2 B3 B4 B5 B6 B7 B8	E_O	$\overline{E_O}$
Pos Full Scale	1 1 1 1 1 1 1 1	-9.920	+10.000
Pos Full Scale -LSB	1 1 1 1 1 1 1 0	-9.840	+9.920
Zero Scale +LSB	1 0 0 0 0 0 0 1	-0.080	+0.160
Zero Scale	1 0 0 0 0 0 0 0	0.000	+0.080
Zero Scale -LSB	0 1 1 1 1 1 1 1	+0.080	0.000
Neg Full Scale +LSB	0 0 0 0 0 0 0 1	+9.920	-9.840
Neg Full Scale	0 0 0 0 0 0 0 0	+10.000	-9.920

Basic bipolar-output DAC

Fig. 6-14 This circuit shows a DAC that is connected for bipolar-output D/A-converter operation. Raytheon Linear Integrated Circuits, 1989, p. 6-14.

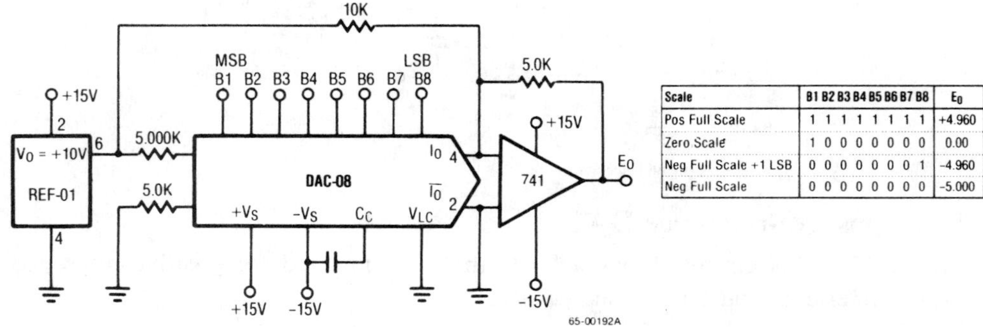

Scale	B1 B2 B3 B4 B5 B6 B7 B8	E_O
Pos Full Scale	1 1 1 1 1 1 1 1	+4.960
Zero Scale	1 0 0 0 0 0 0 0	0.00
Neg Full Scale +1 LSB	0 0 0 0 0 0 0 1	-4.960
Neg Full Scale	0 0 0 0 0 0 0 0	-5.000

Offset binary D/A-converter operation

Fig. 6-15 This circuit shows a DAC/op-amp combination that is used to provide offset-binary operation. Raytheon Linear Integrated Circuits, 1989, p. 6-14.

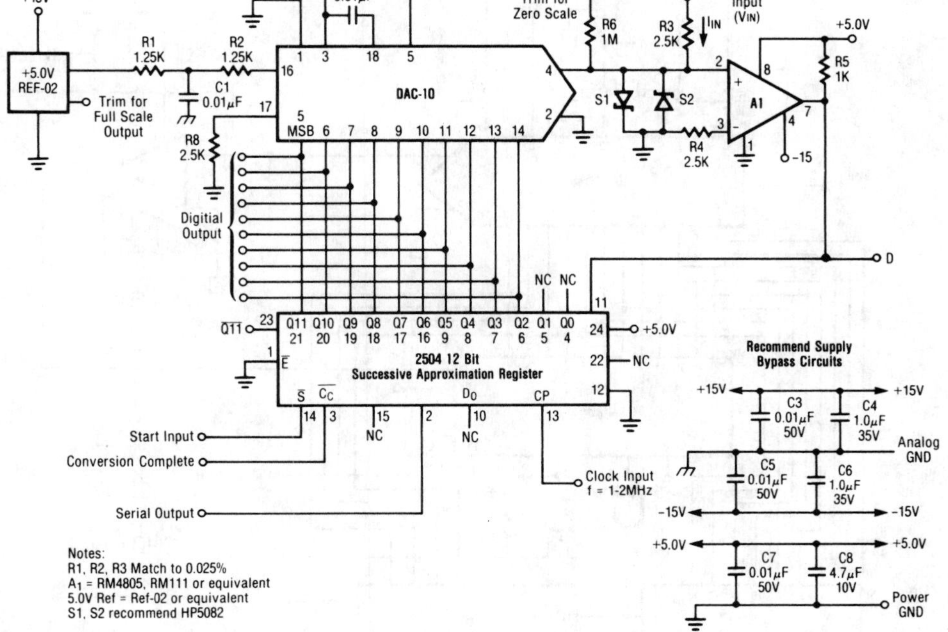

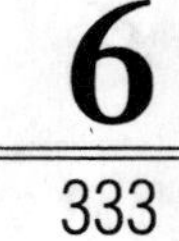

A/D converter

Fig. 6-16 This circuit shows a DAC and comparator that is used with a 12-bit SAR to form a 10-bit A/D converter. The analog input (0 to +10 V) is applied at R3. Both parallel and serial digital outputs are available. R7 is used to trim for zero. Full-scale is trimmed by REF-02 (Fig. 4-1C). Raytheon Linear Integrated Circuits, 1989, p. 6-28.

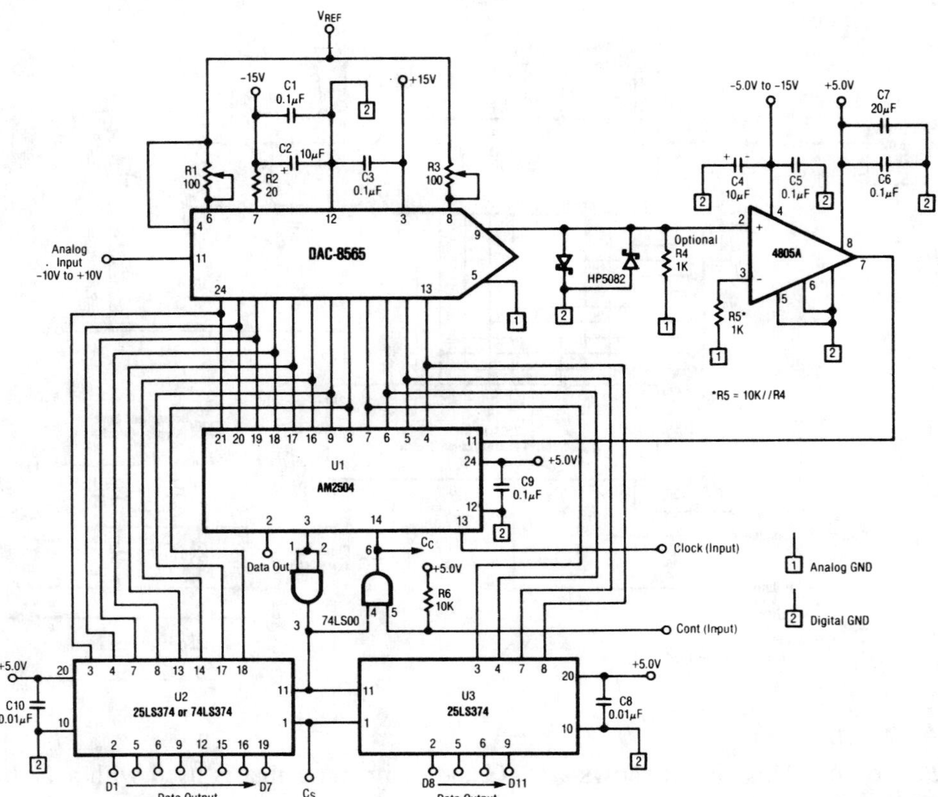

A/D converter with latch

Fig. 6-17 This circuit shows a DAC and comparator that is used with an SAR to form a 12-bit A/D converter. Latched output is provided by the 25LS374. The reference is set by R1 and the bipolar offset is controlled by R3. Raytheon Linear Integrated Circuits, 1989, p. 6-96.

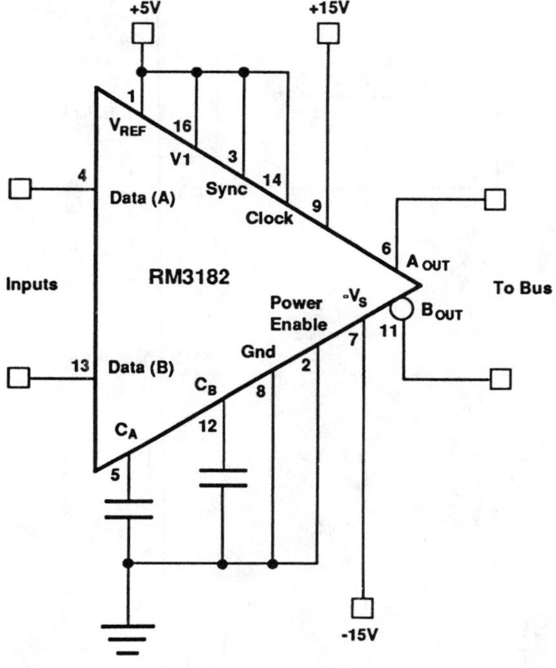

Differential line driver (to drive an ARINC 429 bus)

Fig. 6-18 With this circuit, the output impedance is 75 Ω. When data-A is high and data-B is low, A_{OUT} swings to $+V_{REF}$ and B_{OUT} swings to $-V_{REF}$, and vice versa. The slew rate is set by C_A/C_B. Typical values are 75 pF for 100 Kb/s and 500 pF for 12.5 to 14 Kb/s. Raytheon Linear Integrated Circuits, 1989, p. 11-6.

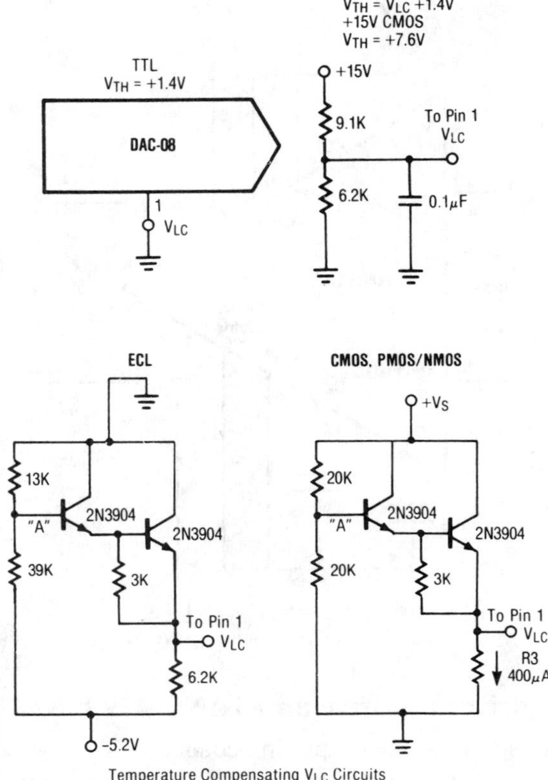

Temperature Compensating V_{LC} Circuits

Interfacing D/A converters with digital-logic families

Fig. 6-19 This circuit shows the interface circuits that are required to interface a typical TTL DAC with ECL, CMOS, PMOS, and NMOS digital devices. Raytheon Linear Integrated Circuits, 1989, p. 6-15.

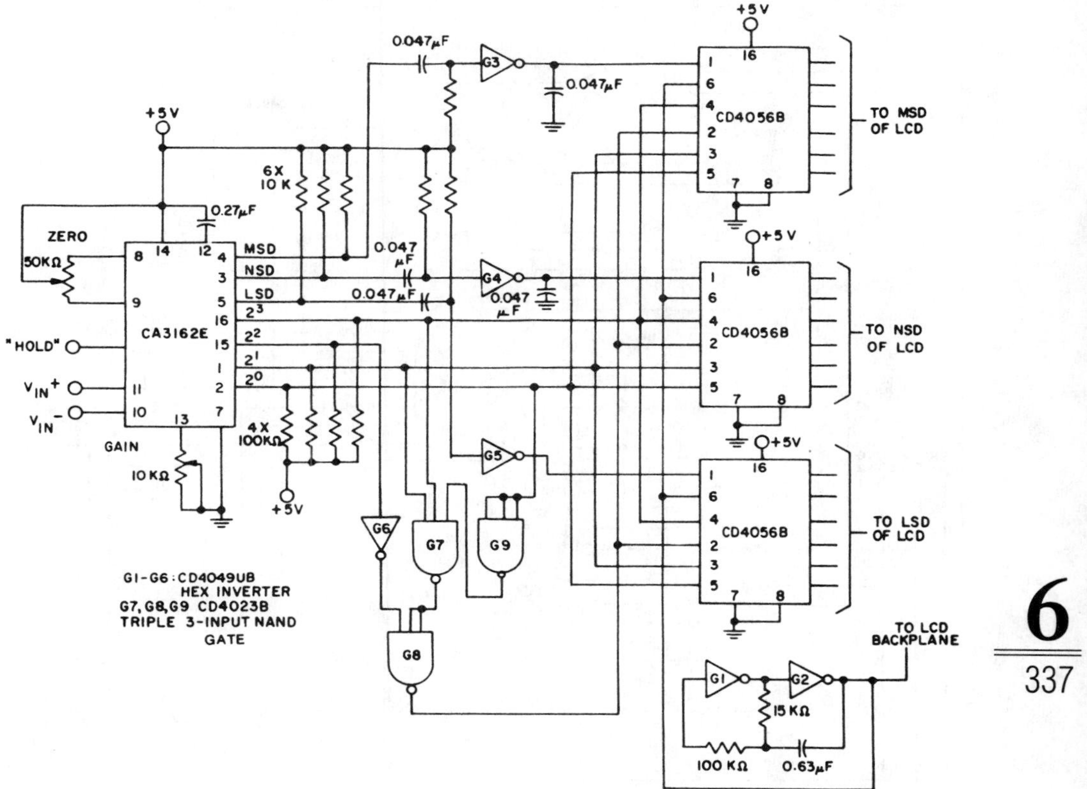

A/D converter for 3-digit LCD display

Fig. 6-20 This circuit shows a CA3162E and three CD4056B LCD decoder/drivers connected in a typical LCD application. With pin 6 of the CA3162E floating (as shown), the sampling rate is 4 Hz. The negative sign (−) is decoded as a negative sign (−), and the positive overload indicator (E) as an "H". To adjust, short terminals 10 and 11 of the CA3162E, and adjust the zero control for a reading of 000 on the display. Then, apply 900 mV at terminals 10 and 11, and adjust the gain control for a reading of 900 on the display. Harris Semiconductors, Data Acquisition, 1991, p. 2-5.

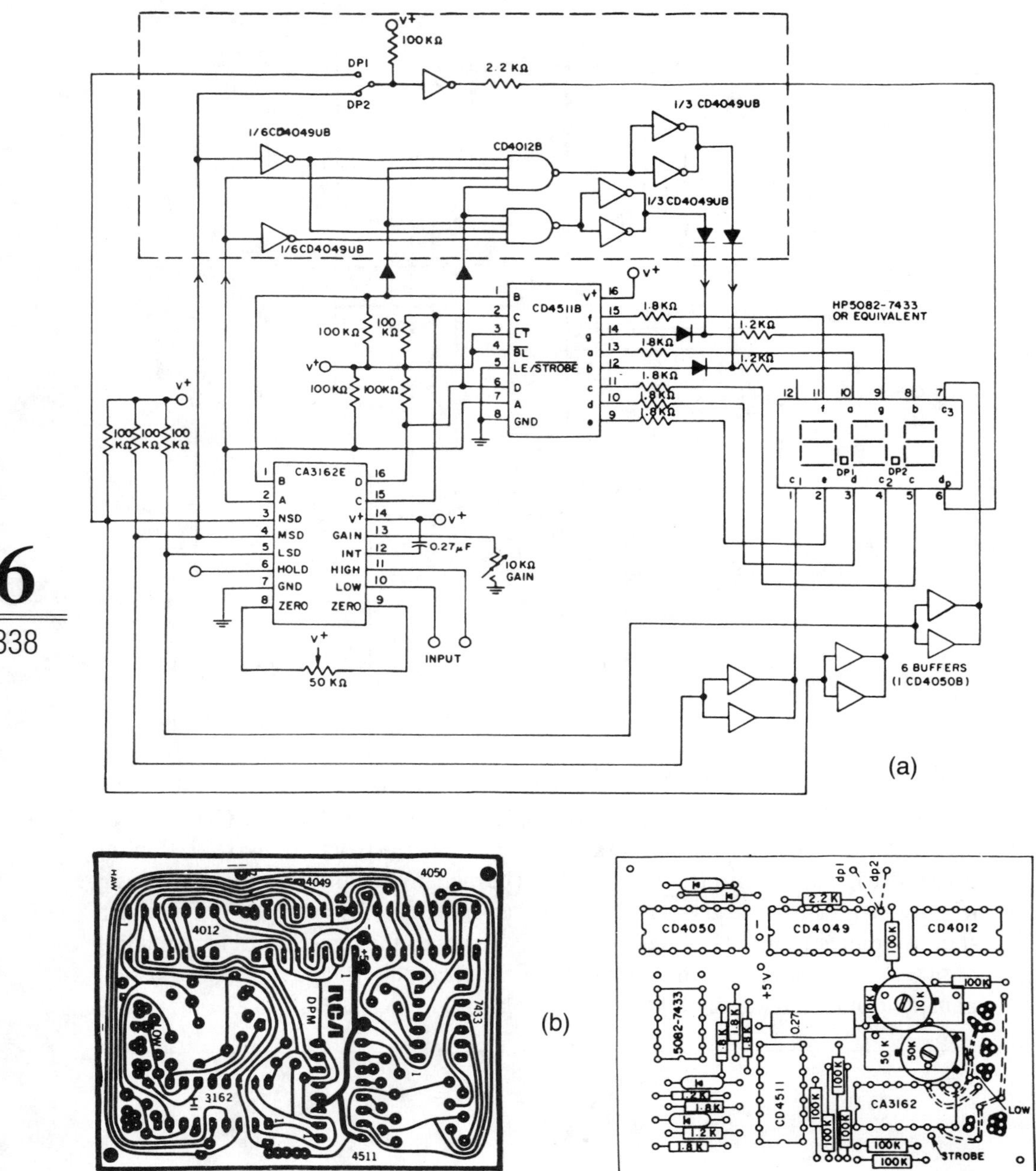

A/D converter for 3-digit LED display

Fig. 6-21 This circuit shows a CA3162A connected to a CD4511B decoder/driver to operate a common-cathode LED display. Figure 6-21B shows the PC template and component layout. The additional logic shown within the dotted area allows the display of negative numbers as low (as −99 mV), depending on the position of the DP1/DP2 switch. The sampling rate can be changed from 4 to 96 Hz when pin 6 of the CA3162A is connected to +5 V. To adjust, follow the procedure described for Fig. 6-20. Harris Semiconductors, Data Acquisition, 1991, p. 2-6, 2-7.

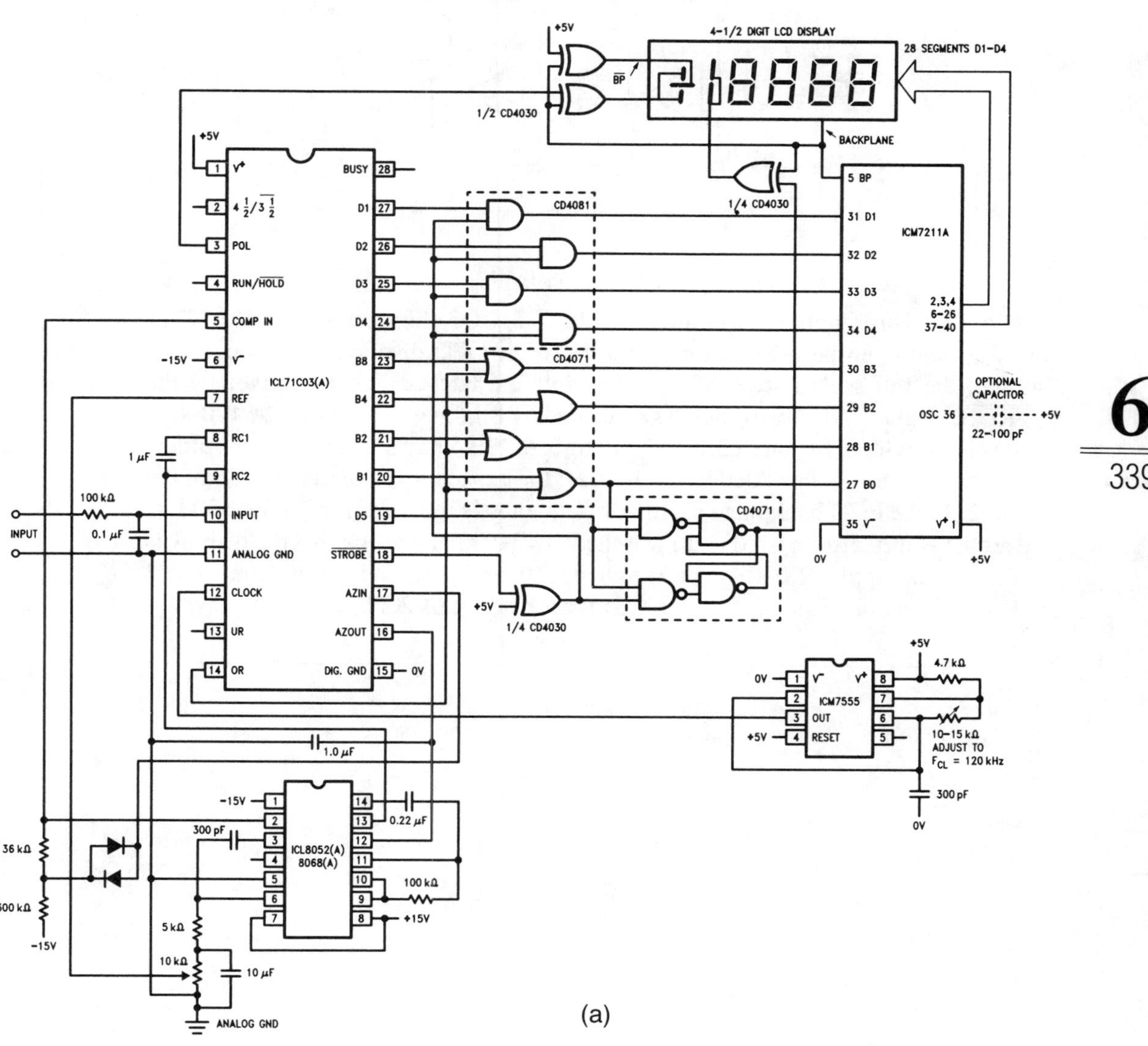

(a)

6

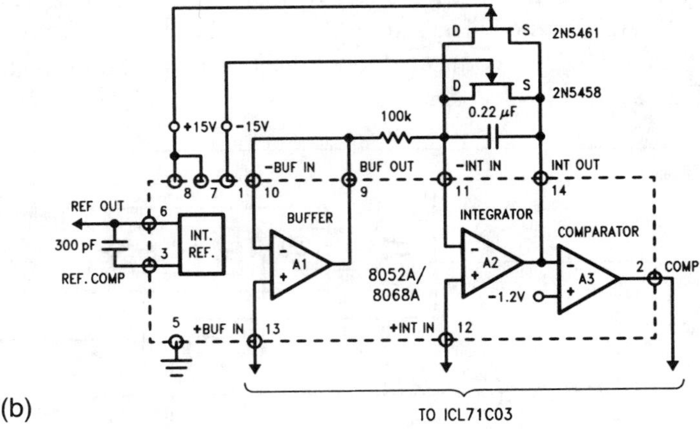

(b)

A/D converter for a 4½-digit LCD display

Fig. 6-22 This circuit shows an ICL8052/ICL71CO3 chip pair with an ICM7211 driver, which can be used to operate a 4½-digit LCD display. With the values shown, the full-scale is ±2.000 V. Data bits are latched into the driver by the $\overline{\text{STROBE}}$ signal and "Overrange" is indicated by blinking the 4 digits. A 120-kHz clock is provided by an ICM7555. Figure 6-22B shows a gross overvoltage protection circuit (to be added if required). If the capacitor is added to pin 36 of the ICM7211, the LCD backplane oscillator frequency is reduced from about 150 Hz. To adjust, short terminals 10 and 11 of the ICL71CO3 and check that the display is 0000. Then, apply 2.000 V at terminals 10 and 11, and adjust the 10-kΩ potentiometer (pin 7 of the ICL71CO3) for a reading of 2000 on the display. Harris Semiconductors, Data Acquisition, 1991, p. 2-25, 2-26.

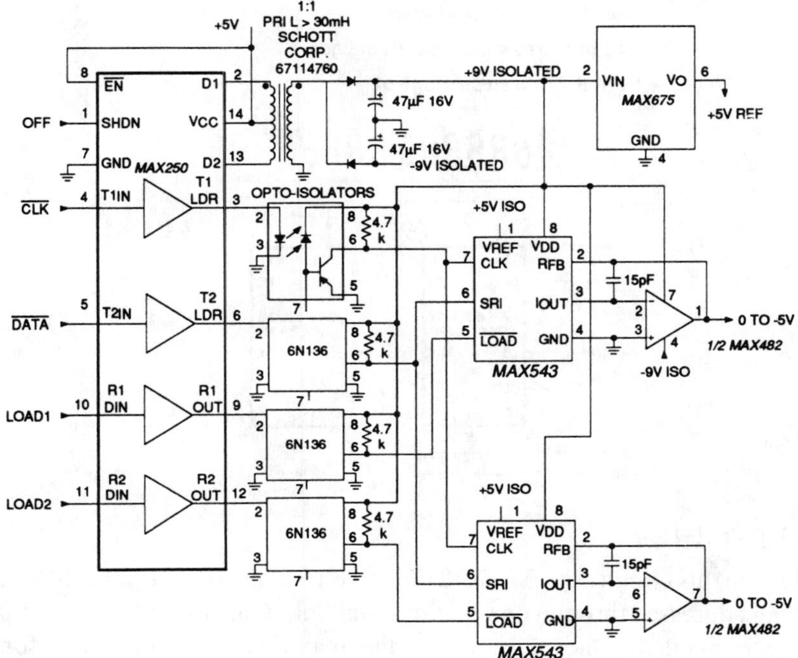

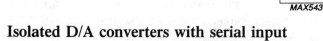

Isolated D/A converters with serial input

Fig. 6-23 In this circuit, two MAX543 serial-input 12-bit D/A converters (DACs) receive power and data from a MAX250 receiver/transmitter. Clock and data lines are shared by both DACs, but separate LOAD1 and LOAD2 inputs latch the data in. The MAX250 also has a shutdown input (SHDN) that powers down the circuit and cuts supply current to less than 10 µA. Maxim, 1992, Applications and Product Highlights, p. 2-9.

- **Fast Serial μP Interface**
- **24 Pin Narrow DIP, SO Packages**
- **Digital Brightness Control**

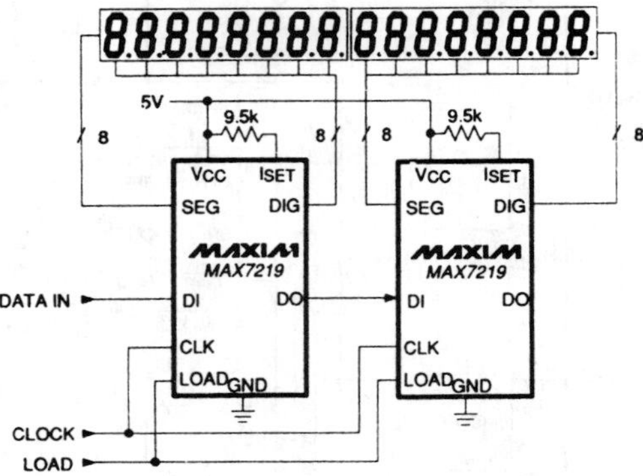

8-digit LED driver

Fig. 6-24 In this circuit, two MAX7217s drive 16 digits from a single 5-V supply, while being addressed through one 3-wire serial link. Only one external component (per chip) is needed. One resistor sets the maximum current (I_{SET}) for all 8 segments, while a fast (up to 10 MHz) cascadable serial interface loads data directly from most microprocessors. Maxim, 1992, Applications and Product Highlights, p. 2-9.

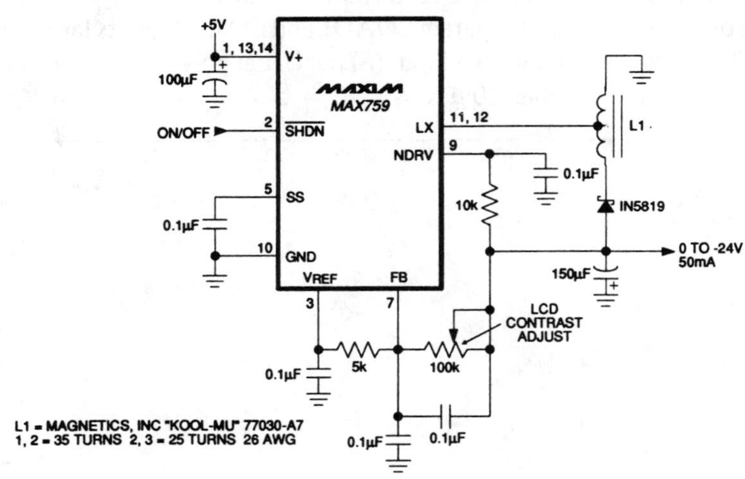

Adjustable LCD supply with autotransformer

Fig. 6-25 This circuit provides a variable output voltage to adjust the contrast of LCDs. The autotransformer L1 steps up the output to prevent overvoltage to the LX transistor. The total voltage from V_{IN} to LX must be limited to 22 V (with a 5-V input), the voltage at LX cannot exceed −17 V. With L1 (a miniature 0.2-in diameter toroid), the voltage at LX is reduced from the output voltage (determined by the turns ratio). Maxim, 1992, Applications and Product Highlights, p. 4-5.

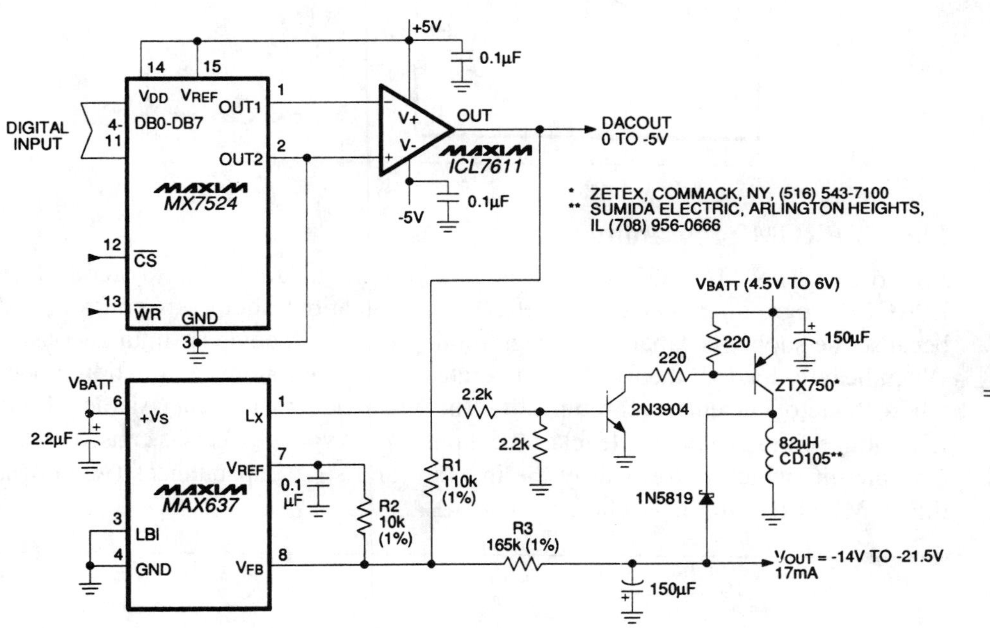

LCD supply with digital adjust

Fig. 6-26 If the expense of a customized autotransformer (L1, Fig. 6-25) is too high, this low-cost pulse-skipping circuit can be used. In this case, the relatively high output voltage is accommodated via an external switch transistor that has 40-V BV_{CEO} specifications. A simple D/A scheme adjusts the output for contrast control and temperature compensation of the LCD. Maxim, 1992, Applications and Product Highlights, p. 4-5.

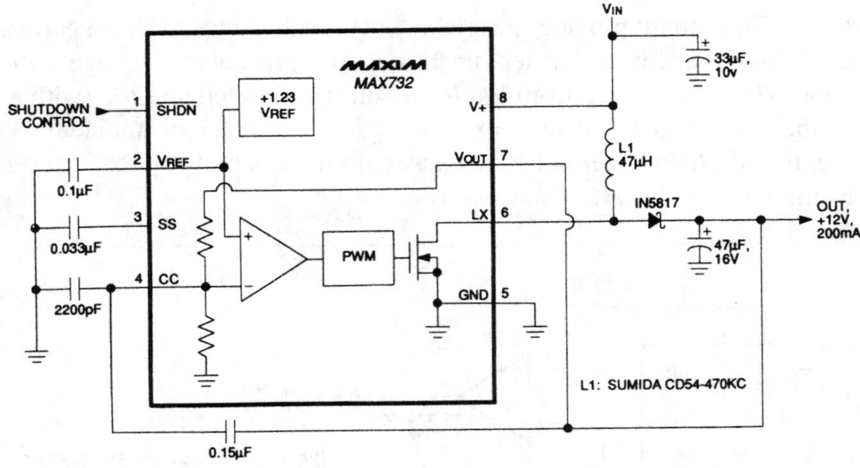

Flash EPROM programmer

Fig. 6-27 The MAX732 current-mode PWM regulator makes a good flash EPROM programmer because of the logic-compatible shutdown control, and because the supply is capable of programming four flash devices simultaneously. When heavily loaded, the MAX732 operates in the continuous-conduction mode, where inductor current never returns to zero. When inductor current is depleted, the LX voltage rings around the level of the input supply voltage. This is caused by the tank circuit, which is formed by the inductor and stray capacitances (windings, diode, MOSFET within the chip, etc.). Maxim, 1992, Applications and Product Highlights, p. 4-7.

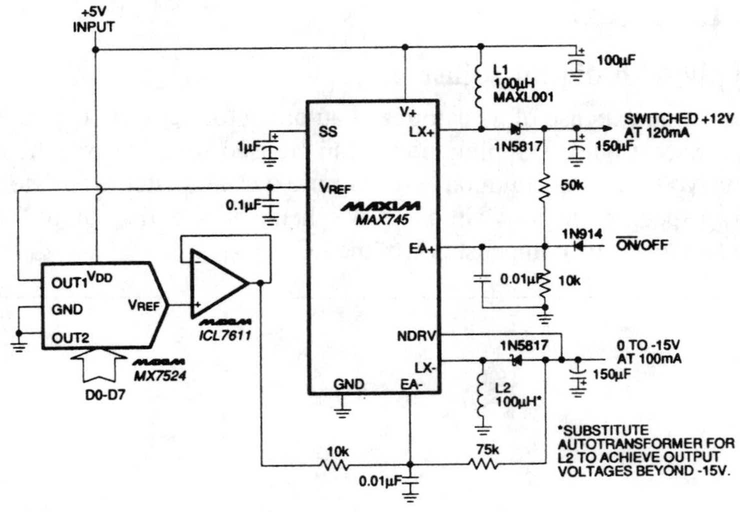

Flash EPROM programmer plus adjustable LCD supply

Fig. 6-28 The MAX745 generates both positive and negative boosted outputs from a +5-V supply. This circuit is configured for small microprocessor systems, and contains a DAC to adjust the negative voltage for LCD contrast control. The circuit also provides for switching on and off the +12-V output to program flash EPROMs. Maxim, 1992, Applications and Product Highlights, p. 4-9.

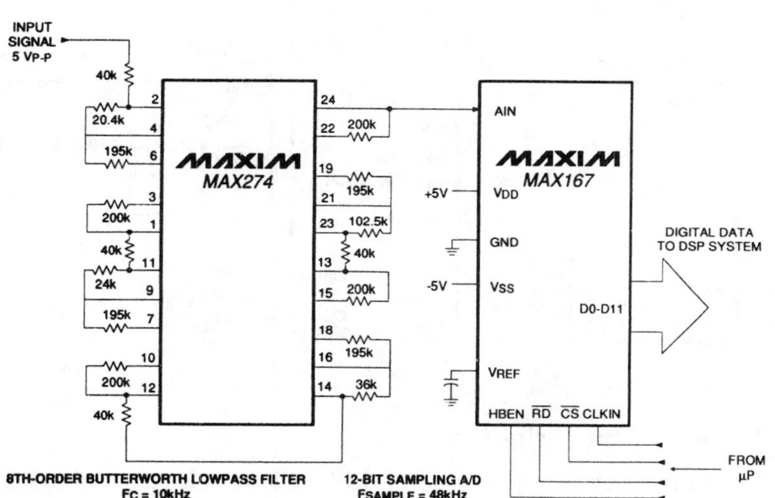

Digital signal-processing system with anti-aliasing

Fig. 6-29 In this circuit, a MAX274 provides a front-end 10-kHz low-pass filter to a digital signal-processing (DSP) system. Signals between dc and 10 kHz are digitized by a 12-bit sampling MAX167 A/D. The DSP processor receives samples at a rate of 48 kHz. The input signal is filtered by a MAX274 that is configured as a 10-kHz 8th-order Butterworth low-pass filter. Signals above 24 kHz (the Nyquist frequency) are attenuated 60 dB. Additional rolloff in the incoming signal reduces potential alias frequencies below the theoretical 74-dB noise floor of the 12-bit system. Maxim, 1992, Applications and Product Highlights, p. 7-4.

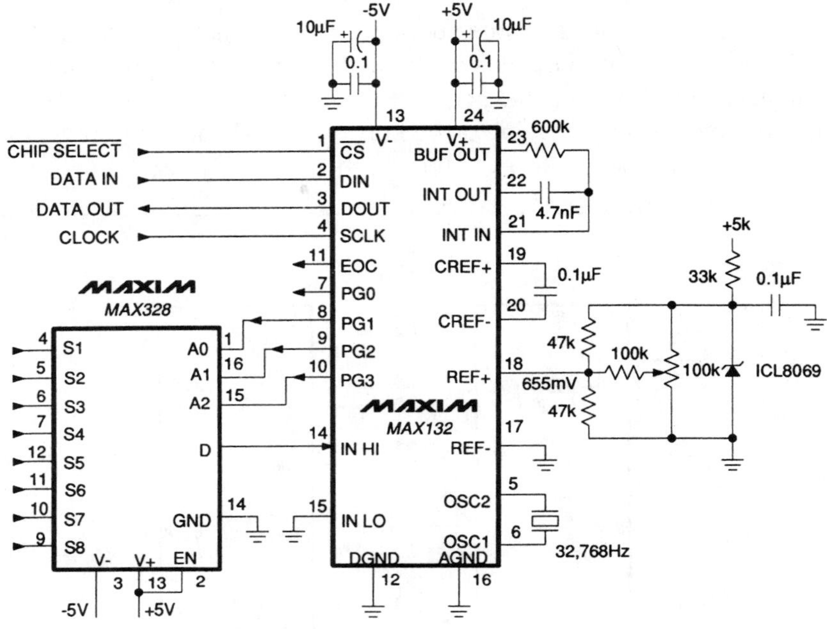

A MUX with 16-bit serial interface

Fig. 6-30 In this circuit, a 16-bit MAX132 A/D provides a serial interface for a MAX328 MUX, with both serial data input and output. The A/D is a 16-bit modified dual-slope converter. In addition to the main 16-bit output, the A/D also provides 3 guard-band bits with add resolution in the A/D noise region, but it can raise resolution if software averaging is used. The MAX328 MUX is addressed through four programmable bits PG0-PG3, which are set through the MAX132 serial input. Maxim, 1992, Applications and Product Highlights, p. 8-4.

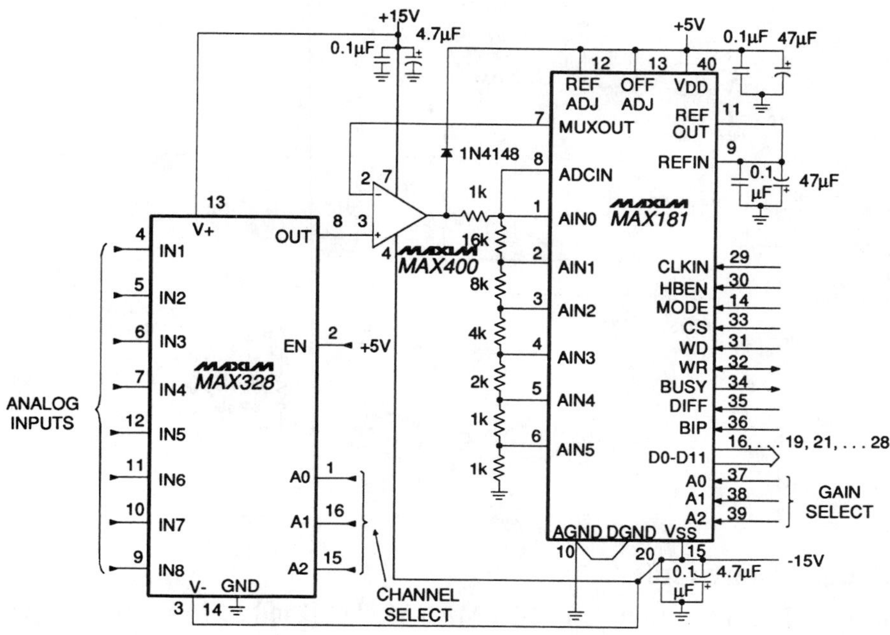

Fault-protected 8-channel 18-bit dynamic-range A/D converter

Fig. 6-31 In this circuit, the MAX181 internal MUX forms a programmable-gain amplifier (PGA) with digitally-selected gains of 1, 2, 4, 8, 16, 32, 64, and 128. An external fault-protected MUX, the MAX378, selects from 8 input channels and protects the input for up to ±60-V voltage levels. Maxim, 1992, Applications and Product Highlights, p. 8-9.

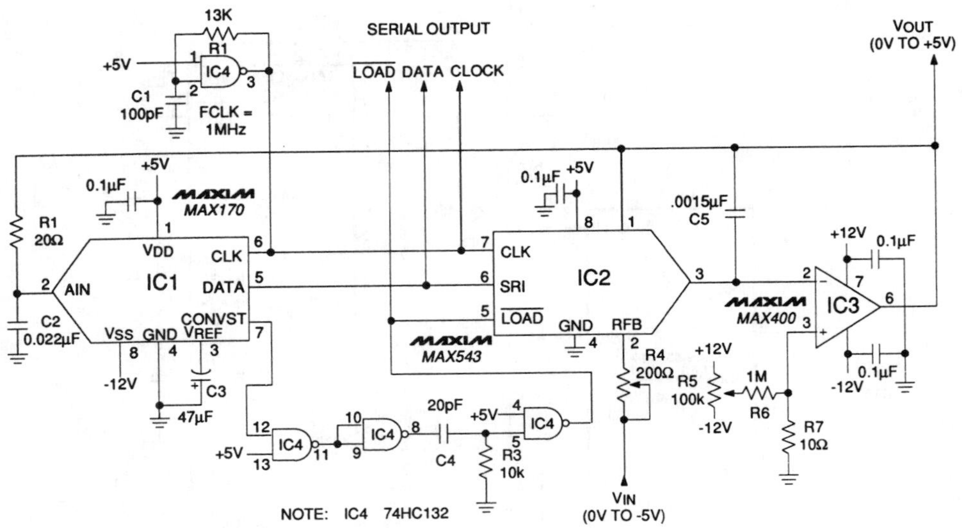

Square-root circuit

6

Fig. 6-32 In this circuit, a 12-bit MAX543 DAC and a MAX170 A/D operate in the feedback loop of a single op amp to determine the square-root of a 0- to −5-V input signal. Both analog (V_{OUT}) and digital outputs (serial DATA output) are provided. Accuracy surpasses 0.1% for −5-mV to 5-V inputs. The MAX543 squares the op-amp input, with an output that is proportional to $I_{OUT} = V_{REF}$(DAC code) + V_{IN}. One DAC multiplying input is digital. The MAX170 A/D digitizes the op-amp output and transmits 12-bit data to the DAC. Consequently, the DAC V_{REF} input and digital input carry the same value. The DAC output is then: $I_{OUT} = \sqrt{5} \times \sqrt{-V_{IN}}$. Typical input/output examples are: $V_{IN} = -5.000$ V, $V_{OUT} = +5.000$ V; $V_{IN} = -0.5$ mV, and $V_{OUT} = +50$ mV. Maxim, 1992, Applications and Product Highlights, p. 8-14.

• 10μA Shutdown
• 5V Operation

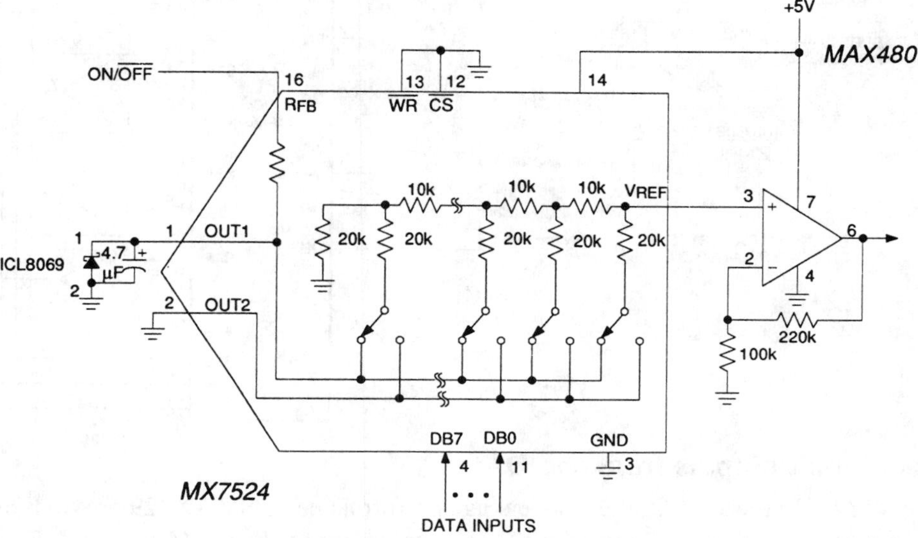

Voltage-output DAC

Fig. 6-33 In this circuit, an MX7524 8-bit current output DAC is connected in an inverted configuration (voltage mode) by driving the I_{OUT} pin with a 1.23-V reference (ICL8069). The DAC output drives a MAX480 where the gain is set so that full-scale output is 4 V. The power consumption is below 1 mW (operating) and much less shut down. The circuit operates when the On/Off input is at 5 V. When On/Off is at 0 V, supply current falls to 15 μA, and the amplifier output falls to 0 V. The DAC input data can be retained if the $\overline{\text{WR}}$ input is used. Digital input lines must be driven with CMOS logic levels (rail-to-rail) to minimize supply current. Maxim, 1992, Applications and Product Highlights, p. 8-15.

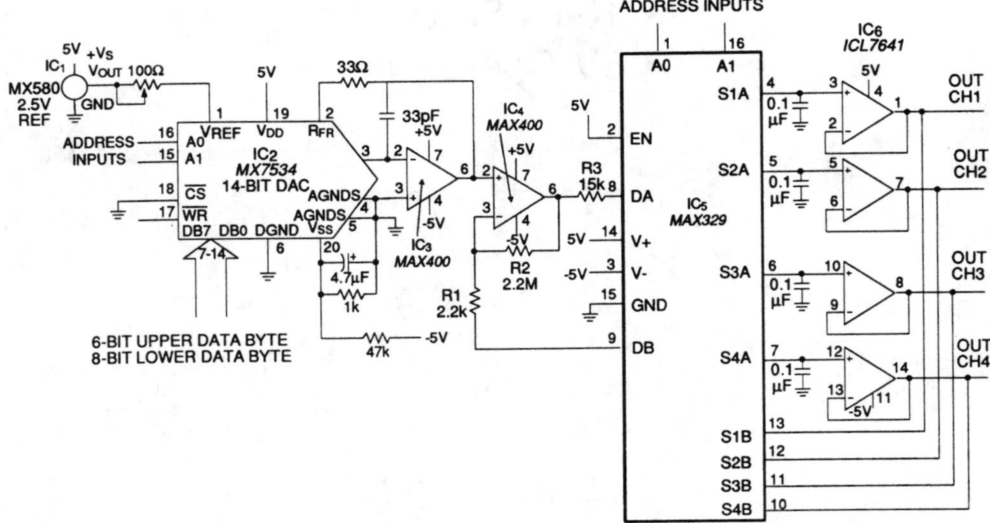

Four 14-bit outputs from one DAC

6

Fig. 6-34 In this circuit, the low-leakage performance of a MAX329 provides an economical way to create four 14-bit DAC outputs from only one 14-bit DAC. Four 0.1-µF capacitors and an ICL7641 quad op amp sample the DAC voltage that is presented by the MUX. The A-section of the MAX329 scans the hold capacitors, while the B section closes a feedback loop, which connects the appropriate buffer within IC6 to a precision op amp (IC4). This loop effectively removes the buffer offset voltage. If a typical MUX charge injection of the 4 pC and a typical leakage (plus op-amp bias) of 1 pA are assumed, the voltage stored on each capacitor drops no more than 1 bit (out of 2^{14}) over 2.5 seconds at 25°C. Maxim, 1992, Applications and Product Highlights, p. 9-5.

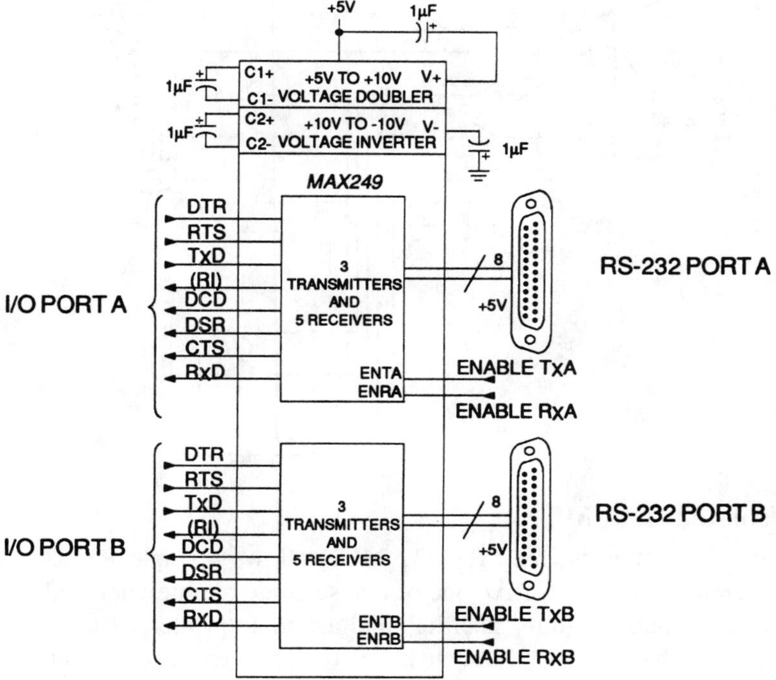

Two RS-232 serial ports with one chip

Fig. 6-35 In this circuit, a MAX249 includes a full complement of drivers and receivers for two complete DTE (Data Terminal Equipment) serial ports. The circuit also saves power with separate shutdown inputs for the inputs and outputs of both DTE ports. When both ports are off, supply current is only 25 μA. The dual serial ports are powered from a single +5-V supply. The guaranteed data rate is 64 kb/s, which ensures compatibility with EIA-232D/232E, V.28/V.24, and EIA-562. Maxim, 1992, Applications and Product Highlights, p. 2-6.

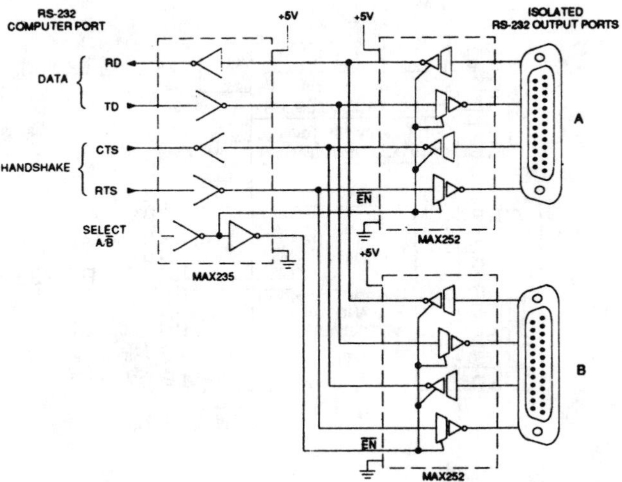

Fully isolated data MUX

Fig. 6-36 This fully-isolated RS-232 MUX allows a single RS-232 computer port, powered from 5 V, to drive one of two separate remotely located peripherals (typically a data-acquisition "satellite" and a remote printer). Isolation is useful here because both devices are likely to have ground potentials different from that of the computer. Also, the connection isolates the peripherals from each other as well as the computer. The MAX235 takes RS-232 data and handshake lines and a "Select A/B" output from the computer and converts them to logic levels. These levels are passed along to both MAX252 isolated transceivers. The RTS line controls which port is selected. Maxim, 1992, Applications and Product Highlights, p. 2-8.

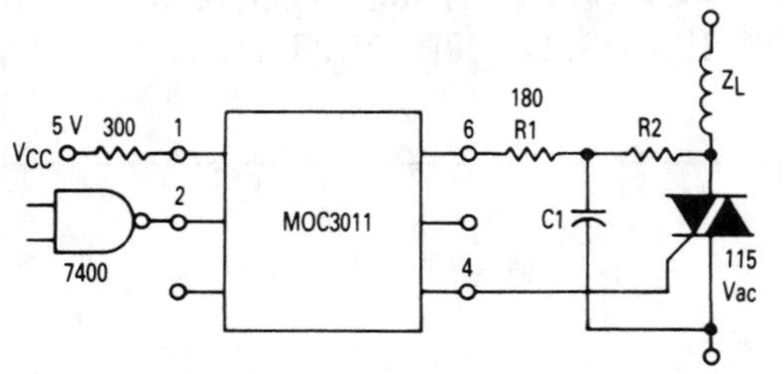

NOTE: CIRCUIT SUPPLIES 25 mA DRIVE TO GATE OF TRIAC AT V_{in} = 25 V AND $T_A \leq 70°C$.

TRIAC		
I_{GT}	R2	C
15 mA	2400	0.1
30 mA	1200	0.2
50 mA	800	0.3

Logic-to-inductive load interface

Fig. 6-37 This circuit shows interface between conventional 7400 TTL logic and a triac, using an MOC3011 optoisolator with triac-driver output (chapter 9). The values of C and R_2 depend on the triac current. Motorola Thyristor Device Data, 1991, p. 1-3-15.

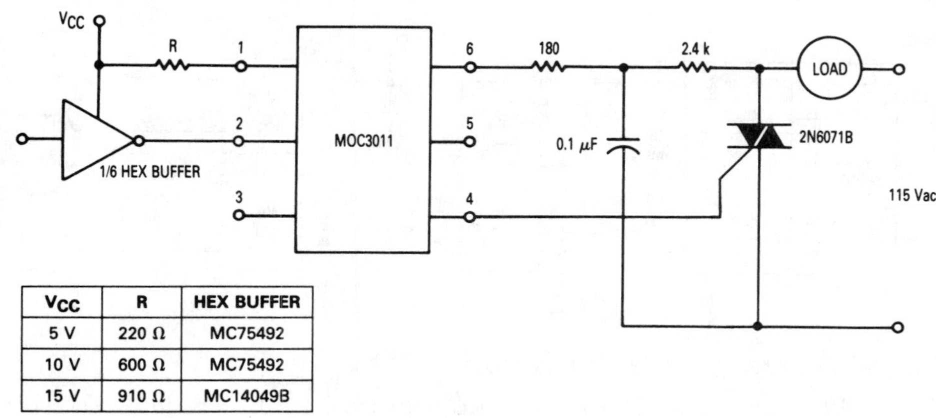

V_{CC}	R	HEX BUFFER
5 V	220 Ω	MC75492
10 V	600 Ω	MC75492
15 V	910 Ω	MC14049B

MOS-to-ac load interface

Fig. 6-38 This circuit shows interface between MOS logic and an ac load, using an MOC3011 optoisolator with triac-driver output (chapter 9). The value of *R* depends on the type of MOS buffer used. Motorola Thyristor Device Data, 1991, p. 1-3-16.

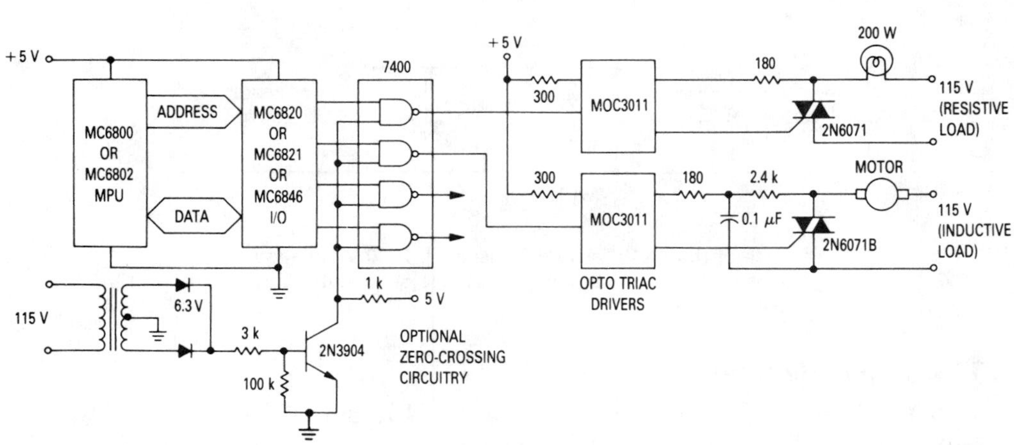

M6800 microcomputer interface

Fig. 6-39 This circuit shows interface between an M6800 or M6802 and resistive or inductive loads. If the second input of the gates is tied to the optional timing circuit shown, the triacs will fire only at the zero crossing of the ac line voltage. This extends life of incandescent lamps, reduces surge-current strains on the triac, and reduces EMI generated by load switching—all without special software or programming within the processor. Motorola Thyristor Device Data, 1991, p. 1-6- 14.

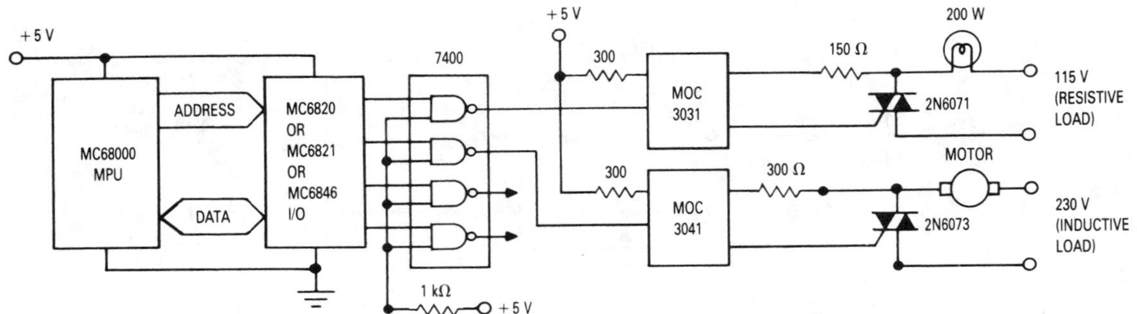

M68000 microcomputer interface

Fig. 6-40 This circuit shows interface between an M68000 and resistive or inductive loads, using MOC3031 and MOC3041 optoisolators. The zero-crossing feature of these optoisolators extends life of incandescent lamps, reduces triac surge-current strains, and reduces EMI, as described for the circuit of Fig. 6-39 (but without the need for a timing circuit at the 7400 gate inputs). Motorola Thyristor Device Data, 1991, p. 1-6-16.

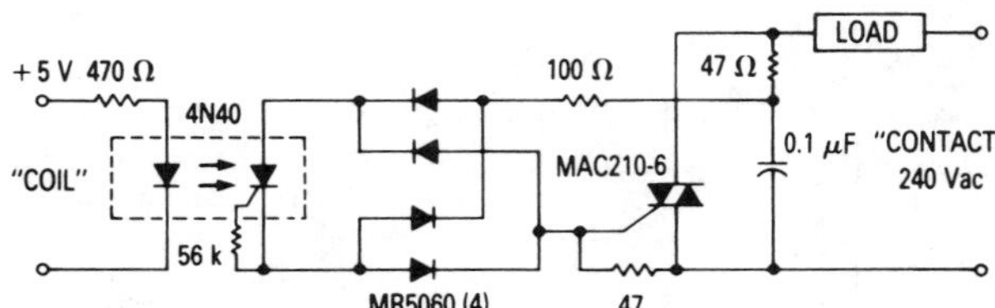

Use of the 4N40 for high sensitivity, 7500 V isolation capability, provides this highly reliable solid state relay design.

TTL-compatible solid-state relay

Fig. 6-41 This circuit is compatible with 74, 74S, and 74H series TTL-logic inputs, and can handle 240 Vac loads up to 10 A. Motorola Optoelectronics Device Data, 1989, p. 6-20.

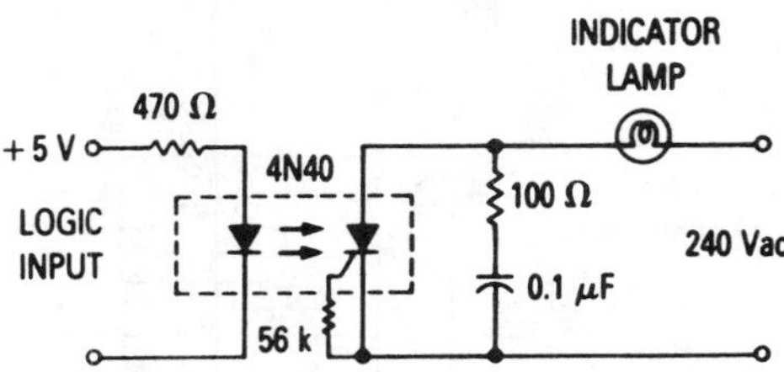

25-W logic-indicator lamp driver

Fig. 6-42 This circuit makes it possible to directly couple (without buffers) TTL and DTL inputs with either indicator or alarm devices. The optoelectronic coupling prevents both logic glitches and the introduction of noise. Motorola Optoelectronics Device Data, 1989, p. 6-20.

In order to interface positive logic to negative-powered electromechanical relays, a change in voltage level and polarity plus electrical isolation are required. The H11Gx can provide this interface and eliminate the external amplifiers and voltage divider networks previously required. The circuit below shows a typical approach for the interface.

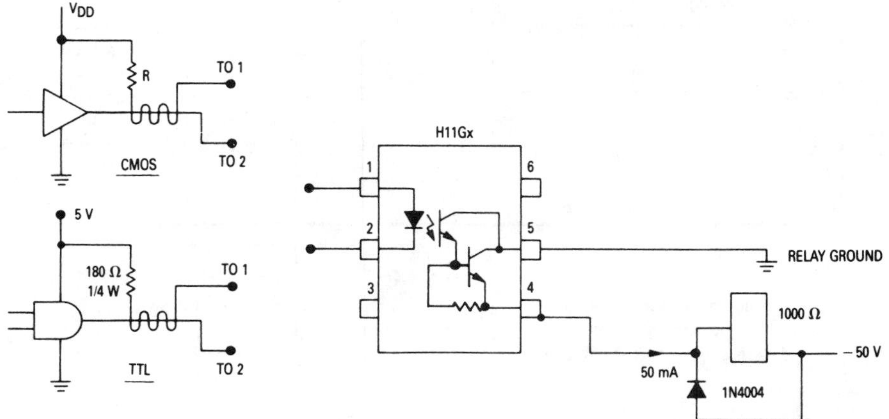

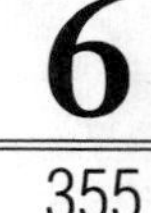

TTL/DTL interface to telephone equipment

Fig. 6-43 This circuit provides interface of positive-logic circuits to negative-powered telephone equipment. In addition, the circuit provides for a change in voltage level (+5 and −50 V), as well as isolation, using an HllG optoisolator. Motorola Optoelectronic Device Data, 1989, p. 6-54.

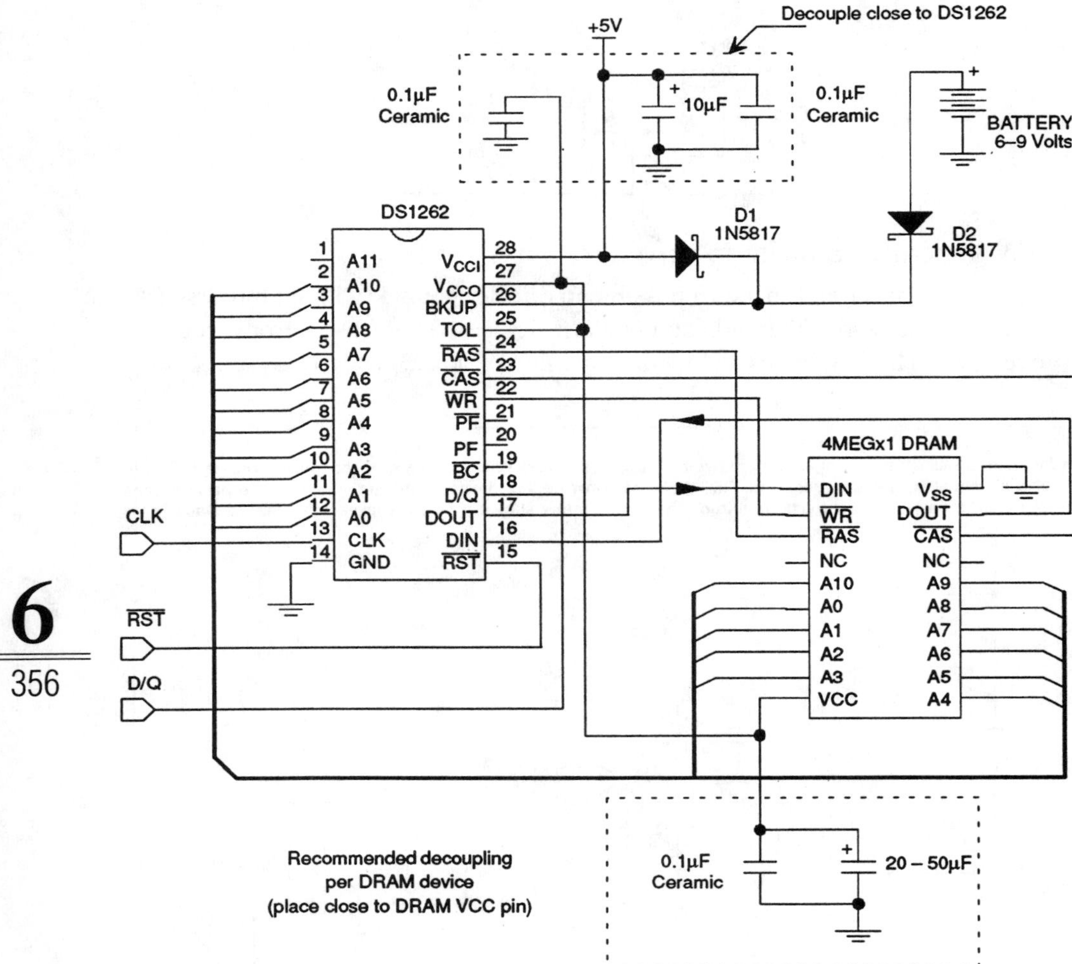

Serial DRAM nonvolatizer

Fig. 6-44 This circuit shows a DS1262 Serial DRAM nonvolatizer chip that is used to enable read/write access of DRAM from a simple 3-wire serial port. Refresh and RAS/CAS timing for the DRAM are performed automatically, transparent to operation of the serial port. In addition, the DS1262 performs all power switching and refresh duties that are necessary to retain DRAM data when the primary power supply fails. The backup supply input accepts a wide voltage range, suitable for use with rechargeable batteries. The DS1262 also provides an electronic "gas gauge" that can predict the condition of the backup supply. When the V_{CC1} power begins to drop, an internal circuit senses the changes, and the data-retention mode begins.

The V_{CC1} input is then disconnected from the V_{CCO} output and the backup supply connected to the BKUP pin is switched in. The DS1262 also monitors the backup supply condition. If the backup supply is below V_{CC1}, the backup-condition output pin $\overline{BC}$ is driven active low, and remains in this state until the backup supply voltage is restored to a level above V_{CC1}. Dallas Semiconductors, Product Data Book, 1992/1993, p. 9-59.

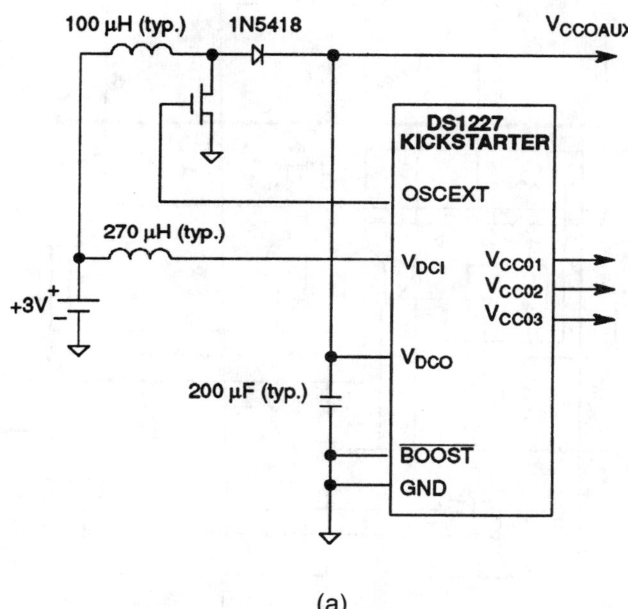

(a)

6

357

Continued

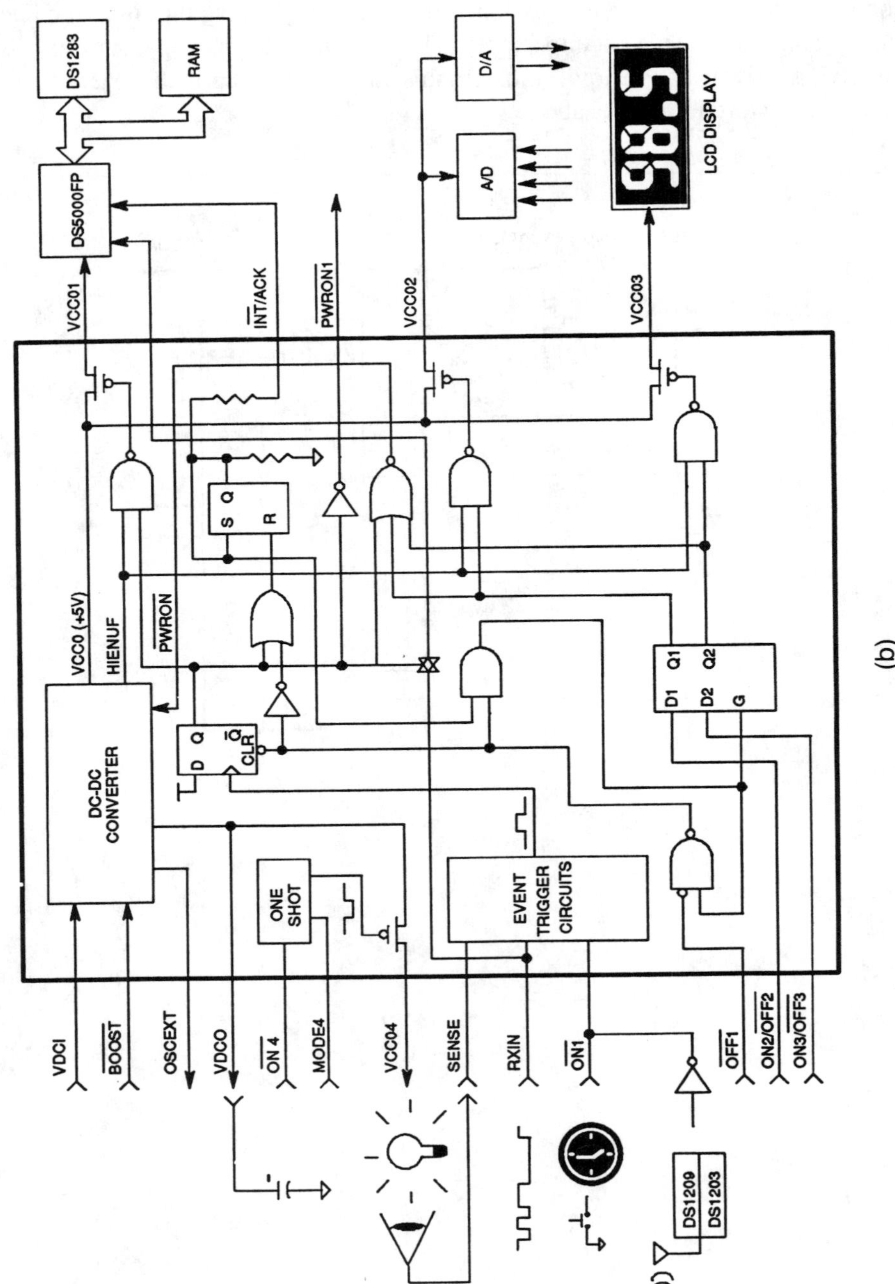

DS1283
RAM
DS5000FP
D/A
A/D
98.5
LCD DISPLAY
VCC01
INT/ACK
PWRON1
VCC02
VCC03
VCCO (+5V)
HIENUF
PWRON
DC-DC CONVERTER
S
Q
R
D
Q
Q̄
CLR
D1
Q1
D2
Q2
G
ONE SHOT
EVENT TRIGGER CIRCUITS
VDCI
BOOST
OSCEXT
VDCO
ON 4
MODE4
VCC04
SENSE
RXIN
ON1
OFF1
ON2/OFF2
ON3/OFF3
DS1209
DS1203

(b)

Digital auxiliary boost supply (kickstarter)

Fig. 6-45 This circuit shows a DS1227 kickstarter chip connected to provide an auxiliary boost supply for digital systems. The DS1227 supplies 5 V on demand from either a 3- or 6-V battery input. The primary 5-V outputs, typically tied to a microcontroller V_{CC} pin, is "kickstarted" on in response to any one of several possible momentary, external signal transitions. Examples of such signals include a clock/calendar alarm from a watchdog timer (Fig. 6-51), or an incoming asynchronous serial-data word from a host PC via a line-powered transceiver (Fig. 6-60), or a simple pushbutton switch. Figure 6-45B shows some typical kickstarter interconnections. Dallas Semiconductor, Product Data Book, 1992/1993, p. 10-17, 10-21.

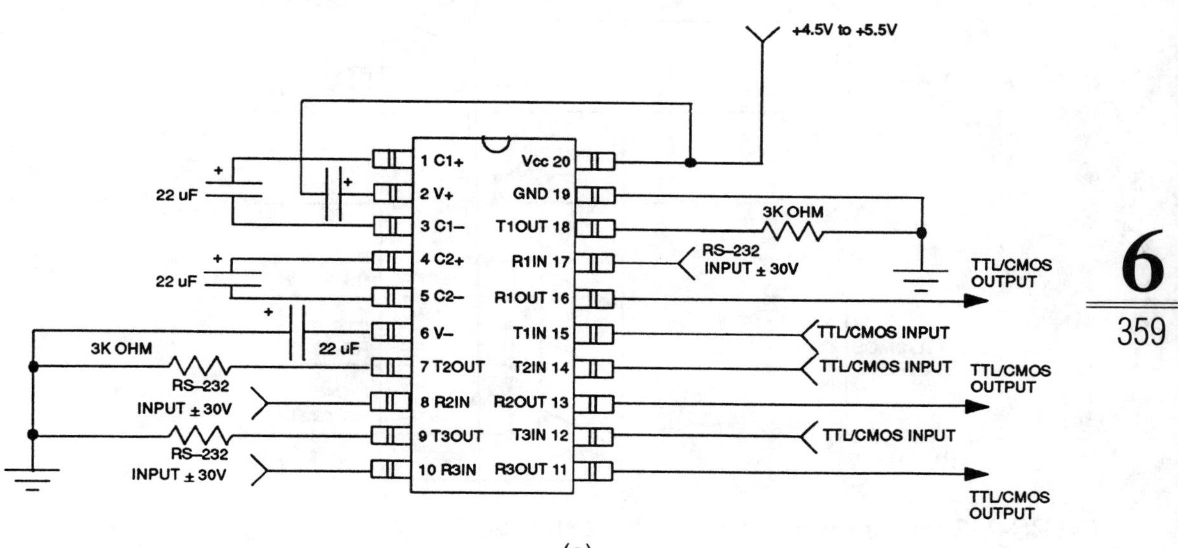

(a)

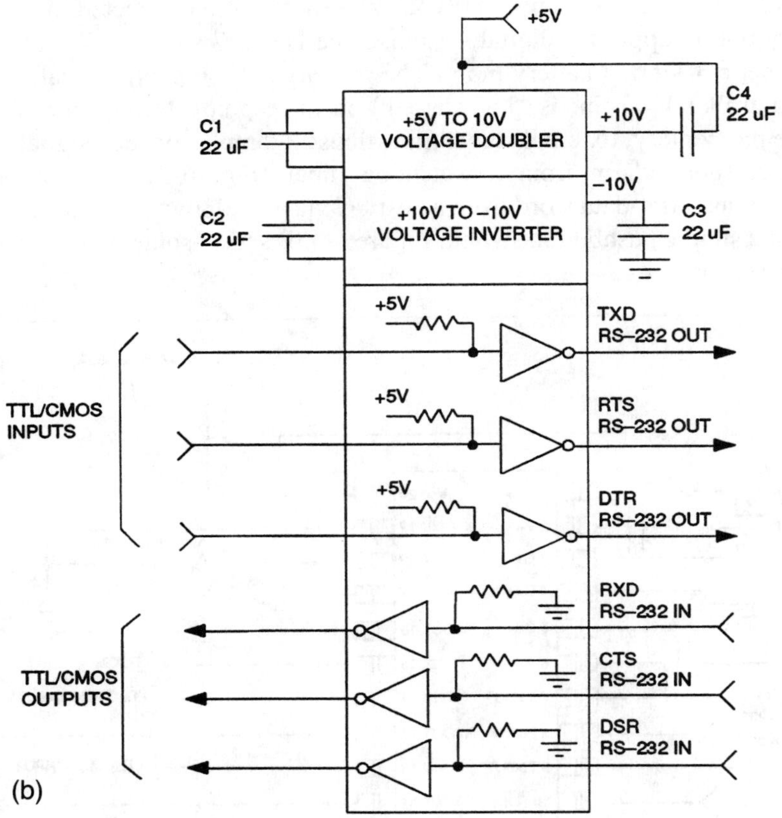

+5V
C1
22 uF
+5V TO 10V
VOLTAGE DOUBLER
+10V
C4
22 uF
C2
22 uF
+10V TO −10V
VOLTAGE INVERTER
−10V
C3
22 uF
+5V
TXD
RS–232 OUT
+5V
RTS
RS–232 OUT
TTL/CMOS
INPUTS
+5V
DTR
RS–232 OUT
RXD
RS–232 IN
CTS
RS–232 IN
TTL/CMOS
OUTPUTS
DSR
RS–232 IN
(b)

RS-232 transmitter/receiver with external power

Fig. 6-46 This circuit shows a DS1229 that is connected as a triple RS-232 transmitter/receiver. The DS1229 has two internal charge pumps that are used to generate ±10 V, and six level translators. Three of the translators are RS-232 that convert TTL/CMOS inputs into +9-V RS-232 outputs. The other three translators are RS-232 receivers that convert RS-232 inputs to 5-V TTL/CMOS outputs. Figure 6-46B shows some typical RS-232 interconnections. Dallas Semiconductor, Product Data Book, 1992/1993, p. 10-33, 10-34.

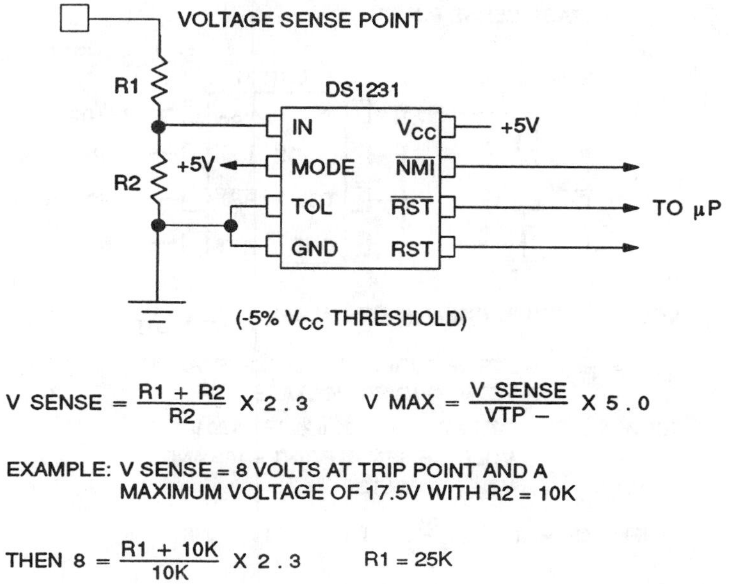

VOLTAGE SENSE POINT
R1
DS1231
R2
+5V
IN
MODE
TOL
GND
V_{CC}
$\overline{NMI}$
$\overline{RST}$
RST
+5V
TO μP
(-5% V_{CC} THRESHOLD)

$$V\ SENSE = \frac{R1 + R2}{R2} \times 2.3 \qquad V\ MAX = \frac{V\ SENSE}{VTP -} \times 5.0$$

EXAMPLE: V SENSE = 8 VOLTS AT TRIP POINT AND A MAXIMUM VOLTAGE OF 17.5V WITH R2 = 10K

$$\text{THEN } 8 = \frac{R1 + 10K}{10K} \times 2.3 \qquad R1 = 25K$$

Digital power monitor

Fig. 6-47 This circuit shows a DS1231 power monitor chip that is used to control a digital-system power supply. The DS1231 uses a temperature-compensated reference circuit to provide both an orderly shutdown and automatic restart of processor-based systems. The DS1231 operates by monitoring the high-voltage inputs to the power-supply regulators (at a voltage sense point), and applying RST or $\overline{RST}$ and $\overline{NMI}$ (nonmaskable interrupt) control signals to the processor. The time for processor shutdown is directly proportional to the available hold-up time of the power supply. Just before the hold-up time is exhausted, the DS1231 unconditionally halts the processor (to prevent spurious cycles) by enabling reset when V_{CC} falls below a selectable 5 or 10% threshold. Notice that the TOL pin is grounded for a 5% threshold. When power returns, the processor is held inactive until well after power conditions have stabilized, safeguarding any nonvolatile memory in the system from inadvertent data changes. Dallas Semiconductor, Product Data Book, 1992/1993, p. 10-40.

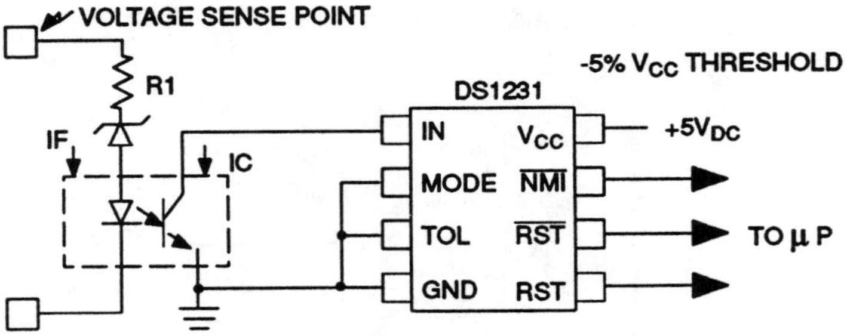

VOLTAGE SENSE POINT

R1

IF

IC

DS1231

-5% V_{CC} THRESHOLD

IN V_{CC} +5V_{DC}

MODE $\overline{NMI}$

TOL $\overline{RST}$ TO μ P

GND RST

VOLTAGE SENSE POINT (TRIP VALUE) $= VZ + \frac{IC}{CTR} \times R1$

$CTR = \frac{IC}{IF}$ CTR = CURRENT TRANSFER RATIO

VZ = ZENNER VOLTAGE

EXAMPLE: CTR = 0.2 IC = 30 μA IF = 150 μA

VOLTAGE SENSE POINT = 105 AND

VZ = 100 VOLTS

THEN $105 = 100 + \frac{30}{0.2} \times R1$ R1 = 33K

Digital power monitor with line isolation

Fig. 6-48 This circuit is similar to that of Fig. 6-47, except that an optoisolator (chapter 9) is used between the voltage sense point and the DS1231 to provide line isolation. Dallas Semiconductor, Product Data Book, 1992/1993, p. 10-41.

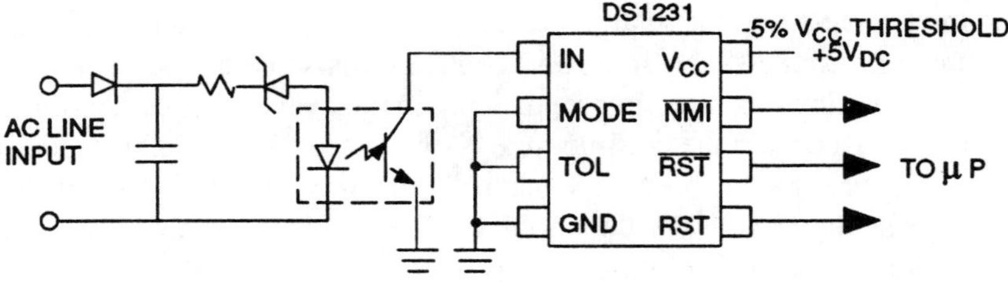

Digital power monitor that operates from the power line

Fig. 6-49 This circuit is similar to that of Fig. 6-47, except that the voltage sense point is taken directly from the ac power line, instead of from the power supply, and an optoisolator (chapter 9) is used between the line and DS1231. Dallas Semiconductor, Product Data Book, 1992/1993, p. 10-42.

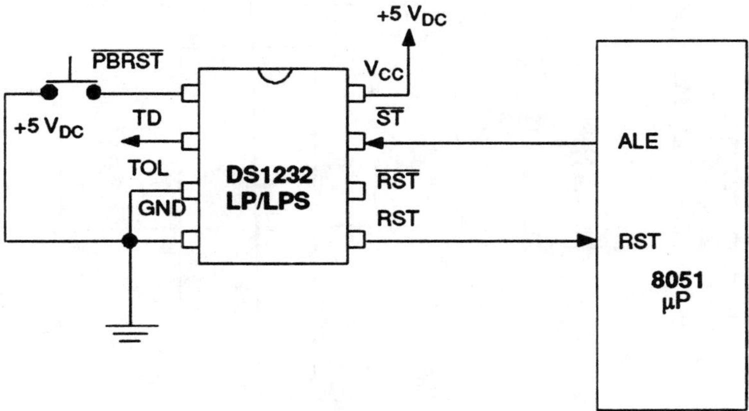

Low-power micromonitor with pushbutton

Fig. 6-50 This circuit shows a DS1232LP/LPS used to monitor and control the power supply and software execution of a processor-based system, and to provide a pushbutton reset. When V_{CC} falls below a preset level (as defined by the TOL pin connection) the DS1232 outputs RST and $\overline{\text{RST}}$ reset signals. With TOL connected to ground, reset becomes active as V_{CC} falls below 4.75 V. When TOL is connected to V_{CC}, reset becomes active as V_{CC} falls below 4.5 V. On power-up, RST and $\overline{\text{RST}}$ are kept active for a minimum of 250 ms to allow the power supply and processor to stabilize. The pushbutton reset input $\overline{\text{PBRST}}$ requires an active low. Internally, this input is debounced and timed so that reset signals of at least 250 ms (minimum) are generated. The 250-ms delay starts as the pushbutton reset input is released from low level. Dallas Semiconductor, Product Data Book, 1992/1993, p. 10-54.

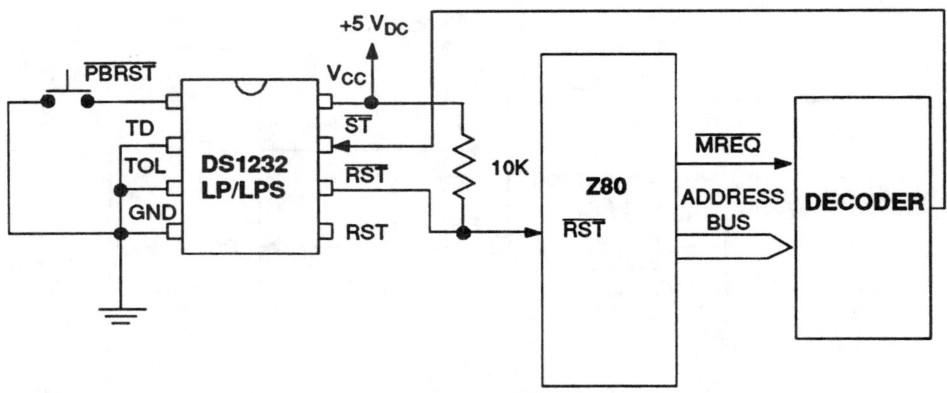

Low-power micromonitor with pushbutton and watchdog timer

Fig. 6-51 This circuit is similar to that of Fig. 6-50, but with the addition of a watchdog-timer function to monitor software execution. The timer function forces RST and $\overline{\text{RST}}$ signals to the active state (shutting down the processor) when the $\overline{\text{ST}}$ input is not stimulated for a predetermined time period (because of some failure in software execution). The time period is set by the TD input to be about 150 ms with TD connected to ground, 600 ms with TD left unconnected, and 1.2 s with TD connected to V_{CC}. The timer starts timing out from the set-time period as soon as reset signals are inactive. The $\overline{\text{ST}}$ input can be taken from address, data and/or control signals. When the processor is executing software, these signals are present, which causes the watchdog to be reset prior to time-out. Dallas Semiconductor, Product Data Book, 1992/1993, p. 10-54.

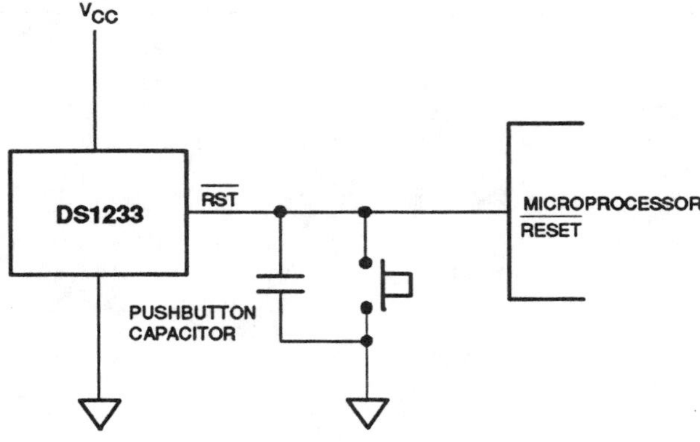

Automatic reset

Fig. 6-52 This circuit shows a DS1233 EconoReset chip that is used to monitor the power supply of a processor-based system and to provide automatic reset. The

DS1233 contains a precision temperature-compensated reference and comparator circuit that is used to monitor the power-supply (V_{CC}) status. When an out-of-tolerance condition is detected, $\overline{\text{RST}}$ becomes active, and shuts off the processor. When V_{CC} returns to an in-tolerance condition, $\overline{\text{RST}}$ is kept in the active state for about 350 ms to allow the power supply and processor to stabilize. The DS1233 also debounces a pushbutton closure (when used), and generates a 350-ms reset pulse upon release. Notice that when an external pushbutton is used, a capacitor between 100 pF and 0.01 μF must be connected between $\overline{\text{RST}}$ and ground. Where additional reset current is required, a minimum capacitance of 500 pF should be used, along with a parallel external pull-up 1-kΩ resistor (minimum). Dallas Semiconductor, Product Data Book, 1992/1993, p. 10-59.

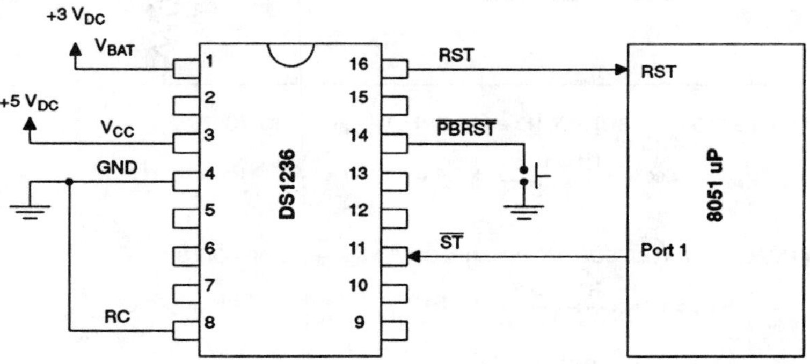

Digital power monitor with watchdog timer and pushbutton

Fig. 6-53 This circuit shows a DS1236 that is used to monitor and control the power-supply and software execution of a processor-based system, and to provide a pushbutton reset. When an out-of-tolerance condition occurs (when V_{CC} is below 4.5 V for 10% operation, or below 4.75 V for 5% operation), the RST and $\overline{\text{RST}}$ outputs are driven to the active state. On power-up, RST and $\overline{\text{RST}}$ are held active for a minimum of 25 ms (100 ms typical) after 4.5 V (or 4.75 V) is reached to allow the power supply and processor to stabilize. The pushbutton input $\overline{\text{PBRST}}$ is debounced and timed so that reset signals are driven to the active state for 25 ms minimum. The watchdog-timer function forces RST and $\overline{\text{RST}}$ to the active state (shutting down the processor) when the $\overline{\text{ST}}$ input is not stimulated for a predetermined time period (because of some failure in software execution). The watchdog time period is 400-ms typical (600-ms maximum). The $\overline{\text{ST}}$ input can be taken from address, data, and/or control signals. When the processing is executing software, these signals are present, and cause the watchdog to be reset prior to time-out. Dallas Semiconductor, Product Data Book, 1992/1993, p. 10-73.

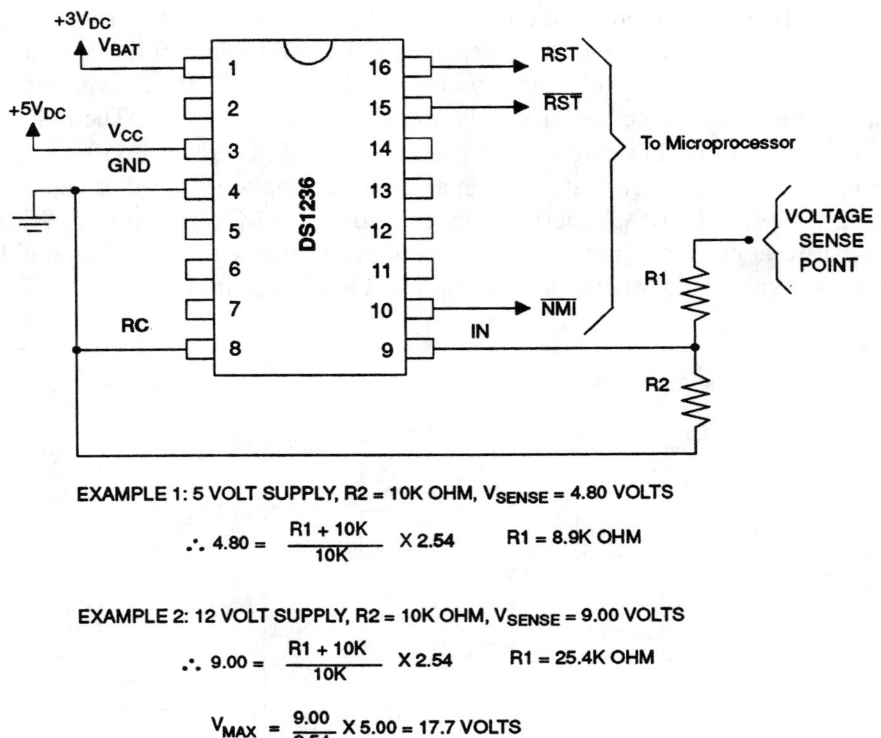

Digital power monitor with early warning

Fig. 6-54 This circuit shows a DS1236 that is used to control a digital-system power supply. The DS1236 generates a nonmaskable interrupt ($\overline{\text{NMI}}$) for early warning of power failure to a microprocessor. A precision comparator monitors the voltage level at the In pin, relative to an internal reference. The In pin is a high-impedance input that allows for a user-defined sense point. An external resistor voltage divider interfaces with voltage signals. The sense point can be taken from the 5-V supply or from a higher voltage level that is closer to the main system power input. Because the In trip point V_{TP} is 2.54 V, the proper values for R_1 and R_2 can be determined by the equations. Dallas Semiconductor, Product Data Book, 1992/1993, p. 7-74.

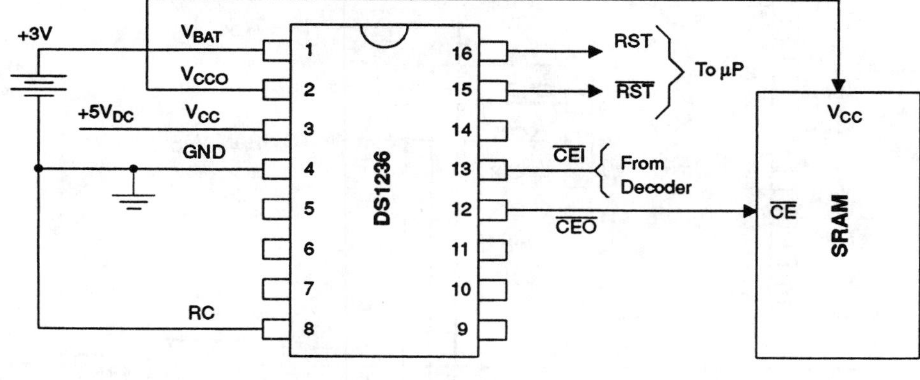

Digital memory backup for SRAMs

Fig. 6-55 This circuit shows a DS1236 that is used to control battery-backup operation for a static RAM. First, the DS1236 contains a switch to direct SRAM power from the 5-V supply (V_{CC}) or from an external battery (V_{BAT}), whichever is greater. The switched supply (V_{CCO}) can also be used to battery-back a CMOS processor. Second, the same power-fail detection (using RST and $\overline{RST}$) that is described for the digital power monitor (Fig. 6-53) is used to hold the chip-enable output ($\overline{CEO}$) to within 0.3 V of V_{CC} or to within 0.7 V of V_{BAT}. This write-protection mechanism occurs as V_{CC} falls below a specified trip point, V_{CCTP}.

Dallas Semiconductor, Product Data Book, 1992/1993, p. 10-74.

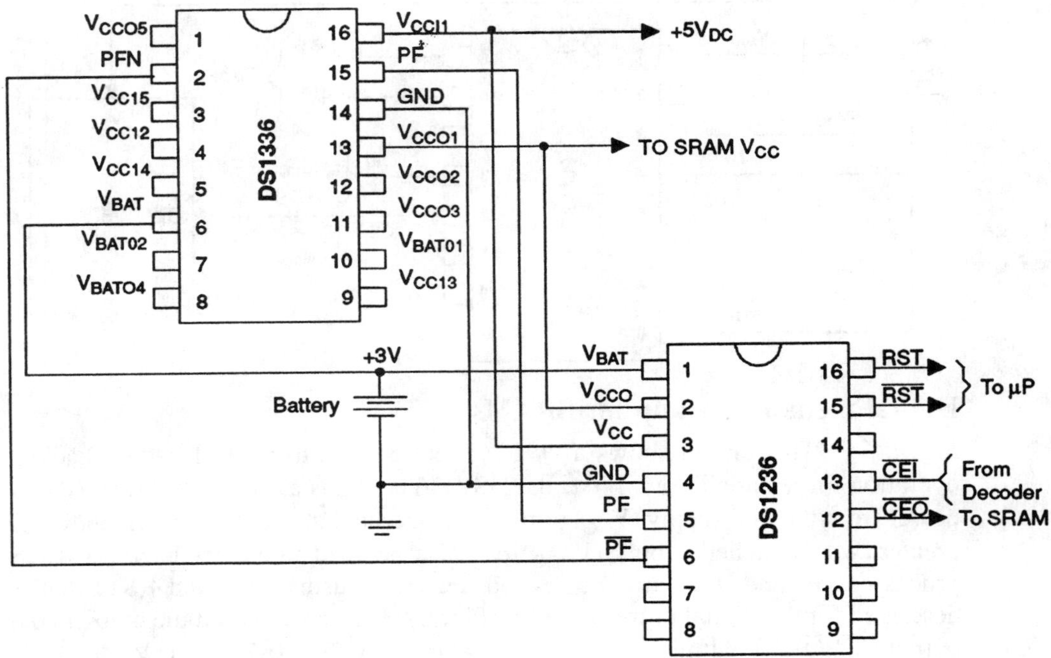

Digital power switching for SRAMs

Fig. 6-56 This circuit shows DS1236 and DS1336 chips that are used to provide power switching. When larger operating currents are required in a battery-backed system, the 5-V supply and battery-supply switches within the DS1236 cannot be large enough to support the required load through V_{CCO} with a reasonable voltage drop. When such large currents are required, the PF and $\overline{PF}$ outputs are provided to gate external power switching devices. Power to the load is switched from V_{CC} to battery on power-down, and from battery to V_{CC} on power-up. The DS1336 uses the $\overline{PF}$ to switch between V_{BAT} and V_{CC}. The load applied to the PF pin from the external switch is supplied by the battery. Thus, if a discrete switch is used, this load should be considered when sizing the battery. Dallas Semiconductor, Product Data Book, 1992/1993, p. 10-77.

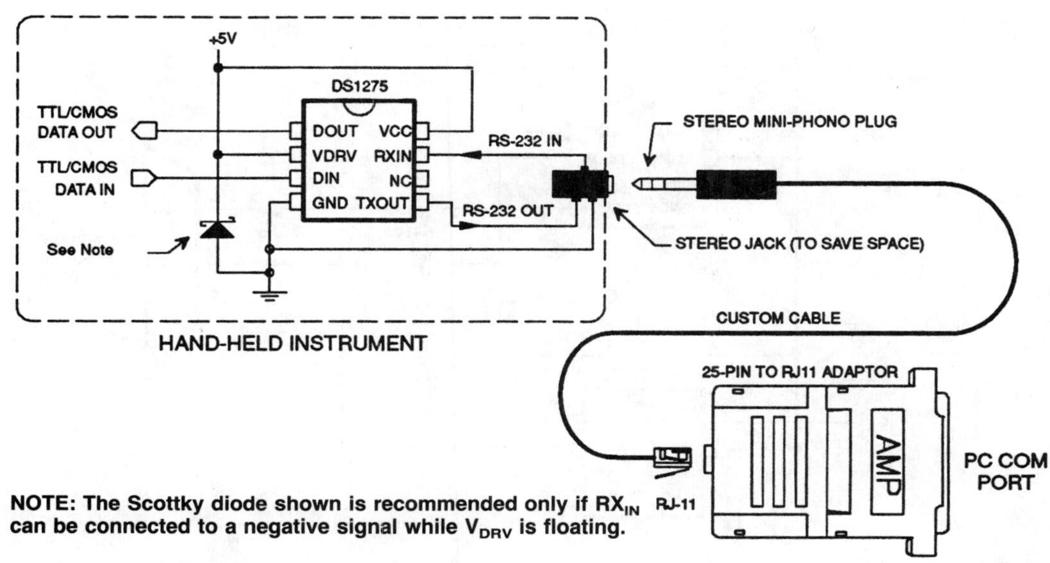

Hand-held line-powered RS-232-C transceiver

Fig. 6-57 This circuit shows a DS1275 as a hand-held IRS-232-C transceiver using a stereo mini-jack and 25-pin to RK11 adaptor. Notice that the DS1275 accommodates both TTL and CMOS data. Dallas Semiconductor, Product Data Book, 1992/1993, p. 10-113.

NOTE:
The capacitor stores negative charge whenever the TXD signal from the PC serial port is in a marking data state (a negative voltage that is typically -10 volts). The top DS1275's TX_{OUT} uses this negative charge reservoir when it is in a marking state. The capacitor will discharge to 0 volts when the TXD line is spacing (and TX_{OUT} is still marking) at a time constant determined by its value and the value of the load resistance reflected back to TX_{OUT}. However, when TXD is marking, the capacitor will quickly charge back to -10 volts. Note that TXD remains in a marking state when idle, which improves the performance of this circuit.

Full-duplex RS-232 that uses negative-charge storage

Fig. 6-58 This circuit shows two DS1275 chips to provide full-duplex operation for both TTL and CMOS. Dallas Semiconductor, Product Data Book, 1992/1993, p. 10-113.

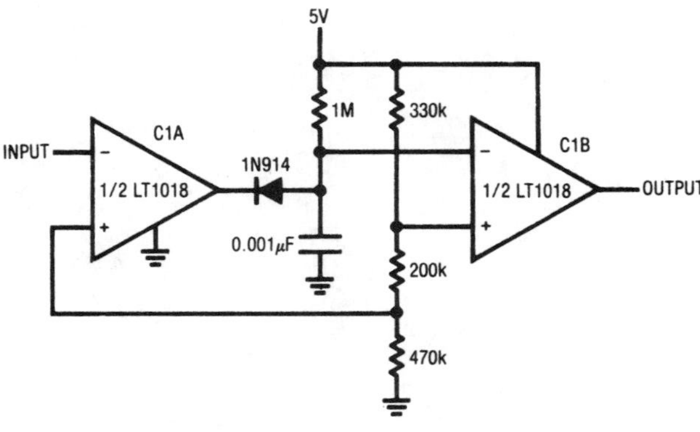

Watchdog timer to prevent lock-up

Fig. 6-59 This circuit uses both sections of an LT1018 to prevent lock-up in processor-based systems. Such lock-up can occur if the system misses an instruction because of transients in hardware or software, and usually results in loss of pulses somewhere in the system. This circuit issues a reset command in response to such a pulse loss. Normally, a pulse train appears at the circuit input and causes the ClA output to pulse low. The diode path discharges the 0.01-μF capacitor each time the ClA output goes low. Interruption of the input pulse train (after the 7th vertical division) allows the capacitor to charge beyond the ClB threshold, and triggers ClB low. This pulse can be used to reset the system (applied to the microprocessor $\overline{\text{RST}}$ input). The ClB negative-input RC values can be adjusted to accommodate various input pulse-repetition rates. Linear Technology, Linear Applications Handbook, 1990, p. AN31-10.

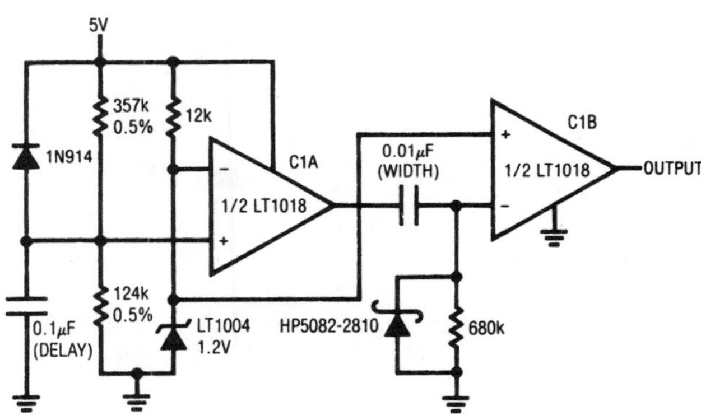

Power-on reset generator

Fig. 6-60 This circuit uses both sections of an LT1018 to reset a digital system after supply turn-on. When supply power is applied, the 5-V rail comes up. The LT1004 clamps at 1.2 V and the ClA noninverting input ramps at a time constant that is determined by the 0.05% resistors and the 0.1-μF capacitor. When the ClA noninverting input ramps beyond the LT1004 potential, the ClA output goes high and delivers a differentiated pulse to the ClB inverting input. The ClB output goes low for a period that is determined by the 0.01-μF/680-kΩ differentiator. The lN914 gives quick reset for the 0.1-μF capacitor, and the Schottky diode clips differentiator-caused negative voltages at the ClB input. The turn-on threshold (4.8 V in this case) is set by the ratio of the 0.5% resistors. The output-pulse delay time is controlled by the 0.1-μF capacitor, which can be varied. Similarly, the RC combination at ClB sets output pulse width, and can be varied. The LT1018 1.2-V minimum supply voltage prevents spurious output during supply power-up. Linear Technology, Linear Applications Handbook, 1990, p. AN31-10.

7

Filter circuits

It is assumed that you are already familiar with filter basics, such as filter responses (low-pass, high-pass, bandpass, and notch), and common filter types (Butterworth, Chebyshev, Elliptic, Bessel). However, the following paragraphs summarize this information, as well as test and troubleshooting for filters. This information is included so that even those readers who are not familiar with electronic procedures can both test the circuits in this chapter and localize problems if the circuits fail to perform as shown.

Filter circuit testing and troubleshooting

The primary purpose of a filter is to discriminate against the passage of certain groups of frequencies while simultaneously passing other groups or portions of the frequency spectrum. Although filter circuits range from the very simple to the very complex, there are only two basic types of filters, active and passive. The *active filters* in this chapter are essentially a frequency-selective amplifier, although no gain is involved, only attenuation. *Passive filters* are either *RC*, used primarily for low- or audio-frequency applications (chapter 1), or *LC* for use at higher frequencies (chapter 2).

Filter frequency response

The four basic filter frequency responses are shown in Fig. 7-A. A *low-pass filter* passes all frequencies below a selected value and attenuates higher frequencies. The low-pass filter is also known as a *high-cut filter*. A *high-pass filter* passes all frequencies above a selected value and attenuates lower frequencies. The high-pass filter is also known as a *low-cut filter*. A *bandpass filter* passes a selected band of

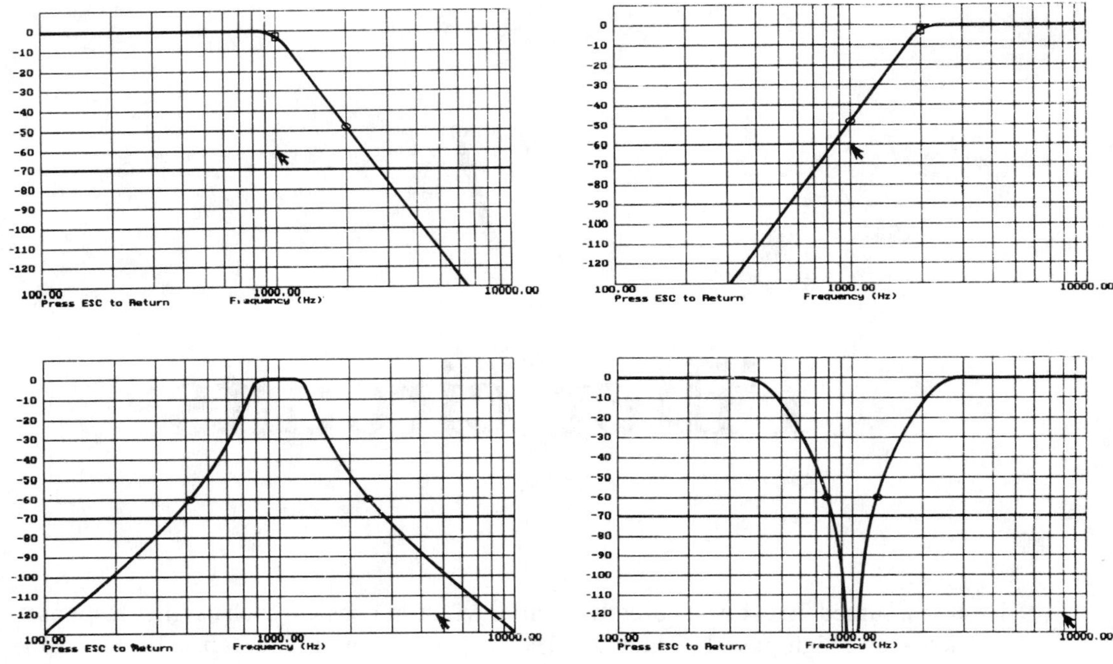

Fig. 7-A Basic filter frequency responses Linear Technology Corporation, Linear Applications Handbook, 1990, p. AN38 1-7.

frequencies. A *notch filter* suppresses a selected band of frequencies while passing all lower and higher frequencies. The notch filter is also known as a *band-elimination, band-stop, band-rejection,* or *band-suppression filter*.

Filter types

Figures 7-B and 7-C show the most common types of filters. Notice the ideal low-pass response in Fig. 7-B. This ideal response is not possible with any of the filter types.

The *Butterworth filter* has the optimum flatness in the passband, but has a slope that rolls off more gradually after the cutoff frequency than the other types. The *Chebyshev filter* can have a steeper initial rolloff than Butterworth, but at the expense of more ripple in the passband. The *Elliptic filter* has the steepest rolloff of all, but shows ripple in both the passband and the stopband (after the cutoff point). The *Bessel filter* has a sloping rolloff, but much steeper than Butterworth.

Basic filter tests

The obvious test for the filter circuits in this chapter is frequency response. Use the procedure for finding the frequency response of amplifiers from chapter 1. Of

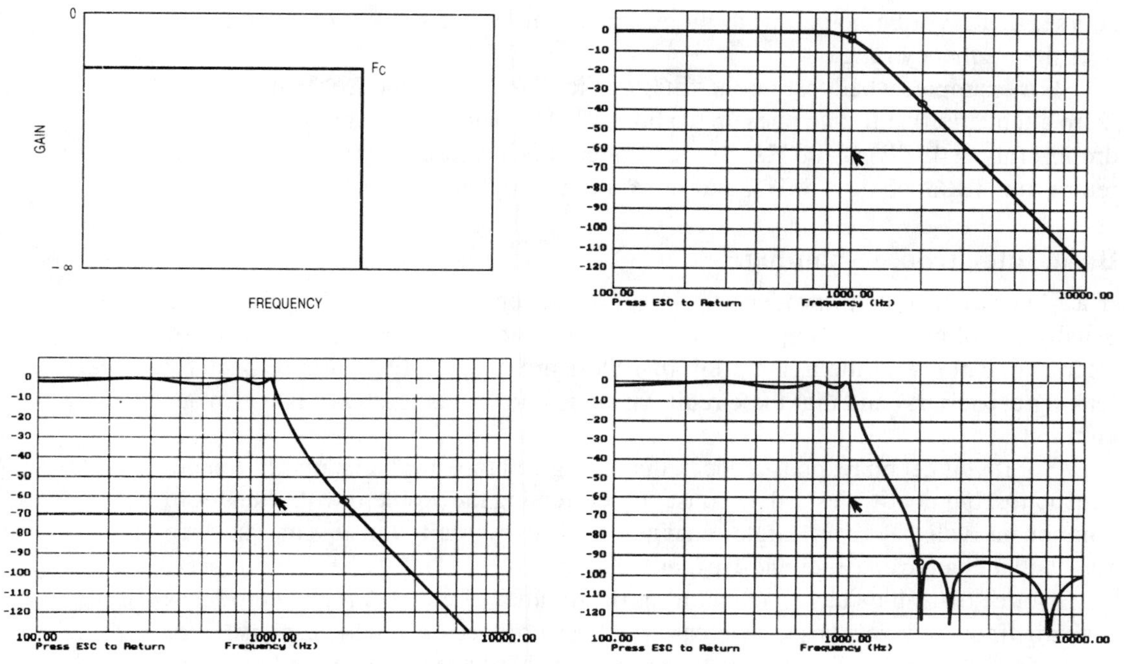

Fig. 7-B Basic Butterworth, Chebyshev, and Elliptic filter responses Linear Technology Corporations Linear Applications Handbook, 1990, P. AN38 1-8.

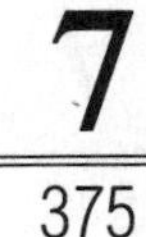

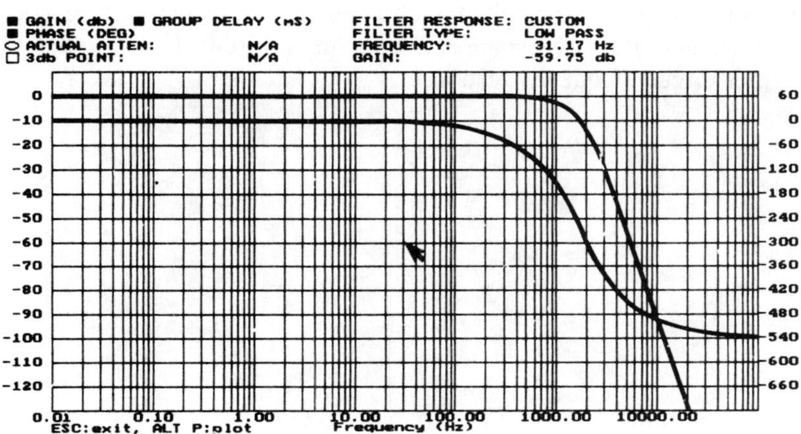

Fig. 7-C Basic Bessel filter response Linear Technology Corporation, Linear Applications Handbook, 1990, p. AN38 3-8.

course, there will be no gain, but the response curve will show filter attenuation over the frequency range.

For example, when testing the filter of Fig. 7-22, the response should be flat (zero attenuation) at frequencies up to about 375 Hz. Then, there should be a sharp dropoff (about 45 dB) at 400 Hz. As the frequency is increased, the response should return to 0 dB, at about 450 Hz, and remain flat at higher frequencies.

Basic filter troubleshooting

If a filter circuit fails to produce the desired frequency response, the problem is usually one of incorrect component values, particularly components that are out of tolerance. This is on the assumption that there are no defective parts (especially leaking capacitors) and that the circuit wiring is correct (as it always is with your circuits!).

So, if you get no response, check the wiring and look for bad parts. If you get a response, but not the desired response, try correcting the problem with changes in component values. Even slight changes in filter circuit values can produce substantial changes in frequency response.

Notice that filters have a *Q* factor, as do all tuned circuits. A high *Q* produces a sharp frequency response, whereas a low *Q* produces a broad response. For example, the circuit of Fig 7-8 has a *Q* of 5, and should produce a broad response with a notch at 3-kHz. The circuit of Fig. 7-50 has a *Q* of 25 and produces a sharp response bandpass at 1 kHz.

Switched-capacitor and instrumentation filters Notice that some of the filter circuits in this chapter require a clock signal (Figs. 7-14 through 7-25, for example). The frequency response of the filter circuit is directly affected by the clock frequency. Therefore, any deviation of the clock from the desired frequency can change frequency response. So, always check clock frequency before you change components to get a desired response. Even a small shift in clock frequency can change the filter-circuit frequency response.

Filter circuits titles and descriptions

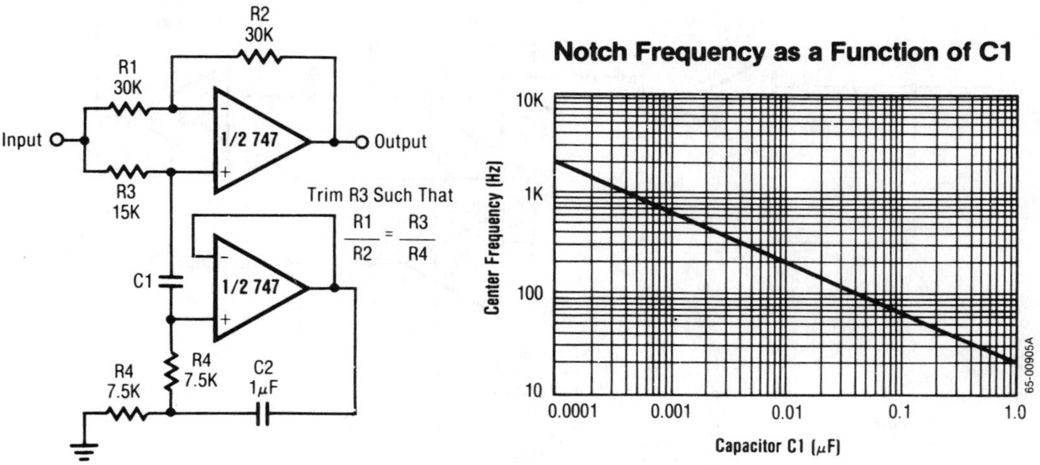

Notch filter using an op amp as a gyrator

Fig. 7-1 The center or notch frequency of this circuit is determined by the value of C_1. Raytheon Linear Integrated Circuits, 1989, p. 4-150.

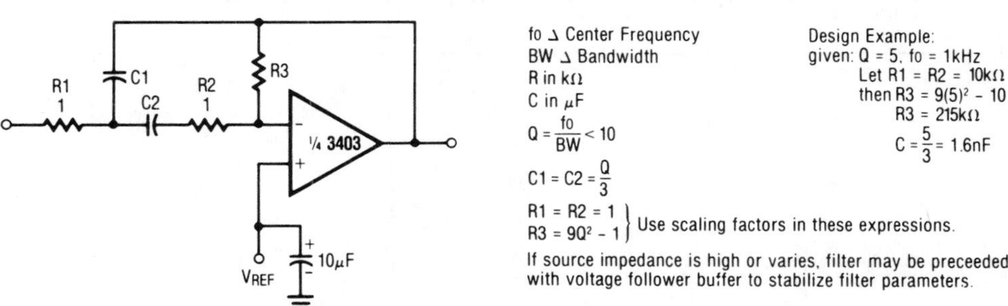

R1
1
C1
C2
R2
1
R3
¼ 3403
VREF
10μF
fo ⩟ Center Frequency
BW ⩟ Bandwidth
R in kΩ
C in μF
Q = fo/BW < 10
C1 = C2 = Q/3
R1 = R2 = 1
R3 = 9Q² - 1
Use scaling factors in these expressions.
If source impedance is high or varies, filter may be preceeded with voltage follower buffer to stabilize filter parameters.
Design Example:
given: Q = 5, fo = 1kHz
Let R1 = R2 = 10kΩ
then R3 = 9(5)² - 10
R3 = 215kΩ
C = 5/3 = 1.6nF

Multiple feedback bandpass filter

Fig. 7-2 This circuit uses one section of a 3403 op amp with multiple feedback to form a bandpass filter. Raytheon Linear Integrated Circuits, 1989, p. 4-160.

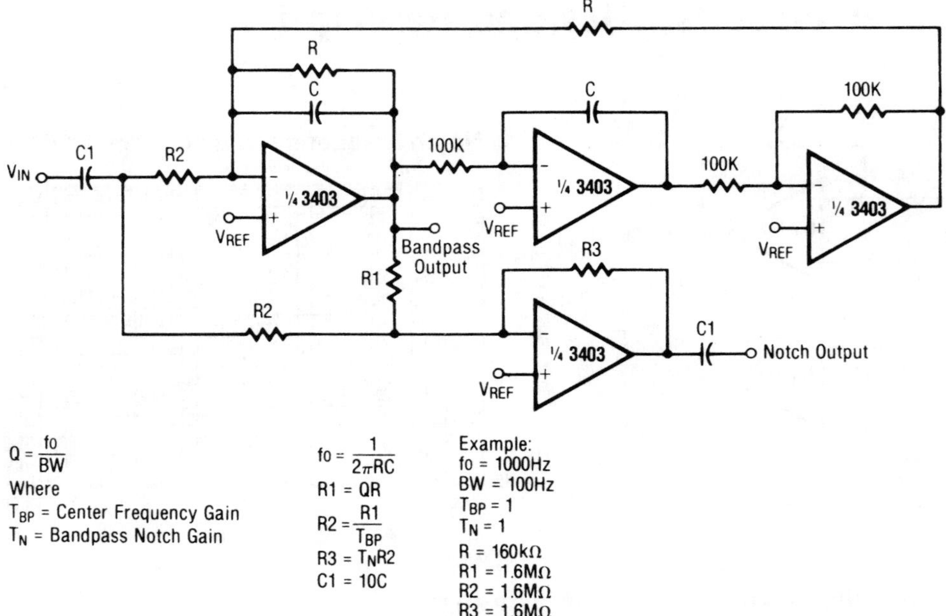

$Q = \frac{fo}{BW}$

Where

T_{BP} = Center Frequency Gain

T_N = Bandpass Notch Gain

$fo = \frac{1}{2\pi RC}$

$R1 = QR$

$R2 = \frac{R1}{T_{BP}}$

$R3 = T_N R2$

$C1 = 10C$

Example:

fo = 1000Hz

BW = 100Hz

$T_{BP} = 1$

$T_N = 1$

$R = 160k\Omega$

$R1 = 1.6M\Omega$

$R2 = 1.6M\Omega$

$R3 = 1.6M\Omega$

$C = 0.001\mu F$

Bi-quad filter

Fig. 7-3 This circuit provides both a notch output and bandpass output. Raytheon Linear Integrated Circuits, 1989, p. 4-161.

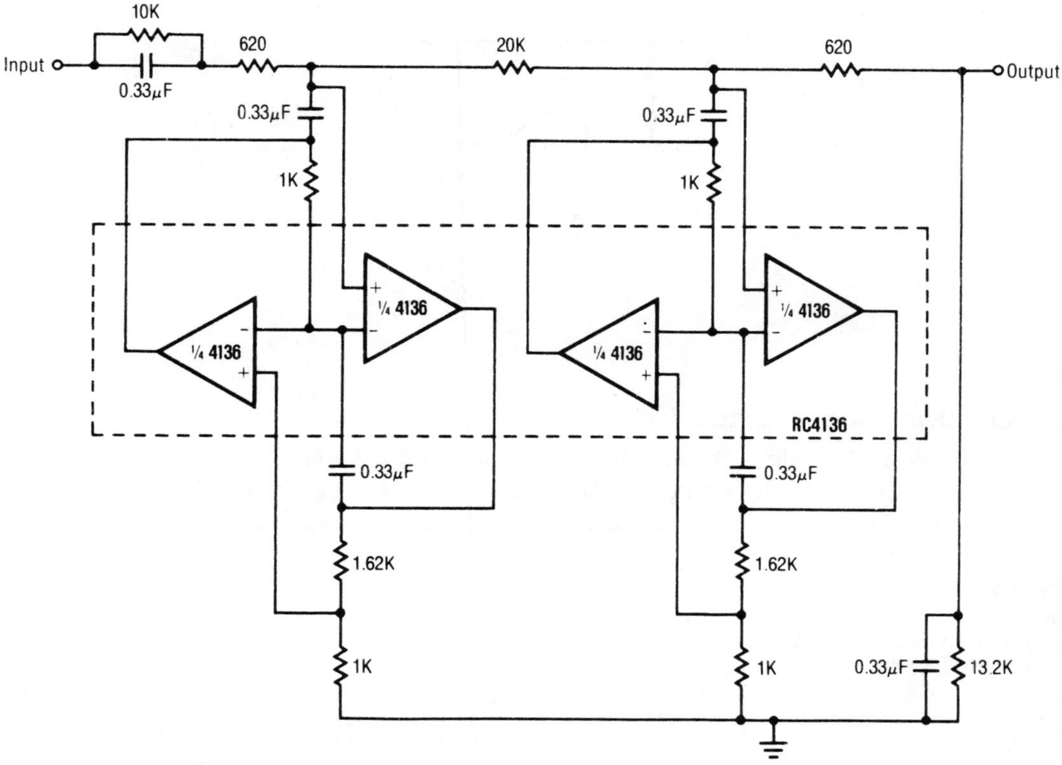

Low-pass Butterworth active filter

Fig. 7-4 With the values shown, this circuit provides a 400-Hz low-pass filter function. Raytheon Linear Integrated Circuits, 1989, p. 4-172.

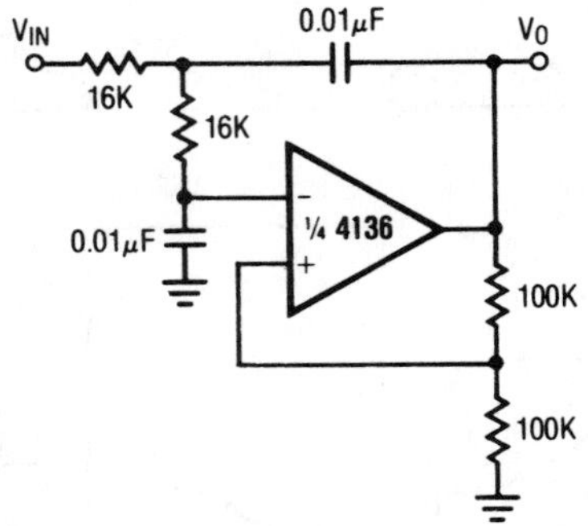

Dc-coupled low-pass active filter

Fig. 7-5 With the values shown, this circuit provides a 1-kHz low-pass filter function, with a direct-coupled input and output. Raytheon Linear Integrated Circuits, 1989, p. 4-174.

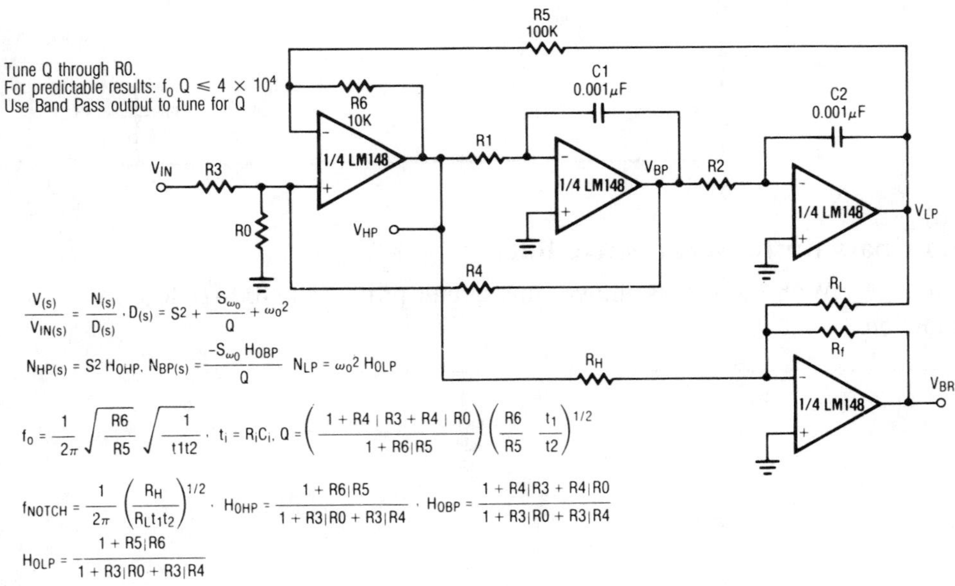

Universal state-space filter

Fig. 7-6 This circuit uses all four sections of a quad op-amp to form a universal state-space filter. Raytheon Linear Integrated Circuits, 1989, p. 4-265.

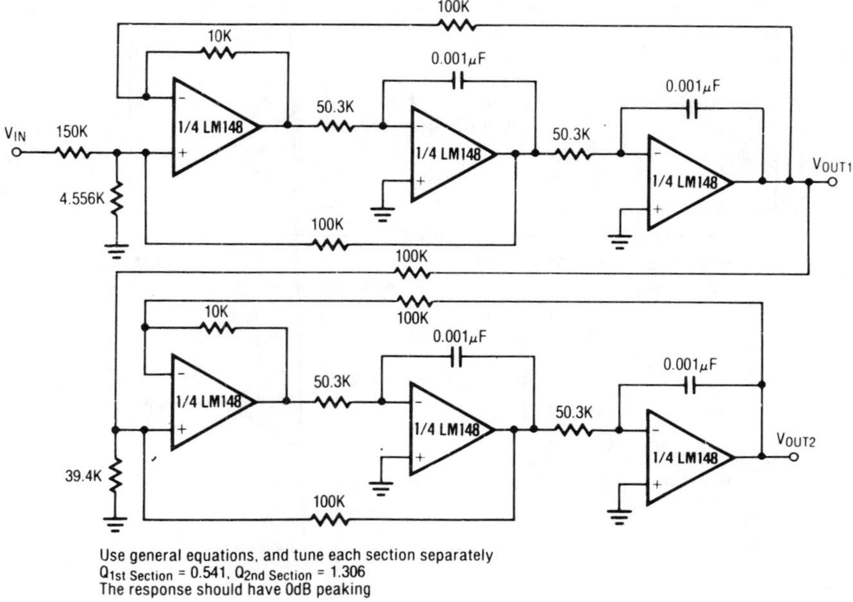

4-pole Butterworth filter

Fig. 7-7 With the values shown, this circuit provides a 1-kHz 4-pole Butterworth filter function, with a direct-coupled input and outputs. Raytheon Linear Integrated Circuits, 1989, p. 4-265.

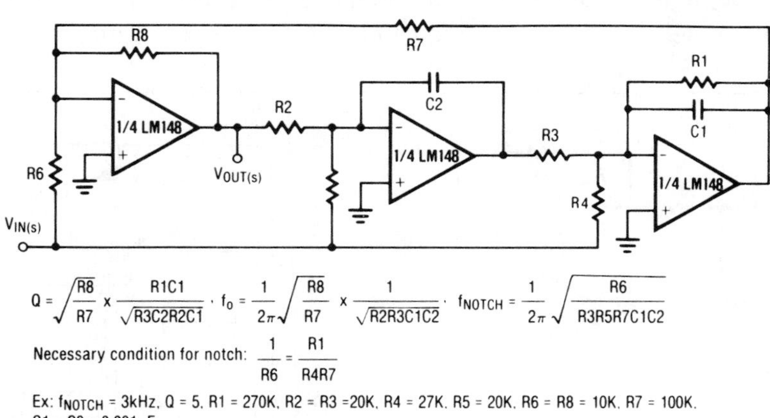

Bi-quad notch filter

Fig. 7-8 This circuit uses three sections of a quad op amp to form a notch filter. Circuit values are given for a 3-kHz notch with a *Q* of 5. The equations are given for other frequencies. Raytheon Linear Integrated Circuits, 1989, p. 4-266.

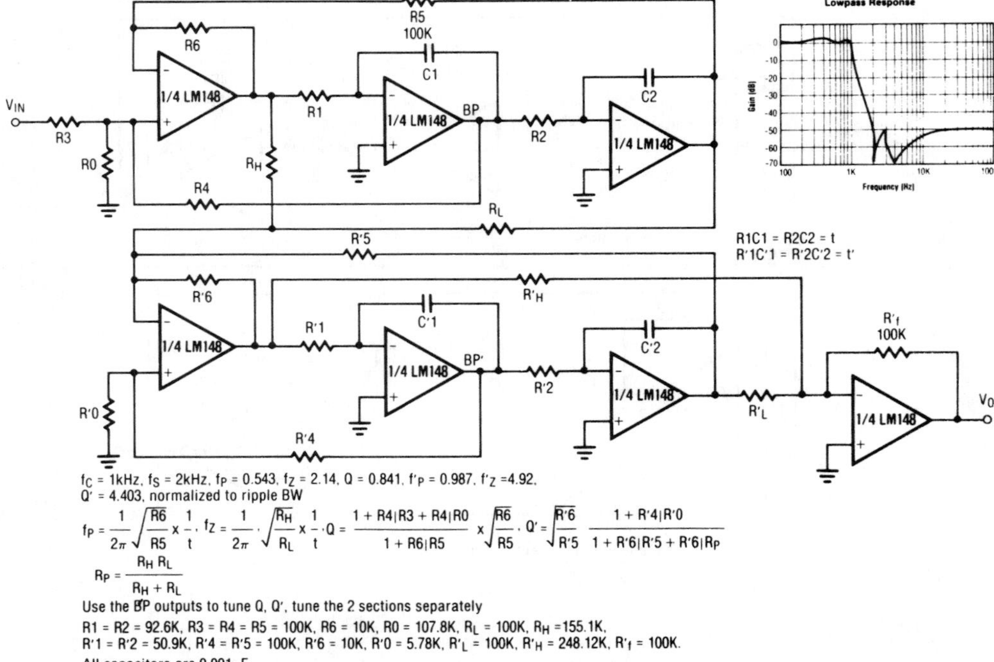

Elliptic filter

Fig. 7-9 With the values shown, this circuit provides a 4th-order 1-kHz elliptic filter (with 4 poles and 4 zeros). The equations are given for other frequencies. Raytheon Linear Integrated Circuits, 1989, p. 4-266.

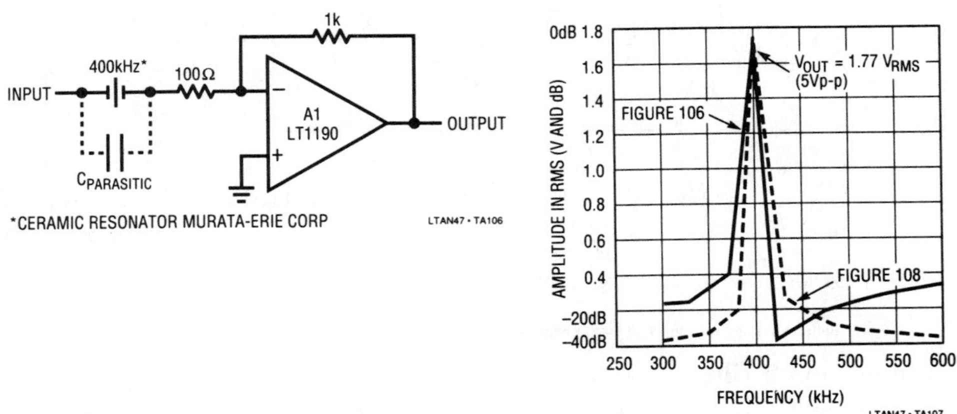

Basic piezo-ceramic based filter

Fig. 7-10 This circuit is a highly selective bandpass filter that uses a resonant ceramic element and a single amplifier. The ceramic element looks like a high impedance off the resonant frequency (400 kHz, in this case). At resonance, the ceramic element has a low impedance, and A1 responds as an inverter with gain. As shown, the ceramic element has stray or parasitic capacitance that causes a slight rise in output at frequencies above 425 kHz. Typically, the A1 output is down about 20 dB at 300 kHz and 40 dB at 425 kHz. The parasitic capacitance can be minimized with a differential network (Fig. 7-11). Linear Technology Corporation, 1991, AN47-48.

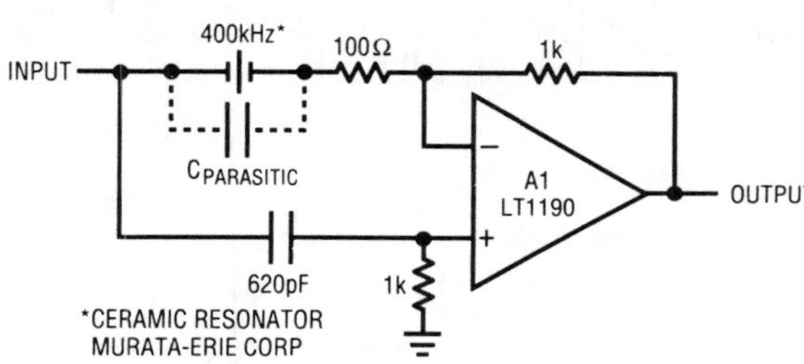

Piezo-ceramic based filter with differential network

Fig. 7-11 This circuit is an improved version of the Fig. 7-10 circuit. With the Fig. 7-11 circuit, a portion of the input is fed to the noninverting input of A1. The RC network at this input looks like the ceramic-resonantor impedance when it is off null. As a result, A1 "sees" similar signals for out-of-band inputs. The high-frequency roll-off of Fig. 7-11 is smooth and about 20 dB deeper than the Fig. 7-10 filter at 475 kHz. The low-frequency side of resonance has similar characteristics at 375 kHz and below. Linear Technology Corporation, 1991, AN47-48.

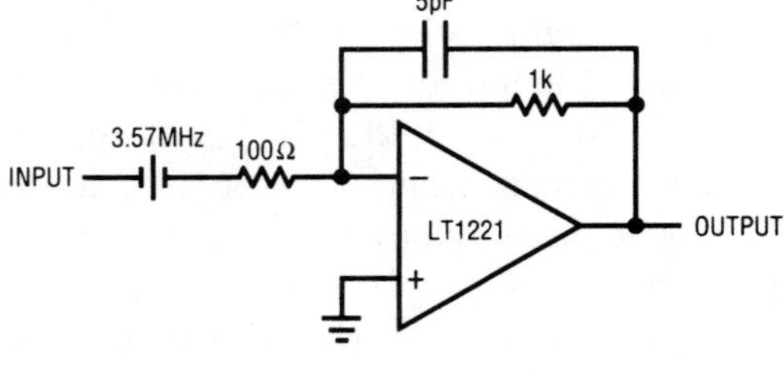

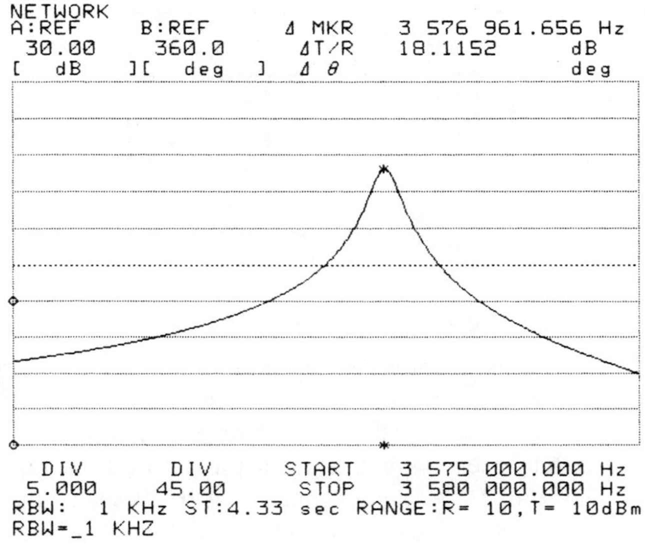

7

Basic crystal filter

Fig. 7-12 This circuit replaces the ceramic element of Fig. 7-10 with a 3.57-MHz quartz crystal. Figure 7-12B shows almost 30-dB attenuation only a few kHz on either side of resonance. Linear Technology Corporation, 1991, AN47-48/49.

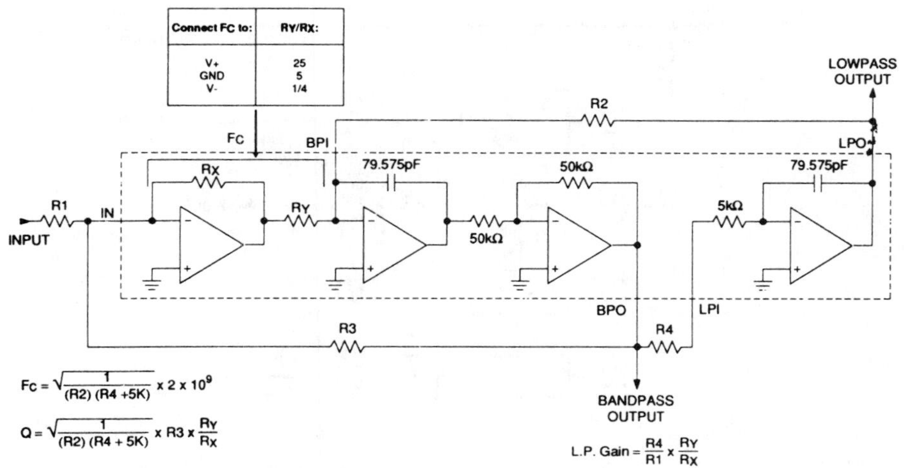

Single 2nd-order filter section

Fig. 7-13 The MAX274/275 continuous-filter architecture uses a four-amplifier state-variable design. The on-chip capacitors and amplifiers, together with external resistors, form cascaded integrators with feedback to provide simultaneous low-pass and bandpass filtered outputs. The low-pass and bandpass frequencies, as well as filter Q, are determined by external resistor values, using the equations shown. No external capacitors are needed. On-chip capacitors are factory trimmed to provide better than 1% pole-frequency accuracy over the temperature range. ±1%-tolerance resistors provide ±2%-accurate pole frequencies. Accurate filter Qs can also be obtained by compensating for amplifier bandwidth limitation using the graphs that are provided on the data sheet. Maxim, 1992, Applications and Product Highlights, p. 7-3.

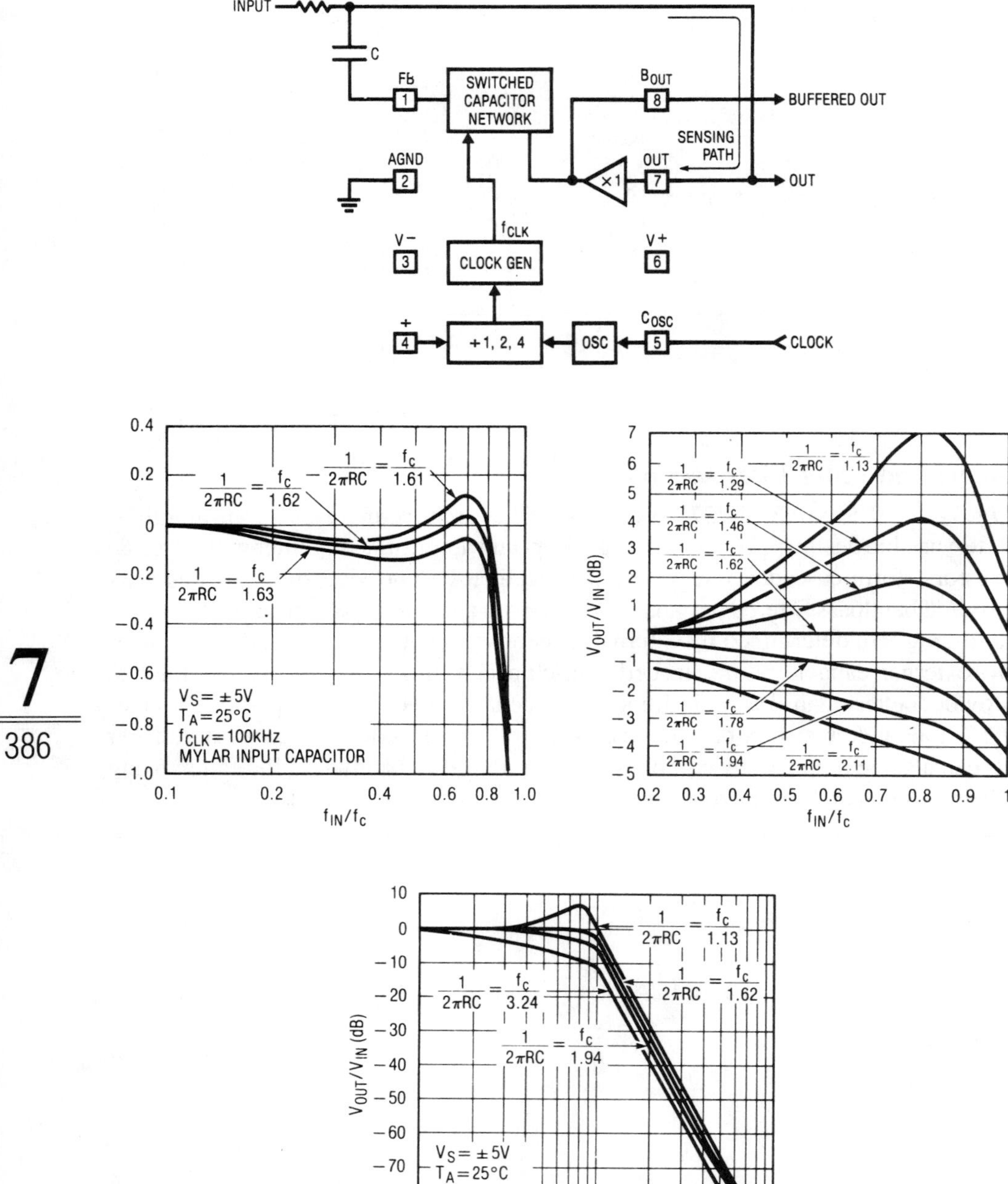

INPUT
R
C
FB
1
SWITCHED CAPACITOR NETWORK
AGND
2
V−
3
CLOCK GEN
f_{CLK}
+
4
÷1, 2, 4
OSC
C_{OSC}
5
CLOCK
V+
6
B_{OUT}
8
BUFFERED OUT
SENSING PATH
OUT
7
×1
OUT
$\frac{1}{2\pi RC} = \frac{f_c}{1.62}$
$\frac{1}{2\pi RC} = \frac{f_c}{1.61}$
$\frac{1}{2\pi RC} = \frac{f_c}{1.63}$
$V_S = \pm 5V$
$T_A = 25°C$
$f_{CLK} = 100kHz$
MYLAR INPUT CAPACITOR
f_{IN}/f_c
V_{OUT}/V_{IN} (dB)
$\frac{1}{2\pi RC} = \frac{f_c}{1.13}$
$\frac{1}{2\pi RC} = \frac{f_c}{1.29}$
$\frac{1}{2\pi RC} = \frac{f_c}{1.46}$
$\frac{1}{2\pi RC} = \frac{f_c}{1.62}$
$\frac{1}{2\pi RC} = \frac{f_c}{1.78}$
$\frac{1}{2\pi RC} = \frac{f_c}{1.94}$
$\frac{1}{2\pi RC} = \frac{f_c}{2.11}$
f_{IN}/f_c
V_{OUT}/V_{IN} (dB)
$\frac{1}{2\pi RC} = \frac{f_c}{1.13}$
$\frac{1}{2\pi RC} = \frac{f_c}{3.24}$
$\frac{1}{2\pi RC} = \frac{f_c}{1.62}$
$\frac{1}{2\pi RC} = \frac{f_c}{1.94}$
$V_S = \pm 5V$
$T_A = 25°C$
$f_c = 100Hz$
$f_{CLK} = 10kHz$
f_{IN}/f_c

Continued

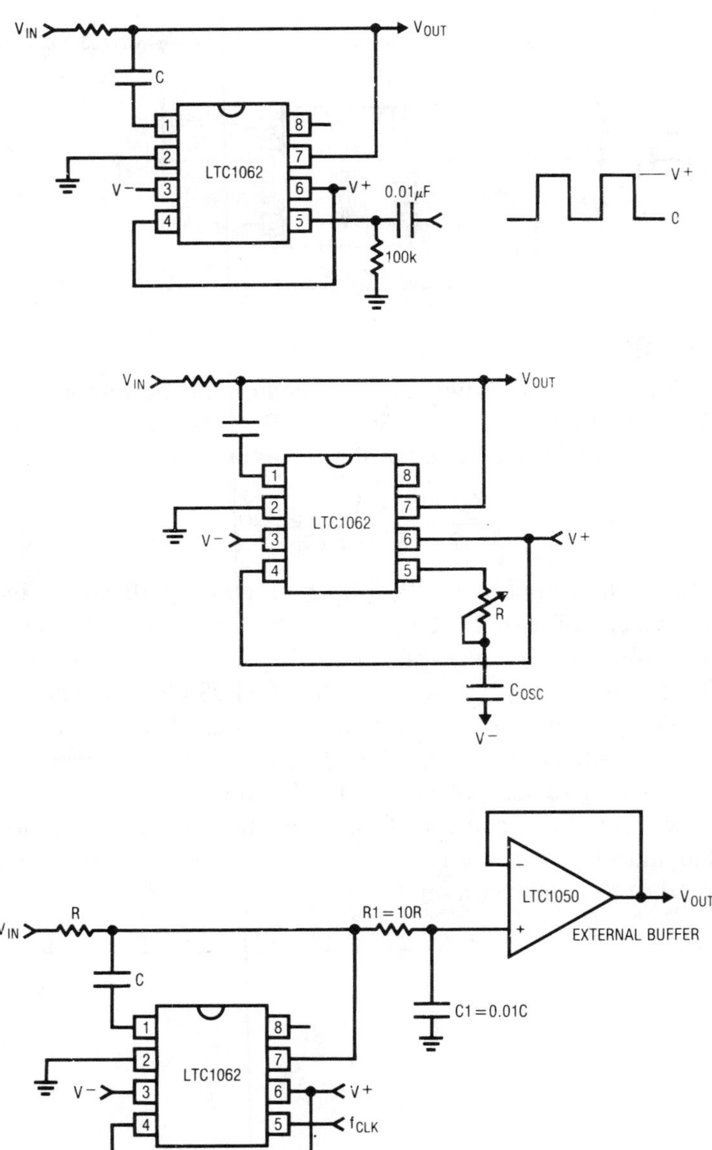

VIN
VOUT
C
LTC1062
V−
V+
0.01μF
100k
V+
C
VIN
VOUT
LTC1062
V−
V+
R
COSC
V−
LTC1050
VOUT
R
R1 = 10R
EXTERNAL BUFFER
VIN
C
C1 = 0.01C
LTC1062
V−
V+
fCLK

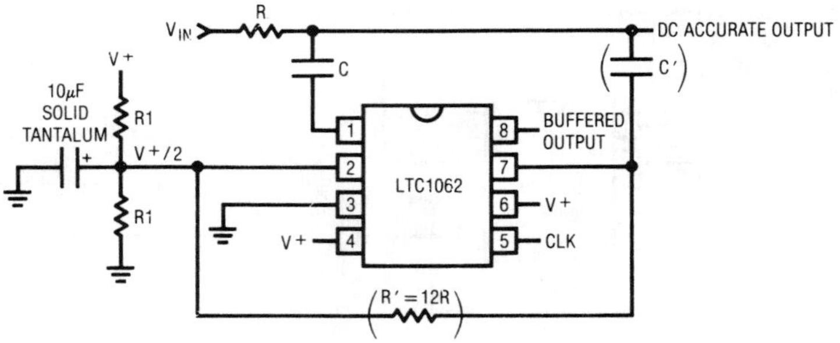

Continued

IC low-pass filter

Fig. 7-14 Figure 7-14A shows the architecture and basic connections for an LTC1062 low-pass filter (similar to a 5th-order Butterworth). The cutoff frequency f_c (at the −3-dB point) is determined by the values of R and C, with the relationship:

$$\frac{f_c}{1.62} \leqslant \frac{1}{6.28\,RC} \leqslant \frac{f_c}{1.63}\,.$$

The clock frequency should be 100 times f_c. Figure 7-14B shows the passband response for values of $1/(6.28RC)$ near $f_c/1.62$. Figures 7-14C and 7-14D are similar, but with a wider range of values. Figure 7-14D shows the LTC1062 operated from a single supply and an external CMOS clock signal. Figure 7-14F shows the LTC1062 operated with the internal clock. If C_{OSC} is 8500 pF and a 50-kΩ pot is connected to pin 5, the clock frequency can be adjusted from 500 Hz to 3.3 kHz (providing for an f_c of 5 to 33 Hz). Figure 7-14G shows how an external buffer can be connected to the LTC1062 to eliminate clock feedthrough and improve the high-frequency attenuation floor. Figure 7-14H shows how the LTC1062 can be operated from a single supply. Linear Technology Corporation, Linear Applications Handbook, 1990, p. AN20-1,-2,-3,-4.

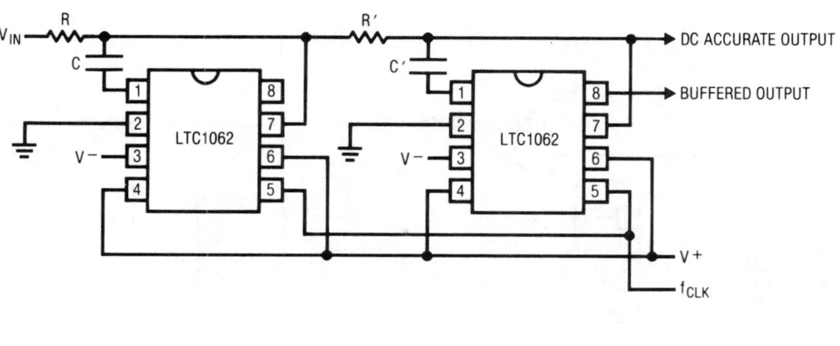

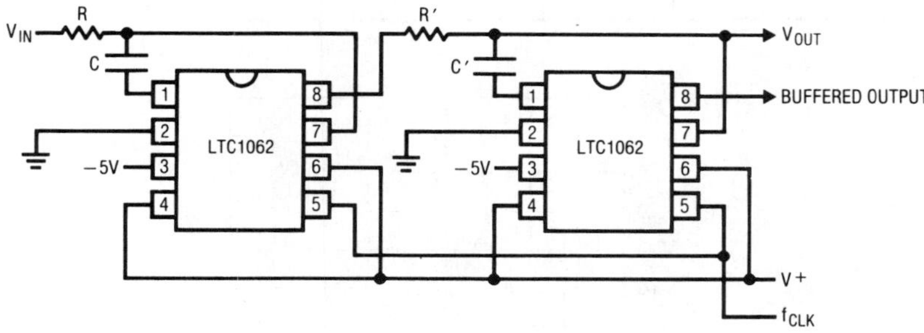

Cascaded IC low-pass filters

Fig. 7-15 Figure 7-15A shows two LTC1062 filters where the second input is taken directly from the dc-accurate output of the first filter. The recommended ratio of R'/R is about 117/1, $1/(6.28RC)$ should equal $f_c/1.57$, and $1/(6.28R'C')$ should equal $f_c/1.6$ when the filters are cascaded (Fig. 7-15A). For example, for an f_c of 4.16 kHz, the clock should be 416 kHz, and $R = 909\ \Omega$, $R' = 107\ \text{k}\Omega$, $C = 0.066$ pF, and $C' = 574$ pF. Figure 7-15B shows cascaded filters where the second input is taken from the buffered output of the first filter. The recommended values for R and C are determined by: $1/(6.28RC) = f_c/1.59$ and $1/(6.28R'C') = f_c/1.64$ when the filters are cascaded as shown in Fig. 7-15B. For example, for an f_c of 4 kHz, the values should be: $R = 97.6\ \text{k}\Omega$, $C = 676$ pF, $R' = 124\ \text{k}\Omega$, and $C' = 508$ pF. Linear Technology Corporation, Linear Applications Handbook, 1990, p. AN20-7.

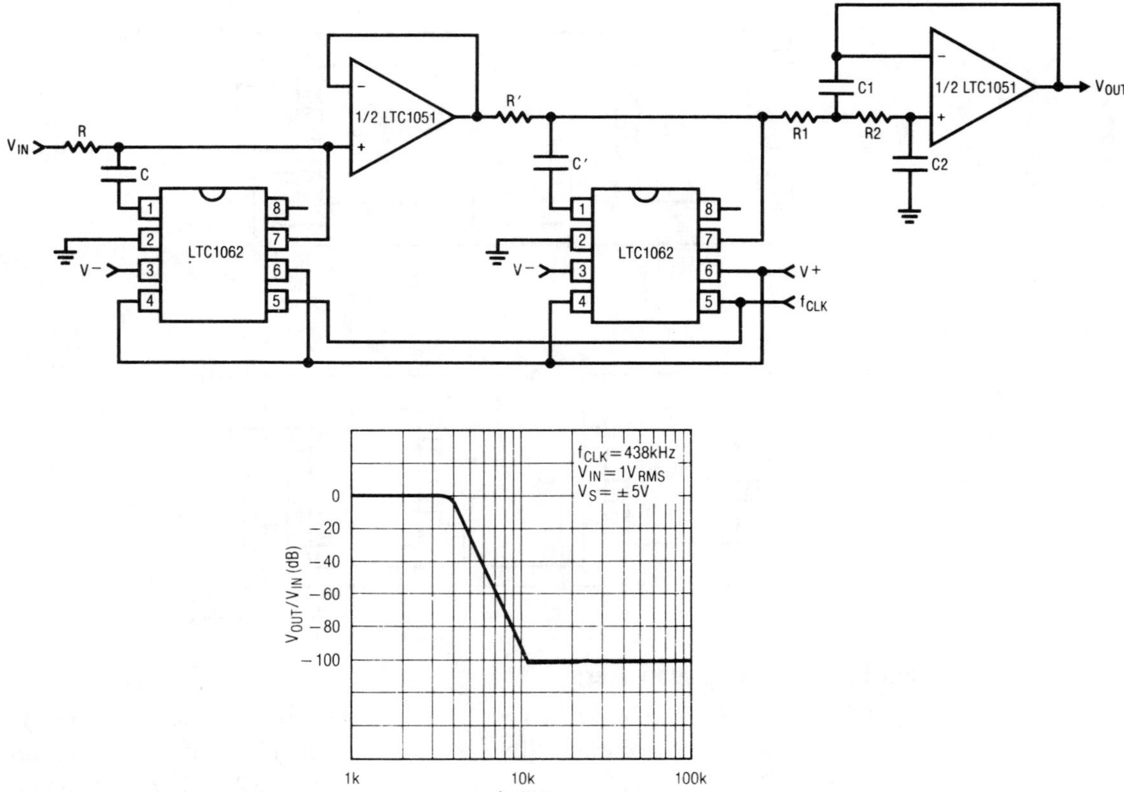

7

390

Low-offset, 12th-order max-flat low-pass filter

Fig. 7-16 This circuit shows a 12th-order filter that uses two LTC1062s and a precision dual op amp. Figure 7-16B shows the frequency response for the following values: f_c = 4 kHz, R = 59 kΩ, C = 0.001 μF, R' = 5.7 kΩ, C' = 0.01 μF, $R_1 = R_2$ = 39.8 kΩ, C_1 = 2000 pF, C_2 = 500 pF, f_{clock} = 438 kHz. Linear Technology Corporation, Linear Applications Handbook, 1990, p. AN20-8,-9.

V_{IN} R R1 R4 R2 R3 C C2 LTC1062 LT1056 OPTIONAL V− V+ CLK IN V_{OUT}

$f_{CLK} = f_{notch} \times 118.3$

$\frac{1}{2\pi RC} = 0.726\, f_{notch}$

AND $R1 = R4 = (R2 + R3)$

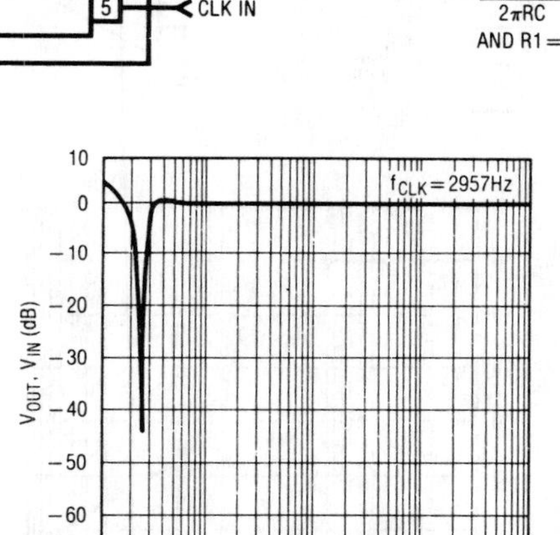

IC notch filter

Fig. 7-17 This circuit shows an LTC1062 and an LT1056 connected to form a notch filter. Figure 7-17B shows the frequency response for a 25-Hz notch filter using values based on the equations of Fig. 7-17A. The optional R2/C2 at the LTC1062 output is used to minimize clock feedthrough. The $1/(6.28R_2C_2)$ value should be 12 to 15 times that of the notch frequency. Linear Technology Corporation, Linear Applications Handbook, 1990, p. AN20-9,-10.

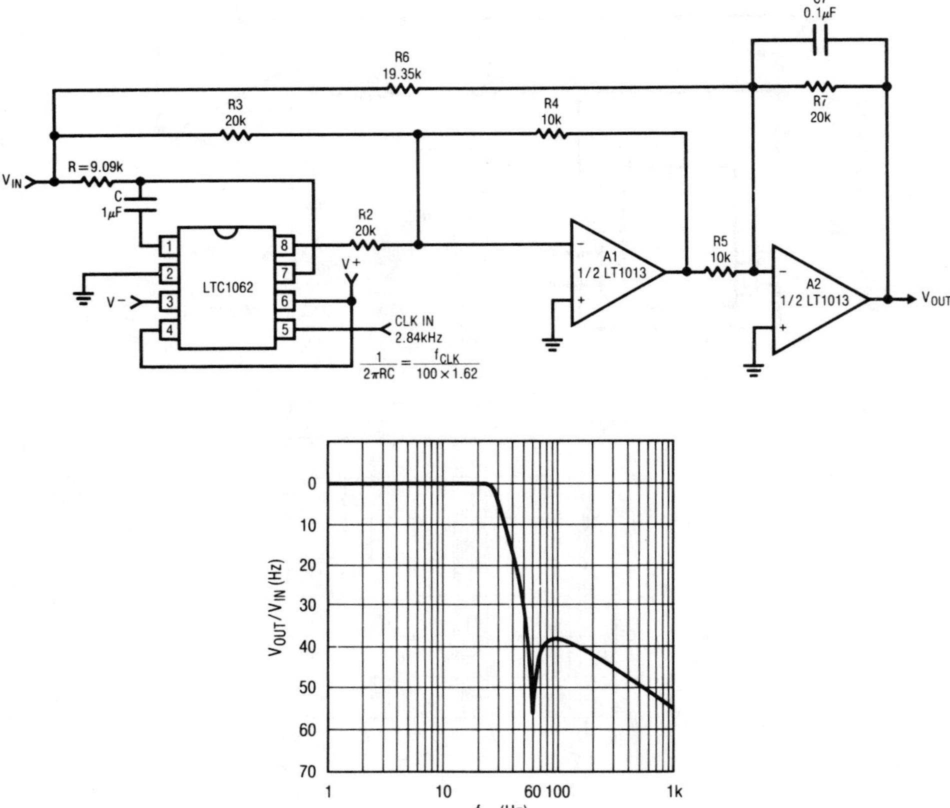

7

Extended notch filter

Fig. 7-18 This circuit is similar to that of Fig. 7-17, but uses two sections of an LT1013 following the LTC1062. Figure 7-18B shows the frequency response for a 60-Hz notch filter that uses the values and equations of Fig. 7-18A. Linear Technology Corporation, Linear Applications Handbook, 1990, p. AN20-10,-11.

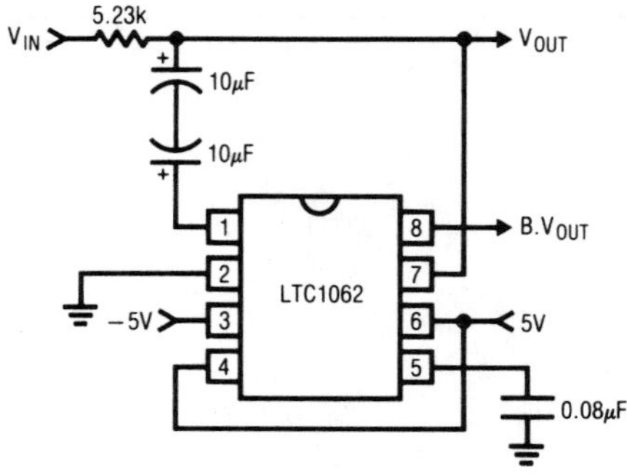

Simple 5-Hz filter that uses back-to-back capacitors

Fig. 7-19 This circuit shows an LTC1062 connected with solid-tantalum capacitors to form a 5-Hz low-pass filter. Notice that no external clock is required. Linear Technology Corporation, Linear Applications Handbook, 1990, p. AN20-11.

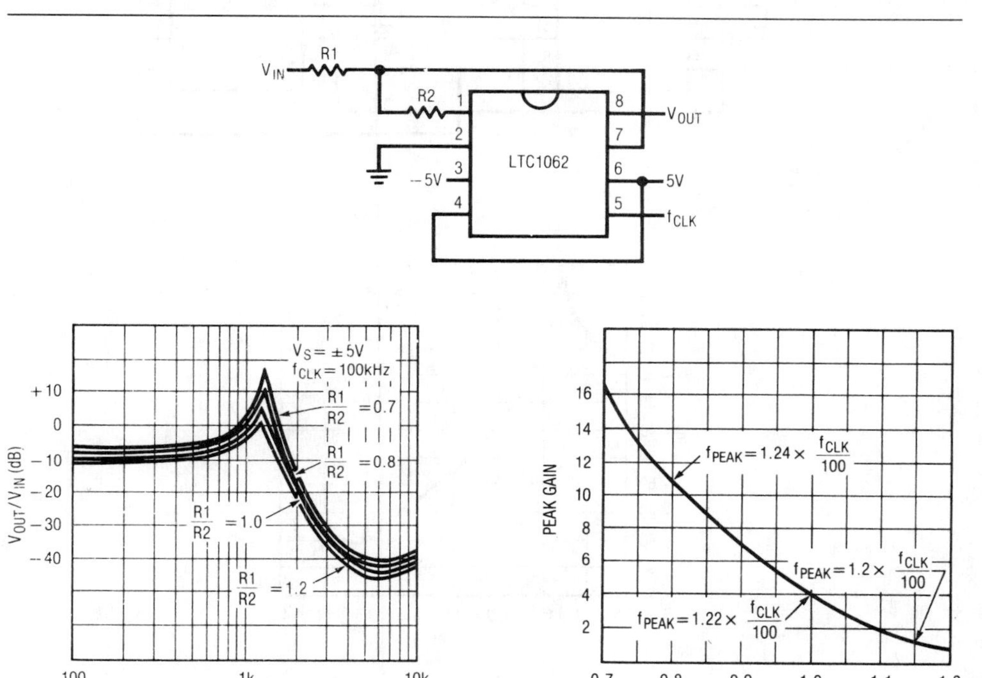

7

Clock-sweepable pseudo-bandpass/notch filter

Fig. 7-20 This circuit shows an LTC1062 connected as a simple clock-sweepable bandpass/notch filter. Figure 7-20B shows the frequency response for a clock frequency of 100 kHz and the various ratios of R_1/R_2. Figure 7-20C shows the variation of peak gain (H_{op}) and peak frequency (f_p) versus different values of the R_1/R_2 ratio. Linear Technology Corporation, Linear Applications Handbook, 1990, p. AN24-4,-5.

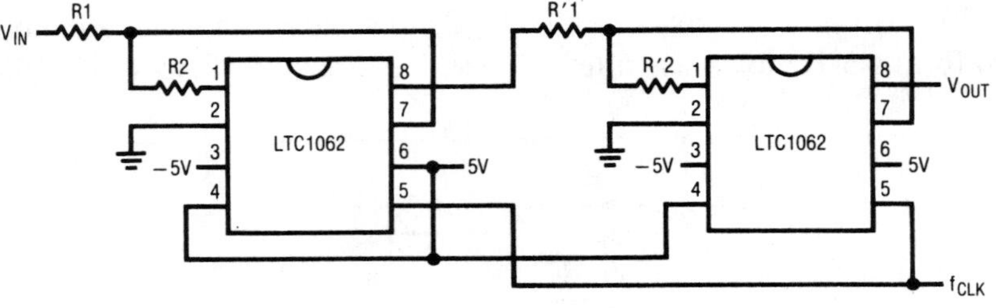

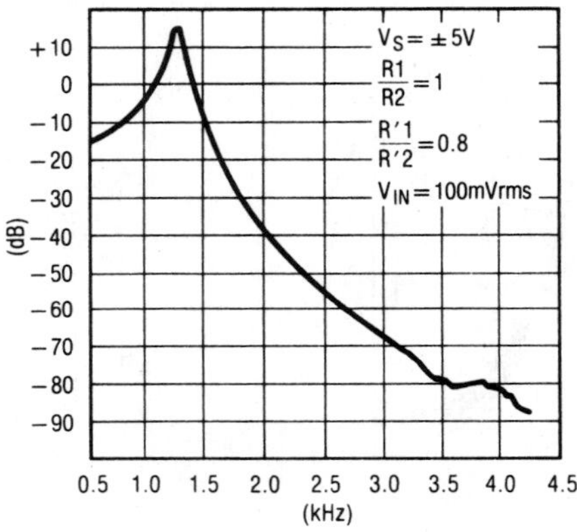

7

394

Selective clock-sweepable bandpass filter

Fig. 7-21 This circuit shows two LTC1062s connected to form a clock-sweepable bandpass filter. Figure 7-21B shows the frequency response for the values given in Fig. 7-21A. Linear Technology Corporation, Linear Applications Handbook, 1990, p. AN24-5.

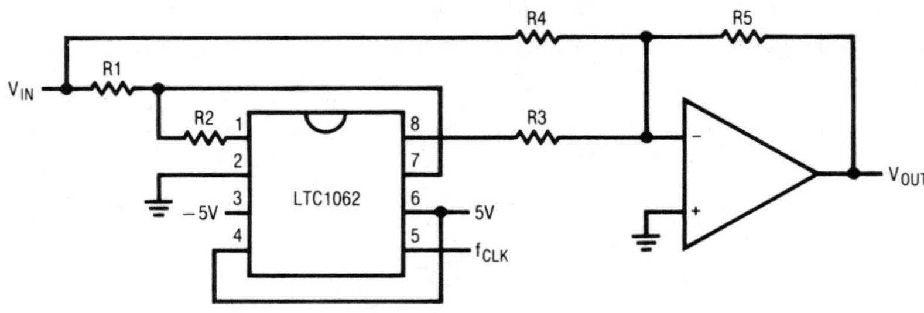

For simplicity use R3 = R4 = R5 = 10k;

$$\frac{R1}{R2} = 1.234, \frac{f_{CLK}}{f_{notch}} = \frac{79.3}{1}$$

Continued

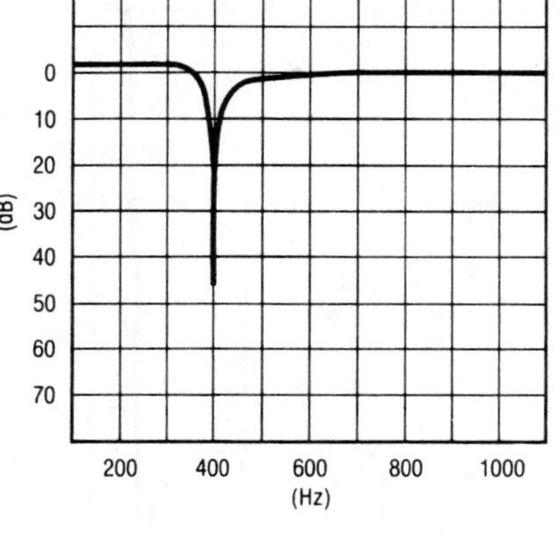

Clock-tunable notch filter

Fig. 7-22 This circuit shows an LTC1062 and an op amp (such as an LT1056) connected to form a notch filter. Figure 7-22B shows the frequency response for a 400-Hz notch filter that uses the following values and equations: $R_3 = R_4 = R_5 = 10\,\mathrm{k}\Omega, R_1/R_2 = 1.234\,\Omega, f_{CLK}/f_{notch} = 79.3/1$. Linear Technology Corporation, Linear Applications Handbook, 1990, p. AN24-6.

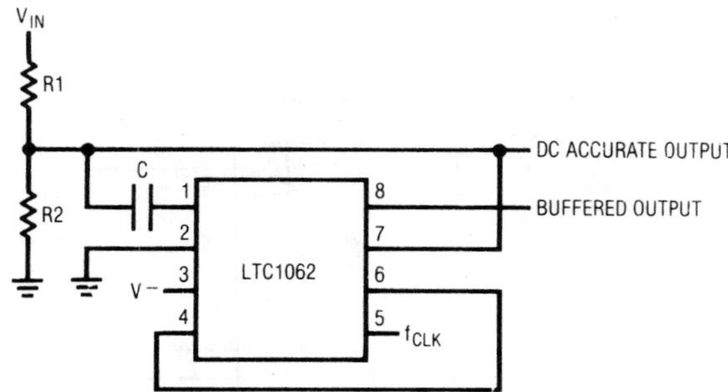

High input-voltage IC filter

Fig. 7-23 This circuit shows how an LTC1062 can be connected to accommodate high input voltages outside the normal input common-mode range. The dc gain of the low-pass filter is: $R_2/(R_1 + R_2)$. For maximum passband flatness, the paralleled combination of R_1/R_2 should be chosen as:

$$\frac{1}{6.28\,(R1 \| R2) \times C} = \frac{f_{cutoff}}{1.63};$$

$R_1//R_2 \geq 5\ \mathrm{k}\Omega$. Notice that there is no need for an external op amp to buffer the divided-down input voltage. The internal buffer input (pin 7) performs this function. Linear Technology Corporation, Linear Applications Handbook, 1992, p. AN24-7.

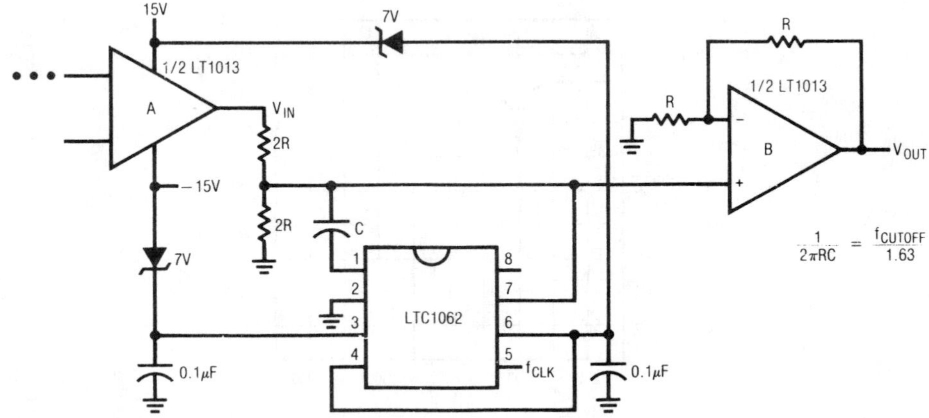

IC filter-operated (from ±15-V) supplies

Fig. 7-24 This circuit shows an LTC1062 interfaced with an LT1013 op amp and operated from ±15-V supplies. The desired cutoff frequency is determined by the equation shown. A typical dc output is 300 mV. Linear Technology Corporation, Linear Applications Handbook, 1992, p. AN24-7.

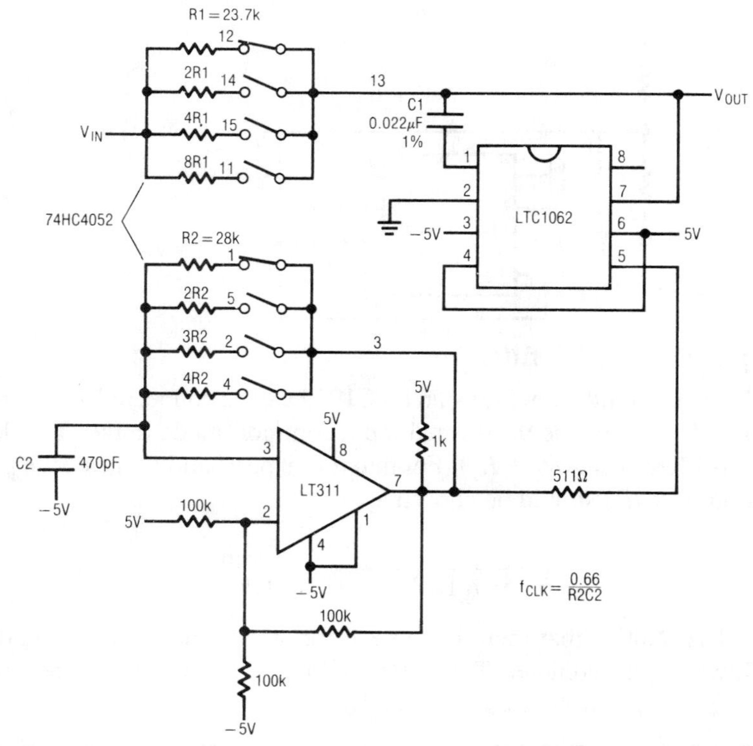

IC filter with programmed cutoff frequencies

Fig. 7-25 This circuit shows an LTC1062 used with a dual four-channel MUX (74HC4052) and an LT311 to provide four different cutoff frequencies (500, 250, 125, and 62.5 Hz). Notice that the clock frequency and the external $R \times C$ product are varied simultaneously so that: $1/(6.28RC) = f_c/1.64 = f_{clock}/164$. Linear Technology Corporation, Linear Applications Handbook, 1992, p. AN24-8.

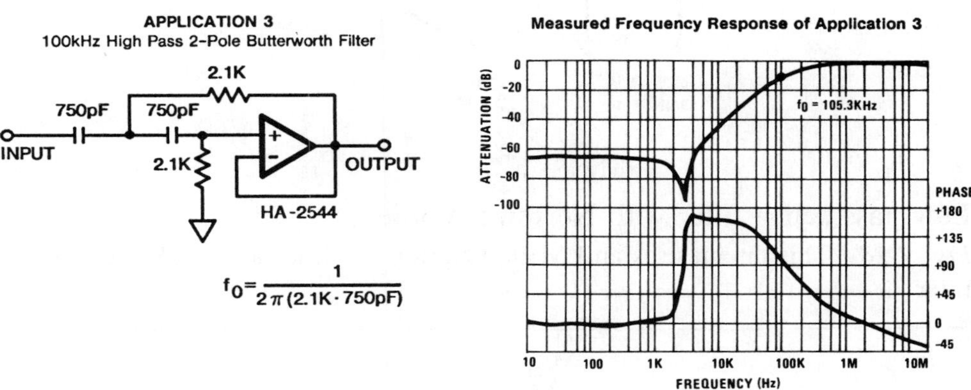

100-kHz high-pass 2-pole Butterworth filter

Fig. 7-26 This circuit uses an HA-2544 video op amp (Fig. 3-29) to form an active 100-kHz filter. Other frequencies can be selected using the equation. Figure 7-26B shows the filter characteristics. Harris Semiconductor, Linear & Telecom ICs, 1991, p. 3-312.

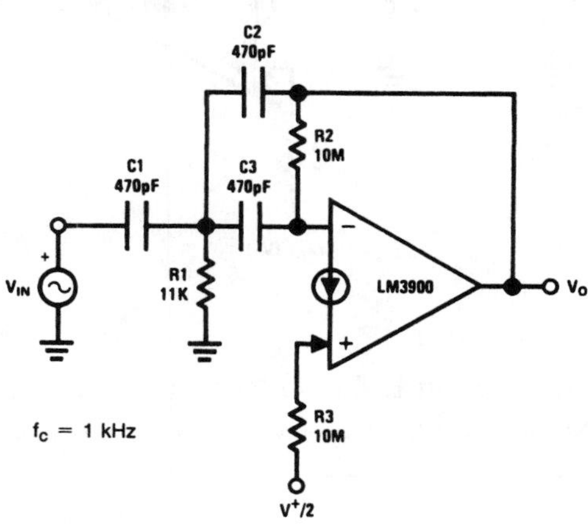

High-pass active filter with Norton amplifier

Fig. 7-27 This circuit uses an LM3900 Norton amplifier as a 1-kHz high-pass filter. National Semiconductor, Linear Applications Handbook, 1991, p. 227.

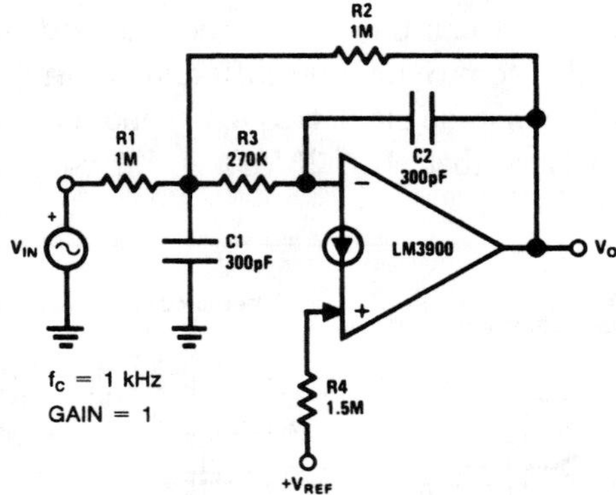

Low-pass active filter with Norton amplifier

Fig. 7-28 This circuit uses an LM3900 Norton amplifier as a 1-kHz low-pass filter. National Semiconductor, Linear Applications Handbook, 1991, p. 227.

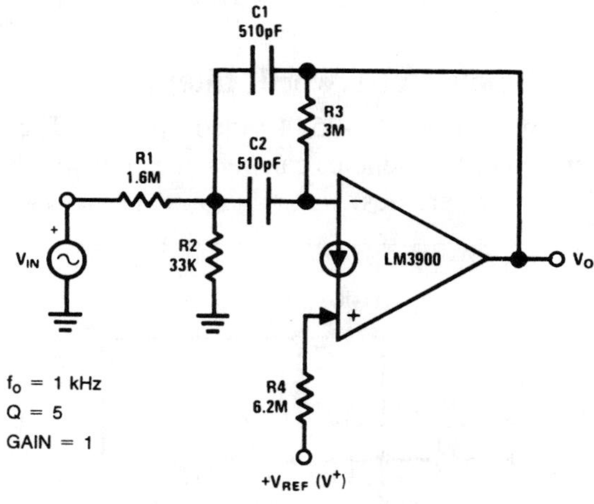

One op-amp bandpass filter

Fig. 7-29 This circuit uses an LM3900 Norton amplifier as a 1-kHz bandpass filter with a Q of 5 and unity gain. National Semiconductor, Linear Applications Handbook, 1991, p. 228.

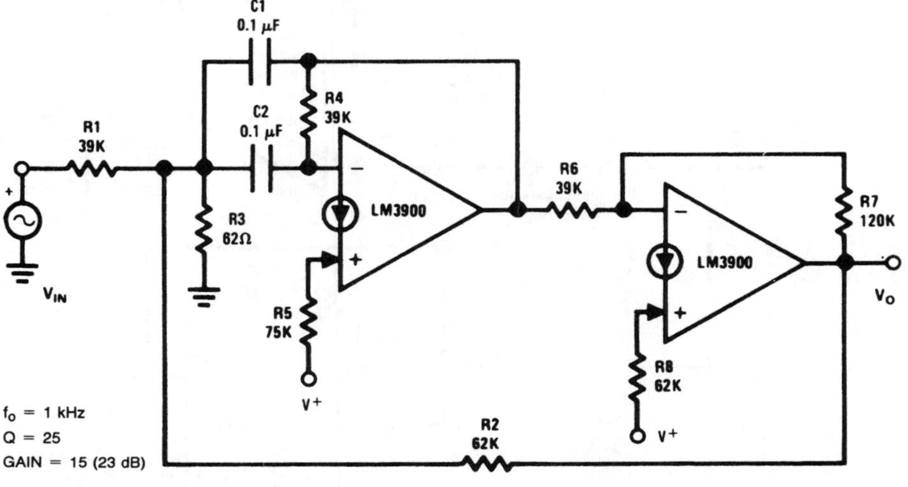

Two op-amp bandpass filter

Fig. 7-30 This circuit uses two LM3900 Norton amplifiers as a 1-kHz bandpass filter with a Q of 25, and a gain of 15 (23 dB). National Semiconductor, Linear Applications Handbook, 1991, p. 229.

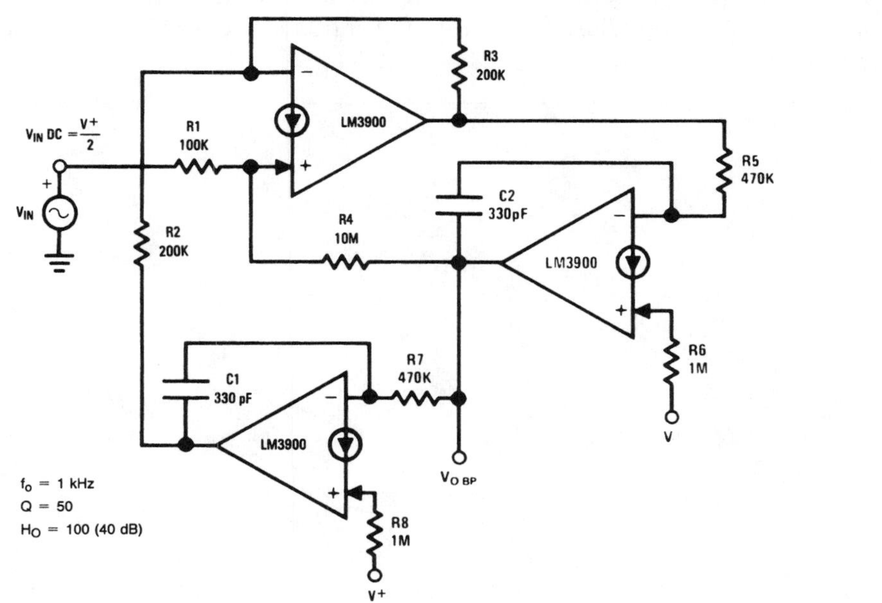

Bi-quad RC active bandpass filter

Fig. 7-31 This circuit uses three LM3900 Norton amplifiers as a 1-kHz bandpass filter with a Q of 50, and a gain of 100 (40 dB). National Semiconductor, Linear Applications Handbook, 1991, p. 230.

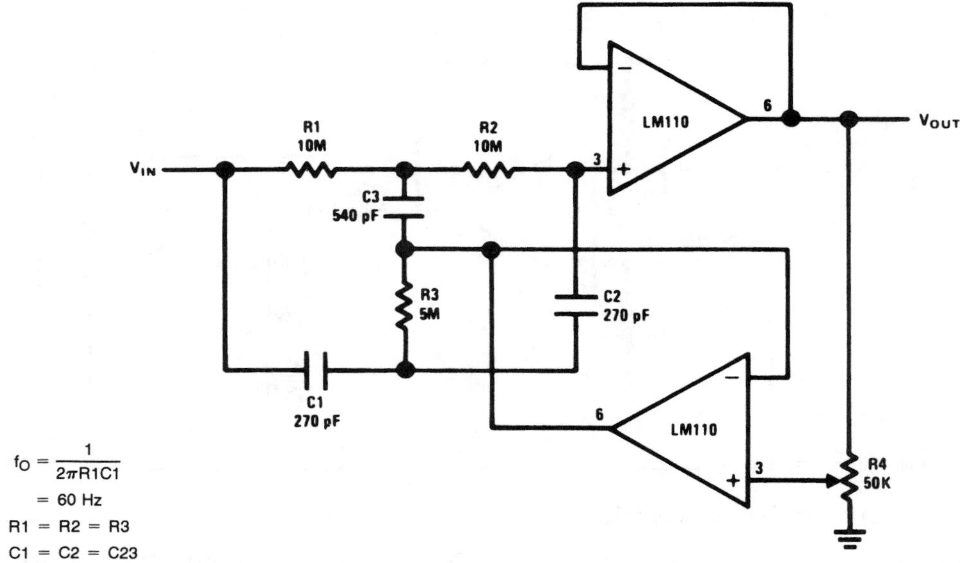

Adjustable-Q notch filter

Fig. 7-32 This circuit uses two LM100s to form a 60-Hz notch filter with adjustable Q (set by adjustment of R4). Other frequencies can be selected using the equation. National Semiconductor, Linear Applications Handbook, 1991, p. 94.

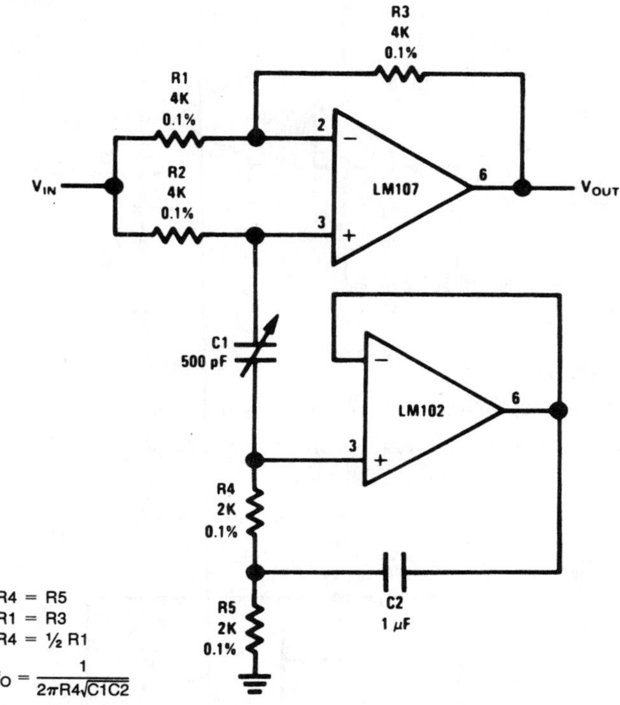

Easily tuned notch filter

Fig. 7-33 This circuit uses two LM107s to form a tunable notch filter. The notch frequency is set by adjustment of C1. The frequency range is determined by the values of R_4, C_1, and C_2, as shown by the equation. National Semiconductor, Linear Applications Handbook, 1991, p. 95.

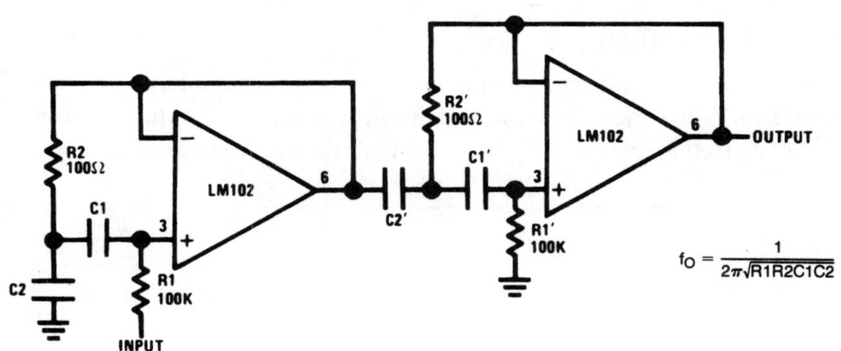

Two-stage tuned circuit

Fig. 7-34 This circuit uses two LM102s to form a two-stage tuned circuit. National Semiconductor, Linear Applications Handbook, 1991, p. 95.

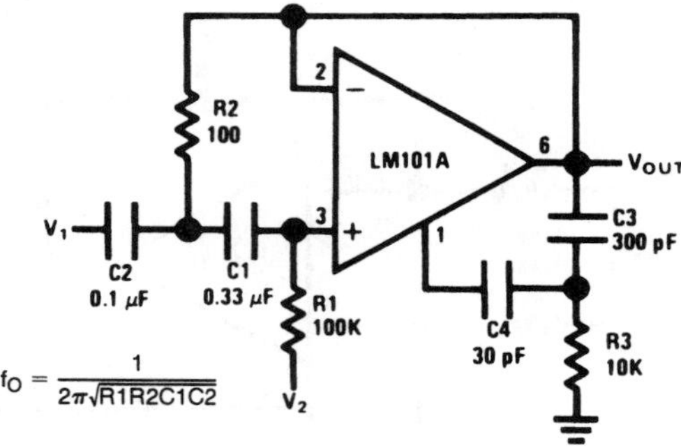

Basic tuned circuit

Fig. 7-35 This circuit uses an LM101A to form a single-stage tuned circuit. The output frequency is determined by the values of R_1, R_2, C_1, and C_2, as shown by the equation. National Semiconductor, Linear Applications Handbook, 1991, p. 95.

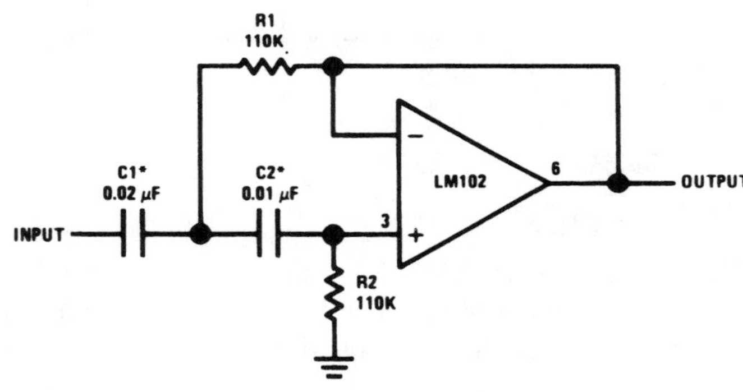

*Values are for 100 Hz cutoff. Use metalized polycarbonate capacitors for good temperature stability.

High-pass active filter

Fig. 7-36 This circuit uses an LM102 connected to form a basic high-pass filter (input capacitors in series and input resistor in parallel). The values are for 100-Hz cutoff. Other frequencies can be selected with different values. National Semiconductor, Linear Applications Handbook, 1991, p. 98.

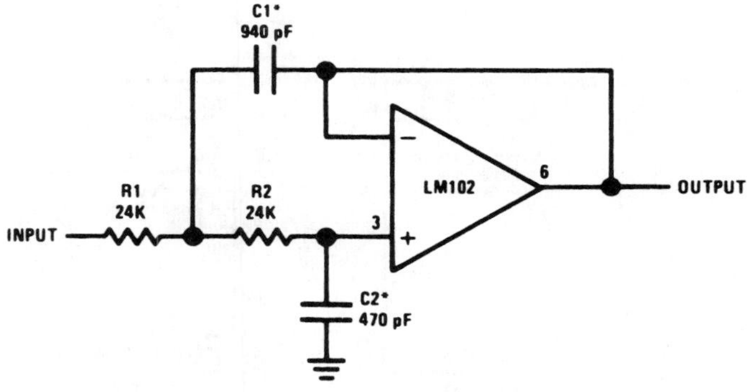

*Values are for 10 kHz cutoff. Use silvered mica capacitors for good temperature stability.

Low-pass active filter

Fig. 7-37 This circuit uses an LM102 connected to form a basic low-pass filter (input capacitors in parallel and input resistors in series). The values are for a 10-kHz cutoff. Other frequencies can be selected with different values. National Semiconductor, Linear Applications Handbook, 1991, p. 98.

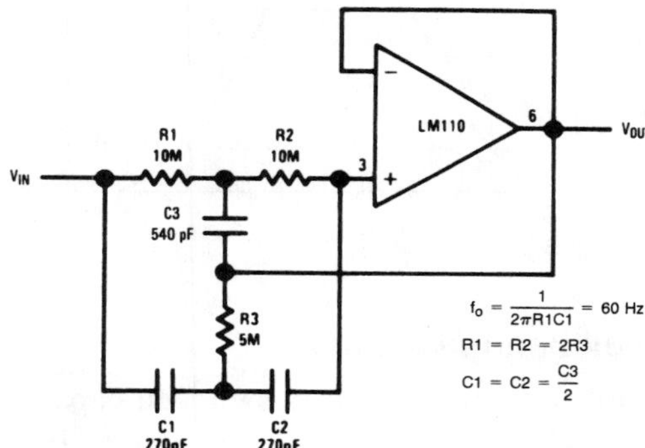

V_{IN} V_{OUT}

$$f_o = \frac{1}{2\pi R1C1} = 60\text{ Hz}$$

$$R1 = R2 = 2R3$$

$$C1 = C2 = \frac{C3}{2}$$

High-Q notch filter

Fig. 7-38 This circuit uses an LM110 connected to form a basic high-Q notch filter. The frequency range is determined by the values of R_1 and C_1, as shown by the equation. National Semiconductor, Linear Applications Handbook, 1991, p. 1202.

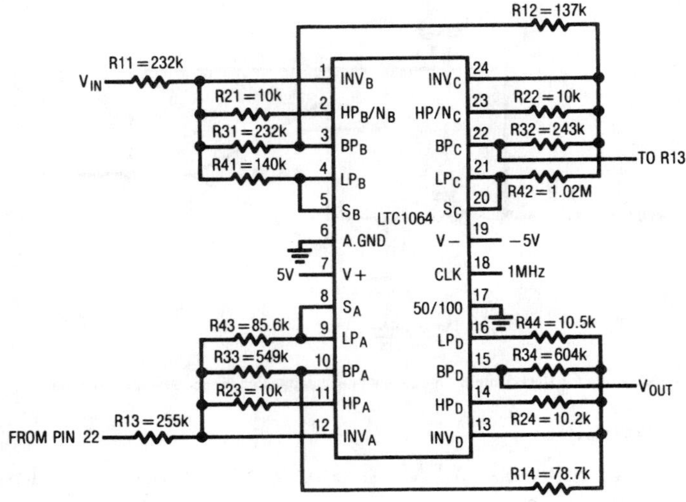

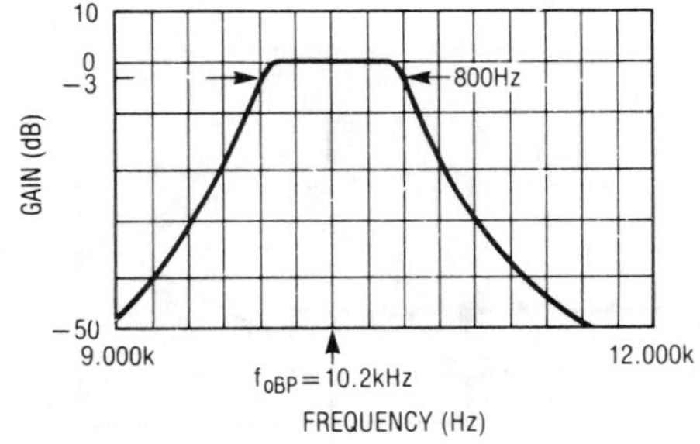

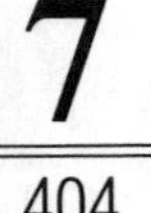

8th-order Chebyshev bandpass filter

Fig. 7-39 This circuit uses an LTC1064 switched-capacitor filter to form an 8th-order Chebyshev bandpass filter, with a center frequency of 10.2 kHz and a bandwidth of 800 Hz. Figure 7-39B shows the characteristics. Linear Technology, Linear Applications Handbook, 1990, p. AN27A-13, 15.

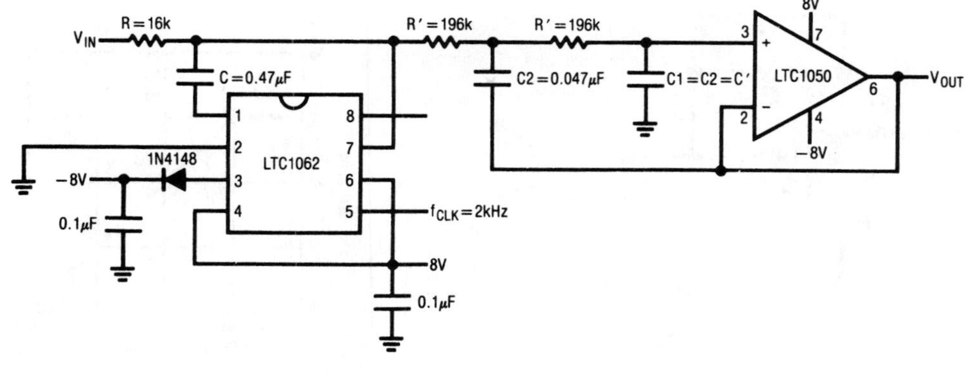

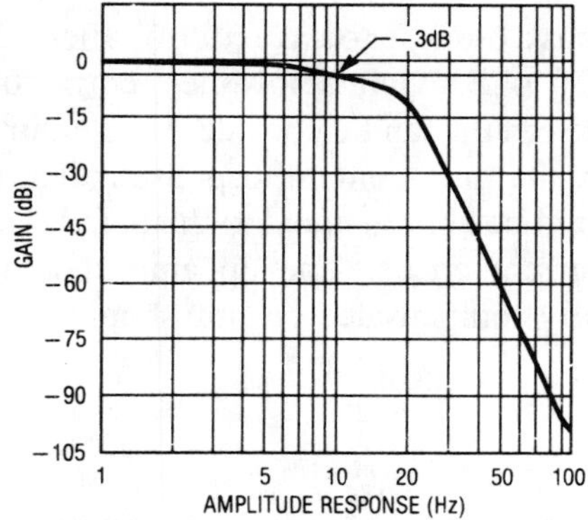

Dc-accurate low-pass Bessel filter

Fig. 7-40 This circuit uses an LTC1050 and LTC1062 to form a low-cost 7th-order 10-Hz low-pass filter, where amplitude and phase response closely approximate a Bessel filter. The required clock frequency is 2 kHz, which yields a clock-to-cutoff frequency ratio of 200:1. Figure 7-40B shows the characteristics.

Linear Technology, Linear Applications Handbook, 1990, p. DN9-1, -2.

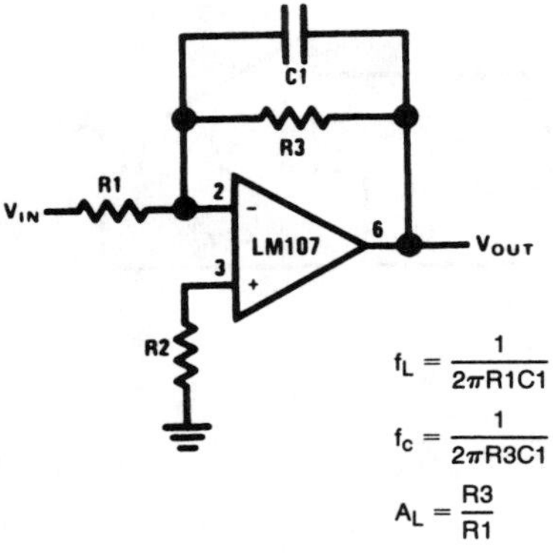

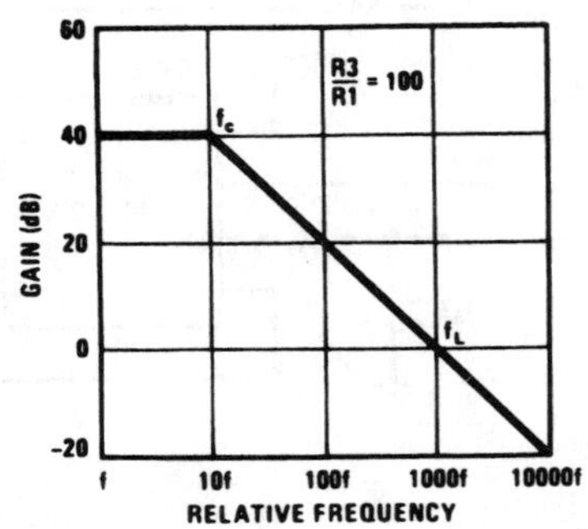

Simple low-pass filter

Fig. 7-41 This circuit has a 6-dB per octave rolloff, after a closed-loop 3-dB point that is defined by f_c (Fig. 7-41B). Gain below the f_c corner frequency is defined by the ratio of R_3 to R_1. The circuit can be considered as an integrator at frequencies well above f_c. However, the time-domain response is that of a single RC, rather than an integral. R2 should be chosen equal to the parallel combination of R1 and R3 to minimize bias-current errors. The op amp should be compensated for unity-gain, or an internally compensated op amp should be used. National Semiconductor, Linear Applications Handbook, 1991, p. 23, 24.

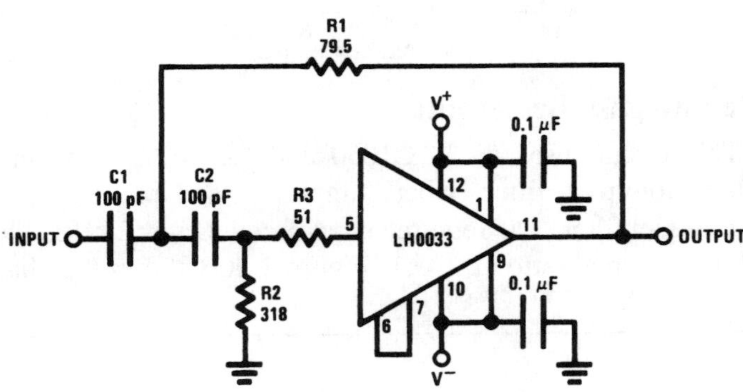

Wideband two-pole high-pass filter

Fig. 7-42 This circuit shows an LH0033 buffer connected to form a basic two-pole high-pass filter with a 10-Hz cutoff frequency. A low-pass filter with the same frequency can be obtained by interchanging R1, R2, C1, and C2. National Semiconductor, Linear Applications Handbook, 1991, p. 542.

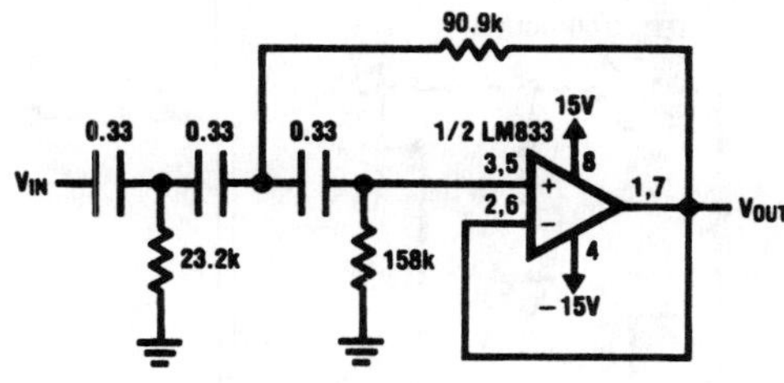

Infrasonic filter

Fig. 7-43 This circuit shows an LM833 connected as a filter for rejecting undesired infrasonic signals. The filter characteristic is 3rd-order Butterworth with the −3-dB frequency at 15 Hz. Resistors and capacitors should be 1%-tolerance components (although 5%-tolerance components can be substituted in less critical applications). National Semiconductor, Linear Applications Handbook, 1991, p. 944.

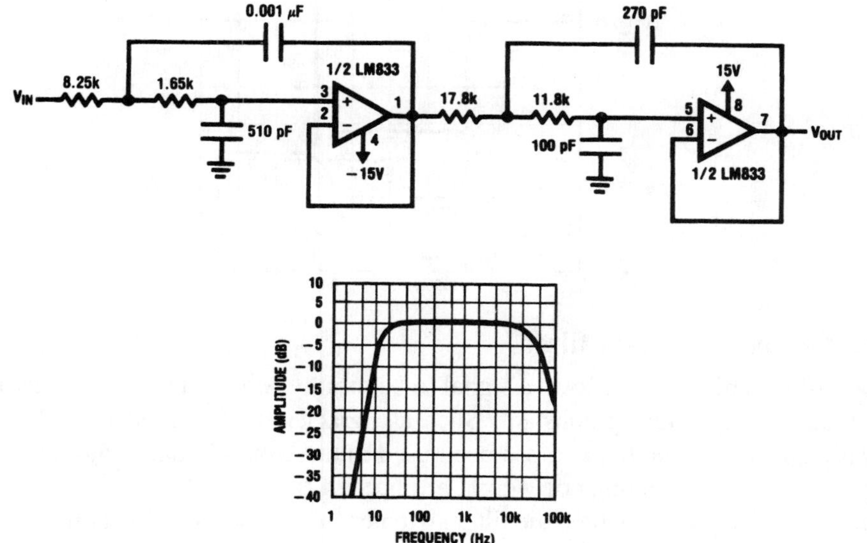

Ultrasonic filter

Fig. 7-44 This circuit shows both sections of an LM833 that is connected as a filter to reject undesired ultrasonic signals. The filter characteristic is 4th-order Bessel with the −3-dB frequency at about 40 kHz. Resistors and capacitors should be 1%-tolerance components. This circuit can be cascaded with that of Fig. 7-43 (with the low-pass preceding the high-pass) to produce a response such as shown in Fig. 7-44B. National Semiconductor, Linear Applications Handbook 1991, p.944, 945.

A1, C1 = LM392 amplifier-comparator dual

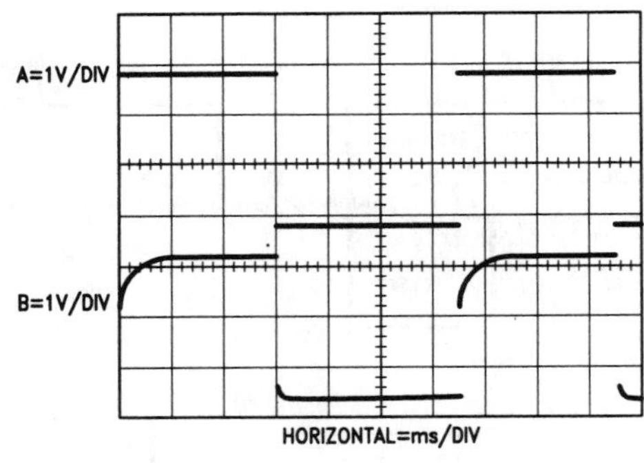

Fed-forward low-pass filter

Fig. 7-45 This circuit allows a signal to be rapidly acquired to final value, but it provides a long filtering constant. Such a characteristic is useful in multiplexed data-acquisition systems, and has been used in electronic infant-weighing scales, where fast, stable readings of weight are needed (in spite of motion on the scale platform). The point at which the filter switches from a short to a long time constant is set by the 1-kΩ pot. Normally, this is adjusted so that switching occurs at 90% to 98% of final value. However, the waveforms shown in Fig. 7-45B were obtained at a 70% trip point to show circuit operation. National Semiconductor, Linear Applications Handbook. 1991, p. 835.

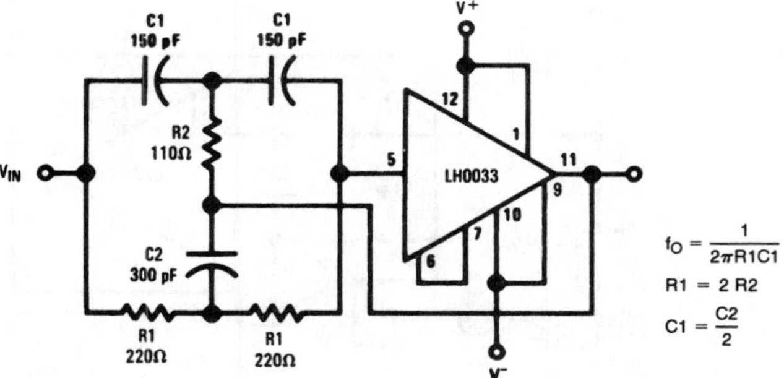

4.5-MHz notch filter

Fig. 7-46 This circuit shows an LH0033 buffer connected to form a notch filter. The frequency range is determined by the values of R_1 and C_1, as shown by the equation. National Semiconductor, Linear Applications Handbook, 1991, p. 161.

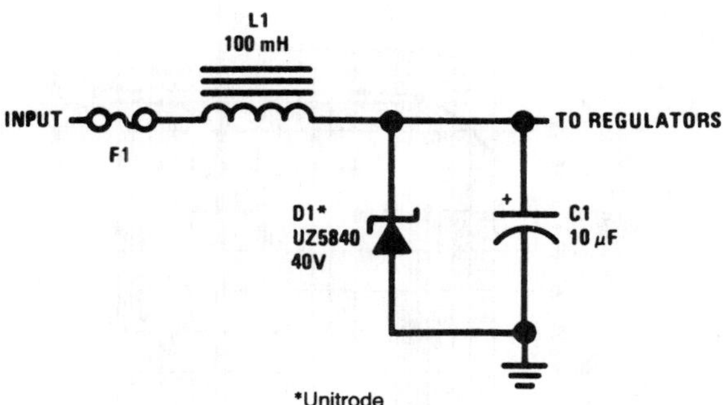

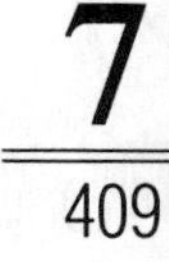

Spike suppressor for unregulated power supplies

Fig. 7-47 This circuit suppresses transients in unregulated supplies. Zener D1 clamps the input voltage to the regulator, and L1 limits the current through D1 during the transient. This circuit will clamp 70-V 4-ms transients. The value of $L_1 = d_V d_t / I$, where d_V is the voltage by which the input transient exceeds the breakdown voltage of D1, d_t is the duration of the transient, and I is the peak current of D1. National Semiconductor, Linear Applications Handbook, 1991, p. 47.

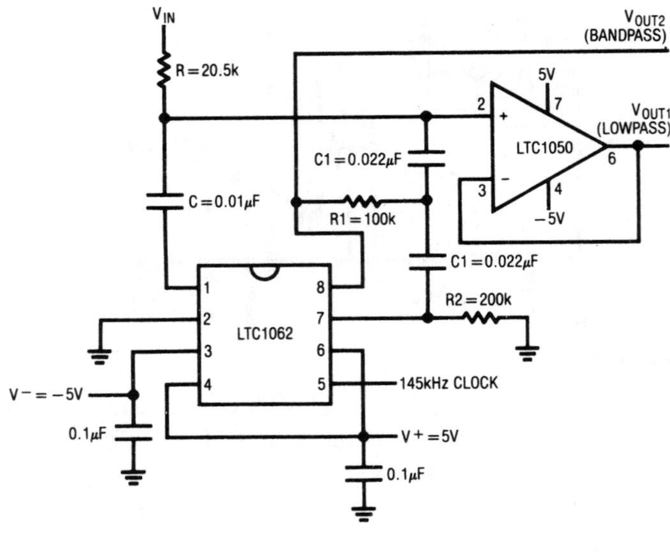

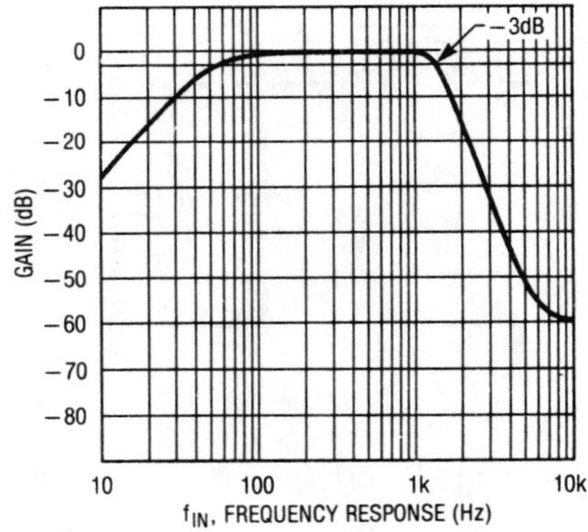

Dc-accurate low-pass/bandbass filter

Fig. 7-48 This circuit uses an LTC1050 and LTC1062 to form a filter that extracts ac information from a dc + ac signal. Figure 7-48B shows the bandpass-filter output characteristics. Linear Technology, Linear Applications Handbook, 1990, p. DN9-2.

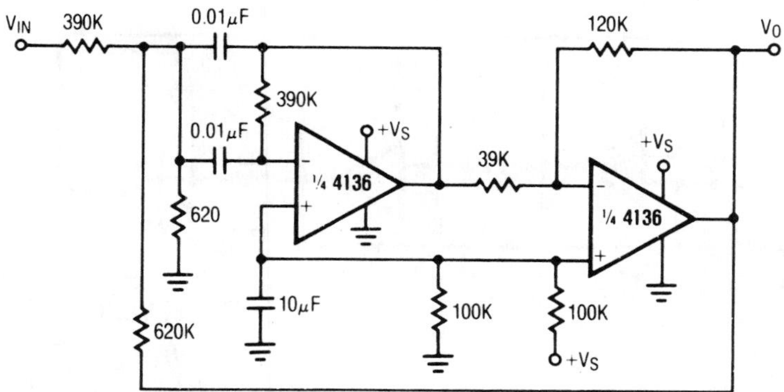

1-kHz bandpass filter

Fig. 7-49 This circuit uses two sections of a 4136 op amp to form a bandpass filter. Raytheon Linear Integrated Circuits, 1989, p. 4-175.

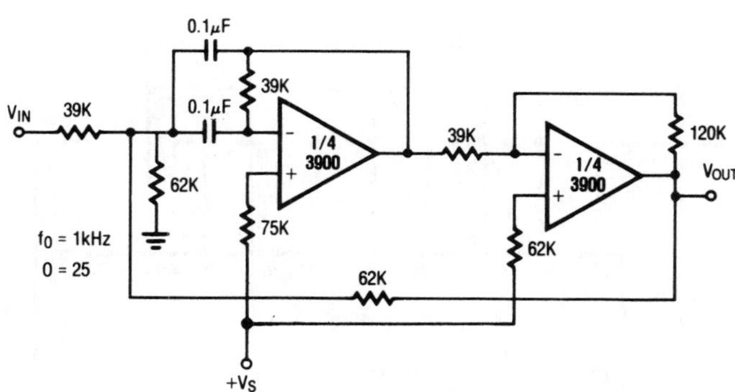

1-kHz bandpass filter with high Q

Fig. 7-50 This circuit uses two sections of a 3900 op amp to form a high-Q (Q = 25) bandpass filter. Raytheon Linear Integrated Circuits, 1989, p. 4.274.

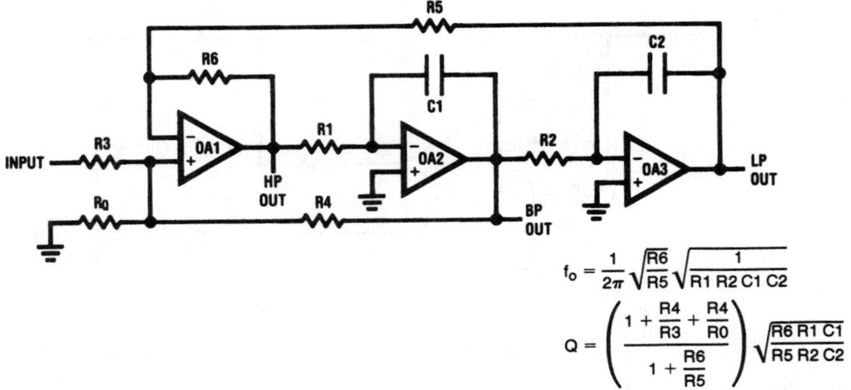

$$f_o = \frac{1}{2\pi}\sqrt{\frac{R6}{R5}}\sqrt{\frac{1}{R1\,R2\,C1\,C2}}$$

$$Q = \left(\frac{1 + \frac{R4}{R3} + \frac{R4}{R0}}{1 + \frac{R6}{R5}}\right)\sqrt{\frac{R6\,R1\,C1}{R5\,R2\,C2}}$$

Universal state-variable filter

Fig. 7-51 This circuit uses three op amps to form a universal state-variable filter. Compare this circuit and equations to that of Fig. 7-6. National Semiconductor, Linear Applications Handbook, 1991, p. 903.

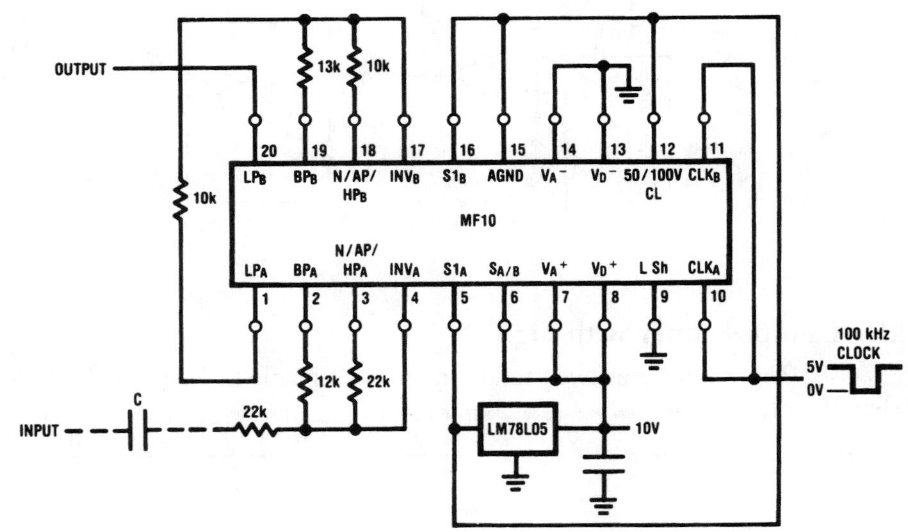

4th-order 1-kHz Butterworth-switched filter

Fig. 7-52 Only six resistors are required for this low-pass filter. However, a 100-kHz clock is required. National Semiconductor, Linear Applications Handbook, 1991, p. 910.

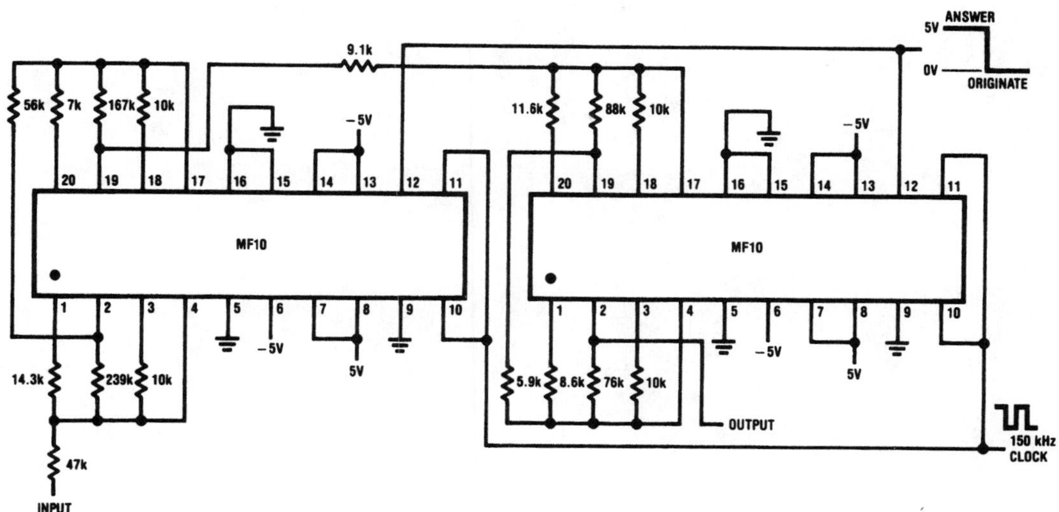

Full-duplex 300-baud modem filter

Fig. 7-53 This circuit is an 8th-order 1-dB ripple Chebyshev bandpass, which functions as both an 1170-Hz originate filter, and a 2125-Hz answer filter. Only one clock is required, and overall gain is 22 dB. National Semiconductor, Linear Applications Handbook, 1991, p. 912.

8

Switching and electronic-control circuits

It is assumed that you are already familiar with electronic-control basics, particularly operation of SCRs and triacs, as well as the various trigger devices. However, the following paragraphs summarize this information, as well as test and troubleshooting for control circuits. This information is included so that even those readers who are not familiar with electronic procedures can both test the circuits in this chapter and can localize problems if the circuits fail to perform as shown.

Control circuit testing and troubleshooting

Figure 8-A shows the symbols and operating characteristics for the two most commonly used electronic-control devices, the *SCR* and the *triac*. These devices can be considered as control rectifiers or switches. The SCR is normally used to control ac, but it can be used to control dc. The triac conducts in both directions and is thus most useful for controlling devices operated by ac power (such as ac motors). Both SCRs and triacs are triggered on by gate trigger signals. Usually, the trigger signal is synchronized with the ac power source being controlled (Fig. 8-A).

With ac power, the control of the power applied to the load is determined by the relative phase of the trigger signal versus the load voltage. Because the trigger control is lost once the SCR is conducting, an ac voltage at the load permits the

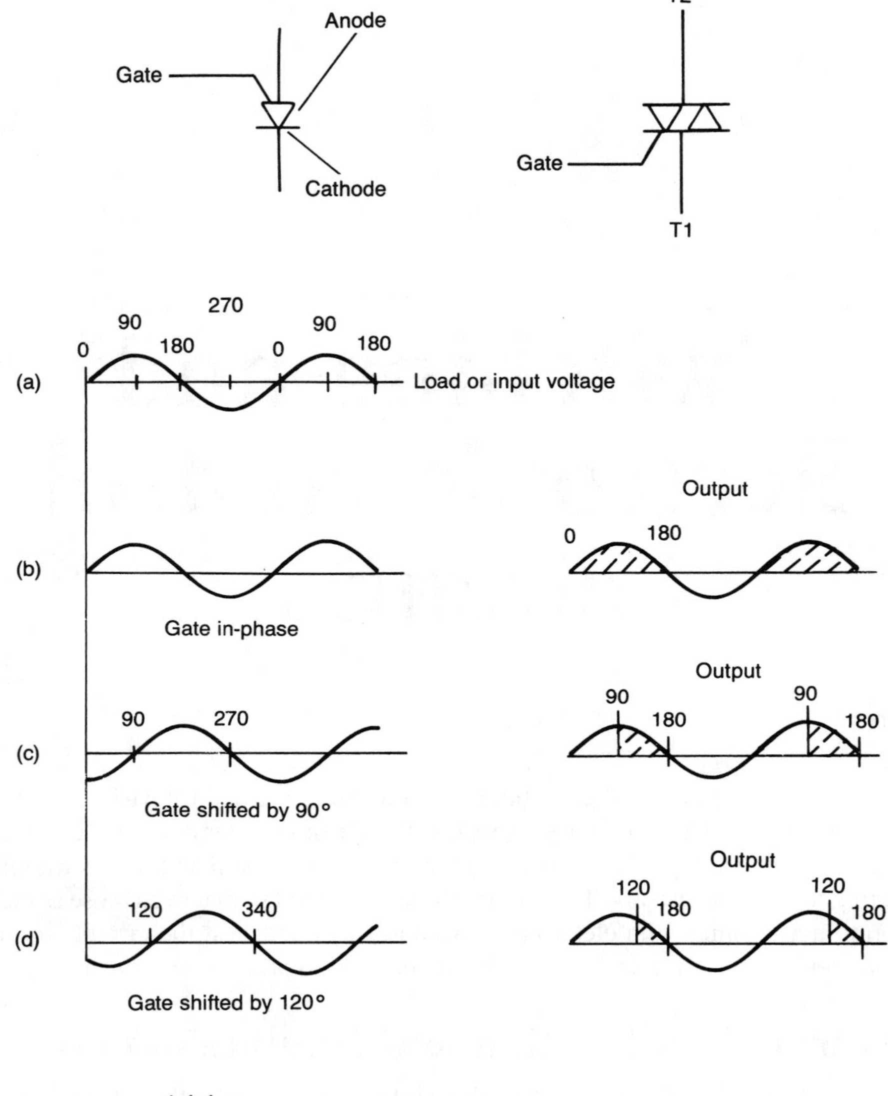

Fig. 8-A Symbols and operating characteristics for SCRs and triacs.

trigger to regain control. Each alternation of ac through the load causes conduction to be interrupted (when the ac voltage drops to zero between cycles), regardless of the polarity of the trigger signal.

If the trigger voltage is in phase with the ac power input signal (Fig. 8-Ab), the

SCR conducts for each successive positive alteration at the anode. When the trigger is positive-going at the same time as the load or anode voltage, load current starts to flow as soon as the load voltage reaches a value that will cause conduction. When the trigger is negative-going, the load voltage is also negative-going, and conduction stops. The SCR acts as a half-wave rectifier, whereas the triac acts as a full-wave rectifier.

If there is a 90° phase difference (Fig. 8-Ac) between the trigger voltage and load voltage (for example, the load voltage lags the trigger voltage by 90°), the SCR does not start conducting until the trigger voltage swings positive—even though the load voltage is initially positive. When the load voltage drops to zero, conduction stops—even though the trigger voltage is still positive.

If the phase shift is increased between trigger and load voltages (Fig. 8-Ad), conduction time is even shorter, and less power is applied to the load circuit. By shifting the phase of the trigger voltage in relation to the load voltage, it is possible to vary the power output—even though the voltages are not changed in strength.

Basic control-circuit tests

The obvious test for the electronic-control circuits in this chapter is to see if power is applied to the load. If the circuit is adjustable (which is usually the case), see that power to the load is changed when the adjustment is made. For example, in the dimmer circuit of Fig. 8-21, power is applied to the load (a lamp) through the 2N6343A triac. During each alternation of the ac input, the capacitor charges through the 500-kΩ pot. When the voltage across the capacitor is equal to the breakover voltage of the MBS4991, the capacitor is discharged through the MBS4991, which applies a trigger to the triac. With the triac on, power is applied to the lamp. The rate at which the capacitor charges is set by the 500-kΩ pot. Thus, the point at which the trigger occurs during each cycle is set by the pot, as is the amount of current through the lamp. In this case, it should be possible to dim the lamp from full-on to full-off with the pot.

If the lamp does not turn on at any setting of the pot, or if the lamp stays on and does not dim, look for defective parts (or improper wiring). The MBS4991 could short, the capacitor could be leaking, or the triac could go into conduction before the trigger point. If the MBS4991 shorts, the triac could remain triggered at all times or could remain not triggered because the capacitor could not charge and discharge properly. If the capacitor is leaking, even a good MBS4991 might not be triggered.

If the triac goes into conduction before the trigger, this is usually a circuit-design fault, or the problem could be in the triac itself. Snubbers are often added to control circuits to prevent such turn-on. For example, in the circuit of Fig. 8-22, the 10-kΩ resistor and 0.1-μF capacitor across Q1 is the snubber that prevents any rapid change of voltage (line transients) from accidentally firing Q1.

Basic control-circuit troubleshooting

When the control circuit fails to turn on (or off), the first step is to check that a trigger is applied to the SCR or triac (right after you have checked all wiring, and have made certain that all components are good). This is a *half-split* and it divides the circuit into two parts.

If the trigger is present, the problem is probably the control device (SCR or triac) or the load/power wiring. If the trigger is absent or abnormal, the problem must be traced back to the trigger source. For example, assume that the pump motor in Fig. 8-7A does not turn on and off at the corresponding water levels. Check for a trigger, either at the triac or at the output (pin 3) of the CA3098 comparator. The pin-3 output should change each time the water level reaches the high and low points.

If there is no change in pin-3 output (no trigger), suspect the CA3096 comparator, the TH1 and TH2 thermistors, or possibly adjustment of the 10-kΩ threshold-adjustment pots. If there is a change, suspect the triac MT1, the pump motor, or the power wiring.

When the control circuit turns on and off, but does not provide proper control, the troubleshooting problem becomes more difficult. Generally, this type of problem means that the trigger is occurring on the wrong part of the conduction cycle. If practical, monitor the phase or time relationship between the trigger and the ac power cycle. Typically, too much power to the load means that the trigger occurs too early in the power cycle (such as in Fig. 8-Ab), whereas late triggering (Fig. 8-Ad) reduces power to the load.

For example, assume that the 100-kΩ pot in Fig. 8-3 does control power to the load (it is possible to vary the power), but the load never receives full power at any setting of the pot. Use one trace of a dual-trace scope to show the ac power cycle, and the other trace to show the trigger voltage. Notice that in this circuit, the trigger is synchronized with the ac power to the load because both are taken from the ac power line. This is not always true in all electronic-control circuits.

With the symptom of low power, you will probably find that the trigger voltage is applied late in the ac power cycle—even at the extreme setting of the 100-kΩ pot. In that case, look for defective parts. Make certain that the MBS4991 fires at the correct breakover voltage (as shown on the datasheet), and that the capacitor is not leaking. Another point to check is the R_{GK} resistor, which is added to prevent early firing of triacs with sensitive gates (sensitive trigger points). Try the circuit with the R_{GK} resistor omitted or with R_{GK} resistors of different values.

Switching and electronic-control circuits titles and descriptions

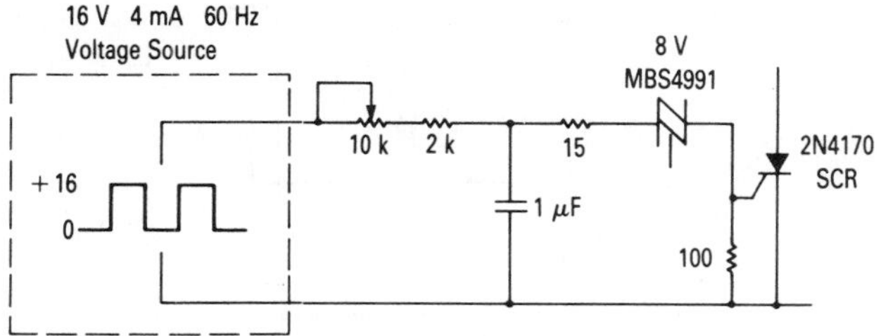

Basic SCR phase control

Fig. 8-1 This circuit shows a basic phase-control scheme using an SCR and relaxation oscillator. The capacitor is charged through the resistor until the breakover voltage of the trigger (an SBS in this case) is reached. The SBS then changes to the on-state, and the capacitor is discharged through the SCR gate. Turn-on of the SCR is accomplished with a short, high-current pulse. In addition to an SBS, commonly-used triggers are UJTs, PUTs, optically coupled thyristors (chapter 9), and SIDACs. Phase control is obtained by varying the RC time constant of the charging circuit so that the trigger turn-on occurs at varying phase angles within a controlled half cycle. With the values shown, the conduction angle can be varied from about 20° to 150°. Motorola Thyristor Device Data, 1991, p. 1-2-10.

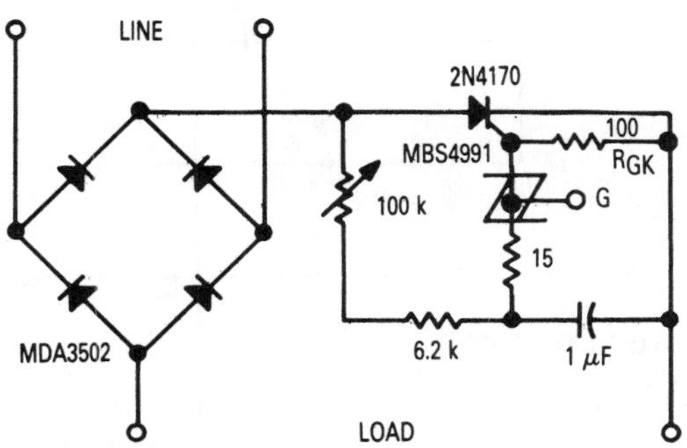

Simple dc power control

Fig. 8-2 This circuit shows a simple dc full-wave control scheme that uses an SCR and an SBS. Control is obtained by varing the 100-kΩ pot. Notice that R_{GK} can be omitted when triacs with nonsensitive gates are used. Motorola Thyristor Device Data, 1991, p. 1-2-13.

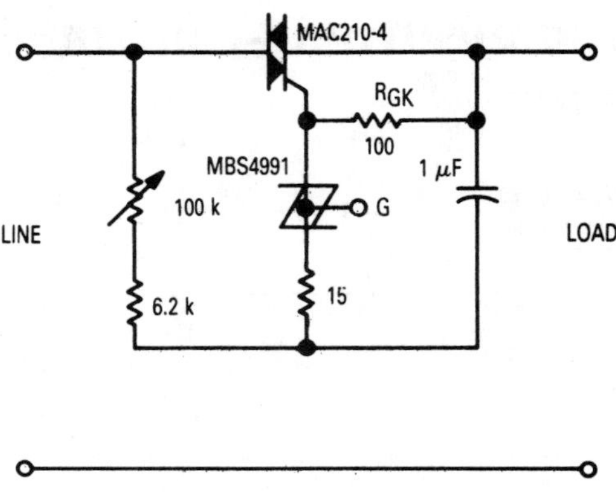

Simple ac power control

Fig. 8-3 This circuit shows a simple ac full-wave control scheme that uses a triac and SBS. Control is obtained by varying the 100-kΩ pot. Notice that P_{GK} can be omitted when triacs with nonsensitive gates are used. Motorola Thyristor Device Data, 1991, p. 1-2-13.

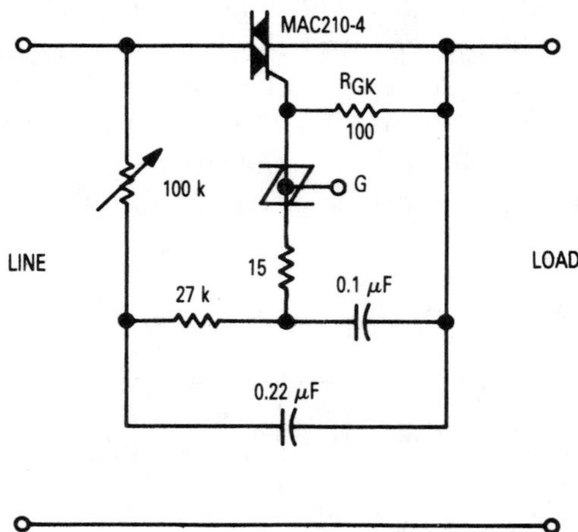

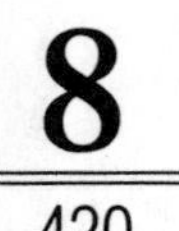

Full-range ac power control

Fig. 8-4 This circuit shows a full-range ac control scheme that uses a triac and SBS. Control is obtained by varying the 100-kΩ pot. Notice that R_{GK} can be omitted when nonsensitive gates are used. The double time constants of this circuit produce low hysteresis, which thus extends the control range. Motorola Thyristor Device Data, 1991, p. 1-2-13.

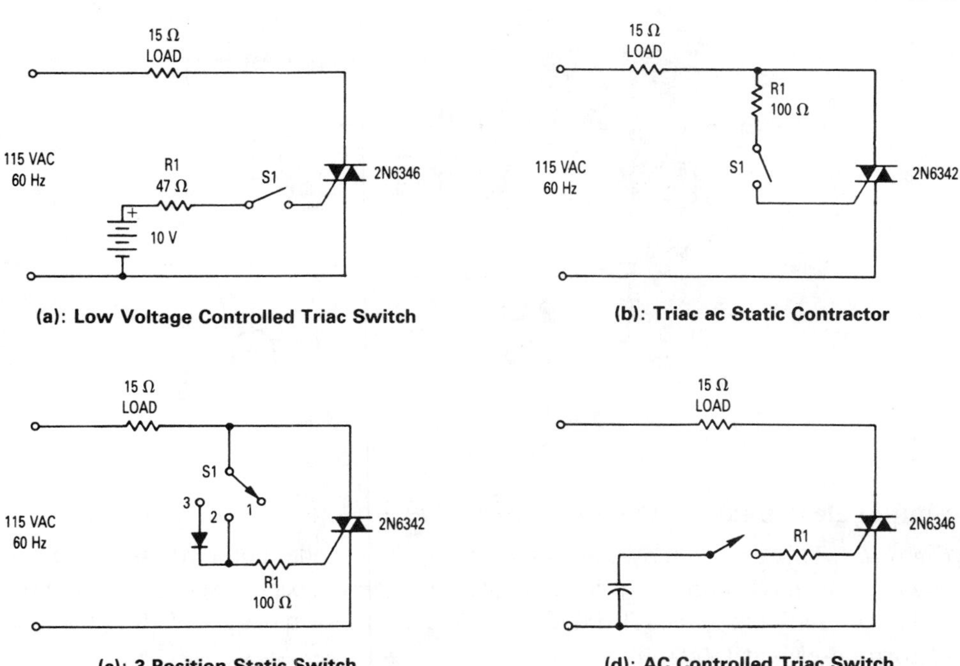

(a): Low Voltage Controlled Triac Switch

(b): Triac ac Static Contractor

(c): 3 Position Static Switch

(d): AC Controlled Triac Switch

Basic triac switches

Fig. 8-5 These circuits show triacs that are used to control a pure resistive load (of about 7 A) across an ac line. In Fig. 8-5A, gate current is supplied to the triac from the 10-V battery when S1 is closed. The triac turns on and remains on until S1 is opened. This circuit switches at zero current, except for initial turn on, so S1 can be a very low-current switch (carring only triac gate current). The triac switch of Fig. 8-5B has the same characteristics as Fig. 8-5A, except that the battery is eliminated. In the circuit of Fig. 8-5C, when S1 is in position 1, the triac receives no gate current and is nonconducting. With S1 in position 2, operation is the same as in Fig. 8-5B. With S1 in position 3, the triac receives gate current only on positive half-cycles, and power to the load is half-wave. Figure 8-5D shows ac control of the triac. Motorola Thyristor Device Data, 1991, p. 1-2-16.

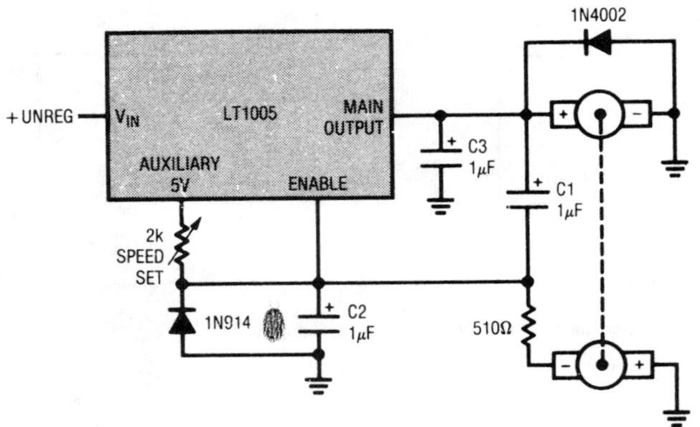

Simple electronic control of motors

Fig. 8-6 This circuit shows how a simple switch-mode motor controller can be made using an LT1005 multifunction regulator. Motor speed is set by the 2-kΩ pot at the Auxiliary pin of the LT1005. The motor-tach shown has a shaft-torque rating of 20 gram-cMs at 3300 rpm. Linear Technology Corporation, Linear Applications Handbook, 1990, p. AN1-7.

(a)

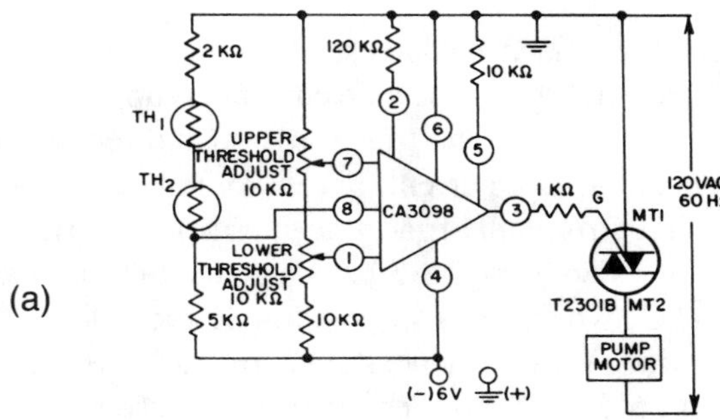

Notes (a) Motor pump is "ON" when water level rises above thermistor TH_2.
(b) Motor pump remains "ON" until water level falls below thermistor TH_1.
(c) Thermistors, operate in self-heating mode.

(b)

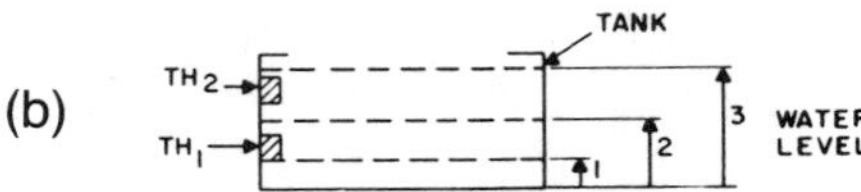

Water-level control

Fig. 8-7 This circuit shows a triac used to provide electronic control of a drain or sump pump. Figure 8-7B shows positioning of the sensor thermistors in the tank that is drained by the pump. The circuit notes explain the operation of the pump control. Harris Semiconductors, Linear & Telecom ICs, 1991, p. 4-14.

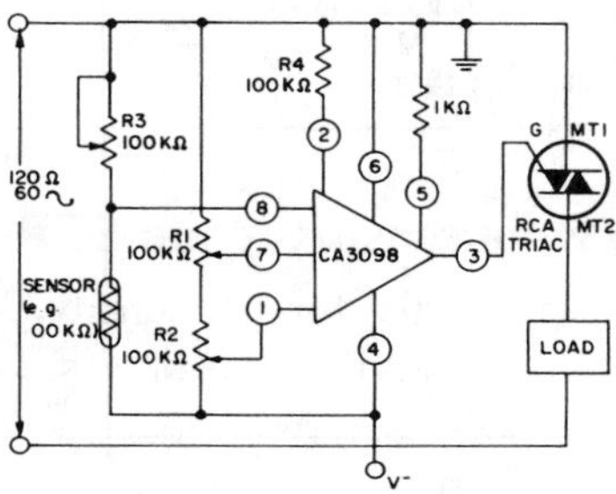

Off/on control of a triac with programmable hysteresis

Fig. 8-8 This circuit shows control of a triac with a CA3098 programmable Schmitt trigger. R1 and R7 set the low and high reference voltages, respectively, while R3 sets the control-signal level from the sensor. Harris Semiconductors, Linear & Telecom ICs, 1991, p. 4-14.

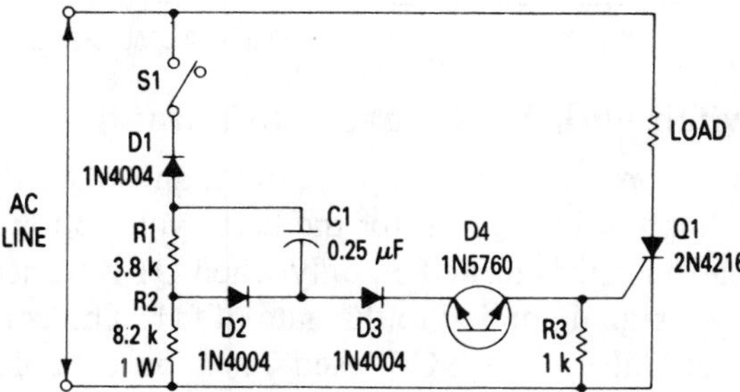

Zero-crossing switch for sensitive-gate SCRs

Fig. 8-9 This circuit is primarily for sensitive-gate SCRs, such as the 2N4216, and loads of 1.5 A or less. Motorola Thyristor Device Data, 1991, p. 1-2-17.

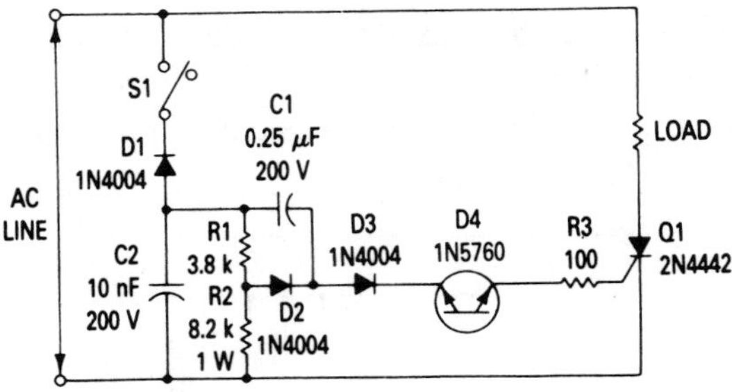

Zero-crossing switch for nonsensitive-gate SCRs

Fig. 8-10 This circuit is primarily for nonsensitive-gate SCRs, such as the 2N4442 (or an MCR218-8), and loads of 8.0 A or less. Motorola Thyristor Device Data, 1991, p. 1-2-17.

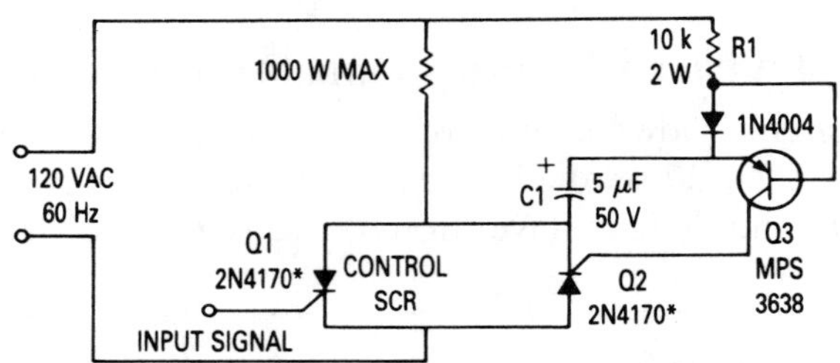

Zero-crossing switch with SCR-slave configuration

Fig. 8-11 This circuit provides a single pulse to the gate of SCR Q2 each time that SCR Q1 turns on, thus turning Q2 on for the half-cycle following the half-cycle during which Q1 was on. Q2 is turned on only when Q1 is turned on, and the load can be controlled by a signal applied to the gate of Q1. The control signal can be either dc or a power pulse. The SCR used must be capable of handling the maximum current requirements of the load to be driven. The 8-A 200-V SCRs shown will handle a 1-kW load. 2N6397s can also be used in this circuit, and will handle loads up to 12 A. Motorola Thyristor Device Data, 1991, p. 1-2-18.

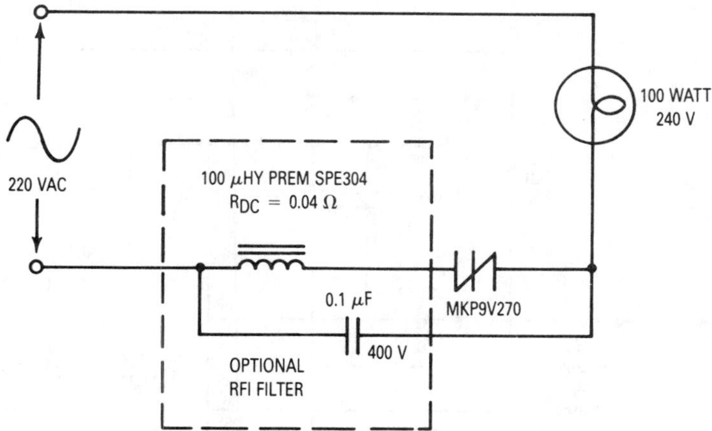

Long-life circuit for an incandescent lamp

Fig. 8-12 This circuit shows an MKP9V270 SIDAC used to phase-control an incandescent lamp, thus lowering the RMS voltage to the filament and prolonging the life of the bulb. This is particularly useful when lamps are used in hard to reach locations, such as in outdoor lighting in signs where replacement costs are high. Bulb life span can be extended by 1.5 to 5 times, depending on the type of lamp, the amount of power reduction to the filament, and the number of times the lamp is switched on from a cold-filament condition. Practical conduction angles run between 110° and 130°, with corresponding power reductions of 10% to 30%. Motorola Thyristor Device Data, 1991, p. 1-4-5.

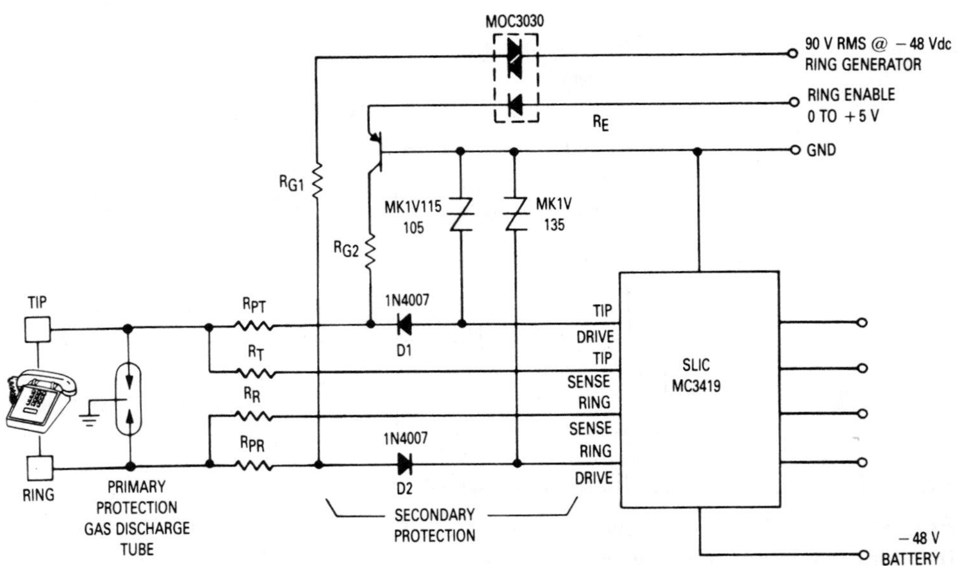

Overvoltage protection for telephone equipment

Fig. 8-13 This circuit shows two SIDACs that are connected to protect the telephone Subscriber Loop Interface Circuit (SLIC) and associated electronics from voltage surges. This is in addition to the primary protection provided by the gas-discharge tube across the tip and ring lines. If a high positive-voltage transient appears on the lines, D1 (with a peak-inverse voltage of 1000 V) blocks the pulse, and the corresponding SIDAC conducts the surge to ground. Conversely, D2 and the related SIDAC protect the SLIC from negative transients. The SIDACs do not conduct when normal signals are present. Motorola Thyristor Device Data, 1991, p. 1-4-6.

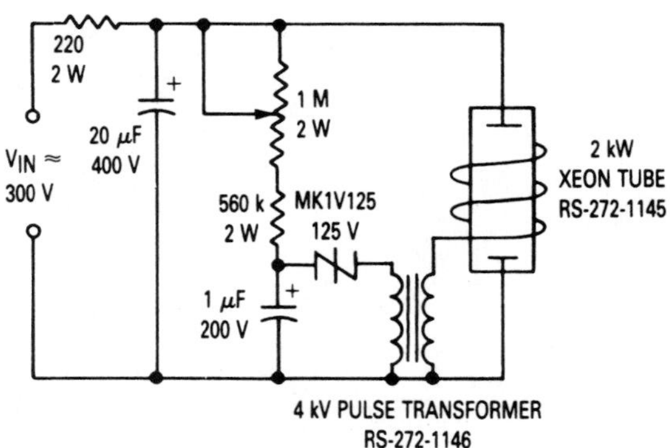

Xeon flasher

Fig. 8-14 This circuit shows an xeon tube triggered by a SIDAC that is connected as a basic relaxation oscillator, where the frequency is determined primarily by the RC time constant (a SIDAC relaxation oscillator is shown in Fig. 5-40). Once the capacitor voltage reaches the SIDAC breakover voltage (as determined by the setting of the series potentiometer), the SIDAC fires, and dumps the charged capacitor. By placing the load (transformer-coupled xeon tube) in the discharge path, the flashing frequency can be controlled. Motorola Thyristor Device Data, 1991, p. 1-4-7.

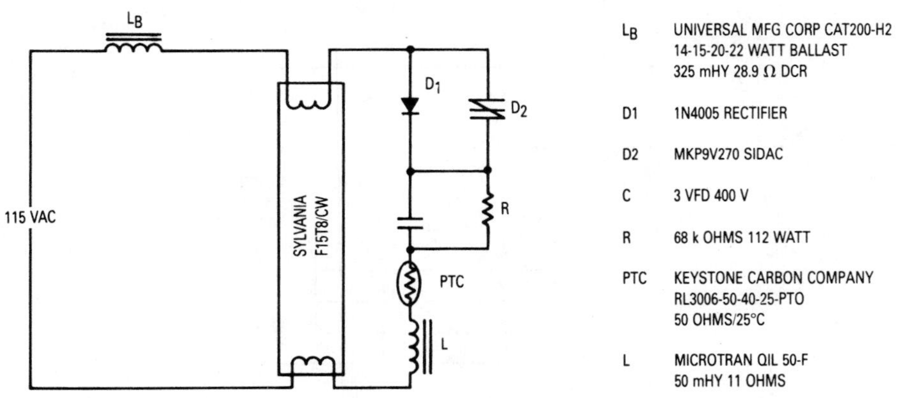

Fluorescent starter

Fig. 8-15 This circuit shows a solid-state fluorescent-lamp starter that uses a SIDAC. In this circuit, the ballast is identical to that used with a conventional glow-tube starter. Motorola Thyristor Device Data, 1991, p. 1-4-8.

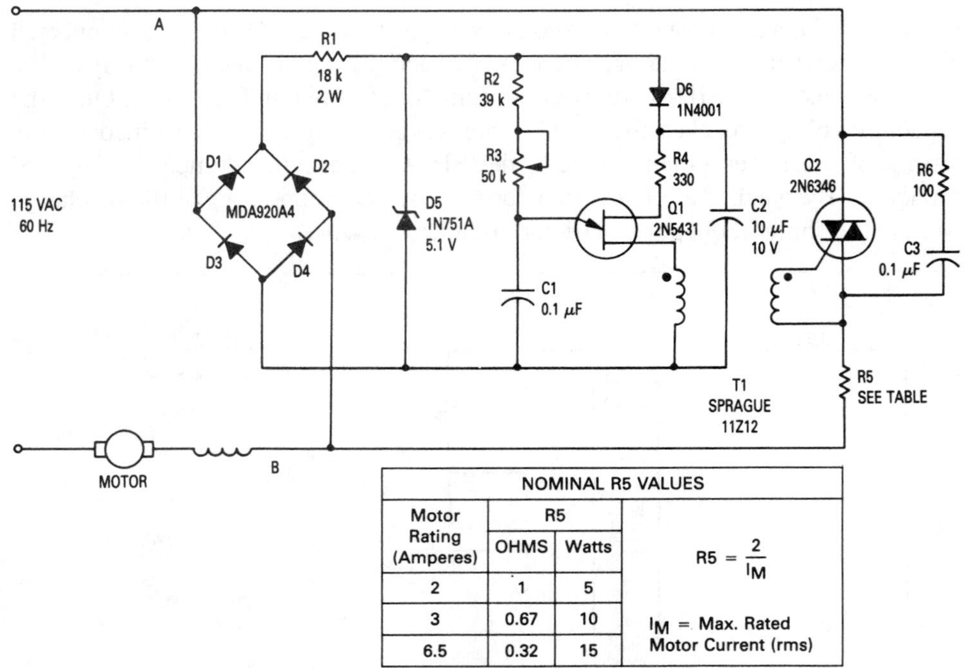

NOMINAL R5 VALUES

Motor Rating (Amperes)	R5 OHMS	R5 Watts
2	1	5
3	0.67	10
6.5	0.32	15

$$R5 = \frac{2}{I_M}$$

I_M = Max. Rated Motor Current (rms)

Motor-speed control with load-current feedback

8

428

Fig. 8-16 This circuit shows a triac motor-speed control that derives feedback from the load current and does not require separate connections to the motor field and armature windings. Thus, this circuit can be conveniently built into an appliance or used as a separate control. When the triac conducts, in response to signals from Q1 through T1, the normal line voltage (less the drop across the triac and R5) is applied to the motor. By delaying the firing of the triac until a later portion of the cycle, the voltage applied to the motor is reduced (or controlled) and speed is reduced proportionally. The use of feedback maintains torque at reduced speeds. The angle at which Q1 fires is proportional to motor current (and the setting of R3) because Q1 is controlled by the voltage across points A and B. As the motor is loaded and draws more current, the firing angle of Q1 is advanced, causing a proportional increase in the voltage applied to the motor, and a consequent increase in the available torque. Motorola Thyristor Device Data, 1991, p. 1-6-6.

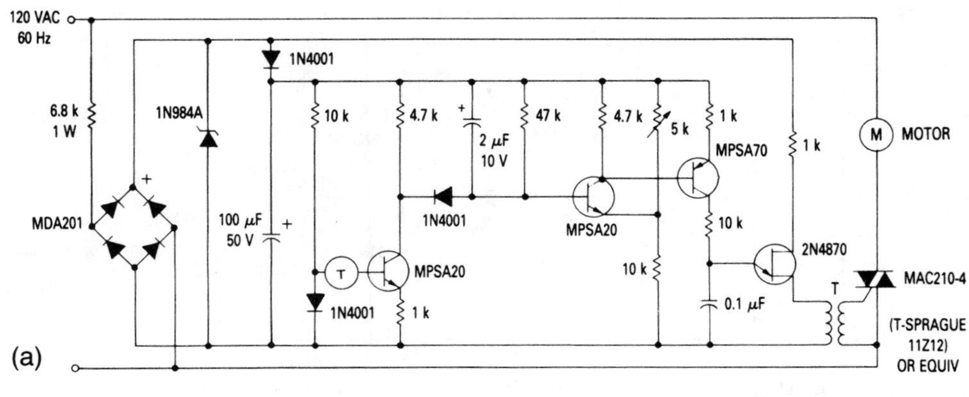

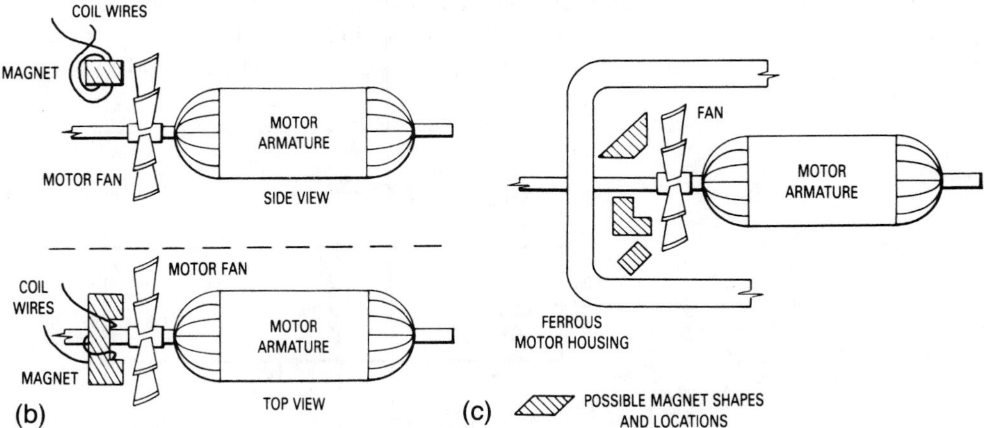

Motor-speed control with tachometer feedback

Fig. 8-17 This circuit shows a triac motor-speed control that derives feedback from a magnet-coil tachometer that is placed near the motor fan (Figs. 8-17B and 8-17C). Motor speed is controlled by the 5-kΩ pot. The MAC210-4 triac is capable of handling motor loads up to 10 A. Motorola Thyristor Device Data, 1991, p. 1-6-7,-9).

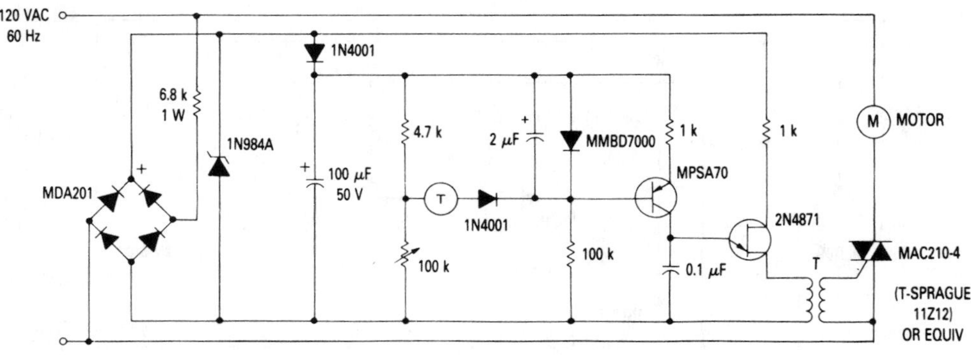

Alternate motor-speed control with tachometer feedback

Fig. 8-18 This circuit is an alternate to that shown in Fig. 8-17, but with a lower parts count. The same magnet-coil tachometer shown in Figs. 8-17B and 8-17C can be used with the circuit of Fig. 8-18. Motor speed is controlled by the 100-kΩ pot.

Motorola Thyristor Device Data, 1991, p. 1-6-9.

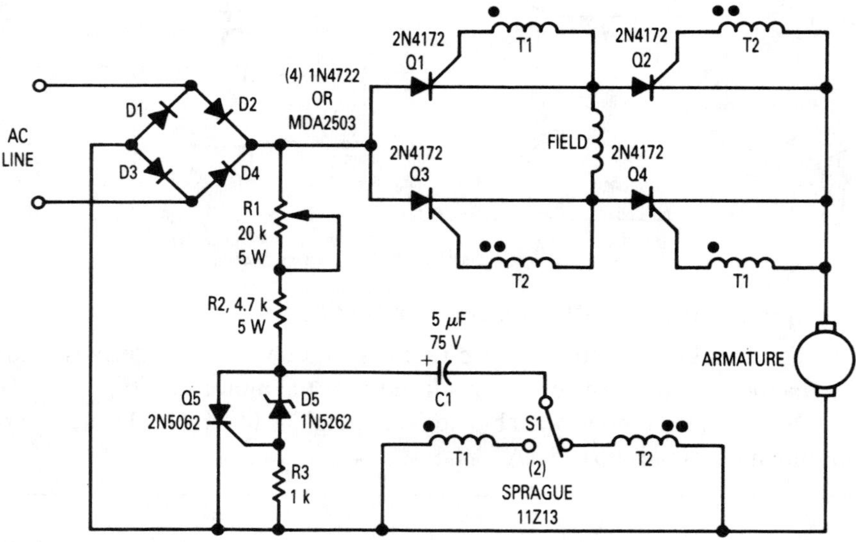

Direction and speed control for series-wound or universal motors

Fig. 8-19 In this circuit, Q1 through Q4 are triggered in diagonal pairs. The pair to be turned on is controlled by S1, which selects either T1 or T2 to receive a pulse from Q5. Current in the motor field is reversed by selecting either Q2 and Q3, or Q1 and Q4, for conduction. Because motor armature current is always in the same direction, the field current reverses in relation to the armature current, which thus reverses the direction of motor rotation. The setting of R1 determines the conduction angle of either Q1 through Q4 or Q2/Q3, thus setting the average motor voltage and thereby the speed. Motorola Thyristor Device Data, 1991, p. 1-6-36.

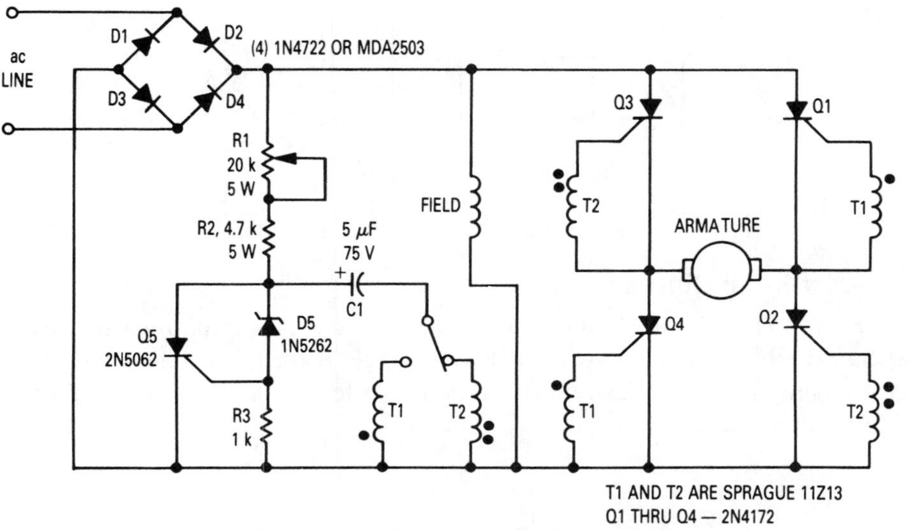

Direction and speed control for shunt-wound motors

Fig. 8-20 This circuit is similar to that of Fig. 8-19, except that the shunt field is placed across the rectified supply and the motor armature is placed in the SCR bridge. Thus, the field current is always in the same direction, but the armature current is reversible, so the direction of rotation is reversible (as determined by which transformer, T1 or T2, is selected through the switch). Again, motor speed is set by R1. Motorola Thyristor Device Data, 1991, p. 1-6-37.

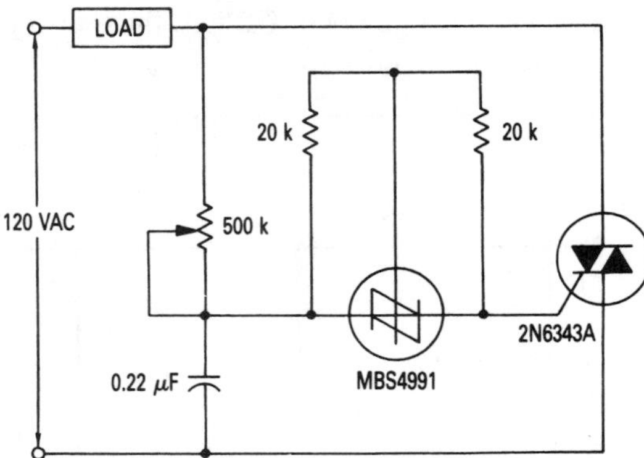

Low-cost lamp dimmer

Fig. 8-21 This circuit shows a full-range, low-cost, lamp dimmer that uses an MBS4991 silicon bilateral switch (SBS) to trigger the triac. The two 20-kΩ shunt resistors minimize the "flash-on" or hysteresis effect of the SBS, and improve the temperature sensitivity. Motorola Thyristor Device Data, 1991, p. 1-6-52.

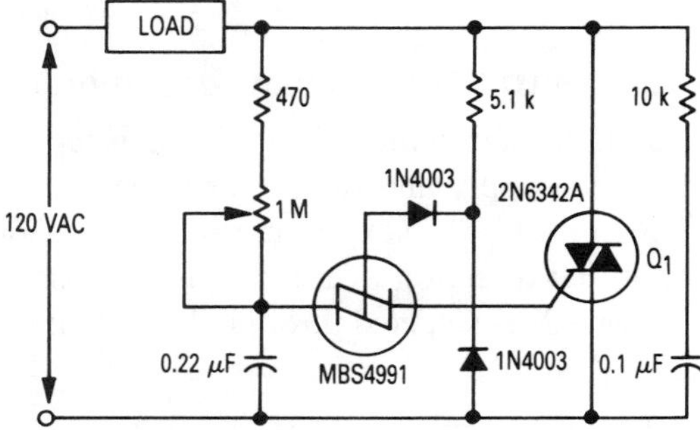

Improved lamp dimmer

Fig. 8-22 This circuit is an improved version of the circuit shown in Fig. 8-21. The hysteresis effect is eliminated by the addition of two diodes and the 5.1-kΩ resistor. The RC network across the triac is a snubber to prevent line transients from accidentally firing the triac. The circuit operates from a 120-V 60-Hz source, and can control up to 1000 W of power to incandescent bulbs (but not fluorescents).
Motorola Thyristor Device Data, 1991, p. 1-6-53.

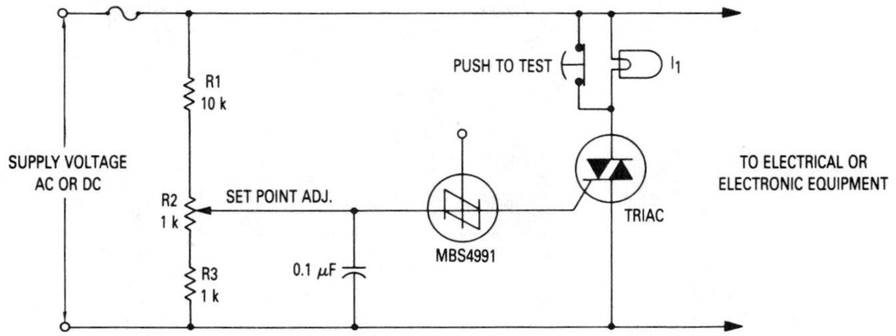

Electronic crowbar

Fig. 8-23 This circuit provides positive protection of expensive electrical or electronic equipment against excessive supply voltage (resulting from improper switching, short circuits, failure of regulators, etc.). The circuit is used where it is economically desirable to shut down equipment, rather than allow the equipment to operate at excessive voltages. The circuit quickly places a short across the power lines (ac or dc), and thereby drops the voltage to the protected device to near zero and blows a fuse. With the values shown, the crowbar operating point (set point) can be adjusted over the range of 60 to 120 Vdc or 42 to 84 Vac. The values of R_1 to R_3 can be changed to cover different supply voltages, but the triac voltage rating must be greater than the highest operating point that is set by R2. Lamp II (with a voltage rating that is equal to the supply) can be used to check the set point and operation of the circuit, by opening the push-to-test switch and adjusting the input or set point to fire the SBS. An alarm unit such as the Mallory Sonalert can be connected across the fuse to provide an audible indication of crowbar action. Notice that this circuit cannot act on short, infrequent power-line transients. Motorola Thyristor Device Data, 1991, p. 1-6-53.

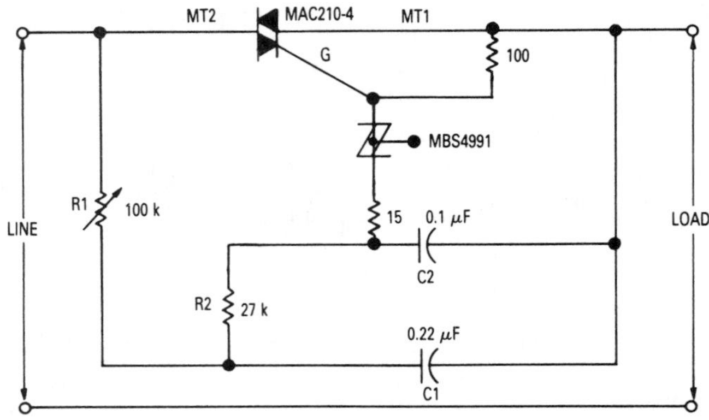

Basic triac control circuit that uses an SBS

Fig. 8-24 This figure shows the basic control circuit for triacs that use SBS triggers. The line voltage and load current depend primarily on the triac characteristics. In this case, the MAC210-4 accommodates loads up to 10 A. Motorola Thyristor Device Data, 1991, p. 1-6-54.

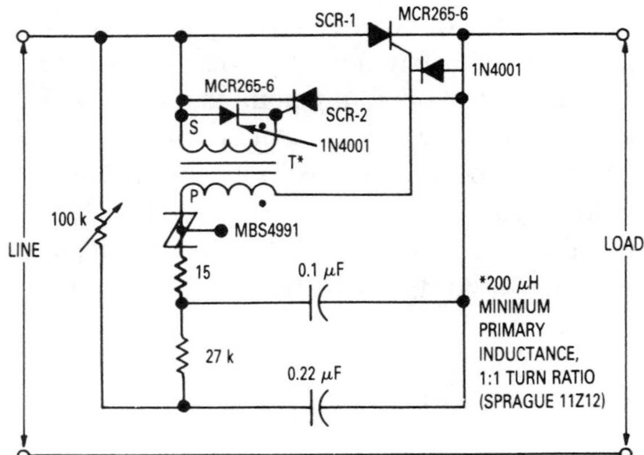

Basic SCR control circuit that uses an SBS

Fig. 8-25 This figure shows the basic control circuit for SCRs that use SBS triggers, and is preferable to that of Fig. 8-24 (triac), where high power must be handled, or where rapidly rising voltages are encountered (high d_v/d_t). Although the circuits of both Figs. 8-24 and 8-25 were designed as incandescent-lamp dimmers, the circuits are well suited to control of universal and shaded-pole motors. Such motors have higher torque at low speeds when open-loop controlled by these circuits, rather than with rheostats or variable transformers (because of the higher voltage pulses applied). Motorola Thyristor Device Data, 1991, p. 1-6-54.

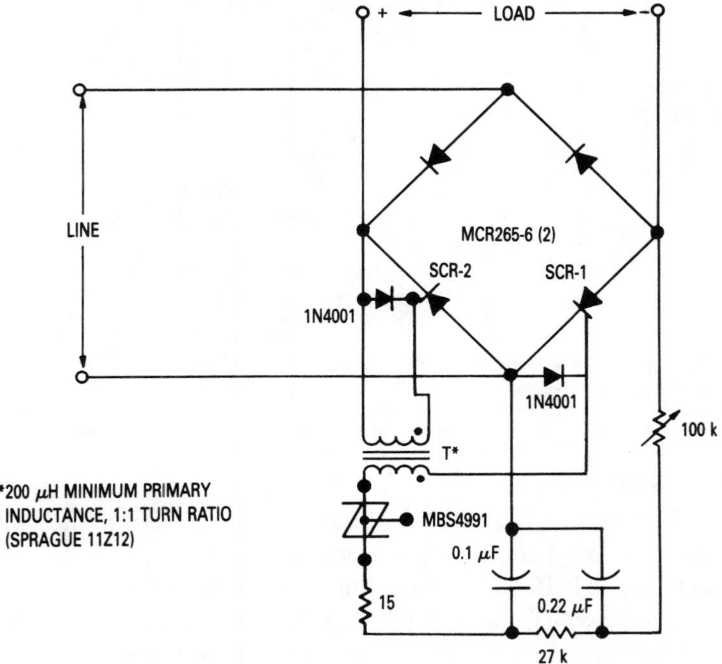

SCR control circuit with dc output that uses an SBS

Fig. 8-26 This circuit is similar to that of Fig. 8-25, except that a dc output is provided to the load. With MCR3818-4 controlled rectifiers mounted on a suitable heatsink, this circuit will control up to 3-kW power from a 120-V line. Motorola Thyristor Device Data, 1991, p. 1-6-55.

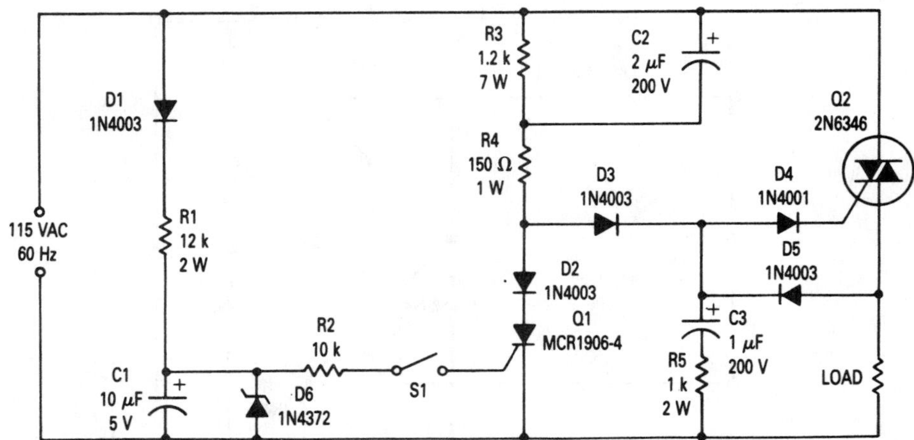

Basic triac zero-point switch

Fig. 8-27 This circuit shows a manually controlled zero-point switch that is useful in power control for resistive loads. Q2 turns on near zero on both the positive and negative half-cycles of the line input. When S1 is closed, Q1 turns on, shunts gate current away from Q2, and keeps Q2 from turning on during either half cycle. The 2N6346 triac shown will handle resistive loads up to 8 A. Motorola Thyristor Device Data, 1991, p. 1-6-55.

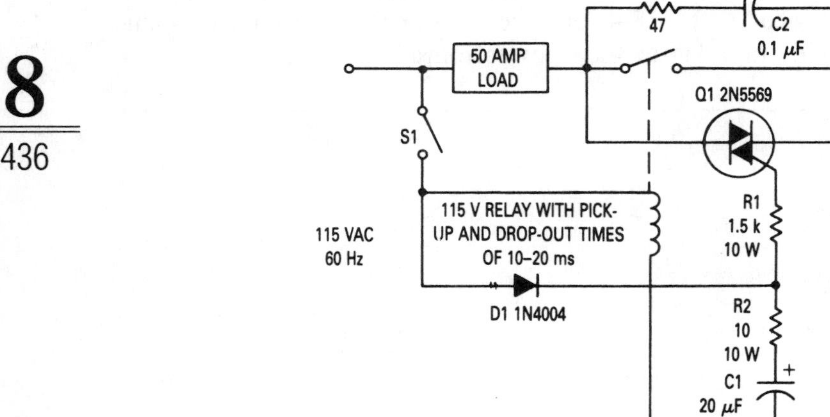

Relay-contact protection that uses a triac

Fig. 8-28 This circuit prevents relay-contact arcing for loads up to 50 A (many 5-A relays can be used for a 50-A load with this circuit). Using the values shown, triac Q1 turns on before the relay closes (when S1 is closed) and remains on after the relay opens. This minimizes arcing (and contact "bounce"), even though the load current passes through the relay contacts. R3 and C1 act as a snubber to reduce d_v/d_t if any other switching element is used on the line (and thus prevents Q1 from being turned on by transients). Motorola Thyristor Device Data, 1991, p. 1-6-57.

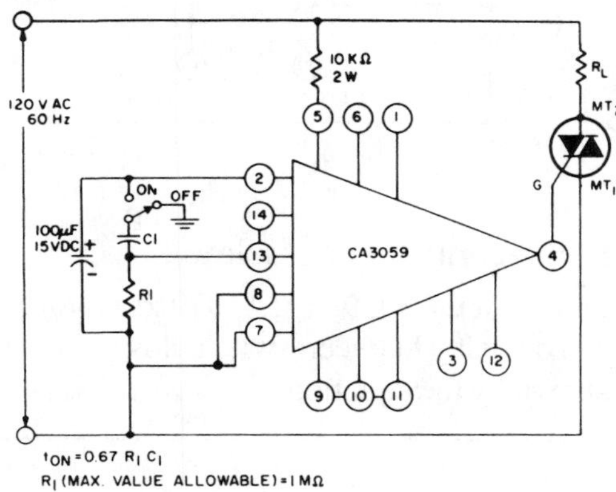

Line-operated one-shot timer

Fig. 8-29 This circuit uses a CA3059 zero-voltage switch to control triac operation. The turn-on time of the triac (and thus the current through load R_L) is set by the values of R and C, as shown by the equations. Harris Semiconductor, Linear & Telecom ICs, 1991, p. 2-35.

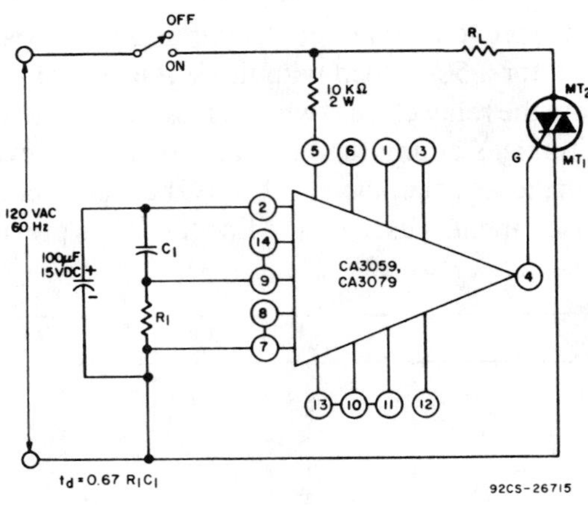

Line-operated thyristor-control time delay

Fig. 8-30 This circuit uses a CA3059 or CA3079 zero-voltage switch to control turn-on time of a triac. The delay between switch closure and turn-on is set by the values of R and C, as shown by the equation. Harris Semiconductor, Linear & Telecom ICs, 1991, 2-35.

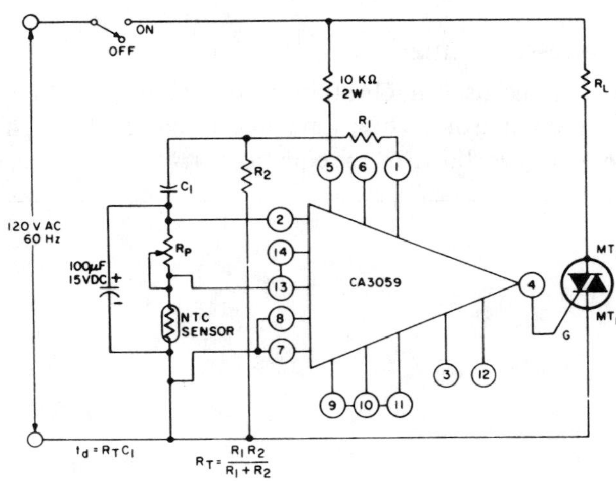

On/off temperature control with delayed turn on

Fig. 8-31 This circuit uses a CA3059 zero-voltage switch to control operation of a triac. The delay between switch closure and turn-on is set by the values of R_1, R_2, and C_1, as shown by the equations. Harris Semiconductor, Linear & Telecom ICs. 1991, p. 2-35.

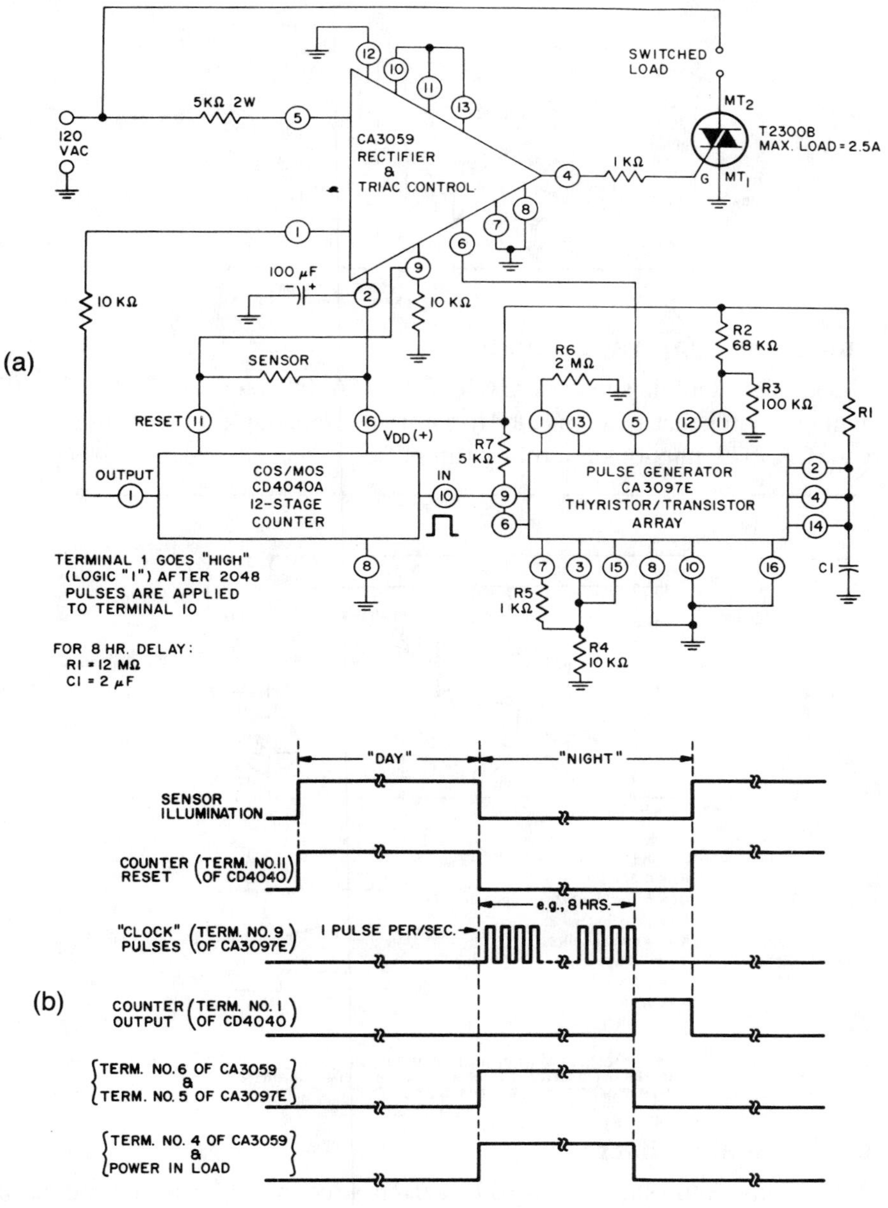

(a)

TERMINAL 1 GOES "HIGH" (LOGIC "1") AFTER 2048 PULSES ARE APPLIED TO TERMINAL 10

FOR 8 HR. DELAY:
R1 = 12 MΩ
C1 = 2 μF

(b)

Line-operated IC timer for long time periods

Fig. 8-32 This circuit provides turn-on to a 2.5-A load after a long time period. The delay is set by the values of R_1 and C_1, as shown by the equations. The timing diagram is shown in Fig. 8-32B. Harris Semiconductor, Linear & Telecom ICs. 1991, p. 2-35.

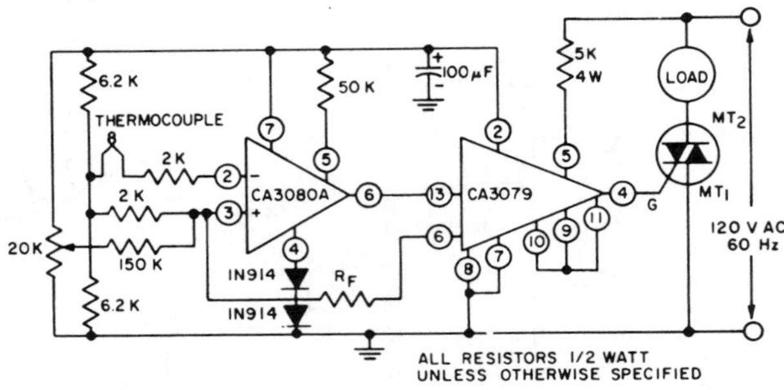

Thermocouple temperature control

Fig. 8-33 This circuit shows a CA3080A OTA (chapter 11) and a CA3079 zero-voltage switch that are connected to form a thermocouple temperature control, where the CA3079 functions as an output amplifier. Harris Semiconductors, Linear & Telecom ICs, 1991, p. 3-69.

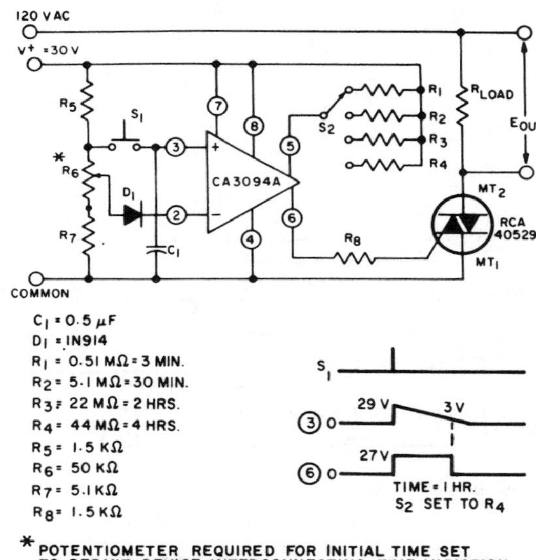

Presettable analog timer

Fig. 8-34 This circuit shows a CA3094A and triac connected to form a presettable timer. The time is preset by the selected values of R_1 through R_4. Harris Semiconductors, Linear & Telecom ICs, 1991, p. 3-78.

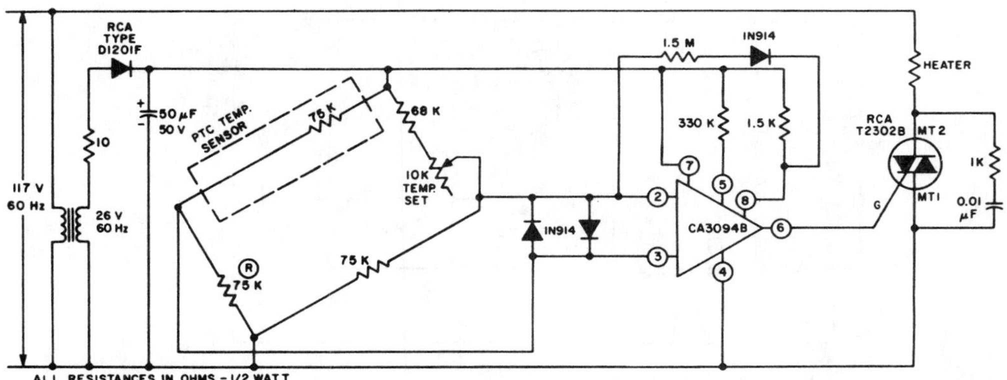

Temperature controller

Fig. 8-35 This circuit shows a CA3094B and triac that are connected to form a temperature controller. Harris Semiconductors, Linear & Telecom ICs, 1991, p. 3-80.

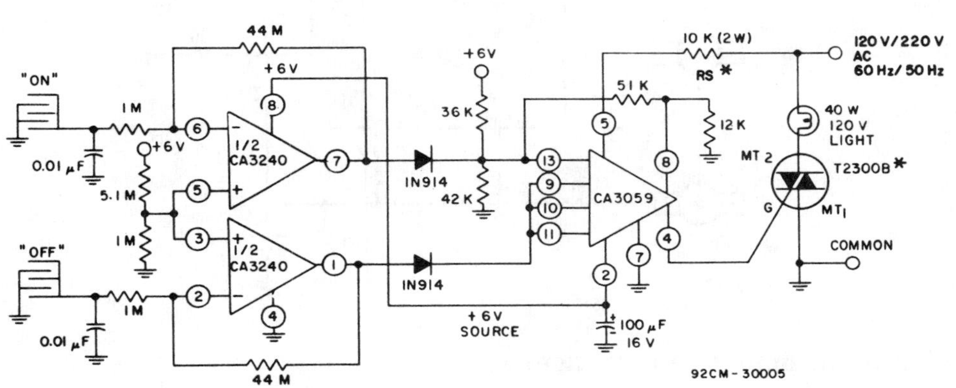

On/off touch switch

Fig. 8-36 This circuit uses a CA3240E to sense small currents flowing between contact points on a touch plate, which consists of a PC board metalization "grid." When the On plate is touched, current flows between the two halves of the grid, and causes a positive shift in the output voltage (pin 7) of the CA3240E. These positive transitions are fed into the CA3079, which is used as a latching circuit and zero-crossing triac driver. When pin 7 of the CA3240 is positive, the triac and lamp are on. The opposite occurs when the Off plate is touched, and pin 1 of the CA3240 is positive. Harris Semiconductors, Linear & Telecom ICs, 1991, p. 3-159.

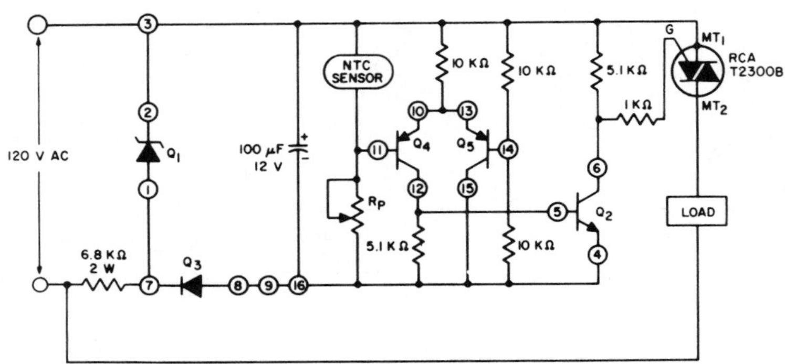

Line-operated level switch

Fig. 8-37 This circuit uses a CA3096 or CA3096A transistor array to control a triac. Harris Semiconductor, Linear & Telecom ICs. 1991, p. 7-41.

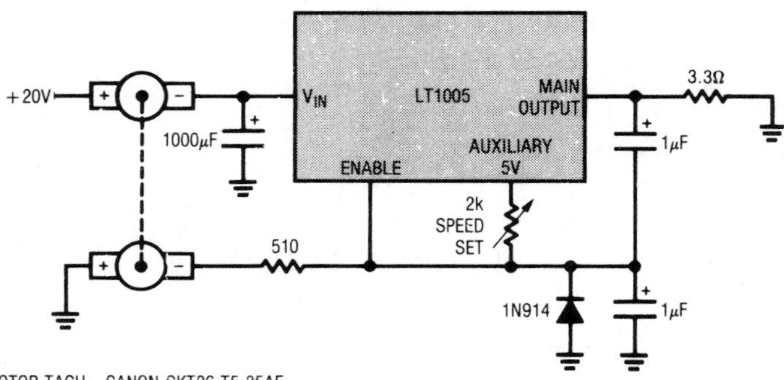

Control of higher-voltage motors

Fig. 8-38 This circuit shows how an LT1005 multifunction regulator can be used to control operation of higher-voltage motors. The circuit is similar to that of Fig. 8-6, except that a 1000-µF capacitor is placed at the regulator to filter transients that are generated by motor switching. When the tach output calls for power, the LT1005 comes on, and allows current to flow through the motor. This forces the LT1005 input toward ground for the duration of the turn-on time. The advantage of this circuit over that of Fig. 8-6 is that higher-voltage motors (up to 20 V) can be used. Linear Technology Corporation, Linear Applications Handbook, 1991, p. AN1-8.

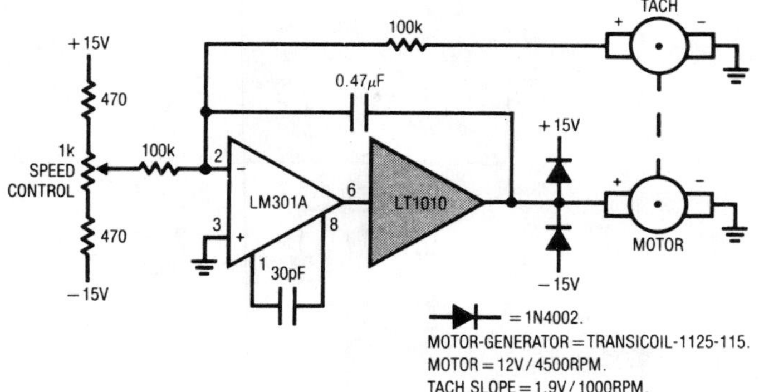

Overload-protected motor-speed controller

Fig. 8-39 With this circuit, the tach signal is fed back and compared to a reference current, with the LM301A closing the control loop. The 0.47-μF capacitor provides stable compensation. Because the tach output is bipolar, the speed is controllable in both directions, with clean transitions through zero. The LT1010 thermal protection prevents device destruction in the event of mechanical overload or malfunction. Linear Technology Corporation, Linear Applications Handbook, 1991, p. AN4-6.

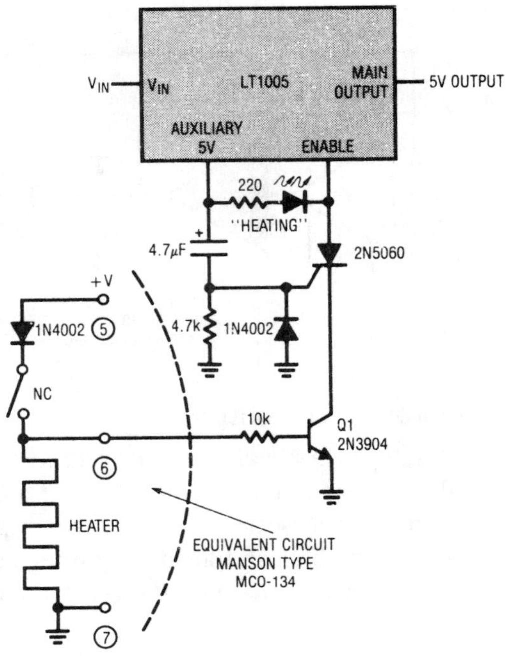

Crystal-oven control circuit

Fig. 8-40 This circuit combines an LT1005 multifunction regulator and a typical commercial crystal oven to prevent equipment use until oven temperature has stabilized (as is often required in precision test equipment). When power is applied, pin 6 of the crystal oven is high, which biases Q1 on. Simultaneously, the SCR gate is triggered by the current flowing through the 4.7-kΩ resistor (the 4.7-μF capacitor charge current). This turns the SCR on and allows the "load" (the LT1005 enable pin) to receive power. The 4.7-kΩ resistor eliminates false SCR triggering, and the diode suppresses reverse gate current when LT1005 input power is removed. Linear Technology Corporation, Linear Applications Handbook, 1991, p. AN1-6.

Programmable ultra-accurate line-operated timer and load control

Fig. 8-41 This circuit provides precise current through a load, as determined by a pulsed triac. The circuit is programmable over the range from 0.5333 s to 2 min and 16 s, in 0.5333-s intervals. Figure 8-41B shows the connections for the various time intervals, and Fig. 8-41C shows the timing diagram. Harris Semiconductor, Linear & Telecom ICs, 1991, p. 2-36, 2-37.

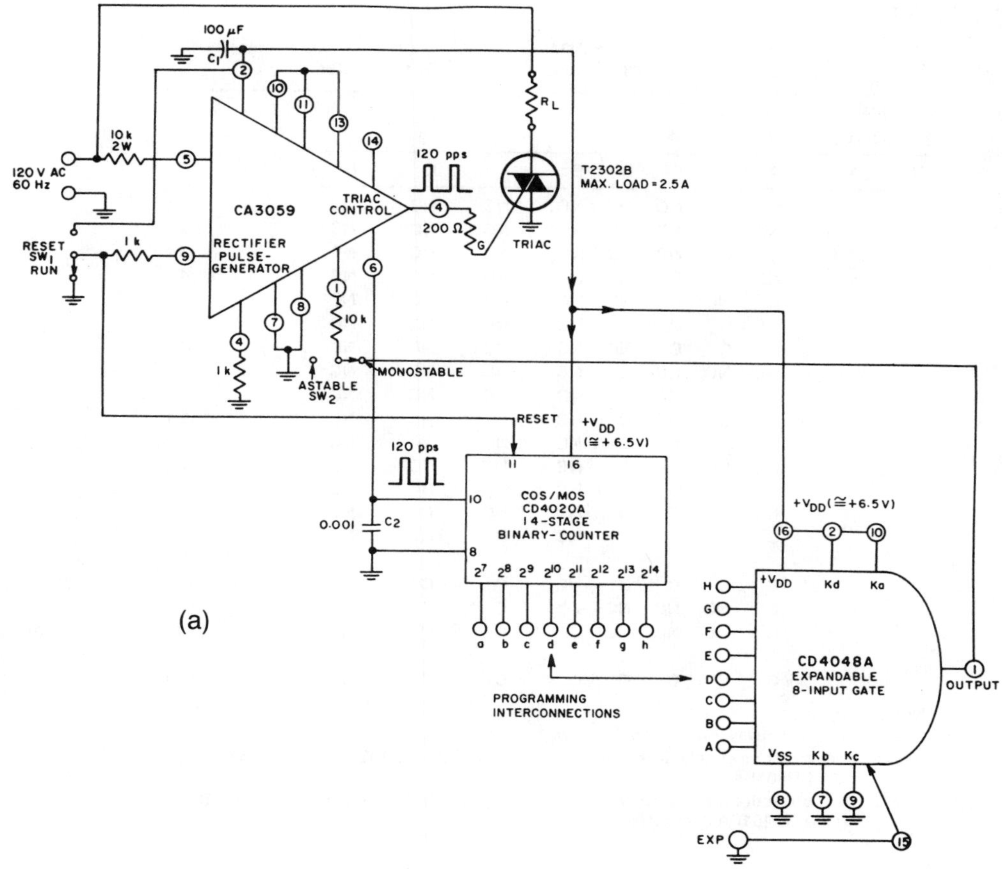

100 µF
C_1
120 V AC
60 Hz
10 k
2 W
RESET
SW_1
RUN
1 k
CA3059
RECTIFIER
PULSE-
GENERATOR
TRIAC
CONTROL
120 pps
200 Ω
G
R_L
T2302B
MAX. LOAD = 2.5 A
TRIAC
10 k
1 k
ASTABLE
SW_2
MONOSTABLE
RESET
$+V_{DD}$
(≅ + 6.5 V)
120 pps
0.001
C2
COS/MOS
CD4020A
14-STAGE
BINARY-COUNTER
2^7 2^8 2^9 2^{10} 2^{11} 2^{12} 2^{13} 2^{14}
a b c d e f g h
PROGRAMMING
INTERCONNECTIONS
$+V_{DD}$ (≅ +6.5 V)
$+V_{DD}$ Kd Ka
H G F E D C B A
CD4048A
EXPANDABLE
8-INPUT GATE
OUTPUT
V_{SS} Kb Kc
EXP

(a)

Continued

(b)

Time Periods (t = 0.5333 s)	1 t	2 t	4 t	8 t	16 t	32 t	64 t	128 t	t_o
Terminals									
CD4020A	a	b	c	d	e	f	g	h	
CD4048A	A	B	C	D	E	F	G	H	
	C	NC	NC	NC	NC	NC	NC	NC	1 t
	NC	C	NC	NC	NC	NC	NC	NC	2 t
	C	C	NC	NC	NC	NC	NC	NC	3 t
	NC	NC	C	NC	NC	NC	NC	NC	4 t
	C	NC	C	NC	NC	NC	NC	NC	5 t
	NC	C	C	NC	NC	NC	NC	NC	6 t
	C	C	C	NC	NC	NC	NC	NC	7 t
	NC	NC	NC	C	NC	NC	NC	NC	8 t
	C	NC	NC	C	NC	NC	NC	NC	9 t
	NC	C	NC	C	NC	NC	NC	NC	10 t
	C	C	NC	C	NC	NC	NC	NC	11 t
	NC	NC	C	C	NC	NC	NC	NC	12 t
	C	NC	C	C	NC	NC	NC	NC	13 t
	NC	C	C	C	NC	NC	NC	NC	14 t
	C	C	C	C	NC	NC	NC	NC	15 t
	C	C	C	C	NC	C	C	NC	111 t
	NC	NC	NC	NC	C	C	C	NC	112 t
	C	NC	NC	NC	C	C	C	NC	113 t
	C	C	C	C	C	C	C	C	255 t

Notes:

t_o = Total time delay = $n_1 t + n_2 t + \ldots n_n t$.

C = Connect. For example, interconnect terminal a of the CD4020A and terminal A of the CD4048A.

NC = No Connection. For example, terminal b of the CD4020A open and terminal B of the CD4048A connected to $+V_{DD}$ bus.

(c)

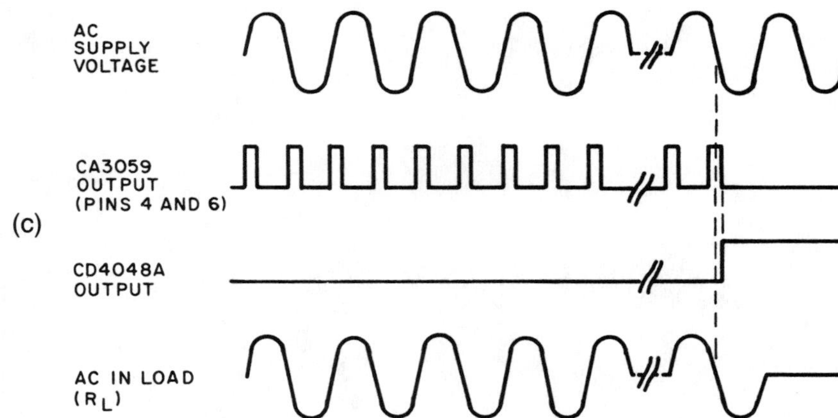

9

UJT, PUT, and optoelectronic circuits

It is assumed that you are already familiar with the devices featured in this chapter. However, the following paragraphs summarize this information, as well as testing and troubleshooting for the related circuits. This information is included so that even those readers who are not familiar with electronic procedures can both test the circuits in this chapter, and localize problems if the circuits fail to perform as shown.

Circuit testing and troubleshooting

Figure 9-2 shows a classic UJT circuit, where the UJT is connected as a relaxation oscillator, which produces a trigger for an SCR (chapter 8). The oscillator frequency is controlled by resistance R_T and capacitance C_T. On each positive half-cycle of the ac line voltage, current flows through C_T, R_T and R_D, charging C_T at a rate determined by the setting of R_T. When the charge on C_T reaches the turn-on voltage, the UJT fires and current passes through R_{B1}, the UJT, R_{B2}, and R_D. This produces a pulse across R_{B1}, and triggers the SCR. Current is then applied to the load through the SCR, on each half cycle.

The UJT and SCR are synchronized since both devices receive power from the same source. When the UJT oscillator frequency is increased (by adjustment of

R_T), the SCR fires earlier in the half cycle, and more power is applied to the load. The opposite occurs when the oscillator frequency is reduced by adjustment of R_T.

Figure 9-22 shows a classic PUT (programmable UJT) circuit where the PUT Q1 is connected as a relaxation oscillator, which produces triggers for two SCRs (connected as an SCR flip-flop). The oscillator frequency is controlled by the values of C_1 and R_1. The firing point of Q1 is programmed by a voltage that is applied to the gate. In this circuit, the programming voltage is supplied by the R3/R4 voltage divider. The programming voltage can also be variable (Fig. 9-23A). Notice that the firing point of the PUT is set by the programming voltage, but the oscillator frequency is set by the timing capacitor C1. Of course, if the programming voltage is set to some higher value, without changing C_1, the oscillator frequency will decrease.

Figures 9-45 through 9-55 show some classic optocoupler or optoisolator circuits. For example, in the circuit of Fig. 9-45, the 4N26 is used to trigger an SCR. When power is applied across pins 1 and 2 of the 4N26, the LED within the sealed unit produces light. This light is applied to the phototransistor in the 4N26 and causes the phototransistor to turn on. This, in effect, causes pin 4 to be connected to pin 5. Current flows through the 1-kΩ resistor at pin 4, which produces a pulse that triggers the SCR. In the circuits of Figs. 9-36 through 9-38, the MOC3009-12 optoisolators contain a driver circuit in addition to the LED and a phototransistor.

Figures 9-30 through 9-35 show some classic fiber-optic receiver and transmitter or driver circuits. Figure 9-34B shows typical fiber-optic termination and assembly. Fiber-optic transmitters (both visible light and infrared) are essentially LEDs that produce light when power is applied. This light is passed through the fiber-optic cable to the receiver, which is essentially a photo-sensitive diode or photodiode. The light can be varied, pulsed, or modulated (as required), and it produces a corresponding output at the receiver. For example, an RF signal applied at C1 in Fig. 9-30 will appear as RF at the output points in Fig. 9-31, after passing through a fiber-optical cable between the two circuits.

UJT testing and troubleshooting

Because most UJTs are used as relaxation oscillators, the obvious test is to check that the circuit is oscillating. For example, in the circuit of Fig. 9-1, you can monitor at the UJT emitter and base-1. The emitter should show a curving ramp pattern (typically), while base-1 (across R_{B1}) should show a pulse. Both the ramp and pulse frequency should change when R_T is varied. If not, look for defective parts and/or improper wiring. Any leakage in timing capacitor C_T impairs operation of the oscillator. If there is excessive leakage, the circuit might not oscillate. If there are pulses at base-1 of the UJT, the problem is likely in the SCR or power wiring. Of course, it is possible that the pulse amplitude is not sufficient to trigger the SCR.

PUT testing and troubleshooting

Testing and troubleshooting for a PUT circuit is essentially the same as for a UJT, except for the programming voltage. For example, in the circuit of Fig. 9-22, you can monitor at the Q1 anode and cathode. The anode should show a curving ramp pattern (typically), while the cathode (across R2) should show a pulse. The frequency of oscillation should be constant because R1 is fixed, as is the programming voltage that is applied to the Q1 gate from R3 and R4.

If Q1 in Fig. 9-22 does not oscillate, or if oscillation is at the wrong frequency, suspect C1. Also, check the programming voltage that is applied to Q1. The programming voltage should be about 2 V (if the circuit is operated from a 3-V battery) with the values shown for R_3 and R_4. If the programming voltage is drastically different from 2 V, the point at which Q1 turns on will be shifted, even though R1 and C1 are good (and at the right value). If the programming voltage is shifted far enough from 2 V, it is possible that Q1 might not oscillate. If pulses are at the cathode of Q1 (across R2), the problem is likely in the SCR circuits. Of course, it is possible that the pulse amplitude is not sufficient to trigger both SCRs.

Optocoupler/optoisolator testing and troubleshooting

The obvious test for optocoupler/optoisolator circuits is to apply power to the LED and see if the phototransistor turns on. For example, in the circuit of Fig. 9-45, apply a 5-mA current to pins 1 and 2 of the 4N26, and check that a signal appears across the 1-kΩ resistor at pin 4 (if a 5-mA pulse is applied to the LED, a pulse should appear across the resistor at pin 4).

If no output is at pin 4, start by checking that pin 5 is connected, to +V (typically 5 V). If the power is present, and a 5-mA signal is applied to pins 1 and 2, but there is no output at 4, suspect the 4N26. If there is an output at pin 4, the problem is likely in the SCR circuits (unless the pulse amplitude is not sufficient to trigger the SCR).

Fiber-optic testing and troubleshooting

Testing and troubleshooting for the fiber-optic circuits of this chapter is similar to that for optocouplers/optoisolators, except that the light is transmitted over a fiber-optic cable (rather than through a sealed optocoupler). For example, you can apply a TTL signal (5-V pulse) to the input of the circuit in Fig. 9-32, and check for a TTL output at U1 in the circuit of Fig. 9-33. The fiber-optic LED of Fig. 9-32 must be connected to the fiber-optic receiver diode of Fig. 9-33 through a fiber-optic cable (assembled as shown in Fig. 9-34B).

If there is no output at U1 in Fig. 9-33, try correcting the problem by adjustment of the 20-kΩ pot at the noninverting input to U1. If this does not cure the problem, recheck the wiring and all parts. If practical, try substituting either the transmitter or the receiver circuit (but not both). Also, make certain that the fiber-optic diodes and cable are properly assembled (cable properly trimmed, core tip seated against the lens, locking nut tightened for a snug fit, etc.).

In the case of the receiver circuits (Fig. 9-33), it is sometimes practical to apply light to the fiber-optic diode and see if an output is produced at U1 (of course, if the diode is infrared, an infrared light must be used). If the diode responds to visible light, and is mounted in an assembly similar to that of Fig. 9-34B, remove the locking nut and cable. Then, expose the diode to light and check if there is any change at the base and collector of Q1, and at the base and emitter of Q2 (alternately cover and uncover the assembly opening).

If there is no change at Q1 and Q2, with the diode alternately exposed to light and dark, suspect the diode. If the emitter of Q2 changes when the diode is exposed to light, but there is no output at U1, suspect U1.

UJT, PUT, and optoelectronic circuits

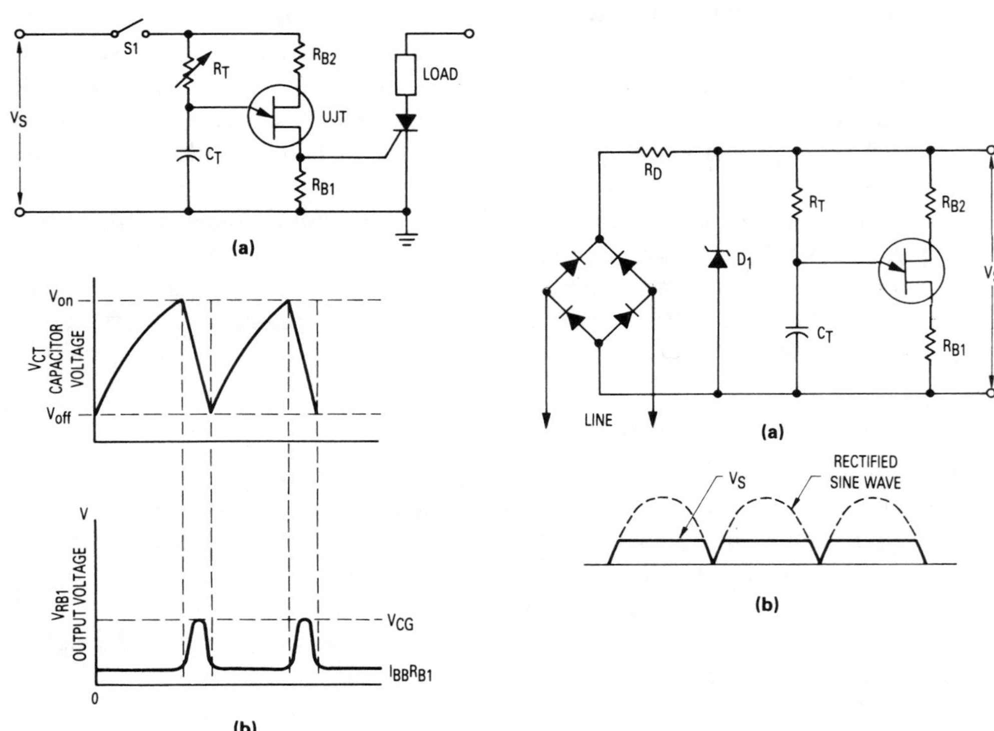

Basic UJT relaxation oscillator

Fig. 9-1 This figure shows the circuit and corresponding waveforms for a unijunction transistor oscillator (Fig. 5-41). When V_S is pure dc, the oscillator is free-running and R_T/C_T determine the frequency of oscillation. Figure 9-1B shows the basic relaxation oscillator, where the output pulses are synchronized to the line-voltage zero-crossing points. Typical values (for a UJT with a base-to-base voltage of 4 to 9 V) are: $R_{B1} = 40$ to $100\ \Omega$, R_{B2} $1\ k\Omega$, $C_T = 0.1\ \mu F$, $R_T = 100\ k\Omega$.
Motorola Thyristor Device Data, 1991, p. 1-6-10,-11.

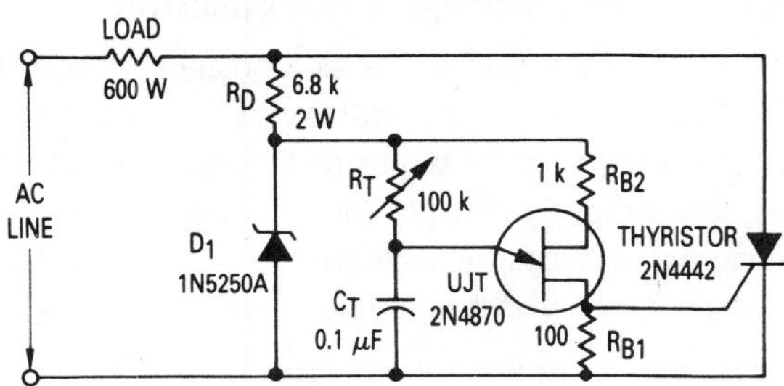

Half-wave thyristor control

Fig. 9-2 This circuit shows a UJT that is a half-wave trigger for a thyristor (an SCR), which controls a 600-W load. Motorola Thyristor Device Data, 1991, p. 1-6-11.

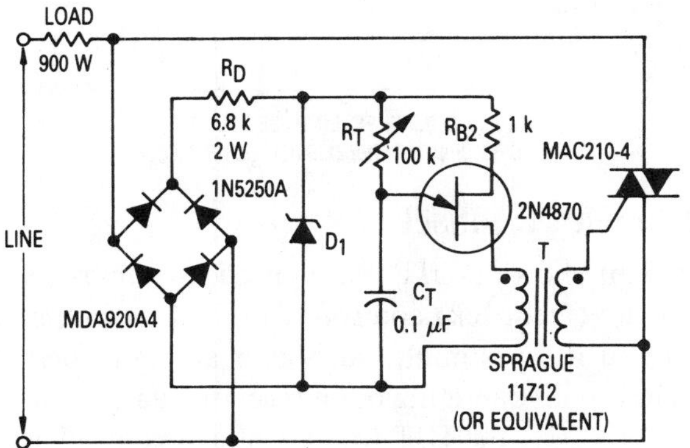

Full-wave thyristor control

Fig. 9-3 This circuit shows a UJT that is used as a full-wave trigger for a thyristor (a triac), which controls a 900-W load. Motorola Thyristor Device Data, 1991, p. 1-6-11.

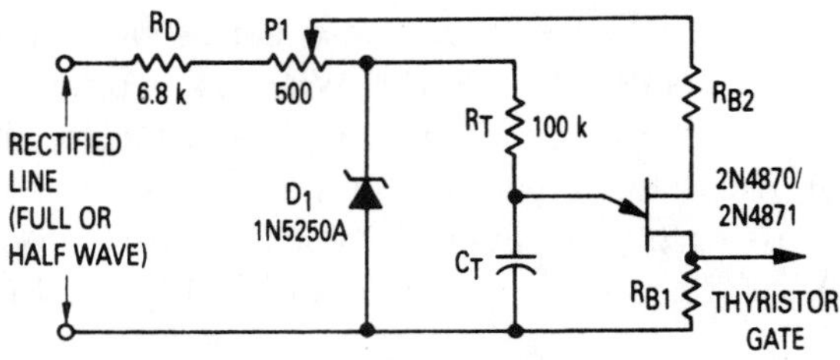

Thyristor control with line-voltage compensation

Fig. 9-4 This circuit shows a UJT used as a trigger, where the thyristor must provide a constant output voltage, regardless of line-voltage changes. As line voltage increases, the voltage on the wiper of P1 increases (as does the UJT peak voltage), which increases the C_T charge. With a higher charge on C_T, the circuit takes more time to trigger, reduces the thyristor conduction angle and maintains the average voltage at a reasonably constant value. Motorola Thyristor Device Data, 1991, p. 1-6-12.

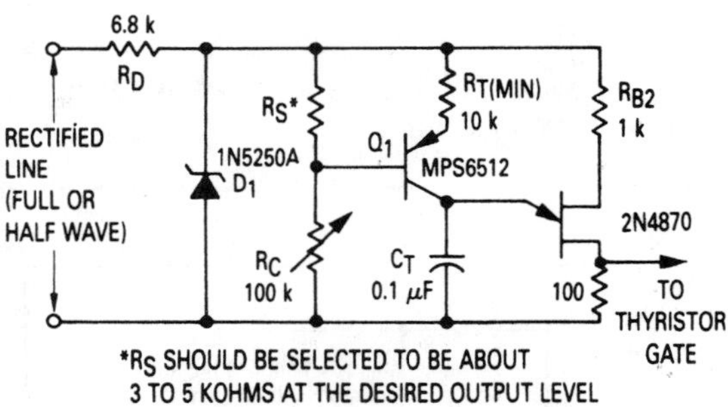

Thyristor control with feedback

Fig. 9-5 This circuit shows a UJT that is used as a thyristor trigger with both manual and automatic (feedback) control. The feedback-sensing resistor R_S can respond to any one of many stimuli, such as heat, light, pressure, moisture, or magnetic fields. The circuit operating point is set manually by R_C. As R_S increases, more current flows into C_T, the UJT triggers at a smaller phase angle, and more power is applied to the load. Thus, for this circuit, R_S must decrease in response to excessive power in the load. If R_S increases with load power, then R_S and R_C must be interchanged. Motorola Thyristor Device Data, 1991, p. 1-6-12.

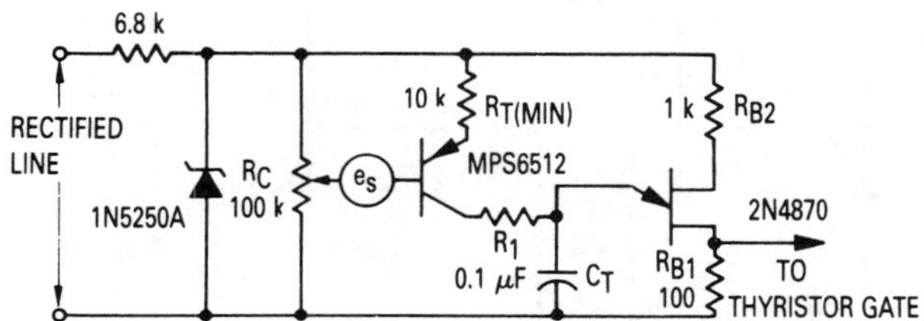

Thyristor control with voltage feedback

Fig. 9-6 This circuit shows a UJT that is used as a thyristor trigger (with feedback), where the quantity to be sensed is in the form of an isolated feedback dc voltage (such as a tachometer output voltage). The feedback voltage is applied between voltage divider R_C and the base of Q1. R_C sets the operating point, and thus provides some manual control. Motorola Thyristor Device Data, 1991, p. 1-6-12.

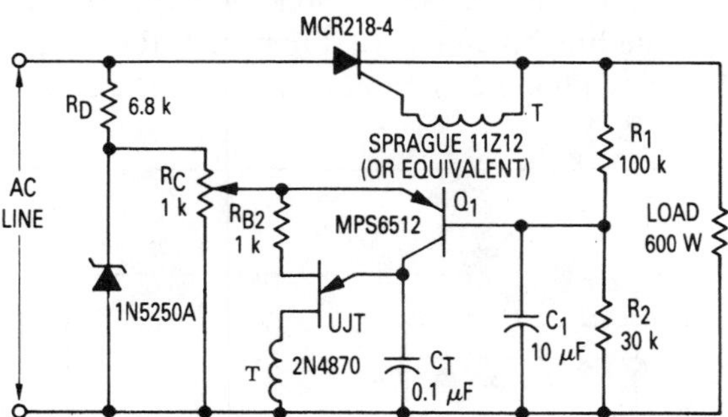

Half-wave thyristor control with average-voltage feedback

Fig. 9-7 This circuit shows a UJT used as a thyristor trigger (with feedback), where the average load voltage is the desired feedback variable. R1, R2, and C1 average the load voltage so that the voltage can be compared with the set point that is determined by R_C. Motorola Thyristor Device Data, 1991, p. 1-6-12.

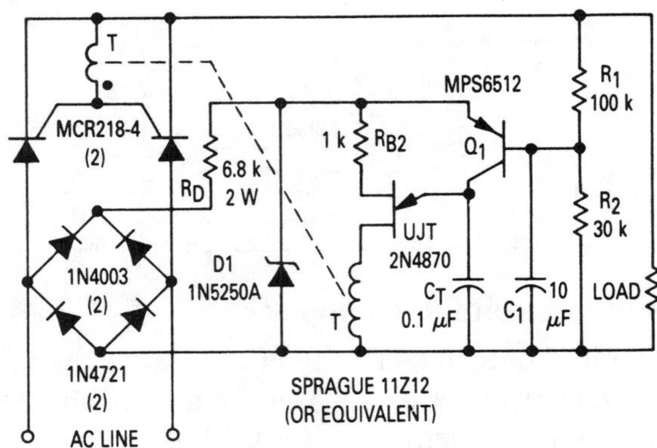

Full-wave thyristor control with average-voltage feedback

Fig. 9-8 This circuit shows a UJT that is used as a thyristor trigger (with feedback), where the average full-wave load voltage is the desired feedback variable. Notice that the line voltage is first rectified. R1, R2, and C1 average the load voltage so that the voltage can be compared with the rectified line voltage. Motorola Thyristor Device Data, 1991, p. 1-6-12.

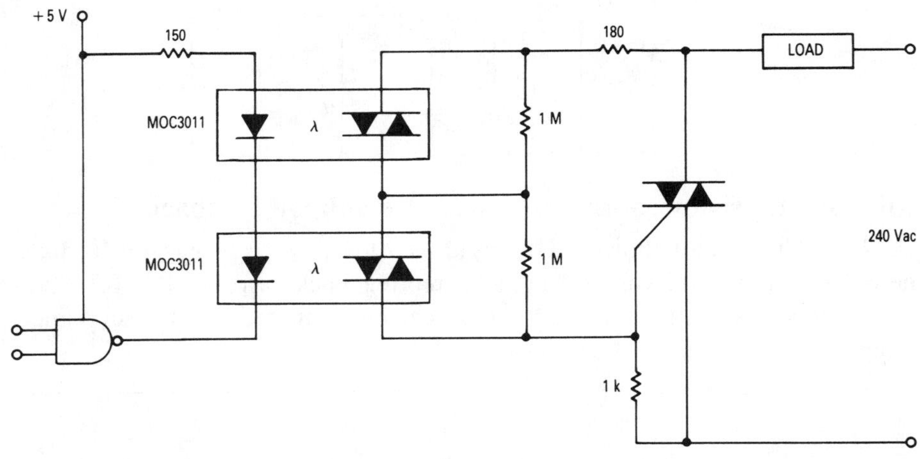

Nonzero-crossing optically isolated triac drivers

Fig. 9-9 This circuit shows two MOC3011 optoisolators used to drive a 240-V triac-controlled load—even though the individual MOC3011 voltage rating is not sufficiently high to be used directly on a 240-V line. The two l-MΩ resistors equalize the voltage on the MOC3011s. Motorola Thyristor Device Data, 1991, p. 1-6-13.

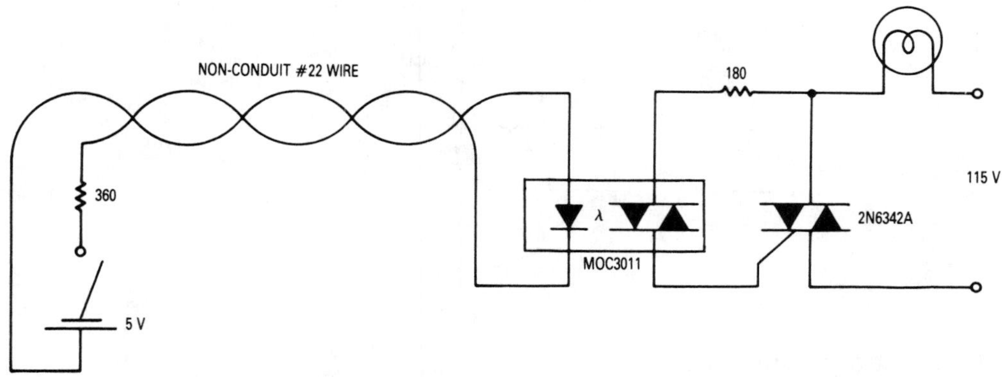

Remote control of ac voltage

Fig. 9-10 This circuit shows an MOC3011 optoisolator used for remote control of ac voltages. Local building codes frequently require that all 115-V light-switch wiring be enclosed in conduit. With this circuit, it is possible to control a large lighting load from a long distance through low-voltage (5 V) signal wiring, which is completely isolated from the ac line. Such wiring usually is not required to be in conduit, so the cost savings (especially in large commercial or residential buildings) can be considerable. Although a lighting load is shown, the load can be a motor, fan, pool pump, etc. Motorola Thyristor Device Data, 1991, p. 1-6-13.

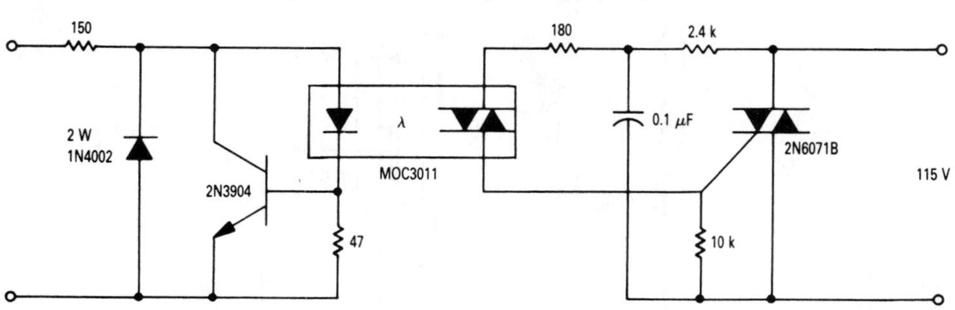

Solid-state relay

Fig. 9-11 This circuit shows a complete, general-purpose, solid-state relay, snubbed for 115-V inductive 5-V loads. The 5-V input is protected by the MOC3011 optoisolator. The output is snubbed by the 2.4-kΩ resistor and 0.1-μF capacitor. Motorola Thyristor Device Data, 1991, p. 1-6-14.

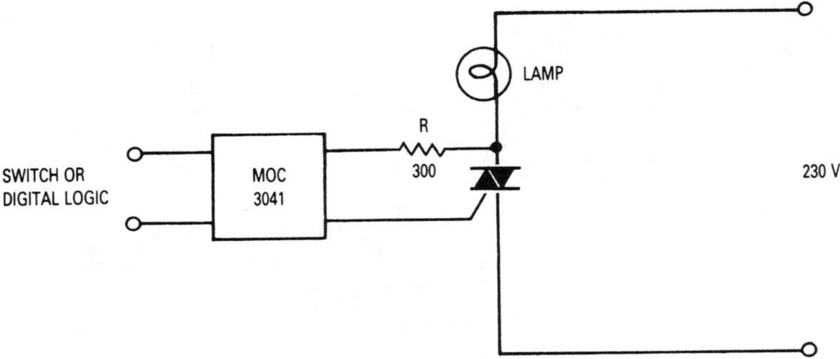

High-wattage lamp control

Fig. 9-12 Many high-wattage incandescent lamps suffer shortened lifetimes when switched on at ac line voltages other than zero. This is because a large inrush of current destroys the filament. A simple solution to this problem is to use an optoisolator and triac, as shown. The triac must be capable of handling the lamp current. The MOC3041 optoisolator can be controlled from a switch or some form of 5-V digital logic. The minimum value of R is determined by the maximum surge current rating of the MOC3041 (I_{TSM}), where: $R_{min} = V_{in(pk)}/I_{TSM}$. With a 230-V line, and a 1.2-A I_{TSM} for the MOC3041, the minimum value of R is: 340 V/1.2A $= 283\ \Omega$. Motorola Thyristor Device Data, 1991, p. 1-6-17.

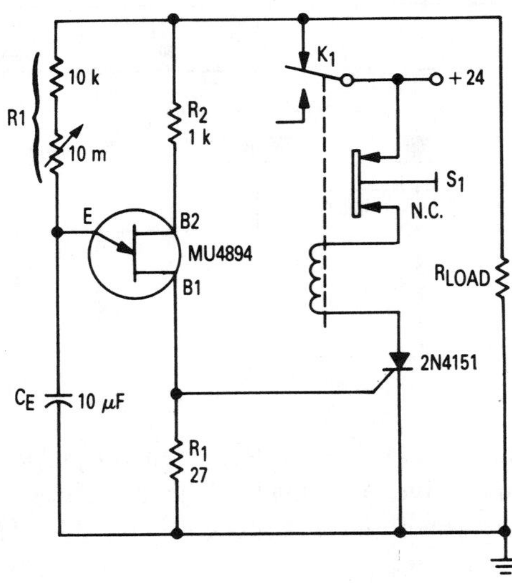

Simple time delay

Fig. 9-13 This circuit shows the basic UJT building block (Fig. 9-1) that is used to provide a simple time-delay function. When normally closed S1 is pushed, the

SCR turns off, K1 is de-energized, and power is applied to the UJT relaxation oscillator and load. After a time delay that varys from less than 1 s to about 2.5 min. (as determined by the setting of the 1-MΩ pot), the UJT fires and turns on the SCR. K1 is energized and power is removed from the UJT and load. K1 stays energized until S1 is pushed again. Motorola Thyristor Device Data, 1991, p. 1-6-39.

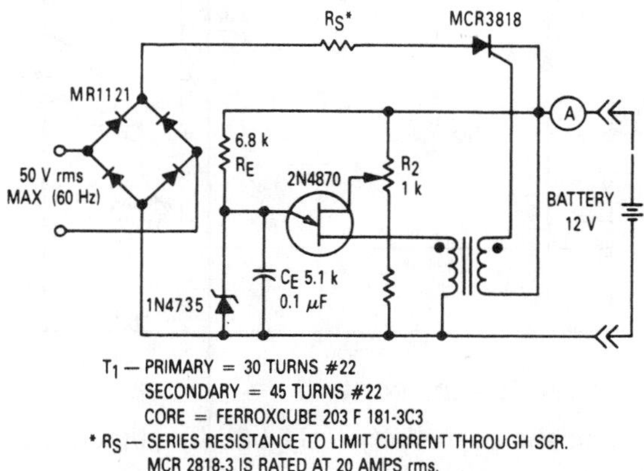

12-V Battery-charger control

Fig. 9-14 This circuit shows the basic UJT building block (Fig. 9-1) that is used to control a battery-charger system. Notice that the battery voltage controls the charger and, when the battery is fully charged, the charger will not supply current to the battery. The setting of R2 determines the amount of charge. Also, the charger will not work if battery polarity is reversed. Motorola Thyristor Device Data, 1991, p. 1-6-42.

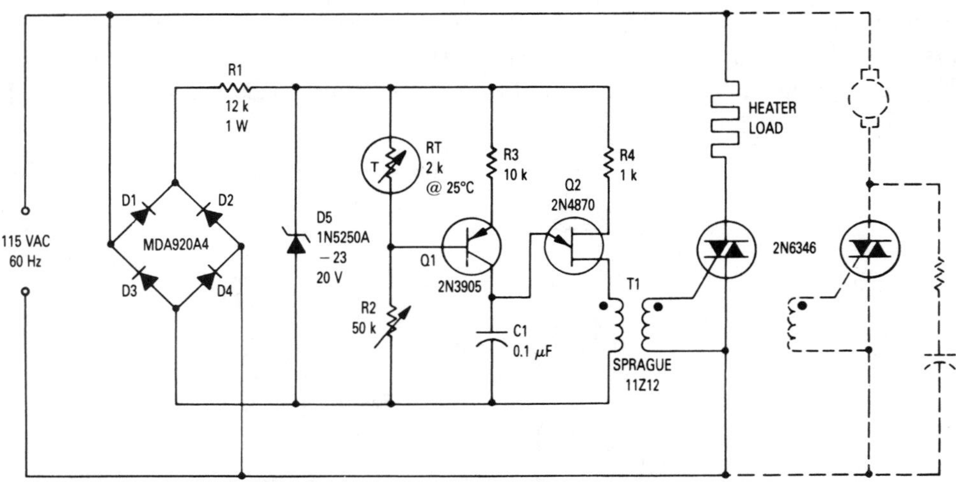

Temperature-sensitive heater control

Fig. 9-15 This circuit shows the basic UJT building block (Fig. 9-1) that is used to control an 800-W heater system. The simple RC circuit of Fig. 9-1 is replaced by Q1, R_T, and R2. Using phase control, the circuit is able to reduce power to the load as the desired temperature is reached, thus eliminating much of the overshoot inherent in mechanical controls. Although the circuit shown is for a heater, the circuit can also be used to control a constant-load motor (such as a blower motor), as indicated by the dotted portion of the diagram. The desired temperature is set by R2. The circuit can also be used for cooling by interchanging R_T and R2. Motorola Thyristor Device Data, 1991, p. 1-6-45.

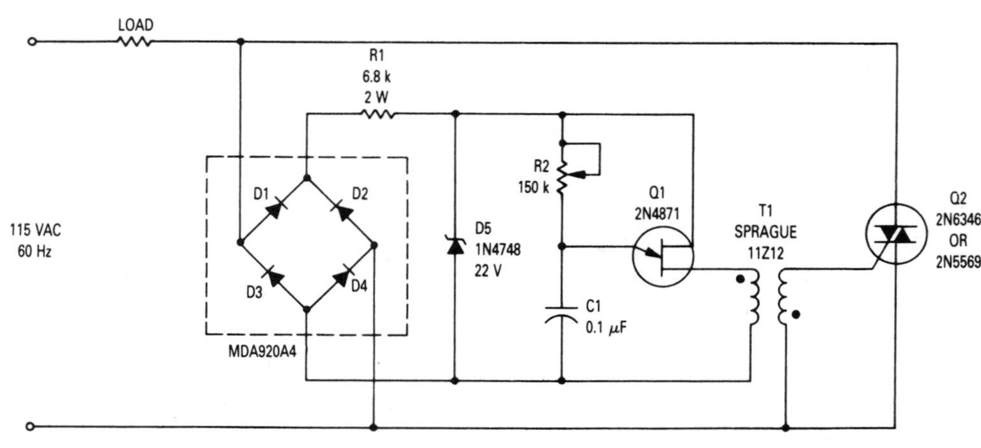

800-W light dimmer

Fig. 9-16 This circuit shows the basic UJT building block (Fig. 9-1) that is used to control power applied to incandescent lights. R2 varies the time constant of the timing circuit, and thus provides phase control for the triac. Power to the lights is controlled by varying the conduction angle of the triac from 0° to 170°. The power available at the 170°-conduction angle is better than 97% of that at the 180° angle.
Motorola Thyristor Device Data, 1991, p. 1-6-45.

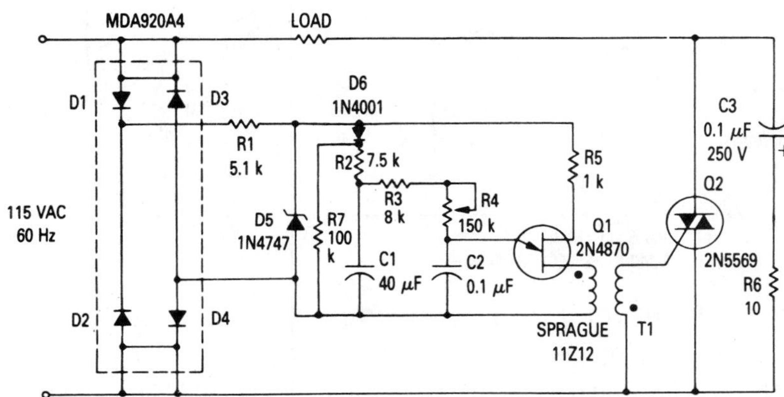

800-W soft-start light dimmer

Fig. 9-17 This circuit shows the basic UJT building block (Fig. 9-1), which is used in a light dimmer with soft-start operation that applies current to the light slowly enough to eliminate high surges (high inrush current). These current surges, found in most cold-filament light dimmers, shorten lamp life. With this circuit, the lamp is heated slowly by a gradually increasing voltage so that inrush current is kept to a minimum. R4 controls the charging rate of C2 and provides the means to control or dim the lamp. Motorola Thyristor Device Data, 1991, p. 1-6-46.

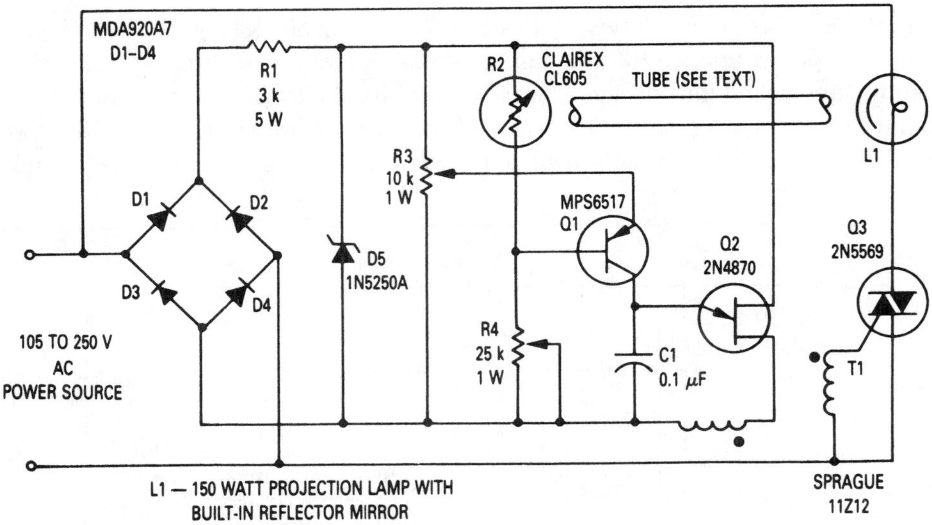

Voltage regulator for a projection lamp

Fig. 9-18 This circuit shows the basic UJT building block (Fig. 9-1), which is used to regulate the voltage across a projection lamp to 100 V (±2%) for an input voltage between 105 and 250 V. This is done by indirectly sensing the light output of lamp L1 and applying the feedback signal to the firing circuit (Q1 and Q2), which controls the conduction angle of triac Q3. The reflector inside the lamp envelope glows red because of the filament heat. Because the reflector has a relatively large mass, the reflector cannot respond to the supply frequency, and the light output provides a form of integration. Photocell R2 is mounted at one end of a black tube, and the other end of the tube is directed at the reflector's back side. R3 and R4 set the lamp voltage to 100 V when the line voltage is 105 and 250 V, respectively. This assures that the lamp voltage will be within the desired tolerance over the operating range of the input voltage. Some interaction occurs between R3 and R4, and the adjustment of each pot might have to be made several times. Because this is an rms voltage regulator, a true rms meter must be used to adjust the load voltages.

Motorola Thyristor Device Data, 1991, p. 1-6-46.

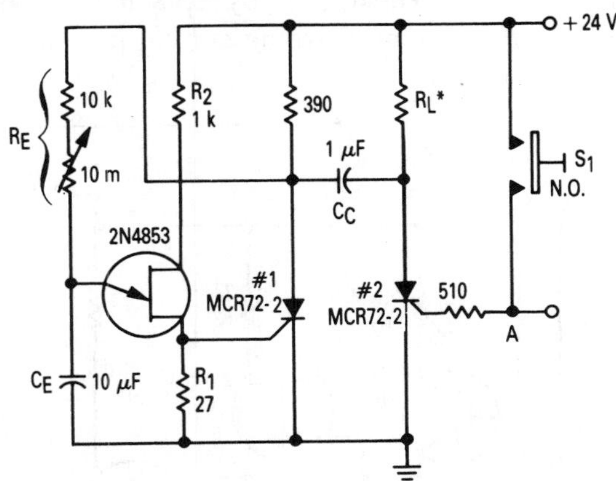

*VALUE OF R_L MUST BE LOW ENOUGH TO ALLOW HOLD CURRENT TO FLOW IN THE SCR.

Solid-state time delay

Fig. 9-19 This circuit shows the basic UJT building block (Fig. 9-1), which is used to provide a time-delay function. The circuit is similar to that of Fig. 9-13, except that the mechanical relay is replaced by an SCR (which generally reduces circuit cost). When S1 is pushed, or when a positive-going pulse is applied at point A, SCR#2 turns on and SCR#1 is turned off by C_C. With SCR#1 off, the supply voltage is applied to R_E and the circuit begins to time. After a period (determined by the setting of R_E), the UJT fires and turns SCR#1 on and SCR#2 off. With the values shown, time delay can be varied from about 1 s to over 2 min. Motorola Thyristor Device Data, 1991, p. 1-6-47.

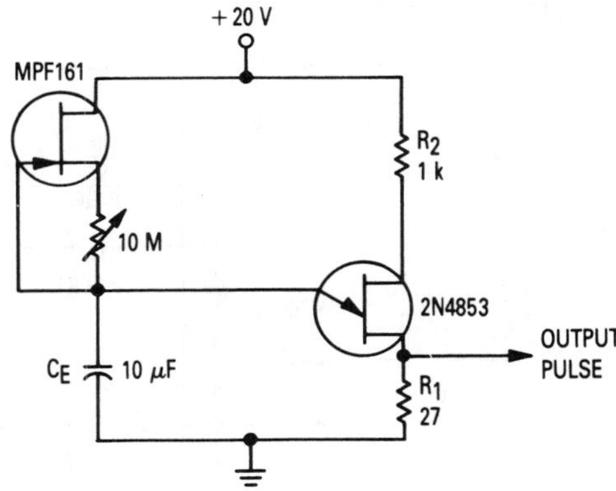

Time delay with constant-charging current

Fig. 9-20 This circuit shows the basic UJT building block (Fig. 9-1), which is used to provide a time-delay function, where the charging current is constant and relatively small (less than 1 μA). This is done by replacing the basic R_E with a JFET and 10-MΩ pot. The 1-μA constant-charging current provides time delays up to about 10 min. Motorola Thyristor Device Data, 1991, p. 1-6-47.

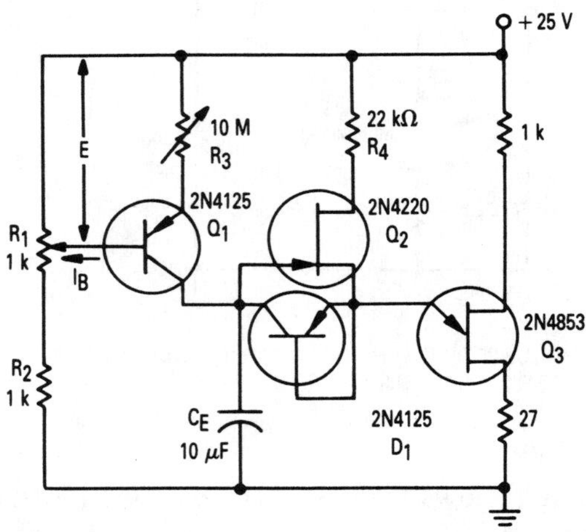

Long-duration time delay

Fig. 9-21 This circuit shows the basic UJT building block (Fig. 9-1), which is used to provide long time delays (up to 10 hours). Q1, R1, R2, and R3 form a constant-current source to charge C_E very slowly (a few nA) to provide the long delay. Q2 and D1 (a 2N4125 connected as a diode) provide a separate discharge path to the emitter of UJT Q3. Notice that there is some interaction between R1 and R3 when setting the time delay. Motorola Thyristor Device Data, 1991, p. 1-6-48.

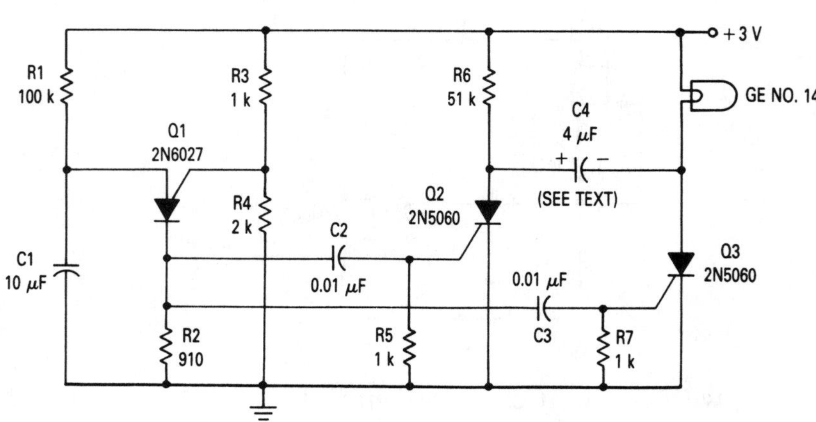

Low-voltage lamp flasher that uses a PUT

Fig. 9-22 This circuit shows a PUT used as a lamp flasher, where the supply is only 3 V. Q1, R1, and C1 form a relaxation oscillator. Q2 and Q3 form an SCR flip-flop. Notice that C4 is not a polarized capacitor. For the values shown, the lamp is on and off for about 0.5 s each. Motorola Thyristor Device Data, 1991, p. 1-6-48.

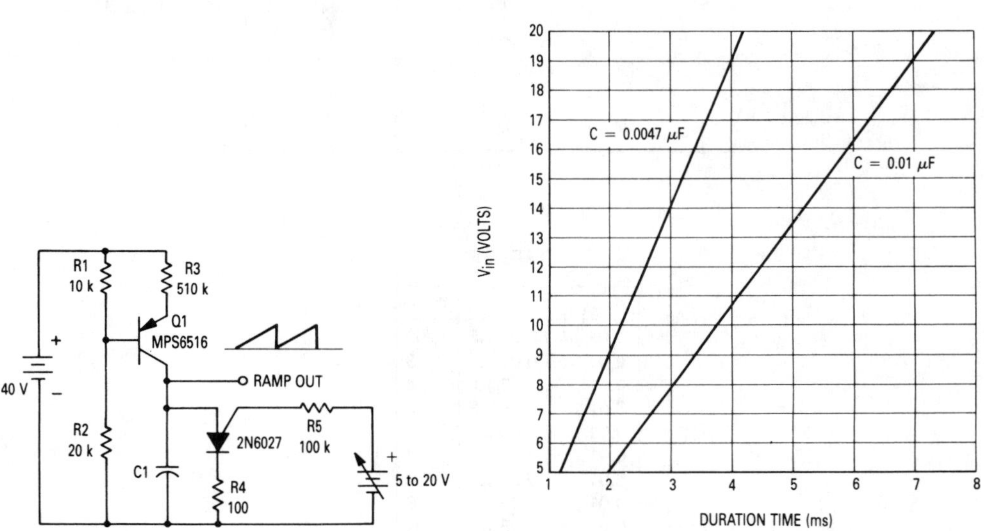

Voltage-controlled ramp generator that uses a PUT

Fig. 9-23 This circuit shows a PUT used as a VCRG. Figure 9-23B shows a plot of voltage-versus-ramp duration for two values of C_1. Motorola Thyristor Device Data, 1991, p. 1-6-49.

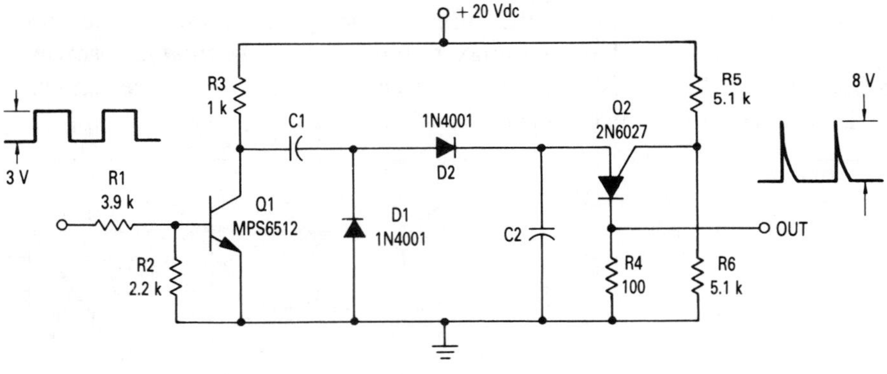

TABLE 6.X

C_1	C_2	Division
0.01 μF	0.01 μF	2
0.01 μF	0.02 μF	3
0.01 μF	0.03 μF	4
0.01 μF	0.04 μF	5
0.01 μF	0.05 μF	6
0.01 μF	0.06 μF	7
0.01 μF	0.07 μF	8
0.01 μF	0.08 μF	9
0.01 μF	0.09 μF	10
0.01 μF	0.1 μF	11

Low-frequency divider

Fig. 9-24 This circuit shows a PUT that is used as a frequency divider, where the ratio of C_1 and C_2 determines the amount of division. For a 10-kHz input frequency, and an amplitude of 3 V, Table 6-X shows the values for C_1 and C_2 that are needed to divide by values from 2 to 11. The division range can be changed by changing the ratio $R_6/(R_6 + R_5)$. With a given C1 and C2, increasing the resistance ratio increases the division range. Motorola Thyristor Device Data, 1991, p. 1-6-49.

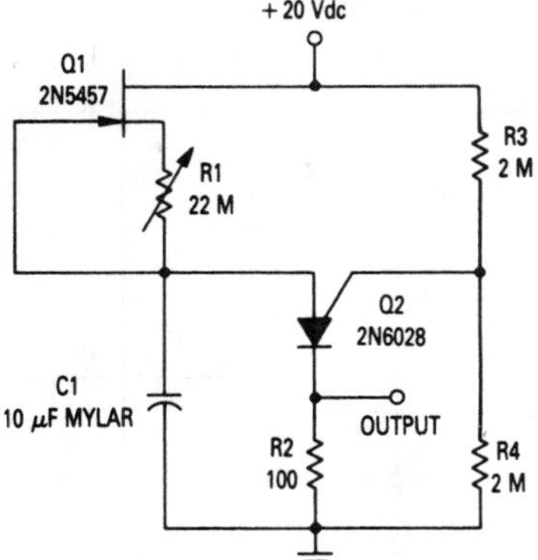

Long-duration timer that uses a PUT

Fig. 9-25 This circuit shows a PUT that is used as a long-duration (up to about 20 minutes) timer. The circuit is similar to that of Fig. 9-20. However, the PUT is superior to the UJT in long-duration timers because the PUT has a lower peak-point firing current (making it possible to charge the capacitor over a longer time). Motorola Thyristor Device Data, 1991, p. 1-6-50.

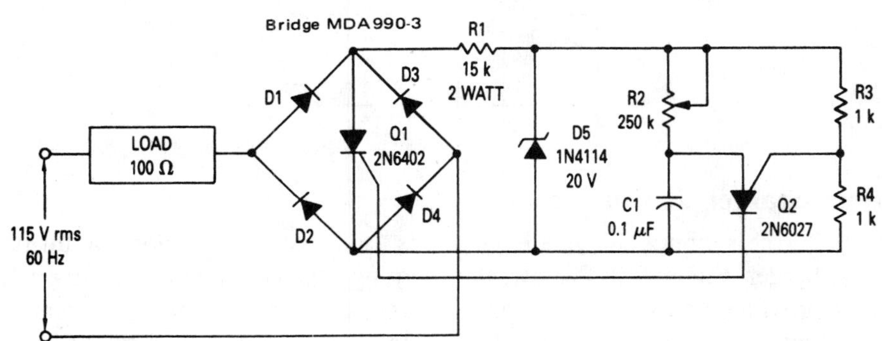

Phase-control circuit that uses a PUT

Fig. 9-26 This circuit shows a PUT that is used to control phase of an SCR. The relaxation oscillator that is formed by Q2 provides conduction control of Q1 from 1 to 7.8 ms (21.6° to 168.5°). This constitutes control of over 97% of the power that is available to the load. Only one SCR is needed to provide phase control of both the positive and negative portion of the sine wave, by putting the SCR across the bridge, which is composed of diodes D1 through D4. Motorola Thyristor Device Data, 1991, p. 1-6-50.

115 V rms
T1
14 V rms
MDA990-1
2N5164
SCR
A
R1 10 k
R4 1 k
2N6027
R2 50 k
12 V
PUT
D1 1N5240 10 V
C1 0.1 μF
T2 11Z12 1:1
R3 47 k
B
SPRAGUE

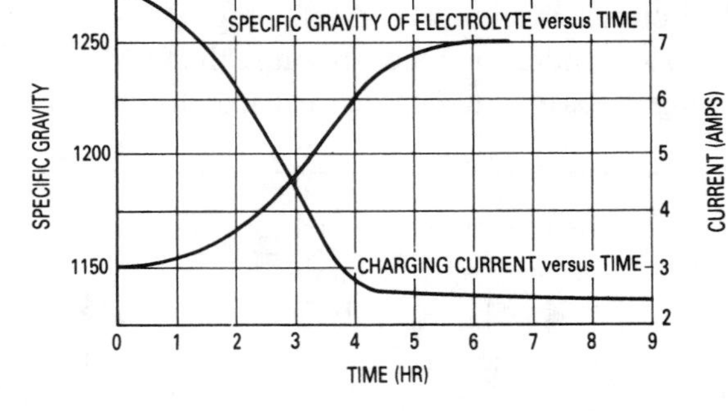

Battery charger that uses a PUT

Fig. 9-27 This circuit shows a PUT that is used to control a short-circuit-proof battery charger. Figure 9-27B shows the charging characteristics of the circuit. The charger provides an average charging current of about 8 A to a 12-V lead-acid storage battery. The circuit will not function or will be damaged by improperly connecting the battery because battery current is used to charge timing capacitor C1 (the SCR remains cut off, and the PUT relaxation oscillator does not function until a battery is connected as shown). The maximum battery voltage (10 to 14 V) is set by R2. Motorola Thyristor Device Data, 1991, p. 1-6-51.

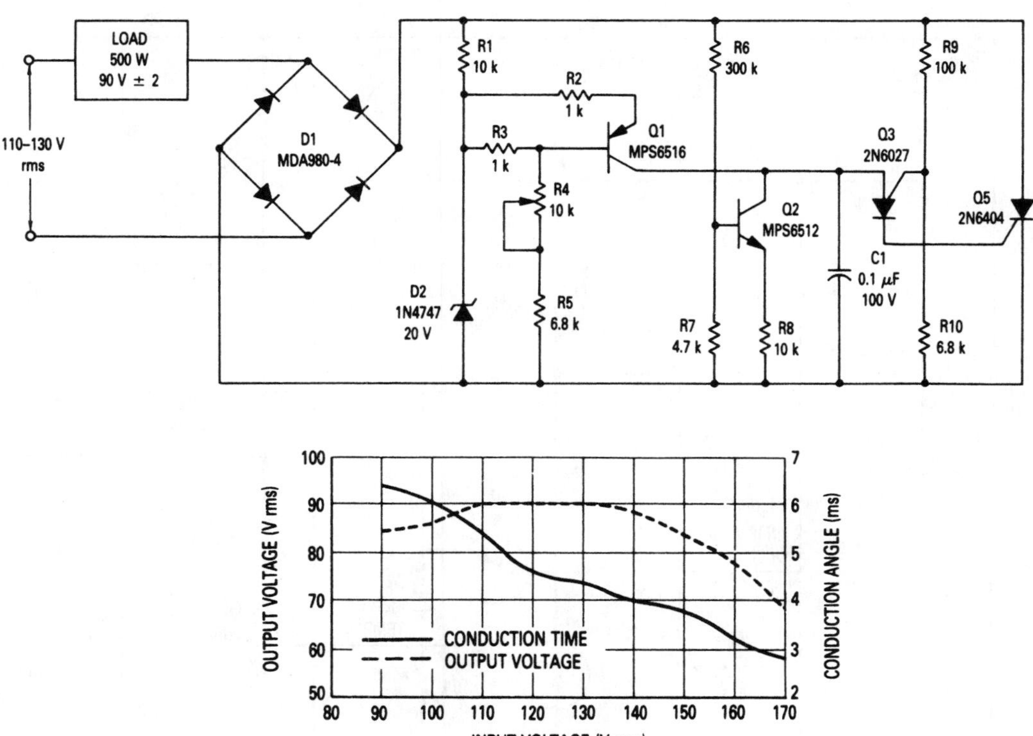

90-V voltage regulator that uses a PUT

Fig. 9-28 This circuit shows a PUT that is used to control a voltage regulator. Figure 9-28B shows a plot of output voltage and conduction angle versus input voltage for the regulator. The output voltage is set by R4. Motorola Thyristor Device Data, 1991, p. 1-6-52.

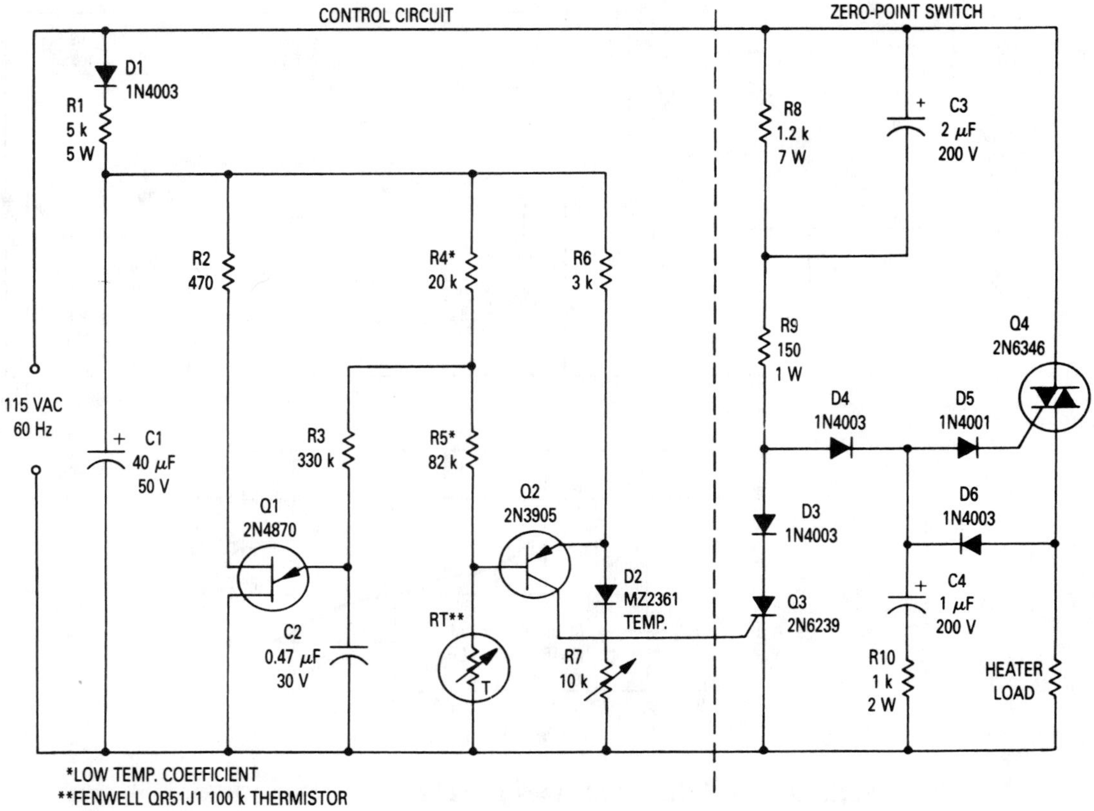

CONTROL CIRCUIT

ZERO-POINT SWITCH

*LOW TEMP. COEFFICIENT

**FENWELL QR51J1 100 k THERMISTOR

Temperature controller that uses zero-point switching

Fig. 9-29 This circuit shows the basic UJT building block (Fig. 9-1), which is used to control a zero-point switching temperature controller. The circuit applies the correct amount of power on a continuous basis at a steady-state duty cycle, depending on the load requirements. Temperature is therefore controlled over a very narrow range and no EMI is generated. The temperature is set by R7. Motorola Thyristor Device Data, 1991, p. 1-6-56.

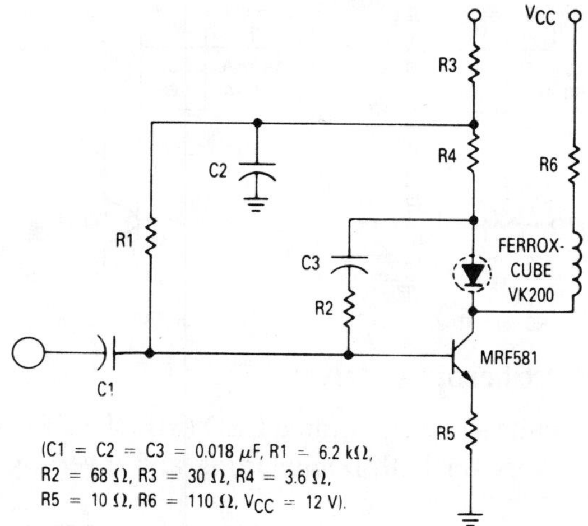

Infrared LED fiberoptic driver

Fig. 9-30 This circuit shows an infrared LED (MFOE3200, 3201, 3202) that is used as a fiberoptic driver at frequencies up to 100 MHz. Motorola Optoelectronics Device Data, 1989, p. 5-60.

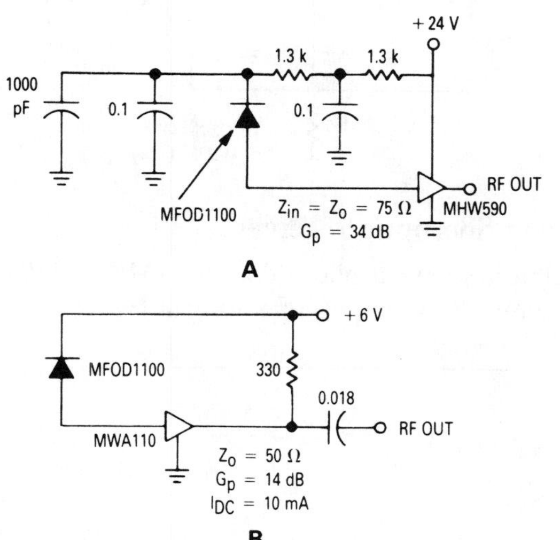

Photodetector fiberoptic receiver

Fig. 9-31 These circuits show a photo detector (MFOD1100) used as a fiberoptic receiver. Motorola Optoelectronics Device Data, 1989, p. 5-60.

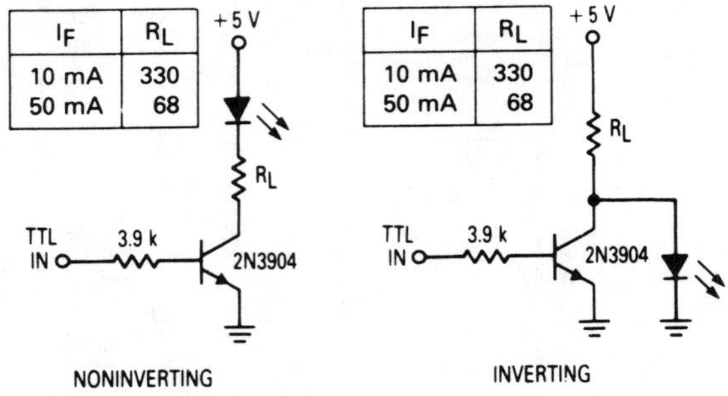

TTL infrared LED fiberoptic driver

Fig. 9-32 These circuits show an infrared LED (MFOE3200, 3201, 3202) that is used as fiberoptic drivers for both noninverting and inverting TTL inputs. Motorola Optoelectronics Device Data, 1989, p. 5-60.

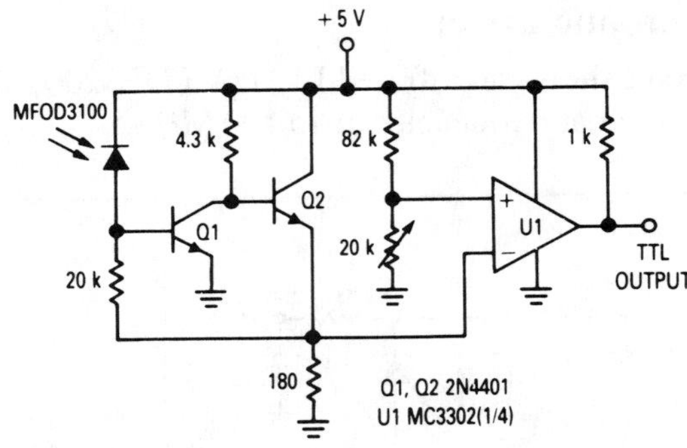

TTL Photodetector fiberoptic receiver

Fig. 9-33 This circuit shows a photo detector (MFOD3100) that is used as a fiberoptic receiver with a TTL output, operating at frequencies up to 1 MHz. Motorola Optoelectronics Device Data, 1989, p. 5-60.

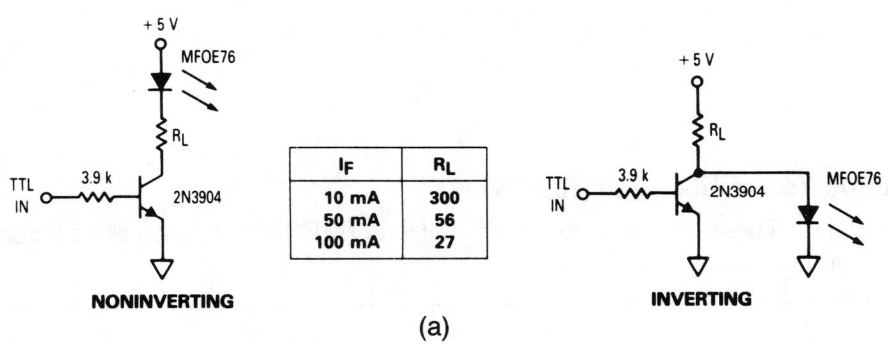

I_F	R_L
10 mA	300
50 mA	56
100 mA	27

Continued

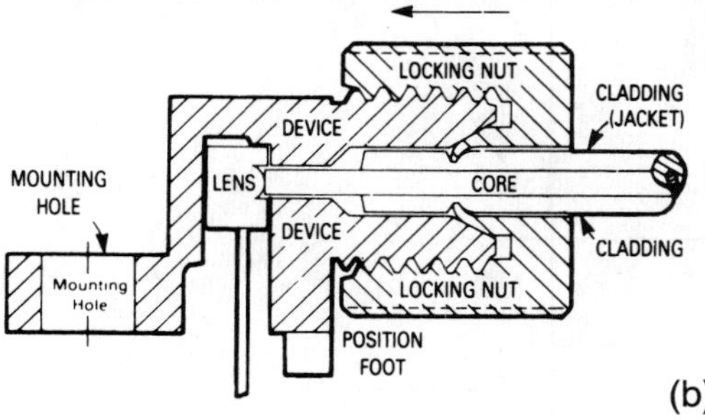

CROSS SECTION OF FLCS PACKAGE

LOCKING NUT
DEVICE
MOUNTING HOLE
LENS
CORE
CLADDING (JACKET)
DEVICE
CLADDING
Mounting Hole
LOCKING NUT
POSITION FOOT

TERMINATION INSTRUCTIONS

1. Cut cable squarely with sharp blade or hot knife.
2. Strip jacket back with 18 gauge wire stripper to expose 0.10–0.18″ of bare fiber core.
   Avoid nicking the fiber core.
3. Insert terminated fiber through locking nut and into the connector until the core tip seats against the molded lens inside the device package.
   Screw connector locking nut down to a snug fit, locking the fiber in place.

(b)

Visible red LED fiberoptic driver

Fig. 9-34 These circuits show a visible red LED (MFOE76) that is used as fiberoptic drivers for both noninverting and inverting TTL inputs. Figure 9-34B shows typical fiberoptic termination and assembly for both receivers and transmitters. Motorola Optoelectronics Device Data, 1989, p. 5-12.

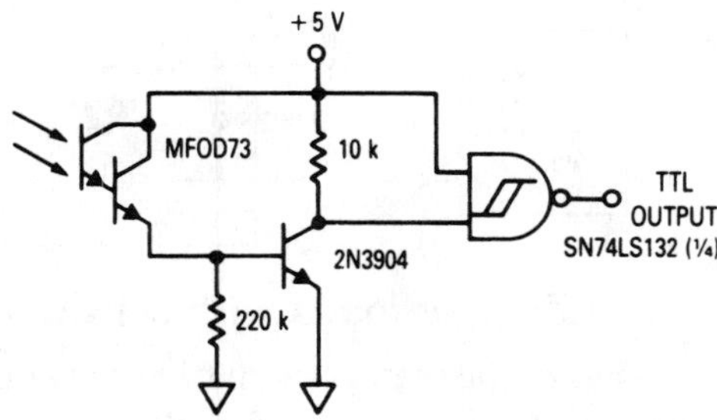

Darlington photodetector fiberoptic receiver

Fig. 9-35 This circuit shows a Darlington photodetector (MFOD73) with a TTL output that operates at frequencies up to 1 kHz. Motorola Optoelectronics Device Data, 1989, p. 5-12.

Note: This optoisolator should not be used to drive a load directly. It is intended to be a trigger device only. Additional information on the use of the MOC3009/3010/3011/3012 is available in Application Note AN-780A.

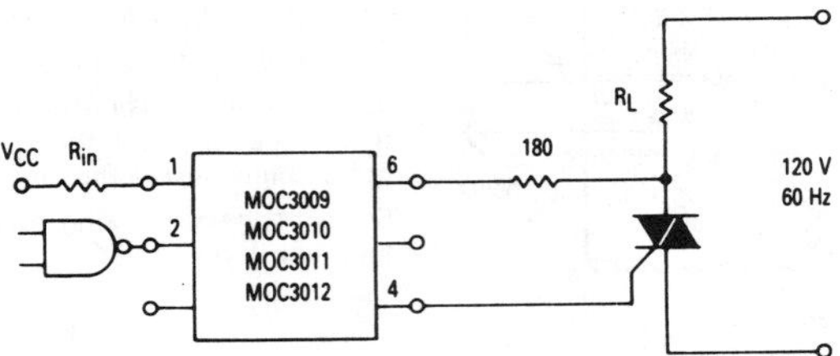

Optoisolator driver for resistive loads

Fig. 9-36 This circuit shows optoisolators (with triac driver outputs) that are used to trigger a triac with resistive load. Motorola Optoelectronics Device Data, 1989, p. 6-92.

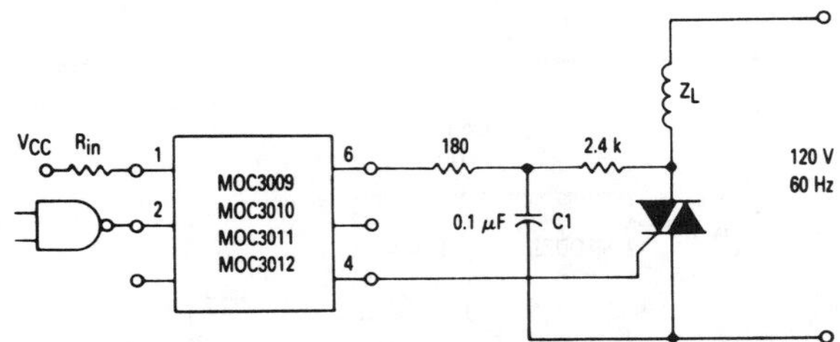

Optoisolator driver for inductive loads with sensitive-gate triacs

Fig. 9-37 This circuit shows optoisolators (with triac driver outputs) that are used to trigger a sensitive-gate triac (I_{GT} less than 15 mA) with inductive load. C1 and the 2.4-kΩ resistor form a snubber for the triac. Motorola Optoelectronics Device Data, 1989, p. 6-92.

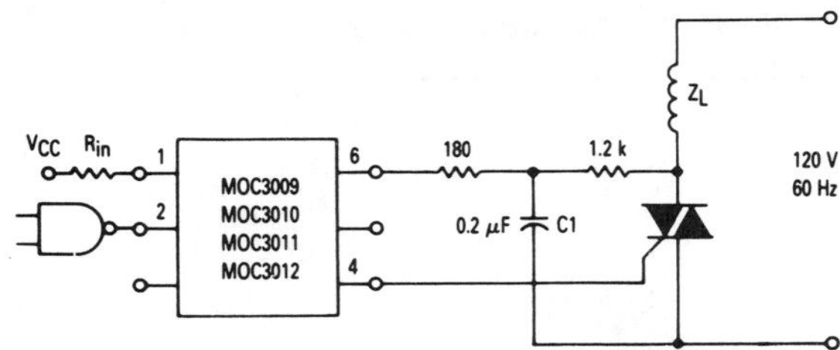

Optoisolator driver for inductive loads with nonsensitive-gate triacs

Fig. 9-38 This circuit shows optoisolators (with triac driver outputs) that are used to trigger a nonsensitive-gate triac (I_{GT} greater than 15 mA but less than 50 mA) with inductive load. C1 and the 1.2-kΩ resistor form a snubber for the triac. Motorola Optoelectronics Device Data, 1989, p. 6-92.

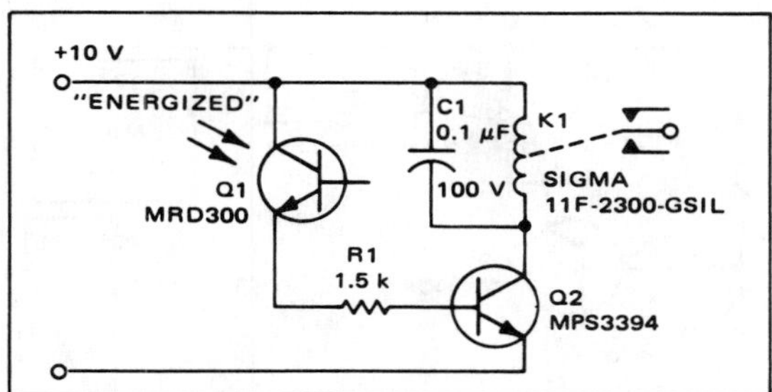

Light-operated relay

Fig. 9-39 This circuit shows an MRD300 phototransistor that is used to control a relay, where the relay is energized when light is applied to Q1. Motorola Optoelectronics Device Data, 1989, p. 9-23.

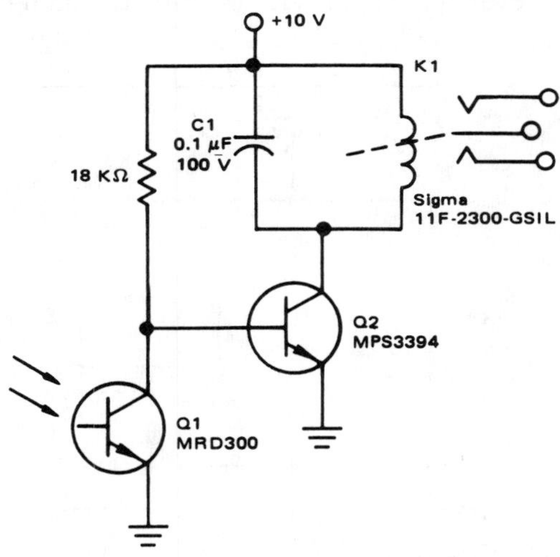

Light-deenergized relay

Fig. 9-40 This circuit shows an MRD300 phototransistor that is used to control a relay, where the relay is de-energized when light is applied to Q1. Motorola Optoelectronics Device Data, 1989, p. 9-24.

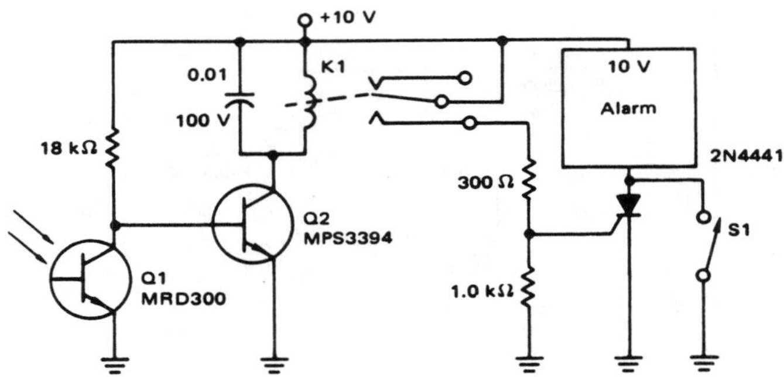

Light-operated alarm

Fig. 9-41 This circuit shows an MRD300 phototransistor that is used to control an alarm (bell, buzzer, etc.). The alarm is turned on when light is removed from Q1. The alarm remains on (even if the light is restored to Q1), until momentary- contact switch S1 is closed. Motorola Optoelectronics Device Data, 1989, p. 9-25.

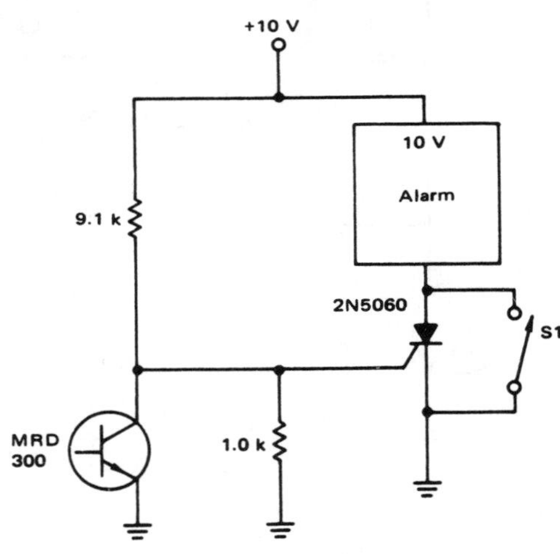

Light-operated alarm with sensitive-gate SCR

Fig. 9-42 This circuit is similar to that of Fig. 9-41, except that the relay is omitted and a sensitive-gate SCR is used. Motorola Optoelectronics Device Data, 1989, p. 9-25.

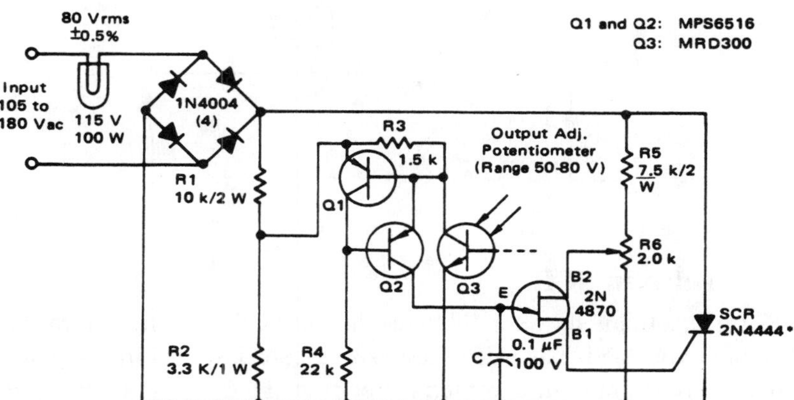

Projection-lamp voltage regulator

Fig. 9-43 This circuit is similar to that of Fig. 9-18. Here, the magnitude of the charging current that is applied to capacitor C and the position of R6 set the firing time of the UJT, which, in turn, sets the firing angle of the SCR. Light from the projection lamp sets the current level in Q3, which diverts current from the timing capacitor. R6 sets the desired brightness level by adjusting the UJT firing time. Motorola Optoelectronics Device Data, 1989, p. 9-25.

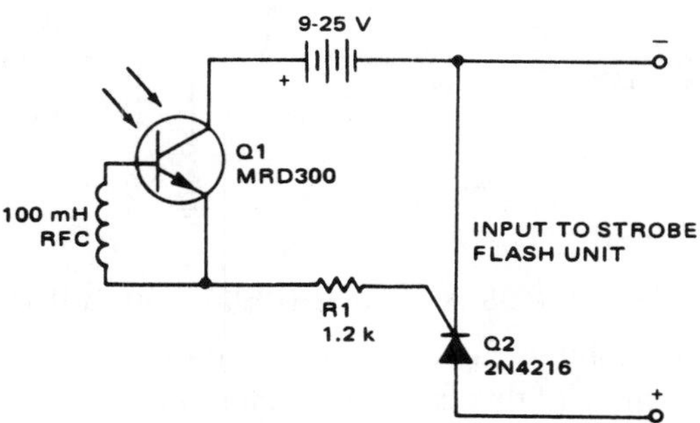

Strobeflash slave adapter

Fig. 9-44 This circuit provides the drive needed to trigger a slave strobeflash, without wires between the master and slave units. When light from the master strobeflash strikes Q1, Q2 is triggered, which applies power to the slave unit. Motorola Optoelectronics Device Data, 1989, p. 9-27.

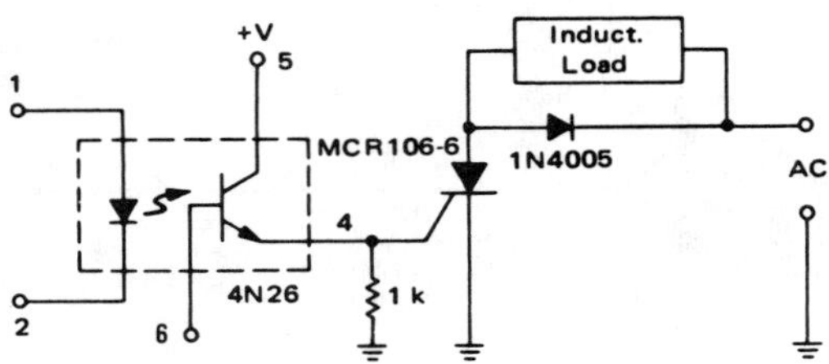

Optocoupler-driven SCR

Fig. 9-45 This circuit shows a 4N26 driving an SCR, which, in turn, is used to control an inductive load. The SCR is a sensitive-gate device (1 mA of gate current) and the 4N26 has a minimum-current transfer ratio of 0.2, so the 4N26 input current (I_F) must be 5 mA. The 1-kΩ resistor prevents the SCR from triggering with small input changes, and the 1N4005 prevents the SCR from triggering with the self-induced voltage when the SCR turns off. Motorola Optoelectronics Device Data, 1989, p. 9-32.

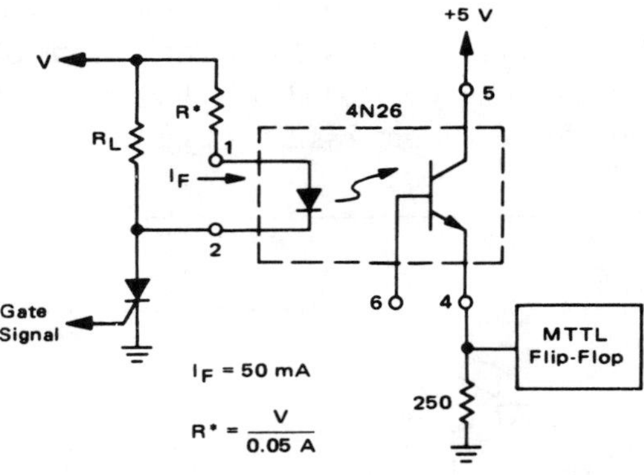

9

476

Optocoupler that is used as a load-to-logic translator

Fig. 9-46 This circuit shows a 4N26 that is used to couple a high-voltage load to a low-voltage logic input. In this case, the value of *R* is chosen so that the output current is greater than 10 mA (to ensure that the 4N26 output voltage exceeds the logic-one level of the flip-flop). Motorola Optoelectronics Device Data, 1989, p. 9-32.

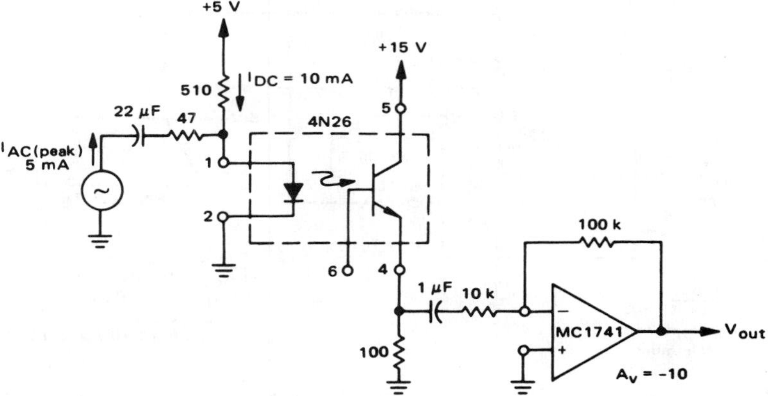

Optocoupler that is used to couple ac to an op amp

Fig. 9-47 In this circuit, the LED portion of the 4N26 is biased with a dc current of 10 mA. The input ac signal is summed with the dc bias so that the 4N26 output has both ac and dc components. Because the op amp is capacitively coupled to the 4N26, only the ac signal appears at the op-amp output. Motorola Optoelectronics Device Data, 1989, p. 9-32.

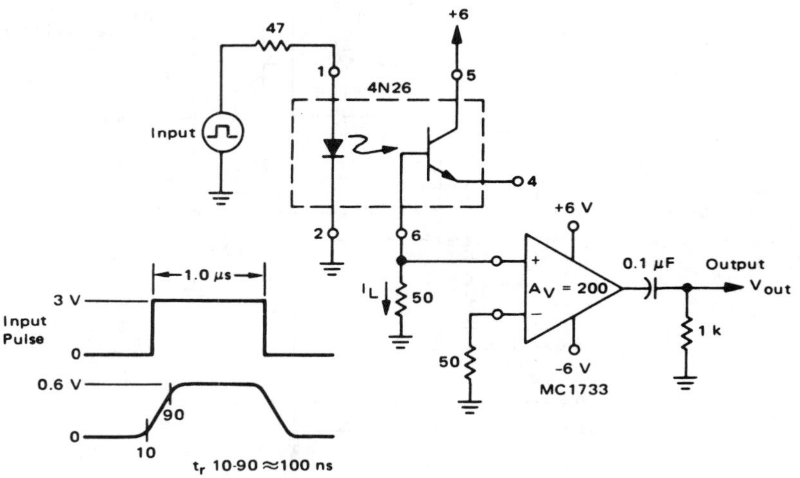

Diode/diode coupler

Fig. 9-48 This circuit shows a 4N26 that is used as a diode/diode coupler. The 4N26 emitter is left open, the load resistor is connected between base and ground, and the collector is tied to the supply. Using the 4N26 in this way reduces the switching time from about 2 or 3 μs to 100 ns. Motorola Optoelectronics Device Data, 1989, p. 9-33.

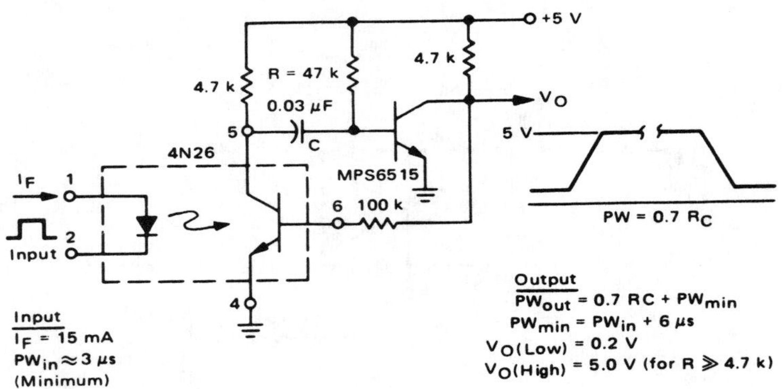

Optocoupler that is used as a pulse stretcher

Fig. 9-49 This circuit shows a 4N26 combined with a standard one-shot to form a pulse stretcher. A pulse of about 3 μs at 15 mA triggers the circuit. The output pulse amplitude is a function of supply voltage on the output side, and is independent of the input. Motorola Optoelectronics Device Data, 1989, p. 9-33.

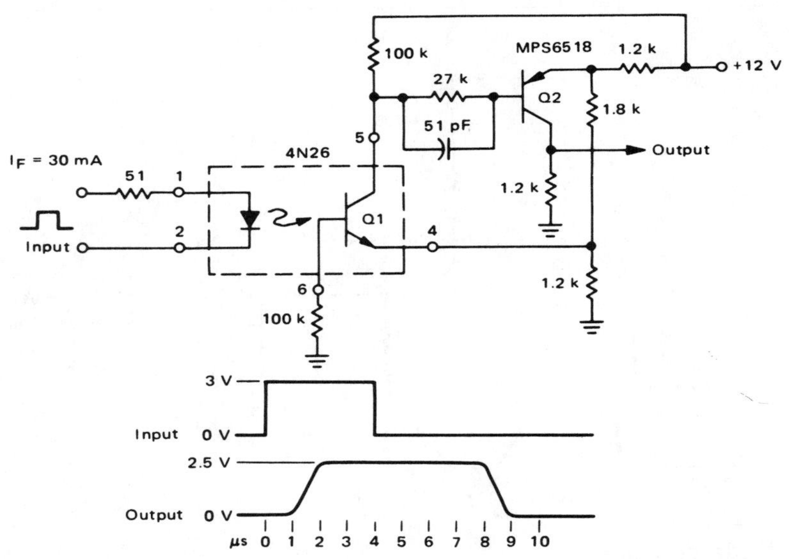

Optically coupled Schmitt trigger

Fig. 9-50 This circuit uses the 4N26 transistor as part of a Schmitt trigger. When Q1 conducts in response to an input pulse, Q2 conducts and the output is high. When the input pulse is removed from the 4N26, Q1 shuts off, and the output is low. With the values shown, the turn-on delay is about 2 μs, and the turn-off delay

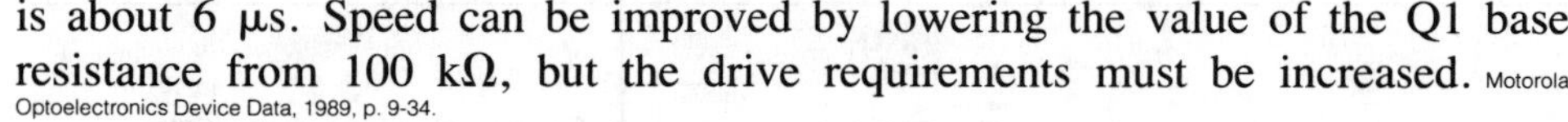

is about 6 μs. Speed can be improved by lowering the value of the Q1 base resistance from 100 kΩ, but the drive requirements must be increased. Motorola Optoelectronics Device Data, 1989, p. 9-34.

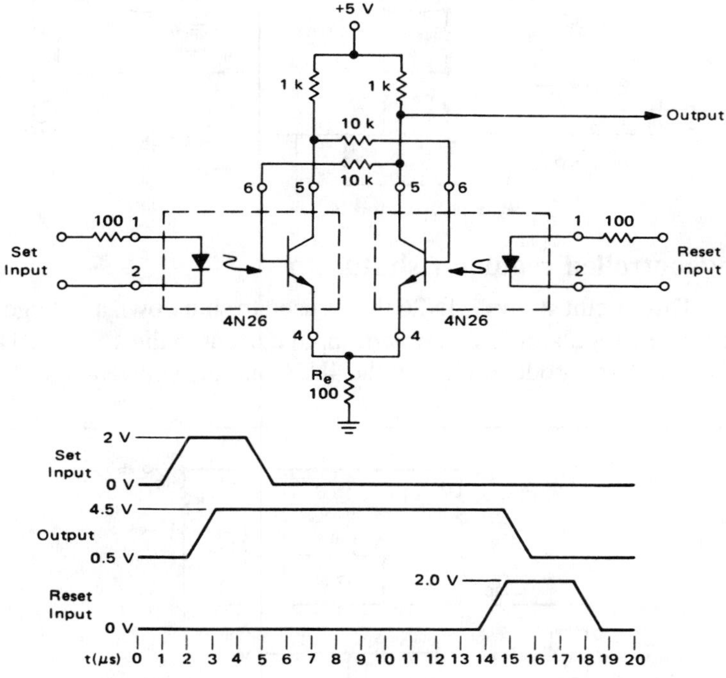

Optically coupled RS flip-flop

Fig. 9-51 This circuit uses two 4N26 couplers to form an RS circuit with two stable states. A +2-V signal at the set input changes the output from low to high, and a +2-V signal at the reset input returns the output to low. The input pulse width must be 3 μs (minimum). Motorola Optoelectronics Device Data, 1989, p. 9-34.

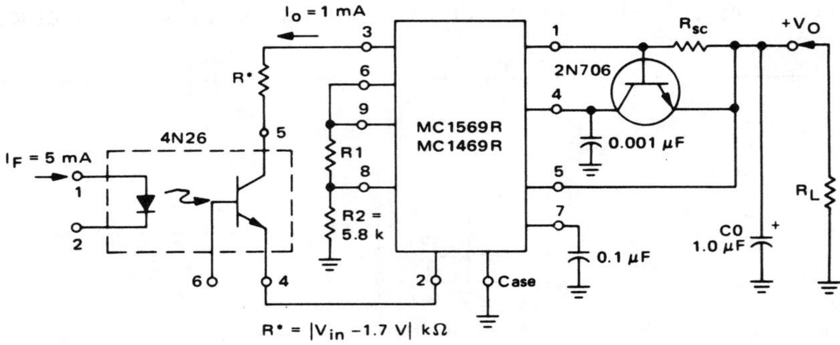

Optically controlled regulator shutdown

Fig. 9-52 This circuit shows a 4N26 that is used to shut down a voltage regulator. To ensure that the regulator is shut down, input current to the 4N26 LED must be a minimum of 5 mA (to produce 1 mA at the 4N26 output). Motorola Optoelectronics Device Data, 1989, p. 9-35.

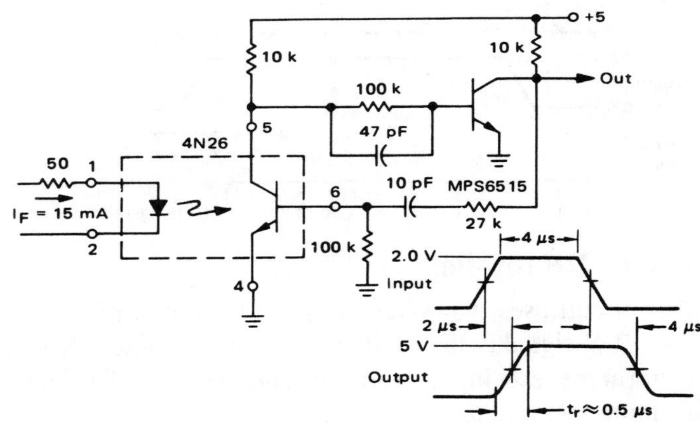

Optocoupler that is used as a pulse amplifier

Fig. 9-53 This circuit shows a 4N26 combined with a transistor to form a simple pulse amplifier that uses positive feedback to the 4N26 base. The feedback decreases pulse rise time from about 2 to 0.5 μs. Motorola Optoelectronics Device Data, 1989, p. 9-35.

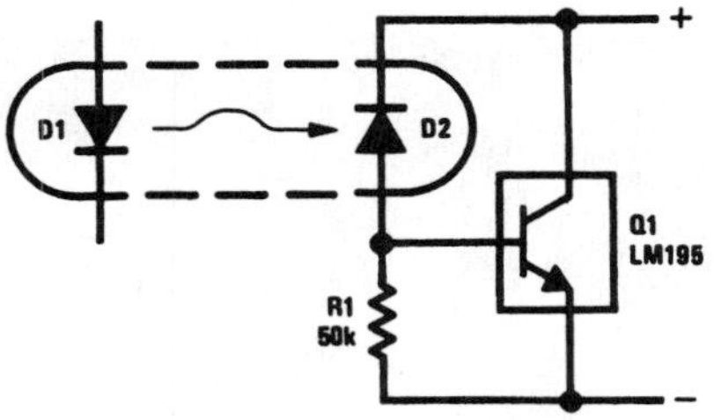

Optically isolated power transistor

Fig. 9-54 In this circuit, D1 and D2 can be part of almost any standard optocoupler or optoisolator. With no drive, R1 absorbs the base current of Q1, holding Q1 off. When power is applied to D1 (or to the LED of an optocoupler), D2 allows current to flow from the collector to base. Less than 20 μA is required from D2 to turn the LM1195 fully on. National Semiconductor, Linear Applications Handbook, 1991, p. 365.

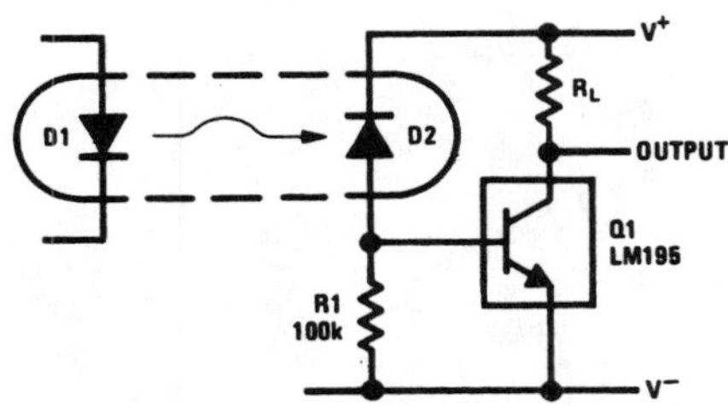

Fast optically isolated transistor or switch

Fig. 9-55 This circuit is similar to that of Fig. 9-54, but it has better ac response. The cathode of D2 is returned to a separate positive supply, rather than to the collector of Q1, which eliminates the added collector-to-base capacitance of D2. With this circuit, a 40-V 1-A load can be switched in 500 ns. Any photosensitive diode can be used for D2, instead of the optoisolator, to make a light-activated switch. National Semiconductor, Linear Applications Handbook, 1991, p. 365.

10

Op-amp circuits

It is assumed that you are already familiar with op-amp basics. However, the following paragraphs summarize this information, as well as test and troubleshooting for op-amp circuits. This information is included so that even if you are not familiar with electronic procedures, you can both test the circuits in this chapter and localize problems if the circuits fail to perform as shown.

Op-amp circuit testing and troubleshooting

The designation *op amp (operational amplifier)* was originally adopted for a series of high-performance direct-coupled amplifiers that were used in analog computers. These amplifiers were used to perform mathematical operations in analog computation (summation, scaling, subtraction, integration, and so on). Today, the availability of inexpensive IC amplifiers has made the packaged op amp useful as a replacement for any amplifier.

Figure 10-1 shows a classic application for an op amp (that of a summing amplifier). The output voltage (V_{OUT}) is the sum of voltages V_1, V_2, and V_3, as shown by the equation in the description for Fig. 10-1. For example, if all three input voltages are 5 V, the output will be 15 V with the values shown.

Notice that the output voltage will be inverted from the input voltage (as shown by the equation). This is because the three voltages to be summed are applied to the inverting input (−) of the op amp, at pin 2.

Also notice that the output (at pin 1) is returned or fed back to the input through R4 (most op amps are operated closed-loop, with feedback). The primary purpose of this feedback is to stabilize gain at some fixed value. Op-amp gain is determined by the ratio of feedback resistance (R_F) to input resistance (R_I) so that gain equals

R_F/R_I. Because R1 through R4 are the same value, there is no gain (or unity gain, or a gain of 1) in this circuit. If R_4 is increased to 100 kΩ, the circuit would have a gain of 10 (100 kΩ/10 kΩ), and V_{OUT} would be 10 times the sum of V_1, V_2, and V_3.

The circuit of Fig. 10-1 shows two power supplies (+15 and −15 V). This is typical for most op amps, and is one of the limitations for op-amp circuits. This is also one of the advantages of a Norton amplifier (chapter 11), which requires only one power source.

The noninverting input of the op amp in Fig. 10-1 is returned to ground through a 2.5-kΩ resistor (R6). This grounded input is typical for op amps that are usually operated with single-input circuits (even though op amps have a differential input). However, there are exceptions. For example, in Fig. 10-6, an op amp is used as a comparator with a dual or differential input and a single-ended output (note that this op amp is operated open-loop, without feedback). As a general guideline, the value for the grounded-input resistor is equal to the parallel resistance of the input and feedback resistors (shown by the equation of Fig 10-26).

Notice that there is no external compensation circuit for most of the op-amp circuits in this chapter. Early op amps often required external compensation circuits (usually capacitors or resistors, or both) to provide a given frequency-response characteristic. Most present-day op amps have internal compensation, and do not require any external components. Again, there are exceptions. For example, in the circuit of Fig. 10-49, compensating capacitors are required at pin 8 of the input amplifiers, and between pin 1 and pin 8 of the output amplifier. From a troubleshooting standpoint, if an op-amp circuit is working, but the frequency response is not as required, look for any compensation circuits.

Op-amp tests

The test procedures for the amplifiers in chapters 1 and 3 can generally be applied to op amps and to the op-amp circuits of this chapter. That is, the tests for frequency response, voltage gain, bandwidth, load sensitivity, input/output impedance, distortion, and background noise that are covered in chapter 1 apply to op amps. However, there are some additional tests that form a basis for troubleshooting op-amp circuits. These tests are summarized in the following paragraphs.

Feedback measurement Because op-amp circuits usually include feedback, it is sometimes necessary to measure feedback voltage at a given frequency with given operating conditions. The basic feedback-measurement connections are shown in Fig. 10-A. Although it is possible to measure the feedback voltage as shown in Fig. 10-Aa, a more accurate measurement is made when the feedback lead is terminated in the normal operating impedance (Fig. 10-Ab). If an input resistance is used in the circuit, and this resistance is considerably lower than the IC input impedance, use the resistance value. If in doubt, measure the input impedance of the IC (as described in chapter 1); then, terminate the feedback lead in that value to

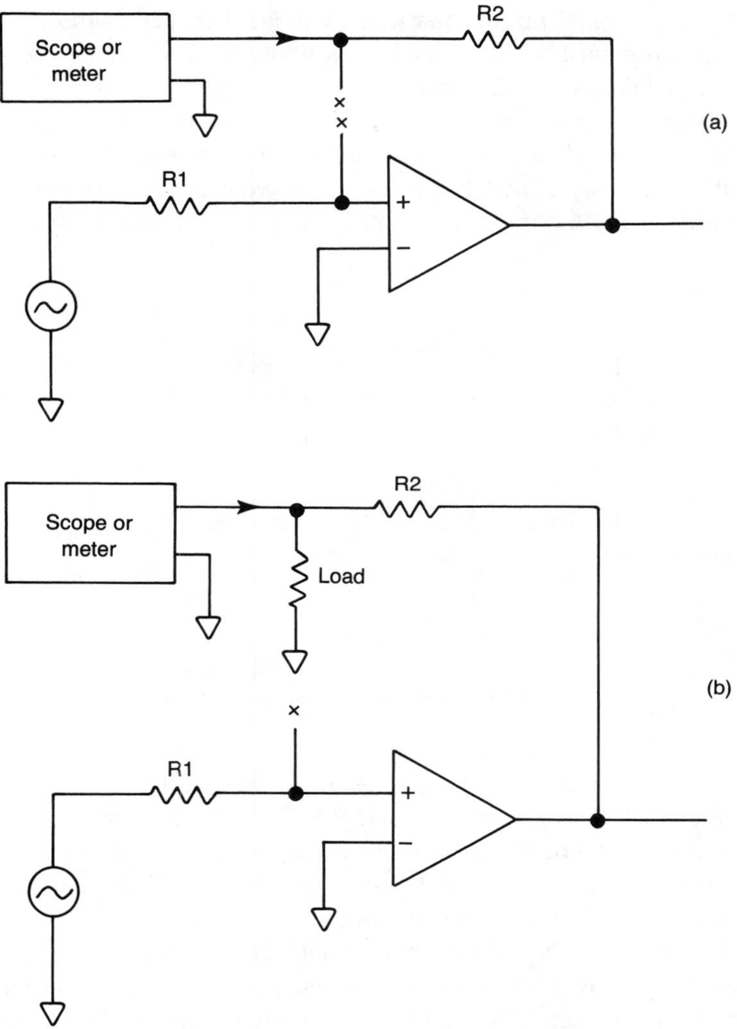

Fig. 10-A Feedback measurement.

measure open-loop feedback voltage. Remember that the open-loop voltage gain must be substantially higher than the closed-loop voltage gain for most op-amp circuits to perform properly.

Input-bias current *Op-amp input bias current* is the average value of the two input-bias currents of the op-amp differential-input stage. In circuit design, the significance of input-bias current is that the resultant voltage drops across input resistors can restrict the input common-mode voltage range at higher impedance levels. The input-bias current produces a voltage drop across the input resistors. This voltage drop must be overcome by the input signal (which can be a problem if the input signal is low and the input resistors are large).

Input-bias current can be measured using the circuit of Fig. 10-B. Any resistance value for R_1 and R_2 can be used, provided that the value produces a measurable voltage drop, and that the resistance values are equal. A value of 1-kΩ, with a tolerance of 1% or better, for both R_1 and R_2 is realistic for typical op amps.

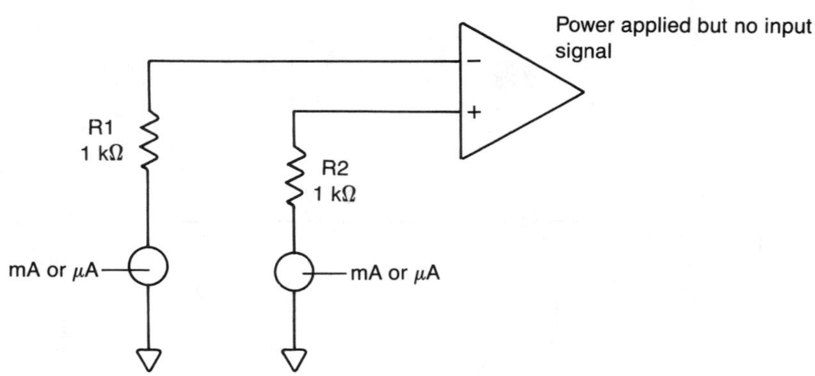

Fig. 10-B Input-bias current measurement.

If it is not practical to connect a meter in series with both inputs (as shown), measure the voltage drop across R1 and R2. Once the voltage drop is found, the input-bias current can be calculated. For example, if the voltage is 3 mV across 1 kΩ, the input bias current is 3 μA. Try switching R1 and R2 to see if any difference is the result of differences in resistor values.

In theory, the input-bias current should be the same for both inputs. In practice, the bias currents should be almost equal. Any great difference in input bias is the result of unbalance in the input differential amplifier of the IC, and can seriously affect circuit operation (and usually indicates a defective IC).

Input-offset voltage and current Op-amp input-offset voltage is the voltage that must be applied at the input terminals to get zero output voltage, whereas input-offset current is the difference in input-bias current at the op-amp input. Offset voltage and current are usually referred back to the input because the output values depend on feedback.

The effect of input-offset on op-amp circuits is that the input signal must overcome the offset before an output is produced. Likewise, with no input, there is a constant shift in output level. For example, if an op amp has a 1-mV input-offset voltage and a 1-mV signal is applied, there is no output. If the signal is increased to 2 mV, the op amp produces only the peaks.

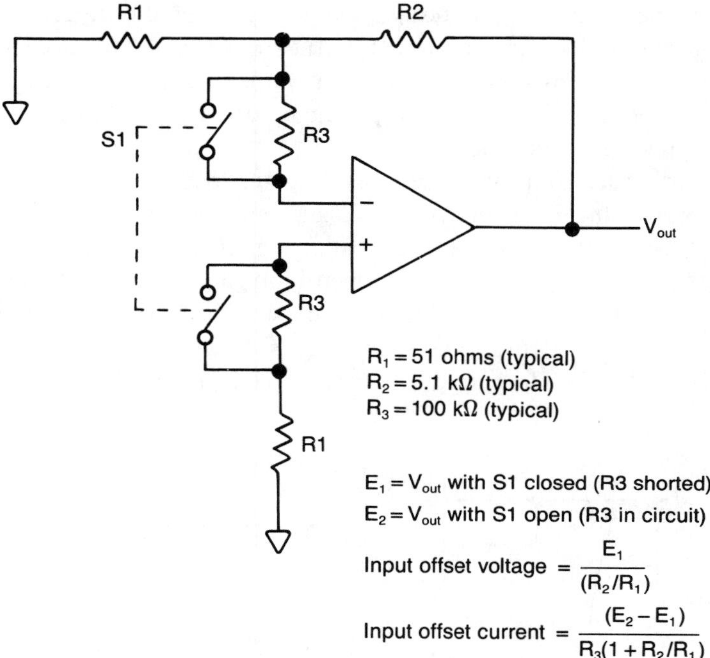

Fig. 10-C Input-offset voltage and current measurements.

Input-offset voltage and current can be measured using the circuit of Fig. 10-C. As shown, the output is alternately measured with R3 shorted and with R3 in the circuit. The two output voltages are recorded as E_1 (S1 closed, R3 shorted) and E_3 (S1 open, R3 in the circuit).

With the two output voltages recorded, the input-offset voltage and input-offset current can be calculated using the equations of Fig. 10-C. For example, assume that $R_1 = 51\ \Omega$, $R_2 = 5.1\ \text{k}\Omega$, $R_3 = 100\ \text{k}\Omega$, $E_1 = 83$ mV, and $E_2 = 363$ mV:

$$\text{Input-offset voltage} = \frac{83\ \text{mV}}{100} = 0.83\ \text{mV}$$

$$\text{Input-offset current} = \frac{280\ \text{mV}}{100\ \text{k}\Omega(1 + 100)} = 0.0277\ \mu\text{A}$$

Common-mode rejection There are many definitions for *common-mode rejection, CMR* (also known as *CMRR, common-mode rejection ratio*). One definition is the ratio of differential gain (usually large) to common-mode gain (usually a fraction). That is, the amplifier might have a large gain of differential signals (different signals at each input terminal, or with one input terminal grounded and the opposite input terminal with a signal), but little gain (or possibly a loss) of common-mode signals (same signal at both terminals). Another definition for CMR is the relationship of change in output voltage to change in input common-mode voltage producing the change, divided by the open-loop gain (amplifier gain without feedback).

No matter what definition is used, the first-step to measure CMR is to find the open-loop gain of the IC at the desired operating frequency (chapter 1). Then, connect the IC in the common-mode circuit of Fig. 10-D. Increase common-mode voltage (at the same frequency used for open-loop gain test) V_{in} until a measurable output (V_{out}) is obtained. Be careful not to exceed the maximum specified input common-mode voltage swing. If no such value is shown, do not exceed the normal input voltage of the IC. Then, find CMR using the equation.

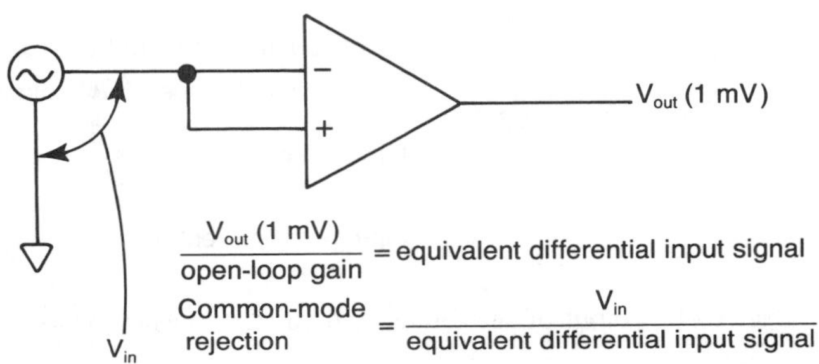

Fig. 10-D Common-mode rejection measurements.

To simplify calculation, increase the input voltage until the output is 1 mV. With an open-loop gain of 100, this provides an equivalent differential input signal of 0.00001 V. Then, measure the input voltage. Move the input-voltage decimal point over five places to find CMR.

Slew rate Op-amp slew rate is the maximum rate of change in output voltage, with respect to time, that the op amp is capable of producing when maintaining linear characteristics (symmetrical output without clipping).

Slew rate is expressed in terms of difference in output voltage divided by difference in time, dV_o/d_t. Usually, slew rate is listed in terms of volts per microsecond. For example, if the output voltage from an op amp is capable of changing 7 V in 1 μs, the slew rate is 7 (which might be listed as 7 V/μs). The major effect of slew rate in op-amp circuits is on power output. All other factors being equal, a lower slew rate results in lower power output.

A simple way to observe and measure op-amp slew rate is to measure the slope of the output waveform of a square-wave input signal (Fig. 10-E). The input square wave must have a rise time that exceeds the slew-rate capability of the op amp. As

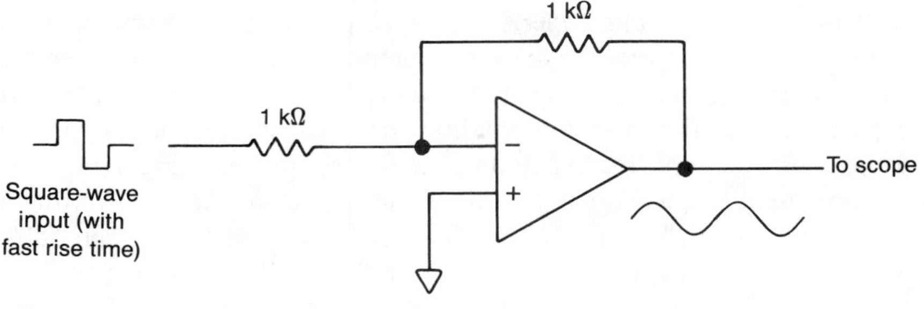

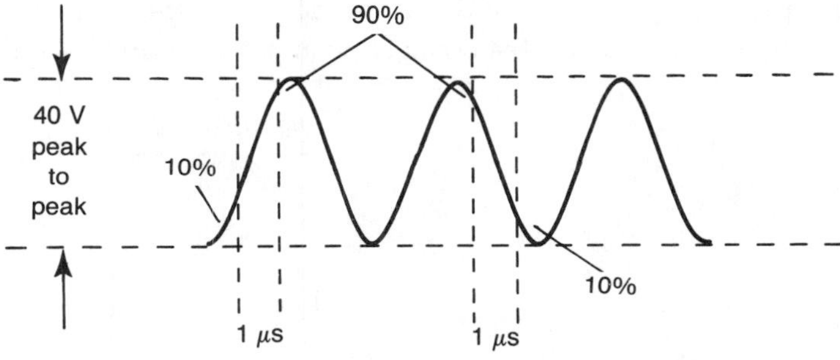

Example shows a slew rate of about 40 (40 V/μs) at unity gain

Fig. 10-E Slew-rate measurements.

a result, the output does not appear as a square wave, but as an integrated wave. In the example shown, the output voltage rises (and falls) about 40 V in 1 μs. Notice that slew rate is usually measured in the closed-loop condition. Also, the slew rate increases with higher gain.

Power-supply sensitivity Op-amp power-supply sensitivity is the ratio of change in input-offset voltage to the change in supply voltage that produces the change, with the remaining supply held constant. The term is expressed in millivolts or microvolts per volt (mV/V or μV/V, respectively), which represents the change of input-offset voltage (in mV or μV, respectively) to a change (in volts) of one power supply.

Power-supply sensitivity can be measured using the circuit of Fig. 10-C (the same test circuit as for input-offset voltage). The procedure is the same as for measurement of input-offset voltage, except that one supply voltage is changed (in 1-V steps) while the other supply voltage is held constant. The amount of change in input-offset voltage for a 1-V change in one power supply is the power-supply sensitivity (or the *input-offset voltage sensitivity*, as it might be called).

Phase shift The phase shift between input and output of an op-amp circuit is usually more critical than with the amplifiers that were described in chapters 1 and

3. This is because an op amp generally uses the principle of feeding back output signals to the input. Under ideal open-loop conditions, the output should be exactly 180° out of phase with the inverting input and in phase with the noninverting input. Any substantial deviation from this condition can cause op-amp circuit problems.

For example, assume that an op-amp circuit uses the inverting input, with the noninverting input grounded, and the circuit output fed back to the inverting input (such as through R2 in Fig. 10-26). If the output is not shifted the full 180° (for example, only a few degrees), the circuit might oscillate. Even if there is no oscillation, the op amp gain will not be stabilized, and the circuit will not operate properly.

A dual-trace scope (Fig. 10-F) is the ideal tool for phase measurement. For the most accurate results, the cables that connect the input and output should be of the

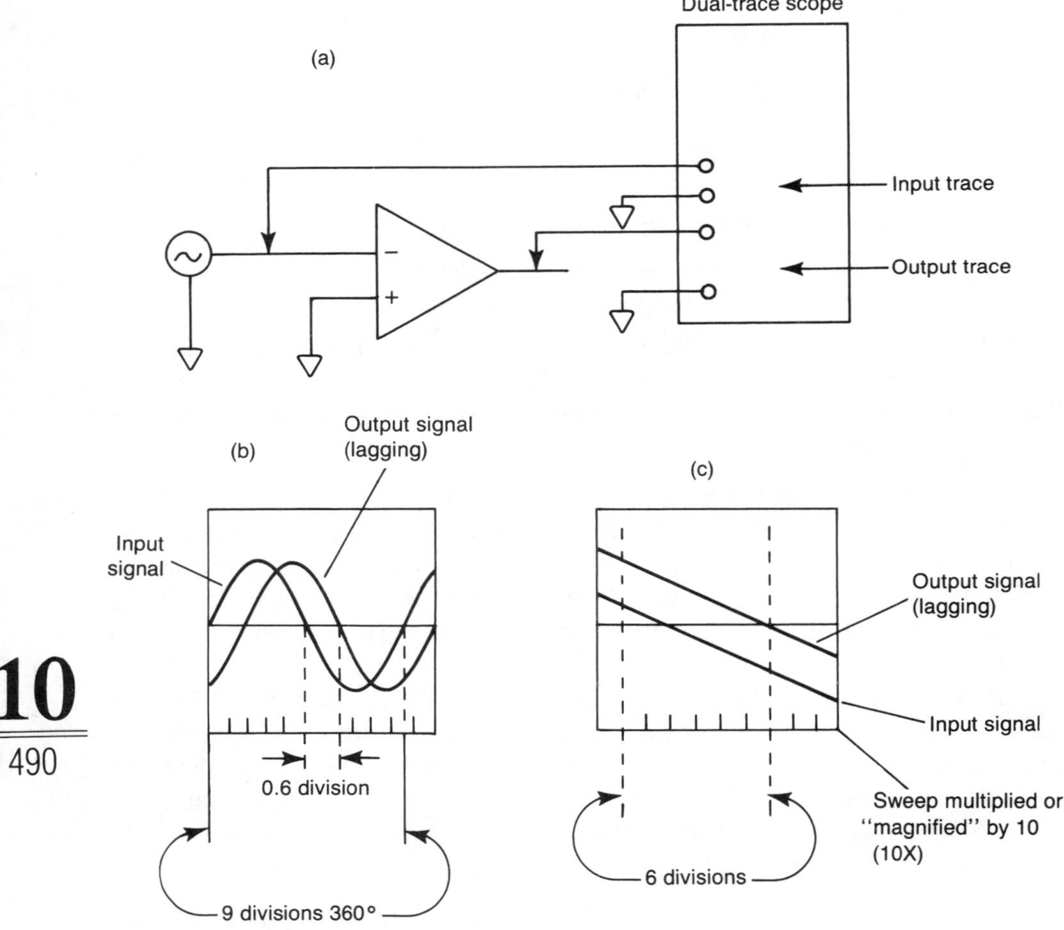

Fig. 10-F Phase-shift measurements.

same length and characteristics. At higher frequencies, a difference in cable length or characteristics can introduce a phase shift.

For simplicity, adjust the scope controls until one cycle of the input signal occupies exactly 9 divisions (typically 9 cm horizontally) of the screen. Then find the phase factor of the input signal. For example, if 9 cm represents one complete cycle (360°), 1 cm represents 40°(360/9 = 40).

With the phase factor established, measure the horizontal distance between corresponding points on the two waveforms (input and output signals). Then, multiply the measured distance by the phase factor of 40°/cm to find the phase difference. For example, if the horizontal distance is 0.6 cm with a 40°/cm phase factor, the phase difference is: 0.6 × 40° = 24°. If the scope has speed magnification, you can get more accurate results. For example, if the sweep rate is increased 10 times, the magnified phase factor is 40°/10 = 4°/cm. Figure 10-Fc shows the same signal that is used in Fig. 10-Fb, but with the sweep rate set by 10 X. With a horizontal difference of 6 cm, the phase difference is 6 × 4° or 24°.

Op-amp circuit troubleshooting

The troubleshooting approach described for the amplifiers of chapters 1 and 3 can be applied to op-amp circuits of this chapter. That is, first test the circuit to see if it performs the desired function. If not, try correcting the problem with adjustments. Then, trace signals using a meter or scope from input to output. If any portion of the circuit has a normal input, and an abnormal output, you have located the area in which the trouble occurs. From that point on, it is a matter of voltage measurements and/or point-to-point resistance measurements. The following are a few examples.

In the circuit of Fig. 10-4A, the output voltage depends on the temperature at the sensor across pins 2 and 6 of the input LT1001. For example, if the sensor temperature is 25°C, the output is 2.5 V at pin 6 of the second LT1001. If the temperature is increased to 30°C, the output voltage should rise to 3.0 V.

If there is no change in output voltage for a change in temperature at the sensor, check the voltage at pin 6 of the input LT1001 and at pin 2 of the output LT1001. If there is no change in temperature at the input LT1001, suspect the sensor, the LT1001 or the associated part. If there is a change at the first LT1001, but not at the input to the second LT1001, suspect the 10-kΩ wirewound resistor. If there is a change at the input to the second LT1001, but not at the output, suspect the LT1001 or the associated circuit parts.

If the output voltage varies when the sensor temperature is changed, but the voltage reading is not correct, the problem is likely one of adjustment. Follow the trim sequence (Fig. 10-4A). Also, look for a leaking capacitor at pin 6 of the second LT1001.

In the circuit of Fig. 10-6, there should be an output voltage when there is a differential voltage at the input. For example, if you ground one input, and apply a signal to the opposite input, there should be an output at the 2N3904 collector.

Likewise, if you ground both inputs, there should be no signal output. However, the output might shift in voltage level because of the input offset voltage.

If there is no output with a differential signal input, check for an output at the junction of the 20- and 4.99-kΩ resistors. If not, suspect the LT1001 or associated parts. If there is an output from the LT1001, but not at the 2N3904 collector, suspect the 2N3904 or 1N914.

In the circuit of Fig. 10-7, there should be an output when the photodiode is exposed to light. The output voltage should be 1 V for each μA of input current produced by the photodiode. Because it is difficult to monitor the input current, measure the voltage across the 500-kΩ resistor (from pin 3 to ground). If there is no change in output voltage when the photodiode is exposed to varying light conditions, suspect the photodiode, the LT1001, or possibly leaking capacitors. If the output does change with light changes, but the output voltage does not correspond to the input current (1 V/μA), check the values of both resistors and both capacitors. It is also possible that the photodiode is defective, and producing a nonlinear output.

Op-amp circuits titles and descriptions

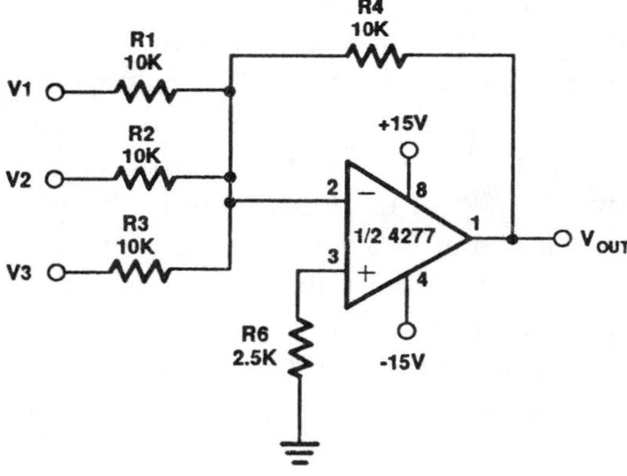

10 Adjustment-free precision summing amplifier

Fig. 10-1 Because R1 through R4 are all the same value, $V_{OUT} = V_1 + V_2 + V_3$. Notice that R_6 equals the parallel resistance of R_1 through R_4, and $V_{OUT} = -R_4$.

$$V_{OUT} = -R_4\left(\frac{V_1}{R_1} + \frac{V_2}{R_2} + \frac{V_3}{R_3}\right).$$

Raytheon Linear Integrated Circuits, 1989, p. 4-33.

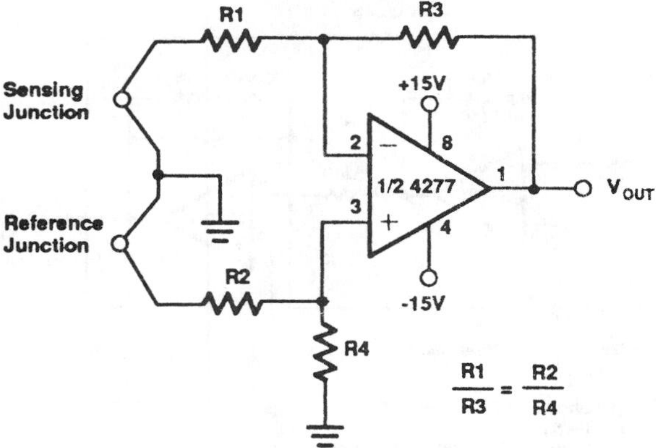

High-stability thermocouple amplifier

Fig. 10-2 Notice that the reference-junction output is applied to the noninverting input of the 4277, and the sensing-junction output is applied to the inverting input.

Raytheon Linear Integrated Circuits, 1989, p.4-33.

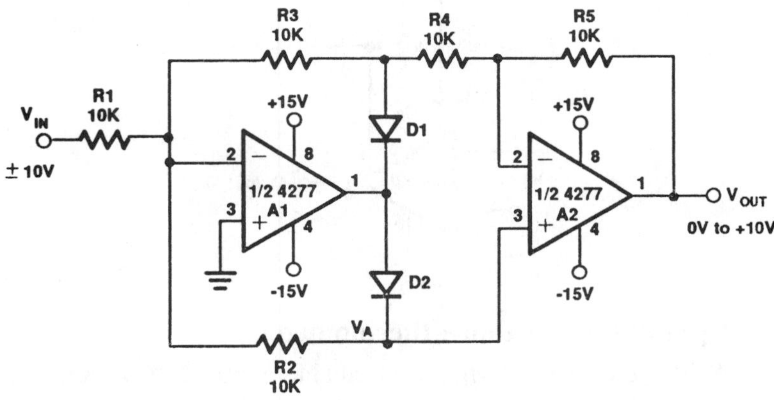

Precision absolute-value circuit

Fig. 10-3 The values and types for D1 and D2 are not critical, but both diodes should be the same, and both must be able to withstand ±15 V (such as a lN914).

Raytheon Linear Integrated Circuits, 1989, p. 4-33.

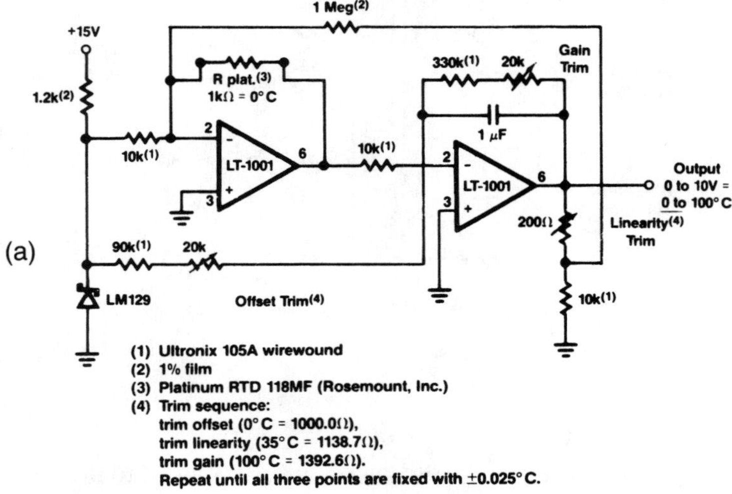

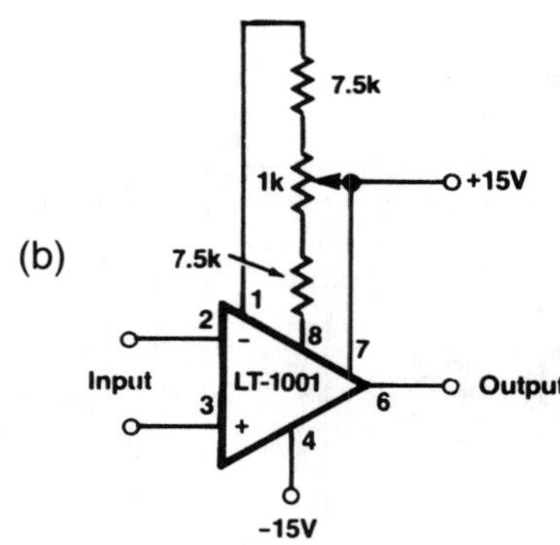

Linearized platinum resistance thermometer

Fig. 10-4 With the values shown, this circuit has a ±0.025°C accuracy over 0° to 100°C. The input offset voltage of the LT-1001, and its drift with temperature, are permanently trimmed at wafer test to a low level. However, if further adjustment of V_{OS} is necessary, nulling with the circuit of Fig. 10-4B will not degrade drift with temperature. The Fig. 10-4B offset circuit has an approximate null range of ±100 μV. Raytheon Linear Integrated Circuits, 1989, p. 4-45.

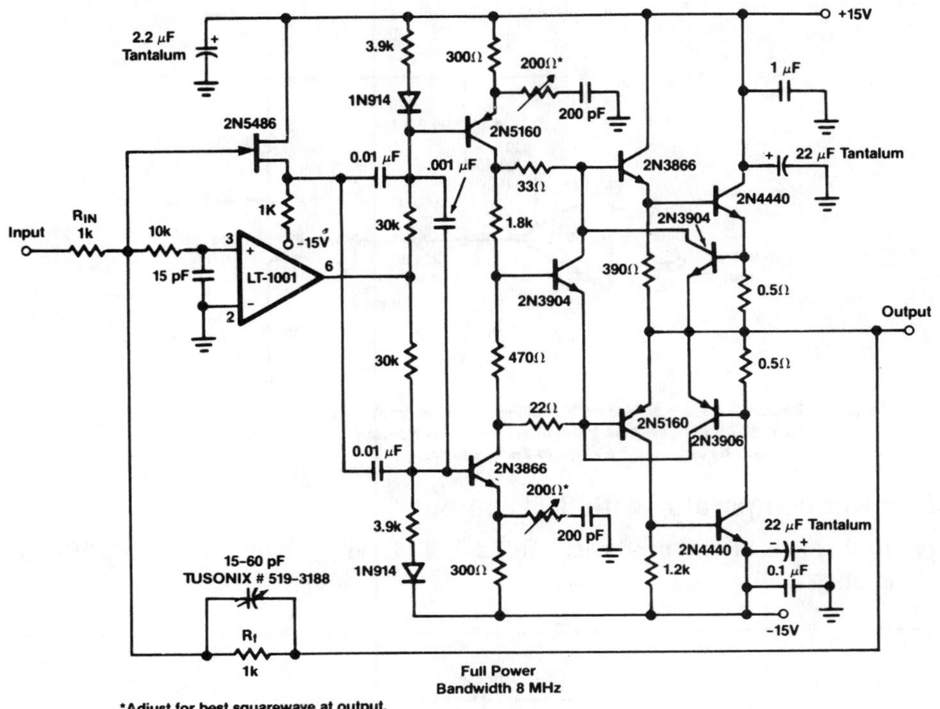

Dc-stabilized 1000-V/μs op amp

Fig. 10-5 As shown, this circuit has a full power bandwidth to 8 MHz, with a slew rate of 1000-V/μs. Raytheon Linear Integrated Circuits, 1989, p. 4-46.

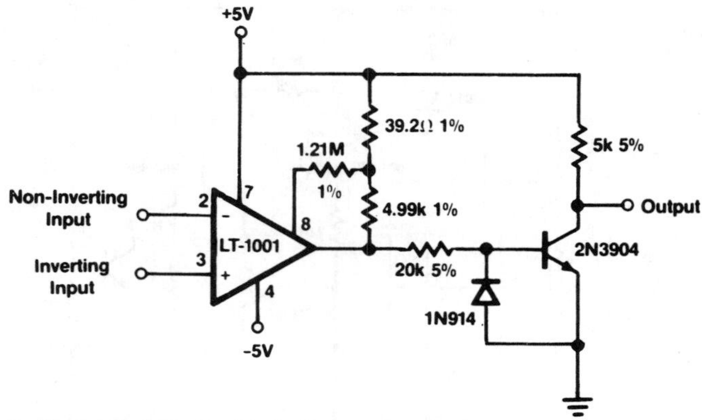

Microvolt comparator with TTL output

Fig. 10-6 This circuit produces a full 5-V TTL output with microvolt (differential) inputs. Raytheon Linear Integrated Circuits, 1989, p. 4-46.

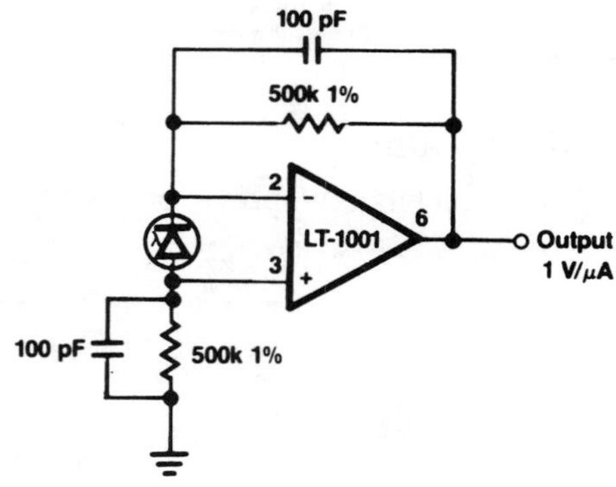

Photodiode amplifier

Fig. 10-7 This circuit produces a 1-V output for a 1-μA input. Raytheon Linear Integrated Circuits, 1989, p. 4-46.

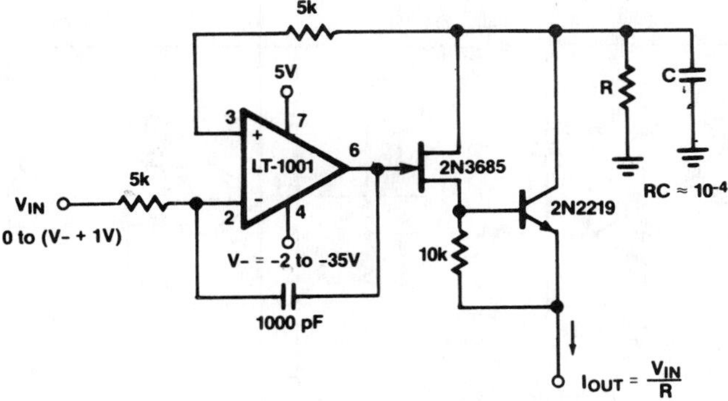

Precision current source

Fig. 10-8 As shown, the output current of this circuit depends on V_{IN} and the value of R. Notice that V_{IN} can be anything from 0 to (V− +1 V), and V− can be −2 to −35 V. Raytheon Linear Integrated Circuits, 1989, p. 4-47.

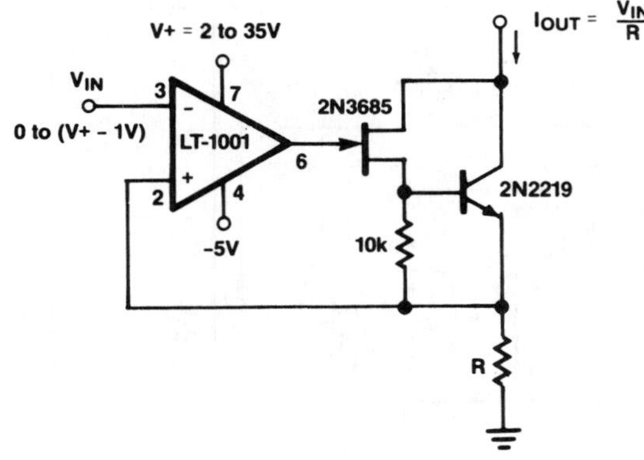

Precision current sink

Fig. 10-9 As shown, the ability of this circuit to sink current depends on V_{IN} and the value of R. Note that V_{IN} can be anything from 0 to (V+ − 1 V), and V+ can be +2 to +35 V. Raytheon Linear Integrated Circuits, 1989, p. 4-47.

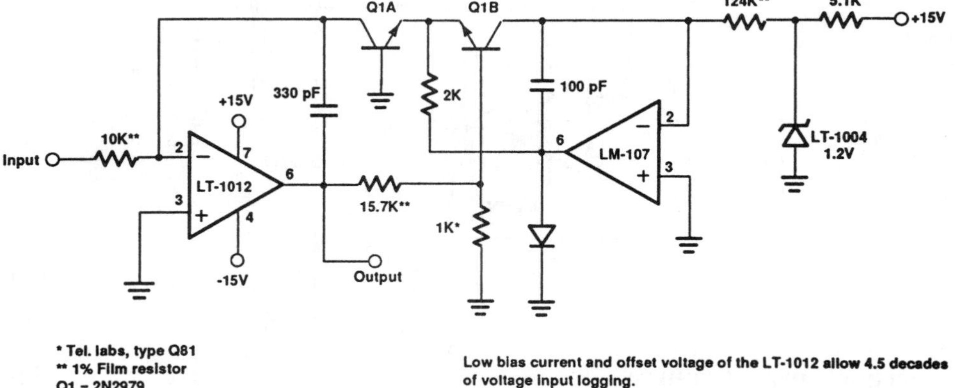

Log amplifier

Fig. 10-10 Compare this op-amp circuit to the log-amplifier circuits of chapters 2 and 3. Raytheon Linear Integrated Circuits, 1989, p. 4-54.

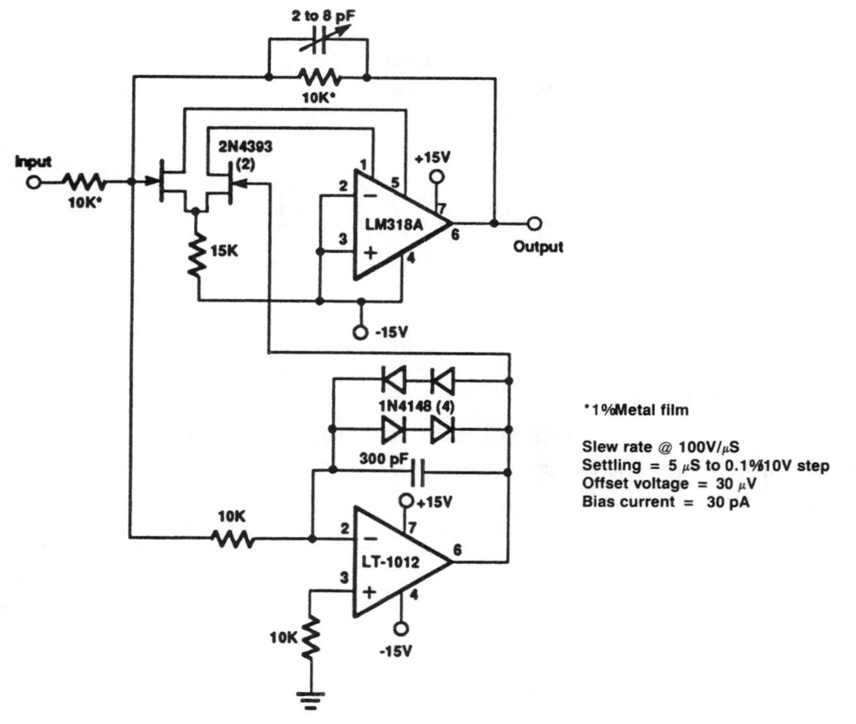

10

Fast precision inverter with 100-V/μs slew rate

Fig. 10-11 This circuit has a setting time of 5 μs to 0.1% with a 10-V step input.

Raytheon Linear Integrated Circuits, 1989, p. 4-55.

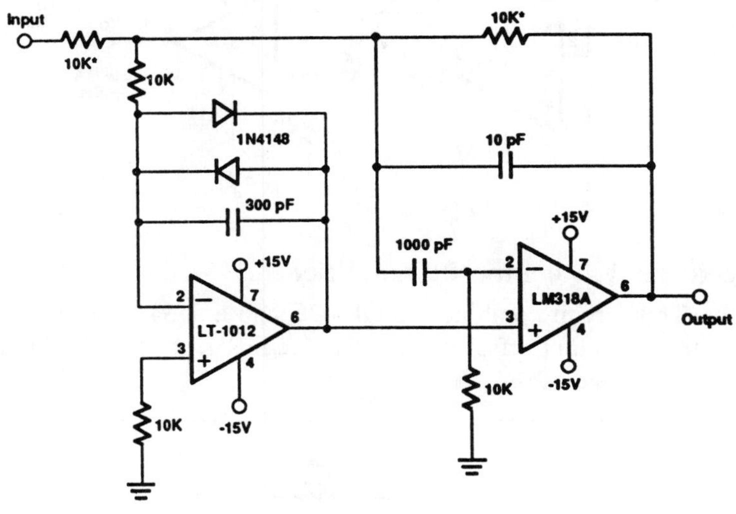

*1%Metal film

Full power bandwidth = 2 MHz
Slew rate = 50V/μS
Settling (10V step) = 12 μS to 0.01%
Bias current dc = 30 pA
Offset drift = 0.3 μV/°C
Offset voltage = 30 μV

Fast precision inverter with 50-V/μs slew rate

Fig. 10-12 This circuit has a settling time of 12 μs to 0.1% with a 10-V step input.

Raytheon Linear Integrated Circuits, 1989, p. 4-55.

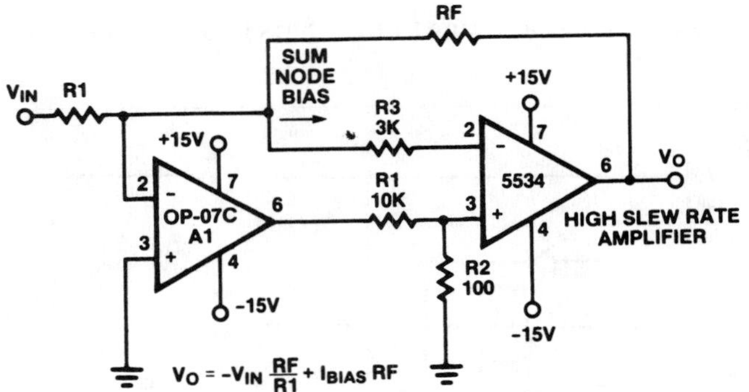

High-speed low V_{OS} composite amplifier

Fig. 10-13 This circuit combines an OP-07C and a 5534 to form a high-speed amplifier with low output offset voltage. Notice that V_O depends primarily on R_F and R_1. Raytheon Linear Integrated Circuits, 1989, p. 4-66.

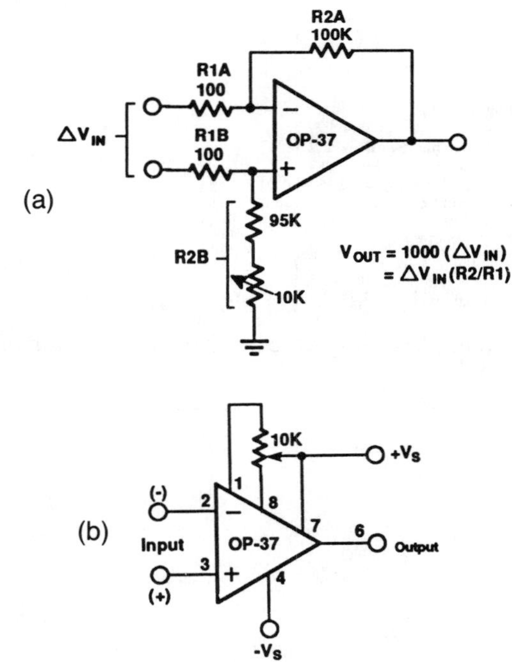

Single op-amp difference amplifier

Fig. 10-14 This circuit provides a voltage gain of 1000 for difference-signal inputs. For example, if the inverting input is 10 μV and the noninverting input is 11 μV, the difference is 1 μV and the output is 1000 μV. Figure 10-14B shows an offset nulling circuit (if required). Raytheon Linear Integrated Circuits, 1989, p. 4-92 and 4-94.

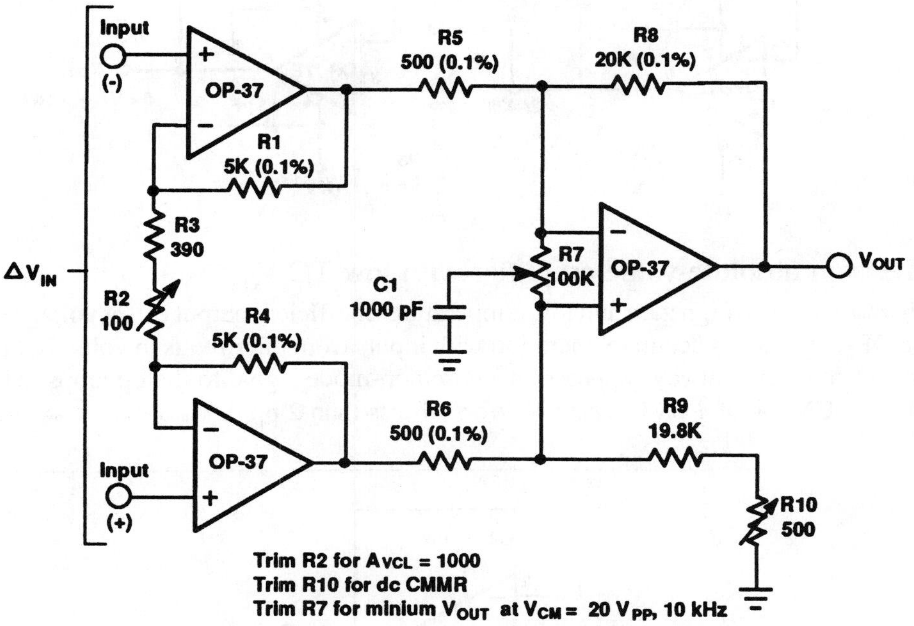

Three op-amp instrumentation amplifier

Fig. 10-15 This circuit uses only OP-37 ICs for instrumentation applications, where low noise, wide bandwidth, low input-offset TC, and low input-bias currents are required. Raytheon Linear Integrated Circuits, 1989, p. 4-95.

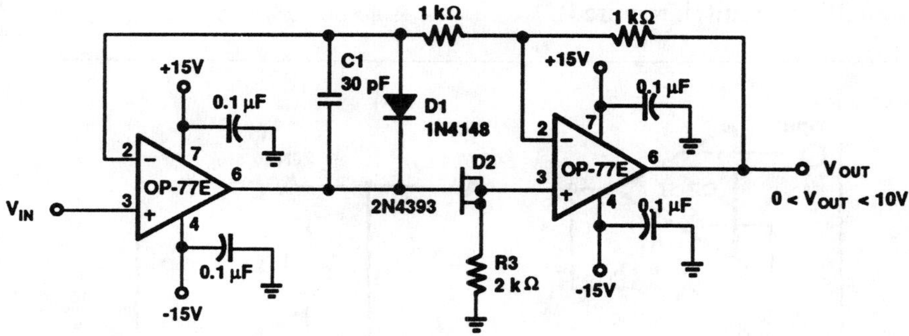

Precision absolute-value amplifier with low TC V_{OS}

Fig. 10-16 The high gain and low temperature-coefficient output offset voltage of the OP-77E assures accurate operation with inputs from microvolts to volts. In this circuit, the signal always appears as a common-mode signal to the op amps. The OP-77E CMRR of 1 μV/V assures errors of less than 2 ppm. Raytheon Linear Integrated Circuits, 1989, p. 4-117.

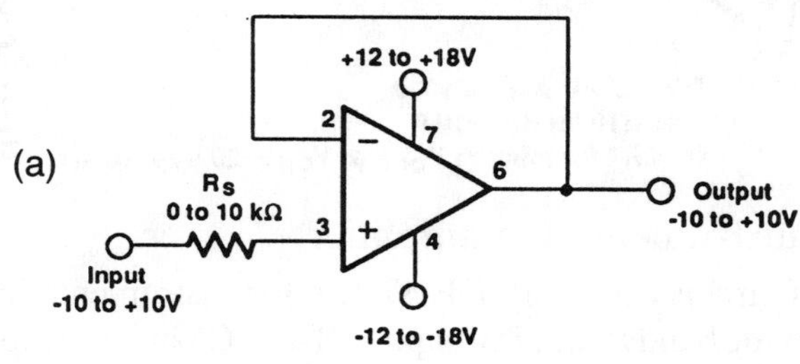

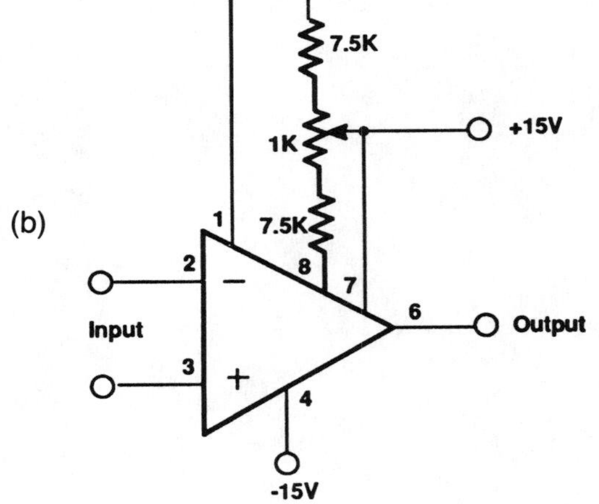

Large-signal voltage follower with 0.00065% worst-case accuracy error

Fig. 10-17 The input offset voltage and drift with temperature of the RC4097 that are shown in this voltage-follower circuit are trimmed at wafer test to a low level. However, if further adjustment of V_{OS} is necessary, nulling with the circuit of Fig. 10-17B will not degrade drift with temperature. The Fig. 10-17B offset circuit has an approximate range of about 4 mV. Raytheon Linear Integrated Circuits, 1989, p. 4-126, 4-127.

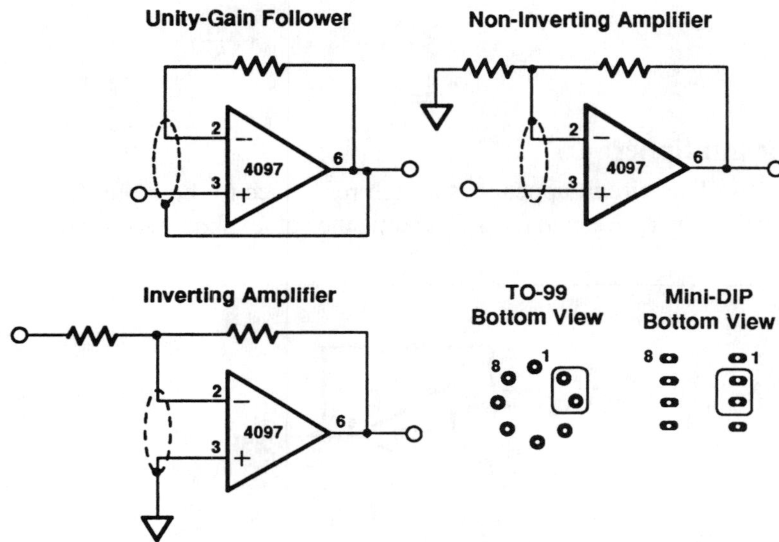

Typical guard-ring layout for op amps

Fig. 10-18 Even with properly cleaned and coated PC boards, leakage currents can limit the circuit performance under severe environmental conditions. These currents can be intercepted before they reach the op-amp inputs by a guard ring or a conductor in the leakage path. The ring should surround the input terminals, and should be at the same potential as the inputs. Notice that the guard ring for the inverting and noninverting configurations connects to the inputs, while the unity-gain circuit connects to the output. Raytheon Linear Integrated Circuits, 1989, p. 4-128.

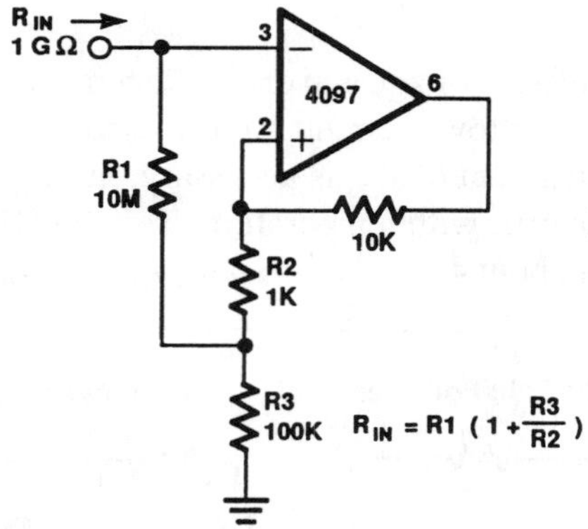

Resistor multiplier

Fig. 10-19 This circuit appears as a high resistance to the input. The resistance value depends on R_1, R_2, and R_3 (1 GΩ using the values shown). Raytheon Linear Integrated Circuits, 1989, p. 4-129.

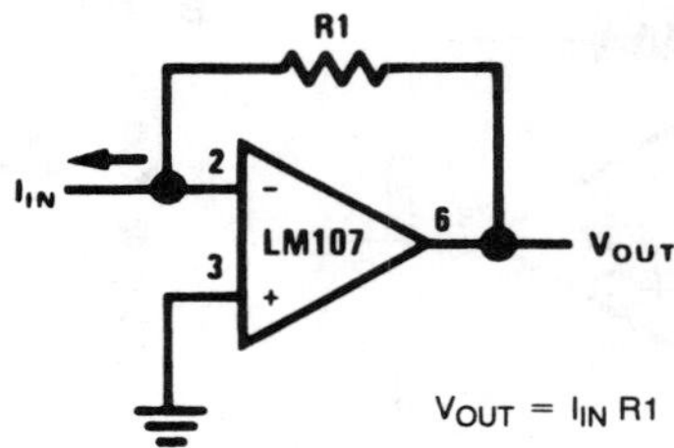

Basic current-to-voltage converter

Fig. 10-20 With this basic circuit, the input current is fed directly into the summing node (pin 2) and the op-amp output changes to extract the same current from the summing node through R1. The scale factor of this circuit is R_1 volts per amp. That is, the output voltage is equal to the input current times R_1. The only conversion error in this circuit is I_{bias}, which is summed algebraically with I_{IN}. The basic circuit can be used to measure current directly because: $I_{IN} = V_{OUT}/R_1$. For example, if V_{OUT} is 1 V (or 1000 mV), and R_1 is 100 Ω, $I_N = 10$ mA. National Semiconductor, Linear Applications Handbook, 1991, p. 24.

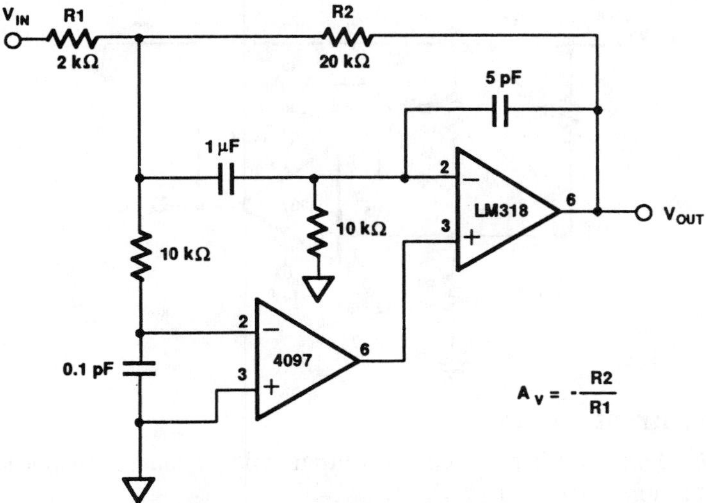

Composite high-speed precision amplifier

Fig. 10-21 This circuit provides a gain of 10 with the values shown. Raytheon Linear Integrated Circuits, 1989, p. 4-129.

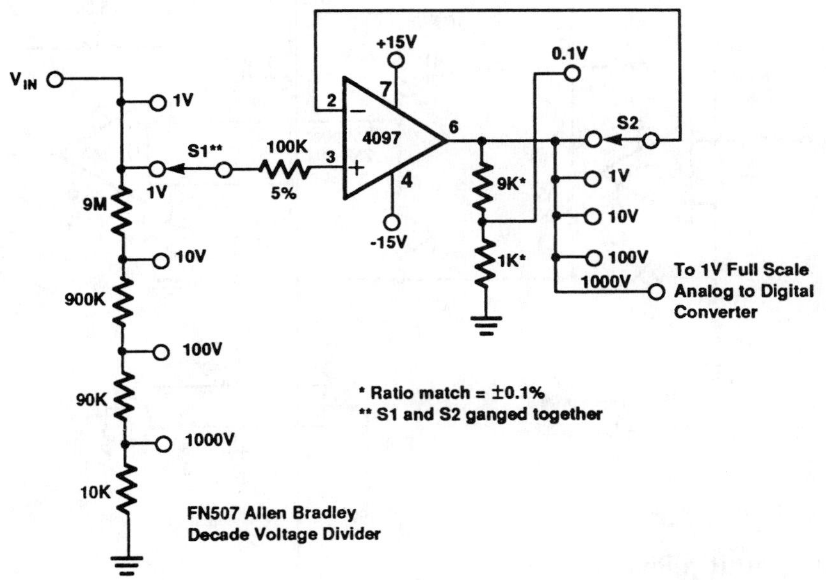

Input amplifier for a 4½-digit voltmeter

Fig. 10-22 This circuit provides an approximate 10-MΩ input impedance, and can be used with any 1-V full-scale A/D converter. Raytheon Linear Integrated Circuits, 1989, p. 4-130.

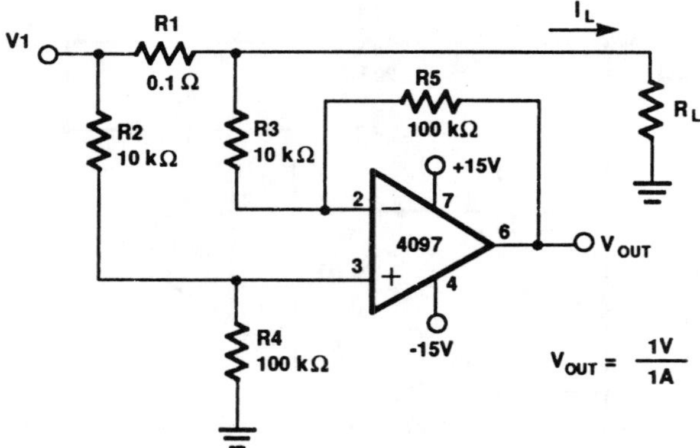

Precision current monitor

Fig. 10-23 This circuit produces an output voltage that is proportional to the current that is applied to the load R_L. Raytheon Linear Integrated Circuits, 1989, p. 130.

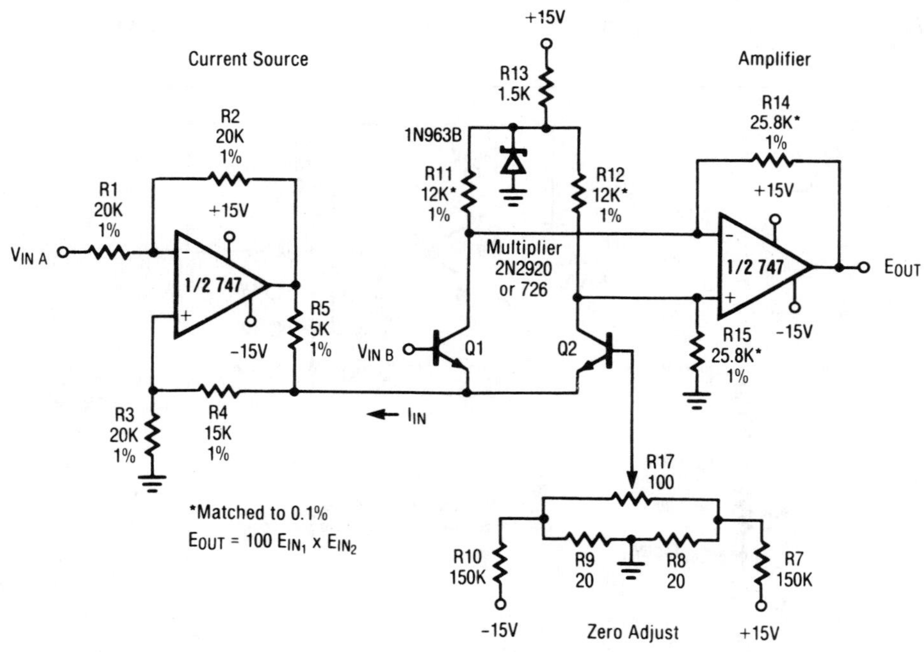

Analog multiplier

Fig. 10-24 The output of this circuit is equal to input A times input B times 100, using the values shown. Raytheon Linear Integrated Circuits, 1989, p. 4-148.

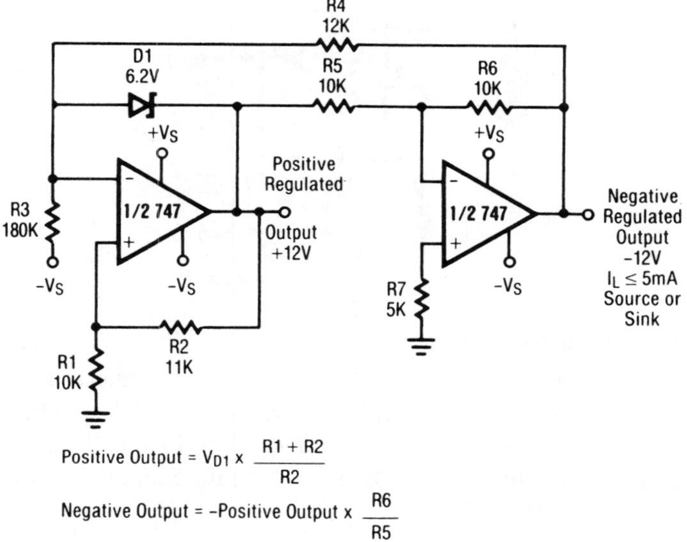

Tracking positive and negative voltage references

Fig. 10-25 This circuit provides both positive and negative voltage references that remain constant, in spite of changes in source voltage or V_S. Raytheon Linear Integrated Circuits, 1989, p. 4-149.

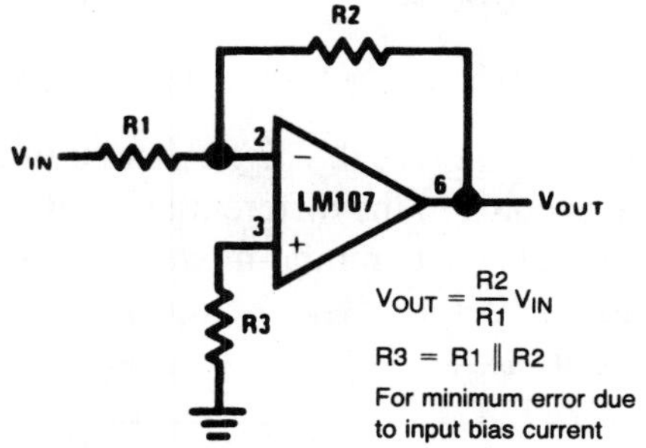

Basic inverting amplifier

Fig. 10-26 This circuit shows an LM107 that is connected in the classic inverting-amplifier configuration, where V_{OUT} is opposite to V_{IN} (if V_{IN} goes positive, V_{OUT} goes negative, and vice versa). The amplitude of the output depends on the ratio of R_1 and R_2 (within the limits of the supply voltage), and R_3 is selected to equal the parallel resistance of R_1 and R_2. National Semiconductor, Linear Applications Handbook, 1991, p. 20.

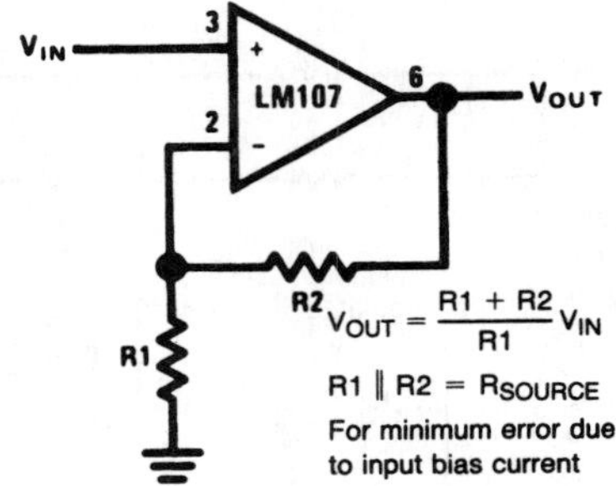

Basic noninverting amplifier

Fig. 10-27 This circuit shows an LM107 that is connected in the classic noninverting amplifier configuration, where V_{OUT} follows V_{IN}. The amplitude of the output depends on the ratio of R_1 and R_2 (within the limits of the supply voltage). The parallel resistance of R_1 and R_2 should equal the source resistance. National Semiconductor, Linear Applications Handbook, 1991, p. 21.

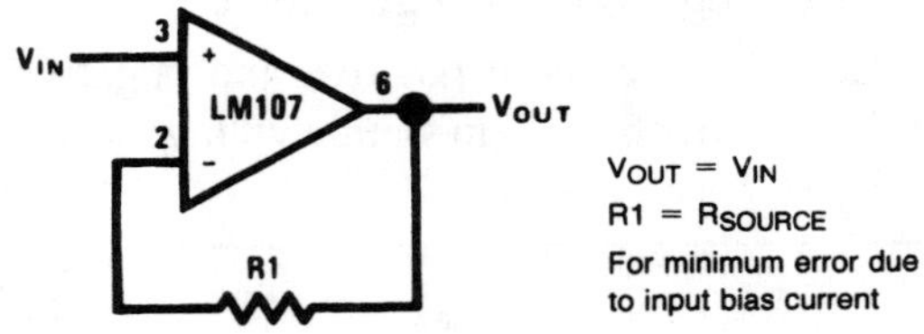

Unity-gain buffer

Fig. 10-28 This configuration gives the highest input impedance of any op-amp circuit. Input impedance is equal to the differential input impedance multiplied by the open-loop gain, in parallel with common-mode input impedance. The gain error of this circuit is equal to the reciprocal of the open-loop gain or to the common-mode rejection, whichever is less. National Semiconductor, Linear Applications Handbook, 1991, p. 21.

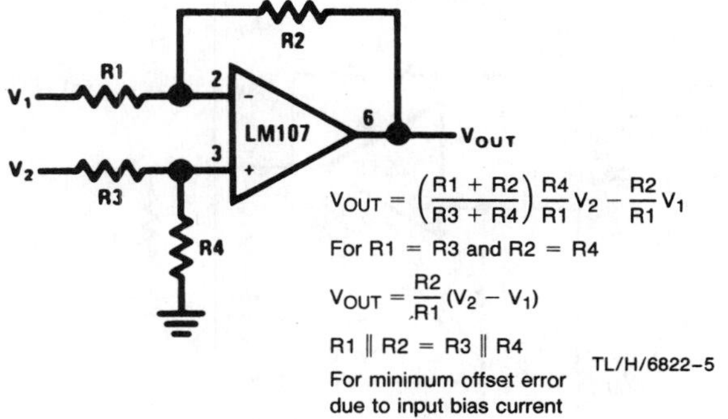

Difference amplifier

Fig. 10-29 This circuit shows an LM107 that is connected in the classic difference-amplifier configuration, where V_{OUT} depends on the difference between V_1 and V_2, as well as the ratios of R_1 through R_4. As shown, the calculations for V_{OUT} are simplified when $R_1 = R_3$ and $R_2 = R_4$. National Semiconductor, Linear Applications Handbook, 1991, p. 22.

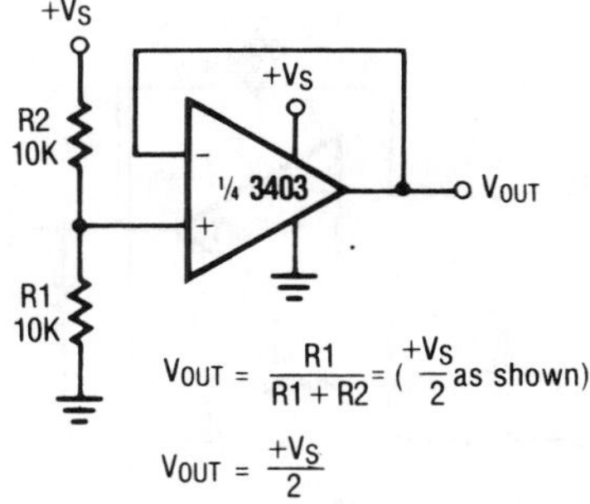

Simple voltage reference

Fig. 10-30 This circuit uses one section of a 3403 op amp as a voltage reference. Compare this basic circuit to the higher-precision circuit of Fig. 4-18. Raytheon Linear Integrated Circuits, 1989, p. 4-159.

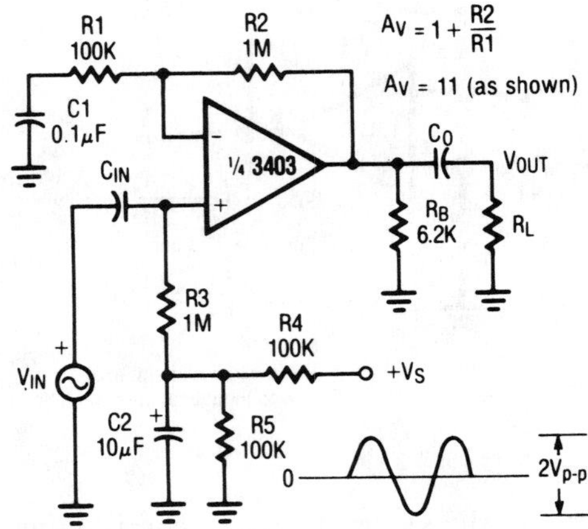

Ac-coupled noninverting amplifier

Fig. 10-31 This circuit is the ac version of the basic dc circuit in Fig. 10-27. Raytheon Linear Integrated Circuits, 1989, p. 4-160.

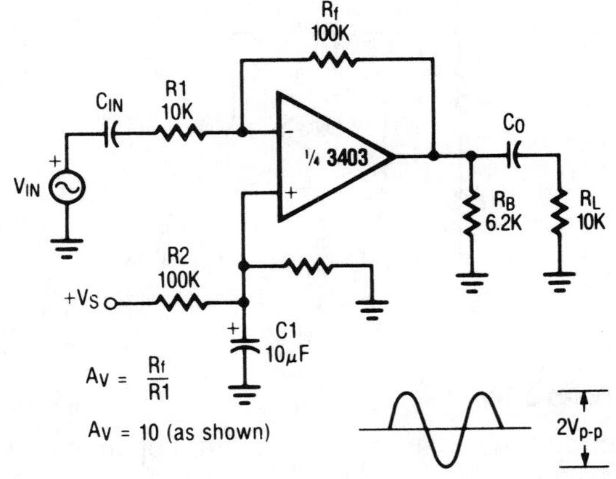

Ac-coupled inverting amplifier

Fig. 10-32 This circuit is the ac version of the basic dc circuit in Fig. 10-26. Raytheon Linear Integrated Circuits, 1989, p. 4-160.

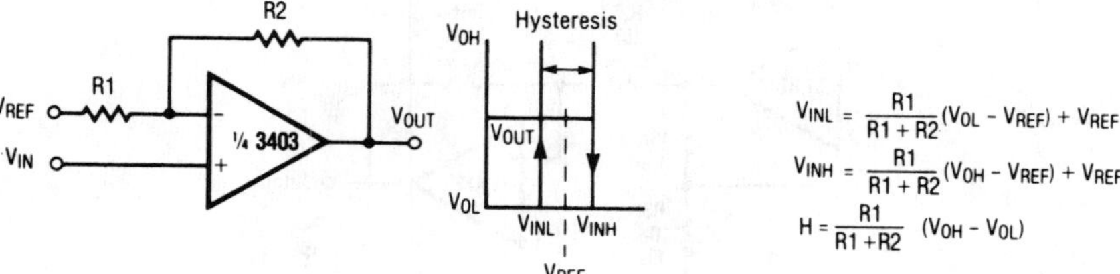

Basic comparator with hysteresis

Fig. 10-33 This circuit uses one section of a 3403 op amp as a comparator (without the use of an IC comparator). Raytheon Linear Integrated Circuits, 1989, p. 4-160.

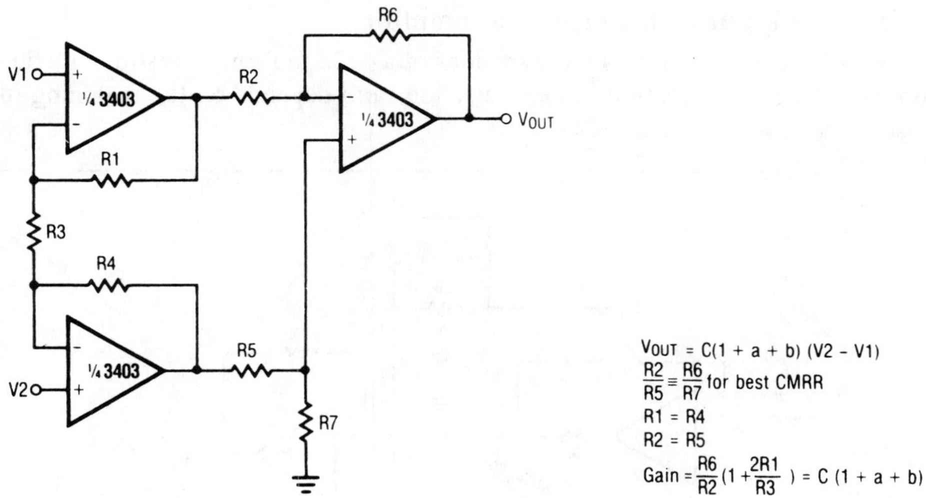

High-impedance differential amplifier

Fig. 10-34 This circuit uses three sections of a 3404 op amp to form a high-impedance amplifier, where the output depends on the difference between V_1 and V_2, as well as the resistance ratios. By eliminating the input resistances (such as shown in Fig. 10-29), input impedance depends on the 3404 alone. Raytheon Linear Integrated Circuits, 1989, p. 4-161.

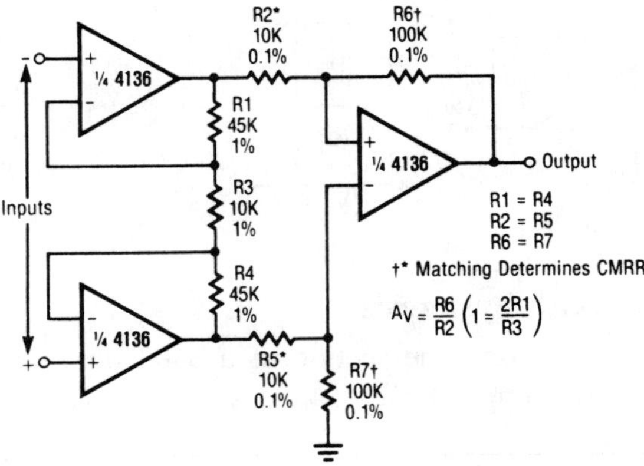

Differential-input instrumentation amplifier

Fig. 10-35 This circuit uses three sections of a 4136 to form an instrumentation amplifier, where the common-mode rejection ratio depends on the matching of resistors. Raytheon Linear Integrated Circuits, 1989, p. 4-177.

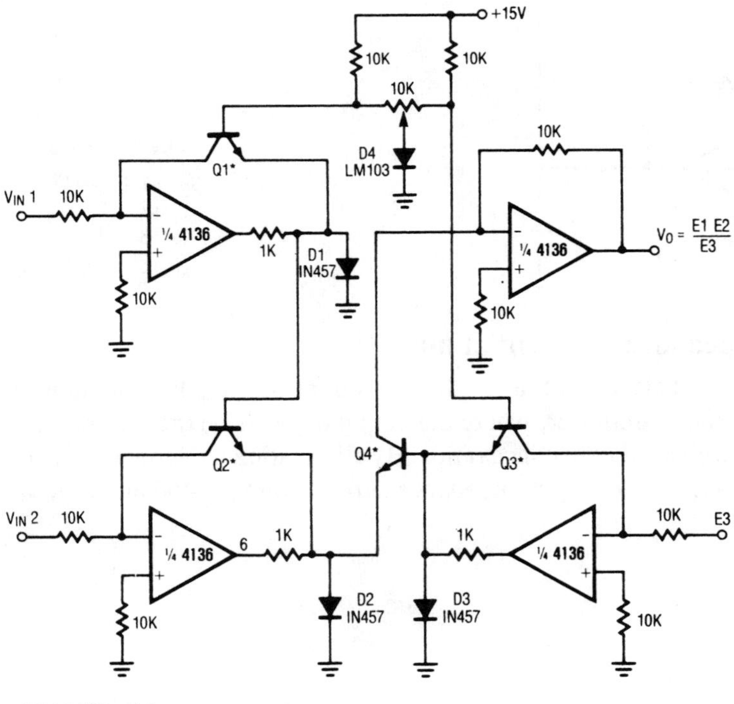

10
512

Analog multiplier/divider

Fig. 10-36 This circuit uses all four sections of a 4136 to form a circuit that both multiplies and divides (simultaneously if required). Virtually any pnp transistors can be used for Q1 through Q4, provided that the transistors are matched. The accuracy of the circuit multiplication and division depends on transistor matching. Raytheon Linear Integrated Circuits, 1989, p. 4-178.

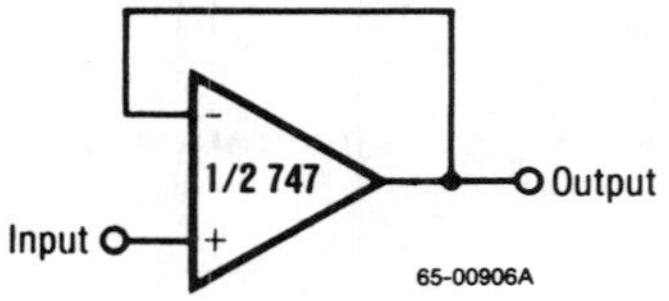

Input
1/2 747
Output
65-00906A

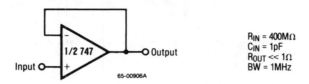

$R_{IN} = 400M\Omega$
$C_{IN} = 1pF$
$R_{OUT} << 1\Omega$
$BW = 1MHz$

Basic unity-gain voltage follower

Fig. 10-37 This circuit uses one-half of a 747 connected in the classic unity-gain voltage-follower configuration, where the input impedance is high and the output impedance is low. Raytheon Linear Integrated Circuits, 1989, p. 4-150.

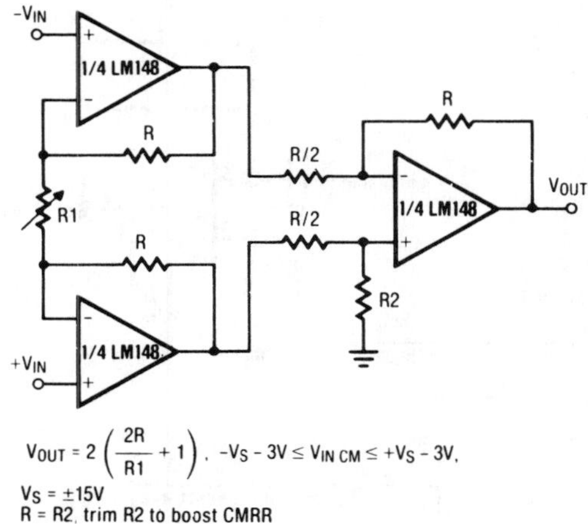

Low-cost instrumentation amplifier

Fig. 10-38 Compare this simple low-cost circuit to that of Fig. 10-15. Raytheon Linear Integrated Circuits, 1989, p. 4-264.

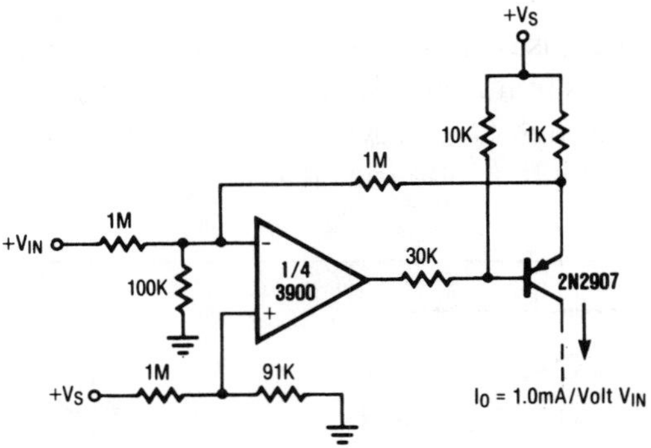

Voltage-controlled current source

Fig. 10-39 This circuit provides a variable current that is controlled by input voltage $+V_{IN}$. The circuit is sometimes called a *transconductance amplifier*, and is similar to the OTAs (operational transconductance amplifiers) that are described in chapter 11. Raytheon Linear Integrated Circuits, 1989, p. 4-273.

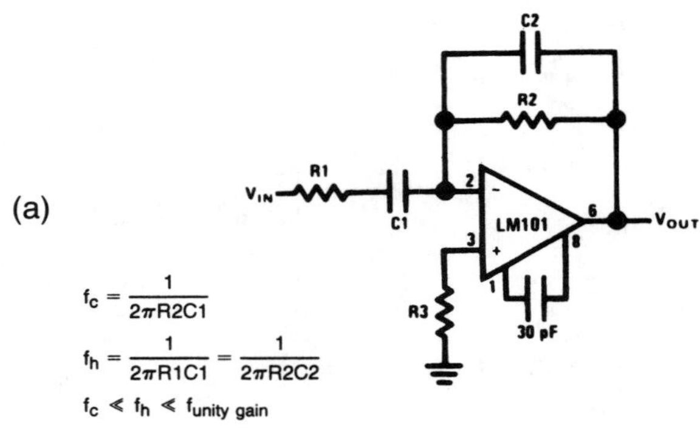

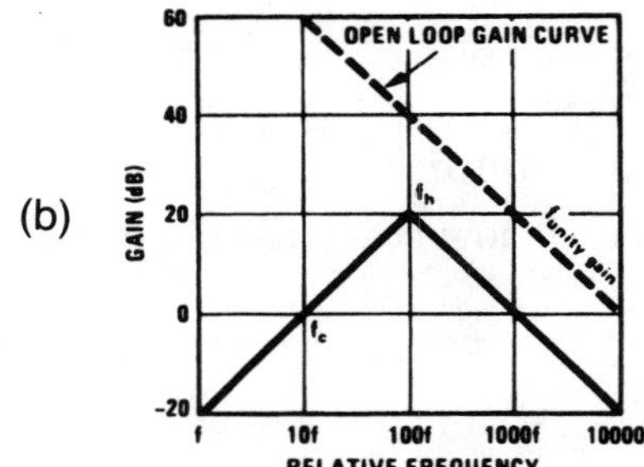

Differentiator

Fig. 10-40 This circuit was originally developed to perform the mathematical operation of differentiation in analog computers. As shown by the frequency-response plot of Fig. 10-40B, a differentiator is actually a form of filter. R2/C2 form a 6-dB per octave rolloff network in the feedback, and R1/C1 form a similar network at the input. National Semiconductor, Linear Applications Handbook, 1991, p. 23.

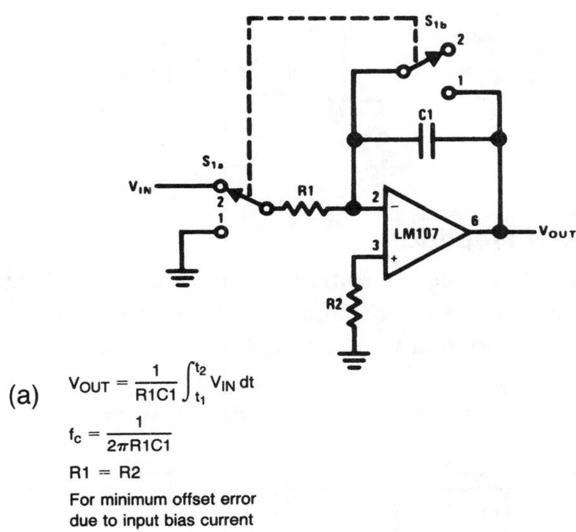

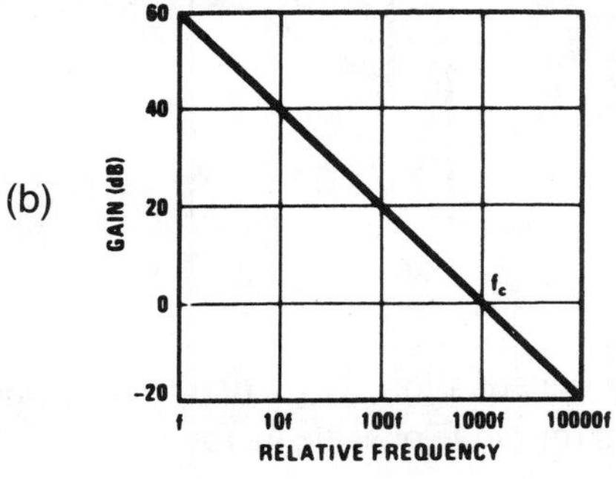

Integrator

Fig. 10-41 This circuit was originally developed to perform the mathematical operation of integration in analog computers. As shown by the frequency-response plot of Fig. 10-41B, an integrator is essentially a low-pass filter with a frequency response that decreases at 6 dB per octave. When S1 is in position 1, the amplifier is connected in unity gain, and C1 is discharged, which sets an initial condition of zero volts. With S1 in position 2, the amplifier is connected as an integrator, where the output changes in accordance with a constant times the time integral of the input. National Semiconductor, Linear Applications Handbook, 1991, p. 23.

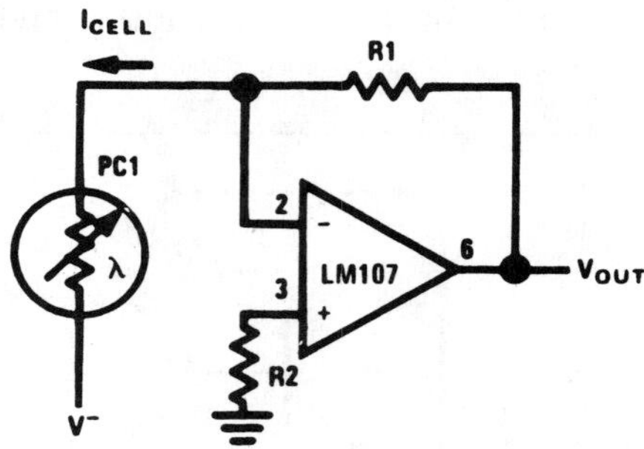

Photoconductive-cell amplifier

Fig. 10-42 R_1 depends on cell sensitivity, and should be chosen for either maximum dynamic range, or for a desired scale factor. R2 is used to minimize output offset, and can be omitted in many applications. National Semiconductor, Linear Applications Handbook, 1991, p. 24.

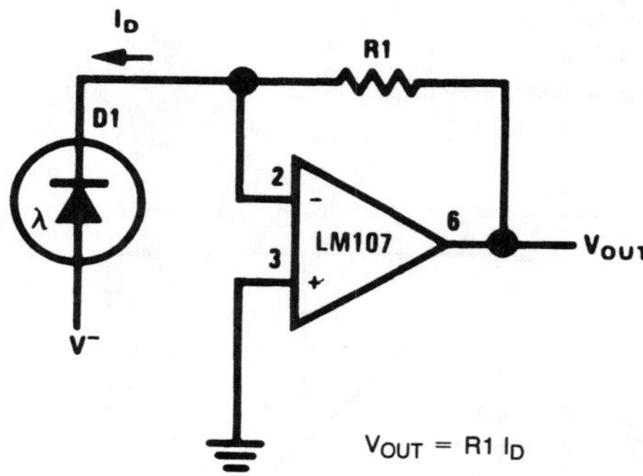

Photodiode amplifier

Fig. 10-43 As shown by the equation, V_{OUT} depends on photodiode current and R_1 (which should be chosen for range or scale factor). National Semiconductor, Linear Applications Handbook, 1991, p. 24.

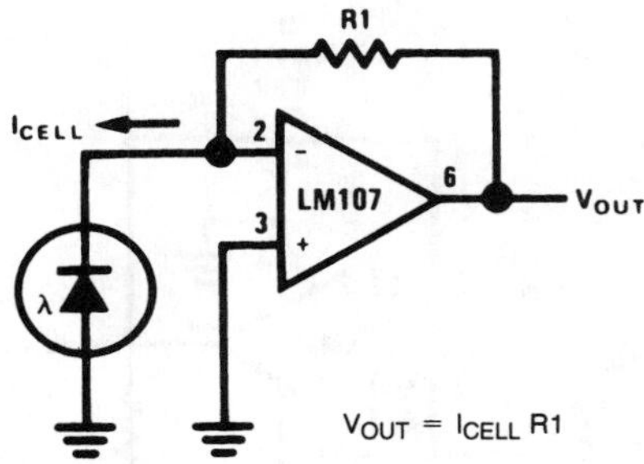

Photovoltaic-cell amplifier

Fig. 10-44 As shown by the equation, V_{OUT} depends on photovoltaic-cell current and R_1 (which should be chosen for range or scale factor). National Semiconductor, Linear Applications Handbook, 1991, p. 24.

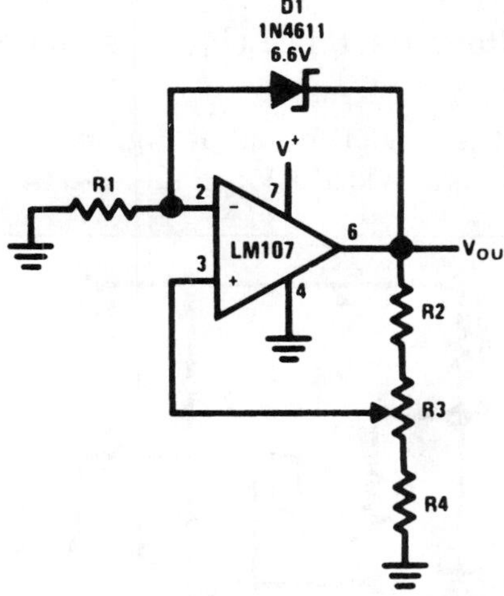

Positive variable voltage reference (V_{OUT} is higher than reference zener)

Fig. 10-45 High-precision extended-temperature applications of this circuit require that the range of adjustment for V_{OUT} is restricted. When this is done, R_1 can be chosen to provide optimum zener current for minimum zener temperature coefficient (*TC*). Because zener current I_Z is not a function of V+, reference *TC* is independent of V+. National Semiconductor, Linear Applications Handbook, 1991, p. 25.

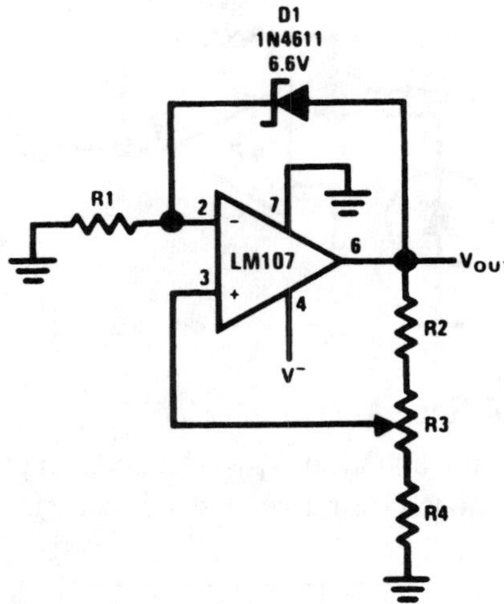

Negative variable voltage reference (V_{OUT} is higher than reference zener)

Fig. 10-46 This circuit is similar to that of Fig. 10-45, except that a variable negative-voltage reference is provided at V_{OUT}. National Semiconductor, Linear Applications Handbook, 1991, p. 25.

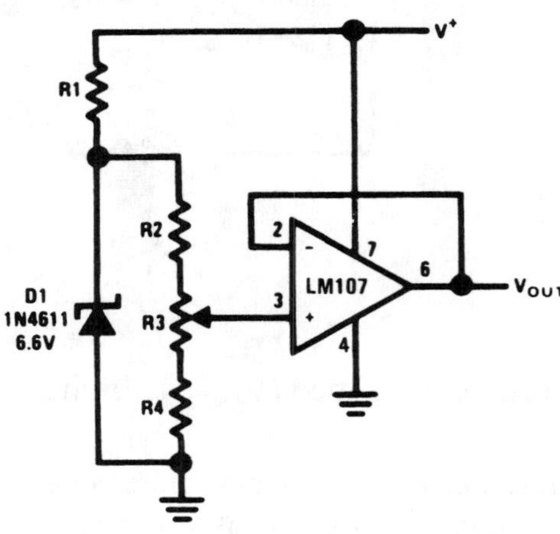

Positive variable voltage reference (V_{OUT} is lower than reference zener)

Fig. 10-47 This circuit is suited for high-precision extended-temperature service, if V+ is reasonably constant (Because I_Z does depend on V+). R1 through R4 are chosen to provide optimum zener current for minimum zener temperature coefficient (*TC*), and to minimize error of I_{bias}. National Semiconductor, Linear Applications Handbook, 1991, p. 26.

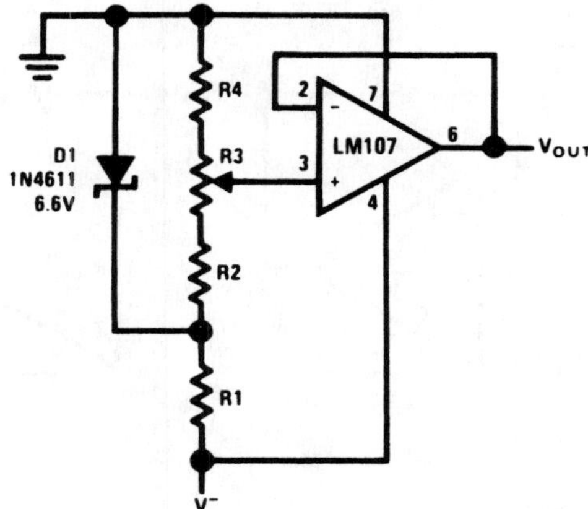

Negative variable voltage reference (V_{OUT} lower than reference zener)

Fig. 10-48 This circuit is similar to that of Fig. 10-47, except that a variable negative-voltage reference is provided for V_{OUT}. National Semiconductor, Linear Applications Handbook, 1991, p. 26.

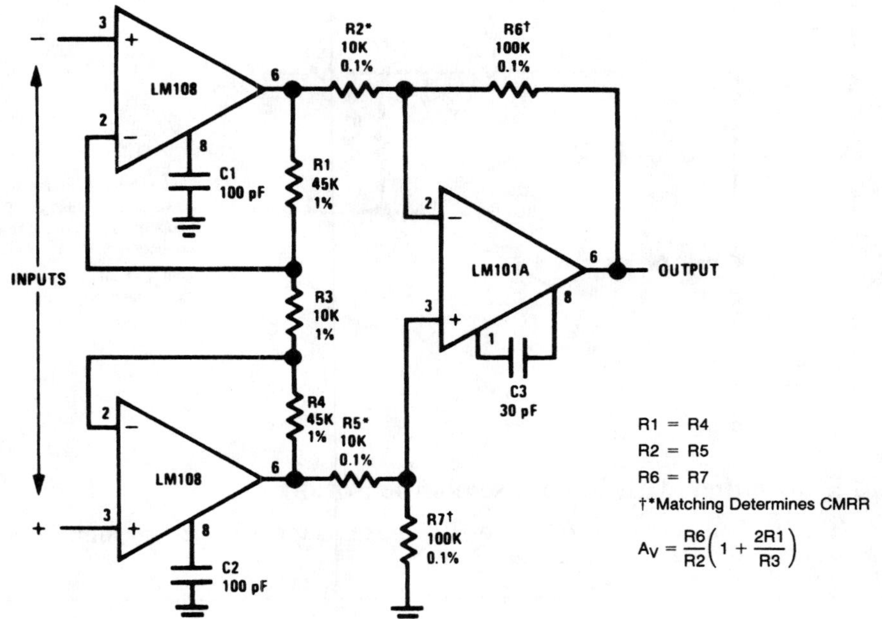

Instrumentation amplifier with ±10-V common-mode range

Fig. 10-49 Compare this circuit to that of Fig. 10-38. National Semiconductor, Linear Applications Handbook, 1991, p. 90.

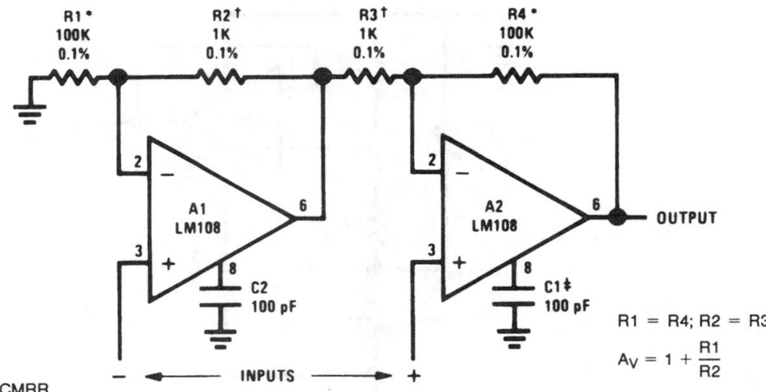

High input impedance instrumentation amplifier

Fig. 10-50 Compare this circuit to that of Fig. 10-38 and 10-49. National Semiconductor, Linear Applications Handbook, 1991, p. 90.

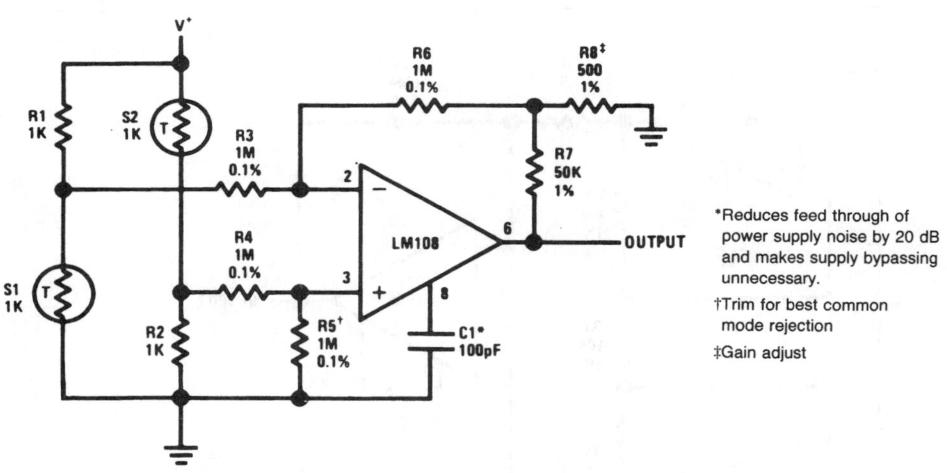

Bridge amplifier with low-noise compensation

Fig. 10-51 Compensating capacitor C1 makes supply bypassing unnecessary. National Semiconductor, Linear Applications Handbook, 1991, p. 90.

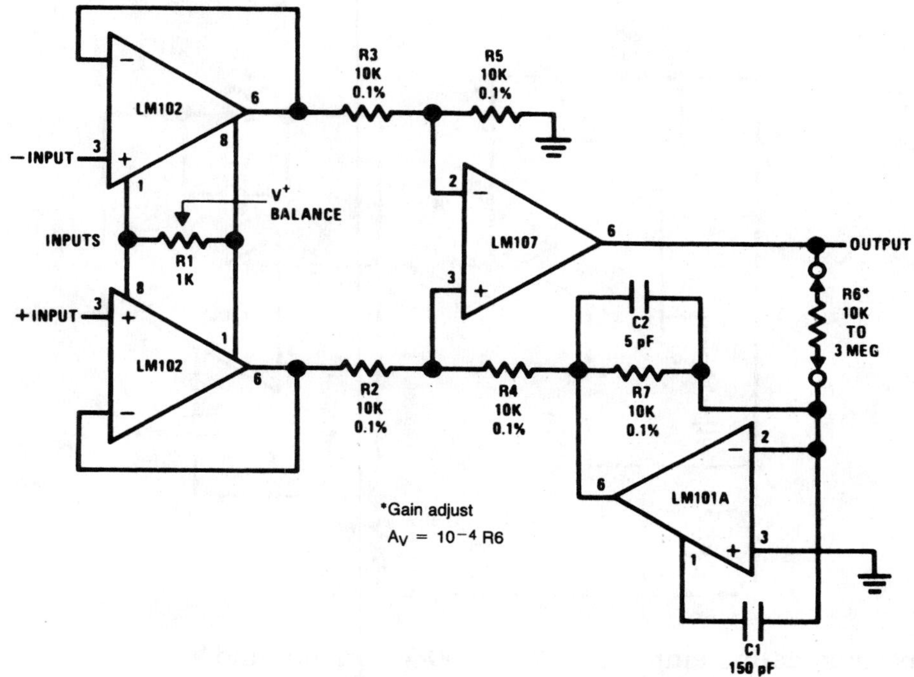

Variable-gain differential-input instrumentation amplifier

Fig. 10-52 The gain of this instrumentation amplifier can be set by selecting R_6. The amplifier can be balanced by adjusting R_1. National Semiconductor, Linear Applications Handbook, 1991, p. 89.

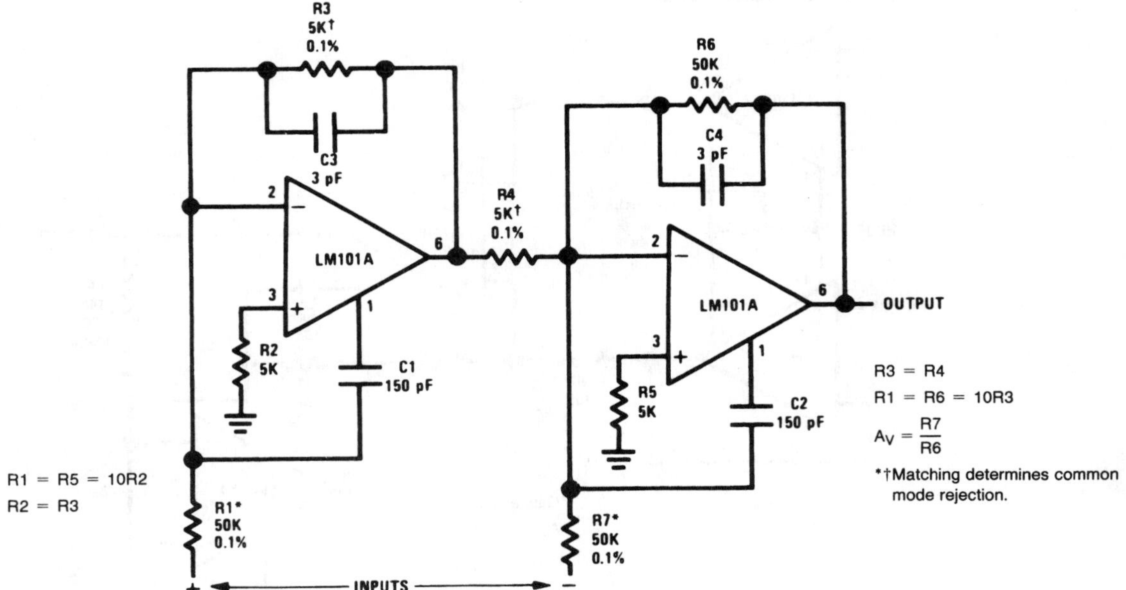

Instrumentation amplifier with ±100-V common-mode range

Fig. 10-53 This circuit provides a much higher common-mode range than that of Fig. 10-49, but has a lower input impedance. National Semiconductor, Linear Applications Handbook, 1991, p. 89.

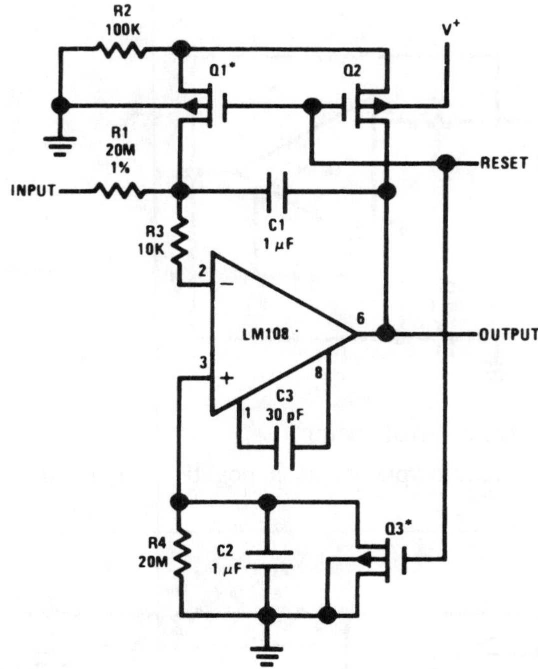

*Q1 and Q3 should not have internal gate-protection diodes.

Worst case drift less than 500 μV/sec over −55°C to +125°C.

Low-drift integrator

Fig. 10-54 Compare this circuit to that of Fig. 10-41. Here, the mechanical switch is replaced by Q1 and Q2. National Semiconductor, Linear Applications Handbook, 1991, p. 93.

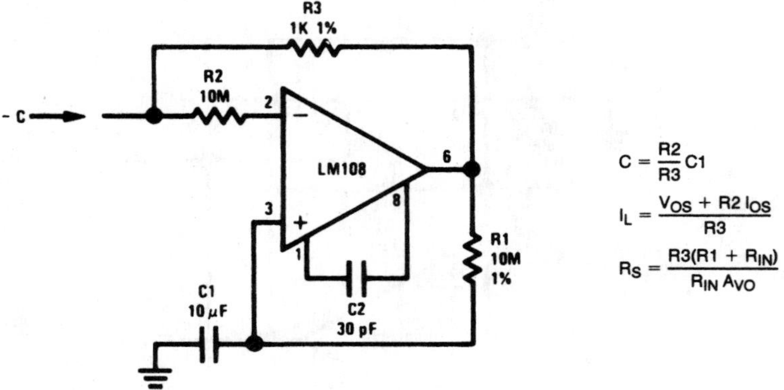

Negative-capacitance multiplier

Fig. 10-55 This circuit appears as a negative capacitance to the input. National Semiconductor, Linear Applications Handbook, 1991, p. 96.

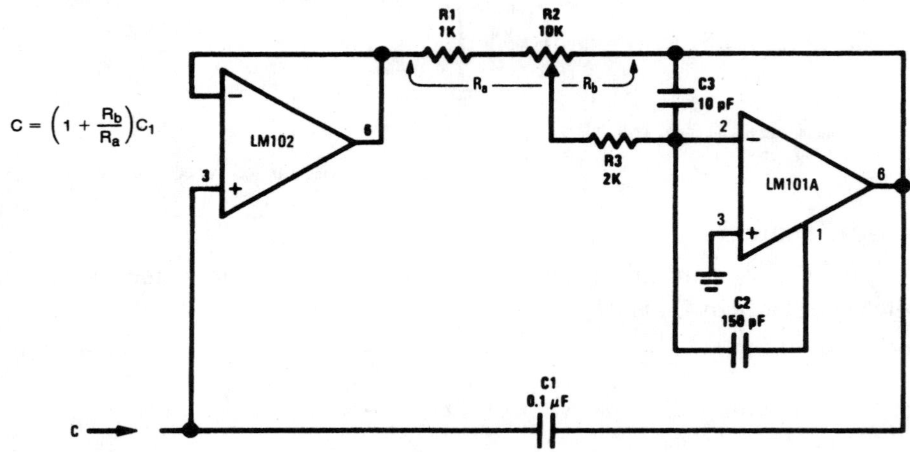

Variable-capacitance multiplier

10

Fig. 10-56 This circuit appears as an adjustable capacitance (set by R2) to the input. National Semiconductor, Linear Applications Handbook, 1991, p. 96.

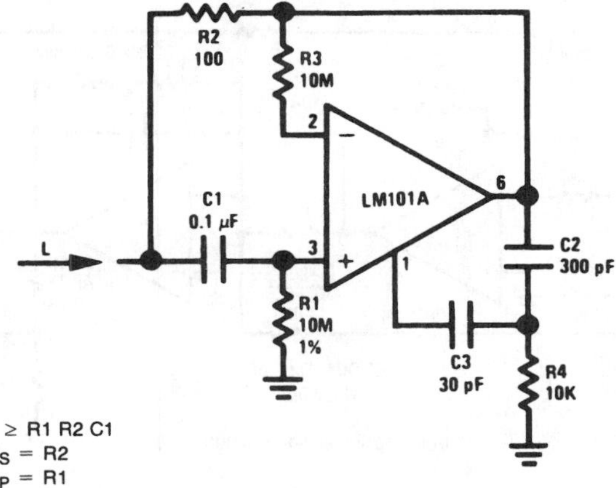

Simulated inductor

Fig. 10-57 This circuit appears as an inductor to the input. The inductance is set by selection of R1, R2, and C1. National Semiconductor, Linear Applications Handbook, 1991, p. 96.

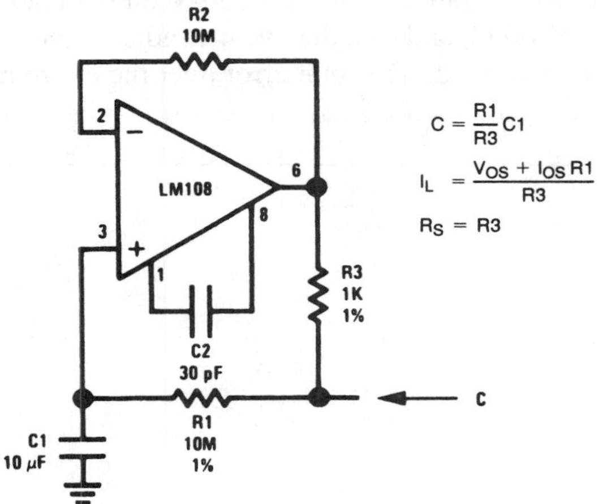

$$C = \frac{R1}{R3} C1$$

$$I_L = \frac{V_{OS} + I_{OS} R1}{R3}$$

$$R_S = R3$$

Capacitance multiplier

Fig. 10-58 This circuit appears as a capacitance to the input. National Semiconductor, Linear Applications Handbook, 1991, p. 96.

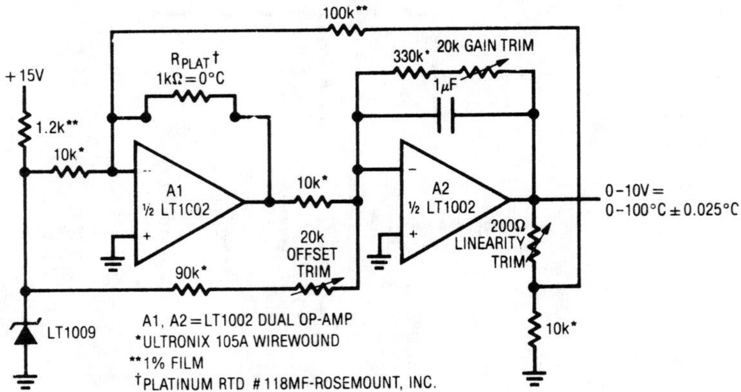

Linear thermometer

Fig. 10-59 This circuit uses both sections of an LT1002 to form a simple (but accurate) thermometer. To calibrate, substitute a precision decade box (General Radio 1432-K) for the sensors. Set the box to the 0°C value (1000.0 Ω) and adjust the offset trim for a 0.000-V output. Then, set the decade box for a 35°C output (1138.7 Ω) and adjust the gain trim for a 3.500-V output reading. Finally, set the box to 1392.6 Ω (100.00°C) and trim the linearity adjustment. Repeat the sequence until all three points are fixed. The total error over the entire range will be within ±0.025°C. Linear Technology Corporation, Linear Applications Handbook, 1990, p. AN6-3.

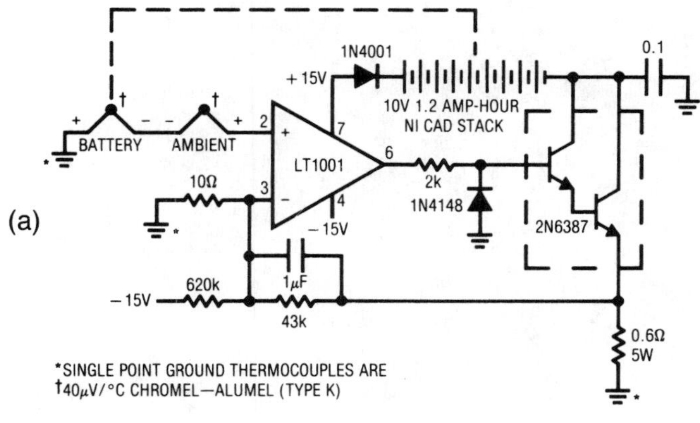

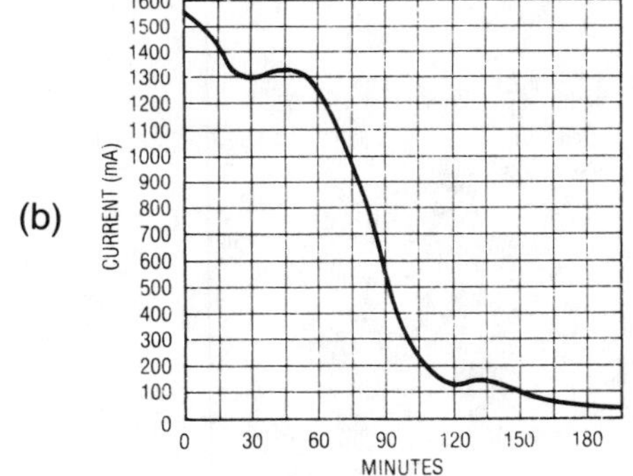

Thermally controlled NiCad battery charger

Fig. 10-60 This circuit uses an LT1001 and a 2N3687 Darlington to form a simple charger for NiCad batteries. The batteries can be charged at a high rate without danger because the cell temperature is measured, and the charging rate is adjusted accordingly. Notice that two thermocouples are used. The battery thermocouple is mounted on one of the cells in the pack, and the ambient thermocouple is exposed to ambient temperature (mounted on a thermal mass that approximates that of the battery pack). The thermocouple voltages cancel and the positive input of the LT1001 is at zero. The temperature difference between the two thermocouples determines the input. As battery temperature rises, this small negative voltage (1°C difference between thermocouples equals 40 μV) becomes larger. The LT1001, operating at a gain of 4300, gradually reduces current through the battery to maintain a balance at the LT1001 input. The effect of this action is shown in Fig. 10-60B. The battery charges at a high rate until heating occurs and the circuit then tapers the charge. With the values shown, the battery-surface temperature rise (above ambient) is limited to the very safe level of 5°C. Linear Technology Corporation, Linear Applications Handbook, 1990, p. AN6-4.

11

OTA and Norton amplifier circuits

Operational transconductance amplifiers (OTAs), including Norton amplifiers, are similar to the op amps of chapter 10, but with certain differences. This introduction summarizes these differences.

OTA and Norton amplifier circuit differences

The characteristics of an OTA are similar to those of an op amp, with two major exceptions. First, the OTA not only includes the usual differential inputs, but it also contains an additional control input in the form of an amplifier bias current, I_{ABC}. This control input increases the OTA's flexibility for use in a wide range of applications. For example, if low power consumption, low input bias, and low offset current, or high input impedance are desired, then low I_{ABC} is selected. On the other hand, if operation into a moderate load impedance is the main consideration, then higher levels of I_{ABC} are used.

The second major difference between an op amp and an OTA is that the OTA output impedance is extremely high (most op amps have very low output impedance). Because of this difference, the output signal of an OTA is best described in terms of current that is proportional to the difference between the voltages of the two inputs (inverting and noninverting).

The output of an OTA is considered to be current, rather than voltage, and the transfer characteristic (or input/output relationship) is best defined in terms of transconductance (g_m), rather than voltage gain. As in the case of a vacuum tube, *transconductance* is the ratio between the difference in current output (I_{OUT}) for a given difference in voltage input (E_{in}).

OTA control current

In the circuit of Fig. 11-30, I_{ABC} at pin 15 is connected to ±6 V through 560-kΩ resistor R_{ABC}. This amount of I_{ABC} produces the desired transconductance to permit a 20-dB gain across a 20-kHz bandwidth. In the circuit of Fig. 11-31, I_{ABC} supplied through a 20-kΩ resistor to all three amplifier sections of the IC to get the required transconductance from each amplifier. In both of these circuits, the IC operates essentially as an op amp, but where the characteristics are tailored to suit circuit requirements by the I_{ABC} current.

Norton amplifiers

The OTA circuits of Figs. 11-1 through 11-29 are also controlled by an external current or voltage (although internal operation of these circuits is quite different from that of an op amp). As shown in Fig. 11-1b, each amplifier is controlled by an external bias (at pins 1 and 16), which sets the transconductance. In this OTA, the transconductance of each amplifier is adjustable over 4 decades.

The amplifiers of Fig. 11-1 through 11-29 are described as a form of Norton amplifier. Notice the modified amplifier symbol, where the output has an adjustable output current (set by the external control bias), and the input has adjustable on-chip predistortion diodes (set by another external bias). On the same chip, but not directly part of the OTA function, are noncommitted Darlington buffers that are used in special circuit applications.

The circuits of Figs. 11-32 through 11-63 are another form of Norton amplifier that require only one power supply. Here, there is no separate pin for control current. Any change in amplifier characteristics is set by current that is applied at the (+) and (−) inputs. Again, notice the modified amplifier symbol. The current-arrow between the inputs implies a current-mode of operation. The symbol also signifies that current is removed from the (−) input, and that the (+) input is a current input (which can control amplifier gain). The signal input can be applied at either the (+) or (−) terminals.

OTA tests

The test procedures for amplifiers that are described in chapters 1, 3, and 10 can generally be applied to the OTAs and Norton amplifiers of this chapter, with certain exceptions.

The output from the amplifiers in this chapter must be considered as a current output (even though the output might produce a voltage across a load). The output can be set or varied by an external voltage or current (either at a separate terminal or at the (+) and (−) inputs. As a result, the external control current (or voltage) must be fixed at some value when making tests.

As an example, if you make a frequency-response check of the circuit in Fig. 11-32A, the $V+$ voltage must remain constant so that a constant current is applied to the (+) input through resistor 2R2.

OTA troubleshooting

The troubleshooting procedures for the amplifiers that are described in chapters 1, 3, and 10 can also be applied to the circuits in this chapter, but remember the effect of the external control voltage or current. Here are some examples.

In the circuit of Fig. 11-16, the frequency of oscillation is set by C, and by the voltage (V_C), which is applied to the op amp through the 30-kΩ resistor. If the circuit is producing an output, but not at the desired frequency, the value of C might be incorrect. If the output frequency cannot be varied, with a variable voltage at the input (either pin 1 or pin 16, Fig. 11-1B), suspect the OTA. If the circuit fails to oscillate, suspect that 1N914 diodes are leaking or that a capacitor is leaking. Check for oscillation signals at the junction of the Darlington buffer input and the feedback capacitor. If there is a signal at the Darlington input, but not at V_0, suspect the Darlington.

In the circuit of Fig. 11-30, the output at pin 16 should follow the input, except that the output should be inverted and with a gain of 20 dB. If not, suspect the op amp. However, the characteristics of the op amp, including gain, are set by the voltage applied at pin 15. Try a different value of R_{ABC} to see if the desired gain can be obtained with all remaining components the same. However, remember that a change in I_{ABC} current also changes other OTA characteristics (such as output impedance).

In the circuit of Fig. 11-37, the output should follow the input (without inversion), with a gain set by the voltage at R5. If there is no output, check for a signal at the (+) input of the op amp. If the signal is present, but there is no signal at V_0, suspect the LM3900. If there is no signal at the (+) input, suspect C_{IN}, CR1, or CR2. If there is an output, but the gain cannot be controlled (or properly controlled), suspect CR2 (which could be leaking). The gain should be about zero when the gain-control voltage is zero, and should be maximum when the gain-control voltage is near 10 V.

OTA circuits titles and descriptions

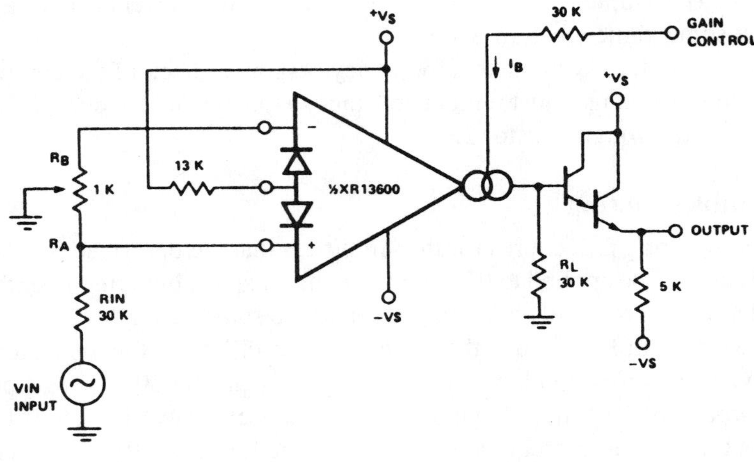

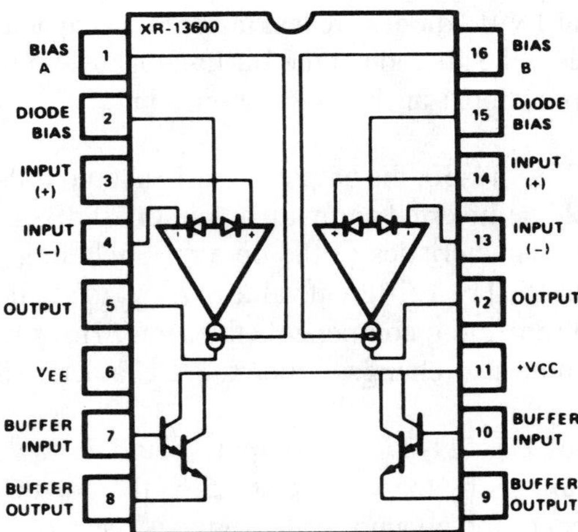

Voltage-controlled amplifier

Fig. 11-1 This circuit uses one section of an XR-13600 dual OTA as a voltage controlled amplifier. Figure 11-1B shows that both sections of the XR-13600 have predistortion diodes and noncommitted Darlington buffer outputs. In the circuit of Fig. 11-1A, the bias current I_B (pins 1 and 16) controls gain of the output. The 1-kΩ potentiometer at the amplifier input is adjusted to minimize the effects (offset) of the gain-control signal at the output. EXAR Corporation Databook, 1990, p. 5-248, 5-252.

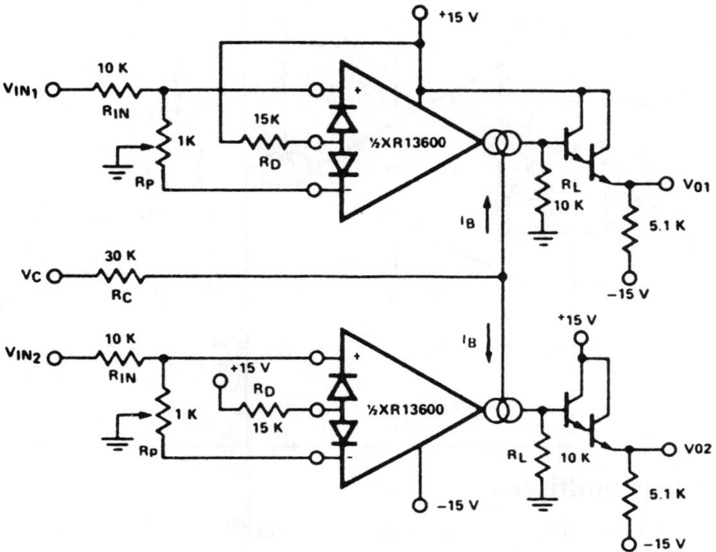

Stereo volume control

Fig. 11-2 This circuit uses both sections of an XR-13600 (Fig. 11-1B) as a stereo volume control with typical channel-to-channel gain tracking of 0.3 dB. The R_P potentiometers are provided to minimize the output offset voltage and they can be replaced with two 510-Ω resistors in ac-coupled applications. With the values shown, amplifier gain is: $V_O/V_{IN} = 940 \times I_B$ (mA). EXAR Corporation Databook, 1990, p. 5-253.

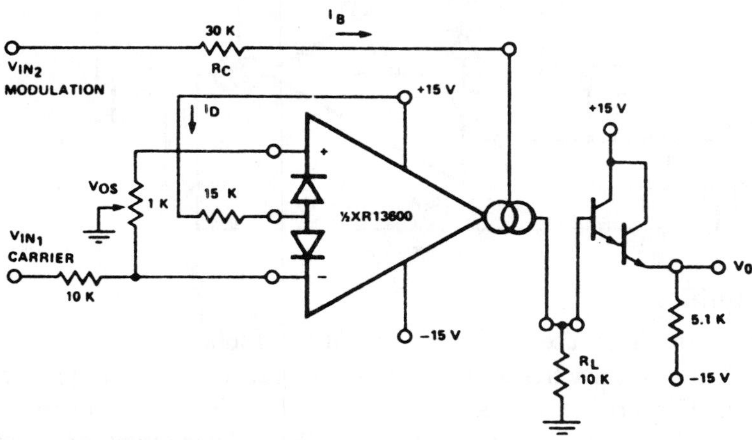

Amplitude modulator

Fig. 11-3 This circuit uses one section of an XR-13600 (Fig. 11-1B) and an amplitude modulator (or two-quadrant multiplier), where $I_O = (-2I_S/I_D) \times (I_B)$. EXAR Corporation Databook, 1990, p. 5-253.

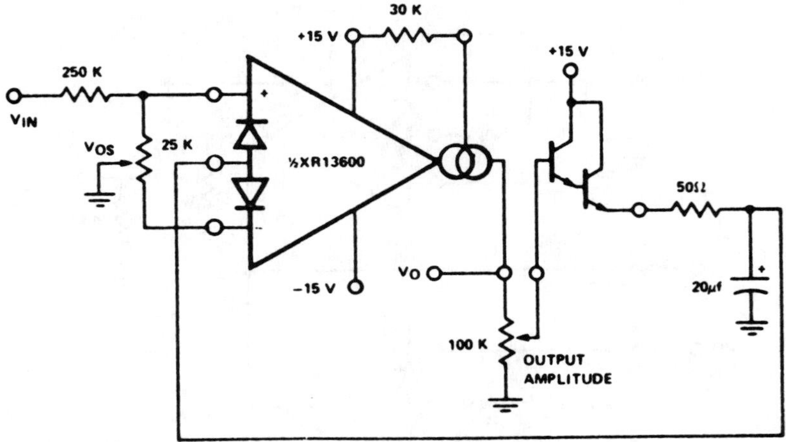

Four-quadrant multiplier

Fig. 11-4 This circuit uses one section of an XR-13600 (Fig. 11-1B) as a four-quadrant multiplier. Notice that the output is taken from the amplifier, and that the buffer output is returned to the amplifier input. The 100-kΩ potentiometer sets the output amplitude and the 25-kΩ potentiometer is used to adjust offset. EXAR Corporation Databook, 1990, p. 5-253.

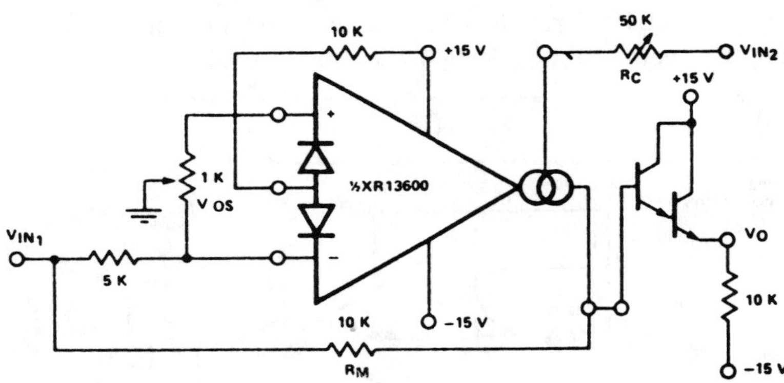

AGC amplifier

Fig. 11-5 This circuit uses one section of an XR-13600 (Fig. 11-1B) as an AGC amplifier. When V_O reaches a high enough amplitude to turn on the Darlington transistors and linearizing diodes, the increase in I_D reduces amplifier gain so as to hold V_O at that level. The control point is set by R_C and the output offset is adjusted by the 1-kΩ potentiometer. EXAR Corporation Databook, 1990, p. 5-253.

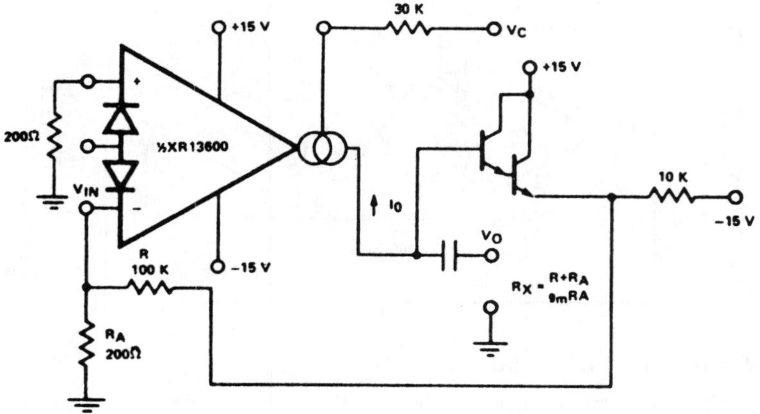

Voltage-controlled resistor

Fig. 11-6 This circuit uses one section of an XR-13600 (Fig. 11-1B) as a single-ended voltage-controlled resistor. A signal applied at R_X generates an input to the XR-13600, which is then multiplied by the *gm* of the amplifier to produce an output current. The resulting "resistor" or $R_X = (R + R_A)/(gm\ R_A)$, where $gm = 19.2\ I_B$ at 25°C. Notice that the attenuation of V_O by R and R_A is necessary to maintain the input within the linear range of the XR-13600 input. EXAR Corporation Databook, 1990, p. 5-254.

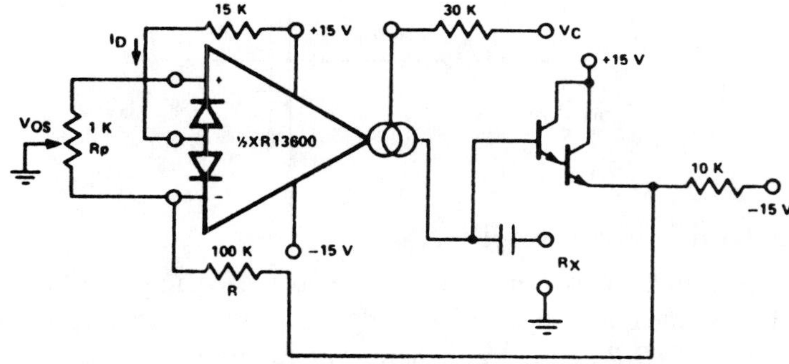

Voltage-controlled resistor with linearizing diodes

Fig. 11-7 This circuit is essentially the same as that of Fig. 11-6, except that the linearizing diodes are included to improve noise performance of the "resistor". EXAR Corporation Databook, 1990, p. 5-254.

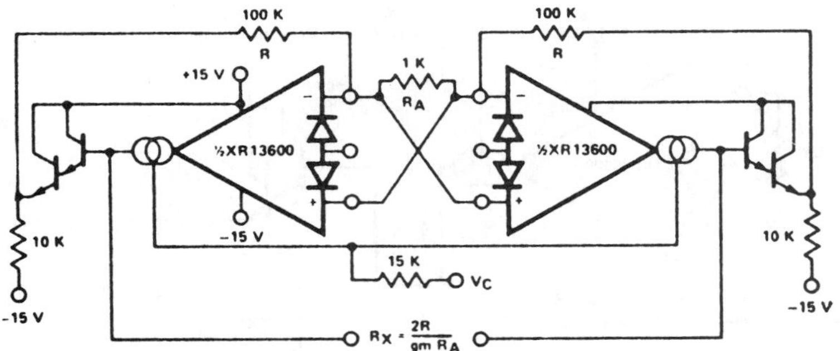

Floating voltage-controlled resistor

Fig. 11-8 This circuit is similar to that of Fig. 11-6, except that both sections of the XR-13600 (Fig. 11-1B) are used. Each "end" of the "resistor" can be at any voltage within the output range of the XR-13600. EXAR Corporation Databook, 1990, p. 5-254.

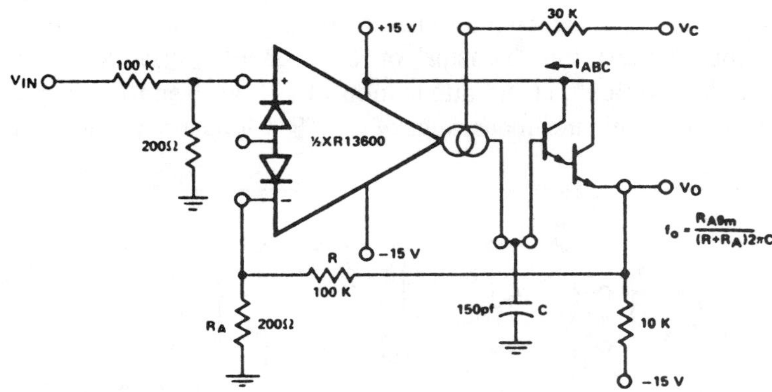

Voltage-controlled low-pass filter

Fig. 11-9 This circuit performs as a unity-gain buffer amplifier at frequencies below cutoff, with the cutoff frequency being the point at which X_C/gm equals the closed-loop gain of (R/R_A). At frequencies above cutoff, the circuit provides a single RC rolloff (6 dB per octave) of the input signal with a −3-dB point that is defined by the given equation, where *gm* (of the XR-13600, Fig. 11-1B) is 19.2 × I_B at room temperature. EXAR Corporation Databook, 1990, p. 5-254.

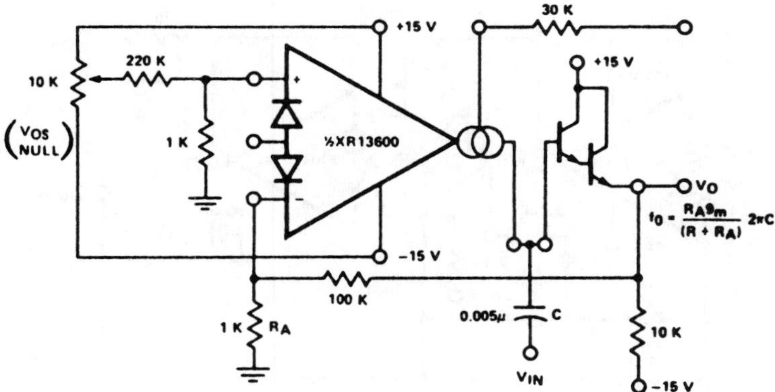

Voltage-controlled high-pass filter

Fig. 11-10 This circuit is similar to that of Fig. 11-9, except that the single RC rolloff is below the defined cutoff frequency. EXAR Corporation Databook, 1990, p. 5-254.

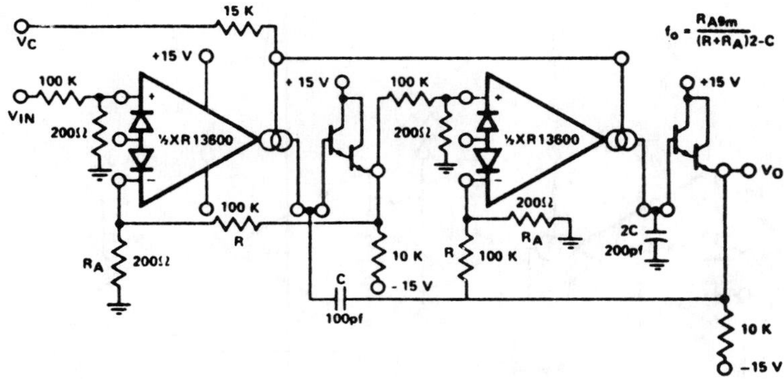

Voltage-controlled Butterworth filter

Fig. 11-11 This circuit uses both sections of an XR-13600 (Fig. 11-1B) to form a 2-pole Butterworth low-pass filter. EXAR Corporation Databook, 1990, p. 5-254.

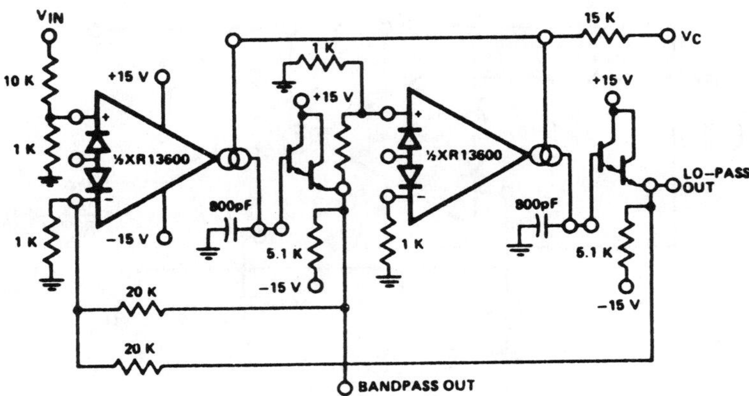

Voltage-controlled state-variable filter

Fig. 11-12 This circuit uses both sections of an XR-13600 (Fig. 11-1B) to form a state-variable filter (with both low-pass and bandpass outputs). EXAR Corporation Databook, 1990, p. 5-255.

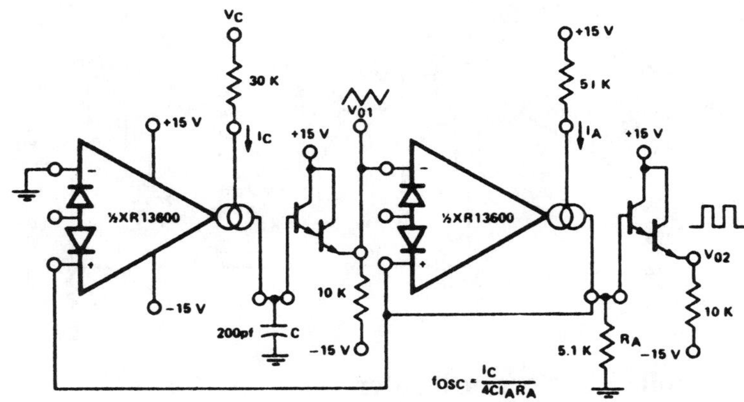

Triangular/square-wave VCO

Fig. 11-13 This classic triangular/square-wave generator (chapter 5) uses both sections of an XR-13600 (Fig. 11-1B). With the values shown, the oscillator provides signals from 200 kHz to below 2 Hz when I_C is varied from 1 mA to 10 nA. The output amplitudes are set by $I_A \times R_A$. Notice that the peak differential-input voltage must be less than 5 V to prevent zenering the inputs. EXAR Corporation Databook, 1990, p. 5-255.

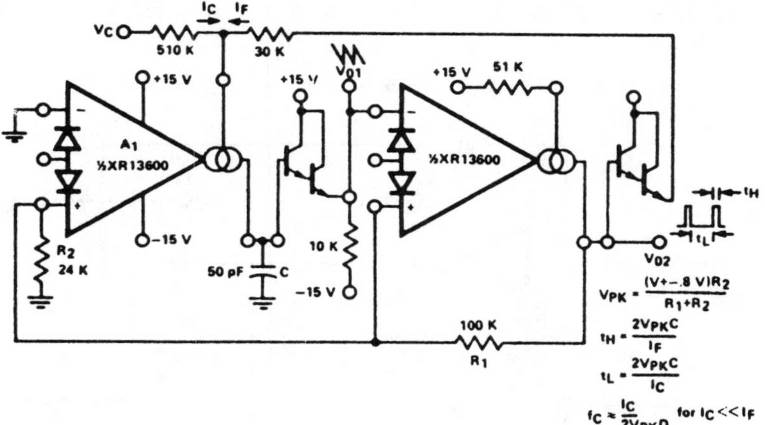

Ramp/pulse VCO

Fig. 11-14 This circuit is similar to that of Fig. 11-13, except that the outputs are ramps and pulses. When V_{02} is high, I_F is added to I_C to increase the bias current of amplifier A_1, and thus to increase the charging rate of capacitor C. When V_{02} is low, I_F goes to zero and the capacitor discharge current is set by I_C. EXAR Corporation Databook, 1990, p. 5-255.

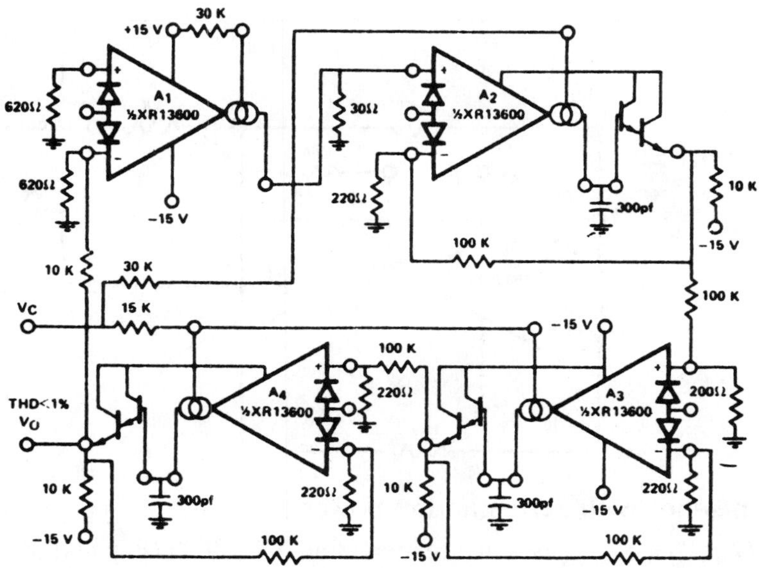

Sinusoidal VCO

Fig. 11-15 This circuit uses two XR-13600 OTAs (Fig. 11-1B), with three of the amplifiers configured as low-pass filters and the fourth as a limiter/inverter. The circuit oscillates at the frequency, where the loop phase-shift is 360° or 180° for the inverter and 60° per filter stage. This VCO operates from 5 Hz to 50 kHz with less than 1% THD. EXAR Corporation Databook, 1990, p. 5-255.

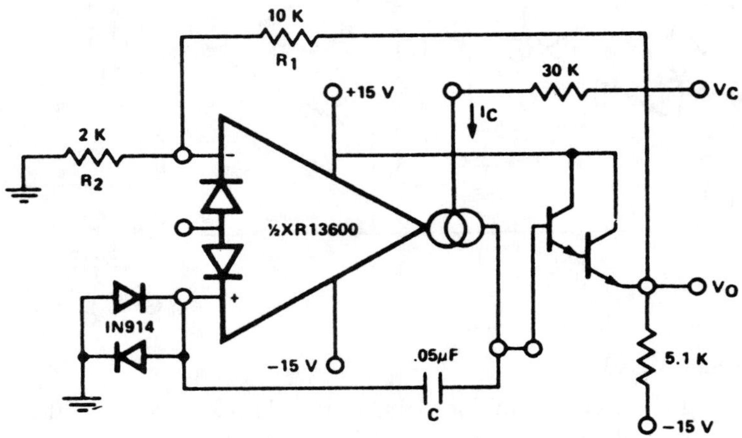

Single-amplifier VCO

Fig. 11-16 This circuit uses only one section of an XR-13600 (Fig. 11-1B) to form a VCO. The remaining section can then be used for another purpose. EXAR Corporation Databook, 1990, p. 5-255.

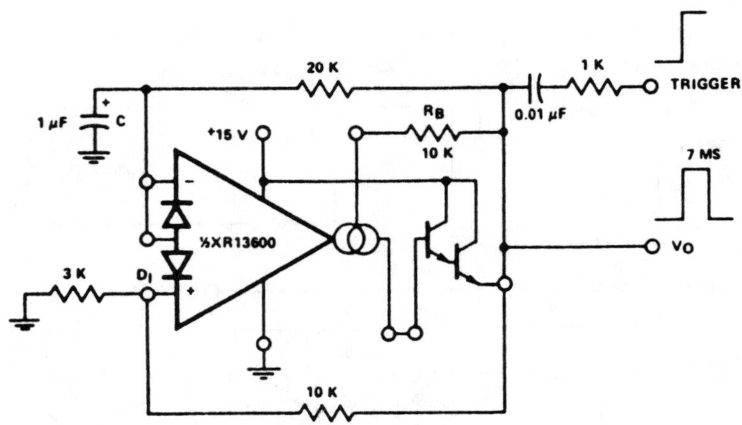

Timer/one-shot with zero standby power

Fig. 11-17 This circuit uses one section of an XR-13600 (Fig. 11-1B) to form a timer or one-shot that draws no current from the supply until it is triggered. The trigger must be at least 2 V, and the output is 7 ms with the values shown. EXAR Corporation Databook, 1990, p. 5-256.

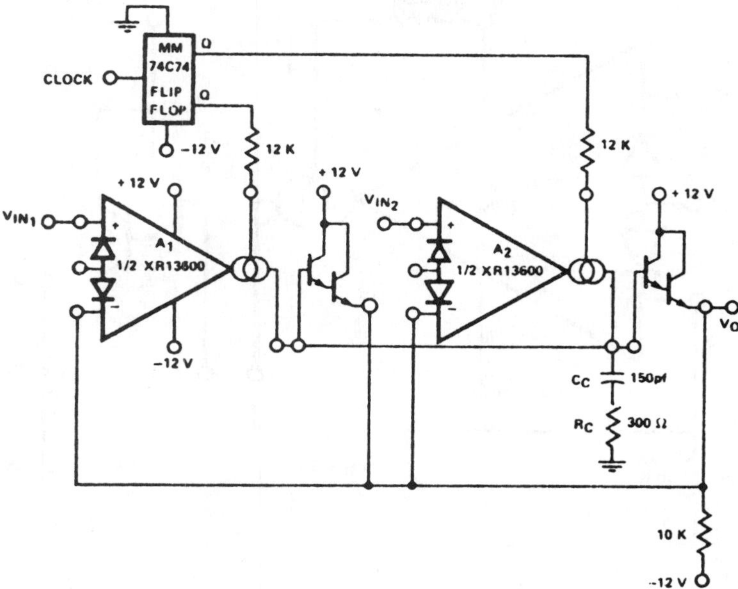

Multiplexer

Fig. 11-18 This circuit uses two sections of an XR-13600 (Fig. 11-1B) and a standard flip-flop to form a multiplexer. The maximum clock rate is limited to about 200 kHz, with a maximum input differential of 5 V. EXAR Corporation Databook, 1990, p. 5-256.

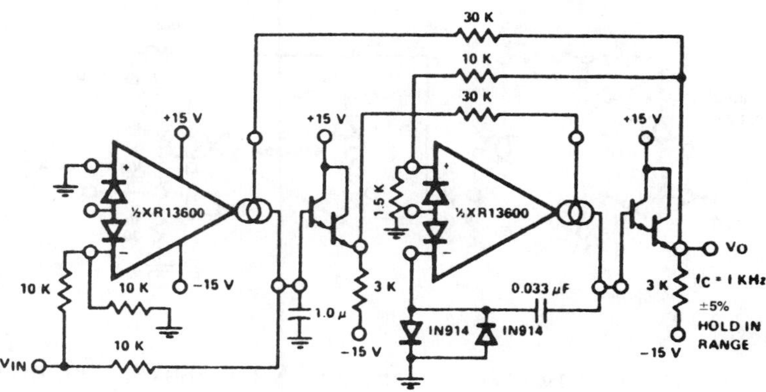

Phase-locked loop

Fig. 11-19 This PLL uses the four-quadrant multiplier (Fig. 11-4) and VCO (Fig. 11-16). The circuit has a ±5% hold-in range and an input sensitivity of about 300 mV. EXAR Corporation Databook, 1990, p. 5-256.

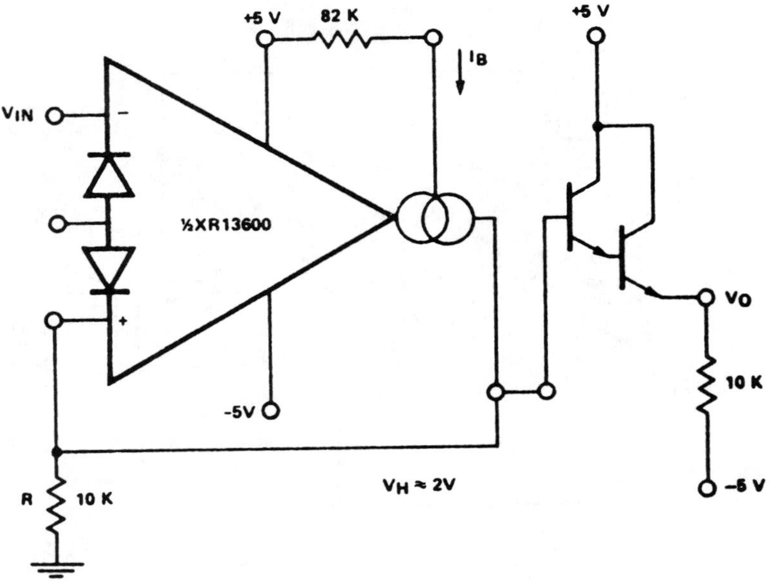

Schmitt trigger

Fig. 11-20 This circuit uses one section of an XR-13600 (Fig. 11-1B) to form a Schmitt trigger. The circuit uses the amplifier output current into R to set the hysteresis of the comparator, thus producing a variable hysteresis, where $V_H = 2 \times R \times I_B$. EXAR Corporation Databook 1990, p. 5-256.

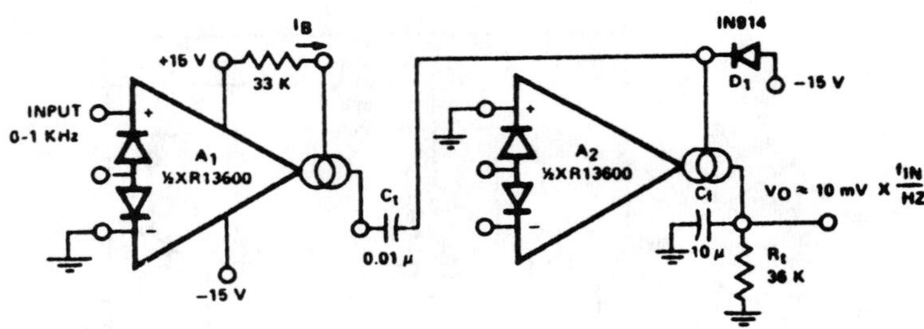

Tachometer

Fig. 11-21 This circuit uses two sections of an XR-13600 (Fig. 11-1b) to form a tachometer or frequency-to-voltage converter. The input is triggered by a positive-going waveform. The maximum F_{IN} is limited by the time required to charge C_t from V_L to V_H with a current of I_B, where V_L and V_H represent the maximum-low and maximum-high output voltage swing of the XR-13600. D1 provides a discharge path for C_t when A_1 switches low. EXAR Corporation Databook, 1990, p. 5-256.

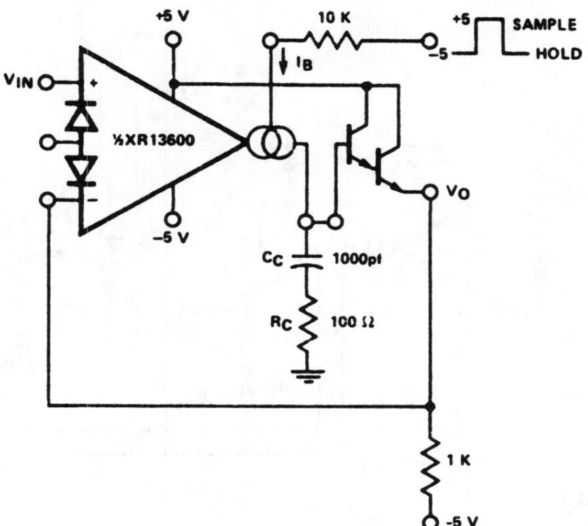

Sample/hold

Fig. 11-22 Although only one section is shown, this sample/hold circuit requires that the Darlington buffer be used from the other half of the package, and that the corresponding amplifier be biased on continuously. EXAR Corporation Databook, 1990, p. 5-257.

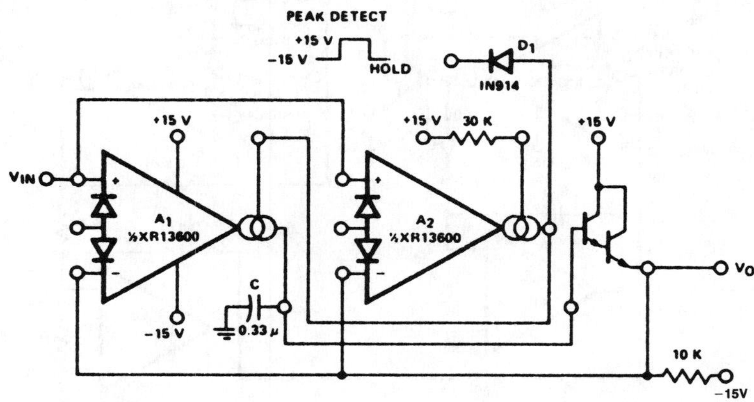

Peak detector and hold

Fig. 11-23 This circuit uses both sections of an XR-13600 (Fig. 11-1B). The peak detector uses A_2 to turn on A_1 whenever V_{IN} becomes more positive than V_O. A_1 then charges C to hold V_O equal to V_{IN} (peak). Notice that the Darlington used must be on the same side of the package as A_2 because the A_1 Darlington is turned on and off with A_1. Pulling the output of A_2 low through D_1 serves to turn off A_1 so that V_O remains constant. EXAR Corporation Databook, 1990, p. 5-257.

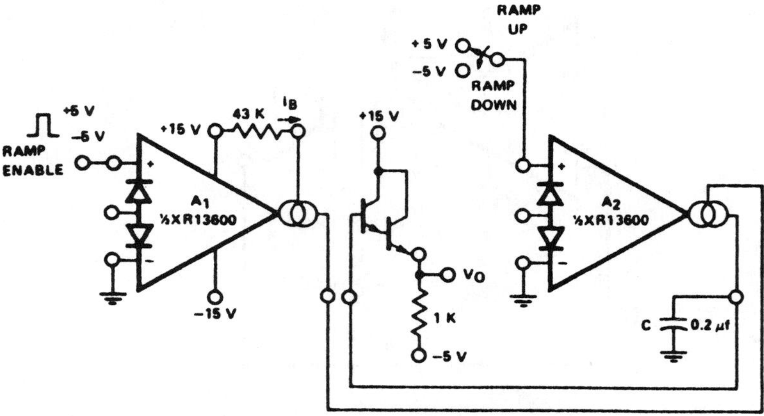

Ramp and hold

Fig. 11-24 This circuit uses both sections of an XR-13600 (Fig. 11-1B). The circuit sources I_B into capacitor C whenever the input to A_1 is made high, giving a ramp rate of about 1 V/ms for the values shown. EXAR Corporation Databook, 1990, p. 5-257.

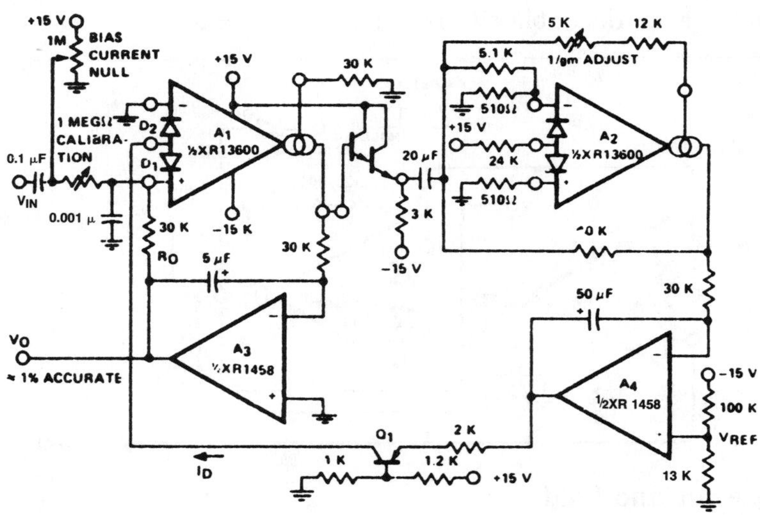

True-rms converter

Fig. 11-25 This circuit uses both sections of an XR-13600 (Fig. 11-1B). The circuit is essentially an AGC amplifier, which adjusts the gain such that the output of A1 is constant. The calibration potentiometer is set such that V_0 reads directly in rms volts. EXAR Corporation Databook, 1990, p. 5-257.

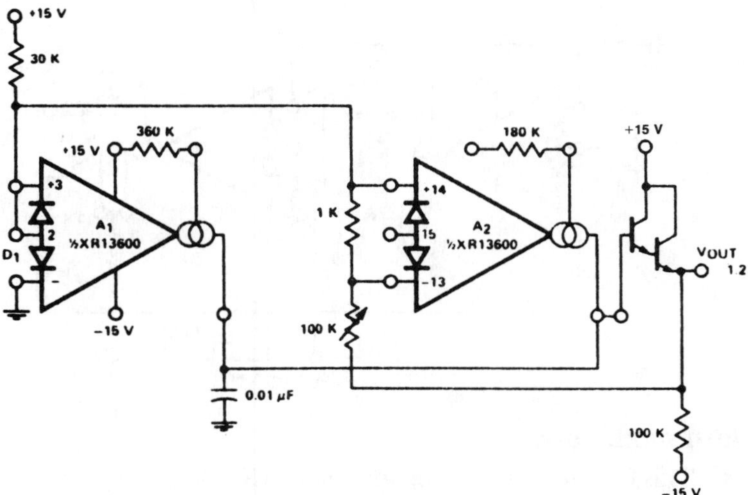

Voltage reference with variable TC

Fig. 11-26 This circuit uses both sections of an XR-13600 (Fig. 11-1B) to form a voltage reference with variable temperature coefficient. The 100-kΩ potentiometer adjusts the output voltage, which has: 1) a positive TC above 1.2 V, 2) zero TC at about 1.2 V, and 3) a negative TC below 1.2 V. This is done by balancing the TC of the A_2 transfer function against the complementary TC of D1. EXAR Corporation Databook, 1990, p. 5-257.

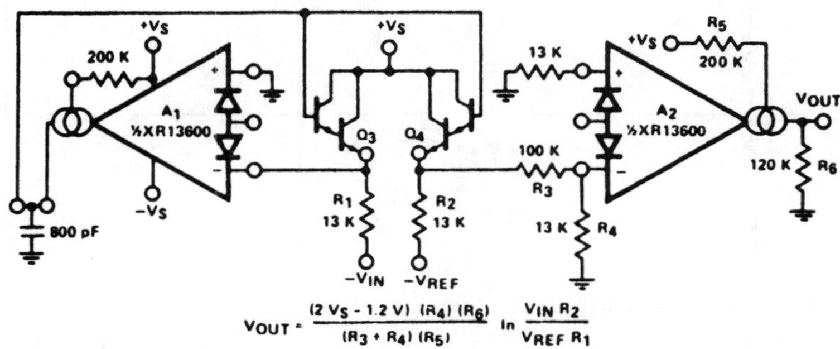

$$V_{OUT} = \frac{(2\,V_S - 1.2\,V)\,(R_4)\,(R_6)}{(R_3 + R_4)\,(R_5)} \ln \frac{V_{IN}\,R_2}{V_{REF}\,R_1}$$

Log amplifier

Fig. 11-27 This circuit uses both sections of an XR-13600 (Fig. 11-1B) to form a log amplifier. The amplifier responds to the ratio of current through Q3/Q4. Zero temperature dependence for V_{OUT} is ensured because the TC of the A_2 transfer function is equal and opposite to the TC of the transistors. EXAR Corporation Databook 1990, p. 5-258.

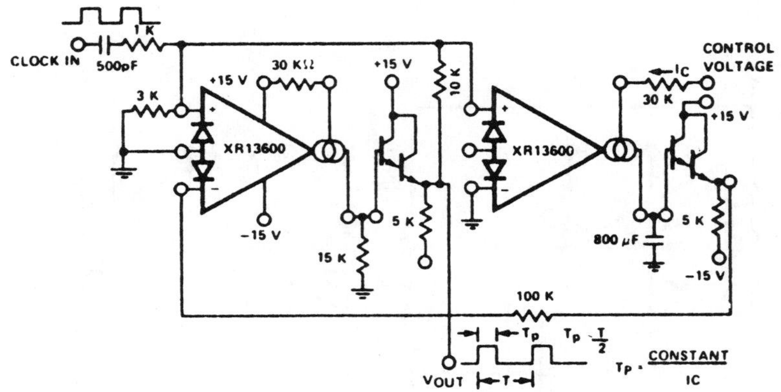

Pulse-width modulator

Fig. 11-28 This circuit uses both sections of an XR-13600 (Fig. 11-1B) to form a pulse-width modulator. EXAR Corporation Databook 1990, p. 5-258.

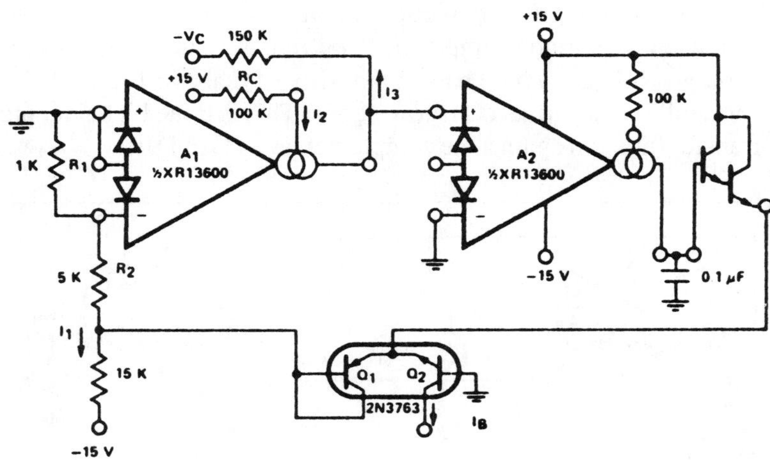

Logarithmic current source

Fig. 11-29 This circuit uses both sections of an XR-13600 (Fig. 11-1B) to form a logarithmic current source. The current can be used for I_B (the bias current) in many OTA applications by providing a logarithmic current output for a linear voltage in. For example, the circuit can be used to bias the stereo volume control (Fig. 11-2) to provide temperature-independent stereo attenuation characteristics. EXAR Corporation Databook, 1990, p. 5-258.

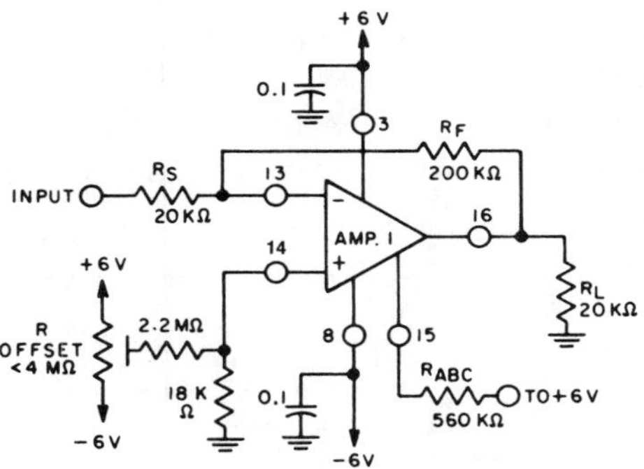

20-dB amplifier

Fig. 11-30 This circuit uses one section of a CA3060 OTA to form a 20-dB amplifier with a typical 20-kHz bandwidth. Notice that the bias input is connected to a fixed +6 V. Harris Semiconductor, Linear & Telecom ICs, 1991, p. 3-49.

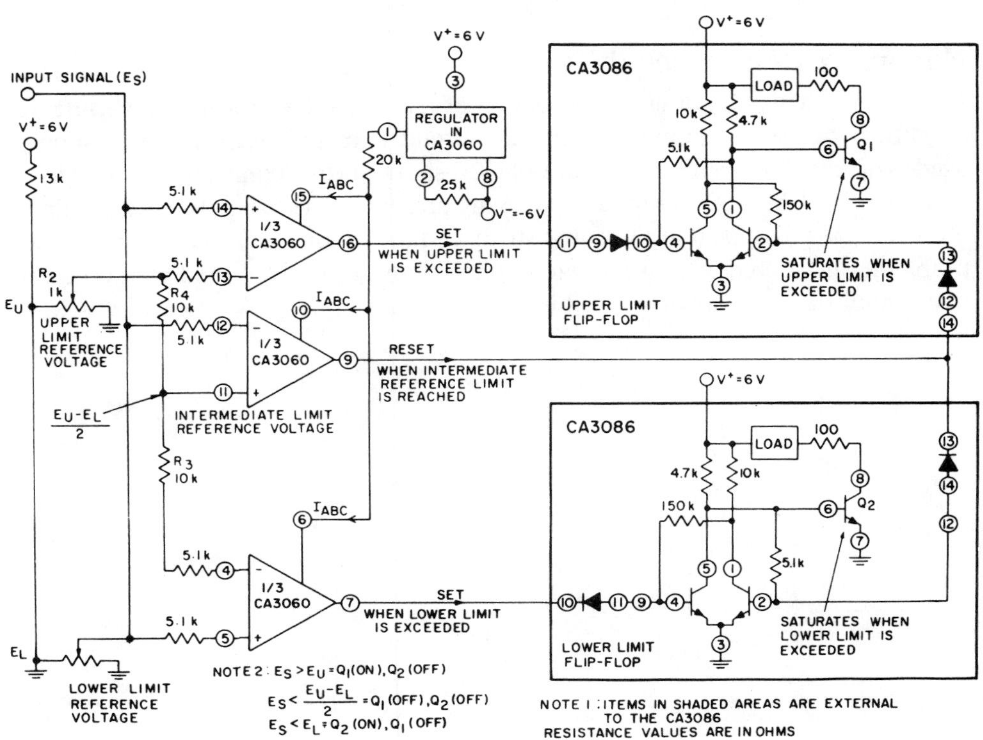

Tri-level comparator

Fig. 11-31 This circuit uses all three sections of a CA3060 to form a tri-level comparator, which has three adjustable limits. If either the upper or lower limit is exceeded, the appropriate output is activated until the input signal returns to a selected intermediate value. Such comparators are used in industrial control applications. The lower and upper limits are set by R1 and R2, respectively. R3 and R4 set the intermediate limit. When both R3 and R4 are the same value, the intermediate limit is midway between the upper and lower limits. The loads shown are 5-V 25-mA lamps. Harris Semiconductor, Linear & Telecom ICs, 1991, p. 3-51.

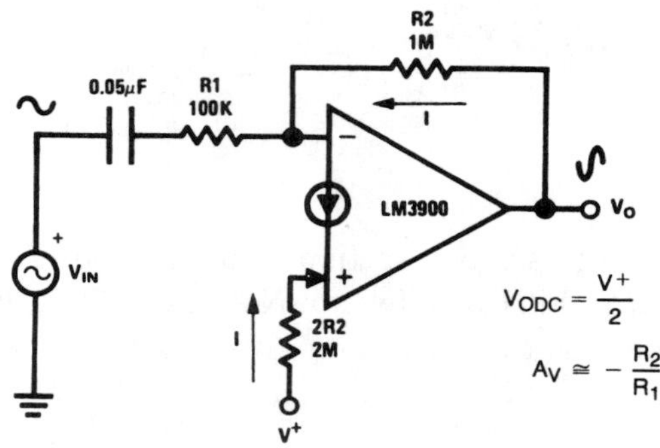

Norton inverting ac amplifier

Fig. 11-32 This circuit uses one section of an LM3900 to form an inverting ac amplifier with single-supply biasing (the circuit is biased from the same supply used to operate the amplifier). Figure 11-32B shows the pin connections for all four sections of an LM3900. If ripple voltages are present on the V+ power-supply line, the voltages will appear at the output with a "gain" of one-half. To eliminate this problem, one source of ripple-filtered voltage can be provided, and then used for many amplifiers, as shown in Fig. 11-33. National Semiconductor, Linear Applications Handbook, 1991, p. 218, 220.

Continued

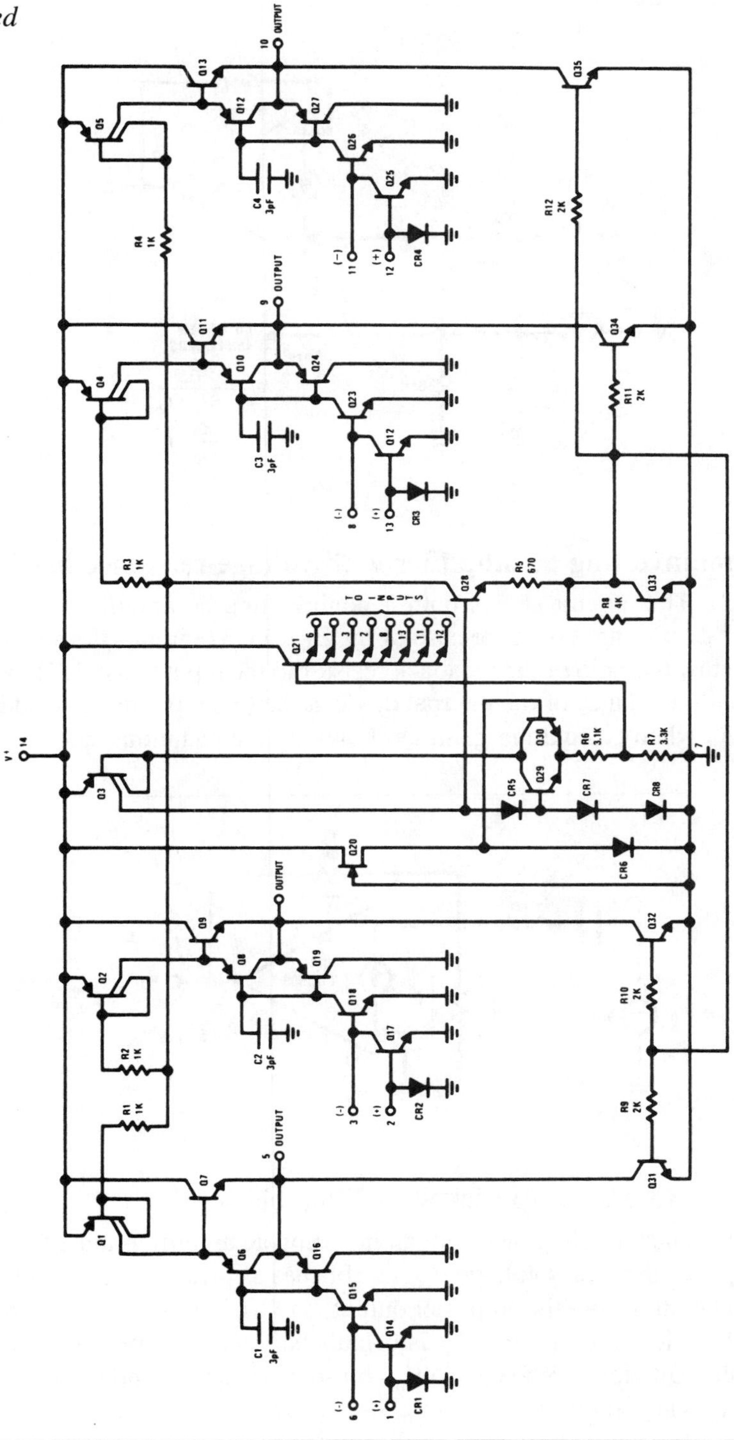

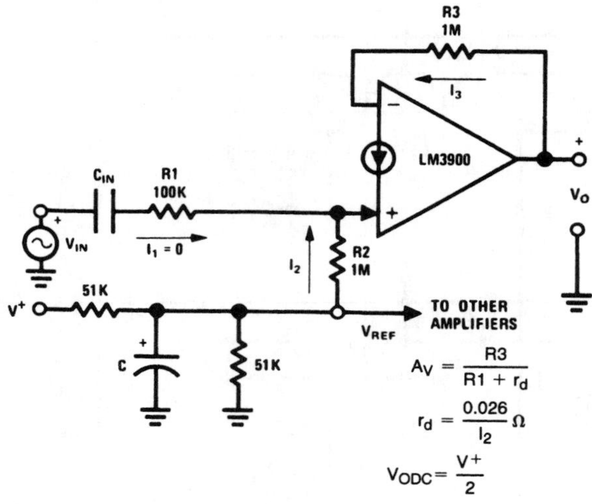

Norton noninverting ac amplifier with voltage-reference biasing

Fig. 11-33 This circuit shows both a noninverting ac amplifier and a second method for dc biasing. As in the case of the Fig. 11-32 circuit, the ac gain of this circuit is set by the ratio of the feedback resistor to the input resistor. However, the small-signal impedance of the internal diode at the (+) input must be added to the values of R_1 when calculating gain, as shown by the equations. National Semiconductor, Linear Applications Handbook, 1991, p. 220.

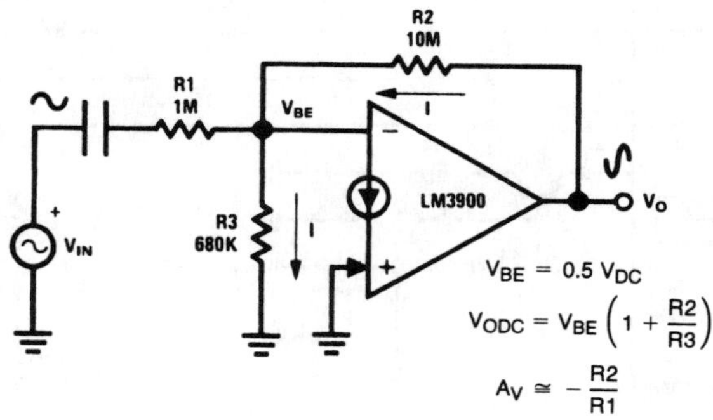

Norton inverting ac amplifier with NV_{BE} biasing

Fig. 11-34 This circuit shows both an inverting ac amplifier and a third method for biasing. The input bias voltage V_{BE} establishes a current through R3 to ground. This current comes from the amplifier output, so V_O must rise to a level that causes the current to flow through R2. V_O is calculated from the ratio of R_2 and R_3, as shown by the equations. When this NV_{BE} biasing is used, R_1 and R_2 are established first, then R_3 is added for the desired V_O. National Semiconductor, Linear Applications Handbook, 1991, p. 220.

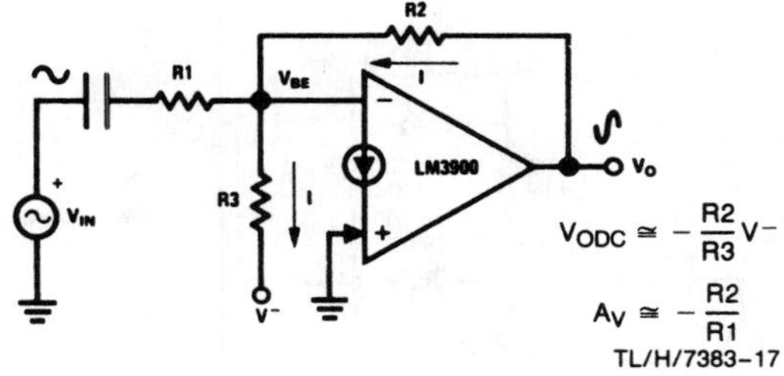

Norton amplifier with a negative supply

Fig. 11-35 This circuit uses one section of an LM3900 to form an ac amplifier with a negative power supply. The dc biasing current, *I*, is established by the negative supply voltage through R3, and provides a very stable output-quiescent point for the amplifier. National Semiconductor, Linear Applications Handbook, 1991, p. 221.

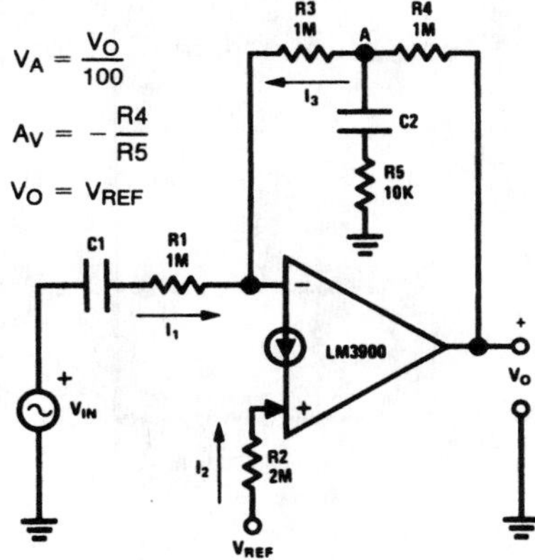

Norton amplifier with high input impedance and high gain

Fig. 11-36 This circuit uses one section of an LM3900 to form an ac amplifier with both high input impedance (1 MΩ) and high gain (100). National Semiconductor, Linear Applications Handbook, 1991, p. 221.

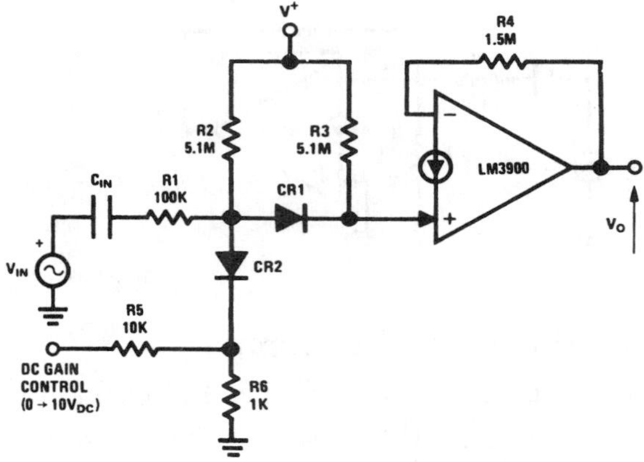

Norton amplifier with dc gain control

Fig. 11-37 This circuit uses one section of an LM3900 to form an ac amplifier with a variable gain control. The dc gain-control input ranges from 0 V for minimum gain to about 10 V for maximum gain. National Semiconductor, Linear Applications Handbook, 1991, p. 221.

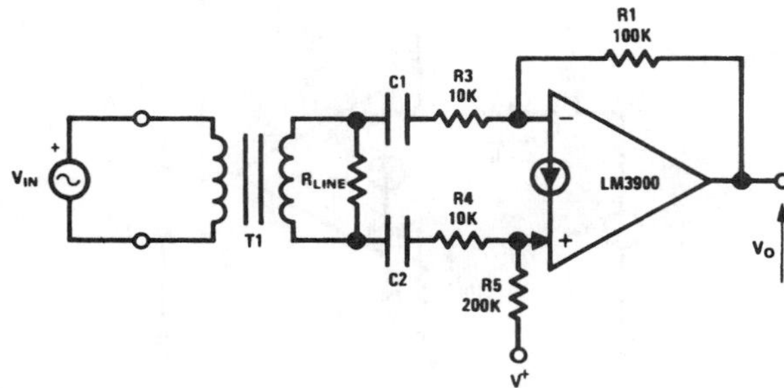

Norton line-receiver amplifier

Fig. 11-38 This circuit uses one section of an LM3900 to form a line-receiver amplifier. The use of both inputs cancels out any common-mode signals. The line is terminated by R_{LINE} and the large input impedance of the amplifier does not affect this matched loading. National Semiconductor, Linear Applications Handbook, 1991, p. 221.

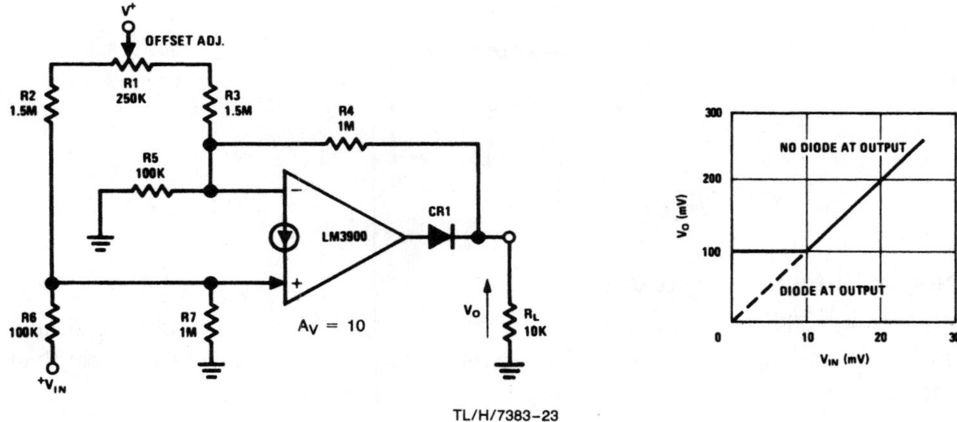

Norton amplifier with zero-volts out for zero-volts in

Fig. 11-39 This circuit uses one section of an LM3900 to form a noninverting direct-coupled amplifier, where zero volts input produces zero volts output. Figure 11-39B shows the voltage transfer function for the circuit, both with and without diode CR1 at the output. National Semiconductor, Linear Applications Handbook, 1991, p. 222, 223.

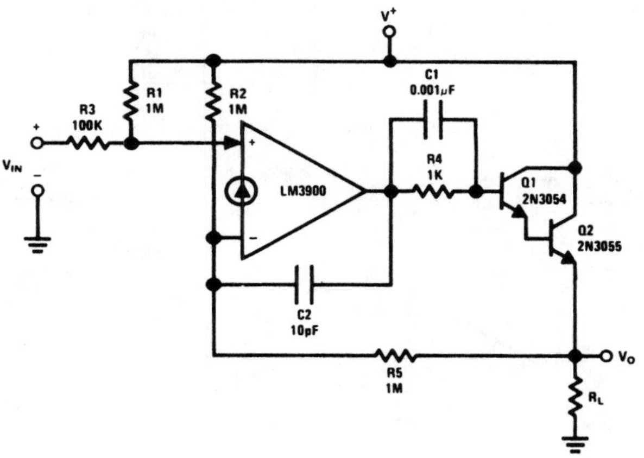

Norton direct-coupled power amplifier

Fig. 11-40 This circuit uses one section of an LM3900 and a Darlington pair to provide an output of 3 A into the load. National Semiconductor, Linear Applications Handbook, 1991, p. 223.

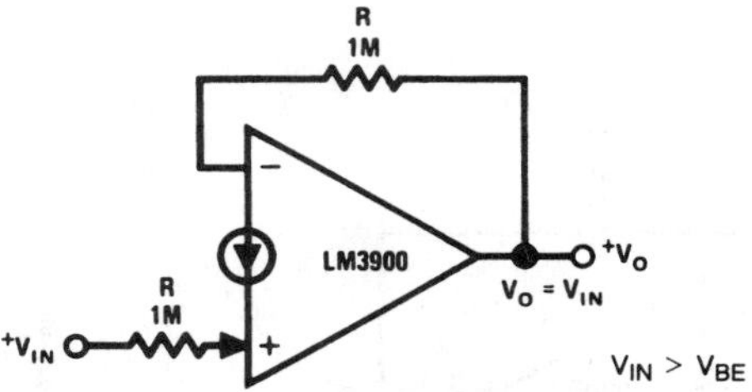

Norton direct-coupled unity-gain buffer

Fig. 11-41 This circuit uses one section of an LM3900 to form a simple direct-coupled unity-gain buffer. Notice that the input voltage must be greater than one V_{BE} (about 0.5 V), but less than the maximum output swing. National Semiconductor, Linear Applications Handbook, 1991, p. 224.

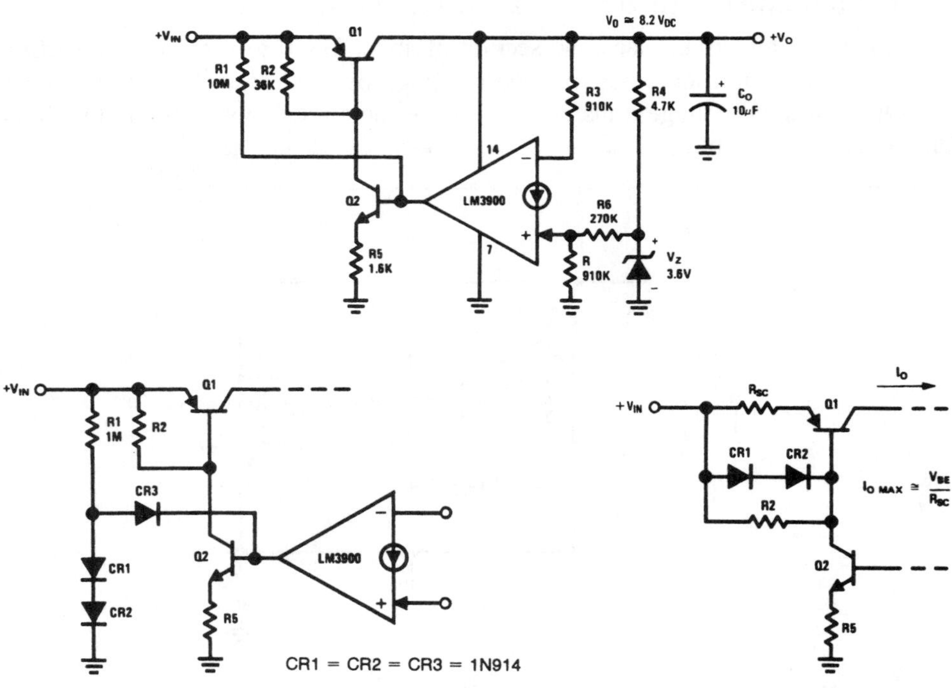

Norton regulator with high-voltage input protection

Fig. 11-42 This circuit uses one section of an LM3900 as the control element in a voltage regulator. Q1 and Q2 absorb any high input voltages, and therefore must be high-voltage transistors. With the values shown, the output is about 8.2 V. Diodes can be added (as shown in Fig. 11-42B) to reduce ripple feedthrough and

Continued

input-voltage dependence. Short-circuit protection can also be added, as shown in Fig. 11-42C. National Semiconductor, Linear Applications Handbook, 1991, p. 225, 226.

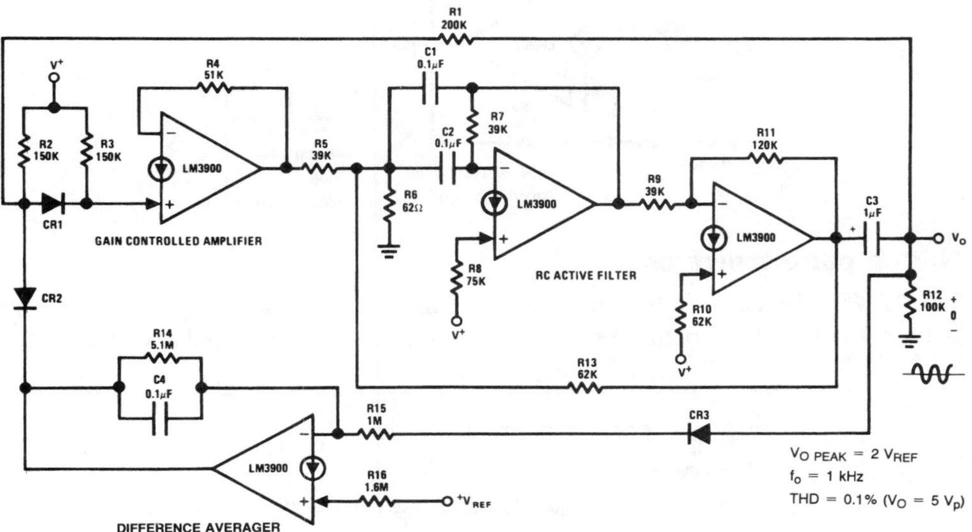

Norton sine-wave oscillator

Fig. 11-43 This circuit uses four sections of an LM3900 to form a sine-wave oscillator. One section is used as a gain-controlled amplifier (to sustain feedback), and another section is a difference averager that maintains a constant output level. The remaining sections are connected as an RC active filter (to produce a 1-kHz output with the values shown). National Semiconductor, Linear Applications Handbook, 1991, p. 231.

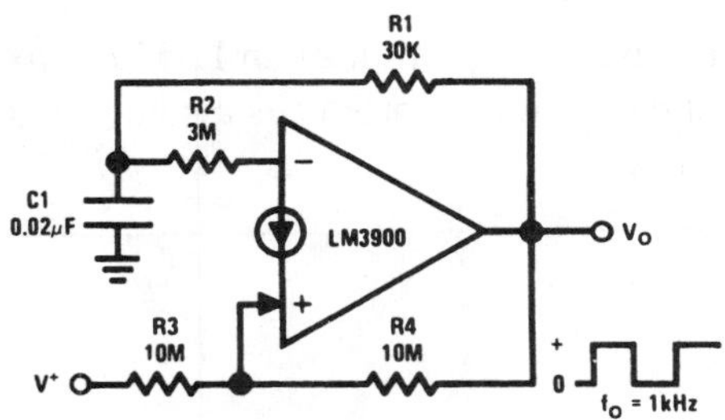

Norton square-wave oscillator

Fig. 11-44 This circuit uses one section of an LM3900 to form a square-wave oscillator (with a 1-kHz output using the values shown). National Semiconductor, Linear Applications Handbook, 1991, p. 232.

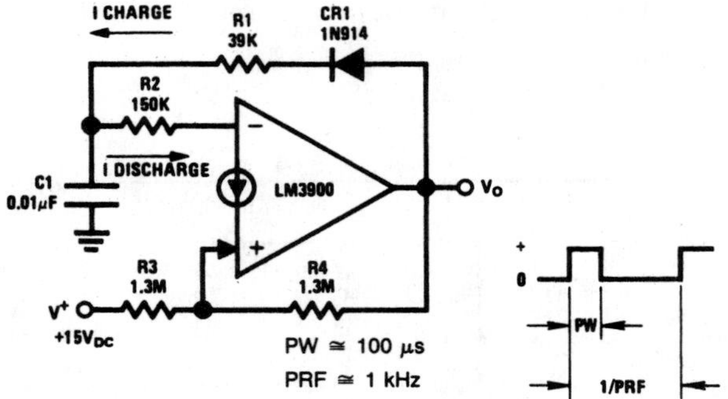

Norton pulse generator

Fig. 11-45 This circuit uses one section of an LM3900 to form a pulse generator (with a 1-kHz pulse repetition rate or frequency, and a 100-μs pulse width). National Semiconductor, Linear Applications Handbook, 1991, p. 232.

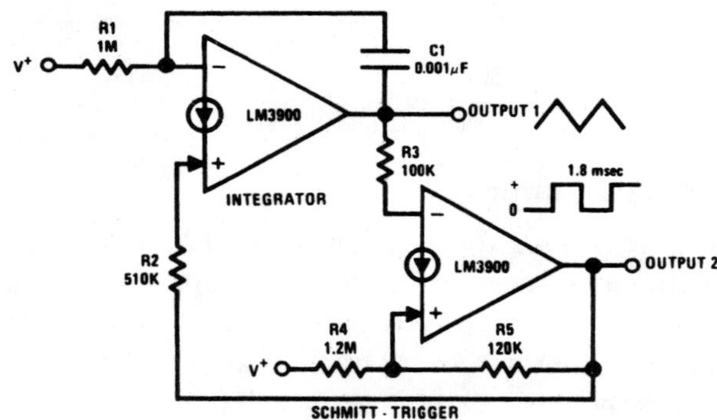

Norton triangle-wave generator

Fig. 11-46 This circuit uses one section of an LM3900 to form a triangle-wave generator. Notice that the circuit also produces a square-wave output at the same frequency. National Semiconductor, Linear Applications Handbook, 1991, p. 233.

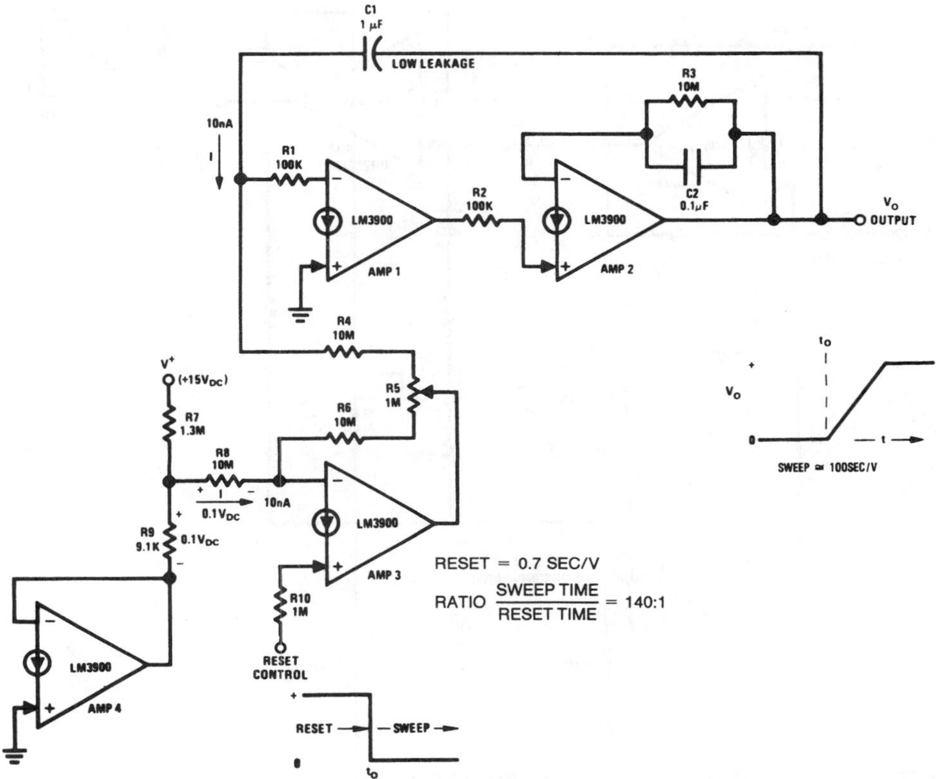

Norton slow-sawtooth waveform generator

Fig. 11-47 This circuit uses four sections of an LM3900 to form a generator with very slow sawtooth waveforms (which can be used to generate long time-delay intervals, as one application). The reset signal is applied to amplifier 3 through R10, and R5 adjusts the sweep. With the values shown, the 10-nA current and 1-μF capacitor establishes a sweep rate of 100 s/V. R4 provides a reset rate of 0.7 s/V.

National Semiconductor, Linear Applications Handbook, 1991, p. 234.

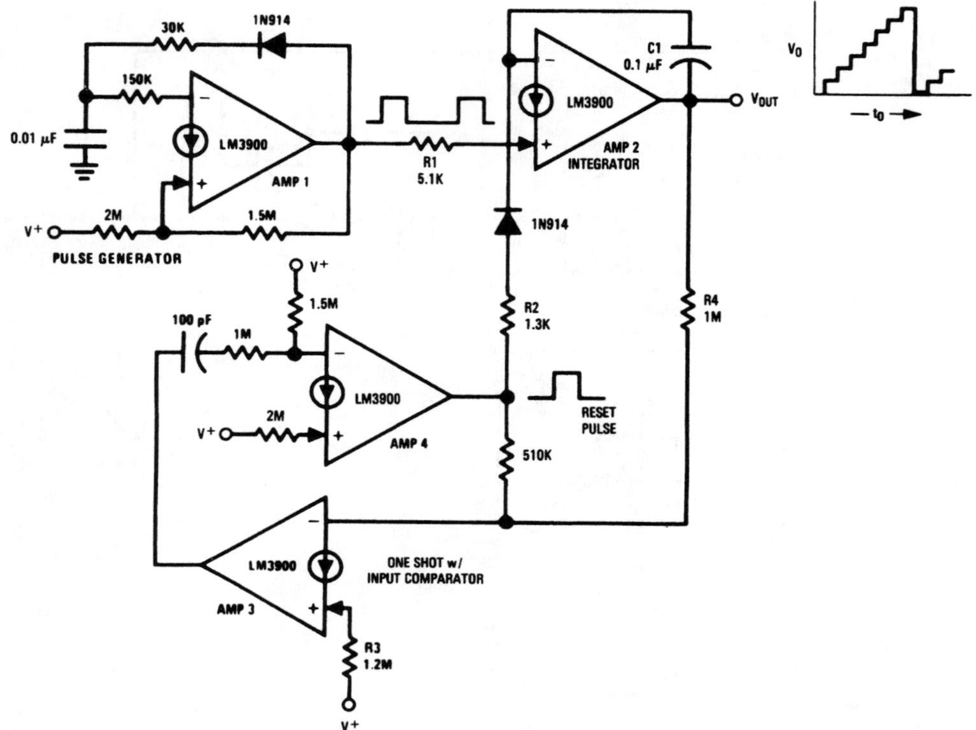

Norton free-running staircase generator

Fig. 11-48 This circuit uses four sections of an LM3900 to form a free-running staircase generator. When the output exceeds about 80% of V+, a 100-μs reset pulse is generated, and the staircase output is caused to fall to about zero volts. The next pulse from amplifier 1 then starts a new stepping cycle. National Semiconductor, Linear Applications Handbook, 1991, p. 236.

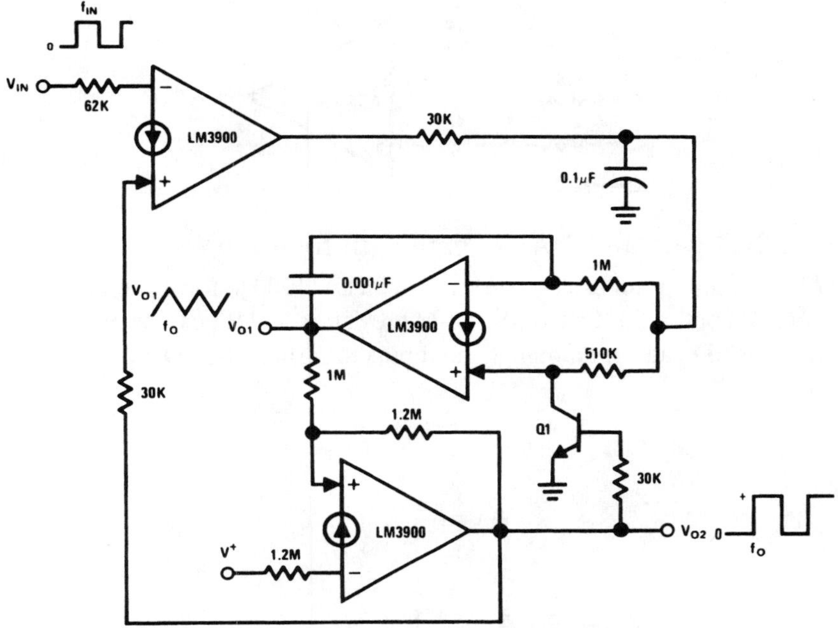

Norton phase-locked loop

Fig. 11-49 This circuit uses three sections of an LM3900 to form a phase-locked loop (with a center frequency of about 3 kHz using the values shown). National Semiconductor, Linear Applications Handbook, 1991, p. 239.

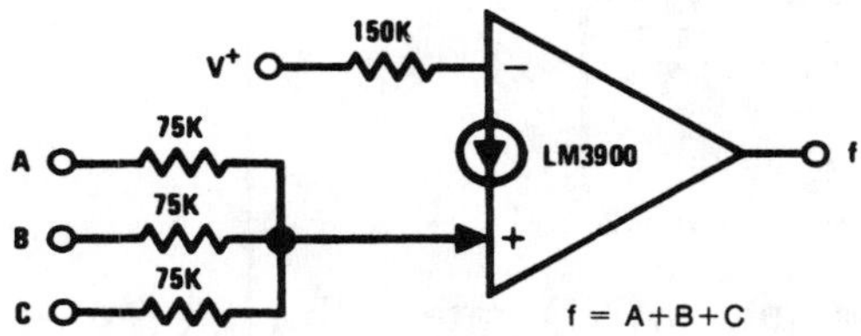

Norton OR gate and NOR gate with high fan out

Fig. 11-50 This circuit uses one section of an LM3900 to form an OR gate with a fan out of 50 gates (if each gate has a 75-kΩ input resistor). More than three inputs can be OR'ed if desired. The circuit can be converted to a NOR gate when the inputs are interchanged. National Semiconductor, Linear Applications Handbook, 1991, p. 240.

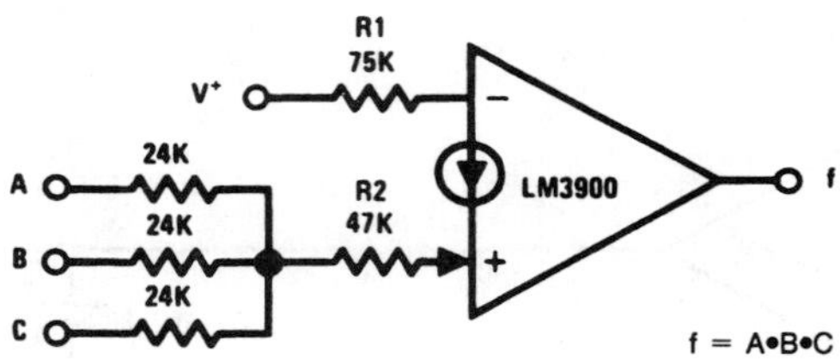

Norton AND gate and NAND gate with high fan out

Fig. 11-51 This circuit uses one section of an LM3900 to form a three-input AND gate with a fan out of 50. One of the 24-kΩ resistors and R2 can be omitted to form a two-input AND gate. Interchange the inputs to form a NAND gate. National Semiconductor, Linear Applications Handbook, 1991, p. 240.

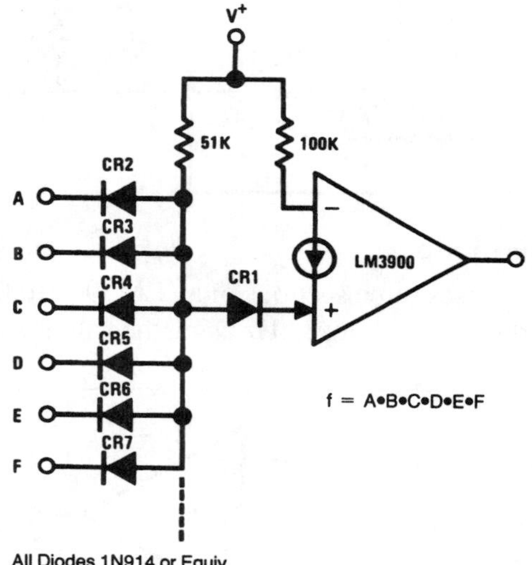

Norton AND gate and NAND gate with high fan in

Fig. 11-52 This circuit uses one section of an LM3900 to form a multiple-input AND gate, with an input-diode network that is similar to that of DTL. Interchange the inputs to form a NAND gate. National Semiconductor, Linear Applications Handbook, 1991, p. 240.

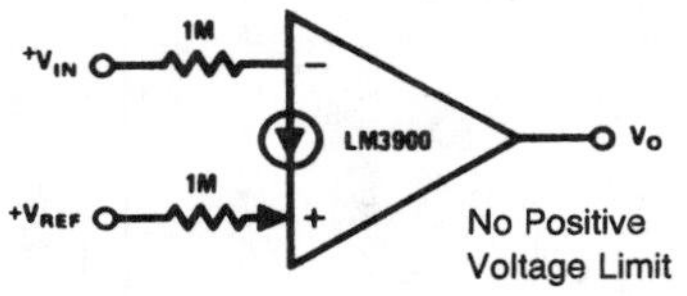

Norton comparator for positive input voltages

Fig. 11-53 This circuit uses one section of an LM3900 to form an inverting comparator. The reference voltage must be larger than V_{BE} (typically 0.5 V), but there is no limit as long as the input resistor is large enough to guarantee that the input current does not exceed 200 μA. National Semiconductor, Linear Applications Handbook, 1991, p. 243.

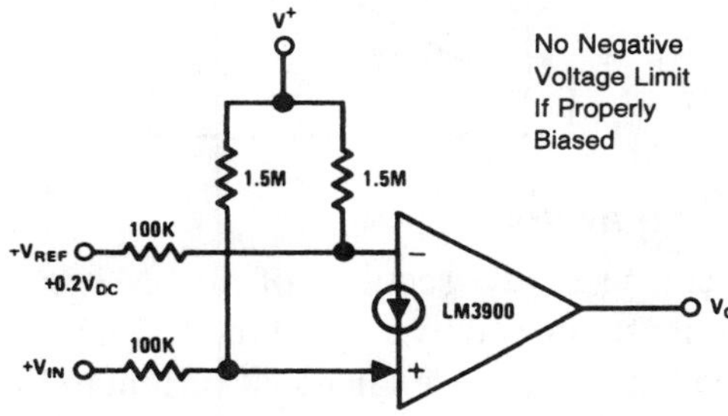

Norton noninverting low-voltage comparator

Fig. 11-54 This circuit uses one section of an LM3900 to form a noninverting comparator. The circuit can compare voltages between zero and 1 V, as well as large negative voltages. When working with negative voltages, the current supplied by the common-mode network must be large enough to satisfy both the current-drain demands of the input voltages, and the bias-current requirements of the amplifier. National Semiconductor, Linear Applications Handbook, 1991, p. 243.

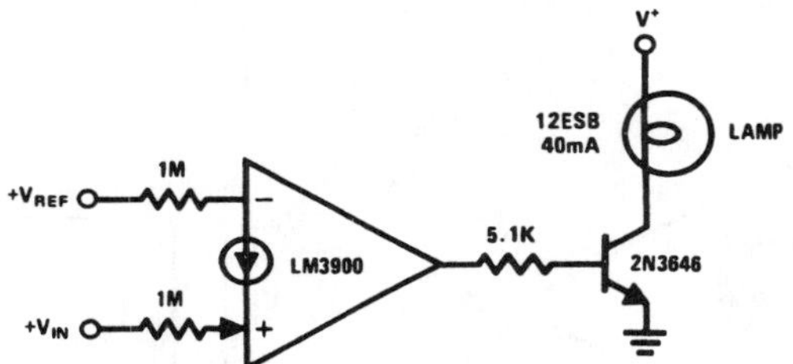

Norton noninverting power comparator

Fig. 11-55 This circuit uses one section of an LM3900 and a 2N3646 to form a noninverting power comparator that is capable of driving a 12-V 40-mA panel lamp. National Semiconductor, Linear Applications Handbook, 1991 p. 243.

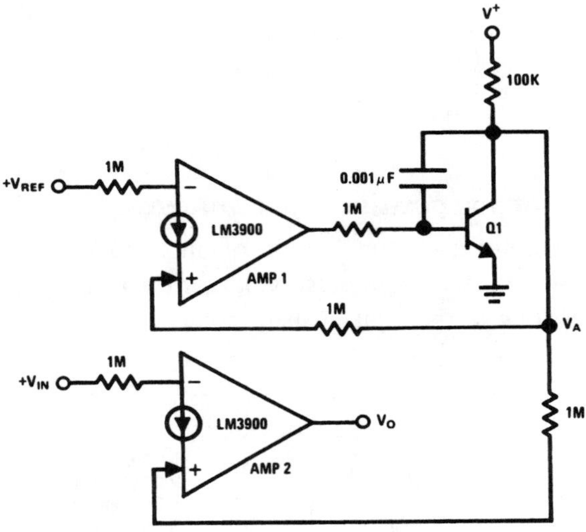

Norton precision comparator

Fig. 11-56 This circuit uses two sections of an LM3900 to form a precision comparator. The current established by V_{REF} at the $(-)$ input of amplifier 1 causes Q1 to adjust V_A. The value of V_A causes an equal current to flow into the $(+)$ input of amplifier 2. This current corresponds more exactly to the reference current of amplifier 1. National Semiconductor, Linear Applications Handbook, 1991, p. 244.

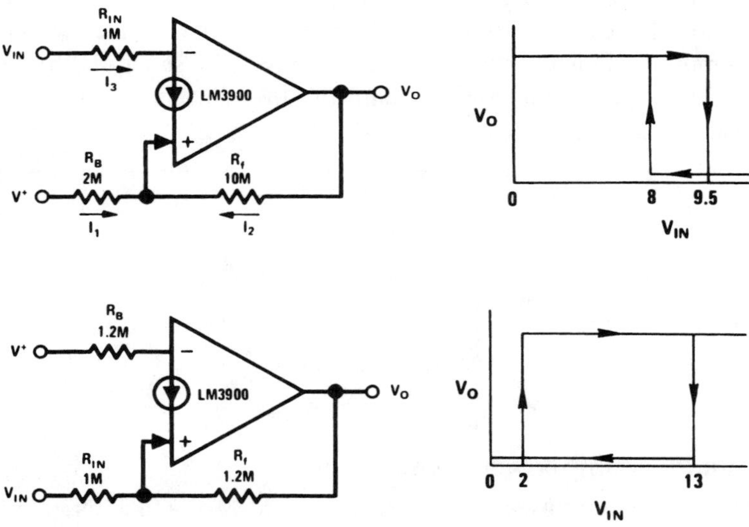

11

Norton Schmitt triggers

Fig. 11-57 This circuit uses one section of an LM3900 connected to form a Schmitt trigger. Both inverting and noninverting versions are shown. By adjusting the values of R_B, R_F, and R_{IN}, the switching values of V_{IN} can be set to any desired levels. National Semiconductor, Linear Applications Handbook, 1991, p. 244.

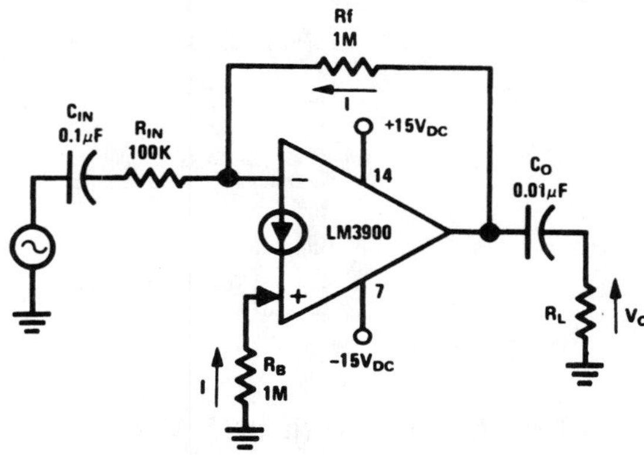

Norton ac amplifier with ±15-V supplies

Fig. 11-58 This circuit uses one section of an LM3900 to form an ac-coupled amplifier, where both inputs bias at one V_{BE} above the $-V_{EE}$ voltage (about -15 V). With $R_F = R_B$, V_O will bias at about 0 V to allow a maximum output swing. Because pin 7 is common to all four amplifiers in the package, the other amplifiers are also biased for operation with ±15-V supplies. National Semiconductor, Linear Applications Handbook, 1991, p. 246.

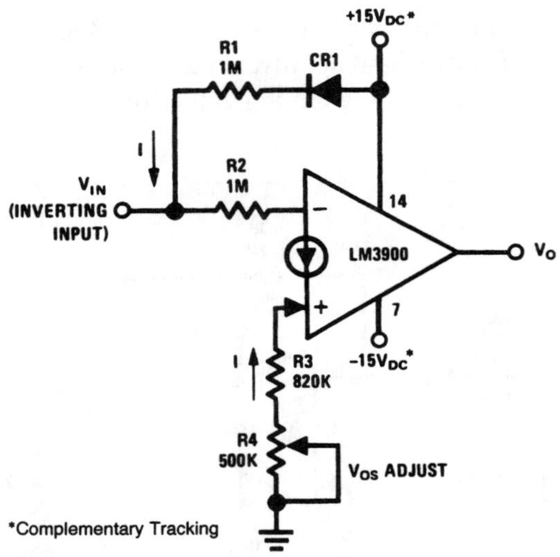

Norton direct-coupled amplifier with ±15-V supplies

Fig. 11-59 This circuit uses one section of an LM3900 to form a direct-coupled inverting amplifier. The ±15-V supplies must have complementary tracking, and R4 adjust for proper offset (or zero offset) at the output. National Semiconductor, Linear Applications Handbook, 1991, p. 246.

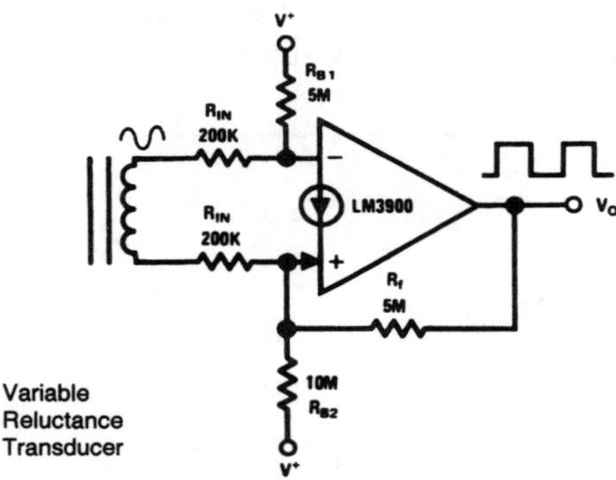

Norton squaring amplifier with hysteresis

Fig. 11-60 This circuit shows one section of an LM3900 that is used to form a squaring amplifier for use with a variable-reluctance transducer. The circuit produces symmetrical hysteresis above and below the zero-output state (for noise immunity), and filters high-frequency input noise disturbances. With the values shown, the trip voltages are about ±150 mV, centered about the zero-output state of the transducer (at low frequencies where the low-pass filter is not attenuating the input signal). National Semiconductor, Linear Applications Handbook, 1991, p. 247.

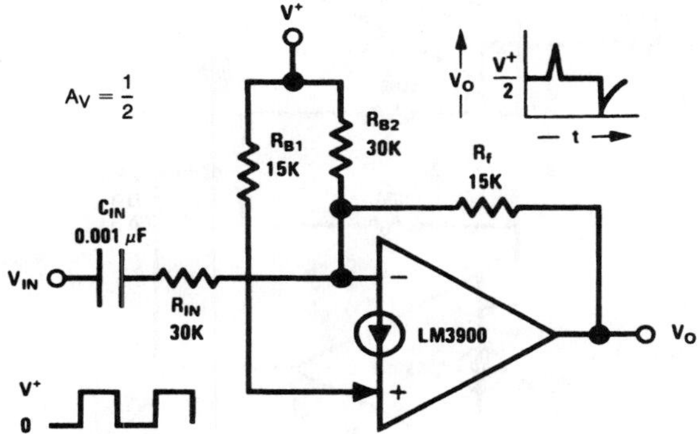

Norton differentiator

Fig. 11-61 This circuit shows one section of an LM3900 that is used to form a differentiator. Notice that the differentiated output is one-half of the square-wave input. National Semiconductor, Linear Applications Handbook, 1991, p. 248.

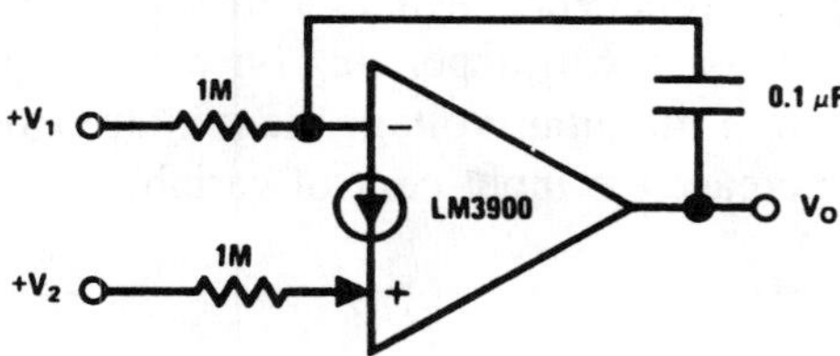

Norton difference integrator

Fig. 11-62 This circuit shows one section of an LM3900 that is used to form a difference integrator. In addition to being the basis for many sweep circuits, this circuit can also provide the time integral of the difference between two input waveforms. This is useful for dc feedback loops because both the comparison to a reference and the integration occur in one amplifier. National Semiconductor, Linear Applications Handbook, 1991, p. 248.

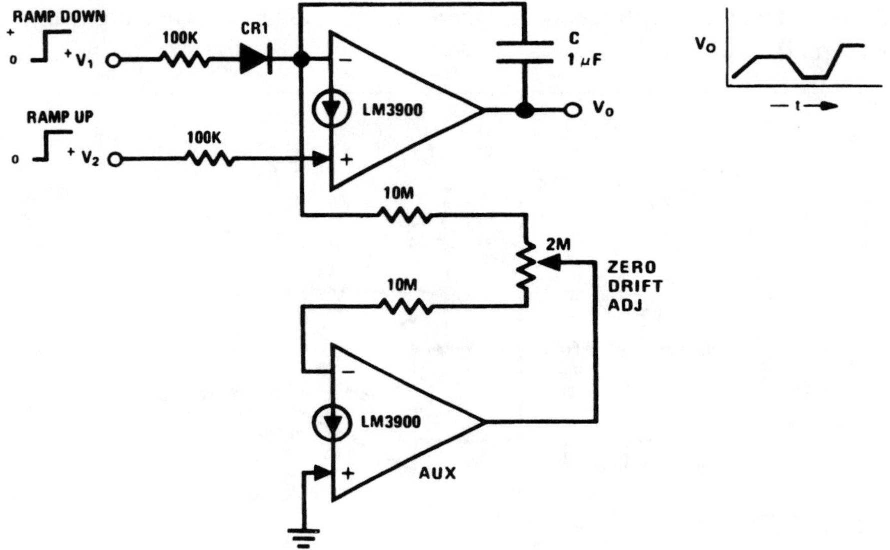

Norton low-drift ramp and hold

Fig. 11-63 This circuit uses two sections of an LM3900 to form a low-drift ramp-and-hold circuit with a zero-drift adjustment. If both inputs are at zero volts, the circuit is in hold (after proper zero-drift adjustment). Raising either input causes the dc output to ramp up or down, depending on which input goes positive. The ramp slope is a function of the input-voltage magnitude. Additional inputs can be placed in parallel to increase the input-control variables. National Semiconductor, Linear Applications Handbook, 1991, p. 249.

12

Special-purpose circuits

The circuits in this chapter cover a variety of functions. Therefore, it is impractical to describe "universal" test and troubleshooting procedures for the circuit group. However, the basic procedures that are covered in chapters 1 through 11 still apply.

For example, in the circuit of Fig. 12-1, the output should be a series of pulses, the frequency of which is set by the input voltages and the values of the components shown in Fig. 12-1B. If you choose the dc to 1.0-kHz operating range (R_O = 6.8 kΩ, C_O = 0.1 μF, R_B = 100 kΩ, C_B= 10 μF), the pulse repetition at R_{OUT} should vary between dc and 1.0 kHz when 0 to +10 V (within about 1%) V is applied at V_{IN}.

Operation of the circuit can be checked with a frequency counter at the output and a digital multimeter at the input. For example, with 5 V at the input, the output should be 500 Hz, and so on. If the output frequency is slightly off, try correcting the problem by adjusting the current-setting resistor at pin 2 of the 4151/4152. Be sure to check across the full input and output range because the resistor setting can affect the frequency at any voltage input.

If there is no output from pin 3 of the 4151/4152 with an input voltage at pin 7, suspect the IC. It is also possible that C_O, C_B, or the 0.01-μF capacitors are leaking or are shorted.

The circuit of Fig. 12-8 can be checked with a voltmeter at the output (pin 6 of the OP-07), and a variable temperature source at the REF-02. Assume that you have chosen a scale factor (resistor values) for a 10-mV/°C readout, and that the REF-02 is in an ambient temperature of 25°C. The output voltage should be 250 mV. Now vary the ambient temperature (with a hair dryer, freon, etc.) and check that the output also varies with hot and cold. You can also check the circuit accuracy by monitoring the REF-02 ambient temperature with a thermometer.

If the output voltage varies with temperature changes, but the scale factor is not correct, try correcting the problem by adjusting R_P and R_{B2}, as described in the circuit description. If there is no variation in output voltage with changes in temperature, check for variation at pins 2 and 3 of the OP-07. If the voltages vary at pins 2 and 3, but not at pin 6, suspect the OP-07. If there is no variation in voltage at the OP-07 input with changes in REF-02 ambient temperature, suspect the REF-02.

The circuit of Fig. 12-19 can be checked by applying pulses to the input (at the CD4001 NOR gate) and by checking the LED display for a readout. Notice that this circuit is essentially a digital circuit, and that all of the test/troubleshooting notes of chapter 6 apply.

For example, if there is no readout whatsoever, suspect the LED display. If only one digit is absent or abnormal, check the signals at pins 25 through 28 of the ICM7217. If only one segment is absent or abnormal, check at pins 15 through 19, 21, and 22.

If the display does not follow the input pulses, check for a corresponding BCD output at pins 4 through 7. With the values and connections as shown, the display reads directly in Hz, and should increase by one for each input pulse that is applied. Notice that the pulses can come from a pulse generator or from a probe. If these are not available, simply charge a capacitor to 5 V by placing the capacitor between ground and V+. Then, discharge the capacitor between ground and the input.

If the BCD output changes when input pulses are applied, suspect the LED display and/or wiring between the display and ICM7217. If the BCD output does not change with input pulses, suspect the ICM7217.

Before you pull the IC, check that there is +5 V at pin 24, that pin 20 is grounded, and that pins 8, 9, and 14 are receiving COUNT, $\overline{\text{STORE}}$, and $\overline{\text{RESET}}$ pulses, respectively. The pulses at pins 8, 9, and 14 are generated by the ICM7207A, and should be at the crystal frequency.

If any one of the pulses are absent or abnormal, check the wiring between pins 2, 13, and 14 of ICM7207A and the ICM7217. Notice that there should still be COUNT pulses at pin 8 of the ICM7217—even with no input pulses. If not, suspect the CD4001 NOR gate.

If all the crystal-frequency signals are absent at the ICM7207A (pins 2, 13, and 14), suspect the IC or the crystal. Check for a 5.24288-MHz signal at pins 5 and 6. Also, check for +5 V at pin 10 and ground at pin 4 of the ICM7207A. If the crystal-frequency signals are absent, suspect the crystal, the two 22-pF capacitors, the 10-MΩ resistor, and the ICM7207A.

Special-purpose circuits

(a)

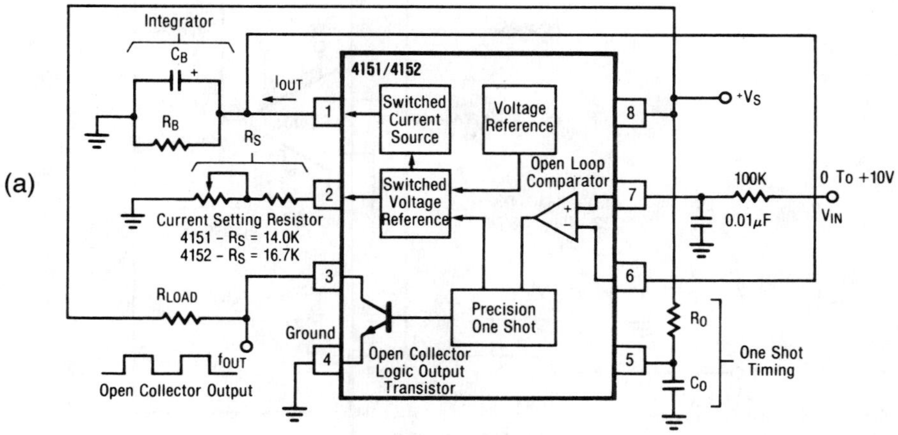

(b)

Operating Range	R_O	C_O	R_B	C_B
DC to 1.0kHz	6.8kΩ	0.1μF	100kΩ	10μF
DC to 10kHz	6.8kΩ	0.01μF	100kΩ	1.0μF
DC to 100kHz	6.8kΩ	0.001μF	100kΩ	0.1μF

Basic voltage-to-frequency converter

Fig. 12-1 Figure 12-1A shows a stand-alone VFC. Figure 12-1B shows the operating range for various component values. This single-supply VFC is recommended for uses, where the dynamic range of the input is limited, and the input does not reach 0 V. Raytheon Linear Integrated Circuits, 1989, p. 7-6, 7-7.

(a)

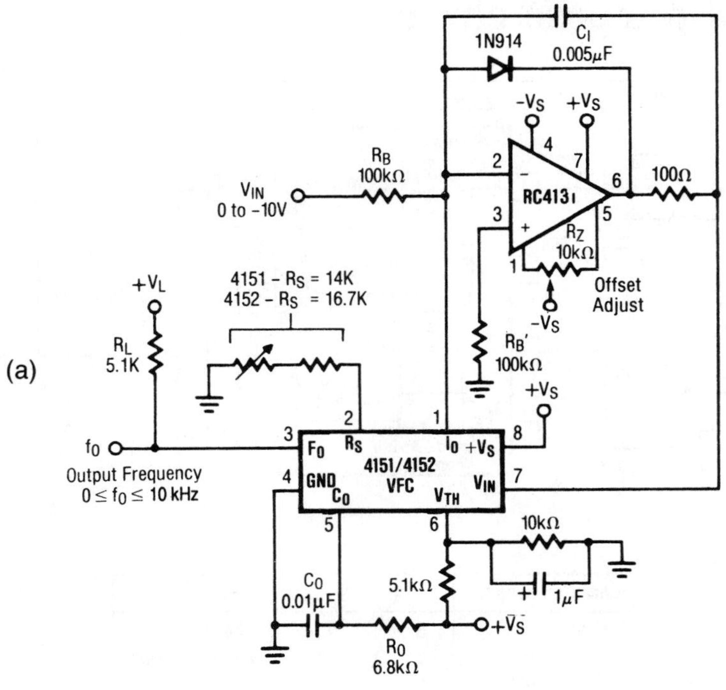

(b)

Range		Scale Factor	R_O	C_O	C_I	R_B
Input V_{IN}	Output F_O					
0 to -10V	0 to 1.0kHz	0.1kHz/V	6.8kΩ	0.1μF	0.05μF	100kΩ
0 to -10V	0 to 10kHz	1.0kHz/V	6.8kΩ	0.01μF	0.005μF	100kΩ
0 to -10V	0 to 100kHz	10kHz/V	6.8kΩ	1000pF	500pF	100kΩ

Precision current-sourced VFC

Fig. 12-2 Figure 12-2A shows a current-sourced VFC that is similar to Fig. 12-1, except that the passive RC integrator is replaced by an active op-amp integrator. This increases the dynamic range down to 0 V, improves response time, and eliminates the nonlinearity error that is introduced by the limited compliance of the switched current source output. Figure 12-2B shows the operating range for various component values. Raytheon Linear Integrated Circuits, 1989, p. 7-7, 7-8.

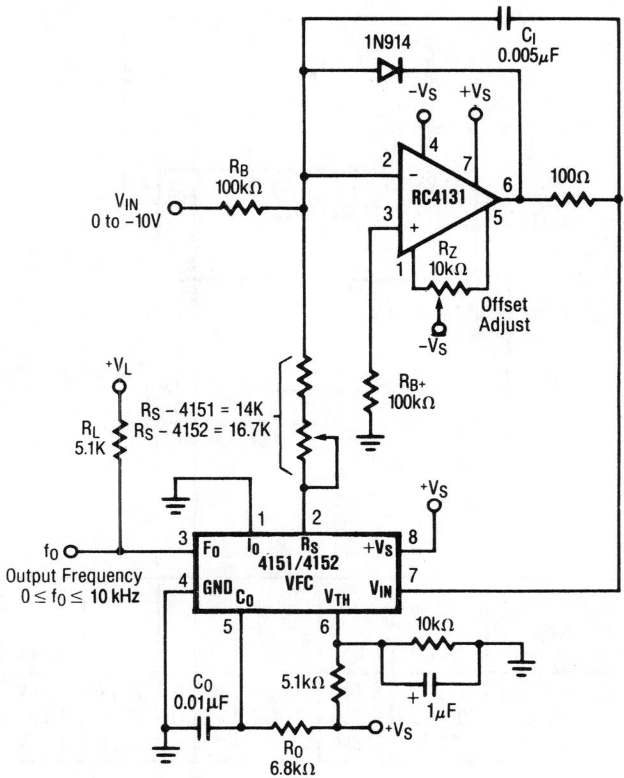

Precision voltage-sourced VFC

Fig. 12-3 Figure 12-3 shows a voltage-sourced VFC that is similar to Fig. 12-2, except that current pulses into the integrator are taken directly from the switched voltage reference. This improves temperature drift at the expense of high-frequency linearity. Raytheon Linear Integrated Circuits, 1989, p. 7-8.

(a)

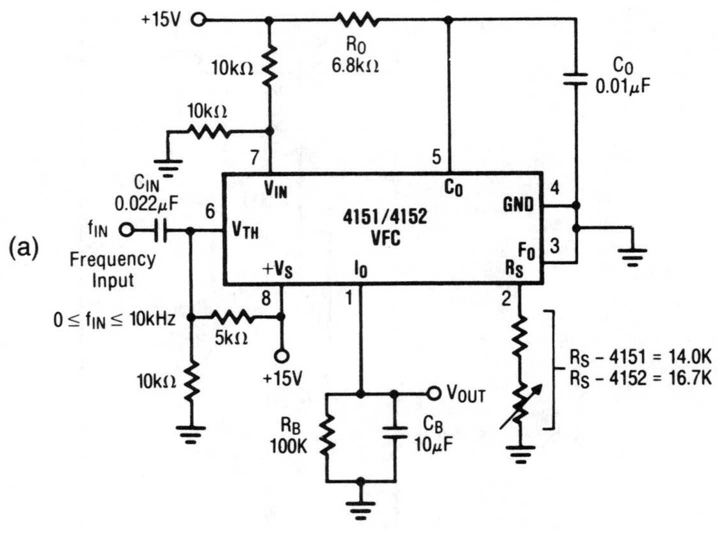

(b)

Input Operating Range	C_{IN}	R_O	C_O	R_B	C_B	Ripple
0 to 1.0kHz	0.02μF	6.8kΩ	0.1μF	100kΩ	100μF	1.0mV
0 to 10kHz	0.002μF	6.8kΩ	0.01μF	100kΩ	10μF	1.0mV
0 to 100kHz	200pF	6.8k8	0.001μF	100kΩ	1.0μF	1.0mV

Basic frequency-to-voltage converter

Fig. 12-4 Figure 12-4A shows a stand-alone FVC. Figure 12-4B shows the operating range for various component values. This circuit performs the opposite of a VFC (the FVC converts an input pulse train into an average output voltage). Raytheon Linear Integrated Circuits, 1989, p. 7-9.

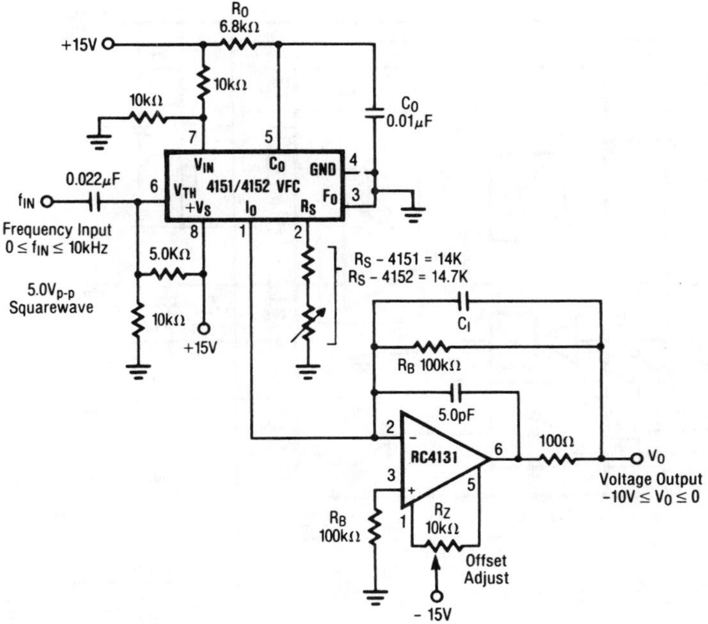

Precision frequency-to-voltage converter

Fig. 12-5 Figure 12-5 shows an FVC that is similar to Fig. 12-4, except that linearity, offset, and response time is improved by adding an op amp to form an active low-pass filter at the output. Raytheon Linear Integrated Circuits, 1989, p. 7-10.

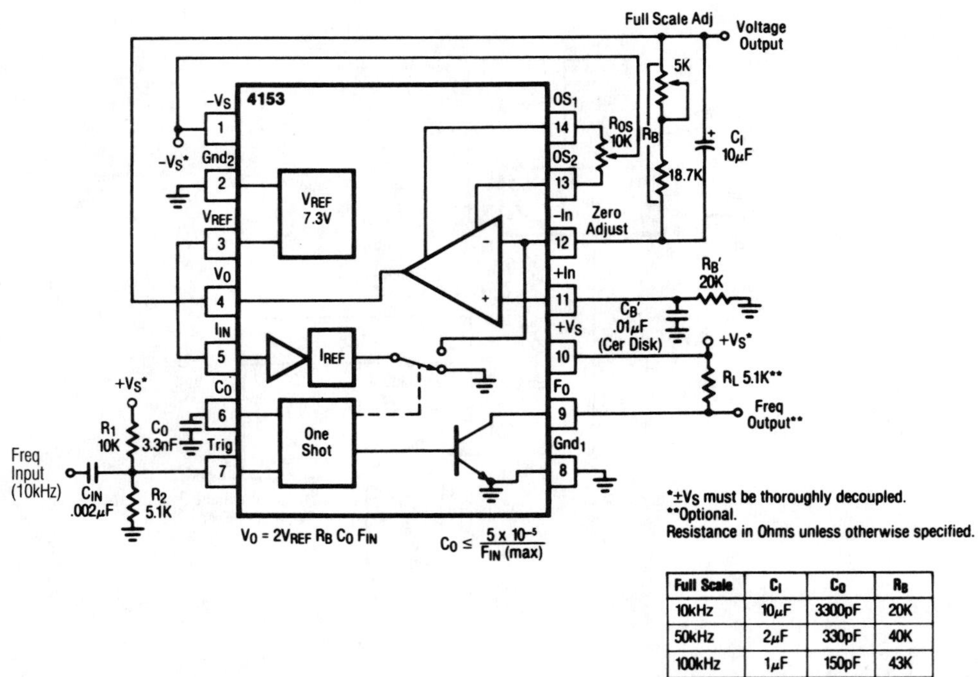

Full Scale	C_I	C_O	R_B
10kHz	10μF	3300pF	20K
50kHz	2μF	330pF	40K
100kHz	1μF	150pF	43K
250kHz	0.2μF	60pF	39K

$$V_{Ripple} = \frac{2V_{REF}\,C_O\,(1 - 1.5 \times 10^4\,C_O\,F_{IN})}{C_I}$$

$$T_{Recovery} = 1.36 \times 10^4\,C_I\,C_O\,R_B\,\Delta F_{IN}$$

Frequency-to-voltage converter with zero and full-scale adjust

Fig. 12-6 This circuit shows an FVC with both zero and full-scale adjustments.

Raytheon Linear Integrated Circuits, 1989, p. 7-19.

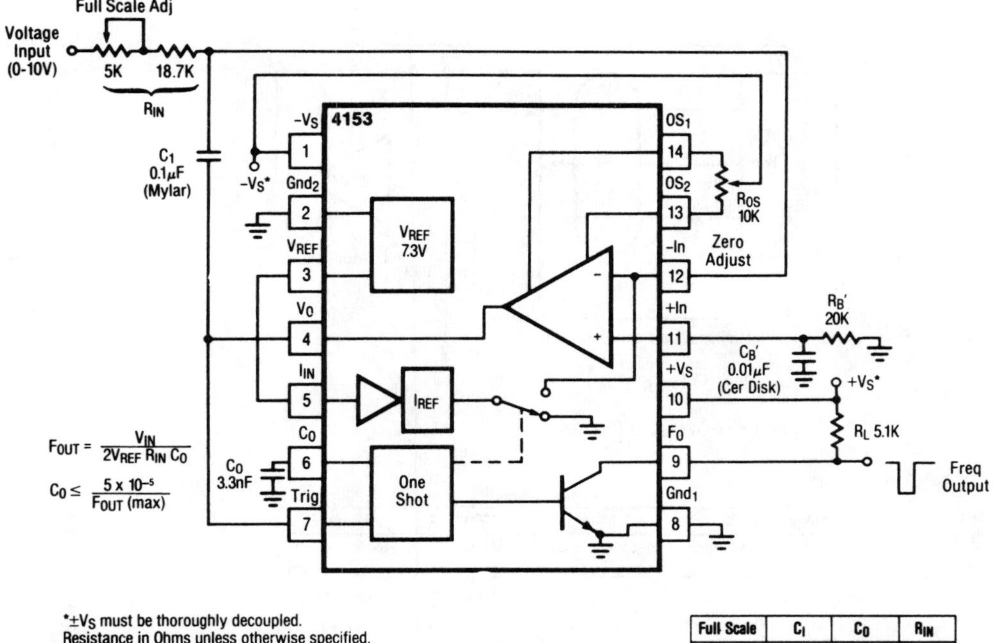

*±V_S must be thoroughly decoupled.
Resistance in Ohms unless otherwise specified.

Full Scale	C_I	C_O	R_{IN}
10kHz	0.1µF	3300pF	20K
50kHz	0.02µF	680pF	20K
100kHz	4300pF	330pF	20K
250kHz	1000pF	130pF	20K

Voltage-to-frequency converter with zero and full-scale adjust

Fig. 12-7 This circuit shows a VFC with both zero and full-scale adjustments.

Raytheon Linear Integrated Circuits, 1989, p. 7-20.

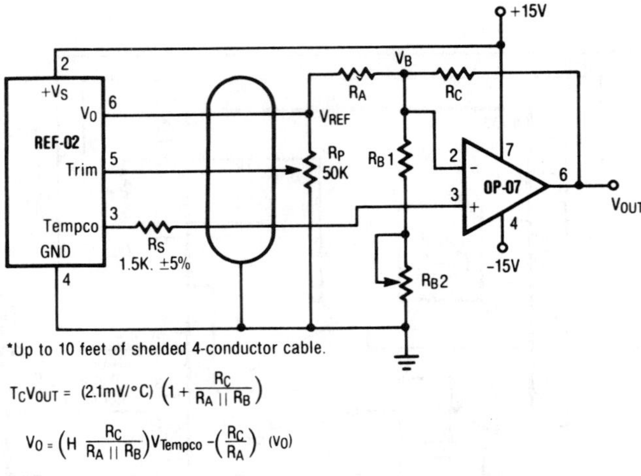

Resistor Values

TCV$_{OUT}$ Slope(s)	10mV/°C	100mV/°C	10mV/°F
Temperature Range	-55°C to +125°C	-55°C to +125°C	-65°F to +257°F
Output Voltage Range	-0.55V to +1.25V	-5.5V to +12.5V	-0.67V to +2.57V
Zero Scale	0V at 0°C	0V at 0°C	0V at 0°F
RA (±1% Resistor)	9.09KΩ	15KΩ	8.25KΩ
RB1 (±1% Resistor)	1.5KΩ	1.82KΩ	1.0KΩ
RB2 (Potentiometer)	200Ω	500Ω	200Ω
RC (±1% Resistor)	5.11KΩ	84.5KΩ	7.5KΩ

Precision electronic thermometer

Fig. 12-8 This circuit shows how a voltage reference can be combined with an op amp to create an electronic thermometer (with ±5% accuracy). To calibrate, measure the voltage at pin 3 of REF-02, and the ambient room temperature (T_A in °C). Then find X as follows:

$$X = \frac{Tempco \text{ (in mV)}}{(S)\ (T_A + 273)}, \text{ where } S = \text{scale factor for values}$$

selected from table in Fig. 12-8. Then, turn off the power, short pin 6 (V_o) of REF-02 to ground, apply exactly 100.00 mV to the op-amp output (pin 6 of OP-07), and adjust R_B2 so that $V_B = X$. Now, remove the short and the 100-mV source, reapply circuit power, and adjust R_P so that the op-amp output voltage equals (T_A) (S). The system is now calibrated. For remote sensor applications, a 1.5-kΩ resistor (R_S) must be connected in series with pin 6 of REF-02. This isolates REF-02 from cable capacitances. Use low temperature coefficient metal-film resistors for R_A, R_B, and R_C. Raytheon Linear Integrated Circuits, 1989, p. 8-16.

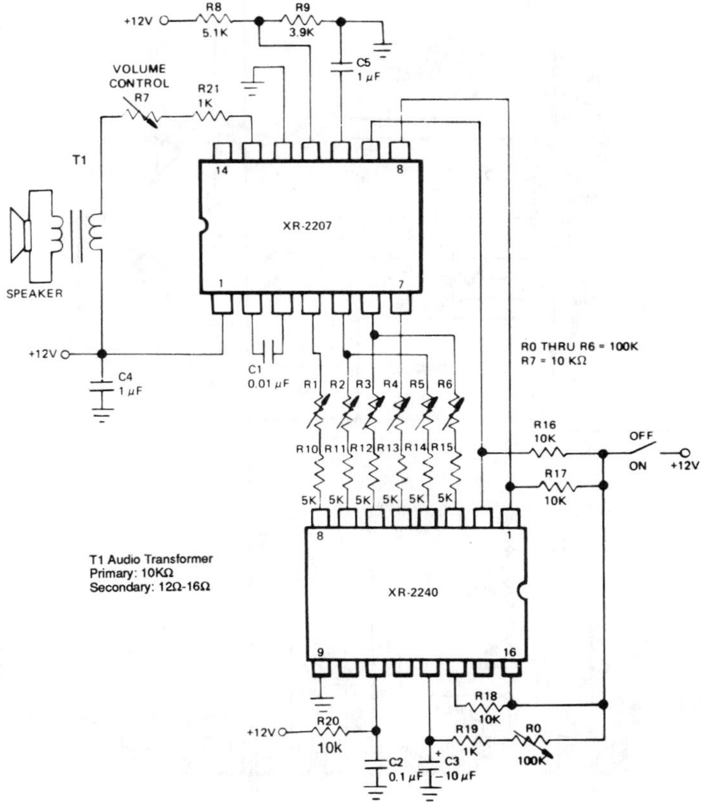

Electronic music synthesizer

Fig. 12-9 In this circuit, the XR-2207 oscillator produces a sequence of tones by oscillating at a frequency set by C1 and resistors R1 through R6 (which set the frequency or "pitch" of the output tone sequence). The XR-2240 counter/timer generates pseudo-random pulse patterns by selectively counting down the time-base frequency. The outputs of XR-2240 (pins 1 through 8) activate the timing resistors R1 through R6 of XR-2207, which convert the binary pulse patterns to tones. C3 and R0 set the "beat" or tempo of the music. The output tone sequence continues for about 1 to 2 minutes (depending on the "beat") and then repeats. The XR-2240 resets to zero when power is applied, so the tone sequence (or music) always starts from the same point when the synthesizer is turned on. EXAR Corporation Databook, 1990, p. 5-326.

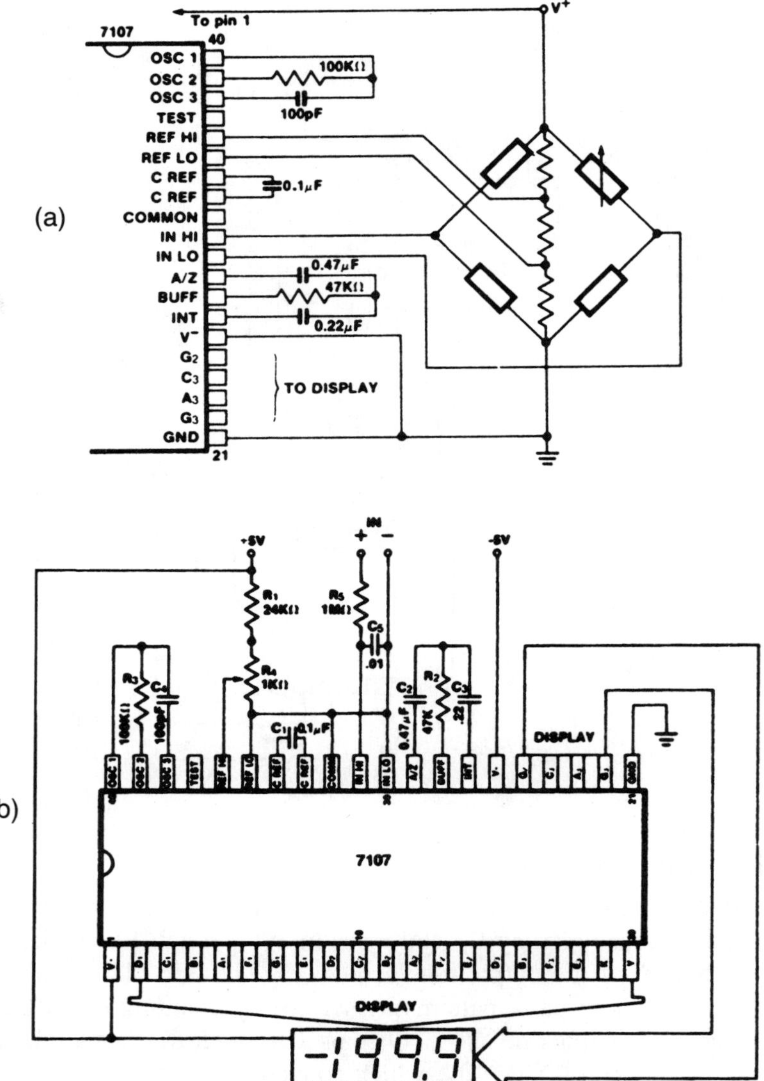

Measuring ratiometric values of quad load cells

Fig. 12-10 This circuit shows an ICL7107 connected to measure ratiometric values of quad load cells, and to display the results on a 3½-digit LED. Direct connections between the ICL7107 and the LED are shown in Fig. 12-10B. The resistor values with the load-cell bridge are determined by the desired sensitivity.

Harris Semiconductors, Data Acquisition, 1991 p. 2-32, 2-39.

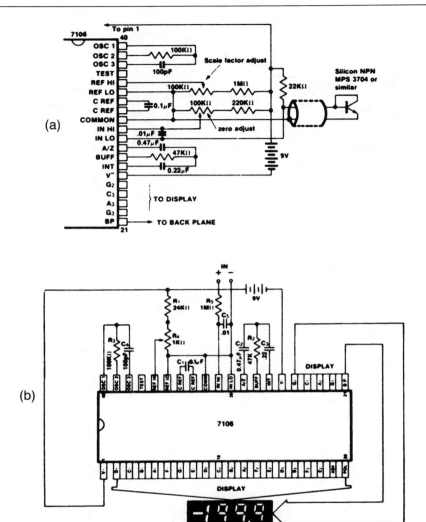

(a)

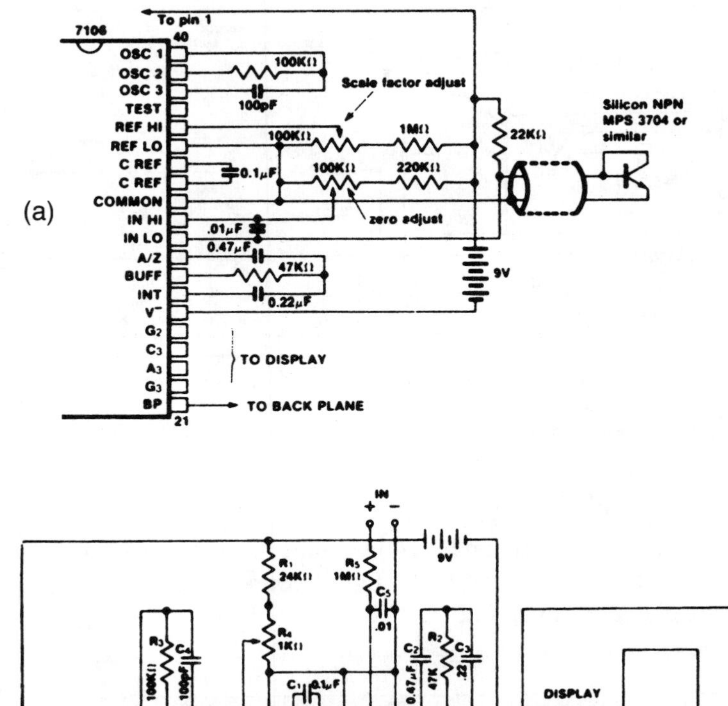

(b)

Digital centigrade thermometer

Fig. 12-11 This circuit shows an ICL7106 connected with a silicon transistor to form a digital thermometer. Direct connections between the ICL7106 and an LCD display are shown in Fig. 12-11B. A diode-connected silicon transistor has a temperature coefficient of about −2 mV/°C. To achieve calibration, place the sensing transistor in ice water and adjust the zeroing potentiometer for a 000.0 reading. Then, place the sensor in boiling water and adjust the scale-factor potentiometer for a 100.0 reading. Harris Semiconductors, Data Acquisition, 1991, p. 2-32, 2-39.

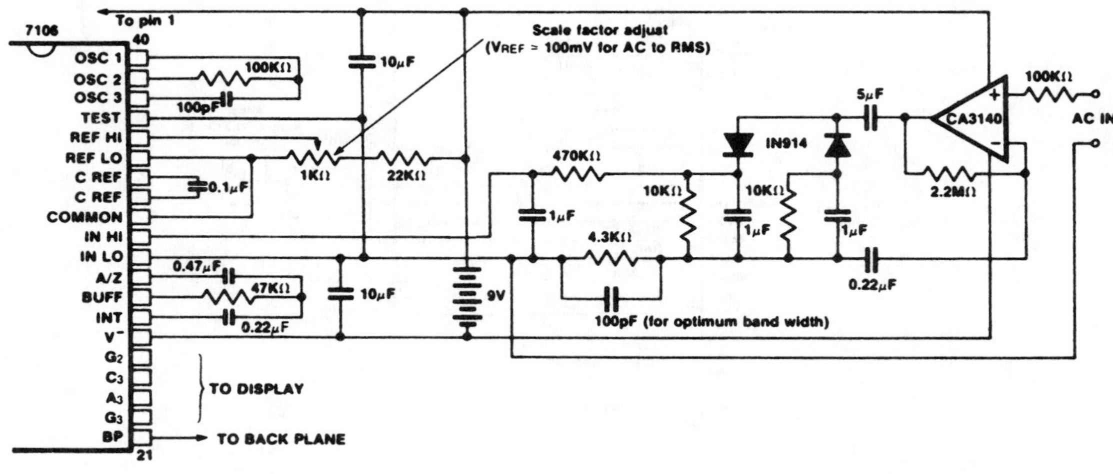

A/D conversion and display for ac signals

Fig. 12-12 This circuit shows an ICL7106 connected to measure ac voltages, and to display the results on a 3½-digit LCD. Direct connections between the ICL7106 and an LCD display are shown in Fig. 12-11B. Adjust the 1-kΩ potentiometer for the desired scale factor. Harris Semiconductors, Data Acquisition, 1991, p. 2-40.

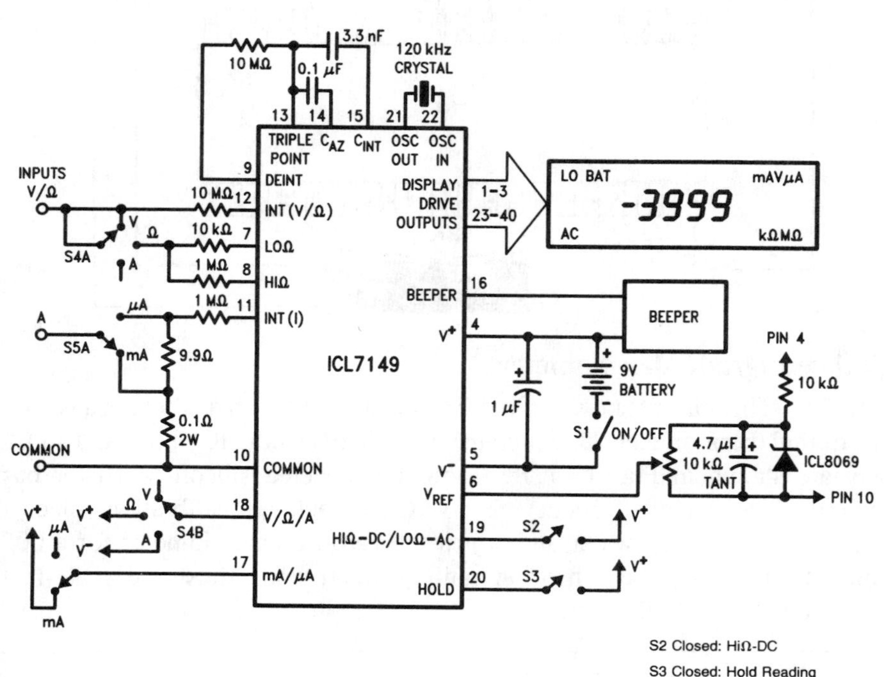

Multimeter (single chip)

Fig. 12-13 This circuit shows an ICL7149 that is connected as a low-power, autoranging digital multimeter. Although the ohms ranges do not need protection, the current ranges should be provided with fast-blow fuses between S5A and the 0.1- and 9.9-Ω shunt resistors. Also, the 10-kΩ resistor at pin 7 must be able to dissipate 1.2 or 4.8 W for short periods during accidental application of 110- or 220-V line voltages, respectively. The suggested crystal is a Statek CX-1V, the display is an LXD part number 38D8R02H, and the beeper is a muRata PKM24-4A0. Harris Semiconductors, Data Acquisition, 1991, p. 2-130.

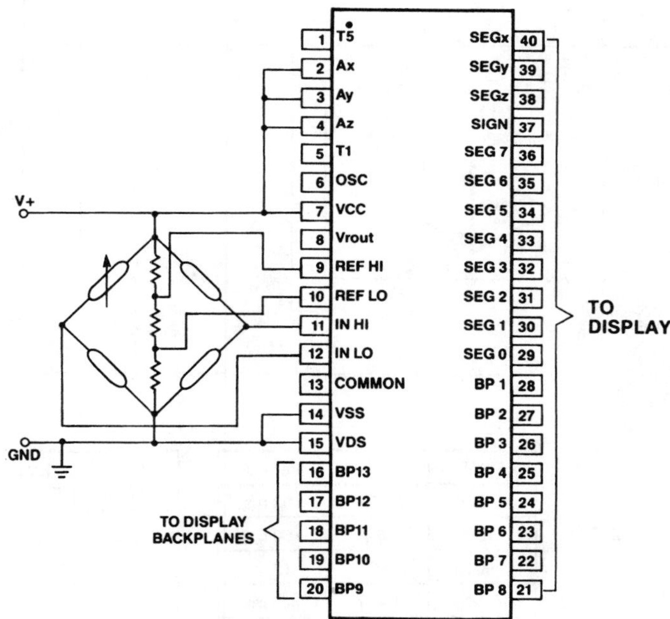

Bargraph display of quad load-cell values

Fig. 12-14 This shows an ICL7182 bargraph converter connected to measure ratiometric values of quad load cells, and to display the results on a bargraph. The resistor values within the load-cell bridge are determined by the desired sensitivity. Harris Semiconductors, Data Acquisition, 1991, p. 2-143.

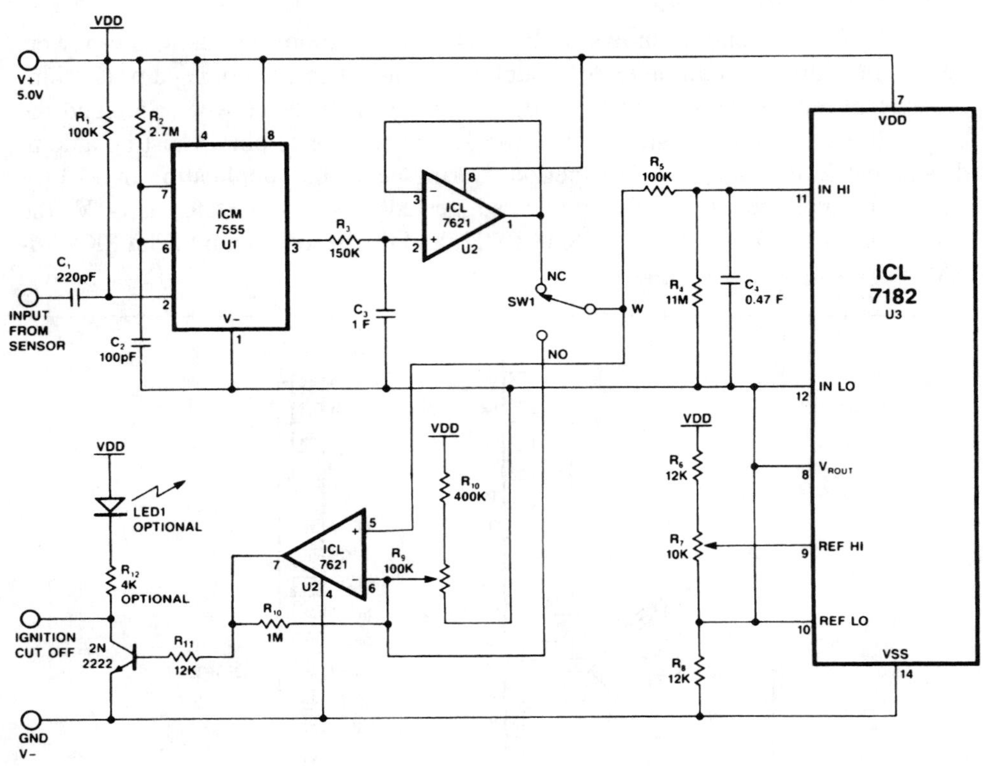

$$\frac{V_O}{V_{IN}} = \frac{1/RC}{S + 1/R_3C_3}$$

$$fc = \frac{1}{2\pi R_3C_3}$$

SW1 Momentary Switch SPST

V_{IN} = 264 mV @ 5000 RPM
4 Stroke V8

RPM	Hz	Period
600	10	100 ms
1000	16.7	60 ms
5000	83	12 ms
10,000	166.7	6 ms

No. of Cylinders	Events Per Cycle	Strokes Per Cycle
1	0.5	4
4	2	4
6	3	4
8	4	4

Tachometer with set point

Fig. 12-15 This circuit shows an ICL7182 bargraph converter that is connected to form a tachometer with set point. The connections between the ICL7182 and bargraph are shown in Fig. 12-14. Harris Semiconductors, Data Acquisition, 1991, p. 2-144.

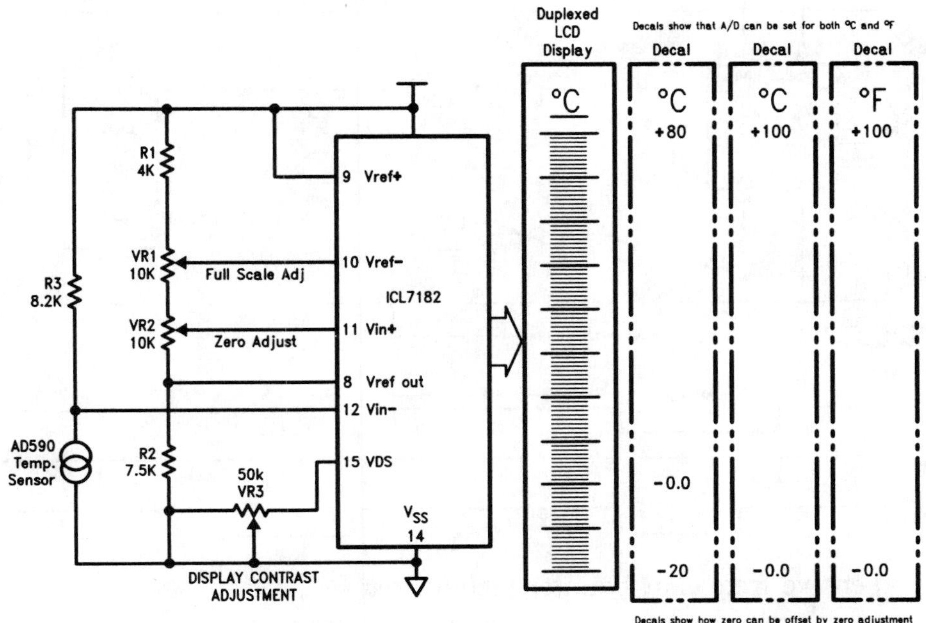

Basic digital thermometer

Fig. 12-16 This circuit shows an ICL7182 bargraph converter that is connected to form a basic digital thermometer, with a bargraph readout. The connections between the ICL7182 and bargraph are shown in Fig. 12-14. Harris Semiconductors, Data Acquisition, 1991, p. 2-145.

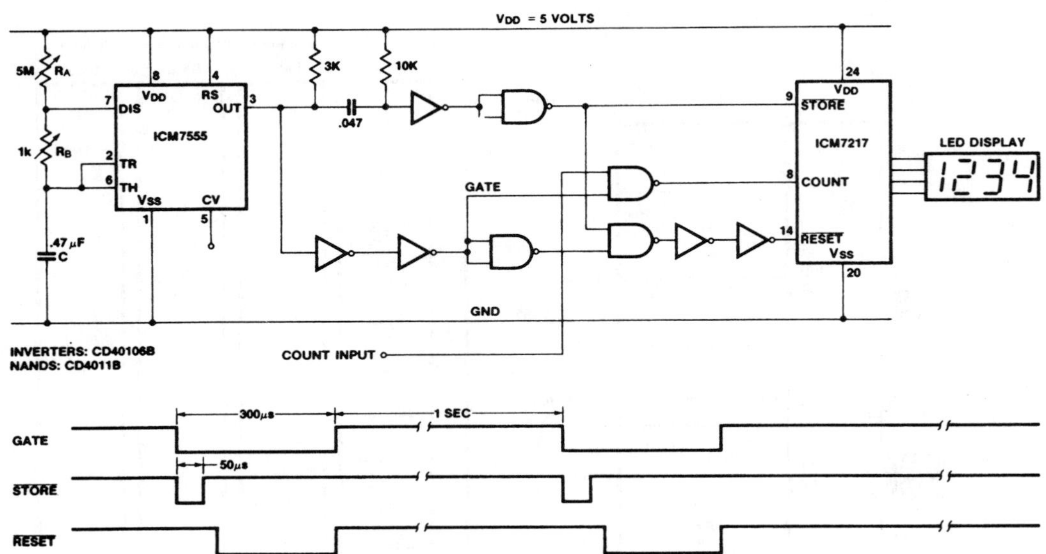

Inexpensive frequency counter/tachometer

Fig. 12-17 This circuit shows an ICM7217 and an ICM7555 that are connected as a basic frequency counter. The connections between the ICM7217 and a common-cathode LED display are shown in Fig. 12-18. The frequency counter is calibrated (against a known standard) using R_A as a coarse control and R_B as a fine control. Notice that the ICM7555 timer is connected as an astable multivibrator. Harris Semiconductors, Data Acquisition, 1991, p. 11-55.

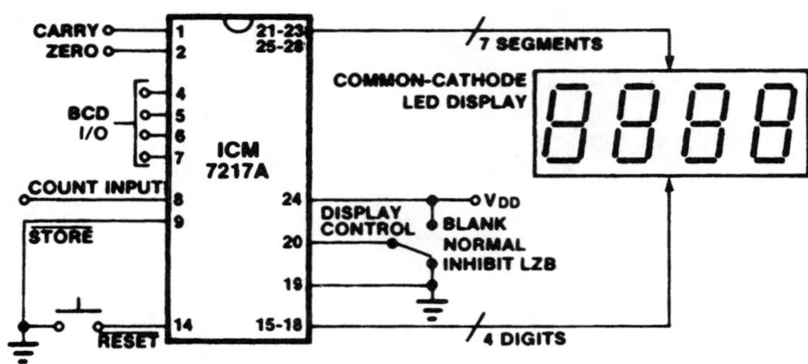

Simple unit counter with BCD output

Fig. 12-18 This circuit shows an ICM7217 and a calculator-type 4-digit display that is connected to form a very simple unit counter and display system. Harris Semiconductors, Data Acquisition, 1991, p. 11-55.

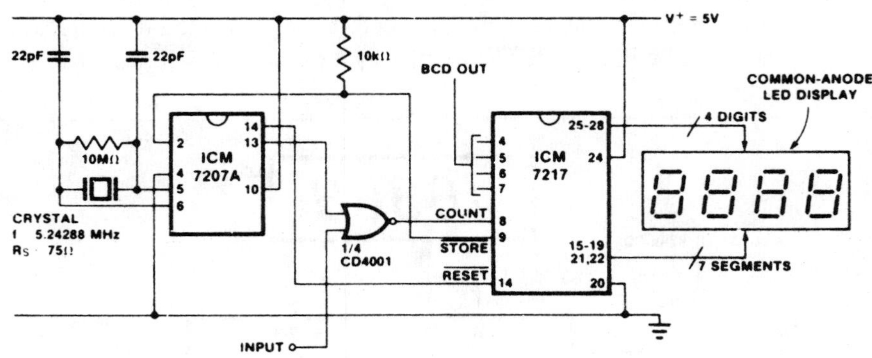

Precision frequency counter

Fig. 12-19 This circuit shows a simple 4-digit frequency counter using an ICM7217 and an ICM7207 (which provides the 1-s gating window and the $\overline{\text{STORE}}$ and $\overline{\text{RESET}}$ signals). The display reads hertz directly, as connected. With pin 11 of the ICM7027 connected to V_{DD}, the gating time is 0.1 s. This displays tens of hertz at the least significant digit. For shorter gating times, use a 6.5536-MHz crystal (0.01 s with pin 11 connected to V_{D}, and 0.1 s with pin 11 open). Harris Semiconductors, Data Acquisition, 1991, p. 11-58.

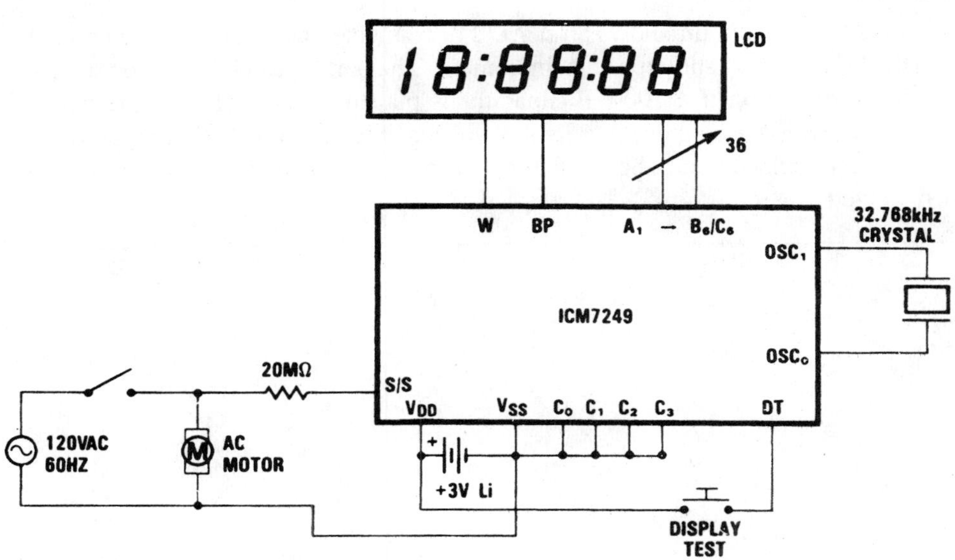

Motor hour meter

Fig. 12-20 This circuit shows an ICM7249 that is connected as an hours-in-use meter and is capable of displaying how many hours that line voltage is applied to the motor. This configuration will operate continuously for 2½ years using a 3-V lithium cell. Without the display (which only needs to be connected when a reading is required), the circuit will operate for about 10 years. Harris Semiconductors, Data Acquisition, 1991, p. 11-89.

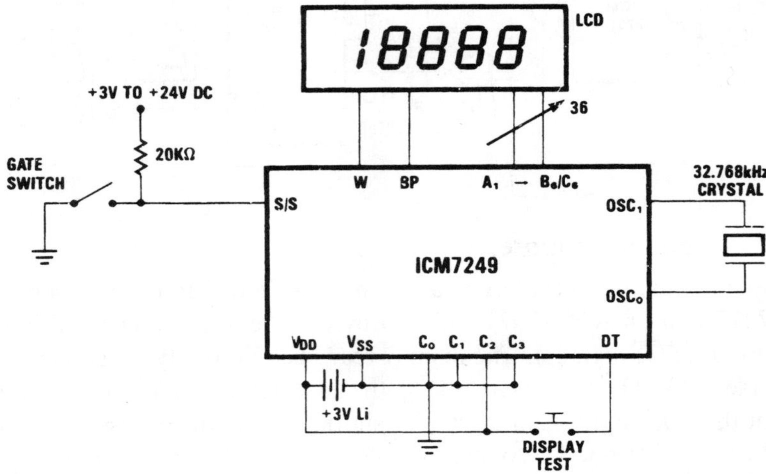

Attendance counter

Fig. 12-21 This circuit shows an ICM7249 connected as an attendance counter, with the LCD display showing each increment. The battery can be replaced without disturbing operation if a 100-μF capacitor is placed in parallel with the battery before removal. Disconnect the display (if possible) during the battery replacement. After replacement, the capacitor can be removed and the display can be reconnected. Harris Semiconductors, Data Acquisition, 1991, p. 11-90.

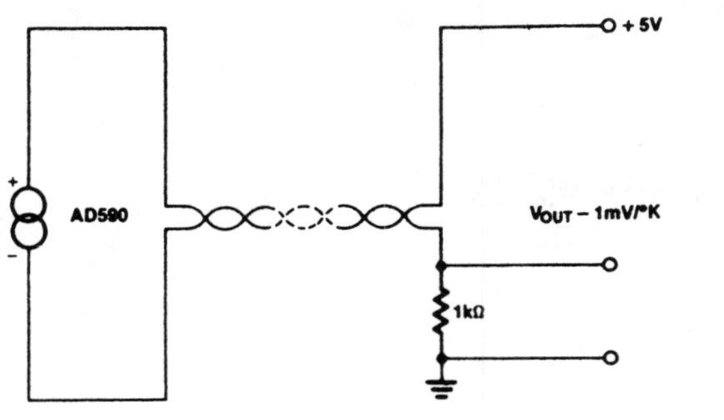

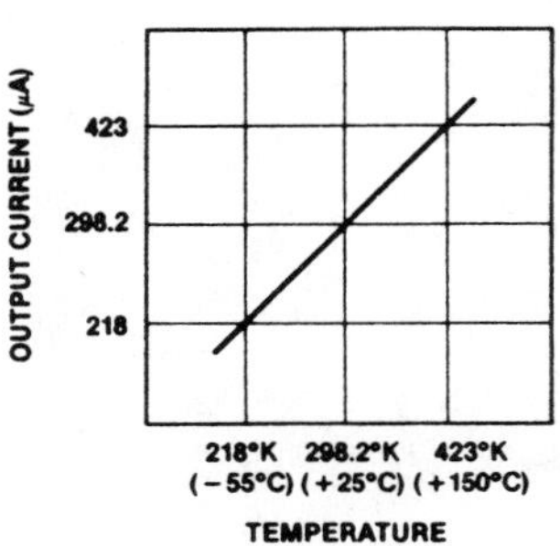

Absolute-temperature sensor

Fig. 12-22 This circuit shows an AD590 that is connected as a basic absolute-temperature sensor. The output is proportional to absolute temperature, as shown by the graph. Harris Semiconductors, Data Acquisition, 1991, p. 12-7.

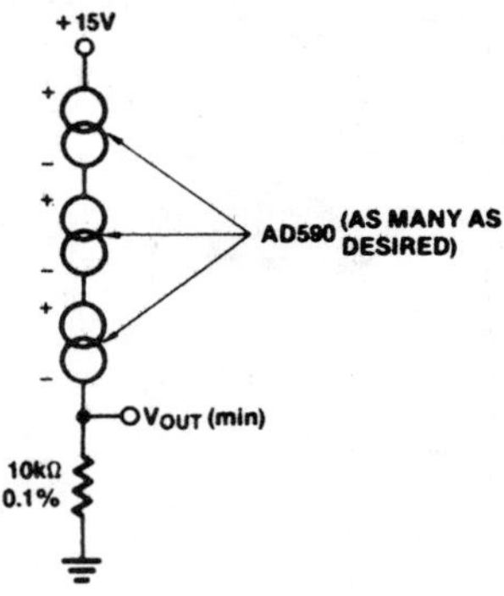

Lowest-temperature sensor

Fig. 12-23 This circuit shows an AD590 connected as a lowest-temperature sensor. The available current is that of the "coldest" sensor. Harris Semiconductors, Data Acquisition, 1991, p. 12-7.

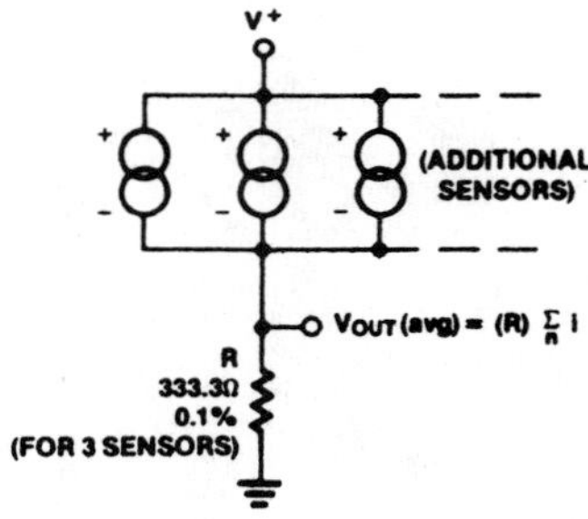

Average-temperature sensor

Fig. 12-24 This circuit shows an AD590 connected as an average-temperature sensor. The sum of the AD590 currents appears across R, which is chosen by $R = (10\ \text{k}\Omega)/n$, where n = the number of sensors. Harris Semiconductors, Data Acquisition, 1991, p. 12-7.

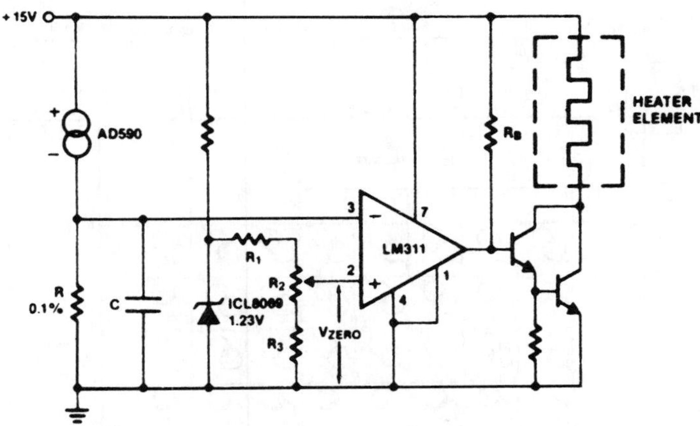

Single-setpoint temperature controller

Fig. 12-25 In this circuit, the AD590 produces a temperature-dependent voltage across R (C is for filtering noise). Setting R2 produces a scale-zero voltage. For the Celsius scale, make $R = 1\ \text{k}\Omega$ and $V_{ZERO} = 0.273$ V. For the Fahrenheit scale, $R = 1.8\ \text{k}\Omega$ and $V_{ZERO} = 0.460$ V. Harris Semiconductors, Data Acquisition, 1991, p. 12-8.

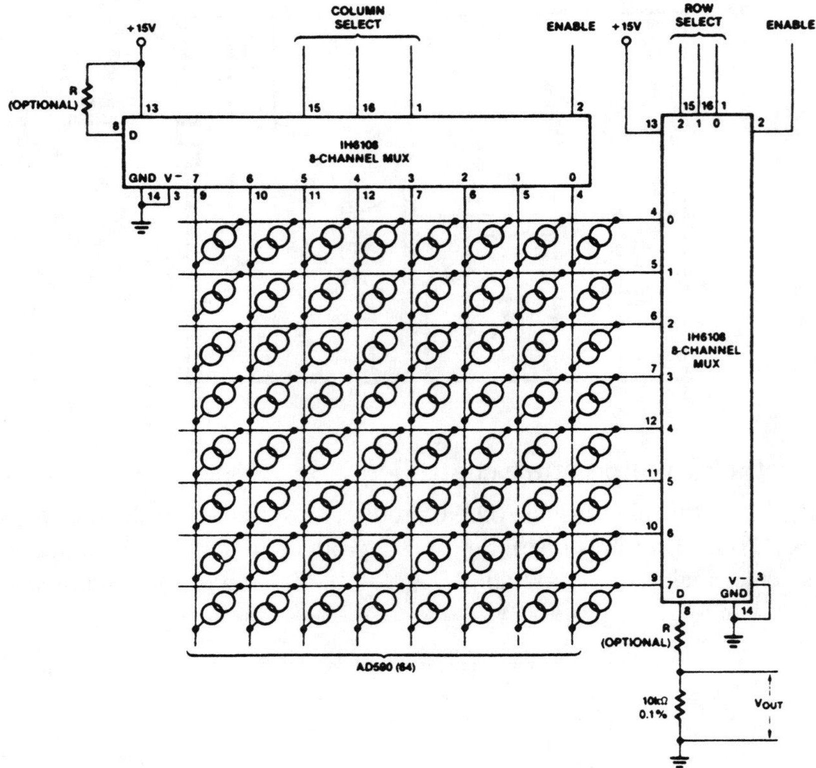

Multiplexed temperature sensors

Fig. 12-26 This circuit shows 64 AD590 transducers that are connected as multiplexed temperature sensors. A 6-bit digital word selects one of the 64 sensors (each at a different location, as desired). Harris Semiconductors, Data Acquisition, 1991, p. 12-8.

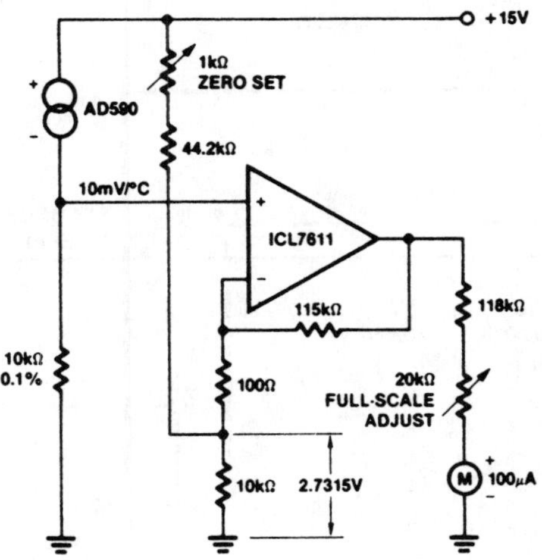

Centigrade thermometer

Fig. 12-27 This circuit shows an AD590 and a low-current op amp that are connected to form a 0 to 100°C thermometer. The readout is a 100-μA meter, which is adjusted by the Zero Set and Full-Scale Adjust potentiometer so that 1 μA indicates 1°C, 10 μA indicates 10°C, and so on. Harris Semiconductors, Data Acquisition, 1991, p. 12-9.

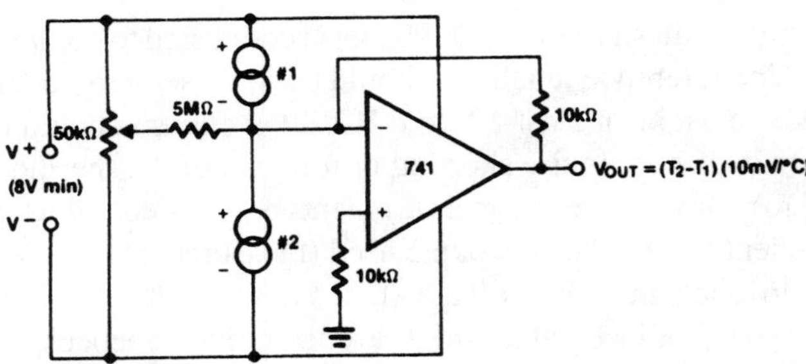

Differential thermometer

Fig. 12-28 This circuit shows two AD590s and an op amp that is connected to measure temperature differential. The 50-kΩ potentiometer trims offsets in both devices, and can be used to set the size of the difference interval. The circuit can be used for liquid-level detection (if there is a measurable temperature difference in the liquid at different levels). Harris Semiconductors, Data Acquisition, 1991, p. 12-9.

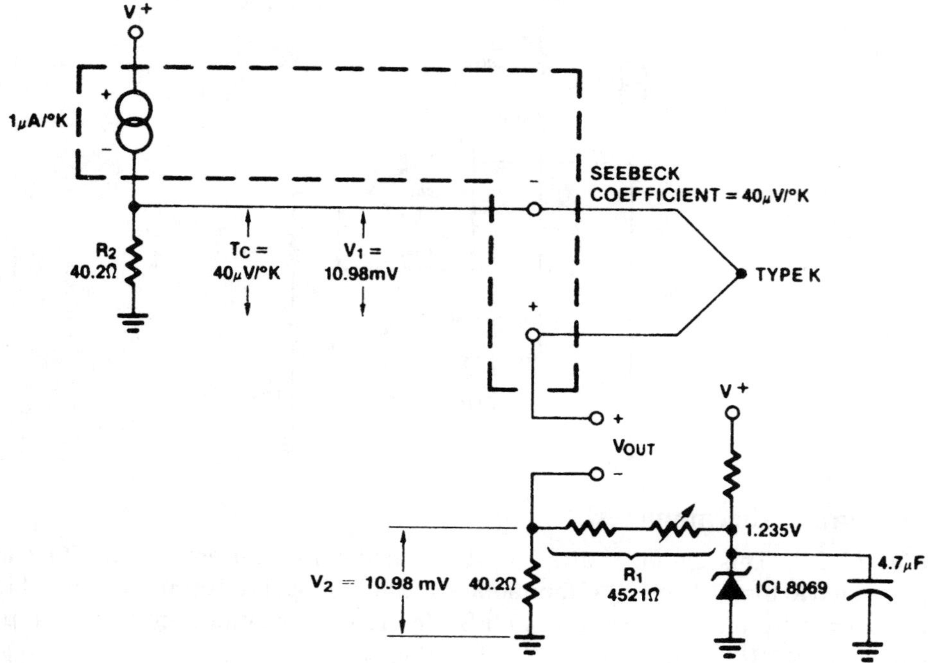

Cold-junction compensation for type-K thermocouple

Fig. 12-29 This circuit shows an AD590 that is connected to compensate a type-K thermocouple. The reference junctions should be in close thermal contact with the AD590 case. *V*+ must be at least 4 V and ICL8069 current should be set at 1 to 2 mA. Calibration does not require shorting or removal of the thermocouple. Set R_1 for $V_2 = 10.98$ mV. If very precise measurements are needed, adjust R_2 to the exact Seebeck coefficient for the thermocouple used (measured or from table), then note V_1 and set R_1 to buck out this voltage (that is, set V_2 to equal V1). For other thermocouple types, adjust values to the appropriate Seebeck coefficient. Harris Semiconductors, Data Acquistion, 1991, p. 12-10.

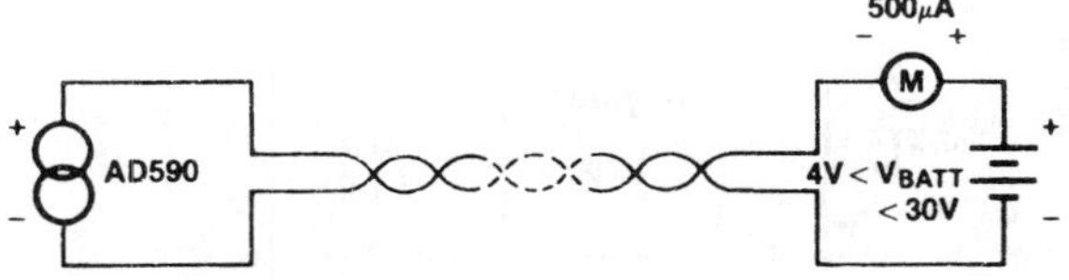

Simple thermometer

Fig. 12-30 With the connections as shown, the meter displays current output directly in degrees Kelvin. Using the AD590J, the sensor output is within ±10° over the entire range. Harris Semiconductors, Data Acquisition, 1991, p. 12-10.

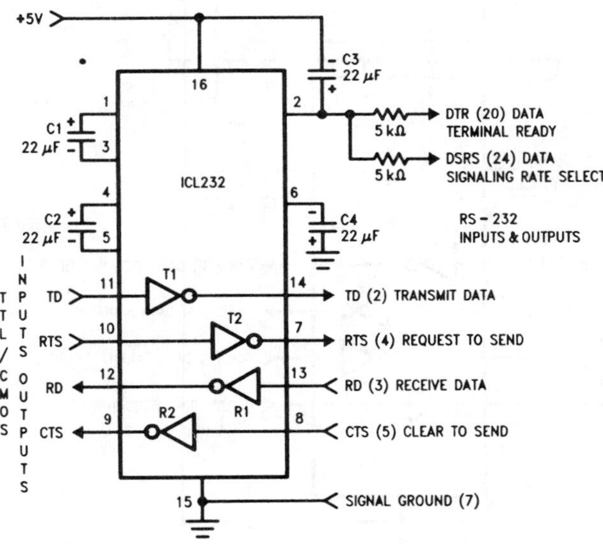

Simple duplex RS-232 port

Fig. 12-31 This circuit shows an ICL232 that is connected as a simple duplex RS-232 port with CTS/RTS handshaking. Fixed output signals, such as DTR (data terminal ready) and DSRS (data signaling rate select), are generated by driving them through a 5-kΩ resistor that is connected to V+. Harris Semiconductor, Data Acquisition, 1991, p. 13-6.

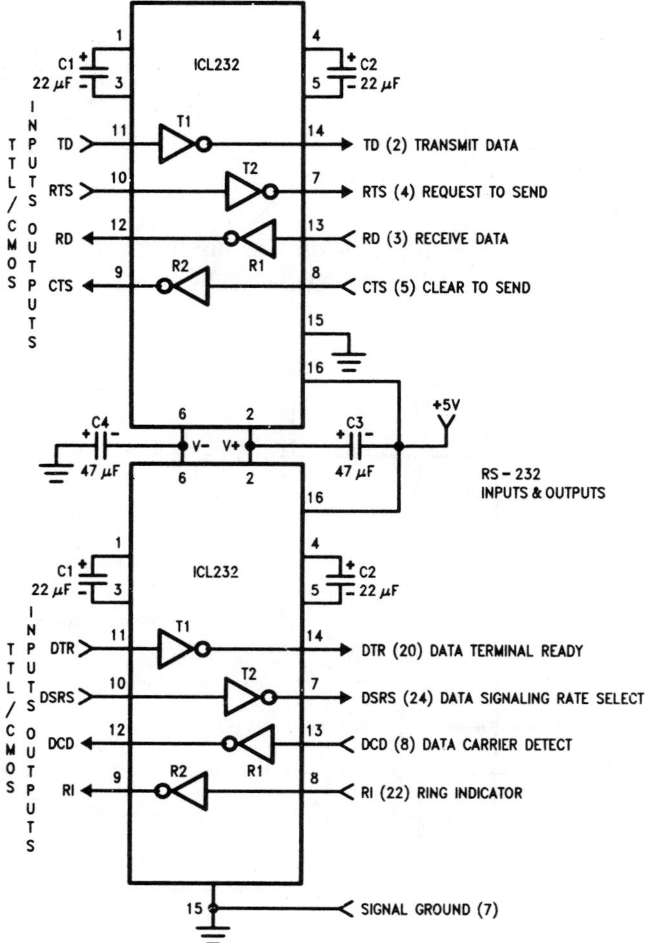

RS-232 port with four pairs of inputs/outputs

Fig. 12-32 This circuit shows two ICL232s that are combined to accommodate four pairs of inputs/outputs. Notice that each circuit requires two charge-pump capacitors C1/C2, but can share common reservoir capacitors C3/C4. Harris Semiconductors, Data Acquisition, 1991, p. 13-7.

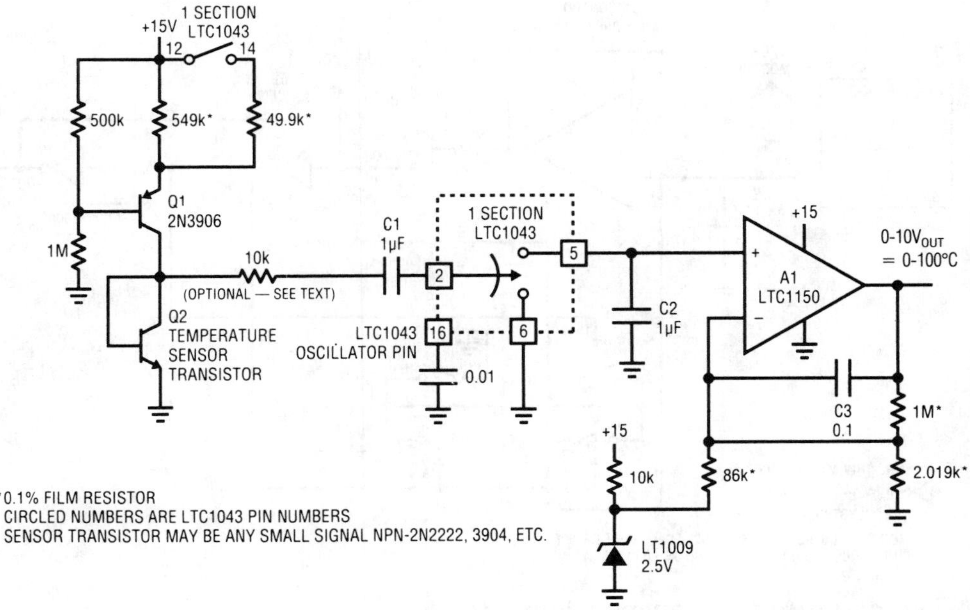

Transistor-based thermometer

Fig. 12-33 This circuit provides a 0- to 10-V output, corresponding to a 0 to 100°C temperature range at the sensor transistor Q2. Accuracy is ±1°C. No calibration is required, and any common small-signal npn can serve as the sensor. The need for calibration is eliminated because Q1 operates as a switched-value current source, alternating between about 10 and 100 μA as the LTC1043 commutates switch pins 12 and 14. The two current values are not important, as long as the ratio remains constant. Linear Technology Corporation, 1991, AN45-7.

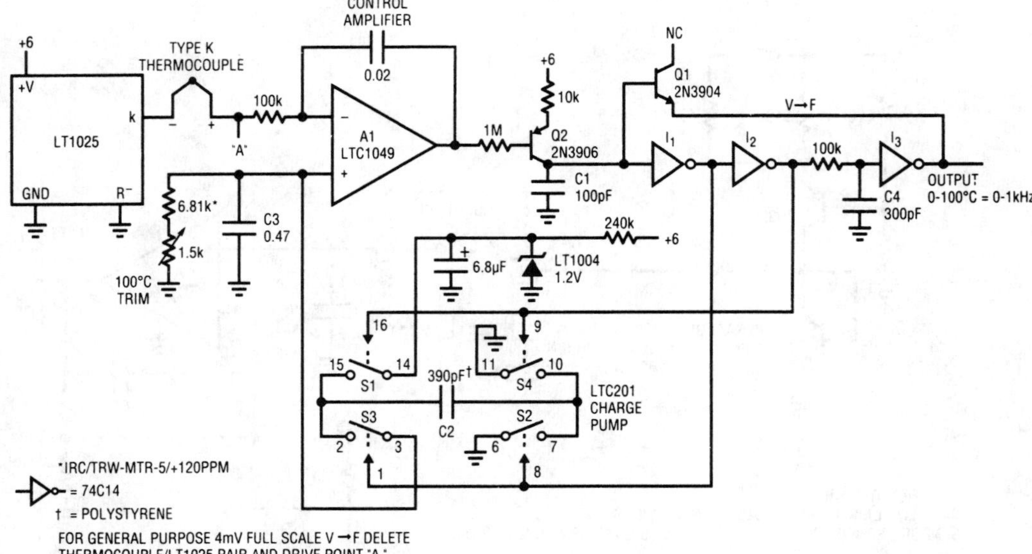

Temperature-to-frequency converter

Fig. 12-34 This circuit produces a 0- to 1-kHz output in response to a sensed 0 to 100°C temperature range. Cold-junction compensation is included, and accuracy is within 1°C with stable 0.1°C resolution. A single 4.75- to 10-V supply is required, with a maximum current consumption of 360 μA. To calibrate this circuit, disconnect the thermocouple and drive point A with 4.06 mV. Then, set the 1.5-kΩ trim for exactly 1000-Hz output. Connect the thermocouple and the circuit is ready for use. Recalibration is not required if the thermocouple is replaced. Linear Technology Corporation, 1991, AN45-8.

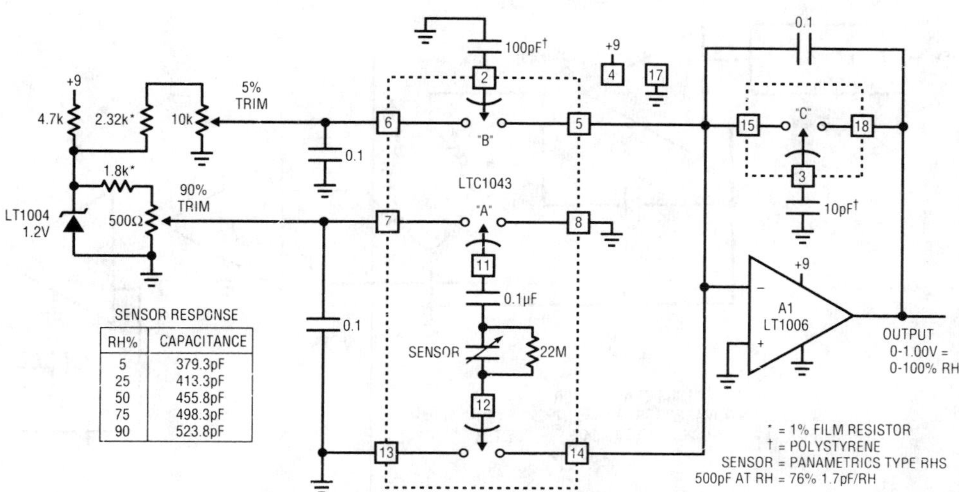

RH%	CAPACITANCE
5	379.3pF
25	413.3pF
50	455.8pF
75	498.3pF
90	523.8pF

Battery-powered relative-humidity signal conditioner

Fig. 12-35 This circuit produces a 0- to 1.00-V output in response to a sensed 0 to 100% relative humidity (RH), and operates from a 9-V battery. To calibrate, place the sensor in a 5% RH environment, and set the 5% RH trim for 50 mV at the output. Then, place the sensor in 90% RH and set the 90% trim for 900-mV output. Repeat as necessary. If the known RH environments are not available, the capacitance-versus-RH table in Fig. 12-35 can be used (although the table applies to an ideal sensor). The capacitor values can be built-up or directly dialed on a precision variable air capacitor (General Radio #722D). Linear Technology Corporation, 1991, AN45-10.

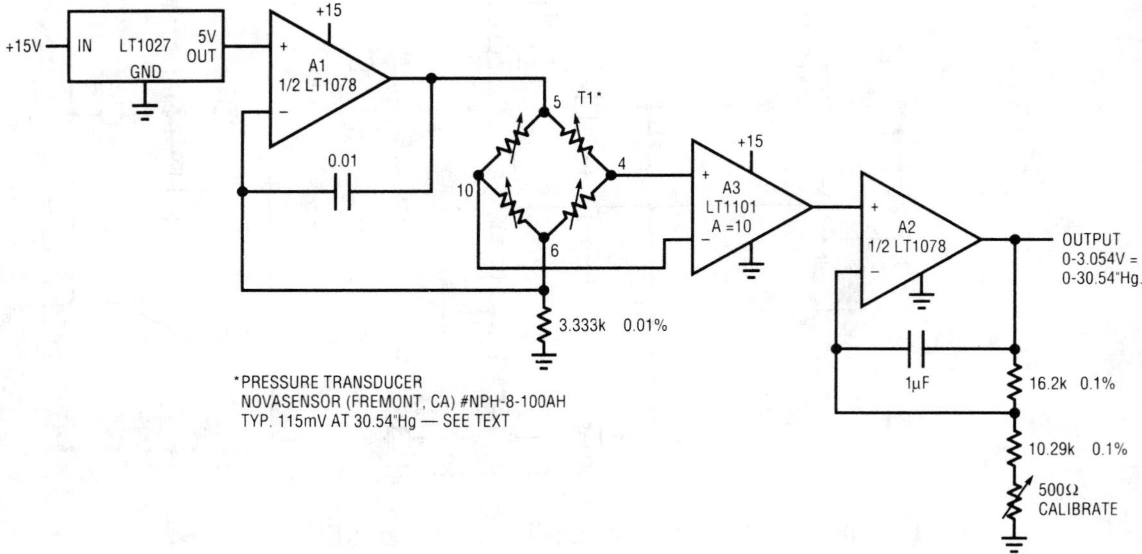

Simple precision barometer

Fig. 12-36 This circuit uses a semiconductor-based pressure transducer to form a low-cost simple barometer, and produces a 0- to 3.054-V output, in response to a sensed 0 to 30.54-in Hg. Transducer T1 specifies a nominal 115 mV at full scale, although each device is supplied with precise calibration data. To calibrate, adjust the potentiometer at A2 until the output corresponds to the scale factor that is supplied with the transducer. Linear Technology Corporation, 1991, AN45-11.

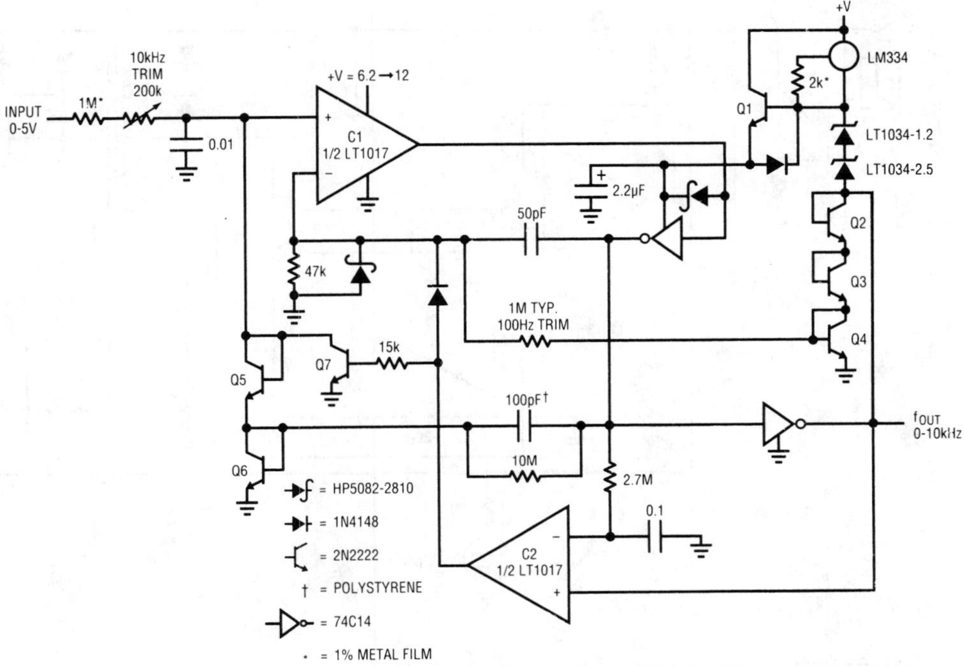

Micropower voltage-to-frequency converter

Fig. 12-37 This circuit produces a 0- to 10-kHz output in response to a 0- to 5-V input, with a maximum current consumption of only 90 μA. To calibrate, apply 50 mV and select the value at the C1 input for a 100-Hz output. Then, apply 5 V and trim the input potentiometer for a 10-kHz output. Linear Technology Corporation, 1991, AN45-15.

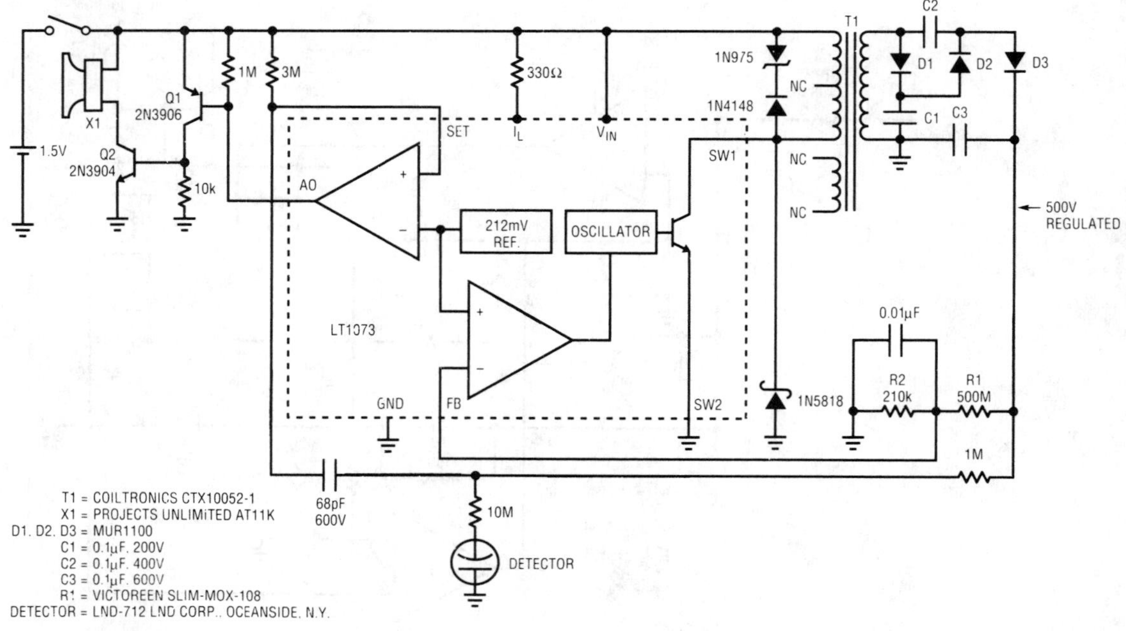

1.5-V-powered radiation detector

Fig. 12-38 This circuit provides an audible "tick" signal each time radiation or a cosmic ray passes through the detector. The LT1073 switching regulator pulses T1 which, in turn, drives a voltage tripler to provide 500-V bias to the detector. R1 and R2 provide feedback to the LT1073, which closes the control loop. No calibration is needed. Linear Technology Corporation, 1991, AN45-11.

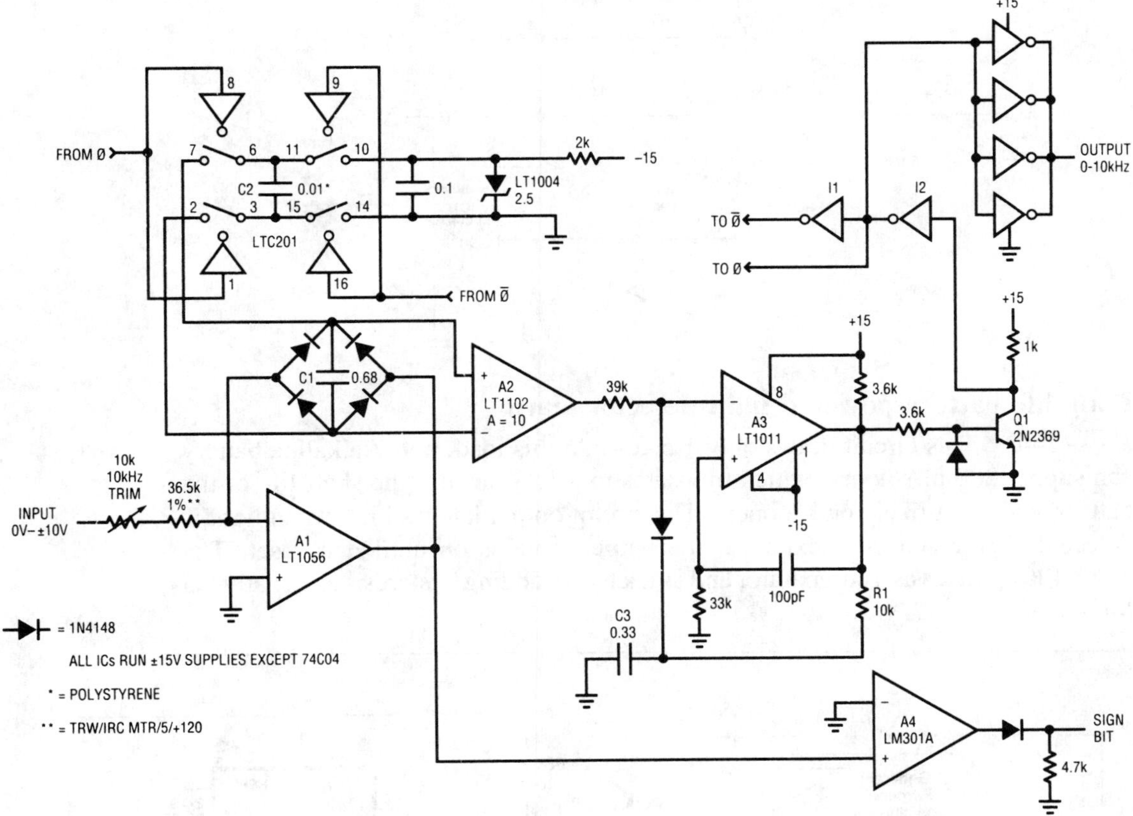

Bipolar voltage-to-frequency converter

Fig. 12-39 This circuit produces a 0- to 10-kHz output in response to a 0- to ±10-V input. The A4 output indicates the sign (+ or −) of the input. To calibrate, apply either a −10- or +10-V input and set the 10-kΩ trim for exactly a 10-kHz output. The low offsets of A1 and A2 permit operation down to a few Hz with no zero trim required. Linear Technology Corporation, 1991, AN45-17.

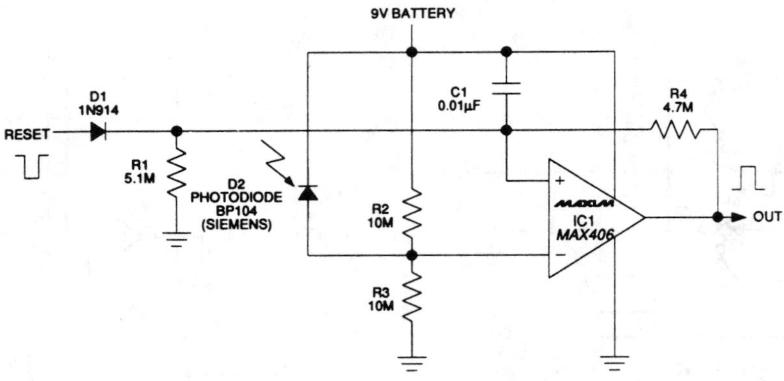

Long-life battery-powered light-detector alarm

Fig. 12-40 This circuit draws only 1.5 µA. At this load, a 9-V alkaline battery can supply 200 mA/hours, which translates to a 15-year life (the shelf life of the battery will most likely end sooner). The circuit output latches high when light is detected. If the sensor is exposed, the output remains on until it is reset. The MAX406 operates as a comparator and as a latch by adding hysteresis externally via R4. Maxim, 1992, Applications and Product Highlights, p. 5-8.

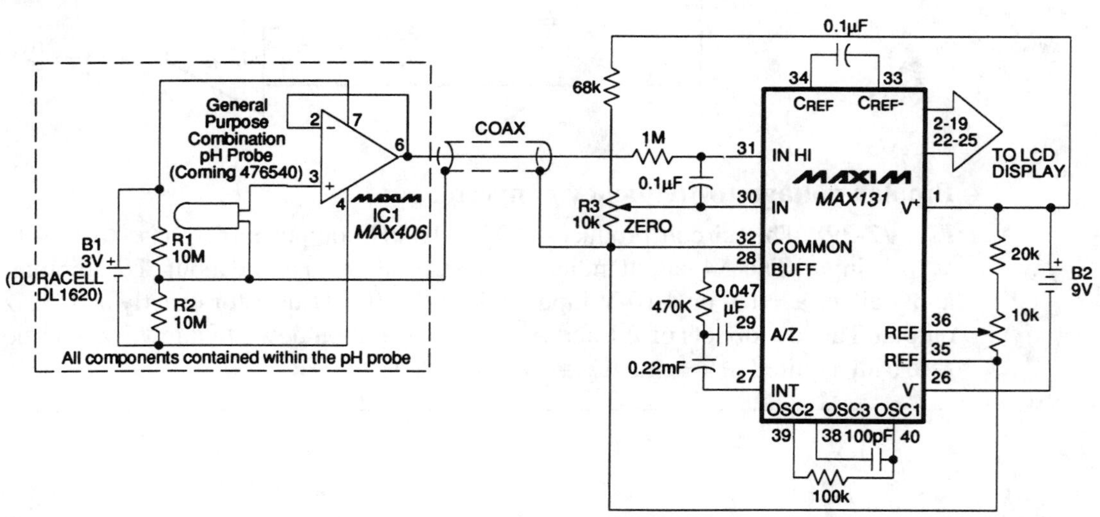

Buffered pH probe allows low-cost cable

Fig. 12-41 The probe circuit eliminates expensive low-leakage cables that often connect pH probes to meters. A MAX406 and a lithium battery are included in the probe housing. A conventional low-cost coax carries the buffered pH signal to the

MAX131 A/D converter. The battery life depends on the dc loading of the amplifier output (MAX131 input current and cable leakage). In most cases, battery life exceeds the functional life of the probe. Maxim, 1992, Applications and Product Highlights, p. 5-9.

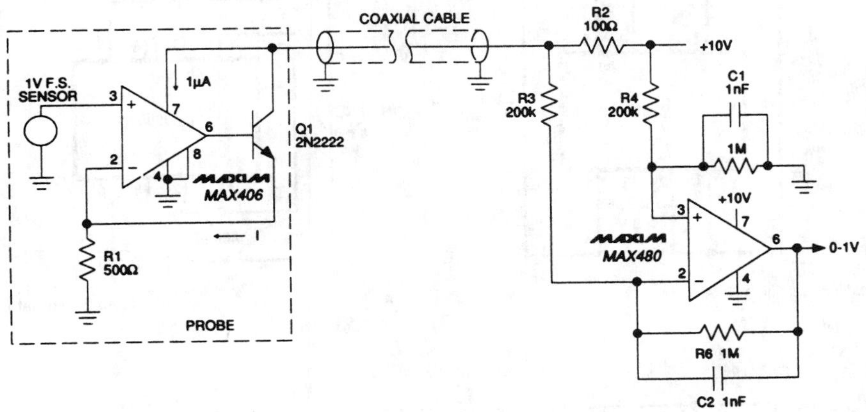

Remotely-powered sensor amp

Fig. 12-42 A simple two-wire current transmitter uses no power at the transmitting end (except that from the transmitted signal). At the transmitter, a 0- to 1-V input drives a MAX406 and npn transistor that is connected as a voltage-controlled current source. Although the MAX406 supply current is taken from the signal, only 1-μA out of 2-mA error is added. The output is sent through the coax to the MAX480, which reconstructs a ground-referenced 0- to 1-V signal. Maxim, 1992, Applications and Product Highlights, p. 5-9.

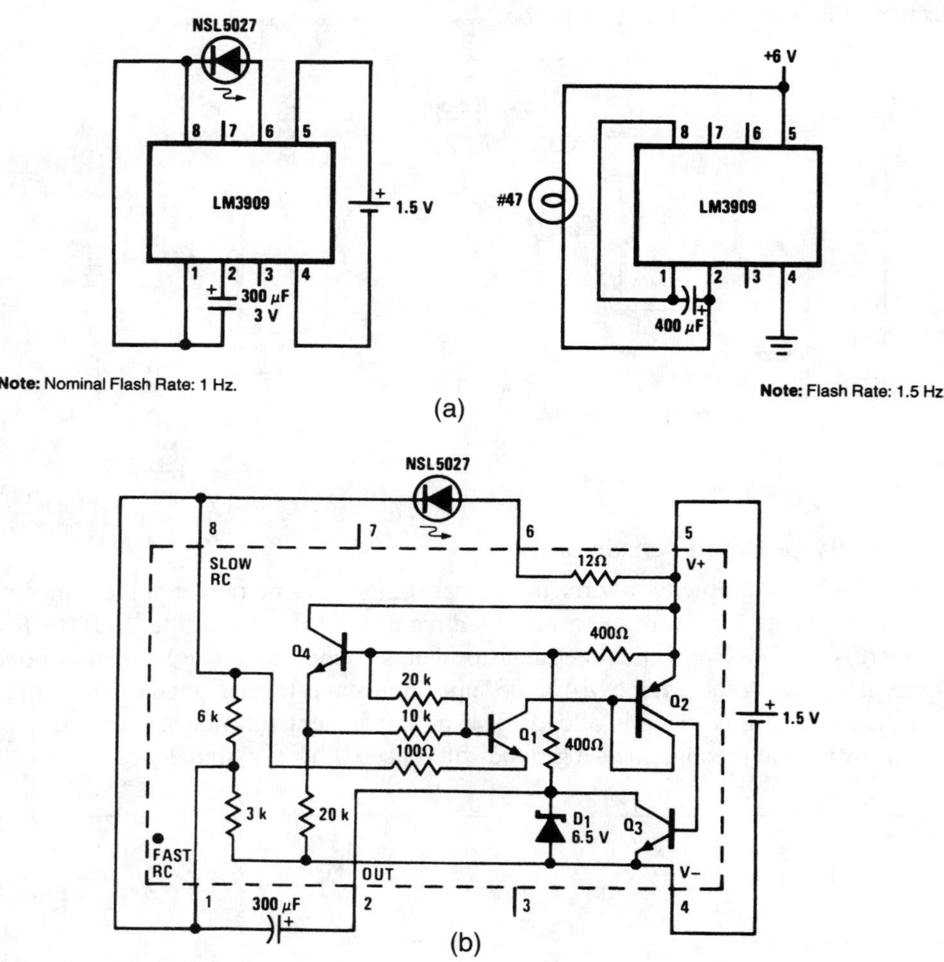

Low-power flashers

Fig. 12-43 This circuit shows two simple flashers that use an LM3909. Figure 12-43B shows the internal circuit. Notice that the LM3909 requires only 1.5 V. However, 6 V is required when the LM3909 must drive a #47 lamp (or any similar load), instead of an LED. The flashing rate is determined by the value of the capacitor between pins 1 and 2. National Semiconductor, Linear Applications Handbook, 1991, p. 394.

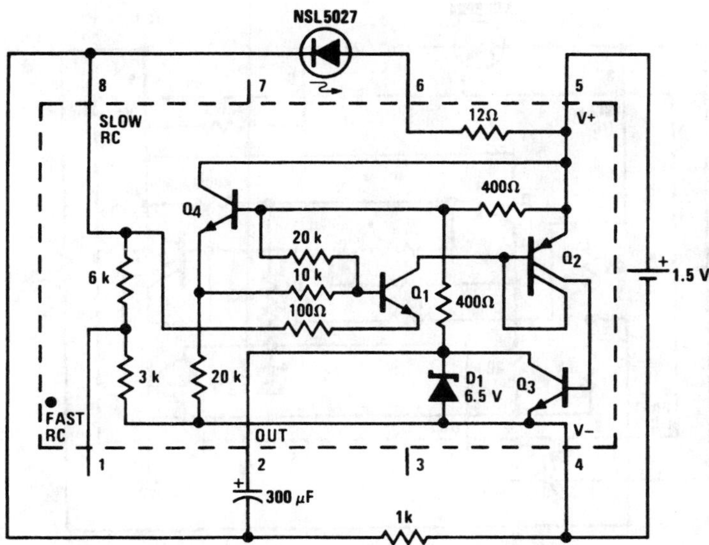

Low-power fast blinker

Fig. 12-44 This circuit shows an LED flasher that uses an LM3909. Notice that the circuit is similar to that of Fig. 12-43B, except that a 1-kΩ resistor is connected between pin 4 and pin 8. This produces a little over three times the flashing rate of the Fig. 12-43B circuit. National Semiconductor, Linear Applications Handbook, 1991, p. 395.

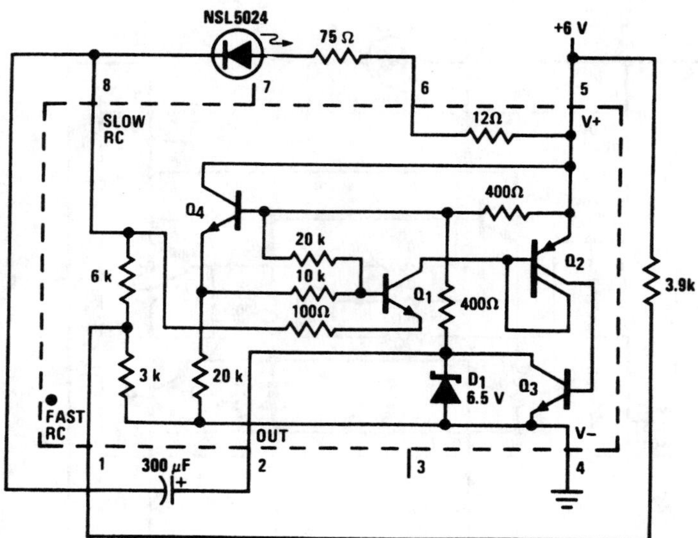

6-V flasher

Fig. 12-45 This circuit shows an LED flasher that uses an LM3909 with a 6-V supply. Notice that the circuit is similar to that of Fig. 12-43B, except that a 75-Ω resistor is required between pin 6 and the LED, a 3.9-kΩ resistor is used between pin 5 and pin 1, the 300-μF capacitor should be at least 3 V, and pin 5 receives +6 V from the external source. The flashing rate remains at about 1 Hz. National Semiconductor, Linear Applications Handbook, 1991, p. 395.

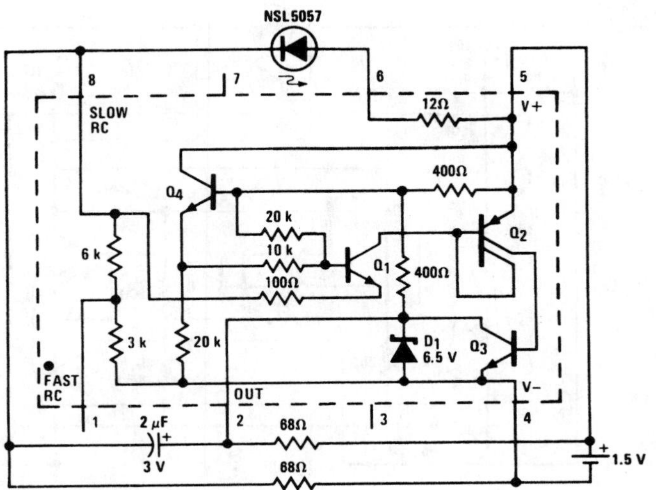

A continuous 1.5-V indicator

Fig. 12-46 This LM3909 circuit shows a continuously appearing indicator light that is powered by 1.5 V. Duty cycle and frequency of the current pulses to the LED are increased (from that of Fig. 12-43B) until the average energy supplied provides sufficient light. At frequencies above 2 kHz, even the fastest movement of the light source or the observer's head will not produce significant flicker. This indicator draws about 12 mA from the 1.5-V battery, and is not intended as a long-life system.
National Semiconductor, Linear Applications Handbook, 1991, p. 396.

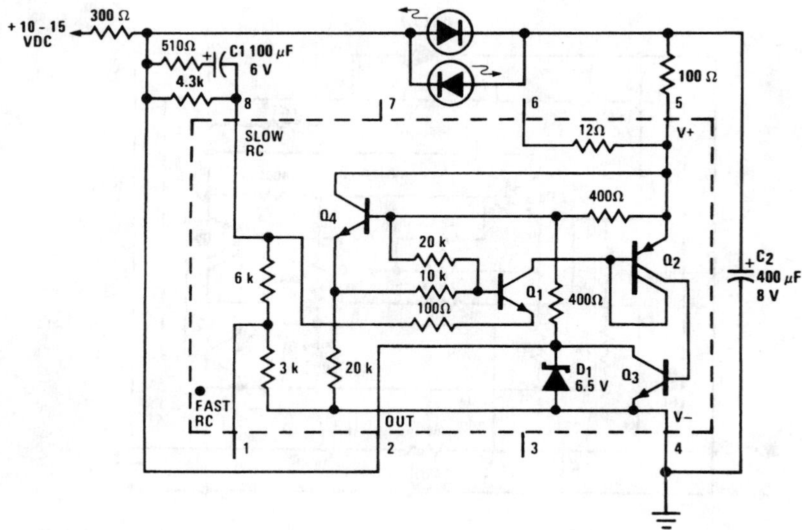

Alternating flasher

Fig. 12-47 This LM3909 circuit is essentially a relaxation-type oscillator (chapter 5) that flashes two LEDs in sequence. With a 12-V supply, the repetition rate is about 2.5 kHz. Timing and storage capacitor C2 alternately charges through the upper LED and is discharged through the other LED by Q3 within the LM3909. If a red/green flasher is desired, the green LED should have its anode or plus lead toward pin 5 (like the lower LED). A shorter, but higher, voltage pulse is available in this position. National Semiconductor, Linear Applications Handbook, 1991, p. 396.

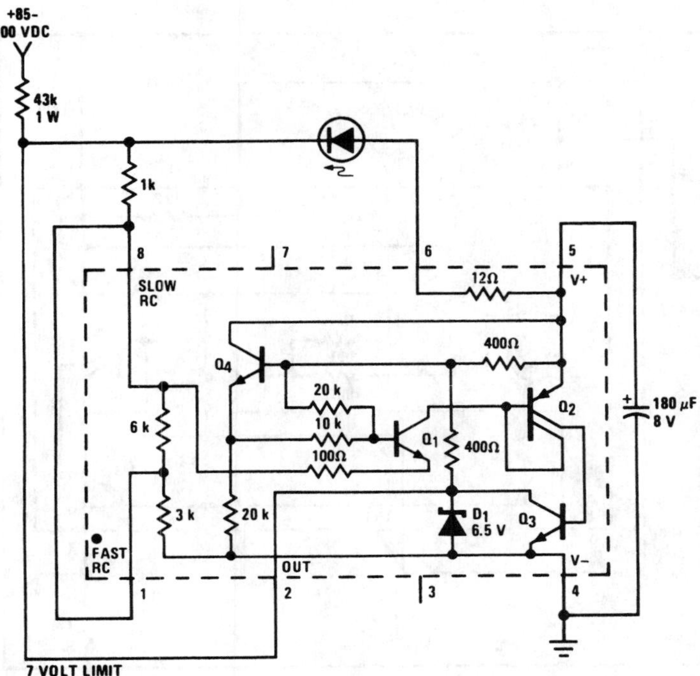

Safe high-voltage flasher or monitor

Fig. 12-48 A high-voltage (85 to 200 V) power supply can be monitored at a remote location safely and with little current drain using this LM3909 circuit. If the 43-kΩ dropping resistor is located at the source end, voltages to the LED and LM3909 are limited to less than 7 V above ground. A chart outlining operation of this circuit at various voltages appears on the LM3909 data sheet. National Semiconductor, Linear Applications Handbook. 1991, p. 397.

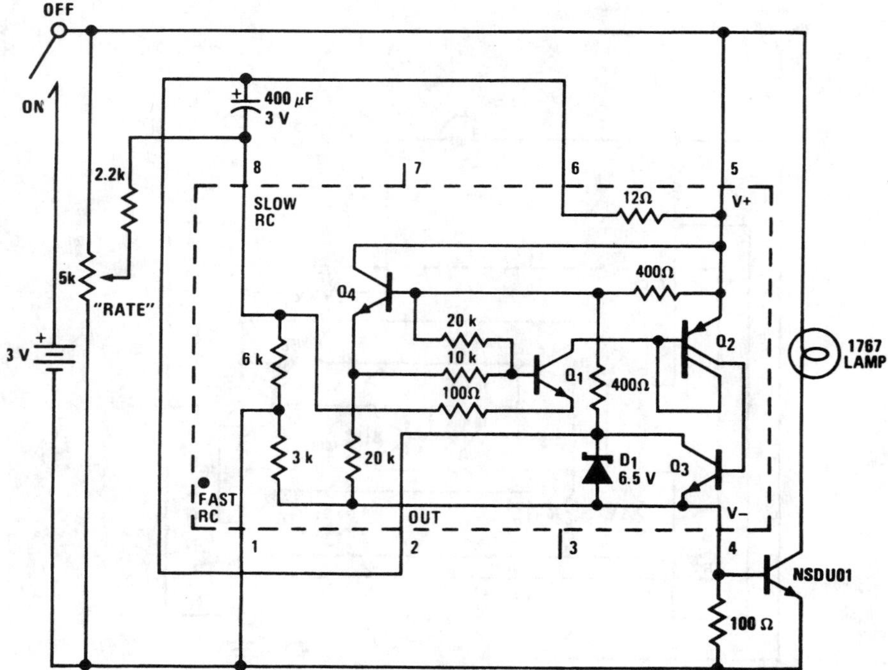

Mini-strobe variable flasher

Fig. 12-49 This LM3909 flasher circuit can be used as a variable-rate warning light, or for advertising or special effects. The rate control can adjust from no flashes to a continuous on. The miniature 1767 lamp can be flashed several times per second. As a toy, the flashing rate can be adjusted to mimic the strobes at rock concerts or the flicker of old-time movies. In a dark room, the flashes are almost fast enough to stop a person's motion. National Semiconductor, Linear Applications Handbook, 1991, p. 397.

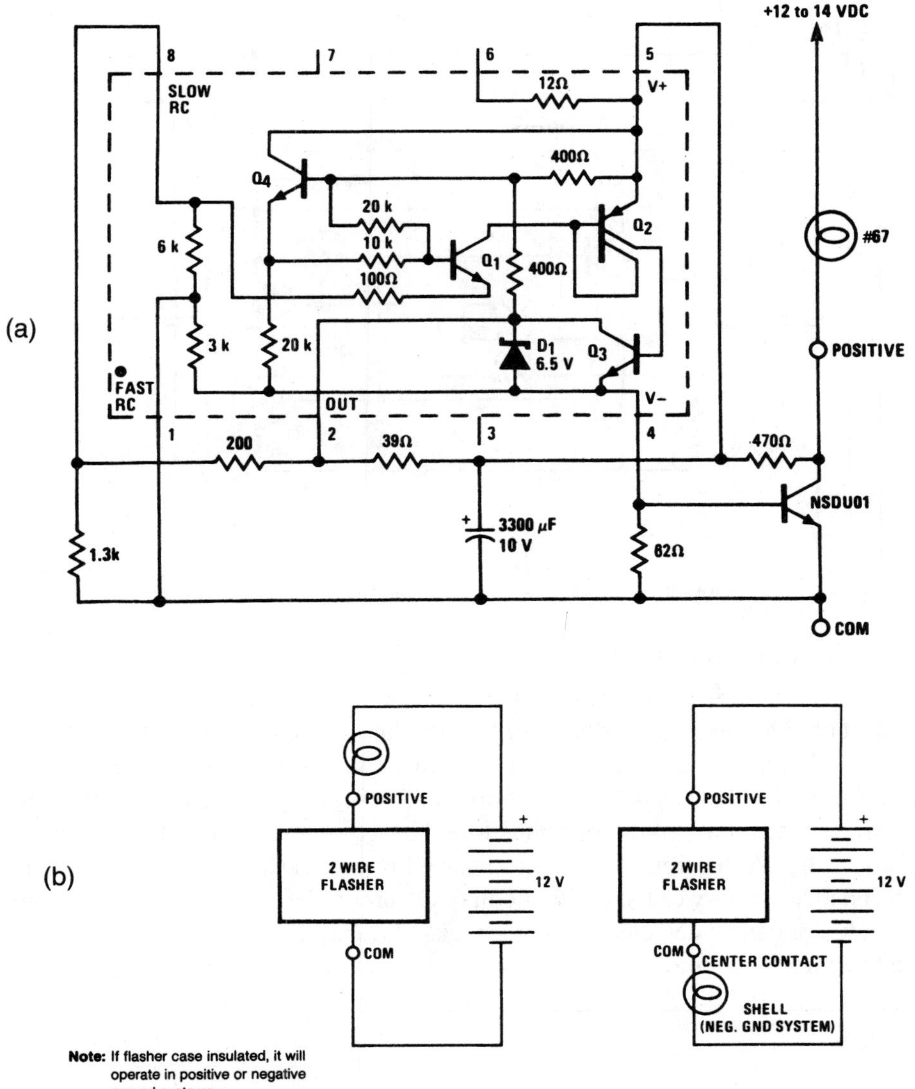

12-V flasher

Fig. 12-50 This LM3909 flasher circuit can be powered from an automotive storage battery, provides a 1-Hz flash, and can power a 600-mA lamp. A particular advantage of this circuit is that only two external wires are required (Fig. 12-50B). Also, no circuit failure can cause a battery drain greater than that of the #67 lamp.

National Semiconductor, Linear Applications Handbook, 1991, p. 398.

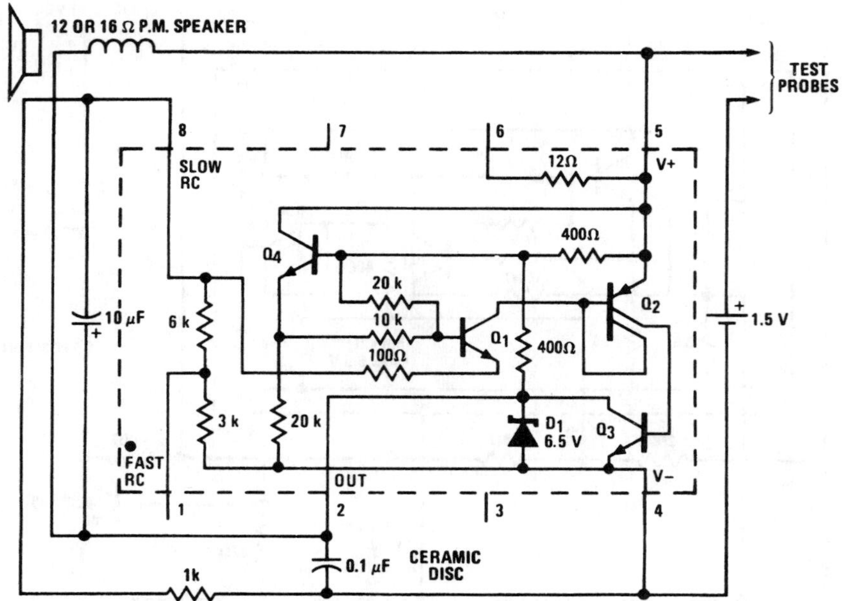

Continuity and coil checker

Fig. 12-51 Using this circuit, a short (up to about 100 Ω) across the test probes provides enough power to produce audible oscillation. In this circuit, the LM3909 is connected as an audio oscillator. By probing two values in quick succession, small differences (such as between a short and 5 Ω) can be detected by differences in the tone. A novel use of this circuit is found in setting the timing of certain types of motorcycles. A different tone can be heard if there is a short (or no short) in the low-resistance primary of the ignition coil (a difference between a 1-Ω resistor and a 1-Ω inductor can be detected). Quick checks of transformers and motors can also be made with this simple circuit. National Semiconductor, Linear Applications Handbook, 1991, p. 399.

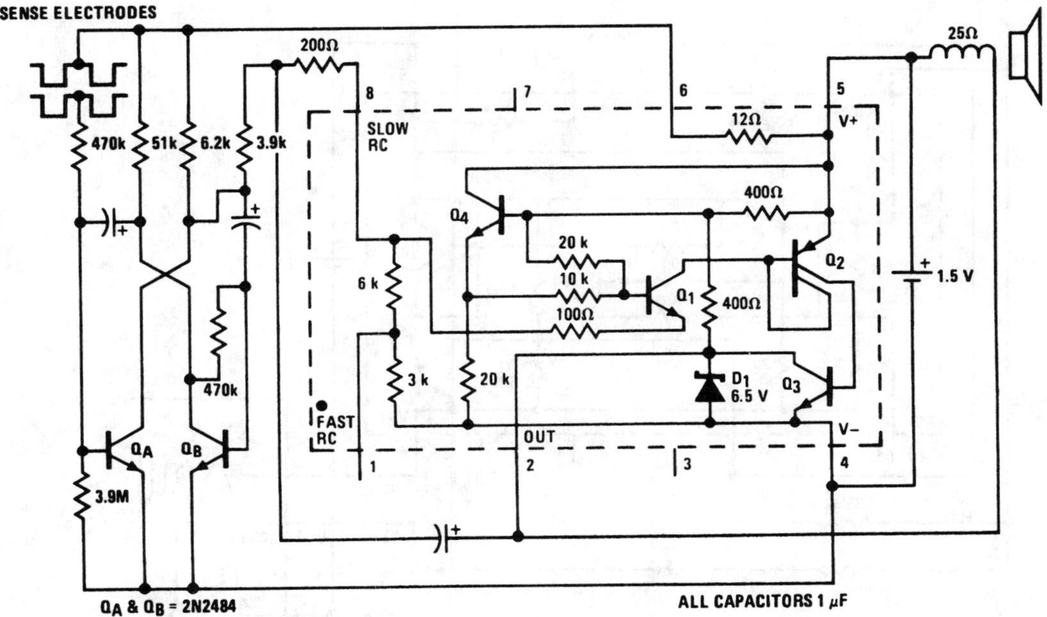

Water-seepage alarm

Fig. 12-52 This LM3909 oscillator circuit can detect water seepage in darkrooms, laundry rooms, etc., and can be used on potentially damp floors with complete safety (because there is no connection to the power line). Also, the standby battery drain of about 100 μA yields a battery life that is close to shelf life. The circuit operates as a multivibrator, which starts at about 1 Hz and oscillates faster as more leakage passes across the sense electrodes.

The sensor consists of two electrodes (6 or 8 inches long) spaced about ⅛″ apart. Two strips of stainless steel on insulators, or the appropriate zig-zag path cut in the copper cladding of a PC board can be used. The sensor should be built into the base of the box in which the circuits are packaged. The bare PC board between the copper-sensing areas should be coated with warm wax so that moisture on the floor (not the moisture absorbed by the board) is detected. The circuit and sensor can be tested by simply touching a damp finger to the electrode gap. National Semiconductor, Linear Applications Handbook, 1991, p. 400.

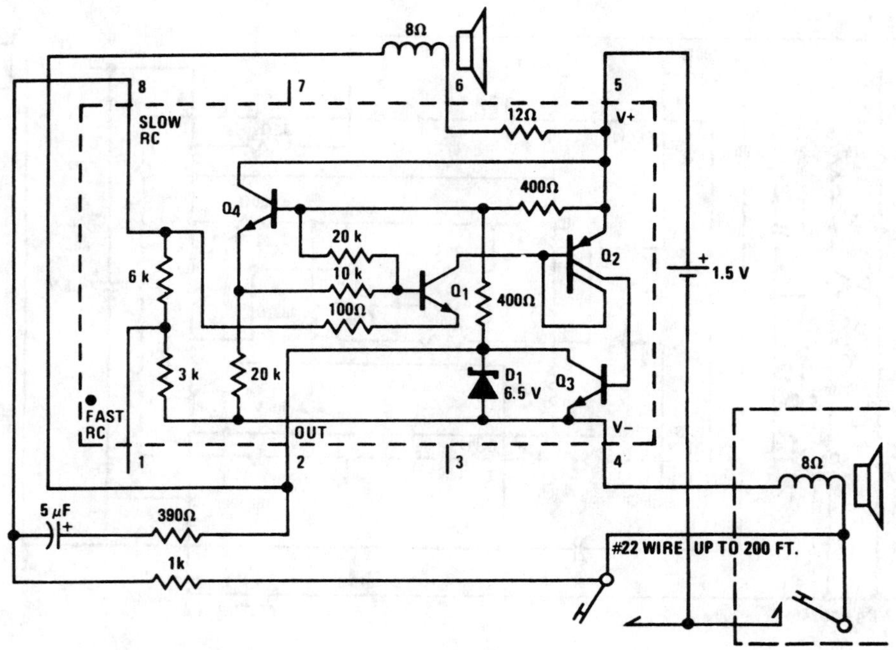

Morse-code set

Fig. 12-53 In this circuit, the LM3909 (connected as an oscillator) drives speakers at both sending and receiving ends, simultaneously. National Semiconductor, Linear Applications Handbook, 1991, p. 400.

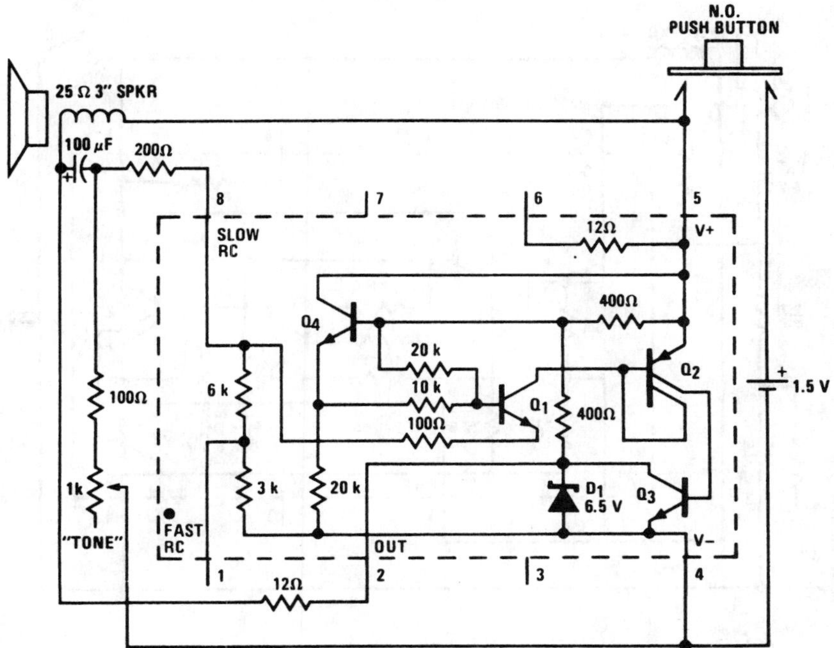

Electronic "trombone"

Fig. 12-54 With this circuit, the LM3909 oscillates at the speaker-cone "free-air" resonance point. That is, if the speaker is in an enclosure with a higher resonant frequency than the speaker, this becomes the frequency at which the circuit oscillates. An educational audio-demonstration device, or simply an enjoyable toy, has been fabricated as follows. A roughly cubical box of about 64″³ was made with one end able to slide in and out like a piston. The box was stiffened with thin layers of pressed wood, etc. Speakers, circuit, battery, and all were mounted on the sliding end, with the speaker facing out through a 2¼″ hole. A tube was provided (2½″ long, 5⁄16″ ID) to bleed air in and out as the piston was moved while not affecting resonant frequency. Minimum volume with the piston in was about 10″³. "Slide tones" can be generated, or a tune played, by properly positioning the piston part and working the push button. National Semiconductor, Linear Applications Handbook, 1991, p. 401.

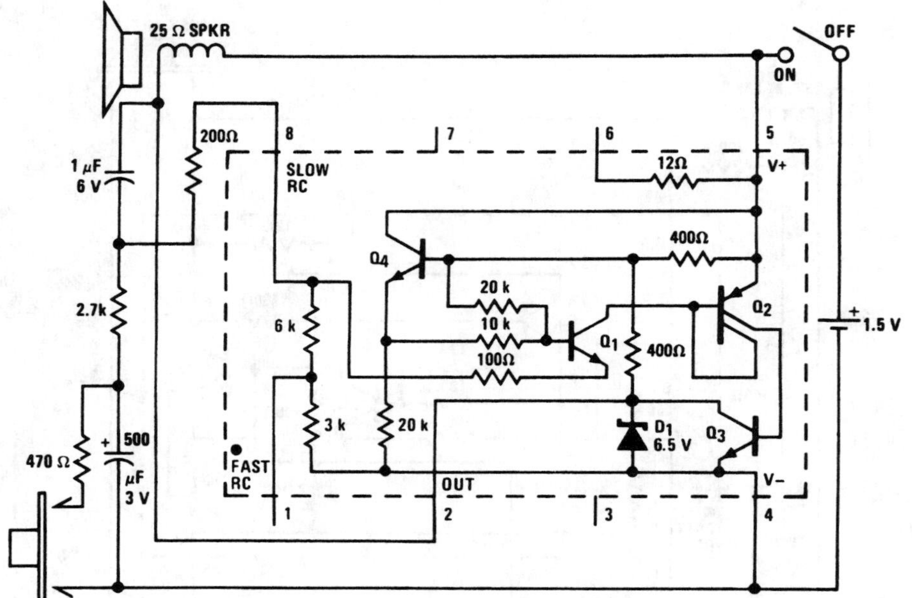

Fire siren

Fig. 12-55 This LM3909 circuit produces a rapidly rising wail upon pressing the button, and a slower "coasting down" sound upon release. If it is desirable to have the tone stop sometime after the button is released, an 18-kΩ resistor can be placed between pins 8 and 6 of the LM3909. The sound is then much like that of a motor-driven siren. National Semiconductor, Linear Applications Handbook, 1991, p. 402.

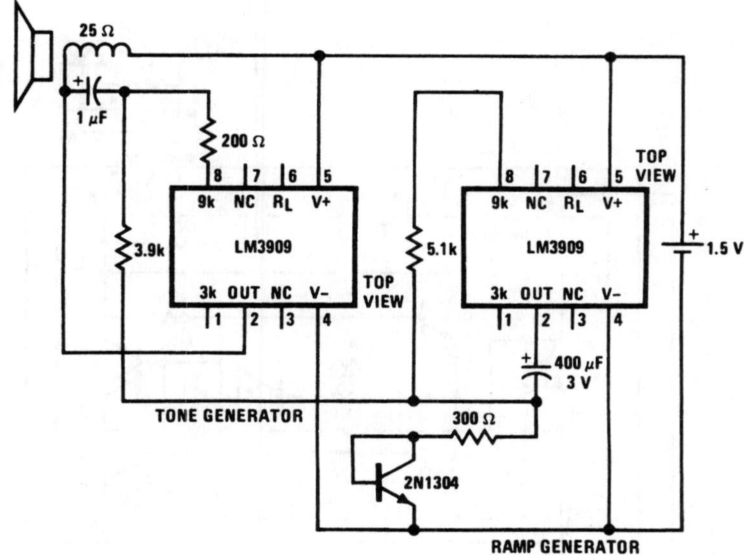

Whooper siren

Fig. 12-56 This circuit uses two LM3909s, one as a tone generator and one as a ramp generator, to produce a "whooper" sound that is somewhat like the electronic sirens that are used on city police cars, ambulances, and airport "crash wagons." The rapid modulation makes the tone seem louder for the same amount of power input. The tone generator is similar to that of the Fig. 12-55 circuit, except that the pushbutton is replaced by a rapidly rising and falling modulating voltage that is produced by the ramp generator. National Semiconductor, Linear Applications Handbook, 1991, p. 402.

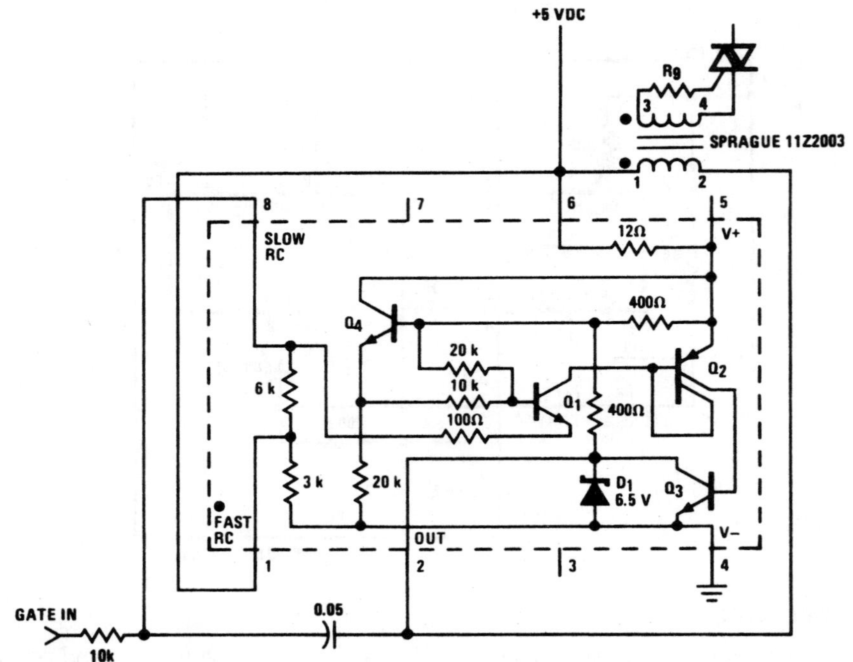

Triac trigger

Fig. 12-57 This circuit uses an LM3909 (operating from a standard 5-V logic supply) to trigger a triac (chapter 8). With no gate input, or a TTL-logic high input, the LM3909 is biased off because pin 1 is tied to V+. With a logic low at the gate input, the LM3909 provides 10-μs pulses at about a 7-kHz rate. A TTL gate loaded only by this circuit is assumed because worst-case voltage swing might be insufficient. The trigger is not of the synchronized zero-crossing type (chapter 8) because the first trigger pulse after gating on could occur at any time. However, the repetition rate is such that after the first cycle, a triac is triggered within 8 V of zero (with a resistive load and a 115-Vac line). National Semiconductor, Linear Applications Handbook, 1991, p. 403.

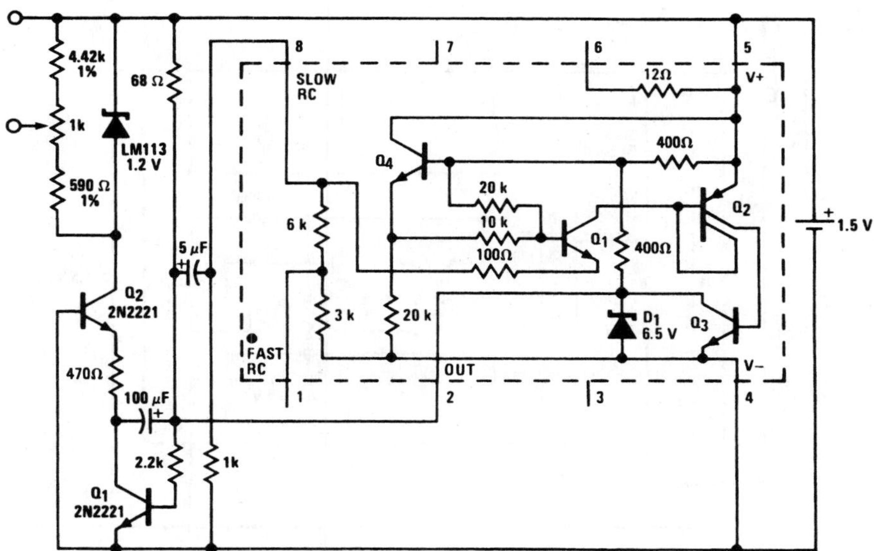

Scope calibrator

Fig. 12-58 This circuit uses an LM3909 to produce a precision square-wave signal that is used to calibrate scopes and scope probes or to check the gain and transient response of audio amplifiers (as described in chapter 1). The output is adjustable (with the 1-kΩ pot) and can be held to within a few tenths percent of 1 V. The output signal is a clean rectangular wave of about 1.5 ms On and 5.5 ms Off. The 0.01% temperature coefficient of the LM113 regulator results in negligible drift of the waveform amplitude under lab conditions. Because the circuit is battery powered, there is no inconvenient line cord, and no noise or hum. National Semiconductor, Linear Applications Handbook, 1991, p. 404.

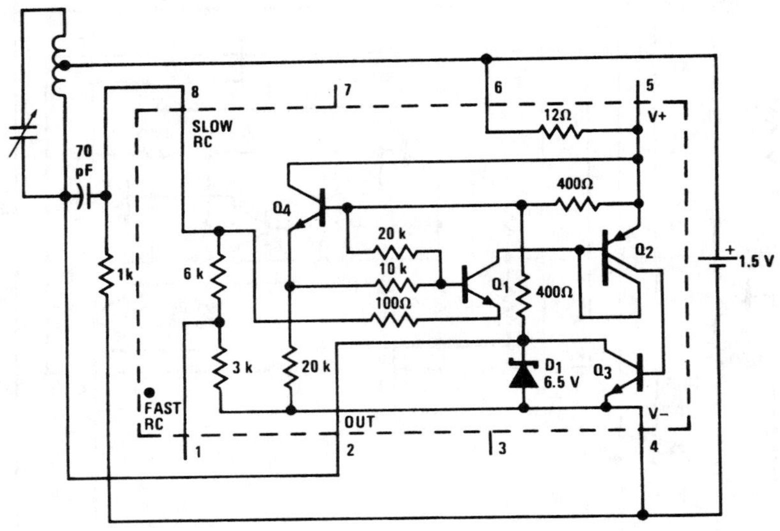

RF oscillator

Fig. 12-59 This circuit uses the LM3909 as a low-power RF oscillator (chapter 5). The tuned circuit is a standard AM-radio ferrite antenna coil (loopstick) with a tap 40% of the turns up from one end. The tuning capacitor is a standard 360-pF AM-radio tuning capacitor. The high-frequency limit is about 800 kHz. National Semiconductor, Linear Applications Handbook, 1991, p. 404.

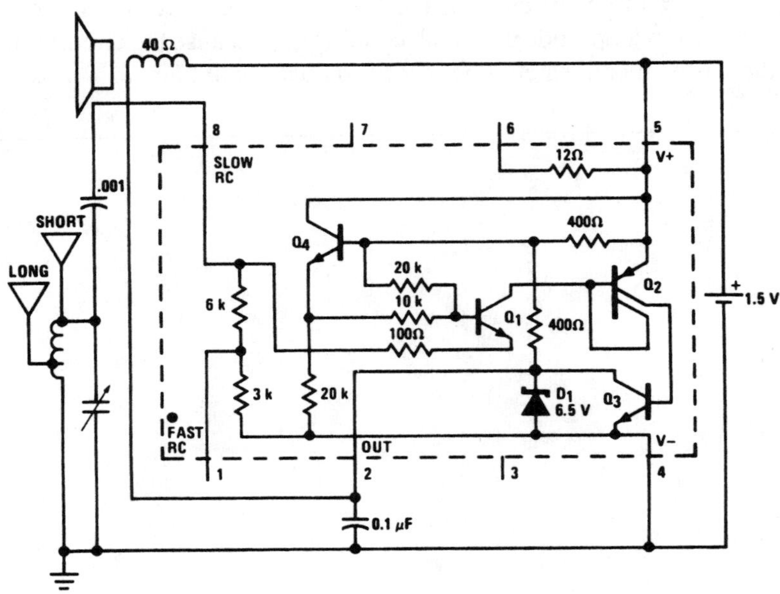

AM radio

Fig. 12-60 This circuit uses the LM3909 as a basic AM radio (chapter 2). The tuned circuit is a standard AM-radio ferrite antenna coil (loopstick with a tap 40% of the turns from one end) and a 360-pF tuning capacitor). The short antenna is 10 to 20′ and the long antenna is 30 to 100′. Notice that this "radio" does not oscillate, and tuning is only as good as a simple crystal set. Because of the low power drain, the circuit will operate continuously for about a month on a single D flashlight cell, and drive a 6″ speaker. National Semiconductor, Linear Applications Handbook, 1991, p. 405.

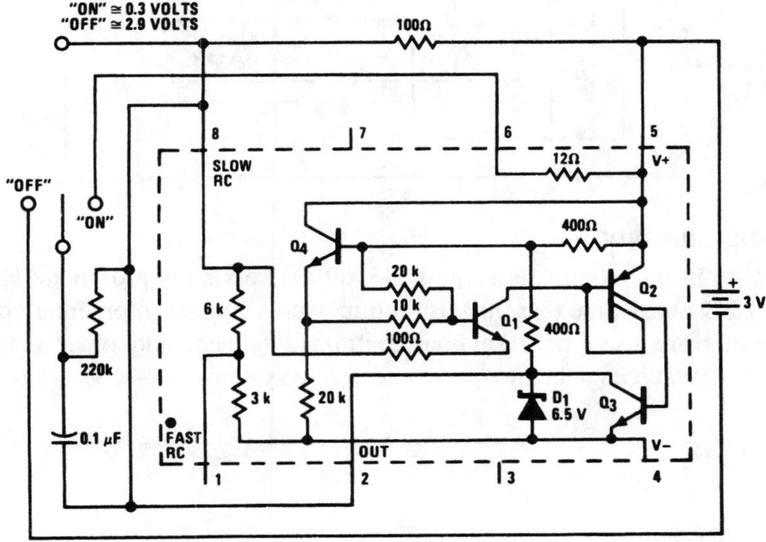

Latch circuit

Fig. 12-61 In this application, the LM3909 switches to, and holds its condition, whenever the switch changes sides—even if the contact is momentary. This results in an output that remains in the 0.3-V (on) or 2.9-V (off) condition. National Semiconductor, Linear Applications Handbook, 1991, p. 405.

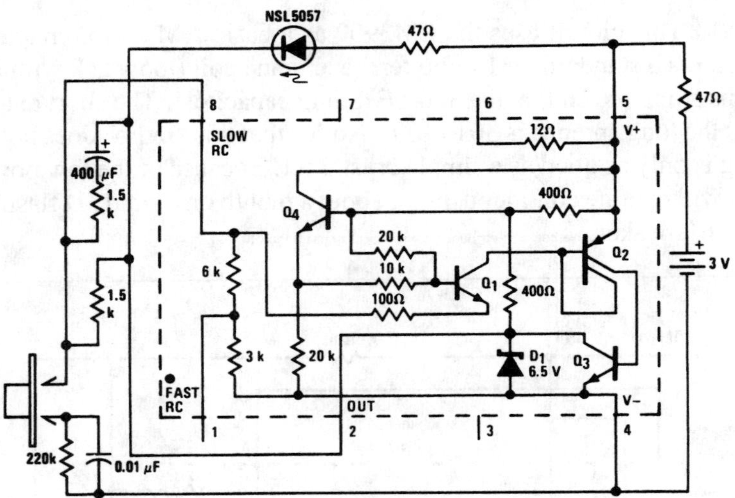

Indicating one-shot

Fig. 12-62 In this application, the LM3909 delivers an approximate ½-s flash from the LED every time that the pushbutton makes contact (momentary or held). Such circuits are used with keyboards, limit switches, and other mechanical contacts that must feed data into electronic digital systems. National Semiconductor, Linear Applications Handbook. 1991, p. 406.

Index

E

F

G

O

T

U

V

W

ABOUT THE AUTHOR

For over 40 years, **John D. Lenk** has been a self-employed consulting technical writer spcializing in practical troubleshooting guides. A long-time writer of international best-sellers in the electronics field, he is the author of 75 books on electronics which together have sold more than 1 million copies and have been translated into eight languages. Mr. Lenk's guides regularly become classics in their fields and his most recent books include *Complete Guide to modern VCR Troubleshooting and Repair, Digital television, Lenk's Video Handbook*, and *Practical Solid-State Troubleshooting*, which sold over 100,000 copies.

ABOUT THE AUTHOR